IB	IIB	IIIA	IVA	VA	VIA	VIIA	0
								4.00260 He 2
			10.81 B 5	12.011 C 6	14.0067 N 7	15.9994 O 8	18.9984 F 9	20.179 Ne 10
			26.9815 Al 13	28.086 Si 14	30.9738 P 15	32.06 S 16	35.453 Cl 17	39.948 Ar 18
58.71 Ni 28	63.546 Cu 29	65.37 Zn 30	69.72 Ga 31	72.59 Ge 32	74.9216 As 33	78.96 Se 34	79.904 Br 35	83.80 Kr 36
106.4 Pd 46	107.868 Ag 47	112.40 Cd 48	114.82 In 49	118.69 Sn 50	121.75 Sb 51	127.60 Te 52	126.9045 I 53	131.30 Xe 54
195.09 Pt 78	196.9665 Au 79	200.59 Hg 80	204.37 Tl 81	207.2 Pb 82	208.9806 Bi 83	(210) Po 84	(210) At 85	(222) Rn 86

157.25 Gd 64	158.9254 Tb 65	162.50 Dy 66	164.9303 Ho 67	167.26 Er 68	168.9342 Tm 69	173.04 Yb 70	174.97 Lu 71
(247) Cm 96	(249) Bk 97	(251) Cf 98	(254) Es 99	(253) Fm 100	(256) Md 101	(254) No 102	(257) Lr 103

Chemistry

CHEMISTRY

Reactions, Structure, and Properties

SECOND EDITION

CLYDE R. DILLARD · DAVID E. GOLDBERG

Brooklyn College of The City University of New York

Macmillan Publishing Co., Inc.
New York
Collier Macmillan Publishers
London

Macmillan Publishing Co., Inc.
866 Third Avenue, New York, New York 10022

Collier Macmillan Canada, Ltd.

Library of Congress Cataloging in Publication Data

Dillard, Clyde R   (date)
    Chemistry, reactions, structure, and properties.

    Includes index.
    1. Chemistry. I. Goldberg, David Elliott,
(date)   joint author. II. Title.
QD31.2.D54 1978   540   76–55768
ISBN 0–02–329580–5

Printing: 1 2 3 4 5 6 7 8   Year: 8 9 0 1 2 3 4

This book is dedicated to
Maurice Goldberg
and to the memory of
Christine Dillard and Adele Goldberg

Preface

In the past three decades, the rapid growth in chemical knowledge and advances in chemical instrumentation have indeed been phenomenal. It has been necessary to change the content and emphasis of upper level chemistry courses to reflect these advances. The first edition of this book was designed to change the general chemistry course to give students a better background for advanced work.

Our objective—to raise the level of instruction in general chemistry considerably above that of the traditional high school course—is unchanged, but this revision reflects our experiences with the first edition as well as the consensus of colleagues who have used the book elsewhere. The book has four specific goals: (1) to develop skills in the fundamentals of stoichiometry, equilibrium calculations, and the applications of thermodynamics to chemical systems; (2) to provide a foundation for upper level courses by presenting a coherent introduction to atomic and molecular structure; (3) to introduce some topics in the classical physical chemistry of gases, liquids, solids, and solutions and to show how the gross properties of matter can be interpreted in terms of molecular theory; and (4) to stimulate further interest in chemistry and provide a basis for open-ended inquiry through presentation of selected items of descriptive chemistry.

For the most part, we have retained the order of topics of the first edition because we believe that all students, regardless of background, must first master the basic concepts of stoichiometry, equilibrium, atomic structure, and molecular structure. Moreover, it is our opinion that a majority of students gain more from discussions of theoretical and structural chemistry *after* they have had some background with chemical reactions. Since quantum chemistry cannot be taught rigorously at the freshman level, the students are required to accept the results on faith. It is reassuring to the student to be well grounded in those aspects of quantitative chemistry which can be readily demonstrated in the laboratory. In like manner, at the freshman level the approach to thermodynamics has to be an intuitive one. Students who have understood equilibrium and who have predicted the spontaneity of redox reactions are better able to grasp the significance of enthalpy, free energy, and entropy.

The order of topics permits from the first week laboratory experiments of immediate relevance to the text. Also a rigorous quantitative approach in early chapters and an early introduction of concepts of thermodynamics are in keeping with our aim of raising the level of the course. However, in the second edition the chapter on thermodynamics has been revised by moving some of the original content to later chapters and by rewriting the remainder for improved clarity. There is also greater emphasis on the use of theoretical "models" as a technique of the scientific method.

The treatment of organic chemistry is now dispersed rather than concentrated in a single chapter. Organic compounds are introduced in Chapter 1 so that organic substances can be included in discussions of stoichiometry and thermochemistry. In Chapter 12 organic molecules are discussed with emphasis on structure. This discussion provides examples of theories of bonding discussed earlier and also provides background for the following chapter on experimental determination of structure.

Finally, organic molecules are discussed in Chapter 23, with emphasis on reactions and practical applications.

Although this text is intended primarily for the science-oriented student, it does not abandon those students whose backgrounds and interests may be less specific. No knowledge of mathematics beyond algebra is required. The Appendix includes instruction on mathematical techniques such as the factor-label method and the use of significant figures. In consideration of the varied backgrounds of the students, the exercises at the end of each chapter are classified in order of difficulty as basic, general, or advanced. It is hoped that the exercises in the last category will both challenge and stimulate the better-prepared student. We believe that students preparing for careers in science should have an early introduction to computers and/or programmable calculators wherever such facilities are available. Accordingly, several problems requiring access to a computer have been included in the advanced exercises.

Implicitly, the text is divided into four sections: Chapters 1–8 deal with stoichiometry and equilibrium, Chapters 9–14 with atomic and molecular structure, Chapters 15–18 with kinetics and the physical chemistry of matter in bulk, and Chapters 19–23 with selected topics of descriptive chemistry. Some instructors may wish to follow a different order of presentation than that given in the text, and these blocks of chapters allow for alternative sequences of topics.

An *Instructor's Supplement,* available from the publisher, contains discussions of the objectives and points of emphasis of each chapter, lists of collateral reading, and suggested lecture demonstrations. A *Student Solution Supplement,* which gives solutions for all the exercises, and a *Student Study Guide* have also been prepared.

We are indebted to many for assistance and suggestions in the preparation of this second edition as well as to those who contributed to the earlier edition. In particular we wish to acknowledge the help and advice of Professors Ted Benfey, Uldis Blukis, F. B. Bramwell, John L. Deutsch, Seymour Dondes, David A. Franz, Ira Levine, Robert Naylor, William M. Ritchey, Andrew Streitwieser, Jr., and Peter E. Yankwich and of Dr. K. Mark Thomas.

C. R. D.

D. E. G.

Contents

Detailed Contents

23 Organic Reactions and Biochemistry   658

Appendix   697

To the Student

This book is designed to provide a foundation in the principles of chemistry which will be useful not only in advanced chemistry courses but also in other subjects where a knowledge of chemistry is required. To aid you in making the most effective use of the book, several comments and suggestions are offered.

Many of the concepts are presented as "tools"; that is, emphasis is placed on the application of principles rather than on mere memorization of facts. You might find it useful to prepare your own review summaries of the chapters by noting the essential concepts and important definitions which are presented in **boldface** type.

Chemistry is a quantitative science; that is, various chemical phenomena can be studied by measuring some effect or property. Also most chemical theories can be expressed in mathematical terms. The numerical problems presented in this text emphasize these facts. Do not approach a problem merely to obtain "the answer." Rather, try to understand the principles involved. In most problems the numbers represent actual experimental magnitudes and are expressed to the proper number of significant figures. Your answers to such problems should also contain the proper number of significant figures. A discussion of how to use significant figures is given in Appendix A–5.

Numerous "worked" examples are used throughout the book. Attempt to solve these before reading the solution presented in the text. Note that the "factor-label" method is used in setting up many of the solutions for the worked examples. That is, the units and conversion factors are expressed in such a way that they necessarily give the appropriate units for the calculated result. This technique is very useful to verify that the correct approach is being used to solve a problem (see the discussion in Appendix A–11). Carefully study the worked examples, and use the factor-label method in solving other problems. It should be noted that the numerical solutions to the worked examples are expressed in the proper number of significant figures.

A table of units and conversion factors is given in Appendix A–3, and numerous other tables of data are provided in the text. While certain physical constants and data which are *defined* as having specific values must be learned, in general one should not attempt to memorize values of constants for individual compounds or reactions. In this text, the tables of data most often used are marked with color panels extending to the edge of the page so that you can locate them easily.

Chapters 1 and 2 contain material which should be familiar to most students. They are intended to provide the essential background for the discussions in the remainder of the book. It is necessary that the content of these chapters be thoroughly understood, even though your instructor may elect to omit this introductory material. Therefore, you should make sure that you understand it by working some of the general exercises in each chapter. Remedy any deficiencies by reviewing the required material either in this book or in other sources.

The exercises listed at the ends of the chapters are classified as "Basic," "General," and "Advanced." The Basic Exercises serve as a review of the particular chapter and should be attempted whether or not they are assigned as homework. The General Exercises require the application of the concepts of the current chapter and may also review material from earlier chapters. The Advanced Exercises are more challenging, and the skills and knowledge required are not necessarily limited to those presented in the text. Always attempt some of the Advanced Exercises. Sometimes they will provoke questions which, when answered, will greatly increase your understanding of the principles involved. Answers to selected basic and general exercises are given at the end of the text. Worked-out solutions to virtually all of the exercises are available in a separate *Student Solution Supplement*.

Alternate explanations and additional discussions of important points are presented in a *Student Study Guide*, by William Ritchey and Robert Naylor. This guide also includes chapter summaries, a self-testing quiz for each chapter, and answers to the quizzes.

1

Essential Concepts

Modern chemistry encompasses a vast array of topics ranging from the chemical processes occurring in cells of living tissues to the chemical compositions of distant stars. Chemical theories are applied to explain such diverse phenomena as heredity and the cosmic origins of the universe. Mankind has always been interested in the material objects in the environment. Ancient philosophers proposed a concept of "elements" to explain the characteristics of various materials and even speculated about the nature of atoms. The alchemists attempted to alter the characteristics of cheaper metals to make them into gold, and although their experiments were unsuccessful, the knowledge they accumulated provided a basis for further investigations. Various artisans' procedures, such as the manufacture of wines, dyes, and ceramics and the tanning of hides, involved chemical reactions. Speculation about such reactions, along with deliberate experimentation, contributed to the growth of chemical knowledge. As experiments and observations became more quantitative, involving careful measurements, chemistry developed into a logical and mathematically rigorous science.

The basic concepts of chemistry are so well established that it is no longer necessary to trace each one from its historical inception. The concepts of chemical elements and the existence of atoms and molecules will be taken as established fact. It is now apparent that most aspects of chemistry will eventually be understood in terms of the atomic theory, molecular structure, and thermodynamics. This is not to say that there is nothing new to be discovered or that every phenomenon has been satisfactorily explained; rather it is to say that the present state of knowledge is such that most new advances and discoveries are made more by design than by trial and error methods. The applications of the known principles of chemistry to create new materials and to predict and demonstrate new phenomena are perhaps the most fascinating aspect of a chemist's career.

Chemistry is an experimental science. The ultimate test of the validity of any chemical principle or theory is that it can be consistently used to explain or predict observable phenomena. Of course there are many well known examples of the successful application of chemical principles. Synthetic fibers, plastics, modern drugs, antibiotics, synthetic rubber, and the widespread use of chemicals in agriculture and industry are dramatic examples of this success. Equally dramatic, but less widely appreciated, is the progress which has been made in delineating the intimate details of the structures of atoms and molecules themselves. The distances between atoms and their geometric arrangements in molecules and crystals have been precisely determined by a variety of experimental techniques. The information thus gained has been applied to predict or explain the physical characteristics and chemical behaviors of various materials.

1

The aim of this general chemistry textbook is to present the essential principles of chemistry and to illustrate them by means of selected facts. This first chapter is a brief survey of concepts and terminology which are essential to the discussion of chemical reactions. These concepts will be refined and more fully developed in later sections of this book. The intent here is to establish a language of terms, such as atoms, molecules, formulas, and bonds, so that the choice of chemical reaction systems to be described starting in Chapter 2 will not be limited to a few examples. As a part of the language of chemistry, it is very important to be able to recognize the names of, and to write formulas for, various chemical substances. The use of the periodic table and the use of simple rules of chemical bonding to derive many formulas will be illustrated.

1-1   The Scope of Chemistry

In the discussions which follow, any material object, whether simple or complex, will be referred to as **matter.** Chemistry is the science which studies matter in terms of its composition, its properties, and its structure. Many apparently simple materials, such as water and air, are indeed complex. On the other hand, complex materials are conveniently described in terms of some rather simple concepts. For example, any sample of matter is made up of one or more chemical elements. An **element** is a form of matter which cannot be further broken down to simpler forms by means of a chemical reaction. There are over 100 known elements, each of which has its own set of characteristics distinguishing it from all other elements. An abbreviation of the English or Latin name of an element is used as a **symbol** for the element. A list of these names and symbols is found inside the back cover of this book. It is important to learn both the names and symbols of the most common elements. The symbol is also sometimes used to represent an atom of the element.

Compounds are chemical combinations of two or more elements in definite ratios by mass in which the set of characteristics of each element is lost. The compound has its own set of characteristics. The term **substance** is used to denote any pure element or any pure compound. Every specimen of matter is either a substance or a mixture of substances.

Those characteristics of a substance which serve to identify it and distinguish it from other substances are called its **properties.** Familiar examples of properties are color, hardness, tendency to vaporize, and tendency to burn or to resist burning. The practical uses of various materials are determined by their properties. For example, houses are built of wood, bricks, or stone, but never of wax, which could melt in hot weather. Both coal and gasoline are used as fuels, but because gasoline is a liquid which vaporizes readily, it is more suitable as a fuel for internal combustion engines than coal is. The characteristic ways in which a substance undergoes chemical reactions, which are reactions in which a change in the composition of the substance takes place, are called its **chemical properties.** Numerous examples of chemical properties will be given throughout this book. Those properties which describe the substance itself but do not involve any chemical reaction are called its **physical properties.**

The most useful properties for identification of a substance are those which can be

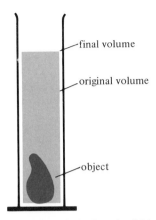

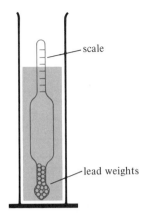

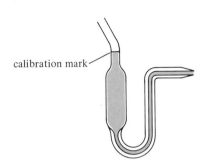

(a) An irregularly shaped solid is first weighed and then placed in a measured volume of liquid in which it does not dissolve. The difference in readings in the cylinder corresponds to the volume of the solid. The density of the solid is equal to the mass divided by the volume.

(b) The hydrometer method for measuring the density of liquids is based on Archimedes' principle that a floating object displaces its own weight of liquid. The hydrometer float is constructed so that intervals on its scale correspond to definite volumes, and since its weight is known, the scale is calibrated in dimensions of density.

(c) The pycnometer is a vessel which has a precisely known volume. Liquid is drawn up to the calibration mark in a weighed pycnometer. The filled pycnometer is re-weighed, and the mass of liquid is obtained by difference. The density may be calculated from the mass difference and the known volume. The volume of the pycnometer is determined by weighing a liquid of known density in it.

Figure 1-1. **Density Measurement**

measured and expressed in numerical terms.[1] For example, one can measure the temperature of a solid substance while it is being heated. The temperature at which the solid changes to a liquid is called its **melting point.** Another conveniently measured property of matter is **density** (mass per unit volume). Three ways of measuring density are illustrated in Figure 1-1.

The number of digits included in a number reported by a scientist reflects the precision with which the measurement was made. The significant digits, or significant figures, are those which are known definitely, plus the final estimated digit. Units, conversion factors, and significant digits are all discussed in the Appendix.

Example

A block of platinum 6.00 cm long, 3.50 cm wide, and 4.00 cm thick has a mass of 1802 grams. What is the density of platinum?

The volume, V, of the block is determined by multiplying its length, l, times its width, w, times its thickness, t:

$$V = lwt = (6.00 \text{ cm})(3.50 \text{ cm})(4.00 \text{ cm}) = 84.0 \text{ cm}^3$$

---

[1] Scientific measurements are usually expressed in the units of the metric system and/or those of the Système Internationale (SI), an expanded version of the metric system approved internationally for scientific use.

The density is the mass per unit volume:

$$d = \frac{m}{V} = \frac{1802 \text{ grams}}{84.0 \text{ cm}^3} = 21.5 \text{ grams/cm}^3$$

Example

The density of aluminum is 2.70 grams/cm³. An irregularly shaped piece of aluminum weighing 40.0 grams is added to a 100 ml graduated cylinder containing exactly 50.0 ml of water. To what height in the cylinder will the water level rise? See Figure 1–1(a).

The water level will change by an amount which reflects the volume of the piece of aluminum. The volume of the piece may be determined from the density of aluminum and its mass by using either the equation for density or the factor-label method (see Appendix A–11):

The equation   *The factor-label method*

$$d = \frac{m}{V}$$

$$V = \frac{m}{d}$$

$$V = \frac{40.0 \text{ grams}}{2.70 \text{ grams/cm}^3} = 14.8 \text{ cm}^3$$

$$\overbrace{40.0 \text{ grams}}^{\text{quantity given}} \underbrace{\left(\frac{1 \text{ cm}^3}{2.70 \text{ grams}}\right)}_{\text{ratio}} = 14.8 \text{ cm}^3$$

The water level in the cylinder is the volume of the water plus that of the aluminum.[2]

total volume = 50.0 ml + 14.8 ml = 64.8 ml

Any material which is not a substance is a mixture. A **mixture** consists of two or more substances—which may be elements or compounds or both—and the relative quantities of the substances may be varied arbitrarily. Most of the characteristics of the original substances are retained by the mixture. For example, a mixture of sand and water will have characteristics of both substances regardless of the relative quantities of each in the mixture. A **homogeneous mixture** appears, even under a microscope, to be composed of only one kind of matter. Such mixtures are called **solutions.** Air, brass, and a homogeneous mixture of sugar in water are examples of solutions. Mixtures with parts which are visibly different are **heterogeneous.** The distinctions among two compounds, the compounds and their constituent elements, and the pure substances and a homogeneous mixture of them are illustrated in Table 1–1, in which the properties of carbon, hydrogen, and oxygen are compared with those of sugar, water, and a solution of sugar in water.

It is possible to identify a particular sample as a (homogeneous) mixture rather than a pure substance by determining that its melting point or boiling point is not sharp, but ranges over several degrees. In a boiling mixture, the first material which evaporates is apt to have a different composition from that which remains; in contrast, a pure substance does not change composition upon evaporation.

One of the more important activities (if not the most important activity) of modern chemists is the explanation of properties of substances in terms of their structures,

---

[2] 1 cm³ = 1 ml.

TABLE 1-1. A Comparison of Elements, Compounds, and a Mixture

	Elements		
	Carbon (Graphite)	Hydrogen	Oxygen
appearance	black solid	colorless gas	colorless gas
composition	100% C	100% H	100% O
melting point	3600°C	−259°C	−218.5°C
boiling point (at 1 atm)	4200°C	−253°C	−183°C
density (at 1 atm)	2.25 grams/cm³	0.0899 grams/liter	1.43 grams/liter

	Compounds		Mixture
	Sugar (Sucrose)	Water	Solution of Sugar in Water
appearance	white solid	clear liquid	clear liquid
composition	42.1% C	11.1% H	arbitrary
	6.4% H	88.9% O	
	51.5% O		
melting point	185–186°C	0°C	below 0°C; depends on composition
boiling point (at 1 atm)	decomposes	100°C	above 100°C; depends on composition
density (at 1 atm)	1.588 grams/cm³	1.00 gram/ml	over 1.00 gram/ml; depends on composition

and the prediction of other properties on the basis of such information. Broadly defined, **structure** is the manner in which the constituent parts of a material are put together. It is difficult to give a more narrow definition of the term *structure* because it has many shades of meaning. For example, the structure of sugar (sucrose) might be described as the regular geometric pattern of a crystal of the substance. On a finer scale, sucrose consists of a combination of two simpler sugars—glucose and fructose. On a still smaller scale, the structures of glucose and fructose could be described by showing how different arrangements of their constituent atoms lead to two different sugars. All three of these sugars are composed of the elements carbon, hydrogen, and oxygen. Ultimately, the structures of the carbon, hydrogen, and oxygen atoms might be considered. Conversely, one might start with the structures of the atoms and predict the structures of compounds and crystals.

The experimental determination of structure includes analysis of the substance to determine the elements present and the relative masses of each. Once the composition is known, it is sometimes possible to infer the structure of a substance from observations of its physical and chemical properties. More direct clues to the structures of materials are obtained by means of polarimetry, X-ray diffraction, infrared spectroscopy, and nuclear magnetic resonance spectroscopy. Details of these experimental techniques will be given later, particularly in Chapter 13.

From the above discussion, it is seen that the scope of chemistry is indeed wide and comprehensive. The interrelationships between the various aspects of chemistry are diagrammed in Figure 1–2. The activities of chemists include determinations of the composition, properties, and structures of substances; the synthesis of new materials

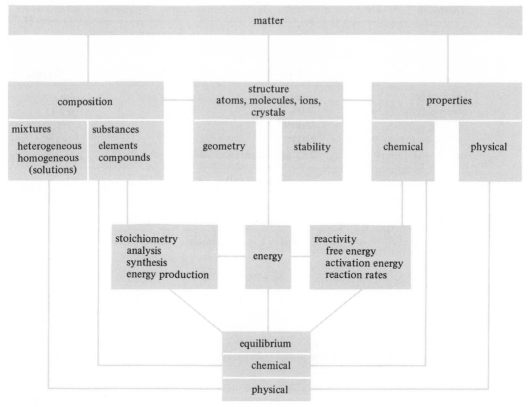

Figure 1-2. **The Scope of Chemistry**

having specific properties and structures; the production of energy from chemical reactions; and the interpretation of the nature of matter on the basis of structure.

From time to time throughout this text the reader should refer to Figure 1-2 so as to place the topic under discussion in perspective against a background of all chemistry.

1-2   Models

The scientific interpretation or explanation of observed phenomena often is best done by means of **models**—mental images proposed to express ideas more easily. In chemistry, models may be physical representations of structures made from balls and sticks, but models may also be merely diagrams or graphs or even mathematical equations. Models of themselves may have no physical reality but they do facilitate the interpretation of real experiments and in some cases models can aid in predicting new phenomena. Among the more familiar models are those dealing with the concept of atoms and atomic structure. For example, in this chapter a model of atomic structure will be proposed which provides just enough details to permit the prediction of some simple chemical formulas. Later, in Chapter 9, the experimental basis for current concepts of atoms will be reviewed, and in Chapter 10, a detailed model of atomic structure will be presented.

Two or more different models may be proposed to explain the same phenomenon. This is not necessarily a contradiction—rather, since both models are inexact views of matter, each is able to explain some aspect of the phenomenon which the other cannot. There is no such thing as a correct model. As more experimental results are obtained, a model may be revised and expanded or it may be abandoned in favor of a more satisfactory alternative.

In addition to the many models which will be presented in this text, the student should attempt to construct others with the view that this is a useful scientific activity. Indeed, by studying how a model behaves, one can learn much about a real phenomenon.

1-3   Atomic Structure

An atom is the smallest particle of an element which, in chemical processes, retains the composition of the element. The results of a number of experiments, which will be described later, indicate that the nature of atoms can be described by the following model: Every atom is composed of a number of subatomic particles called **protons, neutrons,** and **electrons.** The characteristics of these particles are summarized in Table 1–2. The collection of protons and neutrons constitutes the **nucleus** of the atom, and the nucleus is surrounded by a number of electrons. The number of protons within the nucleus is called the **atomic number** of the element. All atoms of a given element have the same atomic number. Since the protons and neutrons are located in the nucleus and since they are heavy compared to the electron, the nucleus contains all of the positive charge and almost all of the mass of the atom. In an electrically neutral atom, there are as many electrons outside the nucleus as there are protons within the nucleus.

It has been demonstrated that the nucleus of an atom has a radius about one hundred-thousandth of the radius of the atom as a whole; therefore the electrons outside the nucleus occupy the bulk of the volume of the atom. The chief factors which determine the properties of a given element are the number and distribution of the electrons in its atoms. Apart from its influence on the number and distribution of electrons in an atom, the nucleus is not involved in ordinary chemical processes.

While the nuclei of the atoms of a given element must contain a fixed number of protons, the number of neutrons may vary. The sum of the number of protons and the number of neutrons in the nucleus is known as the **mass number.** Thus it is possible for two atoms of the same element to have different mass numbers. Atoms with the same atomic number but with different mass numbers are called **isotopes** of each other. Isotopes are denoted with ordinary chemical symbols, with a subscript to the left used to denote the atomic number and a superscript to the left to denote the

TABLE 1-2. Subatomic Particles

Particle	Relative Mass	Charge	Location in the Atom
proton	1	+1	nucleus
neutron	1	0	nucleus
electron	0.0005444	−1	outside the nucleus

TABLE 1-3. Isotopes of Hydrogen and Carbon

Element	Atomic Number, Z	Mass Number, A	Number of Neutrons $(A - Z)$	Name	Symbol
hydrogen	1	1	0	hydrogen-1	^1_1H
hydrogen	1	2	1	deuterium	^2_1H or D
hydrogen	1	3	2	tritium	^3_1H or T
carbon	6	12	6	carbon-12	$^{12}_6C$
carbon	6	13	7	carbon-13	$^{13}_6C$
carbon	6	14	8	carbon-14	$^{14}_6C$

mass number. For example, the symbols $^{12}_6C$ and $^{14}_6C$ represent isotopes of carbon having six protons and having six and eight neutrons in their nuclei, respectively. (Since the atomic number and the chemical symbol both denote the element, the subscript is sometimes omitted.) The heavier isotopes of hydrogen are sometimes given individual names and symbols: ^2_1H, deuterium, D; ^3_1H, tritium, T. The atomic number of an element is customarily represented by the letter Z, and the mass number by the letter A. The number of neutrons in the nucleus is simply the difference, $A - Z$. Table 1-3 summarizes the descriptions of the isotopes of carbon and of hydrogen.

As they occur in nature, most elements consist of a mixture of isotopes. For example, in naturally occurring hydrogen, 99.985% of the atoms are ^1_1H atoms; only one atom in six thousand has a mass number of 2; tritium is radioactive and unstable and is not present in natural hydrogen. In most chemical discussions, the symbol of an element stands for the naturally occurring mixture of isotopes, and the mass number is indicated only when one refers to a specific isotope.

1-4 Molecules and Ions

The smallest neutral particle of a substance capable of independent existence as a gas or as a definite entity in solution is called a **molecule.** Molecules may be monatomic or polyatomic. In polyatomic molecules the constituent atoms are held together by forces called **chemical bonds.** Atoms of the various elements form bonds in characteristic numbers, and the nature of the chemical bonds is extremely important. Some rudiments of chemical bonding are presented in Section 1-8, and the concepts are much more fully developed in Chapter 11. Neon gas is an example of a substance having monatomic molecules. Oxygen gas, on the other hand, is composed of diatomic molecules. Molecules of compounds are necessarily polyatomic. For example, sucrose (table sugar) is a compound of carbon, hydrogen, and oxygen which consists of molecules each containing 45 atoms.

An atom or a group of atoms possessing an electric charge is called an **ion.** If electrons are removed from a neutral atom or molecule, a positive ion results because the protons outnumber the remaining electrons. Similarly, if an atom gains electrons in excess of its atomic number, it becomes a negative ion. To denote ions, the symbols of the respective elements are written with the charge indicated as a superscript to the right, as in Na^+, Al^{3+}, Cl^-, etc. One major class of compounds consists of oppositely

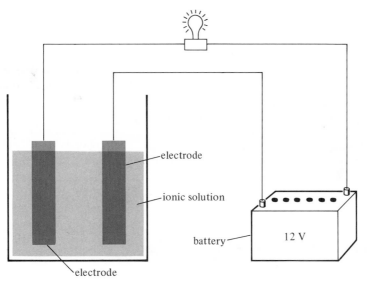

Figure 1-3. **Simple Conductivity Apparatus**
The circuit is complete, and the lamp lights, only when the solution contains
ions and is thereby able to conduct the current. The battery imparts a positive
charge on one electrode and a negative charge on the other.

charged ions. Each compound as a whole is electrically neutral because the relative
numbers of positive and negative ions are such that their charges exactly balance. For
example, ordinary table salt, sodium chloride, consists of equal numbers of Na^+ and
Cl^- ions. If a solution containing ions is placed between electrodes in an apparatus
such as that shown in Figure 1-3, each ion will move toward the electrode which has
a charge opposite to that of the ion. The concerted movement of charged particles
constitutes an electric current. When a battery is connected to a solution of sodium
chloride in water, the solution conducts electricity. Sodium ions move toward the
negative electrode, and chloride ions move toward the positive electrode. In order for
ions to conduct electricity, they must be free to move. Thus melted sodium chloride
conducts electricity, even though solid sodium chloride does not. Substances which
conduct electricity by means of ions are called **electrolytes.**

Some electrolytes conduct electricity only when the pure substance is dissolved in
water or another suitable solvent. The conductivity of these substances stems from
ions formed as a result of a chemical reaction of the substance with the solvent. An
example of this type of compound is hydrogen chloride. Pure hydrogen chloride is a
gas at room temperature, but it can be liquefied by sufficient cooling. However, the
resulting liquid does not conduct electricity. An aqueous (water) solution of hydrogen
chloride is known as hydrochloric acid. Owing to the formation of positive and
negative ions by reaction with water, hydrochloric acid solutions conduct electricity
well. Substances whose solutions conduct electricity well are called **strong electrolytes;** substances whose solutions conduct electricity only poorly are called **weak
electrolytes.** Compounds which in the pure state or in solution do not consist of or
form ions are **nonelectrolytes.** They do not conduct electricity even when melted or
when dissolved in water.

1-5   Metals, Nonmetals, and Metalloids

One very broad classification of the elements includes two main classes, metals and nonmetals, and an intermediate class called metalloids. The general characteristics of **metals** are quite familiar. At room temperature they are opaque solids. (Mercury, a liquid, is an exception.) Metals have a characteristic metallic luster. They are good conductors of heat, and they conduct electricity both in the solid and the molten state. Metals are ductile—that is, they can be drawn into wires; and they are malleable—that is, they can be rolled or hammered into thin sheets. Metals generally have high densities and many have high melting points. In general, metals are insoluble except in other metals. In some solutions metals may appear to dissolve, but actually they undergo chemical reaction. The particles found in the solution are ions rather than atoms of the metal. In compound formation, atoms of metals tend to lose electrons and become positive ions.

Nonmetallic elements are less easily characterized as a class. Some are solids at room temperature, but except for carbon, nonmetals have comparatively low melting points. One nonmetal, bromine, is a liquid at room temperature. Some, notably nitrogen, oxygen, fluorine, chlorine, hydrogen, helium, neon, argon, krypton, and xenon, are gases at room temperature. Nonmetals are poor conductors of heat, and except for one form of carbon—graphite—they do not conduct electricity. In compounds of one nonmetallic element and one metal, atoms of nonmetallic elements tend to exist as negative ions. However, nonmetals also form compounds with other nonmetals, and such compounds are usually nonelectrolytes.

The **metalloids** include the elements boron, silicon, arsenic, tellurium, and astatine. Their properties are intermediate between those of metals and nonmetals. All metalloids are solids, and some have a metallic luster. Some conduct electricity in the solid state and are used in the manufacture of transistors (Section 19-3).

1-6   Formulas of Electrolytes

Since substances have definite compositions, they can be represented by formulas consisting of the symbols of the constituent elements with numerical subscripts written after each symbol to denote the relative number of atoms of each element in the substance. For example, the formula for the substance calcium chloride is $CaCl_2$, which signifies that there are two chlorine atoms present for every calcium atom. (The number 1 is not explicitly written as a subscript.) Formulas for ions are written in a similar manner, but in addition they have a superscript which denotes the charge on the ion, as in NH_4^+, the ammonium ion, CO_3^{2-}, the carbonate ion, and Cl^-, the chloride ion. However, formulas of ionic compounds are usually written without showing the charges on the ions, as, for example, NaCl and $BaCl_2$.

Compounds composed of only two elements are called **binary** compounds. Often one of the elements is a metal and the other a nonmetal. A number of binary compounds of two nonmetals exist, but binary compounds of two metals are encountered less frequently. In formulas for compounds of a metal and a nonmetal, the symbol of the metal is written first.

The names of ionic compounds are simply combinations of the names of the positive ion—the **cation**—given first, followed by that of the negative ion—the **anion.**

TABLE 1-4. Some Common Ions

1+		2+		2+		3+	
ammonium	NH_4^+	barium	Ba^{2+}	magnesium	Mg^{2+}	aluminum	Al^{3+}
copper(I)	Cu^+	calcium	Ca^{2+}	manganese(II)	Mn^{2+}	chromium(III)	Cr^{3+}
hydrogen	H^+	chromium(II)	Cr^{2+}	mercury(II)	Hg^{2+}	iron(III)	Fe^{3+}
potassium	K^+	copper(II)	Cu^{2+}	mercury(I)	Hg_2^{2+}		
silver	Ag^+	iron(II)	Fe^{2+}	tin(II)	Sn^{2+}		
sodium	Na^+	lead(II)	Pb^{2+}	strontium	Sr^{2+}		
				zinc	Zn^{2+}		

3−		2−		1−		1−	
arsenate	AsO_4^{3-}	carbonate	CO_3^{2-}	acetate	$C_2H_3O_2^-$	hydrogen	
arsenite	AsO_3^{3-}	chromate	CrO_4^{2-}	bromide	Br^-	sulfite	HSO_3^-
phosphate	PO_4^{3-}	dichromate	$Cr_2O_7^{2-}$	chlorate	ClO_3^-	hydride	H^-
phosphite	PO_3^{3-}	oxalate	$C_2O_4^{2-}$	chloride	Cl^-	hydroxide	OH^-
		oxide	O^{2-}	chlorite	ClO_2^-	hypochlorite	ClO^-
		sulfide	S^{2-}	cyanide	CN^-	iodate	IO_3^-
		sulfate	SO_4^{2-}	fluoride	F^-	iodide	I^-
		sulfite	SO_3^{2-}	hydrogen		nitrate	NO_3^-
				carbonate	HCO_3^-	nitrite	NO_2^-
				(bicarbonate)		perchlorate	ClO_4^-
				hydrogen		permanganate	MnO_4^-
				sulfate	HSO_4^-		

The names of monatomic (one-atom) cations are simply the names of the elements. Thus, Na^+ is the sodium ion, and Ba^{2+} is the barium ion. Polyatomic cations include the ammonium ion, NH_4^+, and certain oxycations of the metals, such as the uranyl ion, UO_2^{2+}, and the vanadyl ion, VO^{2+}. Thus the oxycations are denoted by the suffix *yl* added to the stem of the element name. Monatomic ions of a given element having different charges are denoted by writing the charge in Roman numerals in parentheses after the name of the element. For example, iron(II) ion and iron(III) ion are the names of Fe^{2+} and Fe^{3+}, respectively. Other examples are found in Table 1–4 and are further explained in Chapter 4.

Negative ions, anions, may be monatomic or polyatomic. All monatomic anions have names ending with *ide*. Two polyatomic anions which also have names ending with *ide* are the hydroxide ion, OH^-, and the cyanide ion, CN^-.

Many polyatomic anions contain oxygen in addition to another element. The number of oxygen atoms in such **oxyanions** is denoted by the use of the suffixes *ite* and *ate*, meaning fewer and more oxygen atoms, respectively. In cases where it is necessary to denote more than two oxyanions of the same element, the prefixes *hypo* and *per*, meaning still fewer and still more oxygen atoms, respectively, may be used. A series of oxyanions is named in Table 1–5. (Other types of polyatomic anions will be discussed in Chapters 11, 14, and 21.)

Compounds which are electrolytes are electrically neutral, even though their constituent ions may have numerically different charges. The relative numbers of ions of each charge type which are present must be such as to achieve a net charge of 0. For example, sodium chloride consists of sodium ions and chloride ions. Hence the formula is written as Na^+Cl^- or simply NaCl. The calcium ion has a charge of 2+,

TABLE 1-5. Names of Oxyanions[a]

Fewest Oxygen Atoms hypo____ite		Fewer Oxygen Atoms ____ite		More Oxygen Atoms ____ate		Most Oxygen Atoms per____ate	
ClO^-	hypochlorite	ClO_2^-	chlorite	ClO_3^-	chlorate	ClO_4^-	perchlorate
BrO^-	hypobromite	BrO_2^-	bromite	BrO_3^-	bromate	BrO_4^-	perbromate
IO^-	hypoiodite	IO_2^-	iodite	IO_3^-	iodate	IO_4^-	periodate (meta)
PO_2^{3-}	hypophosphite	PO_3^{3-}	phosphite	PO_4^{3-}	phosphate		
		NO_2^-	nitrite	NO_3^-	nitrate		
		SO_3^{2-}	sulfite	SO_4^{2-}	sulfate		
				CO_3^{2-}	carbonate		

[a] The acids from which the oxyanions ending in *ite* are derived have names ending in *ous acid*. The acids from which the oxyanions ending in *ate* are derived have names ending in *ic acid*. Thus $HClO_4$ is called perchloric acid.

and in calcium chloride two chloride ions are present per calcium ion. Hence the formula for calcium chloride must be written as $CaCl_2$. Whenever more than one polyatomic ion is required in a formula, the formula for the ion is enclosed in parentheses and the appropriate subscript is written outside the parentheses. Thus ammonium sulfide is written $(NH_4)_2S$.

Example

Name the compounds whose formulas are listed: **(a)** NaCl, **(b)** KBr, **(c)** $MgCl_2$, **(d)** NaOH, **(e)** $Ca(OH)_2$, **(f)** LiCN.
(a) Sodium chloride, **(b)** potassium bromide, **(c)** magnesium chloride, **(d)** sodium hydroxide, **(e)** calcium hydroxide, **(f)** lithium cyanide.

Example

Write formulas for **(a)** barium nitrate, **(b)** aluminum sulfate, and **(c)** iron(II) hydroxide.
(a) Since two negative charges are required to balance the charge on one barium ion, Ba^{2+}, and each NO_3^- ion has only one negative charge, the formula must be written $Ba(NO_3)_2$.
(b) To achieve equal numbers of positive and negative charges, the formula must be $Al_2(SO_4)_3$.
(c) Since iron(II) is specified, it takes two OH^- ions to supply the appropriate number of negative charges—hence the formula is $Fe(OH)_2$.

A useful aid in writing formulas and learning the charges on simple ions is the periodic table, which will be discussed in the following section.

1-7   The Periodic Table

Beginning in the late seventeenth century with the work of Robert Boyle, who proposed the presently accepted concept of an element, numerous investigations produced a considerable knowledge of the properties of elements and their compounds. In 1869, D. Mendeleev and L. Meyer, working independently, proposed the **periodic law.** In modern form, the law states that the properties of the elements are periodic functions of their atomic numbers. In other words, when the elements are listed in order of increasing atomic number, elements having closely similar properties will fall at definite intervals along the list. Thus it is possible to arrange the list of elements in tabular form with elements having similar properties placed in vertical columns. Such an arrangement is called a **periodic table.**

A standard version of the periodic table is shown inside the front cover. Each horizontal row of elements constitutes a **period.** It should be noted that the lengths of the periods vary. There is a very short period containing only 2 elements, followed by two short periods of 8 elements each, and then two long periods of 18 elements each. The next period includes 32 elements, and the last period is apparently incomplete. With this arrangement, elements in the same vertical column have similar characteristics. These columns constitute the chemical **families** or **groups.** The groups headed by the members of the two 8-element periods are designated as **main group** elements, and the members of the other groups are called **transition** or **inner transition** elements. For convenience, the main groups are often denoted by a Roman numeral and the letter A, while the transition groups are labeled with a Roman numeral and the letter B.

In the periodic table, a heavy stepped line divides the elements into metals and nonmetals. Elements to the left of this line (with the exception of hydrogen) are metals, while those to the right are nonmetals. This division is for convenience only; elements bordering the line—the metalloids—have properties characteristic of both metals and nonmetals. It may be seen that most of the elements, including all the transition and inner transition elements, are metals.

Except for hydrogen, a gas, the elements of group I A make up the **alkali metal** family. They are very reactive metals, and they are never found in the elemental state in nature. However, their compounds are widespread. All the members of the alkali metal family form ions having a charge of 1 + only. In contrast, the elements of group I B—copper, silver, and gold—are comparatively inert. They are called the **coinage metals.** Being easily obtained from their naturally occurring compounds, they are among the oldest metals known to man. They are similar to the alkali metals in that they exist as 1 + ions in many of their compounds. However, as is characteristic of most transition elements, they form ions having other charges as well.

The elements of group II A are known as the **alkaline earth metals.** Their characteristic ionic charge is 2 +. These metals, particularly the last two members of the group, are almost as reactive as the alkali metals. The group II B elements—zinc, cadmium, and mercury—are less reactive than are those of group II A, but are more reactive than the neighboring elements of group I B. The characteristic charge on their ions is also 2 +.

With the exception of boron, group III A elements are also fairly reactive metals. Aluminum appears to be inert toward reaction with air, but this behavior stems from the fact that the metal forms a thin, invisible film of aluminum oxide on the surface, which protects the bulk of the metal from further oxidation. The metals of group III A form ions of 3 + charge. Group III B consists of the metals scandium, yttrium, lanthanum, and actinium.

Group IV A consists of a nonmetal, carbon, two metalloids, silicon and germanium, and two metals, tin and lead. Each of these elements forms some compounds with formulas which indicate that four other atoms are present per group IV A atom, as, for example, carbon tetrachloride, CCl_4. However, such compounds are not composed of ions. Lead, tin, and to some extent, germanium also form compounds in which dipositive ions occur. The group IV B metals—titanium, zirconium, and hafnium—

also form compounds in which each group IV B atom is combined with four other atoms; these compounds are nonelectrolytes when pure.

The elements of group V A include three nonmetals—nitrogen, phosphorus, and arsenic—and two metals—antimony and bismuth. Although compounds with the formulas N_2O_5, PCl_5, and $AsCl_5$ exist, none of them is ionic. These elements do form compounds—nitrides, phosphides, and arsenides—in which ions having charges of minus three occur. The elements of group V B are all metals. These elements form such a variety of different compounds that their characteristics are not easily generalized.

With the exception of polonium, the elements of group VI A are typical nonmetals. They are sometimes known as the chalcogens, from the Greek word meaning "ash formers." In their binary compounds with metals they exist as ions having a charge of $2-$. The elements of group VII A are all nonmetals and are known as the **halogens,** from the Greek term meaning "salt formers." They are the most reactive nonmetals and are capable of reacting with practically all the metals and with most nonmetals, including each other. In their binary compounds with metals, they exist as ions with a $1-$ charge. The elements of group 0 are known as the noble gases. Until recently they were thought to be completely inert. However, since 1961, compounds of xenon, krypton, and radon have been prepared.

The elements of groups VI B, VII B, and VIII B are all metals. They form such a wide variety of compounds that it is not practical at this point to present any examples as being typical of the behavior of the respective groups.

The periodicity of chemical behavior is illustrated by the fact that, excluding the first period, each period begins with a very reactive metal. Successive elements along the period show decreasing metallic character, eventually becoming nonmetals, and finally, in group VII A, a very reactive nonmetal is found. Each period ends with a member of the noble gas family. The formulas of the oxides of the elements, shown in Figure 1-4, illustrate another periodic relationship.

Example

Use the periodic table as an aid to write formulas for **(a)** radium sulfide, **(b)** magnesium astatide, and **(c)** potassium phosphide.
(a) RaS, **(b)** $MgAt_2$, **(c)** K_3P.

Example

With the aid of the list of elements (back cover), Table 1-4, and the periodic table, name the following substances: **(a)** Li_2SO_4, **(b)** $KReO_4$, **(c)** Na_2MoO_4.
(a) Lithium sulfate, **(b)** potassium perrhenate, **(c)** sodium molybdate. These names are analogous to sodium sulfate, potassium permanganate, and sodium chromate, respectively.

There are several cases in which the periodic table cannot be used to deduce the formulas and/or the names of compounds. For example, while the formula of the chromate ion, CrO_4^{2-}, can be deduced from its group number as being analogous to the sulfate ion, SO_4^{2-}, the formulas for nitrate ion, NO_3^-, and phosphate ion, PO_4^{3-}, are not similar. Likewise, periodate ion, IO_6^{5-}, does not resemble perchlorate ion, ClO_4^-. Further, permanganate ion, MnO_4^-, and manganate ion, MnO_4^{2-}, differ only in charge. Nevertheless, the periodic table is invaluable as a scheme for the classifi-

IA	IIA	IIIB	IVB	VB	VIB	VIIB	VIIIB	VIIIB	VIIIB	IB	IIB	IIIA	IVA	VA	VIA	VIIA	0
H_2O																	
Li_2O	BeO											B_2O_3	CO_2	N_2O_5		OF_2	
Na_2O	MgO											Al_2O_3	SiO_2	P_2O_5	SO_3	Cl_2O_7	
K_2O	CaO	Sc_2O_3	TiO_2	V_2O_5	CrO_3	Mn_2O_7	Fe_2O_3	Co_2O_3	NiO	Cu_2O	ZnO	Ga_2O_3	GeO_2	As_2O_5	SeO_3	Br_2O_7	
Rb_2O	SrO	Y_2O_3	ZrO_2	Nb_2O_5	MoO_3		RuO_4	Rh_2O_3	PdO	Ag_2O	CdO	In_2O_3	SnO_2	Sb_2O_5	TeO_3	I_2O_7	XeO_3
Cs_2O	BaO	La_2O_3	HfO_2	Ta_2O_5	WO_3	Re_2O_7	OsO_4	IrO_2	PtO	Au_2O	HgO	Tl_2O	PbO_2	Bi_2O_3			
	RaO	Ac_2O_3															

Figure 1-4. Formulas of the Oxides of the Elements

It should be noted that although the formula of only one oxide of each element is presented here, many elements form more than one oxide.

cation of the elements. As will be demonstrated throughout this book, the table is also useful in correlating the properties of compounds. The theoretical basis for the periodic law will be given in Section 10–10.

1-8   Chemical Bonding

Some facts about chemical elements and compounds raise questions about the nature of the bonds which hold their atoms together. For example, (1) certain elements such as hydrogen and chlorine exist in the elementary form as diatomic molecules—H_2 and Cl_2, (2) chlorine combines with sodium to form an electrolyte, NaCl, and (3) chlorine combines with carbon to form a nonelectrolyte, CCl_4. Are the same type of forces involved in all these cases? If not, how do the forces between atoms in substances which are electrolytes differ from the forces between atoms in substances which are nonelectrolytes? Can suitable models be proposed which explain such diverse phenomena in terms of a few general concepts?

Since sodium chloride consists of positive sodium ions and negative chloride ions, the electrostatic attraction between opposite charges is called **ionic bonding.** The attractive force due to a charged particle extends in all directions and a given ion should attract a number of oppositely charged ions. The number of oppositely charged ions closest to a given ion would be limited only by their sizes and the magnitudes of the charges. Rather than simple binary molecules, one would expect a vast array of individual, charged particles. In the case of substances composed of ions, this is indeed the situation. For example, the structure of solid NaCl has been determined by X-ray studies. The results obtained lead to the model—an orderly array of oppositely charged ions—which is pictured in Figure 1–5. The model agrees quite well with the known properties of NaCl. The solid substance does not conduct

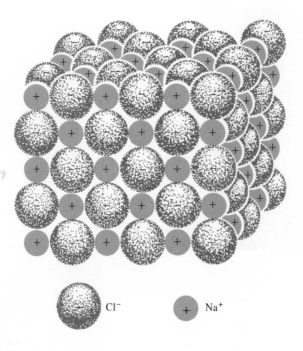

Figure 1–5.
Model of Solid Sodium Chloride

Cl^- Na^+

electricity, but when NaCl is melted or dissolved in water, the constituent ions are free to move under the influence of an applied electrical potential. The physical appearance of NaCl is crystalline, also suggesting an orderly geometric array of ions.

On the other hand, a different model is required to explain the existence of diatomic molecules of hydrogen gas. Since both atoms are identical, formation of oppositely charged ions is not expected. When two hydrogen atoms are brought as close together as they are in the hydrogen molecule, their electrons pair up.[3] The positively charged nuclei are attracted by this pair of electrons (and repelled by each other) so that the electrons are found between the nuclei. The pair of electrons is shared by the two atoms. The bond formed by the sharing of a pair of electrons between two atoms is called a **covalent bond.**

H·   ·H   H:H

separate atoms   covalently bonded atoms

The element helium exists as a monatomic gas and forms no known compounds. A helium atom contains a pair of electrons, and the free atom is obviously the most stable state for helium, since a helium atom does not combine with anything. Thus an electron pair appears to be a stable configuration for the very light elements like helium. In the H_2 molecule two hydrogen atoms assume a configuration somewhat analogous to that of helium. A single hydrogen atom may also achieve the helium configuration by gaining an electron to form the hydride ion, H^-. This ion does exist in compounds like NaH, which has a structure similar to that of NaCl.

The concept of a covalent bond formed by the sharing of a pair of electrons can be extended to other atoms. However, except in hydrogen, not all of the electrons in a given atom are available for bonding. For main group elements in the periodic table, the number of available electrons is equal to the group number. These available electrons are called **valence electrons.** For the first eight-element period the atoms may be diagrammed by using the symbol of the element and a dot for each valence electron, as follows:

Group No.   I   II   III   IV   V   VI   VII   0

Li·   ·Be·   ·B·   ·C·   :N·   :O: :F: :Ne:

The element neon, like helium, is very stable and exists as a monatomic gas. Neon atoms contain eight valence electrons. Other atoms tend to assume the neon configuration—to be associated with eight valence electrons—either by formation of covalent bonds or by the formation of ions. The tendency of atoms to combine so that each atom is surrounded by an octet of valence electrons is called the **octet rule.** (The very light atoms H, Li, and Be form duet structures, having two electrons like helium.) Although the octet rule does not apply to all cases involving covalent bonding, it is a useful guide for formulating the structures of many simple molecules. For example, two fluorine atoms can share a pair of electrons to form the fluorine molecule, which can be represented as follows:

:F:F:

---

[3] The phenomenon of electron pairing will be further described in Section 11-1.

where the sharing of a pair of electrons constitutes a single covalent bond. As the result of forming a covalent bond, both fluorine atoms in the molecule attain an electronic octet configuration like that of neon.

To complete its octet, an atom may have to share more than one pair of electrons, either by combining with several other atoms or by forming **double** or **triple** bonds. It is convenient to indicate a pair of electrons shared between two atoms by a single line drawn between the symbols for the atoms. Double or triple lines indicate double or triple bonds. In such a representation of covalent bonds, the unshared electrons need not be shown.

single bond	double bonds	triple bond	two single bonds
H : H	:Ö : : C : : Ö:	: N : : : N :	:F̈ : S̈ : F̈:
H—H	O=C=O	N≡N	F—S—F

The constituent atoms in polyatomic ions are also linked by covalent bonds. In such cases a net negative charge on the ion is due to an excess of electrons over those provided by the constituent neutral atoms. For example, the chlorite ion, ClO_2^-, contains one electron more than those provided by two oxygen atoms and a chlorine atom.

$$:\ddot{O}:\ddot{C}l:\ddot{O}:^-$$

Similarly, a net positive charge on an ion means that there are correspondingly fewer electrons than would be found on the constituent neutral atoms.

The following procedure may be used to determine the total number of electrons which must be shared by several atoms in order to satisfy the octet rule for all of them. The molecule sulfur dioxide, SO_2, is used to illustrate the several steps in the procedure.

| *Steps* | *Example* |

1. Determine the number of valence electrons available. (For main group elements, from the group number. If the species is a positive ion, the number of available electrons is reduced, if the species is a negative ion, the number is increased.)

$$
\begin{array}{ll}
S & 6 \\
2\,O & 12 \\
\hline
\text{total} & 18
\end{array}
$$

2. Determine the number of electrons necessary to satisfy the octet (or duet) rule with no electron sharing. Use 8 electrons for most atoms, 2 for hydrogen, 0 for monatomic positive ions.

$$
\begin{array}{ll}
S & 8 \\
2\,O & 16 \\
\hline
\text{total} & 24
\end{array}
$$

3. The difference between the numbers obtained in steps 2 and 1 is the number of bonding electrons.

$$
\begin{array}{lr}
\text{electrons required} & 24 \\
-\text{electrons available} & -18 \\
\hline
\text{electrons to be shared} & 6
\end{array}
$$

4. Place the atoms as symmetrically as possible. (Notice that a hydrogen atom

O S O

cannot be bonded to more than one atom, since it is capable of sharing only two electrons.)

5. Place the number of electrons to be shared between the atoms. First place one pair of electrons between each pair of atoms. Then use the remaining pairs of electrons to make double or triple bonds.

$$O::S:O$$

6. Add the remainder of the available electrons to complete the octets (or duets) of all the atoms. There should be just enough.

$$:\ddot{O}::\ddot{S}:\ddot{O}:$$

Example

Write electron dot and line structures for the molecules **(a)** phosphorus trichloride, PCl_3, **(b)** sulfur trioxide, SO_3, **(c)** carbon monoxide, CO, and **(d)** hydroxide ion, OH^-.

(a) $:\ddot{Cl}:\ddot{P}:\ddot{Cl}:$ Cl—P—Cl
   $\quad:\ddot{Cl}:$ $\qquad$ |
   $\qquad\qquad$ Cl

(b) $:\ddot{O}:S::\ddot{O}:$ O—S=O
   $\quad:\ddot{O}:$ $\qquad$ |
   $\qquad\qquad$ O

(c) $:C:::O:$ C≡O

(d) $\left[:\ddot{O}:H\right]^-$ O—H⁻

In example (a) each atom has an octet, formed by sharing one pair of electrons with each neighboring atom in single bonds. In case (b) the sharing of two pairs of electrons forms a double bond, and in case (c) the sharing of three pairs of electrons constitutes a triple bond. In example (d) the hydroxide ion has a negative charge, indicating one electron in excess of those provided by the hydrogen and oxygen atoms. The hydrogen atom does not have an octet, but by sharing a pair of electrons it attains a configuration similar to that of helium.

The electron dot representations are very useful in determining the number of covalent bonds which a given atom may form; they are sometimes helpful in predicting the paths of chemical reactions; and they are aids in keeping track of changes in the numbers of electrons associated with a given atom during certain types of chemical reactions. It must be understood that such representations do not portray the spatial arrangements of atoms in molecules. However, they do aid in the construction of physical models (see Section 11–4).

1-9   Formulas of Compounds Between Nonmetals

Bonding in compounds containing only nonmetals is essentially covalent. In the name of a compound of two nonmetals, the element which lies farther to the left in the periodic table or farther down is given first, without any change in its ending. The other nonmetal is given last, and its ending is changed to *ide*. A prefix is used before the second element to denote the number of atoms of that element (and, rarely, also before the first nonmetal). Some prefixes and examples of their use are given in Table 1–6. These prefixes are not usually used for compounds of a metal and a nonmetal, only for compounds of two nonmetals.

TABLE 1-6. Prefixes for Nomenclature of Binary Compounds of Nonmetals

Prefix	Number	Example	Formula
mon(o)	1	carbon *mon*oxide	CO
di	2	carbon *di*oxide	CO_2
tri	3	sulfur *tri*oxide	SO_3
tetra	4	carbon *tetra*chloride	CCl_4
penta	5	phosphorus *penta*chloride	PCl_5
hexa	6	sulfur *hexa*fluoride	SF_6

Example

Name the following compounds   **(a)** CS_2,   **(b)** PCl_3,   **(c)** SF_2,   **(d)** $AlCl_3$.

(a) Carbon disulfide,   **(b)** phosphorus trichloride,   **(c)** sulfur difluoride,   **(d)** aluminum chloride.

There are several compounds of nonmetals to which the octet rule does not apply. The compounds BF_3, PF_5, and SF_6 have 6, 10, and 12 electrons, respectively, about the central atom. An octet rule structure could be written for BF_3 in which a $B=F$ double bond is included, but atoms of the halogens (fluorine, chlorine, bromine, and iodine) do not form double bonds.

1-10   Formulas of Organic Compounds

The association of the word **organic** with carbon compounds reflects the fact that the compounds in plant and animal tissues are made up chiefly of molecules containing carbon combined with certain other elements. However, living organisms are not the only source of such compounds. Many organic compounds are produced in chemical laboratories and chemical plants; thousands of tons of synthetic organic compounds are produced annually, including fibers, pigments, and plastics.

In modern terms an **organic compound** is one in which there is at least one carbon-to-carbon bond and/or carbon-to-hydrogen bond. In addition to carbon and hydrogen, the elements most commonly encountered in organic compounds are oxygen, nitrogen, phosphorus, sulfur, and the halogens (fluorine, chlorine, bromine, and iodine). Literally millions of organic compounds are composed of just these few elements.

The elements which are commonly part of organic compounds are all located in the upper right corner of the periodic table. They are all nonmetals, and the bonds between their atoms are essentially covalent. (Some organic molecules may form ions, but the bonds *within* each organic ion are covalent. For example, the salt sodium acetate consists of sodium ions, Na^+, and acetate ions, $C_2H_3O_2^-$. The carbon, hydrogen, and oxygen atoms in the acetate ion are covalently bonded; and the entire covalently bonded unit is the negative ion.)

The carbon atom has four valence electrons:

In order to complete its octet, each carbon atom must share a total of four electron pairs. Each shared pair is counted as one in the **total bond order** of the atom; in other words, each carbon atom must have a total bond order of four. Therefore, in organic compounds each carbon atom forms either four single bonds, a double bond and two single bonds, a triple bond and a single bond, or two double bonds. Each of these possibilities corresponds to a total bond order of four.

Since a hydrogen atom can accommodate a maximum of two electrons, its total bond order is limited to one. Thus, in any molecule each hydrogen atom can form but one single bond.

The oxygen atom, with six valence electrons, can complete its octet by forming either two single bonds or one double bond. Thus it will have a total bond order of two. The total bond orders of the other elements usually found in organic compounds can be deduced in a similar manner. The results are given in Table 1-7. With this information, it is possible to write electron dot structures for many organic molecules.

Example

Write an electron dot structure for each of the following: **(a)** CH_4,   **(b)** CH_2O, **(c)** CH_4O,   **(d)** C_2Cl_2,   **(e)** CH_5N.

$$\textbf{(a) } \begin{matrix} H \\ H:C:H \\ H \end{matrix} \qquad \textbf{(b) } \begin{matrix} H \\ H:C: :O: \\ \end{matrix} \qquad \textbf{(c) } \begin{matrix} H \\ H:C:O:H \\ H \end{matrix}$$

$$\textbf{(d) } :Cl:C: : :C:Cl: \qquad \textbf{(e) } \begin{matrix} H \\ H:C:N:H \\ H \ \ H \end{matrix}$$

TABLE 1-7. Total Bond Orders of Elements in Organic Compounds

Element	Symbol	Total Bond Order
carbon	$\cdot C \cdot$	4
hydrogen	$H \cdot$	1
oxygen	$\cdot O \cdot$	2
nitrogen	$:N \cdot$	3
sulfur	$\cdot S \cdot$	2
halogen	$:X \cdot$	1

In the structures given in this example, each element is characterized by a total bond order corresponding to that listed in Table 1–7. Because each pair of electrons shared between two atoms is equivalent to a bond order of one, each shared pair can be represented by a line between the symbols of the elements. Unshared electrons on the atoms are not usually shown in such a representation. The resulting representations of molecules are called **graphic formulas** or **structural formulas.** The structural formulas for the compounds in the preceding example can be written as follows:

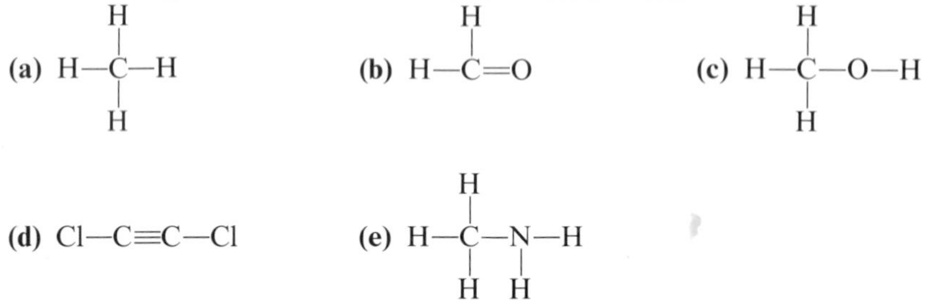

$$
\begin{array}{ccc}
& H & \\
& | & \\
\textbf{(a)} \ H-C-H & & \\
& | & \\
& H &
\end{array}
\qquad
\begin{array}{c}
H \\
| \\
\textbf{(b)} \ H-C=O
\end{array}
\qquad
\begin{array}{c}
H \\
| \\
\textbf{(c)} \ H-C-O-H \\
| \\
H
\end{array}
$$

$$
\textbf{(d)} \ Cl-C\equiv C-Cl
\qquad
\begin{array}{c}
H \\
| \\
\textbf{(e)} \ H-C-N-H \\
| \ \ | \\
H \ \ H
\end{array}
$$

The structures of organic compounds can be represented even more concisely, particularly in printed material, by **line formulas,** so called because they are printed on one line. In line formulas each carbon atom is written as near as possible to the symbol(s) for the element(s) to which it is bonded. Line formulas show the general sequence in which the carbon atoms are attached, but in order to interpret them properly, the permitted total bond orders of all the respective atoms must be kept in mind. The line formulas for the compounds in the preceding example are as follows:

(a) CH_4 **(b)** CH_2O **(c)** CH_3OH **(d)** $ClC\equiv CCl$ **(e)** CH_3NH_2

Some alternative ways of representing the structure of an organic compound are shown in Figure 1–6.

C_3H_8 $CH_3CH_2CH_3$

$$
\begin{array}{ccc}
H & H & H \\
| & | & | \\
H-C- & C- & C-H \\
| & | & | \\
H & H & H
\end{array}
$$

molecular formula   line formula   graphic (structural) formula

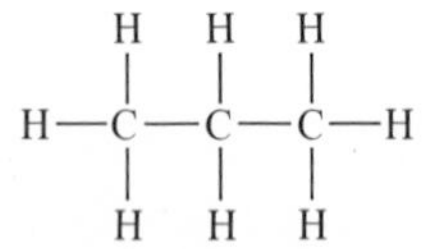

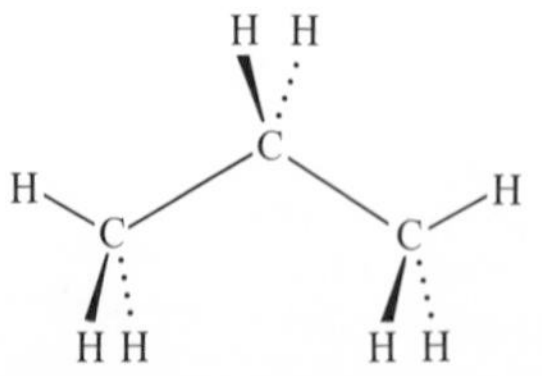

stereographic formula—dotted lines represent bonds directed behind the page; wedges represent bonds directed outward

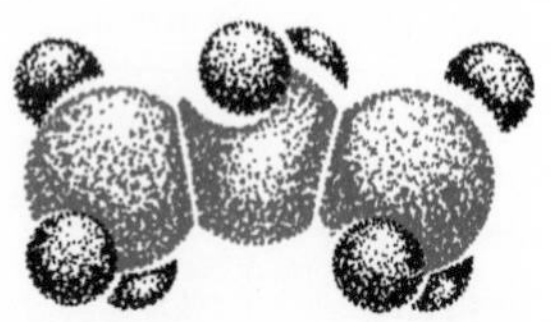

three-dimensional "model" portraying relative dimensions and positions of the atoms

Figure 1–6. **Alternative Representations of Propane**

1-11   Hydrocarbons

Compounds consisting of only carbon and hydrogen have the simplest compositions of all organic compounds. These compounds are called **hydrocarbons.** The hydrocarbons can be classified into four series, based on characteristic structures of the molecules in each series. These series are known as (1) the alkane series, (2) the alkene series, (3) the alkyne series, and (4) the aromatic series. Each series has many subdivisions, and there are molecules which could be classified as belonging to more than one series.

The **alkane series** is also called the **saturated hydrocarbon series** because the molecules of this class have carbon atoms connected by single bonds only. The noncyclic alkanes have the maximum number of hydrogen atoms possible for the number of carbon atoms. These substances may be represented by the general formula C_nH_{2n+2}, and molecules of successive members of the series differ from each other only by a CH_2 unit. The formulas and names of the first ten members of the series, given in Table 1–8, should be learned because these names form the bases for the names of many other organic compounds. It should be noted that the characteristic ending of these names is *ane.* Except for the first four, the names of each of the members of the series begins with a prefix based on the Greek name of the number which indicates the number of carbon atoms in the compound. Names of many other organic compounds are derived from these names by dropping the *ane* ending and adding other endings (see Sections 12–3 and 12–4).

TABLE 1-8. Alkane Series of Continuous Chain Hydrocarbons

The members of the alkane series are constituents of the petroleums, derived from the distillation of crude oil. The first four members of the series are constituents of natural gas. The next four members, including the branched chain molecules with the same molecular formulas, are the constituents of gasoline. Then come, successively, the constituents of kerosine, lubricating oils, Vaseline, and paraffin wax. The compounds are listed according to the boiling points and complexities of the carbon chains.

Formula	Name	Melting Point (°C)	Boiling Point (°C) at 1 atm pressure[a]
CH_4	methane	−182	−161
C_2H_6	ethane	−183	−88
C_3H_8	propane	−190	−42
C_4H_{10}	butane	−135	1
C_5H_{12}	pentane	−131	36
C_6H_{14}	hexane	−94	69
C_7H_{16}	heptane	−90	98
C_8H_{18}	octane	−57	125
C_9H_{20}	nonane	−51	150
$C_{10}H_{22}$	decane	−32	174
$C_{11}H_{24}$	undecane	−27	197
$C_{12}H_{26}$	dodecane	−12	216
$C_{13}H_{28}$	tridecane	−6	234
$C_{14}H_{30}$	tetradecane	6	252
$C_{24}H_{50}$	tetracosane	51	391
$C_{35}H_{72}$	pentatriacontane	75	490
$C_{60}H_{122}$	hexacontane	102	

[a] Pressure is discussed in Section 2–12.

The saturated hydrocarbons are also known as the **paraffins.** The series can be extended indefinitely. Plastics such as polyethylene (see Section 23–10) contain molecules with up to hundreds of thousands of carbon atoms. In addition, it should be noted that if a molecule contains more than three carbon atoms, branched chain molecules and continuous chain molecules having the same molecular formula are possible. Different compounds having the same molecular formula are called **isomers** of one another. For example, it is possible to write formulas for three isomeric compounds having the molecular formula C_5H_{12}. In each compound each carbon atom forms four bonds to other carbon atoms and hydrogen atoms. The structural formulas are

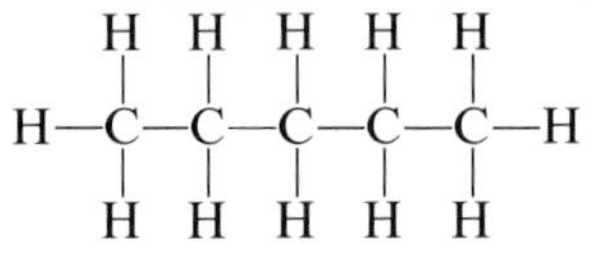

pentane

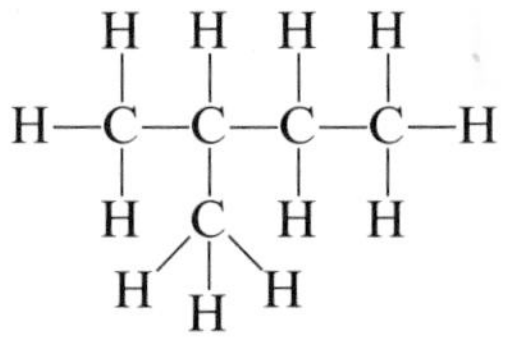

"isopentane"
methylbutane

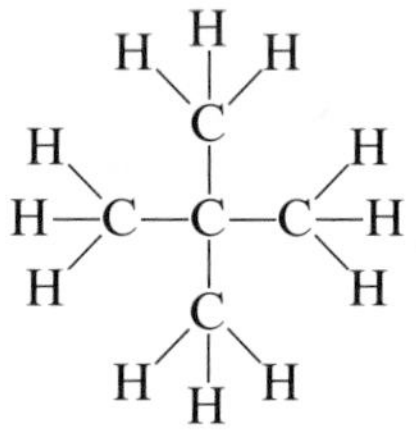

"neopentane"
dimethylpropane

However, graphic formulas such as these do not represent the three-dimensional structures of molecules. These formulas simply show which atoms are connected to each other; they do not show the actual shape of the molecule. The identity of a hydrocarbon is determined by the number of carbon atoms in the longest continuous chain in the molecule and by the position and length of each branch. Formulas represent the same molecules if they have the same number of each kind of atom and (1) the same longest continuous chain of carbon atoms, and (2) the same branches (3) at the same places. If any of the factors (1), (2), or (3) does not hold true, the formulas represent isomers. There are only three compounds with the formula C_5H_{12}. Formulas such as

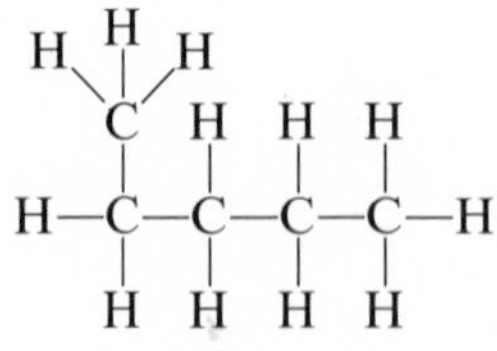

and

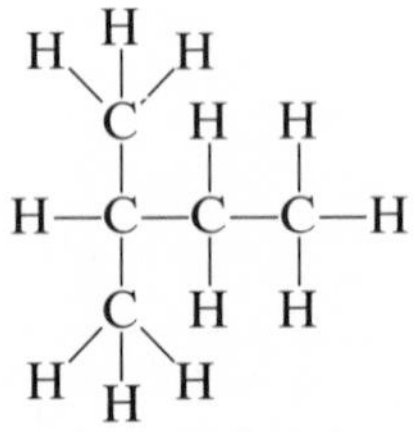

which might be imagined to be different compounds from those listed above, are actually pentane and isopentane, respectively. The different two-dimensional arrangements might be interpreted to be different conformations of the same molecule. A **conformation** is a particular nonpermanent geometric arrangement of the atoms and bonds in a molecule. A molecule in one conformation can be transformed into another conformation by rotation of part(s) of the molecule about one or more single bonds. Models of a molecule in different conformations are shown in Figure 1–7.

Figure 1-7. **A Molecule in Different Conformations**

Example

Which of the following formulas represent different molecules?

The first three are formulas of the same compound. Each has a five-carbon chain with a branch at the second carbon atom from one end. Only the last formula represents a different molecule, since the branch is on the third carbon atom.

Hydrocarbons which contain a carbon to carbon double bond are called **alkenes.** The molecules ethylene and propylene are the first two members of the alkene series.

ethylene   propylene

Hydrocarbons containing triple bonds are known as **alkynes.** Ethyne (popularly called acetylene), propyne, butyne, and pentyne are the first four members of this series.

$$H:C:::C:H \quad \text{or} \quad H—C≡C—H$$

acetylene

Because they have fewer hydrogen atoms than the corresponding alkanes, alkenes and alkynes are said to be **unsaturated.** In general, they are more reactive than the saturated hydrocarbons.

When unsaturated hydrocarbon molecules contain more than three carbon atoms, isomerism due to the position of the multiple bond is possible. For example, there are two isomeric molecules of butyne, $HC{\equiv}C{-}CH_2{-}CH_3$ and $H_3C{-}C{\equiv}C{-}CH_3$, called 1-butyne and 2-butyne, respectively. Details of organic nomenclature will be presented in Section 12–3.

Molecules containing more than one multiple bond are also possible. Butadiene, $H_2C{=}CH{-}CH{=}CH_2$, is such a compound. It and its derivatives are important starting materials in the manufacture of synthetic rubber.

There are also organic molecules in which the carbon atoms are linked together in a ring structure. These **cyclic hydrocarbons** are designated by the prefix *cyclo* in their names. The compound cyclopropane, a saturated hydrocarbon with three carbon atoms, has the following structure:

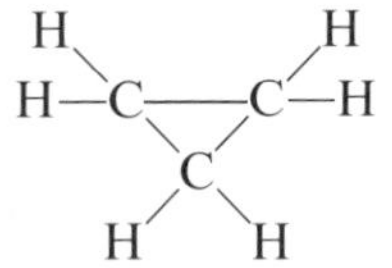

It should be noted that cyclopropane is isomeric with the unsaturated compound propylene. The formula for both compounds is C_3H_6.

The hydrocarbons discussed thus far are often referred to as **aliphatic.** Aliphatic hydrocarbons include the alkanes, the cycloalkanes, the alkenes, and the alkynes.

A special class of cyclic hydrocarbons, the **aromatic** hydrocarbons, are characterized by six-carbon rings, each atom of the ring being bonded to only one atom outside the ring. Two examples are benzene and naphthalene. The simplest electronic structure which can be drawn for these molecules, in which each carbon atom shares four electron pairs with other atoms, has alternating single and double bonds.

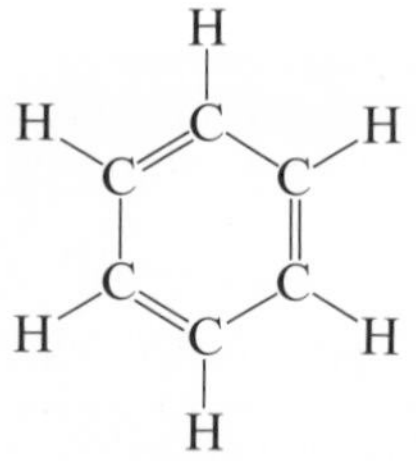

benzene

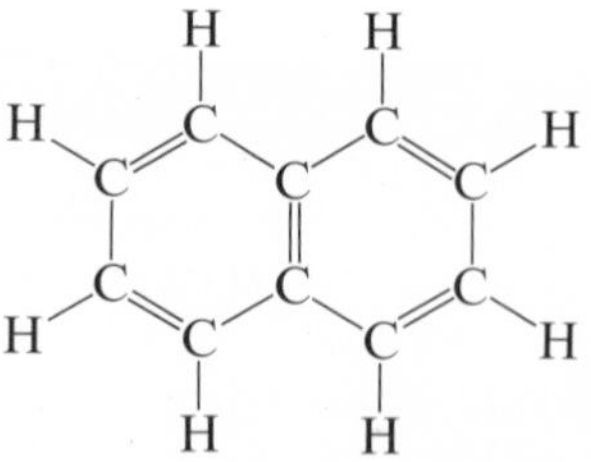

naphthalene

However, this formulation is not entirely satisfactory because these molecules have a much lower reactivity than that expected for molecules with double bonds. Indeed, in some ways they behave more like the saturated hydrocarbons. Theories explaining this behavior are introduced in Section 11–10. Compounds with both aromatic and aliphatic parts, such as ethylbenzene, are considered aromatic.

A hydrogen atom of a hydrocarbon may be replaced by another atom or by a group of atoms, giving a "derived" compound. In these new compounds the hydrocarbon portion is called a **radical,** and the added atom or group of atoms is called the **functional group.** The names of radicals derived from the paraffins come from the names of the parent hydrocarbons, the *ane* ending being replaced by the ending *yl.* For example, $CH_3{-}$ is the methyl radical and $C_2H_5{-}$ is the ethyl radical. The radical derived from benzene, C_6H_5—, is called the phenyl radical. More will be said about radicals and functional groups in later chapters. For the present it is useful to know that this is the basis for naming certain organic compounds.

Example

In organic compounds the functional group —OH is characteristic of alcohols. Write structures and common names for four alcohols derived from the first *three* paraffin hydrocarbons. Which are isomers?

methyl alcohol   ethyl alcohol   propyl alcohol   isopropyl alcohol

The last two are isomers of each other.

1-12   Exercises

Basic Exercises

1. For each pair of materials listed, state several properties which can be used to distinguish between the two: **(a)** steel and aluminum, **(b)** water and a solution of sodium chloride in water, **(c)** water and motor oil, **(d)** FeS and a mixture of iron and sulfur.

2. State the number of protons, neutrons, and electrons in an atom of **(a)** ^{235}U, **(b)** ^{90}Sr, **(c)** D, **(d)** ^{34}S.

3. Define each of the following terms: **(a)** element, **(b)** molecule, **(c)** polyatomic ion, **(d)** electrolyte, **(e)** mass number, **(f)** unsaturated hydrocarbon, **(g)** monatomic molecule.

4. **(a)** List the elements whose symbols start with a letter that is different from the first letter of the name of the element. **(b)** Write the symbol for the element copper. **(c)** Give the group of the periodic table for each of the elements listed in **(a)** and **(b)**.

5. The density of platinum is 21.45 grams/cm³. Calculate its density in kilograms per cubic meter.

6. Distinguish between the terms *isotope* and *isomer*.

7. Fool's gold is so called because it bears a visual similarity to real gold. A block of fool's gold which measures 1.50 cm by 2.50 cm by 3.00 cm has a mass of 56.25 grams. How can this material be distinguished from real gold by means of its physical properties?

8. With the assistance of Table 1–4, if necessary, write formulas for **(a)** ammonium dichromate, **(b)** lead(II) chlorate, **(c)** sodium acetate, **(d)** copper(I) oxide, **(e)** barium arsenate, **(f)** silver sulfate.

9. With the assistance of Table 1–4, if necessary, name the following compounds: **(a)** $Cu(ClO_2)_2$, **(b)** $(NH_4)_2C_2O_4$, **(c)** Hg_2Cl_2, **(d)** $Al_2(CO_3)_3$, **(e)** $Ba(CN)_2$.

10. What is the maximum number of other atoms to which an atom of each of the following can be bonded in organic compounds? **(a)** hydrogen, **(b)** carbon, **(c)** nitrogen, **(d)** oxygen, **(e)** chlorine.

11. For each of the following pairs of elements, write the formula for and name the binary compound which they form, and state whether the bonding in each compound will be

ionic or covalent: **(a)** potassium and phosphorus, **(b)** carbon and fluorine, **(c)** hydrogen and sulfur, **(d)** potassium and hydrogen, **(e)** fluorine and nitrogen.

12. The density of a salt solution is 1.13 grams/cm^3. The solution contains 17.0% sodium chloride. What volume of solution will contain 35.0 grams of NaCl?

13. Write electron dot structures for **(a)** CH_2Cl_2, **(b)** C_2Cl_4, **(c)** H_2O, **(d)** CCl_4, **(e)** NaCl, **(f)** both isomers of $C_2H_4Cl_2$.

14. Draw an electron dot diagram for **(a)** fluoride ion, **(b)** calcium ion, **(c)** calcium fluoride, **(d)** fluorine gas.

15. Write structural formulas for all the compounds which have the molecular formula C_5H_{10}.

16. Draw electron dot diagrams for **(a)** SO_3, **(b)** SO_3^{2-}, **(c)** Na_2SO_3, **(d)** H_2SO_3. **(e)** Name each of these species.

17. Write the formula of a binary compound of fluorine with each main group element in the fourth period of the periodic table.

18. **(a)** How many isomers are there corresponding to the formula $C_4H_{10}O$? **(b)** Secure a set of "ball and stick" models and construct a model of each.

19. Complete the following table.

Isotope of Element	Atomic Number	Number of Protons	Number of Neutrons	Mass Number
^{13}C				
	17		18	
		26		56
			2	3
	52			128
		50	70	

20. Tabulate the symbols and the values of the atomic number, mass number, number of protons, number of neutrons, and the number of electrons for **(a)** $^{23}Na^+$, **(b)** a species with two electrons, three protons, and four neutrons, **(c)** a species with atomic number 92 and with 88 electrons and 146 neutrons.

21. Indicate whether each of the following statements is true for the element in its uncombined (elemental) state only, in its compounds only, or both. **(a)** The halogens occur in diatomic molecules—two like atoms together. **(b)** The metallic elements conduct electricity well. **(c)** Sulfur occurs naturally as a mixture of isotopes. **(d)** The sodium ion is always monopositive. **(e)** Nitrogen atoms are capable of forming covalent bonds.

22. Draw structural formulas for all compounds with the molecular formula $C_5H_{11}Cl$.

23. Pure liquid H_2SO_4 solidifies below 10.4°C. Neither the pure liquid nor the solid conducts electricity; however, aqueous solutions of H_2SO_4 conduct electricity well. Solid Na_2SO_4, which melts at 884°C, does not conduct electricity, but molten Na_2SO_4 as well as aqueous solutions of Na_2SO_4 conduct electricity well. Explain the difference in properties between pure Na_2SO_4 and H_2SO_4 in terms of the model of electrolytes discussed in this chapter.

24. The density of a metal is 9.50 grams/cm^3. Calculate the number of **(a)** kilograms/meter^3, **(b)** cm^3/gram.

25. A solid has a volume of 1.23 cm^3. Its mass plus that of a piece of weighing paper is 10.024 grams; the paper weighs 0.03 gram. Calculate the density of the solid to the proper number of significant digits.

26. Name the following compounds: **(a)** $CuSO_4$, **(b)** $Fe(NO_3)_3$, **(c)** FeI_2, **(d)** $Sr(ClO_3)_2$,

(e) $K_2Cr_2O_7$,   (f) Hg_2Cl_2,   (g) $(NH_4)_2SO_3$,   (h) PCl_3,   (i) SO_3,   (j) CCl_4,   (k) H_2O, (l) NO_2.

27. In which of the following compounds is the bonding essentially ionic, in which is the bonding essentially covalent, and in which are both types of bonding represented? (a) PCl_3,   (b) $(NH_4)_2S$,   (c) $Ba(CN)_2$,   (d) $NaBr$,   (e) CH_3CH_2OH.

28. Construct a table comparing metals with nonmetals in terms of (a) the sign of the charges possible on monatomic ions, (b) the possibility of reaction with other elements of the same class, (c) the range of possible numbers of valence electrons, and (d) the ability of the elements to conduct electricity in the elementary state.

29. Using Table 1–4 and the periodic table, note that for the main group elements, if the periodic group of the element is even, the monatomic ion and the oxyanions of the element have charges that are even numbers, while if the periodic group number is odd, the respective charges are odd. Is this generalization true for elements which are not in a main group of the periodic table? Point out the exceptions in Table 1–4.

30. Which main group element(s) has (have) a different number of outermost electrons than its (their) group number?

31. A 10 gram sample of material A was placed in water, whereupon 4 grams of it, B, dissolved. The remaining material, C, was placed in a second sample of water, but no change took place. The sample of B had a sharp melting point, and after melting was electrolyzed to yield 1.5 grams of D and 2.5 grams of E, neither of which could be further broken down. When the sample of C was heated in air, it reacted completely to give 22 grams of a gas, F. Upon cooling to $-100°C$, material F solidified, and upon warming it sublimed at precisely $-78°C$. Identify each of the lettered materials as either definitely or probably an element, a compound, or a mixture.

32. Write electron dot diagrams for chlorine, Cl_2, and calcium chloride, $CaCl_2$. Explain the difference between the method of satisfying the octet of chlorine in the two cases.

33. Referring to Appendix A–5, if necessary, determine the answers to the following arithmetic operations to the proper number of significant digits:
    (a) $(1.0042 - 0.0034)(1.23) =$
    (b) $(1.0042)(0.0034)(1.23) =$
    (c) $(1.0042)(-0.0034)/1.23 =$

34. Explain why it is necessary to use the prefixes listed in Table 1–5 more often for covalent compounds than for ionic compounds.

35. When 10.0 grams of A was heated, 4.4 grams of B was given off, leaving 5.6 grams of E. The same quantity of B can also be prepared by combination of 1.2 grams of C and 3.2 grams of D. The E can be electrolyzed after melting to yield 4.0 grams of F and 1.6 grams of G, neither of which can be further decomposed by ordinary chemical means. E combines with water to give 1.3 grams of J per gram of E. A combination of E with 1.5 times its mass of water yields a homogeneous material, L. Identify as far as possible each lettered material as element, compound, or mixture. If definite identification cannot be achieved, state what probabilities exist and what possibilities cannot exist.

36. Explain why the alkene and the cycloalkane series of hydrocarbons both have the same general formula, C_nH_{2n}. Explain why the dienes and the alkynes have the same general formula, C_nH_{2n-2}. What is the number of carbon atoms in the smallest member of each group?

General Exercises

37. Using the data of Table 1–4 and the periodic table as guides, if necessary, write formulas for the following compounds: (a) ammonium perrhenate, (b) lithium selenate, (c) copper(I) arsenide, (d) strontium iodate.

38. Write all possible structural formulas for compounds with the molecular formula $C_5H_{10}I_2$.

39. In its elementary form, oxygen exists as diatomic molecules. Look up the formulas of all the other nonmetals in their elementary forms in a reference book such as the *Handbook of Chemistry and Physics* (CRC Press, Cleveland, Ohio). Which, if any, occur(s) as monatomic molecules?

40. Explain why in calculations involving more than one arithmetic operation, rounding off to the proper number of significant figures may be done once at the end if all the operations are multiplications and/or divisions or if they are all additions and/or subtractions, but not if they are combinations of additions or subtractions with multiplications or divisions.

41. Draw a two-dimensional formula for each of the conformations in Figure 1–7.

42. Draw electron dot structures for (a) CH_3OCH_3, (b) CH_3CH_2OH, (c) CH_3COCH_3, (d) C_2H_4, (e) C_2H_2, (f) HCO_2H.

43. Write line formulas for

(a) $CH_3\overset{\displaystyle O}{\underset{\displaystyle \|}{C}}-O-CH_3$,   (b) $CH_3\overset{\displaystyle O}{\underset{\displaystyle \|}{C}}-NH_2$,   (c) $\overset{\displaystyle CH_3}{\underset{\displaystyle CH_3}{>}}CHCH_3$,   (d) $CH_3\overset{\displaystyle O}{\underset{\displaystyle \|}{C}}-CH_3$.

44. Write the structural formulas for all isomers of the aromatic compounds having the molecular formula $C_6H_4Cl_2$. (Four such formulas may be written despite the fact that only three such isomers exist. This disparity will be explained in Section 11–10.)

45. Write structural formulas for all the isomers of cyclobutane.

Advanced Exercises

46. Look up and tabulate the following properties of silicon, tin, gallium, and arsenic: color, density, melting point, ability to conduct electricity. Write the formula for the chloride of each element. Look up and tabulate the density, melting point, and boiling point of each of these compounds. Using these data, predict analogous properties of the element germanium and those of the compound of germanium with chlorine.

47. Write electron dot formulas for each of the following substances, none of which follows the octet rule: (a) BF_3, (b) PF_5, (c) ICl_3, (d) SF_6.

48. A certain homogeneous material has a melting point of 94.6°C. When 10 grams of the material is placed in 20 ml of water, only 2 grams of the material dissolves. Suggest at least two further experiments which could be used to determine whether the material is a mixture or a pure substance.

49. Secure a set of "ball and stick" models, and construct models of methane, ethane, methyl alcohol, dimethyl ether (CH_3OCH_3), ethyl alcohol, and water. (a) Which one(s) of these substances can be considered as "derived" from methane? (b) Which one(s) can be considered as being "derived" from water? (c) Which ones are isomers of each other? (d) Which pairs are in the same class of organic compounds?

50. The statement is often made in elementary texts that all nitrates, acetates, and chlorates are soluble, as well as all sodium and potassium salts. Using a table of solubilities, such as may be found in the *Handbook of Chemistry and Physics* (CRC Press, Cleveland, Ohio), find at least one exception to this statement. Is the example a common chemical?

2
Stoichiometry

One of the fundamental laws of nature is the law of conservation of mass: *Mass can be neither created nor destroyed.* Applied to a chemical reaction, this law requires that the total mass of the products formed must be equal to the total mass of the substances which reacted. It is absolutely imperative to understand the relationship between the quantities of reactants and products, because it is useful in (1) predicting the quantities of reacting materials required to produce a given quantity of products, (2) interpreting the results of a chemical analysis, and (3) choosing the most economical method of carrying out a chemical reaction on a commercial scale. It is also necessary to be able to determine which one of a set of reactants is in excess, because in some cases, depending on the relative quantities used, different products may be produced from the same reactants. For example, when carbon reacts with an excess of oxygen, the compound carbon dioxide is formed, but if there is a limited supply of oxygen, the combination of these two elements produces carbon monoxide. Breathing the latter compound can be fatal; a number of deaths have resulted from burning coal or charcoal in poorly ventilated rooms or inside closed vehicles.

That aspect of chemistry which deals with mass relationships in chemical reactions is called **stoichiometry.** In addition to masses, the quantities of energy involved, whether as heat, light, or electricity, and the volumes of gaseous reactants or products, if any, can all be related to the quantity of chemical change. Stoichiometry is based on three concepts:

1. Conservation of mass.
2. The relative masses of atoms—the atomic weight scale.
3. The concept of the mole.

Atomic weights provide a means of expressing the relative masses of the elements in a compound. The concept of a mole permits expressions of quantity which depend on the *numbers* of atoms or molecules rather than on their masses. The mole concept is useful in describing all aspects of chemical reactions, and it must be learned thoroughly.

2-1 Dalton's Atomic Model

In the eighteenth century chemists demonstrated that in a pure compound, the relative masses of the constituent elements were always the same, regardless of the source of the compound or of the quantity taken for analysis. The fact became known as the **Law of Definite Proportions** or the Law of Constant Composition. The law provided a way of identifying compounds and of distinguishing between different compounds which happened to have the same constituent elements.

Around 1805, John Dalton proposed an explanation of the Law of Definite Proportions in terms of a simple model as follows:

1. The ultimate particles of any element are atoms, and all atoms of a given element are identical.
2. In particular all atoms of a given element have the same mass, and atoms of different elements have different masses.
3. Only whole atoms are involved in chemical processes; compounds are composed of whole atoms in definite ratios.

This model neatly explained the Law of Definite Proportions. Since every compound is made of atoms combined in a definite ratio, and each kind of atom has a characteristic mass, the masses of each element in the compound must be in a fixed ratio.

The model also predicted a second law which became known as the **Law of Multiple Proportions:** When the same two elements combine to form more than one compound, the masses of one element which combine with a fixed mass of the other in all the compounds are in the ratios of small whole numbers. This law is illustrated in Table 2–1 for the case of three compounds of lead and oxygen. It is seen that the masses of oxygen which combine with a fixed mass of lead in the three compounds are simple multiples of a common mass. This result is in accordance with the atomic model because in the three compounds, different numbers of atoms of one element combine with a fixed number of atoms of the other.

Dalton's model had all the qualities of a true scientific theory. It explained observed phenomena. It predicted new results which then could be confirmed by experiment. Most important, it stimulated further investigations using guidelines established by the theory. Dalton's original theory has been amended by subsequent discoveries. For example, as described in Section 1–3, it is now known that all atoms of any given element are not identical.

Dalton used pictorial symbols to represent formulas of compounds. He chose formulas which were as simple as possible and which, in some cases, turned out to be incorrect. Nevertheless, important questions were raised. What is the nature of atoms? Why do atoms combine with other atoms in certain ratios only? How does the way atoms combine influence the structure of compounds? Several generations of chemists have since been concerned with such questions.

2-2   The Atomic Weight Scale

Atoms are extremely small. The radius of a single atom is about 10^{-10} meter.[1] The lightest atom has a mass about 1.6×10^{-24} gram, and the mass of the heaviest atom is only about 250 times as great. It is not convenient to measure such small masses directly, and a scale of relative masses, called the **atomic weight scale,** has been devised.

---

[1] For a description of exponential notation, see Appendix A–6.

TABLE 2-1. Illustration of the Law of Multiple Proportions

Name of Compound	Percent Lead	Percent Oxygen	Mass of oxygen / Mass of lead
litharge	92.83	7.17	$0.0772 = 3 \times 0.0257$
lead dioxide	86.62	13.38	$0.1545 = 6 \times 0.0257$
red lead oxide	90.67	9.33	$0.1029 = 4 \times 0.0257$

The **atomic weight** of an element is the mass of a naturally occurring mixture of isotopic atoms of the element relative to the mass of the same number of atoms of some element chosen as a standard. For many years, chemists used the naturally occurring mixture of isotopes of oxygen as the standard, with its atomic weight defined as exactly 16. The atomic weights of all other elements were determined relative to that standard.

Since 1961, by international agreement, the **atomic mass unit** (amu)—recently given the name **dalton** (D)—has been defined as $\frac{1}{12}$ the mass of the carbon isotope ^{12}C. On this scale, the lightest atom, hydrogen, is approximately 1 atomic mass unit. A table of atomic weights based on this scale may be found inside the back cover of this book. It is worth noting that on the 1961 scale, oxygen, the former standard, has an atomic weight of 15.9994 D; for most purposes the difference from its previous value of exactly 16 D is insignificant.

One need not memorize the values of atomic weights. When used in calculations, the listed values of the atomic weights may be rounded off to an appropriate number of significant figures.[2]

It is important to distinguish between the terms atomic number, mass number, and atomic weight. The first two terms are whole numbers (see Section 1–3). The atomic weight is a relative mass. For a given element, a mass number and the atomic weight may be of the same numerical magnitude, but the mass number refers to a specific isotope while the atomic weight is a weighted average of the masses of all of the isotopes of the element.

The ^{12}C scale of atomic weights was made possible by the development of the **mass spectrometer**, by means of which the relative masses of atoms are precisely determined and the relative abundances of the isotopes of a given element are also obtained. The operation of a mass spectrometer is diagrammed and described in Figure 9–7.

When the relative abundances of the isotopes of a given element and their precise masses are known relative to ^{12}C, the atomic weight of the element may be calculated.

Example

Carbon occurs in nature as a mixture of atoms of which 98.89% have a mass of 12.0000 D and 1.11% have a mass of 13.00335 D. Calculate the atomic weight of carbon.

atomic weight = $(12.0000 \text{ D} \times 0.9889) + (13.00335 \text{ D} \times 0.0111) = 12.011 \text{ D}$

[2]See Appendix A–5 for a discussion of significant figures.

2-3   Formulas and Percent Composition

All compounds may be represented by formulas. Compounds which exist as molecules have formulas which state the number of each kind of atom in the molecule. For example, each molecule of the sugar glucose contains six carbon atoms, twelve hydrogen atoms, and six oxygen atoms. Therefore the formula for glucose is $C_6H_{12}O_6$. Ionic compounds, such as sodium chloride, do not exist as molecules; hence their formulas represent the ratio(s) of each kind of atom present. The collection of atoms represented by the formula is called the **formula unit.**

The masses of molecules or formula units relative to ^{12}C are expressed by their **molecular weights** or **formula weights,** respectively. These are determined by simply adding the atomic weights of each (not merely each kind of) atom in the molecule or formula unit.

Example

What is the formula weight of Na_2S?

$$2 \times \text{atomic weight of Na} = 45.98 \text{ D}$$
$$\text{atomic weight of S} = 32.06 \text{ D}$$
$$\text{formula weight} = \text{total} = 78.04 \text{ D}$$

Example

What is the molecular weight of glucose, $C_6H_{12}O_6$?

$$6 \times \text{atomic weight of C} = 72.06 \text{ D}$$
$$12 \times \text{atomic weight of H} = 12.10 \text{ D}$$
$$6 \times \text{atomic weight of O} = 96.00 \text{ D}$$
$$\text{molecular weight} = \text{total} = 180.16 \text{ D}$$

Some elements exist as polyatomic molecules when they occur in the free state, as, for example, H_2, N_2, O_2, F_2, Cl_2, Br_2, I_2, P_4, and S_8. The formula weights of these elements differ from their atomic weights.

Knowledge of the atomic weights of the elements allows calculation of the percent composition of the elements in any compound whose formula is known. One merely needs to divide the total mass of the atoms of each element in the formula by the total mass of all the atoms in the formula and to multiply by 100 to convert to percent.

Example

Calculate the percent composition of ethyl alcohol, CH_3CH_2OH.
Each molecule contains two carbon atoms, six hydrogen atoms, and one oxygen atom. Hence

$$\text{molecular weight} = (2 \times 12.01 \text{ D}) + (6 \times 1.008 \text{ D}) + (16.00 \text{ D}) = 46.07 \text{ D}$$

$$\% \text{ carbon} = \frac{24.02 \text{ D}}{46.07 \text{ D}} \times 100 = 52.14\% \text{ C}$$

$$\% \text{ hydrogen} = \frac{6.048 \text{ D}}{46.07 \text{ D}} \times 100 = 13.13\% \text{ H}$$

$$\% \text{ oxygen} = \frac{16.00 \text{ D}}{46.07 \text{ D}} \times 100 = 34.73\% \text{ O}$$

2-4   The Mole

Quantities may be expressed in terms of mass, volume, or number, as illustrated by such examples as a pound of butter, a quart of milk, or a dozen eggs. Atomic weights, molecular weights, and formula weights are chemical terms denoting mass. A chemical expression of number is the **mole.** It has been determined that a mole contains 6.02×10^{23} items. This number is called Avogadro's number, after the scientist who first proposed the principle upon which it is based. This number is the number of atoms in that quantity of an element which has a mass in grams numerically equal to the atomic weight of the element. Thus, since the atomic weight of hydrogen is 1.008 D, 1.008 grams of hydrogen contains 6.02×10^{23} atoms of hydrogen. Avogadro's number can be determined experimentally in several ways, two of which will be illustrated in exercise 60, Chapter 5, and in exercise 37, Chapter 18. One mole of a compound contains Avogadro's number of formula units of the compound. Since hydrogen molecules are diatomic, 2 moles of hydrogen atoms are present in each mole of hydrogen molecules. A mole of hydrogen gas will contain 6.02×10^{23} hydrogen molecules and will have a mass of 2.016 grams. For substances which do not exist as molecules, such as sodium chloride, a mole refers to Avogadro's number of formula units of the material. For example, a mole of NaCl contains 6.02×10^{23} sodium ions and the same number of chloride ions. The mass of a mole of NaCl is 58.443 grams.

A chemical formula may be used to refer to *one* atom, molecule, or formula unit or to a *mole* of atoms, molecules, or formula units.

The number of moles in a given mass of a pure substance is easily calculated by dividing the number of grams of substance by the number of grams per mole, that is, the atomic, molecular, or formula weight:

$$\text{number of moles} = \frac{\text{mass in grams}}{\text{number of grams/mole}}$$

Example

How many moles of nitrogen gas, N_2, are there in 35.7 grams of nitrogen?

$$\text{number of moles} = \frac{35.7 \text{ grams}}{28.0 \text{ grams/mole}} = 1.28 \text{ moles}$$

Example

How many hydrogen atoms are present in 0.235 gram of NH_3?

$$\text{number of moles of } NH_3 = \frac{0.235 \text{ gram } NH_3}{17.0 \text{ grams/mole}} = 0.0138 \text{ mole } NH_3$$

$$0.0138 \text{ mole } NH_3 \left(\frac{3 \text{ moles H}}{1 \text{ mole } NH_3}\right)\left(\frac{6.02 \times 10^{23} \text{ H atoms}}{\text{mole H}}\right) = 2.49 \times 10^{22} \text{ H atoms}$$

from the formula   Avogadro's number

In the laboratory it is sometimes convenient to use the unit millimole (mmole), corresponding to 0.001 mole or to 6.02×10^{20} items. The number of millimoles in a sample is the mass in milligrams (mg) of sample divided by the number of milligrams per millimole (which is numerically equal to the number of grams per mole).

Example

How many millimoles of iron are there in 0.500 gram (500 mg) of iron?

$$\text{number of millimoles} = \frac{500 \text{ mg Fe}}{55.85 \text{ mg Fe/mmole Fe}} = 8.95 \text{ mmole Fe}$$

2-5   Empirical Formulas

The results of the chemical analysis of a compound are usually expressed in terms of the percent by mass of each constituent element. If the identity of the compound is unknown, an empirical (simplest) formula may be deduced from its experimentally determined percentage composition. The empirical formula merely shows the relative numbers of atoms of each element in the compound. It may or may not correspond to the actual formula of a molecule of the compound.

The first step in determining the empirical formula of a compound is to determine the relative numbers of moles of atoms of each element present in a given mass of the compound. The ratio of numbers of moles of each element present is the same as the ratio of numbers of atoms of each element present.

Example

What is the empirical formula of a compound which contains 60.0% oxygen and 40.0% sulfur by mass?

Since no specific mass of sample has been given, one may choose any total mass. A 100 gram sample is a convenient choice, because then the mass of each of the elements is numerically equal to the percentage of that element in the compound. Then

$$\frac{60.0 \text{ grams O}}{16.0 \text{ grams O/mole O}} = 3.75 \text{ moles O}$$

$$\frac{40.0 \text{ grams S}}{32.0 \text{ grams S/mole S}} = 1.25 \text{ moles S}$$

Since formulas are always expressed in terms of whole numbers of atoms, the mole ratio must be expressed as a ratio of integers. Therefore to obtain an integral ratio, divide the numbers of moles of the elements present by the number present in smallest amount.

$$\frac{3.75 \text{ moles O}}{1.25 \text{ moles S}} = \frac{3.00 \text{ moles O}}{1.00 \text{ mole S}}$$

The ratio of 3 moles of oxygen atoms to 1 mole of sulfur atoms corresponds to the formula SO_3.

In actual practice, small experimental errors in the analysis may lead to results such that the ratio of atoms obtained is not precisely integral. In these cases the observed results are rounded off to give integers. However, in no case should the numbers be rounded off by more than a few percent.

Example

The insecticide DDT has the following composition by mass: 47.5% C, 2.54% H, and 50.0% Cl. Determine the empirical formula of DDT.

$$\frac{47.5 \text{ grams C}}{12.0 \text{ grams/mole}} = 3.95 \text{ moles C}$$

$$\frac{2.54 \text{ grams H}}{1.008 \text{ grams/mole}} = 2.52 \text{ moles H}$$

$$\frac{50.0 \text{ grams Cl}}{35.5 \text{ grams/mole}} = 1.41 \text{ moles Cl}$$

Chlorine is present in the smallest number of moles. Per mole of chlorine, there are

$$\frac{3.95 \text{ moles C}}{1.41 \text{ moles Cl}} = 2.80 \text{ moles C/mole Cl}$$

$$\frac{2.52 \text{ moles H}}{1.41 \text{ moles Cl}} = 1.79 \text{ moles H/mole Cl}$$

Analytical data are not likely to be 10% in error, therefore the ratio of 1.79/1 should not be merely rounded off to 2/1. Instead, the number of moles of each element per mole of Cl is multiplied by the same small integer to obtain nearly integral ratios:

$$\frac{\text{moles C}}{\text{moles Cl}} = \frac{2.80}{1.00} = \frac{2.80 \times 5}{1.00 \times 5} = \frac{14.0 \text{ moles C}}{5.00 \text{ moles Cl}}$$

$$\frac{\text{moles H}}{\text{moles Cl}} = \frac{1.79}{1.00} = \frac{1.79 \times 5}{1.00 \times 5} = \frac{8.95 \text{ moles H}}{5.00 \text{ moles Cl}}$$

The last ratio is close enough to round off to 9/5, and the resulting 14:9:5 ratio corresponds to the formula $C_{14}H_9Cl_5$.

2-6   Molecular Formulas

In contrast to the empirical formula, which gives only relative numbers of atoms of each element in a compound, a molecular formula gives the actual number of atoms of each element in a molecule. Of course, this information is sufficient to determine the relative numbers of atoms of each element, so molecular formulas yield all the information that empirical formulas give and more. Molecular formulas can be used to distinguish between different substances which have the same empirical formula.

For example, acetylene and benzene have the same empirical formula, CH. However, the molecular formula of acetylene is C_2H_2 and that of benzene is C_6H_6.

One cannot tell merely by inspection whether a formula is a molecular formula. Some formulas, such as Hg_2Cl_2 and $K_2C_2O_4$, are not molecular formulas even though they are multiples of empirical formulas. In these cases the substances are composed of ions rather than neutral molecules. The ions $C_2O_4^{2-}$ and Hg_2^{2+} happen to contain even numbers of each kind of atom.

When a substance does exist in the form of molecules, its molecular formula can be deduced from a knowledge of its empirical formula and its molecular weight. Molecular weights may be found by determining the number of moles present in a given mass of substance, since

$$MW = \frac{\text{number of grams}}{\text{number of moles}}$$

There are several ways of determining the number of moles in a given quantity of substance. For example, at a given temperature and pressure the volume occupied by a given quantity of a gas depends on the number of moles present (see Section 2–16). Other ways of determining molecular weights will be described in Section 16–19 and Figure 9–7.

Example

A compound consisting of 82.66% carbon and 17.34% hydrogen has a molecular weight 58.1 D. Determine its molecular formula.

The empirical formula is determined first:

$$\frac{82.66 \text{ grams C}}{12.01 \text{ grams/mole}} = 6.883 \text{ moles C}$$

$$\frac{17.34 \text{ grams H}}{1.008 \text{ grams/mole}} = 17.20 \text{ moles H}$$

$$\frac{17.20 \text{ moles H}}{6.883 \text{ moles C}} = \frac{2.50 \text{ moles H}}{1.00 \text{ mole C}} = \frac{5 \text{ moles H}}{2 \text{ moles C}}$$

The empirical formula is C_2H_5. This formula has a formula weight of 29.06 D. The number of empirical formula units per molecule is

$$\frac{58.1 \text{ D}}{29.06 \text{ D}} = 2 \text{ units}$$

The molecular formula is $(C_2H_5)_2$ or C_4H_{10}.

2-7   Equations

Chemical reactions are described in a shorthand notation called a chemical **equation** in which the number and kind of formula units of reactants and products are stated. Since no atoms can be created or destroyed in a chemical reaction, the numbers of atoms of each element on each side of the equation must be the same. The required number of each kind of atom is obtained by writing **coefficients** in front of the respective formulas. This procedure is called **balancing** the equation. For example, the equation which describes the reaction of zinc with hydrochloric acid to produce zinc chloride and hydrogen must have a coefficient 2 before the formula HCl:

$$\text{Zn} + 2\,\text{HCl} \rightarrow \text{ZnCl}_2 + \text{H}_2$$

Note that the formula of the HCl is not changed; the number of HCl units is merely adjusted.

The physical state of each of the reactants and products may also be indicated by use of an abbreviation in parentheses after the formula: (s) for solid, (l) for liquid, (g) for gas, and (aq) for aqueous solution. For example, the equation shown above may be rewritten

$$\text{Zn(s)} + 2\,\text{HCl(aq)} \rightarrow \text{ZnCl}_2\text{(aq)} + \text{H}_2\text{(g)}$$

The balanced chemical equation states that for every mole of H_2 gas that is produced, 1 mole of solid zinc and 2 moles of HCl in aqueous solution are used up.

The coefficient before each substance in the balanced equation denotes the relative number of moles of that substance involved in the reaction. The mole ratios and formula weights permit calculation of the relative mass of each substance which reacts or is produced. For example, if it is desired to produce 10.00 grams of hydrogen gas by the above reaction, one may calculate the masses of reactants as follows:

$$10.00 \text{ grams } H_2 \left(\frac{1.000 \text{ mole } H_2}{2.016 \text{ grams } H_2} \right) = 4.960 \text{ moles } H_2$$

According to the balanced equation, each mole of H_2 produced requires 1 mole of Zn and 2 moles of HCl. Therefore, 4.960 moles of zinc and 9.920 moles of HCl are required to produce 4.960 moles of H_2. The mass of each reactant may be determined as shown here for the case of zinc:

$$4.960 \text{ moles Zn} \left(\frac{65.37 \text{ grams Zn}}{\text{mole Zn}} \right) = 324.2 \text{ grams Zn}$$

What mass of HCl would be required?

Example

The equation for the reaction of sucrose (sugar) with oxygen is as follows:

$$\text{C}_{12}\text{H}_{22}\text{O}_{11} + 12\,\text{O}_2 \rightarrow 12\,\text{CO}_2 + 11\,\text{H}_2\text{O}$$

How many grams of CO_2 is produced per gram of sucrose used? How many moles of oxygen gas is needed to react with 1.00 gram of sucrose?

Since the equation states that 12 moles of CO_2 is produced per mole of sucrose which reacts, the number of moles of sucrose in 1.00 gram of sucrose must be calculated.

$$\frac{1.00 \text{ gram sucrose}}{342 \text{ grams sucrose/mole}} = 0.00292 \text{ mole sucrose}$$

$$0.00292 \text{ mole sucrose} \left(\frac{12 \text{ moles } CO_2}{1 \text{ mole sucrose}}\right) = 0.0350 \text{ mole } CO_2$$

$$0.0350 \text{ mole } CO_2 \left(\frac{44.0 \text{ grams } CO_2}{\text{mole } CO_2}\right) = 1.54 \text{ grams } CO_2$$

According to the balanced chemical equation, for every mole of CO_2 produced 1 mole of O_2 is needed, and in this case 0.0350 mole of O_2 will be used up.

Chemical analyses are based on balanced chemical equations involving the substances to be analyzed. For example, to analyze a mixture of $CaCO_3$ and sand, the mixture may be treated with hydrochloric acid. The sand will not react, but the $CaCO_3$ will react according to the equation

$$CaCO_3 \;+\; 2\,HCl \;\rightarrow\; CO_2 \;+\; CaCl_2 \;+\; H_2O$$

The quantity of $CaCO_3$ in a weighed sample of the mixture can be obtained by measuring the carbon dioxide produced.

Example

Four grams of a mixture of calcium carbonate and sand is treated with an excess of hydrochloric acid, and 0.880 gram of CO_2 is produced. What is the percent of $CaCO_3$ in the original mixture?

$$\frac{0.880 \text{ gram } CO_2}{44.0 \text{ grams } CO_2/\text{mole}} = 0.0200 \text{ mole } CO_2$$

$$0.0200 \text{ mole } CO_2 \left(\frac{1 \text{ mole } CaCO_3}{1 \text{ mole } CO_2}\right) = 0.0200 \text{ mole } CaCO_3$$

$$0.0200 \text{ mole } CaCO_3 \left(\frac{100 \text{ grams } CaCO_3}{\text{mole } CaCO_3}\right) = 2.00 \text{ grams } CaCO_3$$

$$\% \, CaCO_3 \text{ in mixture} = \frac{2.00 \text{ grams } CaCO_3}{4.00 \text{ grams mixture}} \times 100 = 50.0\%$$

Example

A 5.00 gram sample of a natural gas, consisting of methane (CH_4) and ethylene (C_2H_4), was burned in excess oxygen, yielding 14.5 grams of CO_2 and some H_2O as products. What percent of the sample was ethylene?
The equations for the reactions are

$$C_2H_4 \;+\; 3\,O_2 \;\rightarrow\; 2\,CO_2 \;+\; 2\,H_2O$$
$$CH_4 \;+\; 2\,O_2 \;\rightarrow\; CO_2 \;+\; 2\,H_2O$$

The total number of moles of carbon dioxide formed can be calculated from the observed mass of carbon dioxide:

$$\frac{14.5 \text{ grams CO}_2}{44.0 \text{ grams CO}_2/\text{mole CO}_2} = 0.330 \text{ mole CO}_2$$

The total number of moles of CO_2 can be expressed in terms of the quantities of C_2H_4 and CH_4 which have reacted. Let x equal the number of grams of ethylene in the original mixture. Then $(5.00 - x)$ will be the number of grams of methane.

$$\text{moles CO}_2 \text{ from C}_2\text{H}_4 = \left(\frac{x \text{ grams C}_2\text{H}_4}{28.0 \text{ grams C}_2\text{H}_4/\text{mole}}\right)\left(\frac{2 \text{ moles CO}_2}{\text{mole C}_2\text{H}_4}\right) = \frac{2x}{28.0} \text{ mole CO}_2$$

$$\text{moles CO}_2 \text{ from CH}_4 = \left(\frac{(5.00 - x) \text{ grams CH}_4}{16.0 \text{ grams CH}_4/\text{mole}}\right)\left(\frac{1 \text{ mole CO}_2}{\text{mole CH}_4}\right) = \frac{5.00 - x}{16.0} \text{ mole CO}_2$$

The sum of these expressions must equal the number of moles calculated above. Hence

$$\frac{2x}{28.0} \text{ mole} + \frac{5.00 - x}{16.0} \text{ mole} = 0.330 \text{ mole}$$

Solving for x,

$$(0.0714x) + (0.312 - 0.0625x) = 0.330$$
$$x = 2.02 \text{ grams C}_2\text{H}_4$$

Then

$$5.00 - x = 2.98 \text{ grams CH}_4$$

and

$$\% \text{ C}_2\text{H}_4 = \frac{2.02 \text{ grams C}_2\text{H}_4}{5.00 \text{ grams total}} \times 100 = 40.4\%$$

2-8   Limiting Quantities

In an actual experiment, the relative quantities of reactants present may differ from those required by the balanced chemical equation. If the reaction mixture contains one of the reactants in greater quantity than is required by the equation, the excess reagent simply does not react. The quantities of the products obtained are determined by the reagent(s) not in excess.

Example

Calculate the mass of carbon tetrachloride which can be produced by the reaction of 10.0 grams of carbon with 100.0 grams of chlorine. Determine the mass of excess reagent which is left unreacted.

First the number of moles of each reactant is determined:

$$(10.0 \text{ grams C})\left(\frac{1 \text{ mole C}}{12.0 \text{ grams C}}\right) = 0.833 \text{ mole C}$$

$$(100.0 \text{ grams Cl}_2)\left(\frac{1 \text{ mole Cl}_2}{70.9 \text{ grams Cl}_2}\right) = 1.41 \text{ moles Cl}_2$$

From the balanced chemical equation

$$C + 2\,Cl_2 \rightarrow CCl_4$$

it is seen that each mole of C requires 2 moles of Cl_2.

$$(0.833 \text{ mole C}) \left(\frac{2 \text{ moles } Cl_2}{1 \text{ mole C}} \right) = 1.67 \text{ moles } Cl_2 \text{ required}$$

However, only 1.41 moles of Cl_2 is available; hence the carbon is in excess.

$$(1.41 \text{ moles } Cl_2) \left(\frac{1 \text{ mole C}}{2 \text{ moles } Cl_2} \right) = 0.705 \text{ mole C required}$$

The excess C is

$$0.833 - 0.705 = 0.128 \text{ mole C}$$

In this case Cl_2 is the **limiting reagent,** which determines the quantity of CCl_4 which can be produced:

$$(1.41 \text{ moles } Cl_2) \left(\frac{1 \text{ mole } CCl_4}{2 \text{ moles } Cl_2} \right) = 0.705 \text{ mole } CCl_4$$

$$(0.705 \text{ mole } CCl_4) \left(\frac{153.8 \text{ grams } CCl_4}{\text{mole } CCl_4} \right) = 108 \text{ grams } CCl_4$$

The mass of unreacted carbon is determined from the number of moles in excess:

$$(0.128 \text{ mole C}) \left(\frac{12.0 \text{ grams C}}{\text{mole C}} \right) = 1.54 \text{ grams C unreacted.}$$

2-9   Net Ionic Equations

Electrolytes in solution have properties characteristic of the ions which they contain. For example, a solution of sodium chloride has properties characteristic of sodium ions in solution and properties characteristic of chloride ions in solution. A solution of barium chloride has properties characteristic of barium ions in solution and properties characteristic of chloride ions in solution. Since aqueous solutions of sodium chloride and barium chloride both have properties of chloride ions, they both yield a precipitate of silver chloride when mixed with solutions of silver nitrate, because a characteristic reaction of chloride ions is the combination with silver ions to yield silver chloride, which is insoluble in water.

$$NaCl(aq) + AgNO_3(aq) \rightarrow AgCl(s) + NaNO_3(aq)$$
$$BaCl_2(aq) + 2\,AgNO_3(aq) \rightarrow 2\,AgCl(s) + Ba(NO_3)_2(aq)$$

When a solution of any soluble sulfate is added to a solution of barium chloride, a precipitate of barium sulfate will be produced.

$$BaCl_2(aq) + K_2SO_4(aq) \rightarrow BaSO_4(s) + 2\,KCl(aq)$$

On the other hand, adding a solution containing sulfate ion to a solution of sodium chloride will not produce a precipitate. The reaction with sulfate ion is a property of barium ions but not of sodium ions or of chloride ions.

The three preceding equations could have been written as follows:

$$Na^+ + Cl^- + Ag^+ + NO_3^- \rightarrow AgCl(s) + Na^+ + NO_3^-$$
$$Ba^{2+} + 2\,Cl^- + 2\,Ag^+ + 2\,NO_3^- \rightarrow 2\,AgCl(s) + Ba^{2+} + 2\,NO_3^-$$
$$Ba^{2+} + 2\,Cl^- + 2\,K^+ + SO_4^{2-} \rightarrow BaSO_4(s) + 2\,K^+ + 2\,Cl^-$$

The formulas of the ions which were not involved in the reaction appear on both sides of the equation. Since these ions were not changed by the reaction, they are called **spectator ions.** Recognizing that the substances whose formulas appear on both sides of the equation are not involved in the overall change, the three reactions described above can be represented by the following two equations:

$$Ag^+ + Cl^- \rightarrow AgCl(s)$$
$$Ba^{2+} + SO_4^{2-} \rightarrow BaSO_4(s)$$

Equations from which the formulas of the spectator ions have been omitted are called **net ionic equations.** From a knowledge of properties of a relatively few ions, the chemistry of a tremendous number of electrolytes can be predicted in terms of net ionic equations. Most reactions for which net ionic equations can be written are of the following types:

1. A reaction between ions in solution to produce a precipitate, or the reaction of an insoluble compound to produce ions. The formation of silver chloride described above is an example of the formation of a precipitate. An example of the formation of ions in solution is the reaction of insoluble MgO with hydrogen ion:

$$MgO(s) + 2\,H^+(aq) \rightarrow Mg^{2+}(aq) + H_2O$$

Other examples of this type of reaction may be predicted from information on the solubilities of electrolytes such as is given in Table 2–2.

2. A reaction which produces molecules containing covalent bonds. For example, the reaction of hydrogen ion with carbonate ion yields carbon dioxide and water:

$$H^+ + HCO_3^- \rightarrow CO_2 + H_2O$$

The reaction of hydrogen ion with hydroxide ion is a very important example of this type of reaction, the net ionic equation being

$$H^+ + OH^- \rightarrow H_2O$$

3. A reaction which involves changes in the charges on the ions involved. For example,

$$Ce^{4+} + Fe^{2+} \rightarrow Ce^{3+} + Fe^{3+}$$

Some reactions between ions may involve a combination of these situations.

TABLE 2-2. Solubilities[a] of Common Ionic Compounds in Water at Room Temperature

	ClO_3^- NO_3^- $C_2H_3O_2^-$	Cl^- Br^- I^-	SO_4^{2-}	CO_3^{2-} SO_3^{2-} PO_4^{3-} CrO_4^{2-} BO_3^{3-}	S^{2-}	OH^-	O^{2-}
Na^+, K^+, NH_4^+	s	s	s	s	s	s	d
Pb^{2+}	s	ss-i	i	i	i	i	i
Ag^+	s	i	ss	i	i	i	i
Hg_2^{2+}	s	i	i	i	i	i	i
Hg^{2+}	s	s-i[b]	s	i	i	i	i
Ba^{2+}	s	s	i	i	s	s	d
Ca^{2+}	s	s	i	i	s	s	d
Mg^{2+}	s	s	s	s	s	i	i

[a] s = soluble (greater than about 1 gram solute/100 grams of water).
ss = slightly soluble (approximately 0.1-1 gram solute/100 grams of water).
i = insoluble (less than about 0.1 gram solute/100 grams of water).
d = decomposes in water.
[b] $HgCl_2$ is soluble, $HgBr_2$ is less soluble, and HgI_2 is insoluble.

Example

With the aid of Table 2-2, write net ionic equations for the process which occurs when solutions of the following electrolytes are mixed: **(a)** $AgClO_3(aq)$ and $Na_2S(aq)$, **(b)** $(NH_4)_3PO_4(aq)$ and $HgSO_4(aq)$.

(a) Table 2-2 indicates that Ag_2S is insoluble while $NaClO_3$ is soluble.

$$2\,Ag^+ \;+\; 2\,\cancel{ClO_3^-} \;+\; 2\,\cancel{Na^+} \;+\; S^{2-} \;\rightarrow\; Ag_2S(s) \;+\; 2\,\cancel{Na^+} \;+\; 2\,\cancel{ClO_3^-}$$

Net equation: $2\,Ag^+ \;+\; S^{2-} \;\rightarrow\; Ag_2S(s)$

(b) $Hg_3(PO_4)_2$ is insoluble while $(NH_4)_2SO_4$ is soluble.

$$\cancel{6\,NH_4^+} + 2\,PO_4^{3-} + 3\,Hg^{2+} + \cancel{3\,SO_4^{2-}} \rightarrow \cancel{6\,NH_4^+} + \cancel{3\,SO_4^{2-}} + Hg_3(PO_4)_2(s)$$

Net equation: $3\,Hg^{2+} \;+\; 2\,PO_4^{3-} \;\rightarrow\; Hg_3(PO_4)_2(s)$

To ascertain that a particular net ionic reaction actually occurs it is necessary to perform the experiment or to predict the behavior of the ions theoretically. Methods of predicting the behavior of ions will be presented in later chapters, beginning with Chapter 4. For the present, it may be assumed that the reactions for which net ionic equations are written actually occur as described.

In making calculations using net ionic equations, quantities should be expressed in moles rather than in terms of units of mass. There is no reagent which consists of a single type of ion only. For example, one will never find a bottle containing only chloride ions. However, the mass of any electrolyte which contains the required number of moles of the desired ion may readily be calculated.

Example

Find the number of moles of chloride ion needed to react with sufficient silver nitrate to make 10.0 grams of AgCl. What mass of $CaCl_2$ is required to provide this number of moles of Cl^-?

The net ionic equation is

$$Ag^+ \;+\; Cl^- \;\rightarrow\; AgCl(s)$$

The number of moles of AgCl, which is equal to the number of moles of Cl^-, is given by

$$\text{moles AgCl desired} = \frac{10.0 \text{ grams}}{143.4 \text{ grams/mole}} = 0.0697 \text{ mole}$$

Hence 0.0697 mole of Cl^- is required. Since $CaCl_2$ contains 2 moles of chloride ion per mole of substance,

$$\text{moles } CaCl_2 \text{ required} = \frac{0.0697 \text{ mole } Cl^-}{2 \text{ moles } Cl^-/\text{mole } CaCl_2} = 0.0348 \text{ mole } CaCl_2$$

and

$$0.0348 \text{ mole } CaCl_2 \left(\frac{111 \text{ grams } CaCl_2}{\text{mole } CaCl_2}\right) = 3.86 \text{ grams } CaCl_2 \text{ required}$$

2-10   Stoichiometry in Solution—Molarity

Very many chemical reactions are carried out in solution. It is easy to measure volumes accurately, and when a known mass of reagent is dissolved in a given volume of solution, a measured fraction of the total volume will contain the corresponding fraction of the mass of the reagent. Solutions containing a known number of moles of substance per unit volume can be mixed in volumes such that the proper mole ratio required by the balanced chemical equation is provided. The **molarity** (M) of a solution is the number of moles of solute (substance dissolved) per liter of solution. It should be emphasized that the volume of the dissolving liquid (solvent) need not be specified; it is the volume of the resulting solution which is important in determining molarity. The nature of the solvent also need not be specified. Water is most often used, but other solvents can be employed.

Example

Calculate the molarity of a solution prepared by adding 100.0 grams of NaCl to sufficient water to make 1.00 liter of solution.

$$\frac{100.0 \text{ grams NaCl}}{58.5 \text{ grams/mole}} = 1.71 \text{ moles NaCl}$$

$$\frac{1.71 \text{ moles NaCl}}{1.00 \text{ liter solution}} = 1.71 \; M$$

The utility of the concept of molarity lies in the fact that the mole is the basic unit of chemical reaction, and it is very convenient to use solutions whose concentrations are expressed in moles per unit volume.

Example

What volume of 1.71 M NaCl solution contains 0.20 mole of NaCl?

$$\text{volume} = \frac{0.20 \text{ mole NaCl}}{1.71 \text{ moles/liter}} = 0.117 \text{ liter} = 117 \text{ ml}$$

Example

$$\text{Zn} + \text{H}_2\text{SO}_4 \rightarrow \text{ZnSO}_4 + \text{H}_2$$

What volume of 3.00 M H_2SO_4 is required to react with 10.0 grams of zinc?

$$\left(\frac{10.0 \text{ grams Zn}}{65.4 \text{ grams Zn/mole Zn}}\right)\left(\frac{1 \text{ mole H}_2\text{SO}_4}{1 \text{ mole Zn}}\right) = 0.153 \text{ mole H}_2\text{SO}_4$$

$$\frac{0.153 \text{ mole H}_2\text{SO}_4}{3.00 \text{ moles/liter}} = 0.0510 \text{ liter} = 51.0 \text{ ml}$$

It should be noted that the nature of the solute may be changed by the solution process. Thus H_2SO_4 exists as molecules when pure, but an aqueous solution of H_2SO_4 contains H^+, HSO_4^-, and SO_4^{2-} ions. In expressing a molarity, the specific solute should be given. If no explicit designation is made, the concentration of the whole substance is usually meant. Thus 1 M H_2SO_4 refers to a solution containing 1 mole of H_2SO_4 per liter of solution, without specifying its nature in solution.

Example

How many moles of chloride ion, Cl^-, is present in 50.0 ml of 0.200 M $BaCl_2$?

$$0.0500 \text{ liter} \left(\frac{0.200 \text{ mole BaCl}_2}{\text{liter}}\right) = 0.0100 \text{ mole BaCl}_2$$

$$0.0100 \text{ mole BaCl}_2 \left(\frac{2 \text{ moles Cl}^-}{\text{mole BaCl}_2}\right) = 0.0200 \text{ mole Cl}^-$$

The 0.200 M $BaCl_2$ solution is a solution of 0.200 M Ba^{2+} and 0.400 M Cl^-.

Upon dilution of a solution, the number of moles of solute does not change, but the concentration does change. The new concentration can be calculated easily from the definition of molarity.

Example

How many milliliters of a 0.100 M solution must be added to water to make 2.00 liters of 0.0250 M solution?

$$2.00 \text{ liters} \left(\frac{0.0250 \text{ mole}}{\text{liter}}\right) = 0.0500 \text{ mole required}$$

$$\frac{0.0500 \text{ mole required}}{0.100 \text{ mole/liter}} = 0.500 \text{ liter} = 500 \text{ ml}$$

Hence 500 ml of 0.100 M solution must be diluted to 2.00 liters.

Since milliliter quantities rather than liter quantities are used in most laboratory experiments, it is convenient to use the definition

$$\text{molarity} = \frac{\text{number of millimoles of solute}}{\text{number of milliliters of solution}}$$

Example

If 20.0 ml of 1.00 M CaCl$_2$ and 60.0 ml of 0.200 M CaCl$_2$ are mixed, what will be the molarity of the final solution?

The total number of millimoles of CaCl$_2$ is the sum of the number of millimoles in the two solutions:

$$20.0 \text{ ml} \left(\frac{1.00 \text{ mmole}}{\text{ml}}\right) + 60.0 \text{ ml} \left(\frac{0.200 \text{ mmole}}{\text{ml}}\right) = 20.0 \text{ mmole} + 12.0 \text{ mmole}$$

$$= 32.0 \text{ mmole}$$

The total volume of the final solution is 80.0 ml. Its molarity is therefore

$$\frac{32.0 \text{ mmole}}{80.0 \text{ ml}} = 0.400 \ M$$

2-11   Titration

Many chemical analyses are performed by **titration,** a procedure in which a reagent in a solution of known concentration, called the standard solution, is allowed to react with a sample containing an unknown quantity of substance to be analyzed. The volume of standard solution required to just react completely with the sample is measured, and the number of moles computed. From the number of moles of standard reagent used, the number of moles of the unknown in the sample may be calculated from the balanced chemical equation. Titrations are possible whenever the following conditions exist:

1. The standard reagent reacts completely with the unknown.
2. The reaction is reasonably fast.
3. The point at which exactly equivalent quantities of the reactants are present, the **equivalence point,** is readily detectable.

In titrations the standard solution is added gradually from a buret either to a solution of a weighed quantity of unknown material or to a precisely known volume of solution of the unknown (Figure 2–1). Addition is continued until the exact mole ratio required by the balanced chemical equation is reached. This point is called the equivalence point of the titration and is usually detected through the addition of an **indicator** to the solution being titrated. Indicators are substances which change color when the substance being added is present in very slight excess. For example, the dye phenolphthalein is colorless in the presence of excess hydrogen ion (acid) but turns

Figure 2-1. **Titration Apparatus**

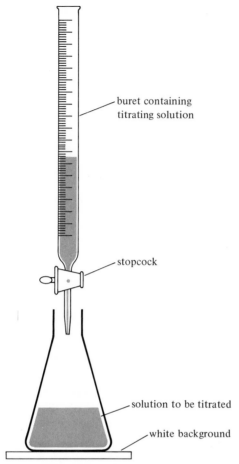

red in the presence of excess hydroxide ion (base); hence it can be used as an indicator in an acid-base titration.

As the equivalence point is approached in a titration, the reagent from the buret is added slowly, drop by drop, in order to ensure that there will be no large excess of reagent after all the unknown has reacted.

Example

Two drops of phenolphthalein solution was added to 40.00 ml of a hydrochloric acid solution, and this solution was titrated with 0.1000 M NaOH. When 30.00 ml of base had been added, part of the solution turned pink, but the color disappeared upon mixing the solution. Addition of NaOH solution was continued dropwise until a one drop addition produced a lasting pink color. At this point the volume of base added was 32.56 ml. What was the concentration of the HCl solution?

$$HCl \; + \; NaOH \; \rightarrow \; NaCl \; + \; H_2O$$

The number of millimoles of added base is

$$(32.56 \text{ ml})(0.1000 \text{ mmole/ml}) = 3.256 \text{ mmole NaOH}$$

Thus 3.256 mmole of HCl was originally present in the 40.00 ml of solution.

$$\frac{3.256 \text{ mmole}}{40.00 \text{ ml}} = 0.08140 \text{ mmole/ml} = 0.08140 \, M$$

The concentration of the HCl solution was 0.08140 M.

Example

The acidic substance in vinegar is acetic acid, $HC_2H_3O_2$. When 6.00 grams of a certain vinegar was titrated with 0.100 M NaOH, 40.11 ml of base had to be added to reach the equivalence point. What percent by mass of this sample of vinegar is acetic acid?

$$HC_2H_3O_2 \; + \; NaOH \; \rightarrow \; NaC_2H_3O_2 \; + \; H_2O$$

$$(40.11 \text{ ml})(0.100 \text{ mmole/ml}) = 4.01 \text{ mmole NaOH}$$

The same number of millimoles of acid must have been present, as is required by the balanced chemical equation.

$$(4.01 \text{ mmole } HC_2H_3O_2)(60.0 \text{ mg/mmole}) = 241 \text{ mg } HC_2H_3O_2$$

$$\frac{241 \text{ mg } HC_2H_3O_2}{6000 \text{ mg vinegar}} \times 100 = 4.01\% \; HC_2H_3O_2$$

Many types of analysis can be performed by means of titration. Additional examples will be described in Sections 4–8, 5–12, and 7–13.

Stoichiometry of Gases

Gases are involved in many chemical reactions. For example, oxygen, a gas, is one of the most important of all chemical reagents. Experimentally, it is not convenient to measure quantities of gas in the same manner that quantities of solids and liquids are measured, because any quantity of gas will occupy the entire volume of its container, regardless of the size of the container. Moreover, gases cannot be quantitatively transferred from one open vessel to another. However, in a container of a given size at a given temperature, the pressure exerted by a gas will depend on the quantity of that gas which is present. It will now be shown how the number of moles of a gas can be determined by measuring its volume, pressure, and temperature.

2-12 Pressure

Pressure is defined as force per unit area. Pressure is exerted by all fluids (liquids and gases) equally in all directions at a given point. In the SI system of measure-

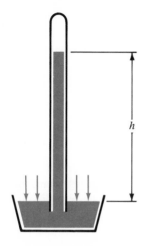

Figure 2-2. Simple Barometer

A glass tube about 1 meter long and closed at one end is completely filled with mercury. The tube is tightly stoppered and then inverted into a dish which contains mercury to a depth of 4–5 cm. The stopper is removed, the end of the tube being kept under the surface of the mercury in the dish. Some mercury runs out of the tube, but most remains in the tube, as shown.

Air pressure acting downward on the surface of the mercury in the dish is responsible for the mercury remaining in the tube instead of falling to the level in the dish under the influence of its own weight. The space above the column of mercury in the tube is under nearly perfect vacuum. Such a device is a crude barometer. The vertical height of the column of mercury held up in the tube, h, will vary as the pressure of the atmosphere varies.

ment,[3] the unit of pressure is the pascal (Pa); in the older metric system, the unit of pressure is dynes/cm². In chemical work, the most widely used measure of gas pressure is the vertical height of a mercury column supported by that pressure. The pressure supporting a column of mercury 1 mm in height is called 1 **torr** (in honor of Torricelli, the scientist who invented the barometer, a device for measuring the pressure of the atmosphere). A simple barometer is diagrammed in Figure 2–2. The **standard atmosphere** is defined as that pressure which will support a column of mercury to a vertical height of 760 mm at a temperature of 0°C. Therefore the pressure of a standard atmosphere is 760 torr. Note that the dimension "1 atmosphere" (1 atm) is not in general the same as "atmospheric pressure." Atmospheric pressure varies widely from day to day and from place to place, whereas the dimension 1 atmosphere is by definition a constant.

If the cross section of a barometer tube of uniform bore is 1 square inch, a 760 mm long column of mercury in the tube will weigh 14.7 pounds. Hence another unit for the standard atmosphere is 14.7 pounds/in.². By the use of appropriate conversion factors,[3] it can be shown that

$$1.00 \text{ atm} = 760 \text{ torr} = 14.7 \text{ pounds/in.}^2 = 101.3 \text{ Pa} = 1.013 \times 10^6 \text{ dynes/cm}^2$$

2-13 Boyle's Law

For a fixed mass of gas at constant temperature, the volume of the gas is inversely proportional to its pressure. This statement is known as **Boyle's law.** This law is approximately true for all gases, but it does not apply to liquids or solids. It is more exact if the pressure of gas is low and the temperature is relatively high. In mathematical form, the law may be represented as

$$PV = k \qquad \text{(constant } T\text{)}$$

Since k is a constant, as P gets larger, V must get correspondingly smaller; thus V varies inversely with P (Figure 2–3).

---

[3] See Appendix A–3 for a tabulation of units and dimensions.

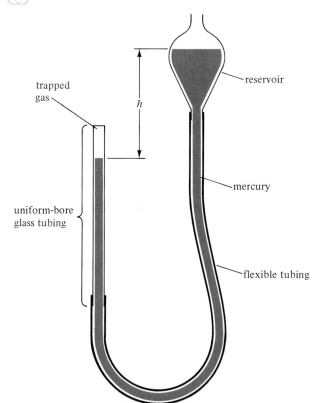

Figure 2-3. Boyle's Law Apparatus
The pressure on a gas in a tube of uniform bore is varied by raising or lowering the mercury reservoir. The pressure on the gas is equal to the atmospheric pressure (expressed in torr) plus the algebraic difference in height, h (in mm), between the levels of mercury in the reservoir and in the tube. Raising the reservoir increases the pressure on the gas, and its volume decreases proportionately. Since the tube has a uniform bore, the volume may be estimated by measurement of the length of the column of gas.

Example

A 1.00 liter sample of gas at 760 torr is compressed to 0.800 liter at constant temperature. Calculate the final pressure of the gas.

$$PV = k = (760 \text{ torr})(1.00 \text{ liter}) = 760 \text{ torr} \cdot \text{liters}$$

But after compression,

$$PV = k$$
$$P(0.800 \text{ liter}) = 760 \text{ torr} \cdot \text{liters}$$
$$P = 950 \text{ torr}$$

It is not necessary to solve numerically for the value of k. A more direct approach is to use the relationship $P_1V_1 = P_2V_2$, where P_1 and V_1 are the initial conditions and P_2 and V_2 are the final conditions. It should be noted that any units, for example, atmospheres for pressure and milliliters for volume, may be used as long as they are used consistently in a given calculation.

2-14   Charles' Law

It may be determined experimentally that the volume of a given mass of gas at constant pressure varies approximately linearly with the temperature, as shown in

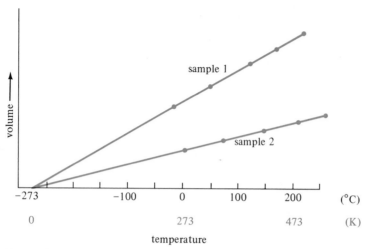

Figure 2–4. Volumes of Two Gas Samples at Constant Pressure as a Function of Temperature

Figure 2–4. The identity of the gas is not important, except, of course, that the gas must not liquefy in the range of temperatures studied. Also, the initial volume of the gas is not important. In Figure 2–4 it may be seen that for two different samples, the volumes are reduced proportionately by reduction in temperature. Assuming that the gas would not liquefy and that the straight line relationship would continue to be valid at such low temperatures, both lines would meet the axis at −273°C. Of course, any real gas would liquefy before that low temperature is reached, but −273°C (more exactly, −273.15°C) is the temperature at which a hypothetical gas which does not condense would have zero volume. This temperature is called the **absolute zero** of temperature.

The Kelvin temperature scale is based on the concept of absolute zero; the zero temperature of the scale is the lowest temperature theoretically possible. The symbol for "degrees Kelvin" is K. A 1° interval (that is, for example, the temperature difference between 57° and 58°) is the same as a 1° interval on the Celsius scale. On the Kelvin scale, therefore, the freezing point of water is 273 K and the boiling point of water at 1 atm pressure is 373 K. Any temperature (t) in degrees Celsius may be converted to the Kelvin temperature (T) by adding 273°:

$$T = t + 273°$$

As shown in Figure 2–4, the volume of a given mass of gas at constant pressure is directly proportional to its *absolute* temperature (T). This is a statement of **Charles' law,** which can be expressed mathematically as

$$V = k'T \quad \text{(constant pressure)}$$

Example

Calculate the volume which 4.00 liters of gas at 0°C will occupy at 100°C and the same pressure.

$$V_1 = k'T_1$$
$$4.00 \text{ liters} = k'(0° + 273°) = k'(273 \text{ K})$$
$$k' = 0.01465 \text{ liter/K}$$

New conditions:

$$V_2 = k'T_2$$
$$V_2 = (0.01465 \text{ liter/K})[(100 + 273) \text{ K}] = 5.46 \text{ liters}$$

Note that temperatures must be expressed in degrees Kelvin. This calculation may be done by combining the separate steps, as follows:

$$\frac{V_1}{T_1} = k' = \frac{V_2}{T_2}$$

$$V_2 = \frac{V_1 T_2}{T_1} = \frac{(4.00 \text{ liters})(373 \text{ K})}{(273 \text{ K})} = 5.46 \text{ liters}$$

2-15 Combined Gas Law

Suppose that it is desired to calculate the volume of a given mass of gas when both its temperature and pressure are changed. The result may be obtained by first applying Boyle's law to calculate the volume at the new pressure, assuming no change in temperature, and then (using that volume) applying Charles' law to calculate the final volume at the new temperature. However, the two steps can be combined to show simultaneously the inverse proportionality of volume and pressure and the direct proportionality of volume and absolute temperature:

$$\frac{P_1 V_1}{T_1} = \frac{P_2 V_2}{T_2}$$

This expression is sometimes referred to as the **combined gas law.**

Example

A sample of gas at 0°C and 1.00 atm pressure occupies 2.50 liters. What change in temperature is necessary to adjust the pressure of that gas to 1.50 atm after it has been transferred to a 2.00 liter container?

$$\frac{P_1 V_1}{T_1} = \frac{P_2 V_2}{T_2}$$

$$T_2 = \frac{P_2 V_2 T_1}{P_1 V_1} = \frac{(1.50 \text{ atm})(2.00 \text{ liters})(0° + 273°)}{(1.00 \text{ atm})(2.50 \text{ liters})} = 328 \text{ K}$$

Hence, an increase in temperature of 55°, from 273 K to 328 K, is required.

It is sometimes convenient to define a set of standard conditions of temperature and pressure. By convention, the standard temperature is chosen as 273 K (0°C) and the standard pressure is chosen as 1 atm (760 torr). Together these conditions are referred to as STP. The volume of a gas at STP may be calculated from its volume at any other temperature and pressure by use of the combined gas law.

2-16   Moles of Gas

The following expression may be derived from the combined gas law:

$$\frac{PV}{T} = k$$

The constant k refers to a given mass of gas and hence a given number of moles of gas. A more general expression would be

$$\frac{PV}{T} = nR \qquad \text{or} \qquad PV = nRT$$

where n is the number of moles of gas in the sample and R is a constant which is characteristic of all gases. The numerical value of R depends, of course, on the units chosen to express P, V, and T.

It is found experimentally that at $0°C$ and 1.00 atm pressure (STP), 1.00 mole of any gas occupies approximately 22.4 liters. Hence the gas constant R can be evaluated as follows:

$$R = \frac{PV}{nT} = \frac{(1.00 \text{ atm})(22.4 \text{ liters})}{(1.00 \text{ mole})(273 \text{ K})} = 0.0821 \text{ liter} \cdot \text{atm/mole} \cdot \text{K}$$

With the equation $PV = nRT$, it is possible to calculate the number of moles of a gaseous sample from pressure, volume, and temperature data. If the volume occupied by a certain mass of a pure gaseous substance is known at some temperature and pressure, it is possible to calculate the mass per mole of gas—the molecular weight.

Example

At 750 torr and 27°C, 0.60 gram of a certain gas occupies 0.50 liter. Calculate its molecular weight.

$$PV = nRT$$

$$\left(\frac{750}{760} \text{ atm}\right)(0.50 \text{ liter}) = n\left(0.0821 \frac{\text{liter} \cdot \text{atm}}{\text{mole} \cdot \text{K}}\right)(300 \text{ K})$$

$$n = 2.0 \times 10^{-2} \text{ mole}$$

$$\text{molecular weight} = \frac{\text{mass}}{\text{number of moles}} = \frac{0.60 \text{ gram}}{2.0 \times 10^{-2} \text{ mole}} = 30 \text{ grams/mole} = 30 \text{ D}$$

Example

What volume of CO_2 at 350°C and 1.00 atm pressure would be produced from the complete thermal decomposition of 1.00 kg of $MgCO_3$?

$$\frac{1000 \text{ grams } MgCO_3}{84.3 \text{ grams/mole}} = 11.9 \text{ moles}$$

$$MgCO_3 \rightarrow MgO + CO_2$$

Since 1 mole of CO_2 is produced per mole of $MgCO_3$ decomposed, 11.9 moles of CO_2 is generated in this case.

$$PV = nRT$$

$$(1.00 \text{ atm})V = (11.9 \text{ moles}) \left(0.0821 \frac{\text{liter} \cdot \text{atm}}{\text{mole} \cdot \text{K}}\right)[(350 + 273)\text{K}]$$

$$V = 609 \text{ liters}$$

2-17 The Law of Partial Pressures

Assuming that no chemical reaction occurs, when two or more different gases are put into the same container, the total pressure is merely the sum of the individual pressures of the different gases. In other words, each gas exerts a pressure independent of the others—as though it were the only gas in the container. Hence the total pressure P is given by the sum of the partial pressures, P_1, P_2, etc.:

$$P = P_1 + P_2 + \cdots$$

This relationship is known as **Dalton's law of partial pressures.** The application of Dalton's law is illustrated by the following examples.

Example

If 2.0 grams of N_2, 0.40 gram of H_2, and 9.0 grams of O_2 are put into a 1.00 liter container at 27°C, what is the total pressure in the container?

The number of moles and the partial pressure of each gas is found:

$$\frac{2.0 \text{ grams } N_2}{28.0 \text{ grams/mole}} = 0.071 \text{ mole } N_2$$

$$P_{N_2} = n_{N_2}RT/V$$

$$P_{N_2} = \frac{(0.071 \text{ mole})(0.0821 \text{ liter} \cdot \text{atm/mole} \cdot \text{K})(300 \text{ K})}{(1.00 \text{ liter})} = 1.7 \text{ atm}$$

$$\frac{0.40 \text{ gram } H_2}{2.0 \text{ grams/mole}} = 0.20 \text{ mole } H_2$$

$$P_{H_2} = \frac{(0.20 \text{ mole})(0.0821 \text{ liter} \cdot \text{atm/mole} \cdot \text{K})(300 \text{ K})}{(1.00 \text{ liter})} = 4.9 \text{ atm}$$

$$\frac{9.0 \text{ grams } O_2}{32.0 \text{ grams/mole}} = 0.28 \text{ mole } O_2$$

$$P_{O_2} = \frac{(0.28 \text{ mole})(0.0821 \text{ liter} \cdot \text{atm/mole} \cdot \text{K})(300 \text{ K})}{(1.00 \text{ liter})} = 6.9 \text{ atm}$$

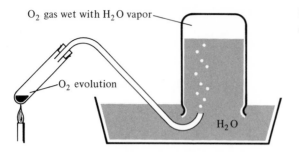

O_2 gas wet with H_2O vapor

O_2 evolution

H_2O

Figure 2-5. **Collection of Oxygen over Water**

The total pressure is

$$P = P_{N_2} + P_{H_2} + P_{O_2} = 13.5 \text{ atm}$$

The same result could have been attained by adding the number of moles of the individual gases and calculating the total pressure from

$$P_{\text{total}} = n_{\text{total}}RT/V$$

In laboratory preparations, gases are often collected over water (Figure 2-5). The collected gas becomes saturated with water vapor, and the total pressure in the collecting vessel is due to the pressure of the gas itself plus the pressure of the water vapor. The pressure exerted by water vapor in the presence of liquid water is called its **vapor pressure** and has a definite value at each temperature. A tabulation of water vapor pressures at various temperatures is given in Appendix A-8. From the data in this table, the pressure exerted by the gas being collected can be calculated.

Example

Oxygen is collected over water at 25°C in a 2.00 liter vessel at a total barometric pressure of 765 torr. Calculate the number of moles of oxygen collected.

The vapor pressure of water at 25°C is 24 torr. Thus the pressure due to the oxygen is 765 torr − 24 torr = 741 torr. The number of moles of oxygen may be calculated:

$$n = \frac{PV}{RT} = \frac{(\frac{741}{760} \text{ atm})(2.00 \text{ liters})}{(0.0821 \text{ liter} \cdot \text{atm/mole} \cdot \text{K})(298 \text{ K})} = 0.0797 \text{ mole}$$

2-18 Ideal Gases

Real gases do not follow Boyle's law or Charles' law exactly. Relatively great deviations are observed when a real gas is under high pressure and/or is at relatively low temperature. An **ideal gas** is defined as a gas which follows the gas laws exactly for all conditions of temperature and pressure. The equation $PV = nRT$ is therefore referred to as the equation of state for an ideal gas. Real gases approach ideal behavior in the limit of very low pressure and relatively high temperature. The nonideal behavior of real gases will be discussed in Sections 15-4 and 15-5.

2-19 Note on the Mole Concept

One mole of any item refers to a specific number of that item regardless of whether one is dealing with atoms, ions, molecules, electrons, or baseball bats. Therefore interpreting chemical reactions in terms of moles is analogous to interpreting them in terms of individual atoms, ions, or molecules. Moreover since the number of moles and mass are related by the following expression

$$\text{number of moles} = \frac{\text{mass}}{\text{formula weight}}$$

knowing any two—number of moles and mass, number of moles and formula weight, or mass and formula weight—permits calculation of the third.

Other relationships discussed in this chapter involving the concept of moles included

$$\text{number of moles} = \text{molarity} \times \text{liters} \quad \text{(for solutions)}$$

$$\text{number of moles} = \frac{PV}{RT} \quad \text{(for gases)}$$

Thus these two relationships can be related to the first one to obtain masses and formula weights from molarities and/or pressure-volume data. Therefore in approaching any problem based on a chemical reaction or related to a chemical formula, the first step should be to THINK MOLES!

2-20 Exercises

kinetic energy
$$K = \tfrac{3}{2} RT \text{ or } \tfrac{1}{2}mV^2$$

Basic Exercises

1. Calculate the percent composition of each of the following compounds: **(a)** H_2O, **(b)** H_2O_2, **(c)** vinyl chloride, C_2H_3Cl, **(d)** $Al_2(SO_4)_3$, **(e)** $NH_4C_2H_3O_2$.
2. Calculate the empirical formula for a compound with the following percent composition: Mg, 23.29%; S, 30.72%; O, 45.99%.
3. An oxide of arsenic was analyzed and found to contain 75.74% As. What is the empirical formula of the compound?
4. Which one of the following, if any, contains the greatest number of oxygen atoms? the greatest number of molecules? 1.0 gram of O atoms, 1.0 gram of O_2, or 1.0 gram of ozone, O_3.
5. How many moles of oxygen atoms are there in each of the following? **(a)** 0.17 mole of O_2, **(b)** 6.02×10^{24} molecules of CO, **(c)** 1.0 mole of $BaS_2O_8 \cdot 4H_2O$, **(d)** 20.0 grams of O_2, **(e)** 1.6 grams of CO_2, **(f)** 1.0 mole of $Ba(NO_3)_2 \cdot H_2O$.
6. How many moles of atoms of each element are there in 1.0 mole of each of the following compounds? **(a)** Fe_3O_4, **(b)** $AsCl_5$, **(c)** $Mg(C_2H_3O_2)_2$, **(d)** $CuSO_4 \cdot 5H_2O$.
7. Convert the following into balanced chemical equations:
   (a) $NCl_3 + H_2O \rightarrow NH_3 + HOCl$
   (b) $PCl_3 + H_2O \rightarrow H_3PO_3 + HCl$
   (c) $SbCl_3 + H_2O \rightarrow Sb(O)Cl + HCl$

8. Complete and balance the following equations using the preceding exercise and the periodic table as aids, if necessary:
   (a) $AsCl_3 + H_2O \rightarrow H_3AsO_3 +$
   (b) $BiCl_3 + H_2O \rightarrow$

9. (a) What mass of P_4O_{10} will be obtained from the reaction of 1.33 grams of P_4 and 5.07 grams of O_2? (b) What mass of P_4O_6 will be obtained from the reaction of 4.07 grams of P_4 and 2.01 grams of O_2?

10. A 0.6000 mole sample of Cu_2S is roasted in excess oxygen to yield copper metal and sulfur dioxide. Calculate the mass of copper metal produced.

11. Calculate the percent sodium in a breakfast cereal which is advertised to contain 110 mg of sodium per 100 grams of cereal.

12. Using Table 2-2 if necessary and assuming that no chemical reaction takes place, state which of the following mixtures would be expected to be completely soluble if a sample containing 1 gram of each was shaken in 100 ml of water. (a) NaCl, AgCl, and $PbCl_2$; (b) $Ba(ClO_3)_2$, NaCl, and KNO_3; (c) $Ba(ClO_3)_2$, $BaCO_3$, and $BaCl_2$.

13. Calculate the mass of mercury in a uniform column 760 mm high and 1.00 cm² in cross-sectional area. Calculate the mass of mercury in a column of equal height but with 2.00 cm² cross-sectional area. Compare the pressures at the bases of the two columns. (The density of mercury is 13.6 grams/cm³.)

14. (a) Calculate the mass of $KClO_3$ necessary to produce 1.23 grams of O_2. (b) What mass of KCl is produced along with this quantity of oxygen?

15. Calculate the mass of $BaCO_3$ produced when excess CO_2 is bubbled through a solution containing 0.205 mole of $Ba(OH)_2$.

16. Calculate the mass of CaO which will react with 6.92 grams of HCl to form H_2O, Ca^{2+}, and Cl^-.

17. When solutions containing 2.00 grams of Na_2SO_4 and 3.00 grams of $BaCl_2$ are mixed, what mass of $BaSO_4$ is produced?

18. Calculate the number of moles of calcium chloride needed to react with excess silver nitrate to produce 6.60 grams of AgCl.

19. Calculate the percent composition of each of the following compounds: (a) benzene, C_6H_6, (b) acetylene, C_2H_2. (c) Compare and explain the results obtained in (a) and (b). (d) What additional data could be used to distinguish between benzene and acetylene?

20. Calcium metal reacts with hydrochloric acid to yield hydrogen and calcium chloride. Write a balanced chemical equation for the reaction. Determine the volume of hydrogen gas at 1.00 atm pressure and 18°C produced from the reaction of 12.2 grams of calcium with excess HCl.

21. Calculate the molecular weight of a gas which has a density of 1.48 grams/liter at 100°C and 600 torr.

22. Write net ionic equations corresponding to each of the following reactions:
    (a) $CoCl_2(aq) + 2\,AgF(aq) \rightarrow 2\,AgCl(s) + CoF_2(aq)$
    (b) $HCl(aq) + LiOH(aq) \rightarrow LiCl(aq) + H_2O(l)$
    (c) $Ag_2O(s) + 2\,HBr(aq) \rightarrow 2\,AgBr(s) + H_2O(l)$
    (d) $Mg(ClO_3)_2(aq) + Ba(OH)_2(aq) \rightarrow Mg(OH)_2(s) + Ba(ClO_3)_2(aq)$
    (e) $Fe_2(SO_4)_3(aq) + Fe(s) \rightarrow 3\,FeSO_4(aq)$

23. How much carbon monoxide is produced from the reaction of 1.00 kg of octane, C_8H_{18}, and 1.00 kg of oxygen?

24. What is the molarity of NaOH in a solution which contains 24.0 grams of NaOH dissolved in 300 ml of solution?

25. How many milliliters of 0.600 M NaOH will be needed to react with 138.0 ml of 4.00 M HCl?

26. Determine the molar concentration of each ionic species in solution after each of the following operations: (a) 200 ml of 2.0 M NaCl is diluted to 500 ml, (b) 200 ml of 2.0 M $BaCl_2$ is diluted to 500 ml, (c) 200 ml of 3.00 M NaCl is added to 300 ml of 4.0 M NaCl, (d) 200 ml of 2.0 M $BaCl_2$ is added to 400 ml of 3.0 M $BaCl_2$ and 400 ml

of water, **(e)** 300 ml of 3.0 M NaCl is added to 200 ml of 4.0 M BaCl$_2$, **(f)** 400 ml of 2.00 M HCl is added to 150 ml of 4.00 M NaOH, **(g)** 100 ml of 2.0 M HCl and 200 ml of 1.5 M NaOH are added to 150 ml of 4.0 M NaCl and 50 ml of water.

27. When 50.00 ml of a nitric acid solution was titrated with 0.334 M NaOH, it required 42.80 ml of the base to achieve the equivalence point. What is the molarity of the nitric acid solution? What mass of HNO$_3$ was dissolved in 90.00 ml of solution?

28. Express the following gas pressures in atmospheres: **(a)** 300 cm Hg, **(b)** 28 lb/in.², **(c)** 380 torr, **(d)** 0.760 torr, **(e)** 7.60 torr.

29. A 15.0 liter vessel containing 5.65 grams of N$_2$ is connected by means of a valve to a 6.00 liter vessel containing 5.00 grams of oxygen. After the valve is opened and the gases are allowed to mix, what will be the partial pressure of each gas and the total pressure at 27°C?

30. Into a 5.00 liter container at 18°C are placed 0.200 mole of H$_2$, 20.0 grams of CO$_2$, and 14.00 grams of O$_2$. Calculate the total pressure in the container and the partial pressure of each gas.

31. A gaseous compound has an empirical formula CH$_3$. Its density is 2.06 grams/liter at the same temperature and pressure at which oxygen has a density of 2.21 grams/liter. Calculate the molecular formula of the gas.

32. If 40.00 ml of 1.600 M HCl and 60.00 ml of 2.000 M NaOH are mixed, what are the molar concentrations of Na$^+$, Cl$^-$, and OH$^-$ in the resulting solution? Assume a total volume of 100.00 ml.

33. How would one prepare exactly 300 ml of 5.00 M HCl solution by diluting sufficient 12.00 M stock solution?

34. Calculate the molarity of each type of ion remaining in solution after 20.0 ml of 6.00 M HCl is mixed with 50.0 ml of 2.00 M Ba(OH)$_2$ and 30.0 ml of water.

35. What volume of 96.0% H$_2$SO$_4$ solution (density 1.83 grams/ml) is required to prepare 2.00 liters of 3.00 M H$_2$SO$_4$ solution?

36. Show that for a mixture of gases $P_{total}V = n_{total}RT$.

37. Assuming the same pressure in each case, calculate the mass of hydrogen required to inflate a balloon to a certain volume, V, at 100°C if 3.5 grams of helium is required to inflate the balloon to half the volume, $0.50V$, at 25°C.

38. Consider several samples of gas undergoing change as described in the table below. Determine the missing value for each sample.

	Initial Conditions			Final Conditions		
	Pressure	Volume	Temperature	Pressure	Volume	Temperature
(a)	760 torr	1.00 liter	25°C	760 torr	_____	200°C
(b)	1.00 atm	500 ml	127°C	_____	200 ml	127°C
(c)	1.23 atm	700 ml	250 K	650 torr	_____	200°C
(d)	600 lb/in.²	3.25 liters	300 K	_____	1.75 liters	100°C
(e)	_____	1.00 liter	300 K	700 torr	3000 ml	400 K
(f)	800 torr	7.50 liters	300 K	_____	2.25 liters	350 K

39. How many milliliters of 0.5000 M KMnO$_4$ solution will react completely with 20.00 grams of K$_2$C$_2$O$_4$·H$_2$O according to the following equation?

$$16\,H^+ \;+\; 2\,MnO_4^- \;+\; 5\,C_2O_4^{2-} \;\rightarrow\; 10\,CO_2 \;+\; 2\,Mn^{2+} \;+\; 8\,H_2O$$

40. How many milliliters of 3.000 M HCl should be added to react completely with 12.35 grams of NaHCO$_3$?

$$HCl \;+\; NaHCO_3 \;\rightarrow\; NaCl \;+\; CO_2 \;+\; H_2O$$

General Exercises

41. Write a balanced chemical equation for the reaction of $ZnCl_2$ with excess NaOH to produce $Na_2Zn(OH)_4$, sodium zincate. What mass of sodium zincate can be produced from 2.00 grams of $ZnCl_2$ with excess NaOH by this reaction?

42. ^{12}C is the standard for the atomic weights of atoms. What is the standard for the molecular weights of molecules? Explain.

43. What mass of solid is produced by treatment of aqueous solutions containing 2.00 grams of $AgNO_3$ and 4.00 grams of KBr, respectively?

44. What mass of AgCl can be obtained from 100 grams of $[Ag(NH_3)_2]Cl$ by means of the following reaction?

$$[Ag(NH_3)_2]Cl \ + \ 2\,HNO_3 \ \rightarrow \ AgCl \ + \ 2\,NH_4NO_3$$

45. Polyethylene may be produced from calcium carbide according to the following sequence of reactions:

$$CaC_2 \ + \ H_2O \ \rightarrow \ CaO \ + \ HC\equiv CH$$
$$HC\equiv CH \ + \ H_2 \ \rightarrow \ H_2C=CH_2$$
$$n\,H_2C=CH_2 \ \rightarrow \ (CH_2CH_2)_n$$

Calculate the mass of polyethylene which can be produced from 20.0 kg of CaC_2.

46. It took exactly 3.00 ml of 6.00 M HCl to convert 1.200 grams of a mixture of $NaHCO_3$ and Na_2CO_3 to NaCl, CO_2, and H_2O. Calculate the volume of CO_2 liberated at 25°C and 760 torr.

47. Calculate the percent of BaO in 29.0 grams of a mixture of BaO and CaO which just reacts with 100.8 ml of 6.00 M HCl.

48. Liquefied natural gas (LNG) is mainly methane. A 10,000 liter tank is constructed to store LNG at -164°C and 1.0 atm pressure, under which conditions its density is 0.415 gram/ml. Calculate the volume of a storage tank capable of holding the same mass of LNG as a gas at 20°C and 1.0 atm pressure.

49. What volume of 3.00 M HNO_3 can react completely with 15.0 grams of a brass (90.0% Cu, 10.0% Zn) according to the following equations?

$$Cu \ + \ 4\,H^+(aq) \ + \ 2\,NO_3^-(aq) \ \rightarrow \ 2\,NO_2(g) \ + \ Cu^{2+} \ + \ 2\,H_2O$$
$$4\,Zn \ + \ 10\,H^+(aq) \ + \ NO_3^-(aq) \ \rightarrow \ NH_4^+ \ + \ 4\,Zn^{2+} \ + \ 3\,H_2O$$

What volume of NO_2 gas at 25°C and 1.00 atm pressure would be produced?

50. A quantity of hydrogen gas occupies a volume of 30.0 ml at a certain temperature and pressure. What volume would half this mass of hydrogen occupy at triple the absolute temperature if the pressure were one-ninth that of the original gas?

51. A certain compound consists of 93.71% C and 6.29% H. Its molecular weight is approximately 130 D. What is its molecular formula?

52. A saturated hydrocarbon contains 82.66% carbon. What is its empirical formula? its molecular formula?

53. A 10.20 mg sample of an organic compound containing carbon, hydrogen, and oxygen only was burned in excess oxygen, yielding 23.10 mg of CO_2 and 4.72 mg of H_2O. Calculate the empirical formula of the compound.

54. (a) Balance the following equations:

$$Na_2CO_3 \ + \ HCl \ \rightarrow \ NaCl \ + \ CO_2 \ + \ H_2O$$
$$NaHCO_3 \ + \ HCl \ \rightarrow \ NaCl \ + \ CO_2 \ + \ H_2O$$

(b) A sample contains a mixture of $NaHCO_3$ and Na_2CO_3. HCl is added to 15.0 grams of the sample, yielding 11.0 grams of NaCl. What percent of the sample is Na_2CO_3?

55. Aqueous copper(II) nitrate reacts with potassium iodide to yield solid copper(I) iodide, potassium nitrate, and iodine. Write a balanced net ionic equation for the reaction.

56. Write one or two overall equations corresponding to each of the following net ionic equations:
   (a) $5 \, Fe^{2+} + MnO_4^- + 8 \, H^+ \rightarrow 5 \, Fe^{3+} + Mn^{2+} + 4 \, H_2O$
   (b) $2 \, I^- + Cl_2 \rightarrow 2 \, Cl^- + I_2$

57. Calculate the concentrations of all species remaining in solution after treatment of 50.0 ml of 0.300 M HCl with 50.0 ml of 0.400 M NH_3.

58. Calculate the change in pressure when 1.04 moles of NO and 20.0 grams of O_2 in a 20.0 liter vessel originally at 27°C react to produce the maximum quantity of NO_2 possible according to the equation

$$2 \, NO + O_2 \rightarrow 2 \, NO_2$$

59. How would one prepare exactly 3.00 liters of 1.00 M NaOH by mixing portions of stock solutions of 2.50 M NaOH and 0.400 M NaOH?

60. When 150.0 ml of 2.000 M NaOH was added to 100.0 ml of a sulfuric acid solution, it required 43.0 ml of 0.5000 M HCl to neutralize the excess base. What was the original concentration of H_2SO_4?

61. Carbon reacts with oxygen to yield carbon monoxide or carbon dioxide, depending on the quantity of oxygen available per mole of carbon. Calculate the number of moles of each product produced when 100 grams of oxygen reacts with   **(a)** 10.0 grams of carbon, **(b)** 100 grams of carbon,   **(c)** 60.0 grams of carbon.

62. Into a 3.00 liter container at 25°C are placed 1.23 moles of O_2 and 2.73 moles of C. **(a)** What is the initial pressure?   **(b)** If the carbon and oxygen react as completely as possible to form CO, what will be the final pressure in the container at 25°C?

63. Calculate the mass of CuS produced and the concentration of H^+ ion produced by bubbling excess H_2S into 1.00 liter of 0.10 M $CuCl_2$ solution. The equation is

$$Cu^{2+} + H_2S(g) \rightarrow CuS(s) + 2 \, H^+$$

64. A refrigeration tank holding 5.00 liters of Freon gas ($C_2Cl_2F_4$) at 25°C and 3.00 atm pressure developed a leak. When the leak was discovered and repaired, the tank had lost 76.0 grams of the gas. What was the pressure of the gas remaining in the tank at 25°C?

65. Complete each line in the table. There are only two naturally occurring isotopes of each element.

| | Isotope A | | | Isotope B | | Atomic |
	Isotope	Percent	Mass (D)	Isotope	Percent	Mass (D)	Weight (D)
(a)	^{191}Ir	37.30	190.9609	^{193}Ir	62.70	192.9633	____
(b)	^{121}Sb	57.25	____	^{123}Sb	42.75	122.9041	121.75
(c)	^{107}Ag	51.82	106.9041	^{109}Ag	48.18	____	107.870
(d)	^{79}Br	____	78.9183	^{81}Br	____	80.9163	79.909
(e)	^{12}C	98.89	____	^{13}C	1.11	____	12.01115

66. Calculate the concentration of each type of ion which remains in solution when each of the following sets of solutions is mixed:
   (a) 100 ml of 0.50 M NaCl + 50 ml of 0.25 M KCl
   (b) 100 ml of 0.50 M NaCl + 50 ml of 0.25 M $AgNO_3$
   (c) 100 ml of 0.50 M NaCl + 50 ml containing 1.00 mmole NaCl + 100 ml of water

67. All the oxygen in $KClO_3$ can be converted to O_2 by heating in the presence of a catalyst. **(a)** What volume of oxygen measured at 20°C and 0.996 atm pressure can be prepared from 450 grams of $KClO_3$? **(b)** If the oxygen were collected over water at 20°C and 0.996 atm barometric pressure, what would be the volume of gas collected?

68. Calcium carbide, CaC_2, reacts with water to produce acetylene, C_2H_2, and calcium hydroxide, $Ca(OH)_2$. Calculate the volume of $C_2H_2(g)$ at 25°C and 0.950 atm produced from the reaction of 128 grams of CaC_2 with 45.0 grams of water.

69. A gaseous compound is composed of 85.7% by mass carbon and 14.3% by mass hydrogen. Its density is 2.28 grams/liter at 300 K and 1.00 atm pressure. Determine the molecular formula of the compound.

70. When sufficient energy is added to a mixture of N_2 and O_2, the gases react to form oxides of nitrogen. Name two sources—one natural and one man-made—which contribute to air pollution from nitrogen oxides.

71. If 0.10 mole of N_2, 0.30 mole of H_2, and 0.20 mole of He are placed in a 10.0 liter container at 25°C, **(a)** what are the volume, temperature, and pressure of each gas? **(b)** what is the total pressure? **(c)** If 0.20 mole of NH_3 and 0.20 mole of He are placed in a similar container at the same temperature, what is the total pressure in the container? **(d)** Would a reaction converting some N_2 and H_2 into NH_3 raise or lower the total pressure in the first container?

72. Balance the following equation, which represents the combustion of pyrites, FeS_2, a pollution-causing impurity in some coals.

$$FeS_2 \ + \ O_2 \ \rightarrow \ SO_2 \ + \ FeO$$

What volume of 6.0 M NaOH would be required to react with the SO_2 produced from 1.0 ton (10^3 kg) of coal containing 0.050% by mass of pyrites impurity?

73. Calculate the mass of chloride ion in 1.00 liter of each of the following solutions: **(a)** 10.0% NaCl (density 1.07 grams/ml), **(b)** 10.0% KCl (density 1.06 grams/ml), **(c)** 1.00 M NaCl, **(d)** 1.00 M KCl. **(e)** Calculate the molarities of the first two solutions.

74. Relative humidity is defined as the ratio of the partial pressure of water in air at a given temperature to the vapor pressure of water at that temperature. Calculate the mass of water per liter of air at **(a)** 20°C and 45% relative humidity, **(b)** 0°C and 95% relative humidity. **(c)** Discuss whether temperature or relative humidity has the greater effect on the mass of water vapor in the air.

Advanced Exercises

75. Calculate the percent composition of poly(vinyl chloride), $(C_2H_3Cl)_n$. Compare the answer to that for vinyl chloride, exercise **1(c)**.

76. To what temperature must a neon gas sample be heated to double its pressure if the initial volume of gas at 75°C is decreased by 15.0%?

77. Using the principles set forth in this chapter, devise a method for determining the molecular weight of a low-boiling liquid.

78. When 0.75 mole of solid "A_4" and 2 moles of gaseous O_2 are heated in a sealed vessel (bomb), completely using up the reactants and producing only one compound, it is found that when the temperature is reduced to the initial temperature, the contents of the vessel exhibit a pressure equal to half the original pressure. What conclusions can be drawn from these data about the product of the reaction?

79. A gas cylinder contains 370 grams of oxygen gas at 30.0 atm pressure and 25°C. What mass of oxygen would escape if first the cylinder were heated to 75°C and then the valve were held open until the gas pressure was 1.00 atm, the temperature being maintained at 75°C?

80. A 10.0 cm column of air is trapped by a column of mercury 8.00 cm long in a capillary tube of uniform bore when the tube is held horizontally in a room at 0.9400 atm pressure (barometric). What will be the length of the air column when the tube is held **(a)** vertically with the open end up? **(b)** vertically with the open end down? **(c)** at a 45° angle from vertical with the open end up?

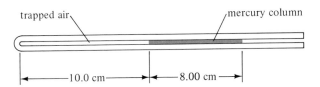

trapped air   mercury column

—10.0 cm—   —8.00 cm—

81. S. Cannizzaro (1858) was able to demonstrate that hydrogen gas consists of molecules with an even number of hydrogen atoms, based on the assumption that Avogadro's hypothesis is true. Avogadro's hypothesis states that equal volumes of gas under the same conditions of temperature and pressure contain equal numbers of molecules. Using the data from the accompanying table, show precisely how he demonstrated this fact.

Gases (100°C)	Density (grams/liter)	Percent Hydrogen
hydrogen	0.0659	100.0
hydrogen chloride	1.19	2.76
water vapor	0.589	11.2
ammonia	0.557	17.7
methane	0.524	25.1

82. What masses of P_4O_6 and P_4O_{10} will be produced by the combustion of 2.00 grams of P_4 in 2.00 grams of oxygen, leaving no P_4?

83. A mixture of Na_2CO_3 and $NaHCO_3$ has a mass of 22.0 grams. Treatment with excess HCl solution liberates 6.00 liters of CO_2 at 25°C and 0.947 atm pressure. Determine the percent Na_2CO_3 in the mixture.

84. In an auto engine with no pollution controls, about 5% of the fuel (assume 100% octane, C_8H_{18}) is unburned. Calculate the relative masses of CO and C_8H_{18} emitted in the exhaust gas. Calculate the relative volumes of these gases.

85. A certain compound was known to have a formula which could be represented as $[PdC_xH_yN_z](ClO_4)_2$. Analysis showed that the compound contained 30.15% carbon and 5.06% hydrogen. When converted to the corresponding thiocyanate, $[PdC_xH_yN_z](SCN)_2$, the analysis was 40.46% carbon and 5.94% hydrogen. Calculate the values of x, y, and z.

86. If the atomic weight of carbon were set at 100 D, what would be the value of Avogadro's number? Is Avogadro's number a fundamental physical constant?

87. (a) What mass of CO would be produced by the reaction of 16.0 grams of O_2 with excess carbon according to the following equation?

$$2\,C \;+\; O_2 \;\rightarrow\; 2\,CO$$

(b) If the atomic weight of carbon were set at 50.00 D, what would be the atomic weight of oxygen? the molecular weight of CO? **(c)** Using the atomic weights and molecular weight of part **(b)**, calculate the mass of CO that would be produced by the reaction of 16.0 grams of O_2 with excess carbon. **(d)** How does this exercise demonstrate the arbitrary nature of the atomic weight scale?

88. The "roasting" of 100.0 grams of a copper ore according to the equations given below yielded 75.4 grams of 89.5% pure copper. If the ore is composed of Cu_2S and CuS with 11.0% inert impurity, calculate the percent of Cu_2S in the ore.

$$Cu_2S \ + \ O_2 \ \rightarrow \ 2\,Cu \ + \ SO_2$$

$$CuS \ + \ O_2 \ \rightarrow \ Cu \ + \ SO_2$$

89. After 11.2 grams of carbon reacts with oxygen originally occupying 21.2 liters at 18°C and 750 torr, the cooled gases are passed through 3.00 liters of 2.50 M NaOH solution. Determine the concentration of NaOH remaining in solution which is not converted to Na_2CO_3. *Note:* CO does not react with NaOH under these conditions.

90. Only gases remain after 18.0 grams of carbon is treated with 5.00 liters of O_2 at 18°C and 5.00 atm pressure. Determine the concentration of $NaHCO_3$ and of Na_2CO_3 produced by the reaction of the CO_2 with 0.500 liter of 2.00 M NaOH. *Note:* CO does not react with NaOH under these conditions.

91. Only gases remain after 18.0 grams of carbon is treated with 25.0 liters of air at 17°C and 5.00 atm pressure. (Assume 19.0% by volume O_2, 80% N_2, and 1.0% CO_2.) Determine the concentrations of $NaHCO_3$ and Na_2CO_3 produced by the reaction of the mixture of gases with 0.750 liter of 1.00 M NaOH.

92. The Rankine scale is an absolute temperature scale which has the same size intervals between degrees as the Fahrenheit scale. What is the freezing point of water in degrees Rankine?

93. What percent of a sample of nitrogen must be allowed to escape if its temperature, pressure, and volume are changed from 220°C, 3.00 atm, and 1.65 liters to 110°C, 0.700 atm, and 1.00 liter, respectively?

94. Under suitable conditions, the reaction

$$C_2H_4(g) \ + \ H_2(g) \ \rightarrow \ C_2H_6(g)$$

occurs at 600 K. Write a computer program to calculate the total pressure in a 1.00 liter reaction vessel for any initial quantities of C_2H_4 and H_2 after the reaction has proceeded from 0% to 100% toward completion. *Note:* Include in your program provision for either C_2H_4 or H_2 being the limiting reagent. Choose any arbitrary number of moles of each reactant and compute the total pressure when 20, 40, 60, and 80% of the limiting reagent has reacted at 600 K.

95. Write a computer program to calculate the mass of gas collected over water in a given volume at any barometric pressure and any temperature between 0 and 80°C. Include in your program a feature which will obtain the necessary vapor pressure of water using the data of Table A–8. Use the program to calculate the masses of the following samples: (a) 2.50 liters of O_2 collected over water at 24°C when the barometric pressure is 763.0 torr, (b) 0.600 liter of He at 45°C and 678 torr, (c) 7.50 liters of N_2 at 0°C and 800 torr.

3

Energy Relationships
in Chemical Reactions

Essentially the universe consists only of matter and energy. Matter has been defined and discussed in Chapter 1. Of course, chemistry is the science which studies matter. Chemistry is also concerned with the study of energy. It is not possible to give a concise definition of energy. Energy is the capacity for effecting change, but it is also the cause of some changes and the result of others. At best, energy is closely associated with change.

Energy, E, and mass, m, are related by Einstein's equation

$$E = mc^2$$

where c is the velocity of light—2.9979×10^{10} cm/sec. The interconversion of mass and energy can be measured in the case of nuclear reactions (Section 20–9), but the change in mass theoretically associated with the energies released or absorbed in ordinary chemical reactions is much too small to be detected by even the most sensitive weighing devices (see exercise 54). Therefore, for ordinary purposes, the laws of conservation of mass and conservation of energy can be considered separate laws.

Energy is involved in chemical reactions in a number of ways. Often energy is supplied from an external source to promote a chemical reaction. For example, aluminum metal is obtained by use of electrical energy to break down aluminum oxide. Some important chemical industries, including aluminum plants, are located near sources of cheap electric power because of the economics of carrying out chemical reactions for profit. On the other hand, many chemical reactions are carried out for the sole purpose of producing a desired form of energy. Some examples are the combustion of fuels, such as gasoline and coal, and the generation of electricity by means of the chemical reactions which occur in a battery.

In addition to practical applications, there are important theoretical relationships between energy and chemical reactions. Why do chemical reactions occur? Why are some substances more reactive than others? Is there a relationship between the energy of a reaction and the composition and structure of each substance involved? Can one predict whether a given reaction will occur spontaneously? How are the quantities of energy involved in a given reaction related to the quantities (number of moles) of substances undergoing reaction? Theories which provide answers to these questions are fundamental to much of modern chemistry.

The branch of science which includes the study of energy transformations is called **thermodynamics.** Basic to thermodynamics are two "laws" derived from experience, which can be stated as follows:

1. Energy can neither be created nor destroyed—the energy of the universe is constant.
2. The entropy of the universe is always increasing.

These generalizations are statements of the first and second laws of thermodynamics. The laws and the meaning of entropy will be discussed and expanded upon in this chapter and in subsequent chapters of this book. It will be shown that energy transformations on a macroscopic scale—that is, between large aggregates of atoms and/or molecules—can be understood in terms of a set of logical principles. Thus thermodynamics provides a model of the behavior of matter in bulk. The power of such a model is that it does not depend on atomic or molecular structure. Furthermore, conclusions about a given process, based on this model, do not require details of *how* the process is carried out. Applied to chemistry, thermodynamics provides criteria for predicting whether a given reaction can occur. If a reaction is feasible, the extent to which it will proceed under a given set of conditions can be predicted. One great value of thermodynamics is that it is possible to use data from experiments which can be conveniently carried out to arrive at conclusions about experiments which are difficult or even impossible to perform.

3-1   Systems, Initial States, Final States

At the outset, it is worthwhile to define some terms which are customarily used in discussing the interactions of energy with matter. That portion of matter which is being investigated is called the **system.** All other objects in the universe which may interact with the system are called the **surroundings.** For example, 1 liter of a 1 M aqueous solution of sodium chloride may be under investigation. The solution's container would be considered part of the surroundings. A system is described by identifying its constituents and their quantities, the temperature, the pressure, and perhaps some other relevant conditions, such as the physical states of the substances involved. A complete description of the system defines its **state.** The initial state of a system is its state before it undergoes a change, and the final state describes the system after a change has occurred. In going from the initial state to the final state, a system may exchange energy with its surrounding, and/or its components may change composition; but there can be no change in the total mass of the system. No matter can be lost to or gained from the surroundings.

The properties of a system which uniquely define the state of the system are called **thermodynamic properties** or **state functions.** For example, consider a system consisting of one mole of an ideal gas. The state of the system is specified by giving any two of the properties pressure, volume, and temperature. As was described in the previous chapter, the pressure, P, volume, V, and temperature, T, of n moles of an ideal gas are related by the equation

$$PV = nRT$$

This equation, expressing the relationship of its state functions, is called the equation of state for an ideal gas. The equations of state for real gases, liquids, solids and solutions are more complicated than that for an ideal gas. Moreover, even for an ideal gas, there are state functions other than P, V, and T. For example, energy is a state function. When the state of a given system is defined in terms of a few of its state functions, all of the other state functions will have definite values. Moreover, when a system undergoes a change in state, the change in value of any state function depends only on the initial and final states of the system and not on how the change is accomplished. Indeed, as will be shown, the importance of state functions lies in the fact that for a given system, the change in their values can be obtained by considering only the initial and final states of the system (see exercise 22).

When a system undergoes a reaction described by a chemical equation, the description of the reactants defines the initial state of the system, and the corresponding description of the products defines the final state. For example, the equation

$$C(s, graphite) \quad + \quad O_2(g, 1\ atm) \quad \xrightarrow[25°C]{} \quad CO_2(g, 1\ atm)$$

describes the reaction occurring at 25°C between 1 mole of solid graphite and 1 mole of oxygen gas at 1 atm pressure to form 1 mole of gaseous carbon dioxide at 1 atm pressure.

The properties of a system which do not depend on the quantity of matter present are called **intensive properties.** For example, density, pressure, and temperature are intensive properties. A teaspoonful of water taken out of a bathtubful has the same temperature as the bathtubful. Properties which are proportional to the quantity of matter in the system are called **extensive properties.** The mass of a sample is an extensive property. Other examples of extensive properties will be described in the following sections.

3-2   Temperature, Heat, and Heat Capacity

Like all other forms of energy, heat is not a substance. In order for energy to be transferred spontaneously in the form of heat, a temperature difference must exist. Temperature is not the same as heat. Temperature is a measure of the relative intensity of heat within a system. One does not speak of a quantity of temperature, and the temperature of an object does not depend on its mass. In contrast, heat is an extensive property; that is, the quantity of heat within an object at a given temperature is proportional to the mass of the object.

When two objects having different temperatures are placed in contact with each other, heat is transferred spontaneously from the object at the higher temperature to the one at the lower temperature. This spontaneous transfer will continue as long as a difference in temperature exists between the two objects. Finally, if left in contact, the two objects will come to the same temperature. If no other processes than heat transfer are involved and no heat is lost to the surroundings, the final temperature will be intermediate between the two original temperatures. The quantity of heat transferred will depend on the masses of the two objects.

The common unit for the expression of quantities of heat is the **calorie,** defined as the heat required to raise the temperature of exactly 1 gram of water 1 degree Celsius. Because electrical measurements can be made more precisely than can calorimetric measurements, the electrical unit of energy, the **joule** (J), has been adopted as the SI unit of energy, and the calorie is defined precisely as 4.184 J. Other energy units and relationships are given in Appendix A–2 and A–3).

The quantity of heat required to raise the temperature of 1 gram of any substance 1 degree Celsius is called the **specific heat capacity** of that substance, or more simply its **specific heat.** The quantity of heat necessary to raise the temperature of 1 mole of a substance 1 degree Celsius is called its **molar heat capacity.**[1] Thus, the quantity of heat involved in any temperature-change process is given by either of the following equations:

spec.   heat = (mass)(specific heat)(temperature change)

molar   heat = (number of moles)(molar heat capacity)(temperature change)

Example

The specific heat of aluminum is 0.214 cal/gram · deg. Calculate the heat necessary to raise the temperature of 40.0 grams of aluminum from 20.0° to 32.3°C.

heat = (mass)(specific heat)(temperature change)

= (40.0 grams)(0.214 cal/gram · deg)(12.3 deg) = 105 cal

Example

The specific heat of silver is 0.0565 cal/gram · deg. Assuming no loss of heat to the surroundings or to the container, calculate the final temperature when 100 grams of silver at 40.0°C is immersed in 60.0 grams of water at 10.0°C.

Let t be the final temperature.

heat = (mass)(specific heat)(temperature change)

heat loss by silver = (100 grams)(0.0565 cal/gram · deg)[(40.0 − t)deg]

heat gain by water = (60.0 grams)(1.00 cal/gram · deg)[(t − 10.0)deg]

Since the heat lost by the silver is gained by the water, the two products are equal:

(100 grams)(0.0565 cal/gram · deg)[(40.0 − t)deg]

= (60.0 grams)(1.00 cal/gram · deg)[(t − 10.0)deg]

$t = 12.6°C$

Some values of mean molar heat capacities of various substances are listed in Table 3–1. For many metals at 25°C, the molar heat capacity is approximately 6 cal/mole · deg.

molar heat capacity = (specific heat)(atomic weight) $\cong$ 6 cal/mole · deg

This fact is often referred to as the law of Dulong and Petit. The law provided a major support of the model of the atom proposed by Dalton, which included the postulate that atoms of a given element had a characteristic "atomic weight." Using the law of Dulong and Petit, it was possible to estimate atomic weights from specific heat data.

---

[1] The heat capacity of a substance depends somewhat on temperature but may be assumed to be constant over reasonable ranges without serious error. For gases, the heat capacities vary somewhat with pressure. The symbol C_p denotes the heat capacity of a gas at constant pressure; the symbol C_v denotes the heat capacity at constant volume (see Section 15–7).

TABLE 3-1. Mean Molar Heat Capacities at Constant Pressure and 298 K

Metallic Elements	C_p cal/mole · deg	J/mole · deg	Other Elements and Compounds	C_p cal/mole · deg	J/mole · deg
Ag	6.1	26	C	2.04	8.5
Al	5.8	24	H_2	6.90	28.9
Au	6.07	25.4	N_2	6.94	29.0
Bi	6.1	26	O_2	7.05	29.5
Cd	6.2	26	Al_2O_3	18.96	79.33
Cr	5.6	23	CH_4	8.60	36.0
Cu	5.85	24.5	C_2H_6	12.71	53.18
Fe	5.9	25	CO	6.97	29.2
Pb	6.4	27	CO_2	8.96	37.5
Sn	6.4	27	Fe_2O_3	24.91	104.2
Zn	6.06	25.4	HBr	6.58	27.5
			$H_2O(g)$	5.92	24.8
			$H_2O(l)$	18.00	75.3
			$H_2O(s)$	8.8	36.8
			NH_3	8.63	36.1
			SnO_2	13.53	56.61

Example

A "new" element, "El," forms a compound with chlorine which contains 1.455 grams of Cl per gram of El. The specific heat of pure El is found to be 0.050 cal/gram · deg. Estimate the atomic weight of El, and deduce the formula of its compound with chlorine.

According to the law of Dulong and Petit,

$$(\text{specific heat})(\text{atomic weight}) \cong 6 \text{ cal/mole} \cdot \text{deg}$$

$$\text{atomic weight} \cong \frac{6 \text{ cal/mole} \cdot \text{deg}}{0.050 \text{ cal/gram} \cdot \text{deg}} \cong 120 \text{ gram/mole} \cong 120 \text{ D}$$

Hence for 1 mole of El atoms, there must be

$$120 \text{ grams El} \left(\frac{1.455 \text{ grams Cl}}{\text{gram El}} \right) = 175 \text{ grams Cl}$$

$$175 \text{ grams Cl} \left(\frac{1 \text{ mole Cl}}{35.5 \text{ grams Cl}} \right) = 5 \text{ moles Cl}$$

There are 5 moles of Cl atoms per mole of El atoms; the formula is $ElCl_5$. These results suggest that the "new" element is antimony, which has an atomic weight of 121.75 D and forms a compound having the formula $SbCl_5$.

3-3   The First Law of Thermodynamics

Any system in a given state will possess a given quantity of energy, called its internal energy, *E*. Internal energy is an extensive property. By either releasing energy or by absorbing energy, a system may change from an initial state where its internal energy is E_1 to a different (final) state where its internal energy is E_2. The change[2] in internal energy is

$$\Delta E = E_2 - E_1$$

It is seldom necessary (and of little practical use) to know the individual values of energies E_1 and E_2. The *difference* in energy between two states is of prime importance and is usually conveniently determined. For example, the transfer of energy in the form of heat was discussed in Section 3–2. The change (increase or decrease) of energy in a given system was determined from its mass, its heat capacity, and the change in temperature. At no time was the total energy of the system considered, only the gain or loss of heat was found.

Energy may be transferred into or out of a system in forms other than heat. For example a chemical system may transfer mechanical energy through expansion of a gaseous product. As will be described in Chapter 5, with proper experimental arrangement, electrical energy may be obtained from a chemical system. It is customary to denote all forms of transferred energy other than heat as work, w. Thus when a system changes from one state to another, the change in its internal energy is given by

$$\Delta E = q + w$$

where q is the heat *absorbed by* the system and w is the work *done on* the system. For both q and w, a positive value is understood to indicate a transfer of energy into the system, while a negative value indicates energy transferred out of the system. A positive value for ΔE thus indicates an increase in the internal energy of a system, and a negative value indicates a decrease.[3] The relationship $\Delta E = q + w$ is a mathematical statement of the first law of thermodynamics—energy can neither be created nor destroyed.

Example

In a certain process, 678 cal of heat is absorbed by a system while the system is doing work corresponding to 294 cal. What is the value of ΔE for the process?

$$q = +678 \text{ cal}$$

Since work is done *by* the system, w has a negative value:

$$w = -294 \text{ cal}$$
$$\Delta E = q + w = 678 \text{ cal} + (-294 \text{ cal}) = 384 \text{ cal}$$

When a system changes from one state to another, the magnitude of the change in

---

[2] The Greek letter Δ always denotes a change in a quantity, and the change is always computed by subtracting the initial value of the quantity from its final value.

[3] In some books the quantity w is given a positive value for work transferred *out* of the system. With this sign convention $\Delta E = q - w$. To avoid confusion when consulting other books, one should note which convention is used for expressing the first law.

internal energy, ΔE, depends only on the initial and final states, and not on the path or process by which the change takes place. If this fact were not true, it would be possible for a system to change from an initial to a final state along one path, then change back to the initial state along another path, and have a net quantity of energy left over. Since the system would then be in its original state, the process could be repeated over and over again. With each repetition some energy would be gained, and the effect would be the creation of energy, which is in contradiction to the first law of thermodynamics. In changing from an initial state to a final state, both q and w could have many different values, depending on the path, but their sum, $q + w$, is invariable and independent of the path. For example, if a heavy box were pushed up a ramp, its energy would be increased by the same quantity whether or not rollers were used. To slide the box without rollers requires that more work be put into the system. The value of w for the process is greater. However, the friction produced by sliding is greater, and more heat is given off in the sliding process. That is, q has a greater negative value. The sum of the greater positive value for w and the greater negative value for q is equal to the sum of q and w for the rolling process. The value of ΔE, the change in potential energy of the box, is the same no matter which process is used. Therefore ΔE depends only on the initial state and the final state, as is expected since E is a state function.

3-4   Enthalpy

In the laboratory, many chemical reactions are carried out in open containers. When a reaction takes place in contact with the atmosphere, the volume of the system will change in such a way that the final pressure of the system equals the atmospheric pressure. Since the atmospheric pressure does not usually change significantly over a period of hours, a reaction occurring in an open vessel may be considered to be a constant pressure process. Any change in the volume of the reaction system would result in work being done against the constant pressure of the atmosphere (see Figure 3–1). As shown in the figure, work done by the system is given by

$$w = -P\Delta V$$

Therefore the change in internal energy for a system undergoing a chemical reaction at constant pressure is given by

$$\Delta E = q_P - P\Delta V \qquad \text{(expansion work only)}$$

where the subscript $_P$ denotes that the heat is absorbed at constant pressure. Rearranging and expanding gives

$$q_P = \Delta E + P\Delta V = E_2 - E_1 + P(V_2 - V_1)$$

Collecting terms for the initial and final states gives

$$q_P = (E_2 + PV_2) - (E_1 + PV_1)$$

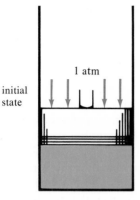

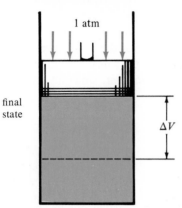

Figure 3–1. **Work of Expansion Against the Constant Pressure of the Atmosphere**

A reaction occurring in a vessel with a movable, piston-like wall liberates a gas, which pushes the piston against the constant pressure of the atmosphere. The volume of the system increases by a quantity, ΔV.

The work done by the system is equal to the pressure times the change in volume. Work can be defined as a force times the distance through which it operates; pressure is force per unit area. The product $P \Delta V$ has the units

$$\left(\frac{\text{dynes}}{\text{cm}^2}\right)(\text{cm}^3) = (\text{dynes})(\text{cm}) = \text{ergs}$$

(See Appendix A–2 and A–3.) In this case work is done *by* the system on the surroundings, and therefore

$$w = -P \Delta V$$

It can be shown that the product PV has the dimensions of energy; thus $(E + PV)$ also has the dimensions of energy. It is convenient to *define* a new state function called **enthalpy,** H, as follows:

$$H = E + PV$$

Then

$$q_P = H_2 - H_1 = \Delta H \qquad \text{(expansion work only)}$$

The change in enthalpy is merely the heat absorbed in a process occurring at constant pressure in which the only possible work is expansion against the atmosphere. Since E, P, and V are state functions, it is apparent that H is also a state function; that is ΔH depends only on the initial and final states of the system involved.

Example

If 1000 cal of heat is added to 1200 ml of oxygen in a cylinder at constant pressure of 1.00 atm (see Figure 3–1), the volume increases to 1500 ml. Calculate ΔE for the process.

The enthalpy change, ΔH, is equal to the quantity of heat added, since the process occurred at constant pressure and only expansion work is involved.

$$\Delta H = 1000 \text{ cal}$$

The following conversion factors may be found in the Appendix:

$$1 \text{ atm} = 1.013 \times 10^6 \text{ dyne/cm}^2$$
$$1 \text{ ml} = 1 \text{ cm}^3$$
$$1 \text{ dyne} \cdot \text{cm} = 1 \text{ erg}$$
$$1 \text{ J} = 1 \times 10^7 \text{ ergs}$$
$$4.184 \text{ J} = 1.000 \text{ cal}$$

The work of expansion, $w = -P\Delta V$, is done by the system on the surroundings, and hence its value is negative. The value is given by

$$-P\Delta V = -(1.013 \times 10^6 \text{ dyne/cm}^2)(1500 \text{ cm}^3 - 1200 \text{ cm}^3)$$
$$= -3.0 \times 10^8 \text{ dyne} \cdot \text{cm} = -3.0 \times 10^8 \text{ ergs} = -30 \text{ J}$$
$$-30 \text{ J} \left(\frac{1 \text{ cal}}{4.184 \text{ J}} \right) = -7.2 \text{ cal}$$

Since $\Delta H = \Delta E + P\Delta V$,

$$\Delta E = \Delta H - P\Delta V = 1000 \text{ cal} - 7.2 \text{ cal} = 993 \text{ cal}$$

The concept of enthalpy is so useful that it is applied to a wide variety of types of reactions and processes. Special names and symbols are often used to denote the enthalpy of certain of these, such as enthalpy of formation (ΔH_f), enthalpy of combustion (ΔH_{comb}), and enthalpy of sublimation (ΔH_{sub}). It must be emphasized that all of the enthalpies of the various types of processes, to be defined below, are merely different examples of specific enthalpy changes.

3-5 Enthalpy Changes in Chemical Reactions

The term **heat of reaction**—more precisely, **enthalpy of reaction**—is often used to denote that in a given chemical process, a definite quantity of energy is absorbed or released per mole of some substance undergoing change. Reactions which occur with the evolution of energy are said to be **exothermic;** reactions in which energy is absorbed are **endothermic.** The quantities of energy evolved in an exothermic reaction or absorbed in an endothermic reaction depend on the quantities of reactants undergoing change. Regardless of the actual form of energy involved, its quantity can be expressed in the same energy units—calories or joules. If heat is liberated in a reaction and not allowed to escape, the products will be at a higher temperature than the reactants; and, as will be shown later, if the heat capacities of the substances involved are known, the quantity of heat produced in the reaction can be accounted for. Finally, if gases are involved, unless the reaction is run at constant volume, some mechanical energy is required to push back the atmosphere (see Figure 3–1), and this energy must also be included in the ΔE associated with the reaction.

The physical states of the reactants and products must be specified to describe a reaction completely. For example, the process represented by the equation

$$2 \text{ HgO(s)} \; + \; 43.4 \text{ kcal} \; \xrightarrow[25^\circ \text{C}]{} \; 2 \text{ Hg(l)} \; + \; O_2(\text{g})$$

is an endothermic reaction in which 2 moles of mercury(II) oxide is decomposed at 25°C to form 2 moles of liquid mercury and 1 mole of oxygen gas (at 1 atm pressure). In this process, 43.4 kcal of heat is absorbed.

Similarly, when carbon is oxidized in air, both heat and light may be produced. The process occurring at 25°C may be described by an equation in which all the energy is expressed in calories, as follows:

$$C(s) \quad + \quad O_2(g) \quad \xrightarrow[25°C]{} \quad CO_2(g) \quad + \quad 94.05 \text{ kcal}$$

In the latter case it should be noted that although gases are involved, there is no net change in the number of moles of gaseous materials; hence in this case the enthalpy of reaction does not include mechanical energy of expansion.

When a chemical reaction occurs at constant pressure and involves no work other than perhaps expansion against the atmosphere, the heat of reaction is equal to the enthalpy change, ΔH. The terms *heat of reaction* and *enthalpy change* are sometimes used interchangeably, but for reactions carried out at other than constant pressure, the heat of reaction is not in general equal to the enthalpy change.

Since energy is given off in an exothermic reaction, the enthalpies of the products total less than the enthalpies of the reactants. Hence the sign of ΔH is negative, indicating a decrease in enthalpy. Conversely, in an endothermic reaction, energy must be provided from outside the system, and there is an increase in enthalpy in going from the initial state to the final state. The sign of ΔH is positive. As two examples, the chemical systems described above can be written with their enthalpy changes as

$$2 \text{ HgO(s)} \quad \rightarrow \quad 2 \text{ Hg(l)} \quad + \quad O_2(g) \qquad \Delta H_{298} = +43.4 \text{ kcal}$$
$$C(s) \quad + \quad O_2(g) \quad \rightarrow \quad CO_2(g) \qquad\qquad\quad \Delta H_{298} = -94.05 \text{ kcal}$$

Example

What quantity of heat is required to decompose 10.8 grams of mercury(II) oxide? According to the equation above, 43.4 kcal is needed to decompose 2 moles of HgO. For 10.8 grams,

$$\left(\frac{10.8 \text{ grams}}{216 \text{ grams/mole}}\right)\left(\frac{43.4 \text{ kcal}}{2.00 \text{ moles}}\right) = 1.09 \text{ kcal required}$$

3-6   Enthalpies of Formation

Knowledge of the enthalpy change for a given reaction permits the evaluation of the reaction as a source of energy. For example, fuels for jet planes or for rockets must yield large quantities of energy per gram of fuel. Even more important, in the design of industrial chemical plants, calculations of the energy requirements of the various processes and their related costs must be made. Often, a process is designed so that the heat liberated in an exothermic step in the process is used to augment the energy requirements for an endothermic step.

From a theoretical point of view, chemical reactions may be regarded as involving the breaking of some bonds and the formation of others. For example, the reaction of 1 mole of propane with oxygen

$$\underset{\substack{|\;\;\;\;|\;\;\;\;| \\ H\;\;H\;\;H}}{\overset{\substack{H\;\;H\;\;H \\ |\;\;\;\;|\;\;\;\;|}}{H\!-\!C\!-\!C\!-\!C\!-\!H}} + 5\,O_2 \rightarrow 3\,CO_2 + 4\,H_2O$$

involves the breaking of 2 moles of carbon-to-carbon bonds, 8 moles of carbon-to-hydrogen bonds, and 5 moles of oxygen-to-oxygen bonds, as well as the formation of 6 moles of carbon-to-oxygen bonds and 8 moles of hydrogen-to-oxygen bonds. The observed enthalpy change must reflect the differences in the strengths of the various bonds. The fact that the reaction is exothermic suggests a model in which, on the average, bonds between oxygen and carbon and between oxygen and hydrogen are stronger than bonds between carbon and hydrogen, between carbon and carbon, and between oxygen and oxygen. Enthalpy data can be used to compare bond strengths in this manner (Section 11-2).

Enthalpy changes for many reactions can be determined experimentally. However, for many others it is either inconvenient or impossible to measure enthalpy changes

TABLE 3-2

TABLE 3-2. Standard Enthalpies of Formation at 298 K

	State	ΔH_f° kcal/mole	ΔH_f° kJ/mole		State	ΔH_f° kcal/mole	ΔH_f° kJ/mole
AgCl	s	−30.36	−127.0	HCl	g	−22.02	−92.13
AgBr	s	−23.8	−99.6	HF	g	−64.2	−269
AgI	s	−14.9	−62.3	HI	g	6.20	25.9
Al_2O_3	s	−399.1	−1670	H_2S	g	−4.815	−20.15
$Au(OH)_3$	s	−100	−418	HgO	s	−21.7	−90.8
B_2H_6	g	7.5	31	KCl	s	−104.18	−435.9
BF_3	g	−265.4	−1110	NF_3	g	−30.4	−127
B_2O_3	s	−305.3	−1277	NH_3	g	−11.04	−46.19
BaO	s	−133.5	−558.6	N_2H_4	g	12.05	50.42
$BaCO_3$	s	−290.8	−1217	NH_4Cl	s	−75.38	−315.4
$BaSO_4$	s	−350.2	−1465	N_2O	g	19.49	81.55
CH_4	g	−17.89	−74.85	NO	g	21.60	90.37
CO	g	−26.41	−110.5	NO_2	g	8.09	33.8
CO_2	g	−94.05	−393.5	NaCl	s	−98.23	−411.0
CaO	s	−151.9	−635.5	$NaHCO_3$	s	−169.8	−710.4
$CaCO_3$	s	−288.5	−1207	Na_2CO_3	s	−341.8	−1430.1
$CaCl_2$	s	−190.0	−795.0	PH_3	g	5.5	23
$Ca(OH)_2$	s	−235.80	−986.6	P_2O_5	s	−365.83	−1531
Cr_2O_3	s	−272.7	−1141	SO_2	g	−70.96	−296.9
CuO	s	−37.6	−157	SO_3	g	−94.45	−395.2
Cu_2S	s	−19.0	−79.5	SiO_2	s	−209.9	−878.2
Fe_2O_3	s	−196.5	−822.2	$SiCl_4$	g	−145.7	−609.6
H_2O	l	−68.32	−285.9	SiF_4	g	−361.29	−1511.6
H_2O	g	−57.79	−241.8	WO_3	s	−201.5	−843.1
H_2SO_4	l	−193.91	−811.3	ZnO	s	−83.2	−348
HBr	g	−8.66	−36.2	ZnS	s	−48.5	−203

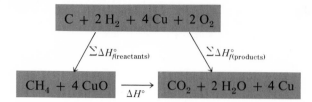

Figure 3-2. **Schematic Diagram Showing**
$\Delta H° = \Sigma\Delta H°_{f(\text{products})} - \Sigma\Delta H°_{f(\text{reactants})}$
To change elements in their standard states (top box) to products in their standard states (right box) requires $\Sigma\Delta H°_{f(\text{products})}$ (right arrow). However, the same conversion can be made by a different path, through the left box to the right box, corresponding to $\Sigma\Delta H°_{f(\text{reactants})} + \Delta H°$. Since the initial and final states are the same, the enthalpy changes along these two paths must be the same. Hence

$$\Sigma\Delta H°_{f(\text{products})} = \Sigma\Delta H°_{f(\text{reactants})} + \Delta H°$$

Rearrangement yields

$$\Delta H° = \Sigma\Delta H°_{f(\text{products})} - \Sigma\Delta H°_{f(\text{reactants})}$$

directly under a given set of conditions. In such cases the enthalpy change must be calculated from other available data.

Since it is the *change* in enthalpy which is important, it is not necessary to know absolute values of the enthalpies of the substances involved in a reaction. Therefore, one can adopt some arbitrary point of reference, or **standard state,** from which the enthalpy changes can be measured. The standard state (not to be confused with STP) of each element and each compound is defined as its most stable physical form at 1 atm pressure and at a specified temperature, usually 298 K. By convention, *each element in its standard state is assigned an enthalpy value of zero*. The enthalpy change that occurs when 1 mole of a compound in its standard state is formed from its elements in their standard states is called the **standard enthalpy of formation** of the compound and is designated by the symbol $\Delta H°_f$. The superscript, °, indicates that all the reactants and all the products are in their standard states. For example, the standard enthalpies of formation of Al_2O_3 and HI are represented as

$$2\,Al(s) \;+\; \tfrac{3}{2}O_2(g, 1\text{ atm}) \;\rightarrow\; Al_2O_3(s) \qquad \Delta H° \text{ at } 298\text{ K} = \Delta H°_f = -399.09 \text{ kcal}$$

$$\tfrac{1}{2}H_2(g, 1\text{ atm}) \;+\; \tfrac{1}{2}I_2(s) \;\rightarrow\; HI(g, 1\text{ atm}) \qquad \Delta H° \text{ at } 298\text{ K} = \Delta H°_f = 6.20 \text{ kcal}$$

Enthalpies of formation of various substances are presented in Table 3-2. It should be noted that the enthalpy change for the dissociation of a compound into its constituent elements is equal in magnitude but opposite in sign to the enthalpy of formation of the compound.

Because ΔH for a reaction is independent of the path of the reaction, the standard enthalpy change for any reaction may be calculated as the difference between the sum[4] of the enthalpies of formation of all the products and the related sum for all the reactants:

$$\Delta H° = \sum \Delta H°_{f(\text{products})} - \sum \Delta H°_{f(\text{reactants})}$$

This principle is illustrated in Figure 3-2 for the reaction

$$CH_4(g) \;+\; 4\,CuO(s) \;\rightarrow\; CO_2(g) \;+\; 2\,H_2O(l) \;+\; 4\,Cu(s)$$

[4] The symbol Σ indicates the "sum of."

Example

Calculate the standard enthalpy change at 298 K for the reaction

$$CH_4(g, 1\ atm)\ +\ 4\ CuO(s)\ \rightarrow\ CO_2(g, 1\ atm)\ +\ 2\ H_2O(l)\ +\ 4\ Cu(s)$$

The sum of the enthalpies of formation of the products (from Table 3-2) is

$$[-94.05 + 2(-68.32) + 4(0)]\ kcal = -230.7\ kcal$$

The related sum for the reactants is

$$[(-17.89) + 4(-37.6)]\ kcal = -168.3\ kcal$$

Then

$$\Delta H_{298}^{\circ} = -230.7 - (-168.3) = -62.4\ kcal$$

The subscript $_{298}$ is used to denote the absolute temperature.

3-7 Hess' Law

The fact that the net enthalpy change is independent of the path permits the calculation of ΔH for reactions which are difficult or impossible to carry out directly. For example, the formation of ethane, C_2H_6, from its elements cannot be carried out in a manner which permits the direct measurement of the enthalpy change. However, it is convenient to measure the enthalpy of reaction of ethane with oxygen, and this result can be combined with other data to allow calculation of ΔH_f° for ethane:

I. $C_2H_6(g, 1\ atm)\ +\ \tfrac{7}{2}O_2(g, 1\ atm)\ \rightarrow\ 2\ CO_2(g, 1\ atm)\ +\ 3\ H_2O(g, 1\ atm)$
$$\Delta H^{\circ} = -368.4\ kcal$$

II. $2\ C(s)\ +\ 2\ O_2(g, 1\ atm)\ \rightarrow\ 2\ CO_2(g, 1\ atm)$
$$\Delta H^{\circ} = 2(-94.05\ kcal)$$

III. $3\ H_2(g)\ +\ \tfrac{3}{2}O_2(g, 1\ atm)\ \rightarrow\ 3\ H_2O(g, 1\ atm)$
$$\Delta H^{\circ} = 3(-57.79\ kcal)$$

Subtracting equation I from the *sum* of equations II and III, and treating the enthalpy changes similarly, one obtains

IV. $2\ C(s)\ +\ 3\ H_2(g, 1\ atm)\ \rightarrow\ C_2H_6(g, 1\ atm)$
$$\Delta H_{f(C_2H_6)}^{\circ} = 3(-57.79) + 2(-94.05) - (-368.4)\ kcal = 6.9\ kcal$$

The relationships among these enthalpy changes are illustrated by an enthalpy diagram (Figure 3-3).

The procedure of algebraically combining equations for several processes and adding the corresponding energies in order to obtain the energy of yet another process is sometimes called **Hess' law.** To use this procedure, it is necessary to collect equations containing all of the species involved as reactants and products, and to

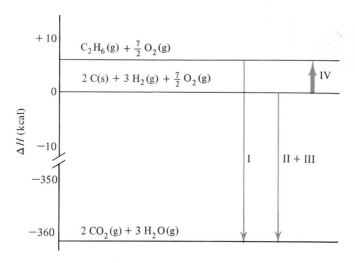

Figure 3-3. Enthalpy Diagram
In this diagram, the scale is drawn so that the elements in their standard states are at zero enthalpy. The enthalpy change for equation IV, represented by the wide arrow, is the difference in enthalpy between equation I and equations II + III.

combine them in such a manner that only the desired reactants and products remain. Since energy and enthalpy are both extensive properties, if other than 1 mole of a given reactant or product is involved, the enthalpy change per mole must be multiplied by the number of moles of that substance. This important principle makes it worthwhile to collect data on the enthalpies of all types of reactions.

3-8 Enthalpies of Combustion

The enthalpy change which occurs when 1 mole of a substance reacts completely with oxygen gas at a given temperature is known as the **enthalpy of combustion,** ΔH_{comb}, of the substance at that temperature. Enthalpies of combustion are conveniently determined in a bomb calorimeter, as described below. Some enthalpies of combustion are listed in Table 3-3. Values of ΔH_{comb} are particularly useful for determining values for ΔH_f for organic compounds. If the values of ΔH_f of the products of the combustion reaction are known, the ΔH_f of the compound may be calculated by means of Hess' law. This procedure is very useful in the case of compounds which cannot be conveniently synthesized by direct combination of their constituent elements.

Example

The enthalpy of combustion of glucose, $C_6H_{12}O_6$, is -673 kcal/mole at 298 K. Calculate the standard enthalpy of formation of glucose. Sketch an enthalpy diagram for the process.

$$C_6H_{12}O_6(s) \quad + \quad 6\,O_2(g) \quad \rightarrow \quad 6\,CO_2(g) \quad + \quad 6\,H_2O(l)$$

Since

$$\Delta H_{reaction} = \sum \Delta H_{f(products)} - \sum \Delta H_{f(reactants)}$$

TABLE 3-3. Some Standard Enthalpies of Combustion at 298 K

			ΔH°_{comb}	
Compound	Formula	State	kcal/mole	kJ/mole
ammonia[a]	NH_3	g	-81.0	-339
cyanamide[b]	NH_2CN	s	-177.2	-741
carbon disulfide[c]	CS_2	l	-246.6	-1032
boron carbide[d]	B_4C	s	-683.3	-2859
carbon monoxide	CO	g	-68.0	-285
methane	CH_4	g	-210.8	-882
ethane	C_2H_6	g	-368.4	-1541
propane	C_3H_8	g	-526.3	-2202
octane	C_8H_{18}	l	-1302.7	-5450
benzene	C_6H_6	l	-782.3	-3273
naphthalene	$C_{10}H_8$	s	-1232.5	-5157
toluene	C_7H_8	l	-934.2	-3909
acetylene	C_2H_2	g	-312.0	-1305
ethylene	C_2H_4	g	-331.6	-1387

[a]Combustion products are N_2 and H_2O.
[b]Combustion products are N_2, CO_2, and H_2O.
[c]Combustion products are CO_2 and SO_2.
[d]Combustion products are B_2O_3 and CO_2.

$$\Delta H^\circ = (6\,\Delta H^\circ_{f(CO_2)} + 6\,\Delta H^\circ_{f(H_2O)}) - (\Delta H^\circ_{f(C_6H_{12}O_6)} + 6\,\Delta H^\circ_{f(O_2)})$$
$$-673\ \text{kcal} = 6(-94.05) + 6(-68.32) - \Delta H^\circ_{f(C_6H_{12}O_6)} - 6(0.0)$$
$$\Delta H^\circ_{f(C_6H_{12}O_6)} = -301\ \text{kcal/mole}$$

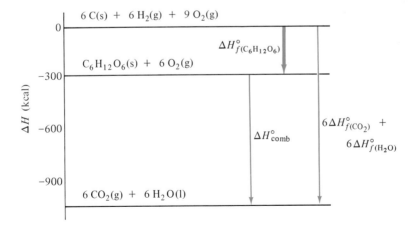

The enthalpy diagram emphasizes that the enthalpy change for the conversion of 6 moles of carbon, 6 moles of hydrogen molecules, and 9 moles of oxygen molecules to 6 moles of carbon dioxide and 6 moles of water is the same whether or not one first "prepares" a mole of glucose and then allows that to react with the rest of the oxygen to give the same products.

3-9 Enthalpy Changes in Other Than Standard States

Few reactions are carried out between reactants in their standard states to yield products in their standard states. However, the enthalpy changes for standard state conditions can be evaluated from the experimental data by accounting for the differences in enthalpies between the actual states and the standard states. These differences are due to such factors as the changes in temperatures and physical states of one or more of the substances involved. For example, it is often desired to estimate the enthalpy change of a reaction at temperatures other than 298 K. Conversely, since many reactions do not take place at 298 K, their enthalpy changes may be measured at an elevated temperature; then it is necessary to determine ΔH_{298}. Since enthalpy change depends on only the initial and final states of the system, it is possible to conceive of the enthalpy change occurring in several steps, each having its respective enthalpy change. The overall ΔH will be the summation of these steps. The process is another application of Hess' law. This procedure is illustrated schematically in Figure 3–4. Assuming that the heat capacities are constant over the entire temperature range[5] the enthalpy change ΔH_{298} can be calculated from the measured value of ΔH_{T_2}:

$$\Delta H_{298} = \Delta H_{\text{reactants}} + \Delta H_{T_2} + \Delta H_{\text{products}}$$

Many of the data of Table 3–2 were obtained by this procedure. Conversely, these data can now be used to calculate enthalpy changes for these reactions at other temperatures.

Example

Using data from Tables 3–1 and 3–2, calculate the enthalpy change for the reaction between carbon and oxygen at 400°C to form 1.00 mole of carbon dioxide at 400°C.

The enthalpy of reaction at 673 K may be obtained from the corresponding enthalpy change at 298 K and the heat capacity of each reactant and product. All the enthalpy changes are identified in Figure 3–4.

$$\Delta H_{673} = \Delta H_{298} - \Delta H_{\text{products}} - \Delta H_{\text{reactants}}$$
$$\Delta H_{298} = -94.05 \text{ kcal}$$

---

[5] For small temperature changes, this assumption is a good approximation. For large changes in temperature, particularly for gases, C_p may be expressed as a function of temperature and the sum evaluated by the methods of calculus.

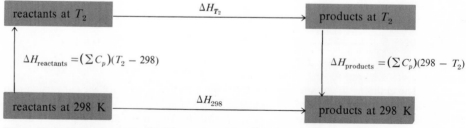

Figure 3-4. Determination of ΔH_{298} from Other Enthalpy Data
ΣC_p and $\Sigma C_p'$ are the sums of the heat capacities of the reactants and of the products, respectively.

$$\Delta H_{\text{products}} = C_{p(CO_2)}(298 - 673) = (8.96 \text{ cal/mole} \cdot \text{deg})(-375 \text{ deg})(1.00 \text{ mole})$$
$$= -3360 \text{ cal} = -3.36 \text{ kcal}$$
$$\Delta H_{\text{reactants}} = (C_{p(O_2)} + C_{p(C)})(673 - 298)$$
$$= [(7.05 + 2.04) \text{ cal/mole} \cdot \text{deg}](375 \text{ deg})(1.00 \text{ mole})$$
$$= +3410 \text{ cal} = +3.41 \text{ kcal}$$
$$\Delta H_{673} = -94.05 \text{ kcal} + 3.36 \text{ kcal} - 3.41 \text{ kcal} = -94.10 \text{ kcal}$$

Reactions involving a given substance in different physical states yield different values of enthalpy change, as is illustrated in Table 3–2 by the case of water. One ΔH_f° value refers to the formation of liquid water at 298 K, and the other refers to the formation of gaseous water in the hypothetical state of 1 atm and 298 K. The difference between the values of ΔH_f° for these reactions gives the enthalpy change for the vaporization of water at 298 K:

$$H_2(g, 1 \text{ atm}) + \tfrac{1}{2}O_2(g, 1 \text{ atm}) \rightarrow H_2O(g, 1 \text{ atm}) \qquad \Delta H_f^\circ = -57.79 \text{ kcal}$$
$$H_2(g, 1 \text{ atm}) + \tfrac{1}{2}O_2(g, 1 \text{ atm}) \rightarrow H_2O(l) \qquad \Delta H_f^\circ = -68.32 \text{ kcal}$$
$$H_2O(l) \rightarrow H_2O(g, 1 \text{ atm}) \qquad \Delta H_{\text{vap}} = 10.53 \text{ kcal}$$

Similarly, any change in state will be accompanied by a definite enthalpy change. ΔH for a phase change for any quantity of a substance is

$$\Delta H = (\text{number of moles})(\Delta H/\text{mole})$$
$$\Delta H = (\text{mass})(\Delta H/\text{gram})$$

Enthalpy changes for processes of changing state are often given descriptive names, some examples of which are listed in Table 3–4.

Example

What is the enthalpy change when 1.00 gram of water is frozen at 0°C?
Freezing is just the reverse of melting; hence the enthalpy change must be -1.435 kcal/mole.

TABLE 3-4. Changes of State

Process	Name of Enthalpy Change		Example	Enthalpy Change[a]		Temperature (K)
				kcal/mole	kJ/mole	
melting	enthalpy of fusion	ΔH_{fus}	$H_2O(s) \rightarrow H_2O(l)$	1.435	6.004	273
boiling	enthalpy of vaporization	ΔH_{vap}	$H_2O(l) \rightarrow H_2O(g, 1 \text{ atm})$	9.72	40.7	373
sublimation	enthalpy of sublimation	ΔH_{sub}	$CO_2(s) \rightarrow CO_2(g, 1 \text{ atm})$	3.87	16.2	183
structure change	enthalpy of transition	ΔH_{trans}	Fe(s, α-form) → Fe(s, β-form)	0.22	0.92	1184
dissolving	enthalpy of solution	ΔH_{soln}	C_2H_5OH + 9 H_2O → 10 mole % solution	-1.678	-7.021	298

[a] At the given temperature.

$$\Delta H = \frac{-1435 \text{ cal/mole}}{18.0 \text{ grams/mole}} = -79.7 \text{ cal/gram}$$

Therefore, 79.7 calories of heat is liberated when 1.00 gram of water is frozen at 0°C and 1.00 atm pressure.

3-10 Experimental Determination of Enthalpy Changes

Enthalpy changes may be determined in an apparatus known as a **calorimeter.** A calorimeter is usually designed for measurement of a specific type of reaction system, and various ingenious devices, which allow determinations of very high accuracy, have been constructed. For reactions carried out in aqueous solution, the calorimeter may consist of a vessel holding a precisely known mass of water. The vessel is equipped with a stirrer and an accurate thermometer (Figure 3–5). The calorimeter is surrounded by some type of insulation so that the exchange of heat with the surroundings is minimized. The insulation may be asbestos, Styrofoam, or a vacuum jacket. When a reaction occurs, the heat liberated raises the temperature of the contents of the calorimeter, and the number of calories liberated may be calculated from the temperature rise and the specific heats of the materials involved. However, it is usually more convenient to determine the "water equivalent" of the calorimeter prior to the actual experiment by adding a known quantity of heat to the system and measuring the resulting temperature rise. The heat required to raise all the parts of the calorimeter is obtained, and this result is expressed as equivalent to the heat capacity of a certain mass of water. When a reaction is run in the calorimeter under

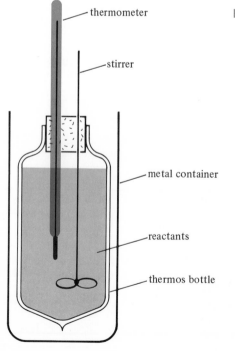

thermometer

stirrer

metal container

reactants

thermos bottle

Figure 3–5. **Simple Calorimeter**

identical conditions, the quantity of heat liberated is calculated from the observed temperature rise as though all the heat were absorbed by water.

Example

When 10,000 J of electrical energy is added to a calorimeter containing 400 grams of water, the temperature of the calorimeter and contents is raised 5.00°C. Calculate the water equivalent of the calorimeter.

$$10,000 \text{ J} \left(\frac{1.000 \text{ cal}}{4.184 \text{ J}} \right) = 2390 \text{ cal}$$

The mass of water which would be raised 5.00°C by this quantity of energy is given by

$$\frac{2390 \text{ cal}}{(1.00 \text{ cal/gram} \cdot \text{deg})(5.00 \text{ deg})} = 478 \text{ grams}$$

Since there is only 400 grams of water present, the calorimeter must absorb the rest of the heat. The calorimeter is equivalent in heat-absorbing capacity to 78 grams of water.

For reactions which involve gaseous materials or which occur at high temperature, enthalpies of reaction may be determined in a bomb calorimeter (Figure 3-6). The "bomb" is a thick-walled vessel which can be filled with gas to pressures of about 30 atm. In a typical experiment, such as the determination of the enthalpy of combustion of a substance in oxygen, a weighed sample of the substance is placed in the bomb, which is then filled with oxygen and sealed. Next the bomb is placed in a

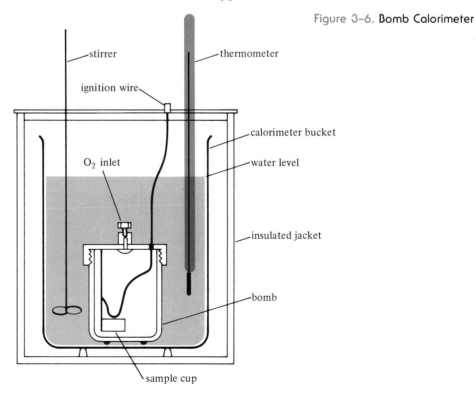

Figure 3-6. Bomb Calorimeter

stirrer

thermometer

ignition wire

calorimeter bucket

O₂ inlet

water level

insulated jacket

bomb

sample cup

weighed quantity of water contained in the calorimeter bucket, and the apparatus is assembled as shown in Figure 3–6. Ignition is brought about by means of an electrical connection through the top of the bomb, and the rise in temperature of the calorimeter system owing to the reaction is measured. The number of calories liberated is obtained by use of the previously determined water equivalent of the calorimeter. In addition, corrections have to be applied for the heat released by the ignition wire and by the friction of the stirrer. Since the reaction is run in a sealed bomb at constant volume instead of at a constant pressure, the observed energy change, ΔE, is different from the enthalpy change, ΔH, which is equal to the heat of reaction for a process occurring at constant pressure. The enthalpy change may be obtained by using the definition of enthalpy and the ideal gas law:

$$H = E + PV = E + nRT$$

where n is the number of moles of *gas* involved. It follows that for a reaction run at a given temperature

$$\Delta H = \Delta E + \Delta n\, RT$$

where $\Delta n = n_2 - n_1$ is the change in the number of moles of gas in going from reactants to products. Note that if ΔE and ΔH are expressed in calories, the gas constant, R, must be expressed as 1.987 cal/mole $\cdot$ K. If ΔE and ΔH are expressed in joules, R is 8.314 J/mole $\cdot$ K.

Example

The reaction of cyanamide, $NH_2CN(s)$, with oxygen was run in a bomb calorimeter, and ΔE was found to be -742.7 kJ/mole of $NH_2CN(s)$ at 298 K. Calculate ΔH_{298} for the reaction.

$$NH_2CN(s) \; + \; \tfrac{3}{2}O_2 \; \rightarrow \; N_2 \; + \; CO_2 \; + \; H_2O(l)$$

The number of moles of gaseous reactant (O_2) is $\tfrac{3}{2}$; the number of moles of gaseous products (N_2 + CO_2) is 2. Therefore

$$\Delta n = 2 - \tfrac{3}{2} = \tfrac{1}{2}$$
$$\Delta H = \Delta E + \Delta n\, RT = -742.7 \text{ kJ} + (0.500 \text{ mole})(8.314 \text{ J/mole} \cdot \text{K})(298 \text{ K})$$
$$= -742.7 \text{ kJ} + 1240 \text{ J} = -741.5 \text{ kJ}$$

3-11   Enthalpy Changes in Reactions Between Ions

When reactions are carried out in solution, the presence of the solvent may influence the enthalpy change. Moreover, many reactions occurring in aqueous solution involve ions rather than neutral molecules. The reaction between silver ion and chloride ion is aqueous solution will have a different enthalpy change from $\Delta H^\circ_{f(AgCl)}$.

$$Ag(s) \; + \; \tfrac{1}{2}Cl_2(g) \; \rightarrow \; AgCl(s) \qquad \Delta H^\circ_f = -30.36 \text{ kcal}$$
$$Ag^+(aq) \; + \; Cl^-(aq) \; \rightarrow \; AgCl(s) \qquad \Delta H^\circ_{298} = -15.67 \text{ kcal}$$

The enthalpy of reaction between the silver ions and chloride ion must be equal to the difference between the enthalpy of formation of the solid silver chloride and the sum of the enthalpies of formation of the aqueous ions. However, there is no direct way to measure the standard enthalpy of formation of a single ion in solution, since a single type of ion is never found alone. A scale of *relative* enthalpies for ions may be established by adopting the convention that *for the hydrogen ion in solution at unit activity,*[6] ΔH_f° *is zero at* 298 K.

Example

Calculate ΔH_f° for chloride ion from the following data:

$$\tfrac{1}{2} H_2(g) \;+\; \tfrac{1}{2} Cl_2(g) \;\rightarrow\; HCl(g) \qquad\qquad \Delta H_f^\circ = -22.1 \text{ kcal}$$

$$HCl(g) \;+\; n\,H_2O \;\rightarrow\; H^+(aq) \;+\; Cl^-(aq) \qquad \Delta H_{298} = -17.9 \text{ kcal}$$

(Here $n\,H_2O$ signifies a large excess of water molecules, and $H^+(aq)$ and $Cl^-(aq)$ represent these ions in the presence of the large excess of water.) Adding the two equations,

$$\tfrac{1}{2} H_2(g) \;+\; \tfrac{1}{2} Cl_2(g) \;+\; n\,H_2O \;\rightarrow\; H^+(aq) \;+\; Cl^-(aq) \qquad \Delta H_{298} = -40.0 \text{ kcal}$$

Remembering that the enthalpy of reaction is the difference between the enthalpies of formation of the products and those of the reactants and considering that the enthalpy of the water on the two sides of the equation cancels out:

$$-40.0 \text{ kcal} = \Delta H_{f(H^+)}^\circ + \Delta H_{f(Cl^-)}^\circ - \tfrac{1}{2}\{\Delta H_{f(H_2)}^\circ + \Delta H_{f(Cl_2)}^\circ\}$$
$$= \quad 0 \quad + \Delta H_{f(Cl^-)}^\circ - \quad 0 \quad - \quad 0$$
$$\Delta H_{f(Cl^-)}^\circ = -40.0 \text{ kcal}$$

Enthalpies of formation of several ions in water, obtained in an analogous manner, are listed in Table 3–5. These may be used to calculate the enthalpies of reactions

[6] Activity is the effective concentration of a species. In solution the activity is equal to the stoichiometric concentration times a factor, called the activity coefficient, which for dilute solutions is usually close to unity. For present purposes, molar concentrations will be used in place of activities. For gases, the activity is approximately equal to the pressure in atmospheres.

TABLE 3-5. Standard Enthalpies of Formation of Aqueous Ions at Unit Activity and 298 K

	ΔH_f°			ΔH_f°	
Ion	kcal/mole	kJ/mole	Ion	kcal/mole	kJ/mole
H^+	0.00	0.00	OH^-	−54.96	−230.0
Ag^+	25.31	105.9	F^-	−78.66	−329.1
K^+	−60.04	−251.2	Cl^-	−40.0	−167
Ca^{2+}	−129.77	−542.96	Br^-	−28.9	−121
Mg^{2+}	−110.41	−461.96	I^-	−13.4	−56.1
Ba^{2+}	−128.67	−538.36	HS^-	−4.22	−17.7
Cu^{2+}	15.39	64.39	S^{2-}	10.0	41.8
Zn^{2+}	−36.34	−152.0	CO_3^{2-}	−161.63	−676.26
Hg^{2+}	41.56	173.9	SO_4^{2-}	−216.9	−907.5

between ions in solution in precisely the same manner as any other enthalpy of formation.

Example

Calculate the enthalpy change of the reaction

$$2 H^+(aq) + CO_3^{2-}(aq) \rightarrow CO_2(g) + H_2O(l)$$

$$\Delta H° = \Delta H°_{f(CO_2)} + \Delta H°_{f(H_2O)} - \Delta H°_{(CO_3^{2-})} - 2\Delta H°_{f(H^+)}$$
$$= -94.0 + (-68.3) - (-161.6) - 2(0) = -0.7 \text{ kcal}$$

3-12   Free Energy and Entropy—Criteria for Spontaneous Change

A major objective of chemists is to understand and control chemical reactions—to know whether or not under a given set of conditions two substances will react when mixed, to predict the extent to which a given reaction will proceed before equilibrium is established, and to determine whether or not a given reaction will be endothermic or exothermic.

The enthalpy change in a chemical reaction is a measure of the difference in energy content of the products and reactants. It is tempting to assume that exothermic reactions will proceed spontaneously upon mixing the reactants and that endothermic reactions do not occur spontaneously. However, there are endothermic reactions which do proceed spontaneously. For example, ammonium chloride decomposes perceptibly at room temperature and appreciably at higher temperatures:

$$NH_4Cl(s) \rightarrow NH_3(g) + HCl(g) \qquad \Delta H°_{298} = +42 \text{ kcal}$$

An endothermic reaction between ions which occurs spontaneously at room temperature is the reaction of magnesium ion with carbonate ion:

$$Mg^{2+}(aq) + CO_3^{2-}(aq) \rightarrow MgCO_3(s) \qquad \Delta H°_{298} = +6 \text{ kcal}$$

Obviously the enthalpy change is not the only factor responsible for the tendency of a system to undergo spontaneous change.

The transfer of heat energy from an object at a higher temperature to one at a lower temperature is a familiar example of a spontaneous process. It must be recalled that heat is a unique form of energy in that at constant temperature, heat cannot be completely converted to any other form of energy. The heat content, or enthalpy, of any system must be considered in two parts:

1. That which is free to be converted to other forms of energy.
2. That which is necessary to maintain the system at the specified temperature and thus is unavailable for conversion.

This concept can be expressed in the form of an equation:

enthalpy = free energy + unavailable energy

Expressed in symbols, the equation becomes

$$H = G + TS$$

where H is the enthalpy, and G is the free energy. The term denoting unavailable energy is expressed as the product of absolute temperature, T, and a factor which is called **entropy,** represented by the letter S. Above the temperature absolute zero, every substance possesses a finite (nonzero) entropy. The nature of entropy will be discussed in the next section. It follows from the preceding equation that all of the enthalpy of a system would be available energy only if the system were at the temperature defined as absolute zero. It turns out that the absolute zero of this thermodynamic temperature scale is identical with the absolute zero defined for ideal gases (see Section 2-14).

In any process ΔH is the difference between the enthalpy of the final state of a system and the enthalpy of the initial state of the system. The following equations can be written to express the enthalpies of two different states:

For state 2: $H_2 = G_2 + T_2 S_2$
For state 1: $H_1 = G_1 + T_1 S_1$

Therefore,

$$\Delta H = H_2 - H_1 = (G_2 - G_1) + (T_2 S_2 - T_1 S_1)$$
$$\Delta H = \Delta G + \Delta(TS)$$

For a process occurring at constant temperature, in which $T_2 = T_1 = T$, the equation becomes

$$\Delta H = \Delta G + T\Delta S \qquad \text{(constant } T\text{)}$$

This equation, which expresses the limitation on the conversion of heat energy to other forms of energy, is a statement of the **second law of thermodynamics.**

Like enthalpy change (ΔH), the free energy change (ΔG) and the entropy change (ΔS) are thermodynamic functions; that is, their magnitudes depend only on the initial and final states of the system and are independent of the manner in which the system changes from its initial state to its final state. In addition, ΔH, ΔG, and ΔS are extensive properties; that is, their numerical magnitudes depend on the number of moles of substances involved in the system. As described in Section 3-10, it is often possible to determine experimentally the enthalpy change, ΔH, and as shown in Sections 5-8 and 6-6, ΔG and ΔS can also be determined experimentally. Generally, free energy changes are expressed in calories (or kilocalories) per mole or joules per mole, and entropy changes are expressed as calories per mole per degree Kelvin (1 cal/mole $\cdot$ K = 1 entropy unit, abbreviated eu) or joules per mole per degree Kelvin.

If a process is to occur spontaneously, it must not require energy from outside the system. (The process might be endothermic, but in such cases energy is provided from the system itself.) In fact, free energy should be released by the system, so that the final state of the system will have less available energy than its initial state. In more concise terms, any spontaneous process must be accompanied by a *decrease* in free energy.

It is customary to solve the equation representing the second law of thermodynamics explicitly for ΔG as follows:

$$\Delta G = \Delta H - T\Delta S \qquad \text{(constant } T)$$

and to apply the following criteria for spontaneous change and for the equilibrium state:

1. If ΔG is negative, the given process may occur spontaneously.
2. If ΔG is positive, the indicated process cannot occur spontaneously; instead the reverse of the indicated process may occur.
3. If ΔG is zero, neither the indicated process nor the reverse process can occur spontaneously. The system is in a state of equilibrium (see Chapter 6). The indicated process is said to be a reversible one because a very small change in conditions can make ΔG either positive or negative.

3-13   The Nature of Entropy

To gain some insight into the nature of entropy, consider the following experiment, which is also diagrammed in Figure 3–7. A vessel containing an ideal gas, A, is connected by means of a valve to a second vessel containing another ideal gas, B, which does not react with A. The system is maintained at a constant temperature.

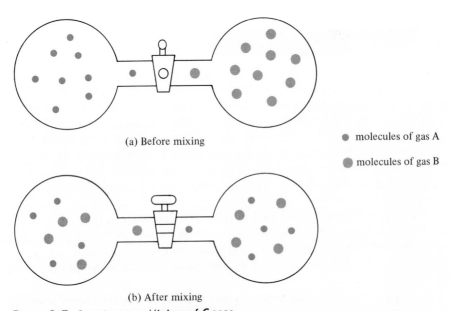

(a) Before mixing

● molecules of gas A

● molecules of gas B

(b) After mixing

Figure 3–7. **Spontaneous Mixing of Gases**
Gases mix because of an entropy increase despite the lack of an enthalpy change.

When the valve is opened, the two gases mix spontaneously, and thereafter both vessels contain the same mixture of the two gases. Since no reaction occurs and the gases are ideal, the enthalpy change is zero. The free energy change for the spontaneous mixing of the gases is given by

$$\Delta G = 0 - T\Delta S = -T\Delta S$$

Since the process is spontaneous, ΔG must be negative; hence ΔS must be positive. In other words, the entropy of the system must have increased.

Entropy is a property which expresses the randomness of a particular state of a system. In the experiment described, once the valve is opened, the most probable situation is one in which there is a random mixture of the two gases. Therefore the entropy of the system is said to have increased relative to the system in which each gas was in a separate vessel. Indeed, it is extremely improbable that the gases will spontaneously separate and return to their original vessels. Such a process would require a decrease in the entropy.

Some processes which increase the randomness of a given system, thus increasing the entropy of the system, include the following:

1. Breaking down large molecules into smaller ones.
2. Increasing the number of moles of gas in the system.
3. Melting a solid.
4. Evaporating a pure liquid.
5. Mixing two or more pure substances, including dissolving one substance in another.

Some processes which generally cause a decrease in the entropy of a system include:

1. Association of small molecules into larger ones.
2. Reaction of a gas to form a liquid or solid.
3. Condensation of a gas.
4. Freezing of a liquid.
5. Crystallization of a dissolved substance.
6. Reaction in solution leading to the formation of a precipitate.

Of course, in any actual system undergoing change, processes tending to increase the entropy and processes tending to decrease the entropy may occur simultaneously. In such cases the overall entropy change will be the net result of the various individual processes.

3-14   Spontaneous Endothermic Reactions

The relationship

$$\Delta G = \Delta H - T\Delta S$$

explains why very exothermic reactions are spontaneous even if there is an accompanying decrease in entropy. For such reactions, ΔH is a sufficiently large negative quantity that ΔG remains negative despite the fact that $-T\Delta S$ is positive. It may also be seen that some endothermic reactions can be spontaneous; if there is a relatively large positive entropy change, the term $-T\Delta S$ will more than compensate for a small positive value of ΔH, and ΔG will be negative. Consider the reaction

$$NH_4Cl(s) \rightarrow NH_3(g) + HCl(g) \qquad \Delta H^\circ_{298} = +42 \text{ kcal}$$

In this case, owing to the formation of 2 moles of gaseous products per mole of solid NH_4Cl reacted, there is a relatively large increase in entropy. The term $-T\Delta S$ more than compensates for the positive value of ΔH, and accordingly, ΔG is negative. Because this is a spontaneous change, the energy used up in the reaction must come from within the system itself. If no energy is added to maintain the temperature, there will be a drop in temperature. That is, thermal energy of the system is used to provide the energy needed for the endothermic process.

3-15   Entropy Changes in Changes of Phase

The transition of a pure substance from one physical state to another (a phase change) which occurs at a fixed temperature is a reversible process. For example, at $0\,°C$ and 1 atm pressure, a system composed of ice and water is in a state of equilibrium. If the pressure on the system is increased very slightly, ice will melt. Conversely, a very small decrease in pressure causes the water to freeze. Applying the criterion that $\Delta G = 0$,

$$\Delta H_{\text{transition}} = T\Delta S$$

Hence the entropy change for such a reversible process can be determined using the relationship

$$\Delta S = \frac{\Delta H_{\text{transition}}}{T}$$

Example

At its melting point, $0\,°C$, the enthalpy of fusion of water is 1.435 kcal/mole. What is the molar entropy change for the melting of ice at $0\,°C$?

$$\Delta S = \frac{1435 \text{ cal/mole}}{273 \text{ K}} = 5.26 \text{ cal/mole} \cdot \text{K}$$

3-16   Standard Free Energy Change

The free energy change of a process in which the reactants in their standard states are converted into products in their standard states is called the **standard free energy change,** $\Delta G°$. By convention, the standard free energy of formation of every element in its standard state at 298 K and 1 atm is taken to be zero. The free energy of

TABLE 3-6. Standard Free Energies of Formation at 298 K

	ΔG_f°			ΔG_f°	
	kcal/mole	kJ/mole		kcal/mole	kJ/mole
$H_2O(l)$	−56.7	−237	ClO_2	29.5	123
$HCl(g)$	−22.7	−95.0	$CH_3Cl(g)$	−19.6	−82.0
$H_2S(g)$	−7.89	−33.0	$CCl_4(l)$	−33.3	−139
$NO_2(g)$	12.4	51.9	$C_6H_{12}O_6(s)^a$	−215	−900
$NH_3(g)$	−3.97	−16.6	$Al_2O_3(s)$	−376.8	−1577
$CO_2(g)$	−94.3	−395	$BaO(s)$	−126.3	−528.4
$CH_4(g)$	−12.14	−50.79	$BaSO_4(s)$	−350.2	−1465
$C_2H_4(g)$	16.28	68.12	$CaO(s)$	−144.4	−604.2
$C_2H_6(g)$	−7.86	−32.9	$CaCO_3(s)$	−269.8	−1129
$C_6H_6(l)$	29.8	125	$CoO(s)$	−30.4	−127
Cl_2O	22.4	93.7	$SiO_2(s)$	−192.4	−805.0

[a] Glucose.

formation of a compound is defined as the free energy change for the reaction of the elements in their standard states to form the compound in its standard state. Standard free energies of formation of several compounds are listed in Table 3-6.

If the standard free energies of formation of all the reactants and products are known, ΔG° for a reaction can be calculated as the difference between the sum of the free energies of formation of the products minus the corresponding sum for the reactants. In Figure 3-8, this generalization is shown in a manner analogous to the use of Hess' law.

Example

Using data from Table 3-6, calculate ΔG° at 298 K for the reaction

$$CH_4(g) \;+\; 2\,O_2 \;\rightarrow\; CO_2(g) \;+\; 2\,H_2O(l)$$

$$\Delta G^\circ = \{\Delta G_{f(CO_2)}^\circ + 2\Delta G_{f(H_2O)}^\circ\} - \{\Delta G_{f(CH_4)}^\circ + 2\Delta G_{f(O_2)}^\circ\}$$
$$= \{-94.3\,\text{kcal} + (-113.4\,\text{kcal})\} - \{(-12.14\,\text{kcal}) + 0\,\text{kcal}\} = -195.6\,\text{kcal}$$

The standard free energy of formation of a compound, ΔG_f°, is a measure of the stability of the substance. For example, if a compound has a large negative value of ΔG_f°, the compound cannot be readily decomposed into its elements. Indeed, that compound is likely to be formed as a product in any reaction in which its constituent

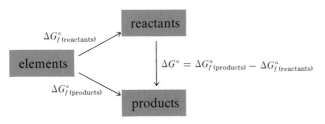

Figure 3-8. **Relationship Between ΔG° of a Reaction and ΔG_f° of Reactants and Products**

$$\Delta G^\circ = \Delta G_{f\,(products)}^\circ - \Delta G_{f\,(reactants)}^\circ$$

elements are involved. On the other hand, a compound for which ΔG_f° is positive will tend to be unstable with respect to decomposition into its elements. Moreover, it will tend to be consumed in almost any reaction, being converted into substances having lower free energies of formation. For example, it might be predicted that methane in its standard state will react with oxygen at 25°C,

$$CH_4(g, 1\ atm)\quad +\quad 2\ O_2(g, 1\ atm)\quad \rightarrow\quad CO_2(g, 1\ atm)\quad +\quad 2\ H_2O(l)$$

since the products are thermodynamically more stable than CH_4. (See the values of ΔG_f° in Table 3–6.) In contrast, the reaction of $CCl_4(l)$ in its standard state with oxygen at 25°C

$$CCl_4(l)\quad +\quad 2\ O_2(g, 1\ atm)\quad \rightarrow\quad CO_2(g, 1\ atm)\quad +\quad 2\ Cl_2O(g, 1\ atm)$$

is not likely to occur because the value of ΔG_f° of Cl_2O, $+22.4$ kcal/mole, is much greater than that for CCl_4. Because CCl_4 is so unreactive, it was formerly used in some types of fire extinguishers. The vapor of CCl_4 has a high density and the fire would be smothered for lack of oxygen. However, ingestion of CCl_4 causes liver damage, and if it is used to extinguish electrical fires, there is a danger that the CCl_4 would be converted to phosgene, $COCl_2$, a highly poisonous compound. Consequently, the use of CCl_4 in fire extinguishers has been prohibited.

It might be assumed that the more the free energy is lowered in a reaction, the faster the reaction will proceed. This assumption is not correct. Predictions based on thermodynamics take into consideration only the initial state and the final state of a system. Nothing is assumed about the rate or path or mechanism for going from one state to another. Applied to chemistry, the useful result of obtaining values for free energy changes for various reactions is the ability to determine whether a given reaction can occur spontaneously. For example, a calculation shows that for the reaction

$$H_2(g)\quad +\quad \tfrac{1}{2} O_2(g)\quad \rightleftharpoons\quad H_2O(l)$$

ΔG° is -56.7 kcal/mole at 298 K. The negative value of ΔG° indicates that the reaction should proceed spontaneously; yet a mixture of hydrogen gas and oxygen gas will remain indefinitely at room temperature without reacting. However, in the presence of a catalyst, such a finely divided palladium, such a mixture would be converted rapidly and smoothly at room temperature into water.

This example points out the difference between the concepts of **kinetic stability** and **thermodynamic stability.** Many systems are thermodynamically unstable at ordinary temperatures and exist only because the rate at which they react is low. Such systems are stable in a kinetic sense only, and if some change which increases the rate of reaction is made, the predicted reaction will occur spontaneously. Some examples of thermodynamically unstable substances are hydrogen peroxide, ozone, trinitrotoluene (TNT), and nitroglycerine. Thermodynamics does not explain *why* one molecule is more stable than another. Also, although a particular reaction may be predicted to occur, there is no guarantee that it will proceed fast enough to be observed; further, some other reaction of the same reactants may be even more thermodynamically favorable. Nevertheless, the calculations of ΔG_f° are important and useful because, as will be shown later, in addition to predicting whether a reaction is feasible, ΔG_f° can also be used to predict the extent to which a chemical reaction may proceed.

3-17 Exercises

Basic Exercises

1. Identify the kind of energy change usually associated with each of the following: **(a)** toaster, **(b)** radio, **(c)** automobile engine, **(d)** automobile battery, **(e)** automobile generator, **(f)** model airplane engine, **(g)** steam engine, **(h)** friction, **(i)** ski jump, **(j)** fluorescent lamp, **(k)** photoelectric cell, **(l)** automobile starter motor, **(m)** furnace.

2. In a certain process, 500 cal of work is done on a system which gives off 200 cal of heat. What is the value of ΔE for the process?

3. Distinguish between a change of state and a phase change. Give an example of each.

4. Contrast the meaning of the word *molar* when used in the term *molar heat capacity* with that in the term *molar solution*.

5. Calculate the number of calories necessary to raise the temperature of 60.0 grams of aluminum from 35 to 55°C.

6. Calculate the heat evolved when 0.75 mole of molten aluminum at its melting point of 658°C is solidified and cooled to 25°. The enthalpy of fusion of aluminum is 76.8 cal/gram.

7. If 2.00×10^3 J of energy raises the temperature of a calorimeter containing 45.0 grams of water from 23.0 to 32.0°C, what is the water equivalent of the calorimeter?

8. Estimate the specific heat of platinum.

9. Determine the final temperature of a system after 100.0 grams of zinc at 95°C is immersed into 50.0 grams of water at 15°C.

10. Determine the specific heat of water in joules per gram per degree.

11. Determine the enthalpy change for the reaction of 96.0 grams of sulfur trioxide with excess barium oxide to yield barium sulfate at 298 K.

12. What is the standard free energy change for the melting of 3.0 moles of water at 0°C? Determine the entropy change for this process. Is the entropy greater for the liquid or the solid?

13. In the equation for expansion work against the atmosphere, $w = -P\Delta V$, which one(s) of the following does P represent? **(a)** the pressure of the system, **(b)** the pressure of the surroundings, **(c)** some constant pressure.

14. Calculate ΔH at 25°C for the reaction of one mole of carbon with excess carbon dioxide to produce carbon monoxide.

15. At 25°C and 1 atm, which one(s) of the following, if any, has a nonzero ΔH_f°? **(a)** Fe, **(b)** O, **(c)** C(s), **(d)** Ne.

16. In Figure 3–3, why is the state $C(s) + 2 H_2(g) + 2 O_2(g)$ assigned a value of 0 kcal? Would it make a difference in the magnitudes of the calculated enthalpy changes if this state were assigned some other value? Explain.

17. Starting with the definition $\Delta H = \Delta E + \Delta(PV)$ and other definitions in Chapter 3, prove that the enthalpy change for a process with no work other than expansion is equal to the heat added at constant pressure.

18. The dimensions of entropy are cal/mole · K while molar heat capacity has dimensions given in cal/mole · deg. **(a)** What is the relationship between these sets of dimensions? **(b)** Which of these functions could use either set of dimensions?

19. A 40.0 gram sample of a metal at 50.0°C is immersed in 100.0 grams of water at 10.0°C. The final temperature of the system is 13.0°C. What is the specific heat of the metal? What is the approximate atomic weight of the metal?

20. The enthalpy change for which of the following processes represents the enthalpy of formation of AgCl? Explain.
    (a) $Ag^+(aq) + Cl^-(aq) \rightarrow AgCl(s)$
    (b) $Ag(s) + \frac{1}{2} Cl_2(g) \rightarrow AgCl(s)$
    (c) $AgCl(s) \rightarrow Ag(s) + \frac{1}{2} Cl_2(g)$
    (d) $Ag(s) + AuCl(s) \rightarrow Au(s) + AgCl(s)$

21. Calculate the enthalpy of condensation of 1.00 gram of water at 100°C.
22. A sample of 0.20 mole of a gas at 44°C and 1.5 atm pressure is cooled to 27°C and compressed to 3.0 atm. Calculate ΔV. Suppose the original sample of gas were heated at constant volume until its pressure was 3.0 atm and then cooled at constant pressure to 27°C. What would ΔV have been? Is volume a state function? If your answer limited to this sample of gas?
23. Given the following information,

$$A \;+\; B \;\rightarrow\; C \;+\; D \qquad \Delta H° = -10.0 \text{ kcal}$$
$$C \;+\; D \;\rightarrow\; E \qquad\qquad\quad \Delta H° = 15.0 \text{ kcal}$$

calculate $\Delta H°$ for each of the following reactions:
(a) $C \;+\; D \;\rightarrow\; A \;+\; B$
(b) $2\,C \;+\; 2\,D \;\rightarrow\; 2\,A \;+\; 2\,B$
(c) $A \;+\; B \;\rightarrow\; E$
24. If 1500 cal of heat is added to a system while the system does work equivalent to 2500 cal by expanding against the surrounding atmosphere, what is the value of ΔE for the system?
25. Calculate the enthalpy of sublimation of 100 grams of carbon dioxide at 183 K (see Table 3–4).
26. Calculate the enthalpy change at 25°C for the reaction of 50.0 grams of CO with oxygen to form CO_2.
27. Calculate the enthalpy change for each of the following reactions at 25°C:
(a) $Ba^{2+}(aq) \;+\; CO_3{}^{2-}(aq) \;\rightarrow\; BaCO_3(s)$
(b) $Ag^+(aq) \;+\; Br^-(aq) \;\rightarrow\; AgBr(s)$
28. The enthalpy of transition of crystalline boron to amorphous boron at 1500°C is 0.40 kcal/mole. What is the enthalpy change accompanying the conversion of 50.0 grams of crystalline boron at that temperature?
29. Using data from Tables 3–2 and 3–6, calculate $\Delta S°_{298}$ for the reaction of 100 grams of nitrogen with oxygen according to the equation

$$N_2(g) \;+\; 2\,O_2(g) \;\rightarrow\; 2\,NO_2(g)$$

30. The enthalpy change for a certain reaction at 298 K is -15.0 kcal/mole. The entropy change under these conditions is -7.2 cal/mole $\cdot$ K. Calculate the free energy change for the reaction, and predict whether the reaction may occur spontaneously.
31. For the reaction

$$2\,A(g) \;+\; B(g) \;\rightarrow\; 2\,D(g)$$

the following data are known: $\Delta E°_{298} = -2.50$ kcal and $\Delta S°_{298} = -10.5$ cal/K. Calculate $\Delta G°_{298}$ for the reaction, and predict whether the reaction may occur spontaneously.
32. Show that the product of pressure times volume, PV, has the dimensions of energy.
33. Which one(s) of the following equations have enthalpy changes equal to (a) $\Delta H_{f(CO_2)}$? (b) $\Delta H_{comb(C)}$? (c) $\Delta H_{comb(CO)}$? (d) $\Delta H_{f(CO)}$?

(i) $C \;+\; O_2 \;\rightarrow\; CO_2$
(ii) $C \;+\; \frac{1}{2}O_2 \;\rightarrow\; CO$
(iii) $CO \;+\; \frac{1}{2}O_2 \;\rightarrow\; CO_2$

General Exercises

34. Calculate the final temperature of the system after a 100 gram piece of lead at 45.0°C is added to a mixture of 2.00 grams of ice in 10.0 grams of water at 0°C.

35. Calculate the final temperature of the system after 10.0 grams of ice at $-10.0°C$ is treated with 2.00 grams of water vapor at $115.0°C$ and 1.00 atm pressure.

36. Assuming no loss of heat to the surroundings, determine the final temperature of the water produced from 2.00 moles of H_2 and 1.00 mole of O_2 initially at $25°C$ which are allowed to react to form $H_2O(g)$.

37. Calculate the mass of mercury which can be liberated from HgO at $25°C$ by the treatment of excess HgO with 10.0 kcal of heat.

38. Explain in terms of thermodynamic properties why heating to a high temperature causes decomposition of $CaCO_3$ to CaO and CO_2.

39. Calculate the heat produced when 1.00 gallon of octane, C_8H_{18}, reacts with oxygen to form carbon monoxide and water vapor at $25°C$. (The density of octane is 0.7025 gram/ml; 1.00 gallon = 3.785 liters.)

40. Calculate ΔH for the process at $25°C$ of dissolving 1.00 mole of KCl in a large excess of water:

$$KCl(s) \rightarrow K^+(aq) + Cl^-(aq)$$

Does this process represent an ionization reaction? Explain.

41. A metallic element whose specific heat is 0.11 cal/gram · deg forms an oxide containing 22.27% oxygen. Identify the element.

42. When 12.0 grams of carbon reacted with oxygen to form CO and CO_2 at $25°C$ and constant pressure, 75.0 kcal of heat was liberated and no carbon remained. Calculate the mass of oxygen which reacted.

43. Calculate the enthalpy of combustion of 100.0 grams of CO at $125°C$.

44. For the reaction at $25°C$

$$X_2O_4(l) \rightarrow 2 XO_2(g)$$

$\Delta E = 2.1$ kcal and $\Delta S = 20$ cal/K. **(a)** Calculate ΔG for the reaction. **(b)** Is the reaction spontaneous as written?

45. The melting point of a certain substance is $70°C$, its normal boiling point is $450°C$, its enthalpy of fusion is 30.0 cal/gram, its enthalpy of vaporization is 45.0 cal/gram, and its specific heat is 0.215 cal/gram · deg. Calculate the heat required to convert 100.0 grams of the substance from the solid state at $70°C$ to vapor at $450°C$.

46. The Solvay process for the industrial production of Na_2CO_3 involves the following reactions:

$$CaCO_3 \longrightarrow CaO + CO_2$$

$$2 CO_2 + 2 NaCl + 2 H_2O + 2 NH_3 \longrightarrow 2 NaHCO_3 + 2 NH_4Cl$$

$$2 NaHCO_3 \xrightarrow{heat} Na_2CO_3 + CO_2 + H_2O$$

$$CaO + H_2O \longrightarrow Ca(OH)_2$$

$$Ca(OH)_2 + 2 NH_4Cl \longrightarrow CaCl_2 + 2 NH_3 + 2 H_2O$$

(a) Calculate ΔH for each step in the process. **(b)** Determine the overall ΔH for the process. **(c)** Write an equation for the net reaction. **(d)** Calculate ΔH for the net reaction, and compare this result to that of part **(b)**.

47. The contact process for the commercial production of sulfuric acid involves the following set of reactions:

$$S_8 + 8 O_2 \longrightarrow 8 SO_2$$

$$2 SO_2 + O_2 \xrightarrow{V_2O_5} 2 SO_3$$

$$SO_3 + H_2O \longrightarrow H_2SO_4$$

(a) Calculate ΔH for each step. (b) For which step is heat evolution most likely to be a problem? (c) Determine the ΔH of the overall reaction per mole of H_2SO_4.

48. A sample of 0.100 mmole of H_2 and 0.0500 mmole of O_2 at 25°C in a sealed bomb is ignited by an electric spark. Calculate the final temperature of the (gaseous) water produced. Ignore the energy of the spark and any heat loss to the surroundings. *Note:* Heat capacity at constant volume, C_v, is given by $C_v = C_p - R$.

49. Calculate ΔH for each of the following reactions:
    (a) $BaCO_3(s) + 2 HCl(aq) \rightarrow BaCl_2(aq) + CO_2(g) + H_2O(l)$
    (b) $AgNO_3(aq) + NaCl(aq) \rightarrow NaNO_3(aq) + AgCl(s)$
    (c) $HNO_3(aq) + NaOH(aq) \rightarrow NaNO_3(aq) + H_2O(l)$
    (d) $HCl(aq) + KOH(aq) \rightarrow KCl(aq) + H_2O(l)$
    (e) $LiOH(aq) + HClO_3(aq) \rightarrow LiClO_3(aq) + H_2O(l)$

50. What quantity of heat is yielded to the surroundings when 0.100 mole of C_8H_{18} at 25°C is completely burned at constant pressure in 1.25 moles of oxygen gas at 25°C, yielding CO_2 and $H_2O(g)$ at 300°C?

51. Calculate the enthalpy change for the reaction of 1.00 mole of carbon dioxide with 1.00 mole of carbon at 600°C to produce carbon monoxide at 600°C.

52. Calculate the molar enthalpy of vaporization of water from the data of Table 3–2. Compare this value with that given in Table 3–4. Why is there a difference?

53. Using the data of Tables 3–2 and 3–3, calculate the enthalpy of formation of each of the following: (a) propane, (b) carbon disulfide, (c) naphthalene.

54. The most exothermic "ordinary" chemical reaction for a given mass of reactants is

$$2 H \rightarrow H_2 \quad \Delta E = -103 \text{ kcal}$$

Calculate the theoretical decrease in mass after the combination of 2.0 moles of hydrogen atoms to form 1.0 mole of hydrogen molecules.

55. Calculate $\Delta S°$ for the combustion of one mole of glucose at 25°C from the enthalpy of combustion at that temperature, −673 kcal/mole, and data from Table 3–6.

56. Using data from Table 3–6, determine which one(s) of the following reactions is(are) feasible at 298 K with all species in their standard states:
    (a) $Cl_2O \rightarrow Cl_2 + \frac{1}{2}O_2$
    (b) $ClO_2 \rightarrow \frac{1}{2}Cl_2 + O_2$
    (c) $Cl_2O + \frac{3}{2}O_2 \rightarrow 2 ClO_2$
    (d) $CCl_4 + 5 O_2 \rightarrow CO_2 + 4 ClO_2$

57. Two 10 gram bars of different metals are heated to 60°C, and then immersed in identical, insulated containers each containing 200 grams of water at 20°C. Will the metal with higher or lower atomic weight cause a greater temperature rise in the water?

58. Calculate the enthalpy change accompanying the freezing of 1.00 mole of water at −10.0°C to ice at −10.0°C.

59. Calculate the enthalpy of combustion of 3.000 grams of coke (carbon) at 100°C.

60. A system is changed from an initial state to a final state by a manner such that $\Delta H = q$. If the change from the initial state to the final state were made by a different path, would ΔH be the same as that for the first path? would q?

61. It is possible for heat to flow from a body at a lower temperature to one at a higher temperature. Explain under what conditions such a process might occur.

62. Outline a practical method by which the specific heat of a solid material may be determined in the laboratory.

63. Propane gas is used as a fuel for household heating. Assuming no heat loss, calculate the number of cubic feet (1 ft³ = 28.3 liters) of propane gas, measured at 25°C and 1.00 atm pressure, which would be necessary to heat 50 gal (1 gal = 3.785 liters) of water from 60 to 150°F.

64. In certain areas where coal is cheap, artificial gas is produced for household use by the "water gas" reaction:

$$C(s) \;+\; H_2O(g) \;\xrightarrow[600^\circ C]{}\; H_2(g) \;+\; CO(g)$$

Assuming that coke is 100% carbon, calculate the maximum heat obtainable at 298 K from the combustion of 1 kg of coke, and compare this value to the maximum heat obtainable at 298 K from burning the water gas produced from 1 kg of coke.

65. From the data for methane, ethane, and propane in Table 3–3, estimate the increase in ΔH_{comb} per added CH_2 group in a hydrocarbon. Predict ΔH_{comb} for octane on this basis, and compare this value to that listed in the table.

66. Calculate the enthalpy changes, ΔH°_{298}, for the following reactions in aqueous solution:

$$HS^- \;\rightarrow\; H^+ \;+\; S^{2-}$$
$$OH^- \;+\; HS^- \;\rightarrow\; S^{2-} \;+\; H_2O$$

Combine these two results to obtain the enthalpy change for the reaction

$$H^+ \;+\; OH^- \;\rightarrow\; H_2O$$

Compare this result to that derived directly from data in Table 3–5.

67. Given the following reactions with their enthalpy changes,

$$N_2(g) \;+\; 2\,O_2(g) \;\rightarrow\; 2\,NO_2(g) \qquad \Delta H_{298} = 16.18 \text{ kcal}$$
$$N_2(g) \;+\; 2\,O_2(g) \;\rightarrow\; N_2O_4(g) \qquad \Delta H_{298} = 2.31 \text{ kcal}$$

calculate the enthalpy of dimerization of NO_2. Is N_2O_4 apt to be stable with respect to NO_2 at 25°C? Explain your answer.

68. Aluminum metal is a very effective reagent for reducing oxides to their elements. An example is the thermite reaction:

$$Fe_2O_3(s) \;+\; 2\,Al(s) \;\rightarrow\; Al_2O_3(s) \;+\; 2\,Fe(s)$$

Calculate the standard free energy change when 1 mole of each of the following oxides is reduced by aluminum at 298 K: **(a)** Fe_2O_3, **(b)** SiO_2, **(c)** CuO, **(d)** CaO.

Advanced Exercises

69. A liter sample of a mixture of methane gas and oxygen, measured at 25°C and 740 torr, was allowed to react at constant pressure in a calorimeter which, together with its contents, had a heat capacity of 1260 cal/deg. The complete combustion of the methane to carbon dioxide and water caused a temperature rise in the calorimeter of 0.667°C. What was the mole percent of CH_4 in the original mixture?

70. Calculate the enthalpy changes for the reactions

$$SiO_2(s) \;+\; 4\,HF(g) \;\rightarrow\; SiF_4(g) \;+\; 2\,H_2O(g)$$
$$SiO_2(s) \;+\; 4\,HCl(g) \;\rightarrow\; SiCl_4(g) \;+\; 2\,H_2O(g)$$

Explain why hydrofluoric acid attacks glass, whereas hydrochloric acid does not.

71. By means of integral calculus, calculate the enthalpy change accompanying the reaction of 1.00 mole of carbon with water in the water gas reaction at 600°C. The heat capacities of hydrogen, carbon monoxide, water vapor, and carbon are given in terms of absolute temperatures as follows:

H_2: $C_p = 6.95 - (0.0001999)T + (4.8 \times 10^{-7})T^2$

CO: $C_p = 6.42 + (0.001665)T - (1.96 \times 10^{-7})T^2$

H_2O: $C_p = 7.256 + (2.298 \times 10^{-3})T + (2.83 \times 10^{-7})T^2$

C: $C_p = 2.04$

72. Assuming that the heat capacities of $H_2(g)$, $N_2(g)$, and $NH_3(g)$ do not vary with temperature and further assuming that ΔS is equal to -47.3 cal/mole $\cdot$ K at 298 K and is independent of temperature for the reaction

$$3\,H_2(g) \;+\; N_2(g) \;\rightarrow\; 2\,NH_3(g)$$

estimate the minimum temperature at which this reaction will occur spontaneously, with all reactants at unit activity.

73. A lead bullet weighing 18.0 grams and traveling at 5.0×10^4 cm/sec is embedded in a wooden block weighing 1.00 kg. If both the block and the bullet were initially at 25.0°C, what is the final temperature of the block containing the bullet? Assume no heat loss to the surroundings. (Specific heat of wood, 0.50 cal/gram $\cdot$ deg; of lead, 0.03 cal/gram $\cdot$ deg.)

74. Only gases remain after 15.50 grams of carbon is treated with 25.0 liters of air at 25°C and 5.50 atm pressure. (Assume 19.0% by volume oxygen, 80.0% nitrogen, 1.0% carbon dioxide.) Determine the heat evolved under constant pressure.

75. List the types of data (e.g., ΔH_f° for NaCl at 298 K) which are necessary to determine the quantity of (electrical) energy required to decompose 1.00 mole of molten NaCl to its elements at its melting point.

76. When 0.100 mole of C_8H_{18} at 25°C is completely burned at constant pressure in some oxygen gas at 25°C, yielding as products gaseous H_2O, CO, and CO_2 at 300°C, the process yielded 90.20 kcal of heat to the surroundings. Calculate the work done and the number of moles of CO and of CO_2 produced.

Oxidation and Reduction

Oxidation-reduction reactions (**redox reactions,** for short) are one of the most important classes of chemical reactions. Practically all chemical reactions which are carried out for the purpose of producing energy are redox reactions. These include combustion of fuels, generation of electricity by means of cells (batteries), and even the metabolism of food. Moreover, the commercial production of metals from their ores is an application of this type of reaction. As will be described in Chapter 5, chemical changes brought about by means of electricity (electrolysis reactions) can also be classified as redox reactions.

Redox reactions may involve several reacting species, and the products obtained from a given set of reactants may depend on whether the reaction takes place in the presence of an acid or in the presence of a base. Despite their apparent complexity, redox reactions may be dealt with systematically through the use of the concept of oxidation number. In this chapter, techniques of balancing redox reactions will be presented. Also some applications of redox reactions in chemical analysis will be described.

4-1 Nature of Oxidation-Reduction Reactions

Examples of redox reactions are illustrated by the following two experiments:

1. When a piece of zinc metal is placed in a solution containing copper(II) ions, the zinc gradually "dissolves" and a precipitate of metallic copper forms. The solution loses the blue color characteristic of copper(II) ion. The observed changes are described by the following equation:

$$Zn \;+\; Cu^{2+} \;\rightarrow\; Cu \;+\; Zn^{2+}$$

2. When a clean copper wire is immersed in a solution containing silver(I) ion, "whiskers" of silver metal grow along the wire, and the originally clear solution turns blue owing to the formation of copper(II) ion according to the following equation:

$$2\,Ag^+ \;+\; Cu \;\rightarrow\; Cu^{2+} \;+\; 2\,Ag$$

In these two experiments, it is apparent that at least two processes are occurring simultaneously—one metal is being consumed while another is being formed. One of the processes, called oxidation, is always accompanied by a simultaneous process, called reduction. Comparison of the equations for the two reactions suggests that the relative reactivities of the metals are as follows: Zn > Cu > Ag. Since the reactions occur spontaneously, they are accompanied by a decrease in free energy. In the next chapter, it will be shown how numerical values for ΔG for redox reactions can be obtained very conveniently. Thus a quantitative measure of relative reactivity is possible.

These two experiments also illustrate the main characteristic of all oxidation-reduction reactions: there are changes in the charges of the species involved. In the cases of atoms and monatomic ions, it is easy to see that these changes are, in effect, the loss and gain of electrons, respectively. However, although no charged species are involved, the reaction described by the equation

$$CH_4(g) + 4\,Cl_2(g) \rightarrow CCl_4(l) + 4\,HCl(g)$$

is also an example of a redox reaction. To explain how all three reactions are similar in nature, a new concept, that of oxidation number, is required.

4-2   Oxidation Number

The **oxidation number** or **oxidation state** (these two terms are synonymous) of an element is a number assigned to its atoms in a molecule or ion to account for the net charge of the molecule or ion.

The assignment of oxidation numbers is done according to some seemingly arbitrary rules, as follows:

1. The algebraic sum of the oxidation numbers of each atom in a formula is equal to the net charge on the group of atoms represented by the formula. Thus the sum of the oxidation numbers in NaCl must be zero, while the oxidation numbers of the five atoms in MnO_4^- must total minus one.
2. In free elements, each atom is assigned an oxidation number of zero, regardless of whether the element exists as monatomic or polyatomic molecules. Therefore, in Hg, Cl_2, and P_4, all of the atoms are assigned an oxidation number of zero.
3. In a monatomic ion, the oxidation number is the same as the charge on the ion. Thus aluminum in the ion Al^{3+} has an oxidation number of plus three; sulfur in S^{2-} has an oxidation number of minus two; and chlorine in Cl^- has an oxidation number of minus one.
4. In their compounds the alkali metals (group I A metals) always exhibit an oxidation number of plus one; the alkaline earth metals (group II A metals) always exhibit an oxidation state of plus two in their compounds.
5. The oxidation number of oxygen in most of its compounds is minus two. The exceptions are the compounds in which there are oxygen to oxygen bonds: the peroxides, such as H_2O_2 and Na_2O_2, and the superoxides, such as KO_2, in which the oxidation numbers are minus one and minus one-half, respectively. Another exception is the compound OF_2, in which oxygen has an oxidation state of plus two.
6. The oxidation number of hydrogen in its compounds is generally plus one, except in the ionic hydrides, compounds of hydrogen with the very active metals. Thus

the hydrides of the elements in groups IA and IIA, as well as the complex hydrides $LiAlH_4$ and $NaBH_4$, have hydrogen in the minus one oxidation state.

7. The oxidation number of fluorine in all its compounds is minus one. The oxidation states of all the other halogens is minus one in all compounds except those with oxygen or halogens having a lower atomic number. Thus the iodine atom in ICl_3 has an oxidation number of plus three and the chlorine atom in ClO_4^- has an oxidation number of plus seven.

8. All atoms of the same element in a given molecule or ion are assigned the same oxidation number.

In using the rules given above, no assumptions are made about the bonding within a molecule. In some cases, fractional oxidation numbers are obtained. Oxidation numbers which cannot be determined by application of these rules might be obtained from the positions of the elements in the periodic table. If there is only one element of unknown oxidation number in a formula, its oxidation number may be determined from rule **1**. It should be noted that since an oxidation number may be positive, negative, or zero, the sign of the oxidation number should be explicitly given.

Example

What is the oxidation number of sulfur in each of the following cases? **(a)** S^{2-}, **(b)** H_2SO_4, **(c)** $S_2O_3^{2-}$, **(d)** CS_2, **(e)** S_8, **(f)** $Na_2S_4O_6$, **(g)** S_2Cl_2.

(a) From rule **3**, the oxidation number is deduced to be minus two.

(b) In H_2SO_4, the oxidation numbers must total zero.

With   4 oxygen at -2
         2 hydrogen at $+1$
            sulfur at x

$$2(+1) + x + 4(-2) = 0$$
$$x = +6$$

(c) In $S_2O_3^{2-}$, the oxidation numbers must total minus two.

     3 oxygen at -2
       2 sulfur at x

$$2x + 3(-2) = -2$$
$$x = +2$$

The oxidation number of sulfur is plus two.

(d) From just the rules stated above any assignment of oxidation numbers which gives a net charge of zero could be made. However, noting that sulfur is in the same periodic group with oxygen, and by analogy with CO_2, the sulfur is usually assigned an oxidation number of minus two.

     2 sulfur at $-2 = -4$
     carbon at $+4 = +4$
         net 0

(e) The oxidation number of any free element is zero.

(f) The oxidation numbers of alkali and alkaline earth metals in their compounds are uniformly plus one and plus two, respectively. The oxidation number of sulfur can be established on this basis:

> 2 sodium at $+1$
> 4 sulfur at $\quad x$
> 6 oxygen at -2

$$2(+1) + 4x + 6(-2) = 0$$
$$x = 2.5$$

(g) The oxidation number of chlorine is minus one (rule 7).

$$2x + 2(-1) = \quad 0$$
$$x = +1$$

Therefore that of sulfur must be plus one.

4-3   Periodicity of the Oxidation States of the Elements

The concept of oxidation numbers makes it possible to systematize much of the chemistry of the elements. For almost all the elements, at least one oxidation state can be predicted from the position of the element in the periodic table. For most elements, the maximum oxidation state is equal to the group number. With the exceptions of the elements of group I B and certain of the group 0 and inner transition elements, no oxidation state exceeds the group number. Only for certain elements of group VIII B are the maximum oxidation states less than the group number. It must be emphasized that the group number does not necessarily give the *only* oxidation state of the element nor even necessarily the most important oxidation state. The elements of groups IV A, V A, VI A, and VII A may have a negative oxidation state which is numerically equal to the group number minus 8. For example, all members of group VII A have an oxidation state equal to $7 - 8 = -1$. As stated explicitly in rule 2, all free elements have an oxidation state of 0. Additional information about the possible oxidation states can be derived from Figure 4-1. In this figure the periodic table is broken up into sections in which the oxidation states may be predicted to some degree. Further discussion of the oxidation states of the elements as related to their atomic structures will be given in Section 10-14.

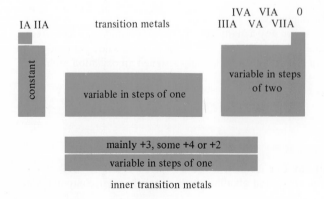

Figure 4-1. **Periodic Variations of Oxidation Numbers**

In the "constant" portion of the periodic table, the elements show only one oxidation state in all their compounds. That oxidation state corresponds to the group number. In each of the other portions of the table, variation in oxidation number is possible. In the portion marked "variable in steps of two," the oxidation states range from a high corresponding to the group number to lower numbers, sometimes to negative numbers, most often in steps of two. In the other portions of the table, the oxidation numbers may vary in steps of one.

Oxidation numbers of atoms or ions which have variable oxidation numbers are indicated by adding Roman numerals, in parentheses, to the names of the atoms. For example, $FeCl_3$ is called iron(III) chloride, and CrO_3 is called chromium(VI) oxide. An element which has the same oxidation number in all its compounds does not require the use of Roman numerals. For example, NaCl is called sodium chloride. "Sodium(I)" would be redundant, since in compounds, the oxidation number of sodium is always $+1$. Nonmetals generally have specific names for each oxidation state, although Roman numerals are sometimes used with them. Nitrogen(I) oxide, nitrogen(II) oxide, and nitrogen(IV) oxide, representing N_2O, NO, and NO_2, respectively, are three examples of the Roman numeral designation. The ending *ide* generally signifies the lowest (negative) oxidation state of an element. Thus chloride signifies chlorine in its -1 oxidation state, and nitride signifies nitrogen in its -3 oxidation state. Positive oxidation states of nonmetals are indicated by names ending in *ite* or *ate,* as in the series hypochlorite, ClO^-; chlorite, ClO_2^-; chlorate, ClO_3^-; and perchlorate, ClO_4^- (Table 1–5).

4-4   Acids and Bases

The terms *acid* and *base* have been used several times previously in this text, with the assumption that their meanings are generally known. However, because acids or bases often influence the course of oxidation-reduction reactions, some discussion of the topic is presented here. Further concepts of acids and bases are given in Chapters 7 and 8.

For present purposes, an acid may be defined as any substance which, when dissolved in aqueous solution, increases the concentration of hydrogen ion in the solution. Acids include hydrogen-containing compounds which dissolve with reaction in water giving hydrogen ions and an equivalent number of negative ions. Conventionally, formulas for this type of acid are written with the ionizable hydrogen first, as, for example, H_2SO_4 and $HC_2H_3O_2$. It should be noted that some acids are capable of providing more than 1 mole of hydrogen ion per mole of compound. In an acidic salt, such as NaH_2PO_4, the anion ($H_2PO_4^-$) is capable of providing hydrogen ions.

Bases are substances which increase the concentration of hydroxide ion in aqueous solution. They include a number of solid, crystalline compounds containing a metal ion and an equivalent number of hydroxide ions (OH^-). Only a few such compounds are appreciably soluble, the most important being NaOH, KOH, and $Ba(OH)_2$. Calcium hydroxide, $Ca(OH)_2$, is sparingly soluble. In addition to these, ammonia, NH_3, and certain of its derivatives dissolve in water with reaction to give some hydroxide ions:

$$NH_3 \;+\; H_2O \;\rightarrow\; NH_4^+ \;+\; OH^-$$

The most significant property of bases is their ability to neutralize acids. If the base and acid are both soluble and are dissociated into ions, the net ionic reaction is simply

$$H^+ \;+\; OH^- \;\rightarrow\; H_2O$$

If the base is insoluble, its reaction with hydrogen ion produces the metal ion in solution, as, for example,

$$Mg(OH)_2(s) \; + \; 2\,H^+(aq) \; \rightarrow \; Mg^{2+}(aq) \; + \; 2\,H_2O(l)$$

Solutions which contain an excess of H^+ over OH^- are **acidic.** Solutions which contain an excess of OH^- over H^+ are **basic** (alkaline). Solutions containing equal concentrations of the two ions are **neutral.**

Certain oxides of nonmetals, for example, CO_2, SO_3, and P_4O_{10}, react with water to yield acid solutions. For this reason they have been called **acid anhydrides.** Their reactions are illustrated by the following equations:

$$CO_2 \; + \; H_2O \; \rightarrow \; H_2CO_3 \; \rightarrow \; H^+ \; + \; HCO_3^-$$
$$P_4O_{10} \; + \; 6\,H_2O \; \rightarrow \; 4\,H_3PO_4 \; \rightarrow \; 4\,H^+ \; + \; 4\,H_2PO_4^-$$

Similarly, certain metal oxides react with water to form bases; hence they have been known as **basic anhydrides:**

$$BaO \; + \; H_2O \; \rightarrow \; Ba^{2+} \; + \; 2\,OH^-$$

Acid anhydrides can react directly with bases, and basic anhydrides can react directly with acids:

$$SO_3 \; + \; 2\,OH^- \; \rightarrow \; SO_4^{2-} \; + \; H_2O$$
$$BaO \; + \; 2\,H^+ \; \rightarrow \; Ba^{2+} \; + \; H_2O$$

4-5   Oxidation-Reduction Reactions

Oxidation-reduction reactions are characterized by changes in the oxidation numbers of some of the elements in the reactants. Any element which increases in oxidation number is said to be **oxidized.** Any element which decreases in oxidation number is **reduced.** In the first of the two reactions described previously,

$$Zn \; + \; Cu^{2+} \; \rightarrow \; Zn^{2+} \; + \; Cu$$

it is obvious that the zinc is oxidized and the copper(II) ion is reduced. The formation of zinc ions can be ascribed to the loss of two electrons by each zinc atom—oxidation. The change from copper(II) ions to copper metal is due to the gain of two electrons by the ion—reduction. Thus the loss of electrons by one species is accompanied by the gain of the electrons by another species. In the second case cited above,

$$Cu \; + \; 2\,Ag^+ \; \rightarrow \; Cu^{2+} \; + \; 2\,Ag$$

the copper is oxidized and the silver ion reduced. Two moles of silver ions is required to accept the electrons from 1 mole of copper atoms.

An **oxidizing agent** causes the oxidation of another species, and in the process the oxidation number of one (or more) of its elements is decreased. A **reducing agent** causes the reduction of another species, and it in turn has one of its elements oxidized. For example, in the reaction between Zn and Cu^{2+} given above, the Cu^{2+} ion is the oxidizing agent, which is reduced; the Zn is the reducing agent, which is oxidized.

oxidizing agent gets reduced

reducing agent gets oxidized

Example

In the following reaction, identify the species oxidized, the species reduced, the oxidizing agent, and the reducing agent.

$$Fe^{2+} + 2 H^+ + NO_3^- \rightarrow Fe^{3+} + NO_2 + H_2O$$

The Fe^{2+} is oxidized to Fe^{3+} by the nitrate ion, which is the oxidizing agent. Nitrogen is reduced from nitrogen(V) in the nitrate to nitrogen(IV) in NO_2, by the reducing agent Fe^{2+}.

Oxidation-reduction reactions need not involve charged species. For example, in the reaction of carbon with oxygen to form carbon dioxide, carbon is the reducing agent and oxygen is the oxidizing agent:

$$C + O_2 \rightarrow CO_2$$

It is not always easy to predict the course of an oxidation-reduction reaction. It is sometimes necessary to know not only the possible oxidation states of each element and the compounds characteristic of each of the oxidation states, but also which compounds and which oxidation states are stable at a given concentration of hydrogen ion or hydroxide ion. For example, when Cr^{3+} is oxidized in acid solution, the product is dichromate ion, $Cr_2O_7^{2-}$. In basic solution, the oxidation of chromium(III) yields chromate ion, CrO_4^{2-}. In both cases the oxidation number of chromium changes from $+3$ to $+6$. On the other hand, the presence of base instead of acid may stabilize a product in a different oxidation state. For example, when permanganate ion, MnO_4^-, is reduced with iron(II) in acid solution, manganese(II) ion is produced. When the reaction is carried out in base, manganese dioxide, MnO_2, is the reduction product. The effect of concentration is illustrated by the reactions of nitric acid, HNO_3, with copper. Both of the following reactions are possible, depending on concentrations and temperature:

$$Cu + 4 H^+ + 2 NO_3^-(conc) \rightarrow 2 NO_2 + Cu^{2+} + 2 H_2O$$
$$3 Cu + 8 H^+ + 2 NO_3^-(dil) \rightarrow 2 NO + 3 Cu^{2+} + 4 H_2O$$

In Chapters 5, 6, and 7, various criteria for predicting the course of oxidation-reduction reactions will be given. In this chapter, however, it will be assumed that the products of such reactions have been determined by experiment, as indeed they can be. For example, when hot concentrated sulfuric acid is treated with copper metal, it is possible to detect the formation of sulfur dioxide by its odor and to note the formation of copper(II) ion by its characteristic blue color. Merely this information may be used to begin writing the equation for the reaction:

$$H_2SO_4(conc) + Cu \rightarrow Cu^{2+} + SO_2 \quad \text{(incomplete)}$$

It is obvious from the law of conservation of mass that other species must be involved in this reaction, even though their presence might not be readily detected. Some negative ions must be present after the reaction is complete in order to form a neutral compound with the positive copper(II) ions. The hydrogen atoms missing from the right side of the equation are in the form of water and HSO_4^-. The complete equation for the reaction is

$$3 \, H_2SO_4(\text{conc}) \; + \; Cu \; \rightarrow \; Cu^{2+} \; + \; 2 \, HSO_4^- \; + \; SO_2 \; + \; 2 \, H_2O$$

The H_2SO_4 which is not reduced provides hydrogen sulfate ion, HSO_4^-, which balances the charge of the Cu^{2+} ion produced, and also provides the hydrogen from which the water is produced.

4-6   Balancing Oxidation-Reduction Equations

Oxidation-reduction equations may be balanced by any one of several methods. Two methods of balancing oxidation-reduction equations will be presented in this chapter—the oxidation number method and the ion-electron half-reaction method. Both methods give the same results, but the ion-electron method is more versatile because

1. It is directly applicable to electrochemical processes.
2. It emphasizes the species actually involved in the net ionic equation.
3. The products can be more easily predicted by its use.
4. Acidic and/or basic conditions are more clearly indicated.
5. A balanced half reaction can be combined with many other half reactions to yield overall equations.

The oxidation number method has the advantage of speed and of being applicable to reactions in which no ions are involved.

The following equation will be balanced as an example of the rules of the oxidation number method. The rules are listed on the left, and each step is illustrated on the right.

$$HNO_3 \; + \; Cu_2O \; \rightarrow \; Cu(NO_3)_2 \; + \; NO \; + \; H_2O$$

Steps	*Example*
1. Write the unbalanced equation and identify the elements which change oxidation number. For convenience, draw connecting lines between the reactant and product species containing the elements changing oxidation number.	$\overset{V}{H}NO_3 + \overset{I}{Cu_2}O \rightarrow \overset{II}{Cu}(NO_3)_2 + \overset{II}{N}O$ (The oxidation numbers are shown for the elements which change oxidation number.)
2. Balance the number of atoms of each element changing oxidation number.	$HNO_3 + Cu_2O \rightarrow 2\,Cu(NO_3)_2 + NO$
3. Indicate the total change in oxidation number for both the oxidation and the reduction.	$1(3 \text{ ox. no. each}) = 3 \text{ decrease}$ $HNO_3 + Cu_2O \rightarrow 2\,Cu(NO_3)_2 + NO$ $2(1 \text{ ox. no. each}) = 2 \text{ increase}$

| *Steps* | *Example* |

4. Multiply the total changes in oxidation number by small integers which make the increase equal to the decrease. Multiply the coefficients of the reacting species by these small integers.

loss of 6 oxidation numbers

$$2\,HNO_3 + 3\,Cu_2O \rightarrow 6\,Cu(NO_3)_2 + 2\,NO$$

gain of 6 oxidation numbers

5. Having established the coefficients of the species changing oxidation number, balance the remainder of the atoms by inspection. If necessary, add acid (or base) and water as needed.

$(12 + 2)\,HNO_3 + 3\,Cu_2O \rightarrow$
$6\,Cu(NO_3)_2 + 2\,NO + 7\,H_2O$
(The 12 additional HNO_3 molecules are needed as reactants to provide NO_3^- ions and H_2O molecules in the products.)

The balanced equation is

$$14\,HNO_3 + 3\,Cu_2O \rightarrow 6\,Cu(NO_3)_2 + 2\,NO + 7\,H_2O$$

The second method, that of ion-electron half reactions, will now be illustrated in a similar manner for the same equation:

$$H^+ + NO_3^- + Cu_2O \rightarrow Cu^{2+} + NO + H_2O$$

| *Steps* | *Example* |

1. Identify the elements oxidized and reduced, along with the oxidation and reduction products.

NO_3^- is reduced to NO
Cu_2O is oxidized to Cu^{2+}

2. Write down the species containing the element oxidized and the oxidation product.

$$Cu_2O \rightarrow Cu^{2+}$$

3. Balance the number of atoms of element oxidized.

$$Cu_2O \rightarrow 2\,Cu^{2+}$$

4. To the side having the higher oxidation number, add the number of electrons equal to the total change in oxidation number.

$$Cu_2O \rightarrow 2\,Cu^{2+} + 2\,e^-$$

5. Balance the numbers of atoms of any other elements, if any, except for hydrogen and oxygen, keeping their oxidation numbers the same on both sides of the equation.

$$Cu_2O \rightarrow 2\,Cu^{2+} + 2\,e^-$$

6. In acid solution, add sufficient hydrogen ions, H^+, to balance the charge. (In basic solution, OH^- is used to balance the charge, as will be illustrated below.)

$$2\,H^+ + Cu_2O \rightarrow 2\,Cu^{2+} + 2\,e^-$$

7. Add water to balance the numbers of hydrogen and oxygen atoms. This step gives the complete half reaction for the element oxidized.

$$2\,H^+ + Cu_2O \rightarrow$$
$$2\,Cu^{2+} + H_2O + 2\,e^-$$

Steps	*Example*

8. Repeat steps 2–7 for the species containing the element reduced to obtain a complete reduction half reaction.

$$4\,H^+ + NO_3^- + 3\,e^- \rightarrow$$
$$NO + 2\,H_2O$$

9. Multiply the results of steps 7 and 8 by the smallest possible integers so that the number of electrons used up in step 8 is equal to that produced in step 7.

(Step 7) × 3 and (step 8) × 2:
$$6\,H^+ + 3\,Cu_2O \rightarrow$$
$$6\,Cu^{2+} + 3\,H_2O + 6\,e^-$$
$$8\,H^+ + 2\,NO_3^- + 6\,e^- \rightarrow$$
$$2\,NO + 4\,H_2O$$

10. Add the two half reactions and subtract from each side any species appearing on both sides of the equation.

$$14\,H^+ + 3\,Cu_2O + 2\,NO_3^- \rightarrow$$
$$6\,Cu^{2+} + 2\,NO + 7\,H_2O$$

The above reaction took place in acid solution, and hydrogen ion was added to balance the charge. To balance an oxidation-reduction equation in *basic* solution, step 6 is amended. To balance the charge, hydroxide ion is added. To illustrate this procedure, the reaction of MnO_4^- with $Bi(OH)_3$ to produce BiO_3^- and MnO_2 is balanced stepwise with the steps for each half reaction written in parallel.

Steps	Oxidation Half Reaction	Reduction Half Reaction
1	$Bi(OH)_3$ oxidized to BiO_3^-	MnO_4^- reduced to MnO_2
2, 3	$Bi(OH)_3 \rightarrow BiO_3^-$	$MnO_4^- \rightarrow MnO_2$
4, 5	$Bi(OH)_3 \rightarrow BiO_3^- + 2\,e^-$	$MnO_4^- + 3\,e^- \rightarrow MnO_2$
6	$Bi(OH)_3 + 3\,OH^- \rightarrow$ $BiO_3^- + 2\,e^-$	$MnO_4^- + 3\,e^- \rightarrow MnO_2 + 4\,OH^-$
7	$Bi(OH)_3 + 3\,OH^- \rightarrow$ $BiO_3^- + 3\,H_2O + 2\,e^-$	$MnO_4^- + 2\,H_2O + 3\,e^- \rightarrow$ $MnO_2 + 4\,OH^-$
9	$3\,Bi(OH)_3 + 9\,OH^- \rightarrow$ $3\,BiO_3^- + 9\,H_2O + 6\,e^-$	$2\,MnO_4^- + 4\,H_2O + 6\,e^- \rightarrow$ $2\,MnO_2 + 8\,OH^-$

The complete balanced equation is

$$OH^- + 3\,Bi(OH)_3 + 2\,MnO_4^- \rightarrow 2\,MnO_2 + 3\,BiO_3^- + 5\,H_2O$$

4-7 Disproportionation

Under certain conditions, a reagent which contains an element in an oxidation state which is intermediate between a higher and a lower oxidation state may react with itself to form products having the element in the higher and lower states. Such a reaction is called **disproportionation** or **autooxidation**. For example, when copper(I) oxide is placed in acid, copper metal and copper(II) ion are produced. In this reaction, the copper(I) species acts as both the oxidizing agent and the reducing agent. The half reactions are

$$2\,H^+ + Cu_2O + 2\,e^- \rightarrow 2\,Cu + H_2O$$
$$2\,H^+ + Cu_2O \rightarrow 2\,Cu^{2+} + H_2O + 2\,e^-$$

and the overall reaction is

$$2\,H^+ \;+\; Cu_2O \;\rightarrow\; Cu \;+\; Cu^{2+} \;+\; H_2O$$

When most nonmetallic elements in the zero oxidation state are treated with strong base, disproportionation occurs. Thus chlorine, bromine, iodine, sulfur, selenium, phosphorus, arsenic, antimony, and other nonmetals disproportionate when treated with strong bases. For example, depending on the temperature and concentration, chlorine can undergo two different disproportionation reactions:

$$Cl_2 \;+\; 2\,OH^- \;\rightarrow\; ClO^- \;+\; Cl^- \;+\; H_2O$$
$$3\,Cl_2 \;+\; 6\,OH^- \;\rightarrow\; ClO_3^- \;+\; 5\,Cl^- \;+\; 3\,H_2O$$

In acid solution, the above-named elements in the zero oxidation state are stable toward disproportionation; in fact, if two compounds of the element, one of which has the element in a positive oxidation state and the other in the negative oxidation state, are mixed in neutral or acid solution, they react to form the free element. For example,

$$2\,H^+ \;+\; Cl^- \;+\; ClO^- \;\rightarrow\; Cl_2 \;+\; H_2O$$

4-8   Redox Titrations

If an oxidation-reduction reaction occurs with sufficient speed, it is possible to titrate one reagent with the other, just as is done for acid-base reactions (Section 2-11). In such redox titrations, the end point is detected by use of a redox indicator, which is a substance present in small concentration which changes color upon oxidation or reduction and which is oxidized or reduced immediately after the reagent being titrated is totally used up. For some oxidation-reduction reactions, such as oxidation by permanganate ion, the very intense color of the reagent itself, together with the faint colors of the reduction products, allows titration without an indicator.

Example

Exactly 40 ml of an acidified solution of 0.4000 M iron(II) ion is titrated with potassium permanganate solution. After addition of 32.00 ml of $KMnO_4$, one additional drop turns the iron solution purple. Calculate the concentration of the permanganate solution.

The reaction which occurs is

$$8\,H^+ \;+\; 5\,Fe^{2+} \;+\; MnO_4^- \;\rightarrow\; Mn^{2+} \;+\; 5\,Fe^{3+} \;+\; 4\,H_2O$$

The number of millimoles of iron(II) ion present is

$$(40.00\ \text{ml})(0.4000\ \text{mmole/ml}) = 16.00\ \text{mmole Fe}^{2+}$$

Hence the number of millimoles of MnO_4^- is

$$16.00 \text{ mmole Fe}^{2+} \left(\frac{1 \text{ mmole MnO}_4^-}{5 \text{ mmole Fe}^{2+}} \right) = 3.200 \text{ mmole MnO}_4^-$$

The concentration is then

$$\frac{3.200 \text{ mmole}}{32.00 \text{ ml}} = 0.1000 \, M \text{ MnO}_4^-$$

4-9   Equivalents

It is sometimes convenient (but not necessary) to use the concept of a chemical **equivalent.** For oxidizing and reducing agents and their reduction and oxidation products, one equivalent is defined as that quantity of substance which reacts with or yields 1 mole of electrons. For other reagents, such as those involved in acid-base reactions, one equivalent is that quantity of substance which reacts with or liberates 1 mole of H^+ or OH^- ion. Consider the equation

$$MnO_4^- \; + \; 8 \, H^+ \; + \; 5 \, e^- \; \rightarrow \; Mn^{2+} \; + \; 4 \, H_2O$$

which states that each mole of MnO_4^- reacts with 5 moles of electrons. Hence, it takes $\frac{1}{5}$ mole of MnO_4^- to react with 1 mole of electrons. Therefore it is apparent that in this reaction, there are 5 equivalents of MnO_4^- per mole of MnO_4^-. For MnO_4^- in this reaction, the equivalent weight—the mass of an equivalent in daltons—is one-fifth of the formula weight of MnO_4^-.

As a consequence of the way equivalents are defined, one equivalent of any oxidizing or reducing agent (or acid or base, in a nonredox reaction) reacts with one equivalent of any other substance capable of reacting with it. However, equivalents must be defined in terms of a particular reaction because some substances undergo more than one kind of reaction. For example, the equivalent weight (the mass of one equivalent) of H_2SO_4 depends on whether it is reduced to H_2S, reduced to SO_2, reduced to H_2, half neutralized to $NaHSO_4$, or completely neutralized to Na_2SO_4, because in each case, a different number of moles of hydrogen ions or moles of electrons may be involved.

Example

What is the equivalent weight of H_2SO_4 when it is reduced to SO_2?

$$2 \, H^+ \; + \; H_2SO_4 \; + \; 2 \, e^- \; \rightarrow \; SO_2 \; + \; 2 \, H_2O$$

$$\left(\frac{98.0 \text{ grams H}_2SO_4}{\text{mole H}_2SO_4} \right) \left(\frac{1 \text{ mole H}_2SO_4}{2 \text{ moles e}^-} \right) = \frac{49.0 \text{ grams H}_2SO_4}{\text{equivalent}}$$

The concept of equivalents is often used in analytical chemistry. However, calculations in which equivalents are used can be done by alternative methods using moles.

4-10   Normality

Just as molarity is useful for expressing the concentration of a solution in terms of moles of solute(s), the concept of normality is used to express the concentration in terms of equivalents of solute(s). The number of equivalents of solute per liter of solution is defined as **normality,** N. From the definition of an equivalent, given in Section 4-9, it is apparent that the normality of a solution depends on the reaction that the solute undergoes. However, the normality of a given substance in a solution is always an integral multiple of its molarity.

Example

What is the normality of a solution of sulfuric acid made by dissolving 2 moles of H_2SO_4 in sufficient water to make 1 liter of solution if **(a)** the solution is to be completely neutralized with NaOH and **(b)** the H_2SO_4 in the solution is to be reduced to H_2S?

(a) H_2SO_4 + 2 NaOH → Na_2SO_4 + 2 H_2O

$$2 \text{ moles } H_2SO_4 \left(\frac{2 \text{ moles } OH^-}{\text{mole } H_2SO_4} \right) = 4.0 \text{ equivalents } H_2SO_4$$

$$\frac{4.0 \text{ equivalents}}{\text{liter}} = 4.0 \ N$$

(b) H_2SO_4 + 8 H^+ + 8 e^- → H_2S + 4 H_2O

$$2 \text{ moles } H_2SO_4 \left(\frac{8 \text{ moles } e^-}{\text{mole } H_2SO_4} \right) = 16 \text{ moles } e^- = 16 \text{ equivalents } H_2SO_4$$

$$\frac{16 \text{ equivalents } H_2SO_4}{\text{liter}} = 16 \ N$$

The concept of normality is useful in calculations for titrations. At the equivalence point of a titration, the number of equivalents of each reagent is the same. Since the normality times the number of liters is equal to the number of equivalents, the normality, N_2, of an unknown solution is calculated from its volume, V_2, and the volume, V_1, and normality, N_1, of the standard solution:

$$N_1 V_1 = N_2 V_2$$

Example

What is the normality of a MnO_4^- solution if 32.00 ml of the solution is required to titrate 40.00 ml of 0.4000 N Fe^{2+}. Compare this result with the molarity of MnO_4^- calculated on page 110, and explain the relationship of the concentration units.

$$(N_{MnO_4^-})(32.00 \text{ ml}) = (0.4000 \ N)(40.00 \text{ ml})$$
$$N_{MnO_4^-} = 0.5000 \ N$$

The concentration may be expressed as 0.1000 M or 0.5000 N. As seen by the balanced equation on page 109, there are 5 equivalents per mole; hence the normality is 5 times the molarity.

4-11   Relative Oxidizing and Reducing Tendencies

Redox half reactions have the following form:

$$\text{oxidizing agent} \;+\; n\,e^- \;\rightleftharpoons\; \text{reducing agent}$$

The oxidizing agent and reducing agent related by such an equation are known as a redox **couple.** Given a suitable oxidizing agent, the reducing agent of the couple can be oxidized; conversely, given a suitable reducing agent, the oxidizing agent can be reduced. Appropriate combination of any two redox couples may result in an oxidation-reduction reaction. Thus it is possible to predict the products of an oxidation-reduction reaction, provided the redox couples are known. However, there are three complications:

1. The fact that a balanced equation can be written does not necessarily mean that the reaction will actually occur.
2. A given reagent may belong to more than one couple. For example, Fe^{2+} ion belongs to both the Fe–Fe^{2+} and Fe^{2+}–Fe^{3+} couples, and when Fe^{2+} is a reagent, it can act either as an oxidizing agent or as a reducing agent.
3. The reaction may proceed so slowly that products are not obtained in a reasonable time.

For a given couple, the stronger the oxidizing agent, the weaker will be the reduction product as a reducing agent. The relative oxidizing and reducing tendencies of several redox couples involving elements in their zero oxidation state are given in Table 4–1. Generalizing from the table, it may be concluded that the nonmetals in their elementary forms are good oxidizing agents and the alkali metals are very good reducing agents. To a degree, these tendencies can be correlated with the periodic table. For example, fluorine, located in the upper right-hand corner of the table, is such a strong oxidizing agent and fluoride ion has correspondingly such a low reducing ability that there is no known chemical reagent which can oxidize fluoride ion to fluorine. The element can be prepared from its compounds only by electrolysis.

TABLE 4-1. Relative Oxidizing and Reducing Tendencies of Some Elements and Their Ions

	Oxidizing Agents	Reducing Agents	
strong	F_2	F^-	weak
	Cl_2	Cl^-	
	Br_2	Br^-	
intermediate	Ag^+	Ag	intermediate
	I_2	I^-	
	Cu^{2+}	Cu	
	H^+	H_2	
	Fe^{2+}	Fe	
	Zn^{2+}	Zn	
weak	Al^{3+}	Al	strong
	Na^+	Na	

The elements located toward the bottom left-hand corner of the periodic table are very good reducing agents; consequently their positive ions are very weak oxidizing agents. Transition metal ions are intermediate in their oxidizing abilities, but within this group there is no general tendency of a periodic nature.

The higher oxidation state compounds of the transition elements are generally strong oxidizing agents, being reduced to simple positive ions in lower oxidation states in many cases. Thus, permanganate and dichromate ions are extremely good oxidizing agents in acid solution. Similarly, the nonmetals in higher oxidation states are generally good oxidizing agents. However, their oxidizing abilities depend strongly on the hydrogen ion concentration. Therefore, it is not easy to make simple predictions of their behavior. For example, the reaction between sulfate ion and iodide ion proceeds readily in concentrated sulfuric acid solution, but in dilute acid or neutral solution the reaction proceeds in the reverse direction:

$$SO_4^{2-} + 2I^- + 4H^+ \underset{\substack{\text{dil acid or} \\ \text{neutral}}}{\overset{\text{conc acid}}{\rightleftharpoons}} SO_2 + I_2 + 2H_2O$$

4-12 Exercises

Basic Exercises

1. What is the oxidation state of nitrogen in each of the following? (a) NH_3, (b) HN_3, (c) N_2H_4, (d) NO_2, (e) N_2O_4, (f) NO_2^-, (g) NH_2OH, (h) NO, (i) HNO_3, (j) N_2O, (k) HCN.

2. Which one(s) of the following involve redox reactions? (a) burning of gasoline, (b) evaporation of water, (c) human respiration, (d) preparation of metals from their ores, (e) production by lightning of nitrogen oxides from nitrogen and oxygen in the atmosphere, (f) production by lightning of ozone (O_3) from O_2, (g) reaction of H_2SO_4 with NaOH.

3. Complete and balance the following:
   (a) $Sn^{2+} + H^+ + NO_3^- \rightarrow Sn^{4+} + NO$
   (b) $Cu + Ag^+ \rightarrow Ag + Cu^{2+}$
   (c) $Cr_2O_7^{2-} + I^- \rightarrow I_2 + Cr^{3+}$
   (d) $Zn + H^+ + NO_3^- \rightarrow Zn^{2+} + NH_4^+$
   (e) $Br_2 + I^- \rightarrow I_2 + Br^-$
   (f) $ClO^- + I^- \rightarrow I_2 + Cl^-$
   (g) $S_2O_3^{2-} + Ag^+ \rightarrow Ag + S_4O_6^{2-}$

4. In the process of drying dishes with a dish towel, what may be considered the wetting agent? the drying agent? What happens to the wetting agent? to the drying agent? In an oxidation-reduction reaction, what happens to the oxidizing agent? to the reducing agent?

5. Which of the following equations represent oxidation-reduction reactions? Identify each oxidizing agent and each reducing agent.
   (a) $K + O_2 \rightarrow KO_2$
   (b) $H_2O_2 + KOH \rightarrow KHO_2 + H_2O$
   (c) $Ca(HCO_3)_2 \xrightarrow{\text{heat}} CaCO_3 + CO_2 + H_2O$
   (d) $Cr_2O_7^{2-} + 2OH^- \rightarrow 2CrO_4^{2-} + H_2O$
   (e) $H_2O_2 \rightarrow H_2O + \frac{1}{2}O_2$

6. State the oxidation number of the underlined element in each of the following:
   (a) $\underline{P}_2O_7^{4-}$, (b) $\underline{C}_3O_2$, (c) $\underline{Mn}O_4^-$, (d) $\underline{Mn}O_4^{2-}$, (e) $\underline{V}O_2^+$, (f) $\underline{U}O_2^{2+}$, (g) $\underline{Cl}O_3^-$.

7. Classify each of the following substances as acid, base, acid anhydride, or basic anhydride: **(a)** H_2SO_3, **(b)** NH_3, **(c)** LiOH, **(d)** Li_2O, **(e)** Cl_2O_3, **(f)** BaO, **(g)** CO_2, **(h)** CrO.

8. H_2SO_4 acts as an oxidizing agent, a dehydrating agent, and an acid. Select equations from the following which illustrate each type of behavior:

   (a) $C_6H_{12}O_6 \xrightarrow{\ H_2SO_4(conc)\ } 6\,C \ + \ 6\,H_2O$
   (b) $5\,H_2SO_4(conc) \ + \ 4\,Zn \ \rightarrow \ H_2S \ + \ 4\,Zn^{2+} \ + \ 4\,SO_4^{2-} \ + \ 4\,H_2O$
   (c) $H_2SO_4(dil) \ + \ Zn \ \rightarrow \ Zn^{2+} \ + \ H_2 \ + \ SO_4^{2-}$
   (d) $H_2SO_4(dil) \ + \ ZnCO_3 \ \rightarrow \ Zn^{2+} \ + \ CO_2 \ + \ SO_4^{2-} \ + \ H_2O$

9. When NaBr is treated with concentrated H_2SO_4, SO_2, HBr, and Br_2 are produced; when NaCl is treated with concentrated H_2SO_4, HCl is produced but no Cl_2 or SO_2 is produced. **(a)** Write balanced chemical equations for all these reactions. **(b)** On the basis of the facts given above, predict which of the following reactions will occur:

$$Br_2 \ + \ 2\,NaCl \ \rightarrow \ Cl_2 \ + \ 2\,NaBr$$
$$Cl_2 \ + \ 2\,NaBr \ \rightarrow \ Br_2 \ + \ 2\,NaCl$$

10. What is the oxidation state of hydrogen in each of the following? **(a)** HCl, **(b)** H^+, **(c)** NaH, **(d)** H_2, **(e)** $LiAlH_4$.

11. Name the compounds in each of the following sets: **(a)** $FeCl_2$, $FeCl_3$; **(b)** VCl_2, VCl_3, VCl_4; **(c)** UCl_3; **(d)** Cu_2O, CuO; **(e)** HNO_2, HNO_3.

12. Write the formulas for sulfide ion, sulfite ion, sulfate ion, and thiosulfate ion (in which one oxygen atom of sulfate is replaced by a sulfur atom). Compare the oxidation number of sulfur in each of these species. Use this exercise to distinguish between the concepts of oxidation number and charge. Is there any direct relationship between the oxidation number of sulfur and the charge on the ion?

13. Both VO_2^+ and VO^{2+} are known as the "vanadyl" ion. **(a)** Determine the oxidation number of vanadium in each. **(b)** Which of the following names corresponds to which ion? oxovanadium(IV) ion, dioxovanadium(V) ion.

14. Zinc metal "dissolves" in aqueous HCl. Solid NaCl dissolves in water. Solid sucrose, ordinary table sugar ($C_{12}H_{22}O_{11}$), dissolves in water. How can it be demonstrated that only the last two of these are actually cases of dissolving, while the first is a chemical reaction?

15. Hydrogen peroxide, H_2O_2, may act as an oxidizing agent or a reducing agent. Explain why this behavior is possible. Write an equation for the disproportionation of H_2O_2.

16. What mass of phosphoric acid, H_3PO_4, is required to make 550 ml of 0.400 N solution **(a)** assuming complete neutralization of the acid? **(b)** assuming reduction to HPO_3^{2-}?

17. What mass of N_2H_4 can be oxidized to N_2 by 24.0 grams of K_2CrO_4, which is reduced to $Cr(OH)_4^-$?

18. Use the data of Table 4–1 to predict the products of each of the following. If no reaction occurs, write "nr."
   (a) $Na \ + \ ZnCl_2 \ \rightarrow$
   (b) $F_2 \ + \ NaCl \ \rightarrow$
   (c) $I_2 \ + \ Fe \ \rightarrow$
   (d) $I^- \ + \ Fe^{2+} \ \rightarrow$

19. How many moles of $FeCl_3$ can be prepared by the reaction of 10.0 grams of $KMnO_4$, 1.07 moles of $FeCl_2$, and 500 ml of 3.00 M HCl? $MnCl_2$ is the reduction product.

20. Which ones of the following are examples of disproportionation reactions? What criteria determine whether a reaction is a disproportionation?
   (a) $Ag(NH_3)_2^+ \ + \ 2\,H^+ \ \rightarrow \ Ag^+ \ + \ 2\,NH_4^+$
   (b) $Cl_2 \ + \ 2\,OH^- \ \rightarrow \ ClO^- \ + \ Cl^- \ + \ H_2O$
   (c) $CaCO_3 \ \rightarrow \ CaO \ + \ CO_2$

(d) $2 HgO \rightarrow 2 Hg + O_2$
(e) $Cu_2O + 2 H^+ \rightarrow Cu + Cu^{2+} + H_2O$
(f) $CuS + O_2 \rightarrow Cu + SO_2$
(g) $2 HCuCl_2 \xrightarrow[\text{dilute with } H_2O]{} Cu + Cu^{2+} + 4 Cl^- + 2 H^+$

21. Determine the oxidation state of chlorine in **(a)** ClO^-, **(b)** ClO_2^-, **(c)** ClO_3^-, **(d)** ClO_4^-. **(e)** Name each of these ions (Table 1–5).

General Exercises

22. Predict the highest and lowest possible oxidation states of each of the following elements: **(a)** Ta, **(b)** Te, **(c)** Tc, **(d)** Ti, **(e)** Tl.
23. Complete and balance the following equations:
    (a) $CrO + H_2O \rightarrow Cr(OH)_2$
    (b) $Cr_2O_3 + HCl \rightarrow CrCl_3 + H_2O$
    (c) $Cr_2O_3 + NaOH \rightarrow NaCr(OH)_4$
    (d) $CrO_3 + NaOH \rightarrow Na_2CrO_4 + H_2O$
    (e) Does the oxidation number of chromium in its oxides have any effect on the tendency of the oxide to act as an acid anhydride or a basic anhydride?
24. Complete and balance the following equations:
    (a) $MnO_4^- + Sn^{2+} \rightarrow Mn^{2+} + Sn^{4+}$
    (b) $Mn^{2+} + Sn^{4+} \rightarrow MnO_4^- + Sn^{2+}$
    (c) Is it possible to predict from the equation alone whether a reaction will go as the equation is written? **(d)** Could a prediction in the above case be made if the following reactions are known to occur?

$$16 H^+ + 2 MnO_4^- + 10 Cl^- \rightarrow 5 Cl_2 + 2 Mn^{2+} + 8 H_2O$$
$$Sn + 2 Cl_2 \rightarrow SnCl_4$$

25. Determine the oxidation number of **(a)** N in HN_3, **(b)** O in KO_2. **(c)** Draw electron dot diagrams for each of these species.
26. Which one(s) of the following are **(a)** very good oxidizing agents? **(b)** very good reducing agents? **(c)** neither? MnO_4^-, I^-, Cl^-, Ce^{4+}, $Cr_2O_7^{2-}$, Na, Na^+, CrO_4^{2-}, HNO_3, Fe^{2+}, F_2, F^-.
27. Complete and balance the following equations:
    (a) $S_2O_8^{2-} + I^- \rightarrow I_3^- + SO_2$
    (b) $[CrCl_6]^{3-} + Zn \rightarrow [ZnCl_4]^{2-} + Cr^{2+}$
    (c) $CH_3CHO + Cr_2O_7^{2-} \rightarrow Cr^{3+} + CH_3CO_2H$
    (d) $H_2SO_4(conc) + Br^- \rightarrow Br_2 + SO_2$
    (e) $Br_2 + OH^- \rightarrow Br^- + BrO_3^-$
    (f) $MnO_4^{2-} + H^+ \rightarrow MnO_4^- + Mn^{2+}$
    (g) $AuCl_4^- + Sn^{2+} \rightarrow Sn^{4+} + AuCl$
    (h) $Au(CN)_2^- + Zn \rightarrow Zn(CN)_4^{2-} + Au$
    (i) $NO_2 + H_2O \rightarrow NO + NO_3^-$
    (j) $CrO_4^{2-} + Cu_2O \rightarrow Cu(OH)_2 + Cr(OH)_4^-$
28. Write balanced equations for the following reactions. In each case assume that the second reagent is in large excess.
    (a) $H_3PO_4 + NaOH \rightarrow$
    (b) $NaOH + H_3PO_4 \rightarrow$
    (c) $H_3PO_4 + Ba(OH)_2 \rightarrow$
    (d) $CrO_4^{2-} + H^+ \rightarrow Cr_2O_7^{2-}$

29. Complete and balance the following:
    (a) $Br^- + BrO_3^- + H^+ \rightarrow$
    (b) $Cu + Cu^{2+} + OH^- \rightarrow$
    (c) $P_4 + OH^- \rightarrow PH_3 + PHO_3^{2-}$
30. H_2O_2 is reduced rapidly by Sn^{2+}, the products being Sn^{4+} and water. H_2O_2 decomposes slowly at room temperature to yield O_2 and water. Calculate the volume of O_2 produced at 20°C and 1.00 atm when 200 grams of 10.0% by mass H_2O_2 in water is treated with 100.0 ml of 2.00 M Sn^{2+}, and then the mixture is allowed to stand until no further reaction occurs.
31. Complete and balance the following equation, in basic solution,

$$Hg_2(CN)_2 + Ce^{4+} \rightarrow CO_3^{2-} + NO_3^- + Hg(OH)_2 + Ce^{3+}$$

    (a) by considering the carbon in $Hg_2(CN)_2$ to be in the -4 oxidation state and the nitrogen to be in the $+3$ oxidation state, (b) by considering C as $+4$ and N as $+5$, (c) by considering Hg to be $+2$ and carbon -4. (d) Explain why the same result is obtained regardless of the choice of oxidation states.
32. Calculate the mass of oxalic acid, $H_2C_2O_4$, which can be oxidized to CO_2 by 100.0 ml of an MnO_4^- solution, 10.0 ml of which is capable of oxidizing 50.0 ml of 1.00 N I^- to I_2.
33. It requires 40.05 ml of 1.000 M Ce^{4+} to titrate 20.00 ml of 1.000 M Sn^{2+} to Sn^{4+}. What is the oxidation state of the cerium in the reduction product?
34. Determine which reagent is in excess and by how much if 100.0 grams of P_4O_6 is treated with 100 grams of $KMnO_4$ in HCl solution to form H_3PO_4 and $MnCl_2$.
35. If 10.00 grams of V_2O_5 is dissolved in acid and reduced to V^{2+} by treatment with zinc metal, how many moles of I_2 could be reduced by the resulting V^{2+} solution, as it is reoxidized to V^{IV}?
36. What mass of $K_2Cr_2O_7$ is required to produce from excess oxalic acid, $H_2C_2O_4$, 5.00 liters of CO_2 at 75°C and 1.07 atm pressure? The reduction product of $Cr_2O_7^{2-}$ is Cr^{3+}.
37. Complete and balance the following, and give the oxidation state of the indicated element in the *missing* product:
    (a) oxygen in: $H_2O_2 + I_2 \rightarrow I^- +$
    (b) oxygen in: $H_2O_2 + Sn^{2+} \rightarrow Sn^{4+} +$
    (c) manganese in: $MnO_4^{2-} + H^+ \rightarrow Mn^{2+} +$
    (d) nitrogen in: $NO_2 + H_2O \rightarrow NO +$
38. Complete and balance the following:
    (a) $MnO_4^- + Cr^{3+} \rightarrow Cr_2O_7^{2-} + Mn^{2+}$
    (b) $CrO_4^{2-} + Fe(OH)_2 \rightarrow Fe(OH)_3 + Cr(OH)_4^-$
    (c) $AuCl_4^- + Zn \rightarrow Au + Zn^{2+} +$
    (d) $Zn + OH^- \rightarrow Zn(OH)_4^{2-} +$
39. Consider the following equations:

$$Cr_2O_7^{2-} + 14\,H^+ + 6\,e^- \rightarrow 2\,Cr^{3+} + 7\,H_2O$$
$$Cr_2O_7^{2-} + 3\,SO_3^{2-} + 8\,H^+ \rightarrow 2\,Cr^{3+} + 3\,SO_4^{2-} + 4\,H_2O$$

    (a) How many moles of dichromate ion are represented in each equation? (b) How many equivalents of dichromate ion are represented in each? (c) What is the normality of a 0.200 M solution of dichromate ion when used in the second reaction?
40. What is the normality of 0.300 M H_3PO_3 when it undergoes the reaction

$$H_3PO_3 + 2\,OH^- \rightarrow HPO_3^{2-} + 2\,H_2O$$

41. Write an equation describing the oxidation of ammonia by oxygen to NO in basic solution. In the commercial Ostwald process, for the production of nitric acid, this reaction

is carried out directly in the gaseous state. Explain why the same equation describes the direct reaction and the reaction in basic solution.

Advanced Exercises

42. Complete and balance the following equations:

(a) $Fe(CN)_6{}^{4-} + H^+ + MnO_4{}^- \rightarrow Fe^{3+} + CO_2 + NO_3{}^- + Mn^{2+}$
(b) $Cu_3P + Cr_2O_7{}^{2-} \rightarrow Cu^{2+} + H_3PO_4 + Cr^{3+}$
(c) $H_2SO_4 + I^- \rightarrow I_2 + H_2S$

43. A 10.00 gram mixture of Cu_2S and CuS was treated with 200.0 ml of 0.7500 M $MnO_4{}^-$ in acid solution, producing SO_2, Cu^{2+}, and Mn^{2+}. The SO_2 was boiled off and the excess $MnO_4{}^-$ was titrated with 175.0 ml of 1.000 M Fe^{2+} solution. Write balanced chemical equations for all the reactions. Calculate the percent CuS in the original mixture.

44. Consider the following balanced equations:

$$H_2O_2 + Sn^{4+} \rightarrow Sn^{2+} + O_2 + 2\,H^+$$

$$3\,H_2O_2 + Sn^{4+} \rightarrow Sn^{2+} + 2\,O_2 + 2\,H_2O + 2\,H^+$$

Does the fact that these two equations are complete and balanced mean that the reaction between H_2O_2 and Sn^{4+} can lead to different sets of products depending on the mole ratio of H_2O_2 to Sn^{4+}, or is some other explanation possible? If necessary, write another equation to support your answer.

45. Write a balanced chemical equation for the oxidation of phosphorus(III) sulfide by nitric acid. The products include NO and SO_2.

46. A 6.000 gram sample contained Fe_3O_4, Fe_2O_3, and inert materials. It was treated with an excess of aqueous KI in acid, which reduced all the iron to Fe^{2+}. The resulting solution was diluted to 50.00 ml, and a 10.00 ml sample of it was taken. The liberated iodine in the small sample was titrated with 5.500 ml of 1.000 M $Na_2S_2O_3$ solution, yielding $S_4O_6{}^{2-}$. The iodine from another 25.00 ml sample was extracted, after which the Fe^{2+} was titrated with 3.20 ml of 1.000 M $MnO_4{}^-$ in H_2SO_4 solution. Calculate the percentages of Fe_3O_4 and of Fe_2O_3 in the original mixture.

47. Cyanide ion is oxidized by powerful oxidizing agents to $NO_3{}^-$ and CO_2 or $CO_3{}^{2-}$, depending on the acidity of the reaction mixture. Nitric acid, a powerful oxidizing agent, is reduced by moderate reducing agents to NO plus other products. Write a complete and balanced equation for the reaction of nitric acid with potassium cyanide. If this reaction were actually carried out, what safety precautions would be necessary?

48. Determine the equivalent weight of bromine in each of the following: **(a)** the reduction half reaction for the disproportionation of Br_2 in base. **(b)** the oxidation half reaction, which yields bromate ion. **(c)** the overall reaction. **(d)** What is the relationship between the answer to **(c)** and the answers to **(a)** and **(b)**? Explain fully.

49. If all the reactants and products are known, the equation for any reaction may be balanced by the procedure described below.

1. Using alphabetic or other symbols, write arbitrary coefficients in front of each formula in the equation.
2. On the basis that the number of atoms of each element must be the same on both sides of the equation, derive the relative magnitudes of the coefficients.
3. Assign one coefficient a numerical value and find the values of the other coefficients as indicated by their relative magnitudes. For example the alphabetic coefficients in the unbalanced equation

$$a\,KMnO_4 + b\,KCl + c\,H_2SO_4 \rightarrow$$
$$d\,Cl_2 + e\,MnSO_4 + f\,H_2O + g\,K_2SO_4$$

must be related as follows: $a = e$, $2b = d$, $c = f = e + g = 4a$, $a + b = 2g$. If a is assigned a value of 10, then $e = 10$, $c = f = 40$, $g = 30$, $b = 50$, $d = 25$. Dividing each of these values by 5 to obtain the smallest integral ratios yields

$$a:b:c:d:e:f:g = 2:10:8:5:2:8:6$$

Hence the balanced equation is

$$2\,KMnO_4 \;+\; 10\,KCl \;+\; 8\,H_2SO_4 \;\rightarrow$$
$$5\,Cl_2 \;+\; 2\,MnSO_4 \;+\; 8\,H_2O \;+\; 6\,K_2SO_4$$

Use this method to balance the following equations:

(a) $KIO_3 \;+\; KI \;+\; H_2SO_4 \;\rightarrow\; KI_3 \;+\; K_2SO_4 \;+\; H_2O$

(b) $Pb(N_3)_2 \;+\; Co(MnO_4)_3 \;\rightarrow\; CoO \;+\; MnO_2 \;+\; Pb_3O_4 \;+\; NO$

(c) $KOH \;+\; K_4Fe(CN)_6 \;+\; Ce(NO_3)_4 \;\rightarrow$
$$Fe(OH)_3 \;+\; Ce(OH)_3 \;+\; K_2CO_3 \;+\; KNO_3 \;+\; H_2O$$

5

Electrochemistry

When a system undergoes change, some of its potential energy[1] may be converted to another form of energy. Conversely, when energy is absorbed by a system, the system may change in such a manner that its potential energy is increased. In the latter case, the system may be regarded as having stored a certain quantity of energy which can be released later. Fuels, such as coal and oil, represent stored chemical energy which can be released in the form of heat when the fuel is burned. Subject, of course to the limitations of the second law of thermodynamics (Section 3–12), this heat may in turn be converted into still another form of energy.

Redox reactions may be applied to convert chemical energy directly into electrical energy. Conversely, electrical energy may be used to promote chemical reactions. The study of these types of interconversions of energy is called **electrochemistry.** Electrochemistry is one of the oldest fields of chemistry, and yet the modern uses of electrochemistry are constantly expanding with the development of many new practical applications. The fact that electricity can be produced by means of chemical reactions allows the storage of electrical energy as internal energy (E) of chemical reagents, which may be used to produce electrical energy at a more convenient time or place. Electric automobiles are being developed and tested as a possible means of avoiding the air pollution caused by products from internal combustion engines. Among the sources of power used in space vehicles are solar batteries, which change the energy from sunlight into electrical energy, and fuel cells, which produce electrical energy from the chemical energy of fuels. Electrochemistry also provides insight into such diverse phenomena as the corrosion of metals, the refining of metals, the transmission of nerve impulses in animals, and the interactions of ions in solution with each other and with the solvent.

In this chapter, the quantitative relationships between electricity and chemical reaction will be presented. Then principles of design and construction of cells for the production of electricity will be presented, and the thermodynamics of the relevant processes will be discussed. Finally, several important applications of electrochemistry will be described.

5-1 Electrical Units

To use the relationships between the various electrical quantities, it is necessary to understand the system of units which is commonly used. To use the relationships between the various electrical quantities, it is necessary to understand the system of units which is commonly used. An electric **current** is a flow of charge. The practical unit of electric **charge** is the **coulomb** (C), which is the charge on 6.25×10^{18} electrons. An electric current is measured in **amperes** (A), defined as

[1] For a discussion of potential energy, see Appendix A–2.

the passage of 1 C of charge per second past a given point. In passing through any medium, moving charges will encounter a **resistance** to their flow. Consequently, charge will flow between two points only if there is a driving "force," called the **potential.** The unit of potential is the **volt** (V), which is the potential needed to cause a current of 1 A in a medium having a resistance of 1 **ohm.**

An important relationship, valid in ordinary electric circuits (but not necessarily in solutions of ions), is Ohm's law. The law relates current, I, potential, ϵ, and resistance, R, and is expressed mathematically as

$$\epsilon = IR$$

The magnitude of electrical energy is the product of the potential times the charge which passes, and the unit is volt · coulomb. Since charge is equal to current times time, the quantity of electrical energy is also equal to the product potential times current times time. Alternatively, since potential times current is defined as **power,** the quantity of energy can be expressed as power times time, and may have the dimensions watts · seconds.

$$E = \epsilon q = \epsilon It = Pt$$

These and other electrical units derived from them are listed in Table 5–1. If time is expressed in seconds, these basic units are interconvertible without numerical coefficients (other than 1). Thus

$$1 \text{ A} = \frac{1 \text{ C}}{1 \text{ sec}}$$

Example

How many joules of energy is expended when a current of 1.00 A passes for 100 sec under a potential of 115 V?

Energy is equal to the product of potential times charge or potential times current times time (Table 5–1).

$$E = \epsilon q = \epsilon It = (115 \text{ V})(1.00 \text{ A})(100 \text{ sec}) = 11{,}500 \text{ J}$$

5-2   Electrolysis

Electrolysis is the production of a chemical reaction by means of an electric current. The basic apparatus for an electrolysis experiment is shown in Figure 5–1. It consists of an **electrolytic cell** containing the electrolyte (either in solution or in the molten state) into which two **electrodes** are placed. The electrodes are connected to the two sides of a source of direct current, such as a battery, so that a complete electrical circuit is established. The current is carried from the battery through the wires by means of electrons (metallic conduction). Within the cell, the current is carried by the positive and negative ions of the electrolyte (electrolytic conduction). The electrodes supply surfaces where the conduction changes from metallic to electrolytic or vice versa. The change in type of conduction at each electrode involves a chemical reaction in which electrons are accepted by a substance at one electrode, while

TABLE 5-1. Some Electrical Quantities and Units

	symbol	Unit	Abbrev.	Relationship to Other Quantities
charge	q	coulomb	C	$q = It$
current	I	ampere	A	$I = q/t$
potential	ϵ (or V)	volt	V	$\epsilon = IR$
power	P	watt	W	$P = \epsilon I$
energy	E	joule	J	$E = Pt = \epsilon It = \epsilon q$
resistance	R	ohm	Ω	$R = \epsilon/I$

simultaneously at the other electrode, electrons are released by another substance and are returned to the battery through the connecting wires. The reactions at the electrodes, involving gain or loss of electrons by the chemical species, are oxidation-reduction reactions. For example, consider the cell shown in Figure 5-1, which consists of two carbon electrodes dipping into a moderately concentrated solution of copper(II) chloride. The battery forces electrons onto one electrode, giving it a negative charge. The positively charged copper(II) ions in the solution migrate toward this electrode, where they accept the electrons in a reduction half reaction:

$$Cu^{2+} + 2 e^- \rightarrow Cu$$

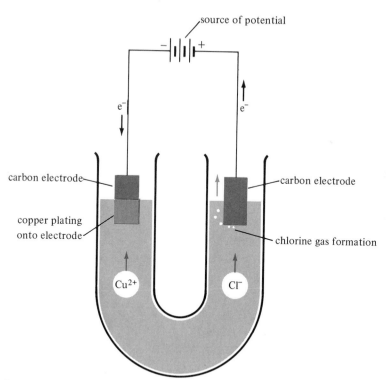

Figure 5-1. Electrolysis of Copper(II) Chloride

anode: $2 Cl^- \rightarrow Cl_2 + 2 e^-$

cathode: $Cu^{2+} + 2 e^- \rightarrow Cu$

cell: $Cu^{2+} + 2 Cl^- \rightarrow Cl_2 + Cu$

Simultaneously, the negatively charged chloride ions migrate to the other electrode, from which electrons are flowing toward the battery. The chloride ions undergo reaction, giving up electrons to the external circuit:

$$2\,Cl^- \rightarrow Cl_2 + 2\,e^-$$

The electrode at which reduction occurs is called the **cathode.** (The positively charged ions, which migrate toward this electrode in an electrolysis reaction, are called **cations.**) The electrode at which oxidation occurs is defined as the **anode.**[2] (Negatively charged ions, which migrate toward the anode during electrolysis, are called **anions.**)

In any electrolysis experiment the actual changes which occur at the electrodes depend on the relative oxidation-reduction tendencies of the substances involved. Carbon is relatively inert toward electrolytic oxidation, and in the cell described above, the electrode was not involved in the oxidation reaction. However, if the anode of the cell were copper metal, the oxidation reaction would have been

$$Cu \rightarrow Cu^{2+} + 2\,e^-$$

because copper metal is more easily oxidized than is chloride ion. Nevertheless, the chloride ions in solution would still migrate toward the anode, where in this case they would function to balance the charge on the copper(II) ions entering the solution from the copper electrode.

The electrolysis of sodium chloride under varying conditions emphasizes the importance of the experimental conditions. If a very dilute solution of sodium chloride is electrolyzed between platinum electrodes, the two electrode reactions are

anode: $\qquad\qquad 2\,H_2O \rightarrow O_2 + 4\,H^+ + 4\,e^-$

cathode: $2\,H_2O + 2\,e^- \rightarrow H_2 + 2\,OH^-$

The number of electrons gained at the cathode must be equal to that lost at the anode. Hence the second equation may be written:

$$4\,H_2O + 4\,e^- \rightarrow 2\,H_2 + 4\,OH^-$$

If the H^+ produced at the anode and the OH^- produced at the cathode are allowed to mix, they form water. The overall process is

anode: $\qquad\qquad\qquad 2\,H_2O \rightarrow O_2 + 4\,H^+ + 4\,e^-$

cathode: $\qquad\quad 4\,H_2O + 4\,e^- \rightarrow 2\,H_2 + 4\,OH^-$

neutralization: $\quad 4\,OH^- + 4\,H^+ \rightarrow 4\,H_2O$

cell: $\qquad\qquad\qquad\qquad 2\,H_2O \rightarrow 2\,H_2 + O_2$

The net result in this case is the decomposition of water by electrolysis. If a solution of moderately concentrated sodium chloride is electrolyzed, however, the electrode reactions are

[2] These definitions of anode and cathode also apply to galvanic cells, to be described below.

anode: $2 \, Cl^- \rightarrow Cl_2 + 2 \, e^-$

cathode: $2 \, H_2O + 2 \, e^- \rightarrow H_2 + 2 \, OH^-$

cell: $2 \, H_2O + 2 \, Cl^- \rightarrow H_2 + Cl_2 + 2 \, OH^-$

Thus the products obtained are chlorine, hydrogen, and sodium hydroxide. The electrolysis of brine (concentrated sodium chloride solution) is an important commercial process for the preparation of these chemicals.

If molten sodium chloride is electrolyzed, the electrode reactions are

anode: $2 \, Cl^- \rightarrow Cl_2 + 2 \, e^-$

cathode: $Na^+ + e^- \rightarrow Na$

cell: $2 \, Na^+ + 2 \, Cl^- \rightarrow 2 \, Na + Cl_2$

In the commercial production of sodium metal, a mixture of sodium carbonate and sodium chloride, which has a melting point of about 600°C, is electrolyzed because less heat is required to maintain the melt at this temperature than is required to maintain pure sodium chloride at its melting point, 801°C. The use of a lower temperature reduces the cost of the process.

The Hall process for the production of aluminum from its oxide, diagrammed in Figure 5–2, is another example of the commercial preparation of a metal by electrolysis.

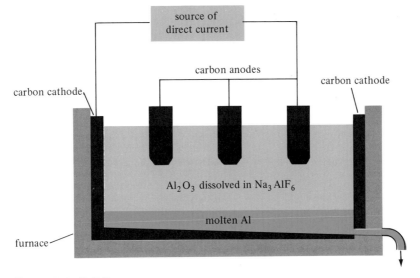

Figure 5–2. **Hall Process**
An undergraduate student named Charles Hall, after being told by his professor about the difficulty of reducing aluminum ores, devised the electrolytic process which bears his name. Purified Al_2O_3 is dissolved in molten cryolite, Na_3AlF_6, and the electrolysis is carried out at 1000°C. Carbon electrodes are used and the oxidation at the anode results in the formation of oxides of carbon. The carbon anodes are thus corroded away and must be continuously fed into the molten solution as the electrolysis progresses. Molten aluminum is drawn off at the bottom of the cell and cast in the form of thick slabs, called "pigs."

5-3   Faraday's Laws

In 1834 Michael Faraday demonstrated that the quantities of chemicals which react at the electrodes of an electrolytic cell are directly proportional to the quantity of charge passed through the cell. He also showed that when a given quantity of charge is passed through several different electrolytes, the same number of equivalents (Section 4–9) of each substance undergoes reaction. For example, if several cells containing aqueous solutions are connected in series, as shown in Figure 5–3, and if 96,487 C of charge is passed through them, the electrode reactions will occur simultaneously and for 7.9997 grams of O_2 produced from H_2SO_4, 1.008 grams of H_2, 107.9 grams of Ag, 31.773 grams of Cu, and 38.27 grams of In will be produced at the respective cathodes. These masses are the equivalent weights of the respective elements. The results of this experiment are generalized in **Faraday's laws:**

1. The masses of products obtained in a given electrolysis experiment are proportional to the quantity of charge which passes through the cell.
2. The relative masses of the products produced from a given quantity of charge are proportional to the respective equivalent weights.

It is concluded that 96,487 C is the charge on 1 mole of electrons. This quantity of charge is designated as **1 faraday** (F).[3] Therefore, since each faraday is equivalent to 1 mole of electrons, the quantity of change which occurs in any electrolysis experiment may be determined from the number of faradays of charge which passes.

Example

A solution of copper(II) sulfate is electrolyzed between copper electrodes by a current of 10.0 A for exactly 1 hour. What changes occur at the electrodes and in the solution? The electrode reactions are

anode: $Cu \rightarrow Cu^{2+} + 2\,e^-$

cathode: $Cu^{2+} + 2\,e^- \rightarrow Cu$

---

[3] For most calculations, the value of the faraday will be taken as 96,500 C.

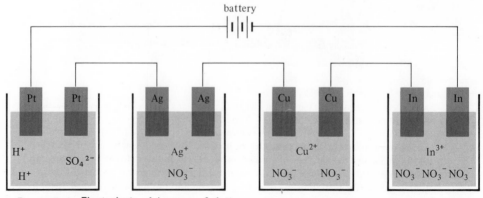

Figure 5-3. **Electrolysis of Aqueous Solutions**

$$\text{number of moles of electrons} = (10.0 \text{ A})(3600 \text{ sec}) \left(\frac{1 \text{ C}}{1 \text{ A} \cdot \text{sec}}\right)\left(\frac{1 \text{ mole e}^-}{96{,}500 \text{ C}}\right)$$

$$= 0.373 \text{ mole e}^- = 0.373 \text{ F}$$

Since it takes 2 moles of electrons to react with or produce each mole of copper metal the number of moles of copper dissolved or deposited is

$$(0.373 \text{ mole e}^-)\left(\frac{1 \text{ mole Cu}}{2 \text{ moles e}^-}\right) = 0.186 \text{ mole Cu}$$

Thus 0.186 mole of copper is dissolved from the anode, 0.186 mole of copper is deposited onto the cathode, and the original copper(II) ion concentration of the solution remains unchanged.

Example

A constant current was passed through a solution of $AuCl_4^-$ ions between gold electrodes. After a period of 10.0 min, the cathode increased in mass by 1.314 grams. How much charge was passed? What was the current, I?
The reaction at the cathode is the reduction of the gold(III) to gold metal:

$$AuCl_4^- + 3 e^- \rightarrow Au + 4 Cl^-$$

$$\text{moles Au} = \frac{1.314 \text{ grams Au}}{197 \text{ grams/mole Au}} = 6.67 \times 10^{-3} \text{ mole Au}$$

$$q = (6.67 \times 10^{-3} \text{ mole Au}) \left(\frac{3 \text{ moles e}^-}{\text{mole Au}}\right) = 2.00 \times 10^{-2} \text{ F}$$

$$I = \frac{q}{t} = \frac{(2.00 \times 10^{-2} \text{ F})(96{,}500 \text{ C/F})}{600 \text{ sec}} = 3.22 \text{ A}$$

5-4 Galvanic Cells

The reaction

$$Zn + Cu^{2+} \rightarrow Cu + Zn^{2+}$$

has been represented (Section 4-1) as occurring by means of the following half-reactions:

$$Zn \rightarrow Zn^{2+} + 2 e^-$$
$$Cu^{2+} + 2 e^- \rightarrow Cu$$

It is assumed that when zinc metal is placed in a copper(II) ion solution, the Cu^{2+} ions accept electrons directly from the zinc atoms. The fact that this reaction occurs spontaneously whenever metallic zinc is put into a solution of copper(II) ions suggests that this reaction occurs with a decrease in free energy. Hence it should be possible to transfer electrons from zinc metal to copper ions whenever a suitable means is provided. An arrangement for doing this is shown in Figure 5-4(a). A piece

(a)

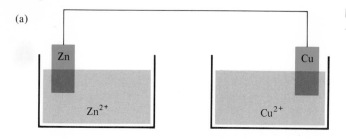

Figure 5-4. **Transfer of Electrons from Zinc to Copper**

(b)

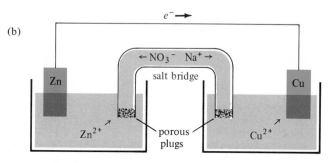

of zinc metal is placed in a beaker containing a solution of a zinc salt, and a piece of copper metal is placed in another beaker which contains a solution of copper salt. Then the two pieces of metal are connected by a wire. It might be expected that zinc atoms would start to lose electrons, thereby entering the solution as Zn^{2+} ions, and that the electrons released would be transferred through the wire to the copper electrode where they would add to Cu^{2+} ions, causing them to deposit onto the electrode as copper atoms. The only thing preventing this process from occurring is the fact that if electrons flowed from the zinc to the copper and onto the copper(II) ions, a net positive charge would very quickly accumulate in the beaker containing the zinc salt and simultaneously a net negative charge would accumulate in the beaker containing the copper salt. Therefore, almost instantaneously, the electrons would cease to flow. To maintain a current, some provision must be made to avoid a charge build up by allowing ions to be transferred between the two solutions. This can be done by connecting the two solutions with an inverted U tube containing a solution of a salt such as sodium nitrate, as shown in Figure 5-4(b). This device is called a **salt bridge.** As zinc ions are added to the left beaker because of the loss of electrons by the zinc atoms, negative ions migrate into the beaker from the salt bridge. Meanwhile, positive ions migrate through the salt bridge into the other beaker and neutralize the accumulated negative charge resulting from the depletion of copper(II) ions. Thus a complete circuit is achieved, and a continuous flow of electrons through the wire is maintained. This arrangement is an example of a **galvanic cell.** Each beaker is said to contain a **half cell,** and the two half-cell reactions are:

$$Zn \rightarrow Zn^{2+} + 2\,e^- \quad \text{(left beaker)}$$
$$Cu^{2+} + 2\,e^- \rightarrow Cu \quad \text{(right beaker)}$$

The overall cell reaction is exactly the same as the reaction which occurs when zinc metal is placed directly in a copper(II) ion solution:

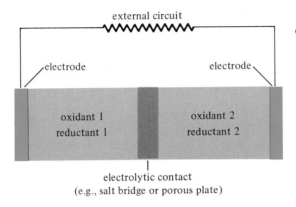

Figure 5-5. **Design of a Galvanic Cell**

external circuit

electrode   electrode

oxidant 1   oxidant 2
reductant 1   reductant 2

electrolytic contact
(e.g., salt bridge or porous plate)

$$Zn \ + \ Cu^{2+} \ \rightarrow \ Zn^{2+} \ + \ Cu$$

Any combination of redox couples can be used to construct a galvanic cell and thereby change chemical energy into electrical energy by forcing electrons through an external conductor from the reducing agent to the oxidizing agent. The basic design of a galvanic cell is shown schematically in Figure 5-5. Essentially each redox couple is made into a half cell, and these are connected by some sort of electrolytic contact, such as a salt bridge or porous partition. This arrangement is necessary in most cases to avoid direct reaction. If the physical states of the constituents in the half cells are such that they cannot mix readily (for example, if the redox couples consist of insoluble substances), both half cells (electrodes) can be placed in the same vessel and no salt bridge is necessary. Examples of actual cell designs are shown in Figures 5-6, 5-7, and 5-8.

In the Daniell cell, shown in Figure 5-6, the zinc metal and the copper metal serve as the reduced forms of the respective couples and also as the electrodes. Electrolytic contact is established through the porous partition by diffusion of the sulfate ions from the copper half cell to the zinc half cell, while the zinc ions pass in the opposite direction.

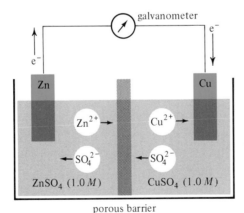

galvanometer

e^-

e^-

Zn   Cu

$Zn^{2+} \rightarrow$ $Cu^{2+} \rightarrow$

$\leftarrow SO_4^{2-}$ $\leftarrow SO_4^{2-}$

$ZnSO_4$ (1.0 M) $CuSO_4$ (1.0 M)

porous barrier

Figure 5-6. **Daniell Cell**

anode: $Zn \ \rightarrow \ Zn^{2+} \ + \ 2 \ e^-$ $\epsilon° = 0.76$ V
cathode: $Cu^{2+} \ + \ 2 \ e^- \ \rightarrow \ Cu$ $\epsilon° = 0.34$ V

$\epsilon°_{cell} = 1.10$ V

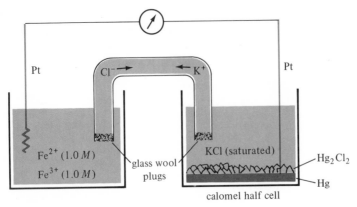

Figure 5–7. Galvanic Cell

anode: $\qquad$ $2\,Hg \;+\; 2\,Cl^- \;\rightarrow\; Hg_2Cl_2 \;+\; 2\,e^-$ $\qquad$ $\epsilon° = -0.244$ V

cathode: $\qquad$ $Fe^{3+} \;+\; e^- \;\rightarrow\; Fe^{2+}$ $\qquad$ $\epsilon° = 0.771$ V

cell: $\quad 2\,Hg \;+\; 2\,Cl^- \;+\; 2\,Fe^{3+} \;\rightarrow\; 2\,Fe^{2+} \;+\; Hg_2Cl_2$ $\quad \epsilon°_{cell} = 0.527$ V

In the cell shown in Figure 5–7, the redox couple on the left consists of dissolved Fe^{2+} and Fe^{3+} in contact with an **inert electrode.** This electrode serves only to conduct electrons and is not otherwise involved in the redox reaction. The other half cell contains mercury metal and an insoluble salt, Hg_2Cl_2. Here, electrolytic contact is established by means of a salt bridge containing a solution of KCl.

In the cell shown in Figure 5–8, one of the electrodes consists of a silver wire coated with silver chloride. The other electrode consists of platinum metal over which hydrogen gas is bubbled. In this type of cell, neither the hydrogen gas nor the insoluble silver chloride can migrate to the opposite electrode, and therefore no salt bridge or porous partition is necessary.

A shorthand notation is often used to describe cells. The substances involved are denoted by their formulas, with their concentrations and/or pressures written in parentheses following the particular formula. A single vertical line indicates contact

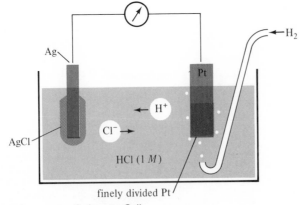

Figure 5–8. Galvanic Cell

anode: $\qquad$ $H_2 \;\rightarrow\; 2\,H^+ \;+\; 2\,e^-$ $\qquad$ $\epsilon° = 0.000$ V

cathode: $\quad AgCl \;+\; e^- \;\rightarrow\; Ag \;+\; Cl^-$ $\qquad$ $\epsilon° = 0.222$ V

cell: $\quad H_2 \;+\; 2\,AgCl \;\rightarrow\; 2\,Ag \;+\; 2\,Cl^- \;+\; 2\,H^+$ $\qquad$ $\epsilon°_{cell} = 0.222$ V

between solid and liquid, or liquid and gas, etc. Two or more substances in the same solution are separated by commas; the salt bridge or other electrolytic contact is indicated by a pair of vertical lines. By convention, the half cell which is the anode is written on the left.

Example

Write line notations for the cells described in Figures 5-6, 5-7, and 5-8.

$$\text{Zn} \mid \text{Zn}^{2+}(1\ M) \parallel \text{Cu}^{2+}(1\ M) \mid \text{Cu}$$
$$\text{Pt} \mid \text{Hg} \mid \text{Hg}_2\text{Cl}_2 \mid \text{KCl}(1\ M) \parallel \text{Fe}^{2+}(1\ M),\ \text{Fe}^{3+}(1\ M) \mid \text{Pt}$$
$$\text{Pt} \mid \text{H}_2(1\ \text{atm}) \mid \text{HCl}(1\ M) \mid \text{AgCl} \mid \text{Ag}$$

5-5   Half-Cell Potentials

The potential which causes electrons to move in the external circuit between the half cells of a galvanic cell is a reflection of the relative oxidizing and reducing tendencies of the respective redox couples. However, it is impossible to measure the potential of a single half cell because it is impossible to have either oxidation or reduction without the other. Hence, only the *difference* in potential between two half cells can be measured. To obtain a numerical scale of potentials, the hydrogen/ hydrogen ion half cell has been chosen as a standard against which all other half cells are measured. A half cell consisting of hydrogen gas at unit activity[4] in contact with a solution of hydrogen ions at unit activity (at a platinum electrode upon which finely divided platinum has been deposited) is assigned a standard potential, $\epsilon°$, of 0.000 V:

$$\text{H}_2(\text{unit activity}) \quad \rightarrow \quad 2\,\text{H}^+(\text{unit activity}) \quad + \quad 2\,\text{e}^- \qquad \epsilon° = 0.000\ \text{V}$$

By convention, this reaction is written as an oxidation, and all other half reactions are written as reductions so that they may be immediately combined with the hydrogen couple.

It is customary to refer to an entire half cell as an "electrode." For example, the hydrogen/hydrogen ion half cell is often called the hydrogen electrode. In any cell which consists of a given half cell and the standard hydrogen half cell, the measured voltage is the electrode potential of the given half cell. If all the species in the given half cell are at unit activity, the measured potential is called the **standard electrode potential** and is denoted $\epsilon°$.

Example

A hydrogen electrode with hydrogen gas at 1 atm and hydrogen ion at unit activity is

---

[4] The activity of an ionic solute is equal to its concentration only for a hypothetical solution in which there is no attractive force between the constituent ions. In real solutions, the activity is equal to the stoichiometric concentration times a factor called the **activity coefficient,** which for dilute solutions is usually close to unity. Therefore, for present purposes, concentrations will be used in place of activities. The error introduced by the use of concentrations could be corrected for theoretically, but the correction is beyond the scope of this text. For gases, the activity is approximately equal to the pressure in atmospheres. Pure liquids and pure solids are defined as being at unit activity.

suitably connected to a half cell consisting of copper metal immersed in 1.00 M copper(II) sulfate solution. The measured potential for the cell

$$Pt \mid H_2(1 \text{ atm}) \mid H^+(1\ M) \parallel Cu^{2+}(1\ M) \mid Cu$$

is 0.34 V. What is the standard electrode potential for the copper/copper(II) ion half cell?

Since all species are in their standard states and since the potential of the hydrogen half cell is zero, the total potential of the cell may be attributed to the copper half cell; therefore the standard electrode potential of the copper/copper(II) ion half cell is $+0.34$ V; that is, $\epsilon° = +0.34$ V.

Some standard electrode potentials are listed in Table 5–2, along with equations for the corresponding half reactions. The half reactions are all written as reduction reactions and therefore the potentials are called reduction potentials.[5] The oxidizing agents in those redox couples whose standard reduction potentials are negative are weaker in oxidizing power at unit activity than is the hydrogen ion at unit activity. (The reducing agents at unit activity are correspondingly stronger reducing agents than is hydrogen gas at 1 atm pressure.) Conversely, the oxidizing agents in those couples whose standard reduction potentials are positive are stronger oxidizing agents than is hydrogen ion at unit activity.

5-6 Cell Reactions

Any combination of two half cells will produce a complete cell. The overall cell reaction is obtained by suitably combining the equations for the two half reactions. If the constituents of each half reaction are in their standard states at unit activity, the cell potential will be the algebraic difference between the respective standard reduction potentials of the two half cells. This potential difference is called the **standard cell potential** and is denoted $\epsilon°_{cell}$. It should be noted that the sign used for the electrode potential is determined by the way the half reaction is written. If half reactions are written as reduction reactions, reduction potentials from Table 5–2 are used. *If the half reactions are written as oxidation reactions, the signs of the reduction potentials are reversed.*

Example

Write the cell reaction and calculate the value of $\epsilon°_{cell}$ for the following cell:

$$Zn \mid Zn^{2+}(1\ M) \parallel Fe^{2+}(1\ M), Fe^{3+}(1\ M) \mid Pt$$

The equations for the half reactions, with the corresponding standard potentials, are

$$Fe^{3+} + e^- \rightarrow Fe^{2+} \qquad \epsilon° = +0.77 \text{ V}$$
$$Zn \rightarrow Zn^{2+} + 2e^- \qquad \epsilon° = +0.76 \text{ V}$$

[5] The convention used here is that recommended by the International Union of Pure and Applied Chemistry (IUPAC). In many U.S. textbooks, half reactions are written as oxidation reactions, and the corresponding potentials are oxidation potentials. The magnitudes of the potentials are not affected by the change in convention, but the signs are reversed. Therefore, before using standard electrode potential data, it is necessary to ascertain which convention is being used.

TABLE 5-2

TABLE 5-2. Standard Reduction Potentials at 25°C

Reaction	Standard Potential, $\epsilon°$ (V)[a]
$F_2 + 2e^- \rightarrow 2F^-$	2.87
$Co^{3+} + e^- \rightarrow Co^{2+}$	1.82
$H_2O_2 + 2H^+ + 2e^- \rightarrow 2H_2O$	1.77
$MnO_4^- + 4H^+ + 3e^- \rightarrow MnO_2 + 2H_2O$	1.70
$PbO_2 + 4H^+ + SO_4^{2-} + 2e^- \rightarrow PbSO_4 + 2H_2O$	1.70
$Ce^{4+} + e^- \rightarrow Ce^{3+}$	1.61
$MnO_4^- + 8H^+ + 5e^- \rightarrow Mn^{2+} + 4H_2O$	1.51
$Au^{3+} + 3e^- \rightarrow Au$	1.50
$Cl_2(g) + 2e^- \rightarrow 2Cl^-$	1.36
$Cr_2O_7^{2-} + 14H^+ + 6e^- \rightarrow 2Cr^{3+} + 7H_2O$	1.33
$MnO_2 + 4H^+ + 2e^- \rightarrow Mn^{2+} + 2H_2O$	1.23
$O_2(g) + 4H^+ + 4e^- \rightarrow 2H_2O$	1.23
$2IO_3^- + 12H^+ + 10e^- \rightarrow I_2 + 6H_2O$	1.20
$Br_2 + 2e^- \rightarrow 2Br^-$	1.09
$OCl^- + H_2O + 2e^- \rightarrow Cl^- + 2OH^-$	0.94
$2Hg^{2+} + 2e^- \rightarrow Hg_2^{2+}$	0.92
$Cu^{2+} + I^- + e^- \rightarrow CuI$	0.85
$Ag^+ + e^- \rightarrow Ag$	0.80
$Hg_2^{2+} + 2e^- \rightarrow 2Hg$	0.79
$Fe^{3+} + e^- \rightarrow Fe^{2+}$	0.771
$O_2 + 2H^+ + 2e^- \rightarrow H_2O_2$	0.68
$Cu^{2+} + Cl^- + e^- \rightarrow CuCl$	0.566
$I_2 + 2e^- \rightarrow 2I^-$	0.54
$Cu^{2+} + 2e^- \rightarrow Cu$	0.34
$Hg_2Cl_2 + 2e^- \rightarrow 2Hg + 2Cl^-$	0.270
$Hg_2Cl_2 + 2e^- \rightarrow 2Hg + 2Cl^-$ (sat KCl)	0.244
$AgCl + e^- \rightarrow Ag + Cl^-$	0.222
$Cu^{2+} + e^- \rightarrow Cu^+$	0.15
$Sn^{4+} + 2e^- \rightarrow Sn^{2+}$	0.13
$2H^+ + 2e^- \rightarrow H_2$	0.000
$Pb^{2+} + 2e^- \rightarrow Pb$	−0.13
$Sn^{2+} + 2e^- \rightarrow Sn$	−0.14
$2CuO + H_2O + 2e^- \rightarrow Cu_2O + 2OH^-$	−0.15
$AgI + e^- \rightarrow Ag + I^-$	−0.151
$CuI + e^- \rightarrow Cu + I^-$	−0.17
$Ni^{2+} + 2e^- \rightarrow Ni$	−0.25
$Co^{2+} + 2e^- \rightarrow Co$	−0.28
$PbSO_4 + 2e^- \rightarrow Pb + SO_4^{2-}$	−0.31
$Cu_2O + H_2O + 2e^- \rightarrow 2Cu + 2OH^-$	−0.34
$Cd^{2+} + 2e^- \rightarrow Cd$	−0.40
$Fe^{2+} + 2e^- \rightarrow Fe$	−0.44
$Cr^{3+} + 3e^- \rightarrow Cr$	−0.74
$Zn^{2+} + 2e^- \rightarrow Zn$	−0.763
$2H_2O + 2e^- \rightarrow H_2 + 2OH^-$	−0.828
$Mn^{2+} + 2e^- \rightarrow Mn$	−1.18
$Al^{3+} + 3e^- \rightarrow Al$	−1.66
$H_2 + 2e^- \rightarrow 2H^-$	−2.25
$Mg^{2+} + 2e^- \rightarrow Mg$	−2.37
$Ce^{3+} + 3e^- \rightarrow Ce$	−2.48
$Na^+ + e^- \rightarrow Na$	−2.71
$Ca^{2+} + 2e^- \rightarrow Ca$	−2.87
$Ba^{2+} + 2e^- \rightarrow Ba$	−2.90
$Cs^+ + e^- \rightarrow Cs$	−2.92
$K^+ + e^- \rightarrow K$	−2.93
$Li^+ + e^- \rightarrow Li$	−3.05

[a] $\epsilon° = \epsilon°_{cell}$ when combined with $H_2(1\ atm) \rightarrow 2H^+(1\ M) + 2e^-$

To obtain the equation for the cell reaction, the equation for the iron(II)/iron(III) half reaction must be multiplied by 2 (to obtain 2 moles of electrons) and combined with the zinc/zinc ion half reaction. The value of the half cell potential is *not* multiplied by 2, however, because potential does *not* depend on the *quantity* of substance undergoing reaction. Addition of the resulting equations and the two potentials results in the following cell reaction:

$$Zn + 2 Fe^{3+} \rightarrow 2 Fe^{2+} + Zn^{2+} \qquad \epsilon^\circ = 1.53 \text{ V}$$

The sign of ϵ_{cell} determines the direction of reaction. A positive sign means that the reaction is spontaneous as written, and a negative sign means that the reverse reaction is spontaneous.

Example

Is 1.0 M H^+ solution under hydrogen gas at 1.0 atm capable of oxidizing silver metal in the presence of 1.0 M silver ion?

The desired reaction is

$$2 Ag + 2 H^+(1.0 M) \rightarrow 2 Ag^+(1.0 M) + H_2(1 \text{ atm})$$

Combining the two half reactions gives

$$2 Ag \rightarrow 2 Ag^+ + 2 e^- \qquad \epsilon^\circ = -0.80 \text{ V}$$
$$2 H^+ + 2 e^- \rightarrow H_2 \qquad \epsilon^\circ = 0.00 \text{ V}$$
$$\overline{2 Ag + 2 H^+ \rightarrow H_2 + 2 Ag^+ \qquad \epsilon^\circ_{cell} = -0.80 \text{ V}}$$

The negative value of the standard cell potential indicates that H^+ solution does not oxidize silver metal under these conditions.

5-7 Measurement of Electrode Potentials

In principle, the standard electrode potential of any redox couple may be obtained by measuring the potential of a cell consisting of a standard hydrogen electrode and the given couple, with all substances involved being present at unit activity. However, since the concentrations of the reactants and products change as the cell reaction proceeds, the measured potential would not long be equal to the standard electrode potential. Therefore cell potentials are measured without drawing significant current from the cell by use of an apparatus called a **potentiometer.** A diagram of a simple potentiometer is given in Figure 5-9. In the potentiometer circuit, a standard cell of known potential is placed in opposition to the unknown cell. The effective potential of the standard cell is varied by means of a variable resistor until no charge flows in either direction. The cell is then said to be operating under **reversible** conditions. A small increase in the applied opposing potential causes the cell reaction in the unknown cell to proceed in the reverse direction, while a small decrease in the opposing potential will cause the unknown cell to discharge and operate at somewhat less than its maximum voltage. Only when the opposing potential *exactly balances* that of the unknown cell is the potential of the unknown cell at its maximum.

Because it requires manipulation of a gas, the use of the standard hydrogen

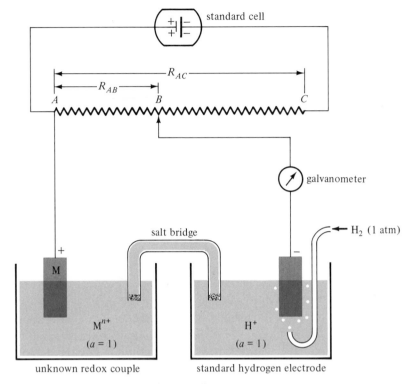

Figure 5-9. **Simplified Potentiometer**
The potential from a standard cell is applied in opposition to the potential from the cell of unknown potential. A movable contact on slide wire R_{AC} is adjusted until no current is registered by the galvanometer. At this point of balance, B, the potential of the unknown cell is just equal to the voltage drop across the portion of the slide wire R_{AB}. The potential of the unknown cell is the fraction of the potential of the standard cell given by R_{AB}/R_{AC}. (If the resistance of the slide wire is uniform along its length, this fraction is equal to the ratio of lengths, $\overline{AB}/\overline{AC}$.) Thus $\epsilon_{unknown} = \epsilon_{std}(R_{AB}/R_{AC})$.

electrode in a potentiometer is rather inconvenient. However, since the *difference* in potential between the two half cells is being measured, any stable reference electrode whose potential is known with respect to the standard hydrogen electrode can be used. One electrode often used for this purpose is the saturated calomel electrode (Figure 5-7). Its potential, ϵ_{cal}, relative to the standard hydrogen electrode is +0.244 V.

Example

The potential of a silver/silver chloride electrode measured with respect to a saturated calomel electrode is −0.022 V. What is the value of the standard reduction potential for the silver/silver chloride electrode?

$$\epsilon^{\circ}_{cell} = \epsilon^{\circ}_{Ag/AgCl} - \epsilon^{\circ}_{cal}$$
$$\epsilon^{\circ}_{Ag/AgCl} = -0.022 \text{ V} + 0.244 \text{ V} = 0.222 \text{ V}$$

5-8   Electrical Work and Free Energy Change

Electrons flowing in the external conductor between the electrodes of a galvanic cell constitute an electric current which can be made to perform various kinds of work, such as running an electric motor, providing heat, providing light, or producing chemical change by electrolysis. The work done by an electric current is equal in magnitude to the product of the charge which flows times the potential through which it flows. Thus electrical work has dimensions of volts · coulombs, or joules (see Table 5–1). For a cell having a potential ϵ_{cell}, the electrical work, w_{el}, is given by

$$w_{el} = nF\epsilon_{cell}$$

where n is the number of moles of electrons which pass and F, the faraday, is equal to the number of coulombs of charge per mole of electrons. If a cell is operated under reversible conditions, the measured voltage, ϵ_{rev}, is its maximum voltage, and the work done under these conditions becomes the maximum work, w_{max}:

$$w_{max} = nF\epsilon_{rev}$$

When a cell is operating reversibly, no charge is actually flowing and no net reaction actually occurs; w_{max} merely represents the maximum work that the cell is *theoretically* capable of producing. However, w_{max} has important thermodynamic significance. Unlike heat energy, electrical energy *can* be quantitatively converted to other forms of energy. For a given cell reaction, w_{max} represents the maximum quantity of available energy, and therefore it is a measure of the change in free energy (Section 3–16). If work is done by a system, its free energy decreases and the following equation applies:

$$-\Delta G = w_{max} = nF\epsilon_{rev}$$

When all substances involved in the reaction are in their standard states, ϵ_{cell}° represents the standard cell potential and ΔG° represents the standard free energy change for the cell reaction. Hence

$$\Delta G^{\circ} = -nF\epsilon_{cell}^{\circ}$$

Example

Calculate the standard free energy change for the reaction

$$Zn + Cu^{2+} \rightarrow Cu + Zn^{2+}$$

ϵ_{cell}° is obtained as follows:

	Zn	$\rightarrow$	Zn^{2+} + 2 e$^-$		$\epsilon^{\circ} = 0.76$ V
Cu^{2+} + 2 e$^-$	$\rightarrow$	Cu			$\epsilon^{\circ} = 0.34$ V
Zn + Cu^{2+}	$\rightarrow$	Zn^{2+} + Cu			$\epsilon_{cell}^{\circ} = 1.10$ V

Since 2 moles of electrons is associated with 1 mole of chemical reaction,

$$\Delta G^{\circ} = -nF\epsilon^{\circ}_{cell} = -(2 \text{ moles e}^-)\left(\frac{96,500 \text{ C}}{\text{mole e}^-}\right)(1.10 \text{ V}) = -212,000 \text{ J}$$

$$= -212,000 \text{ J} \left(\frac{1 \text{ cal}}{4.18 \text{ J}}\right)\left(\frac{1 \text{ kcal}}{10^3 \text{ cal}}\right) = -50.7 \text{ kcal}$$

Reversible cell potentials are relatively easy to measure; hence the free energy change for any reaction which can be arranged to take place in a galvanic cell can be obtained conveniently. Free energy is a state function, and for a given process the magnitude of ΔG depends only on the initial and final states of the system. The ΔG of a reaction which occurs irreversibly can be obtained by measuring the potential of a galvanic cell in which the cell reaction is the desired reaction.

Example

Calculate ΔG°_{298} and ΔS°_{298} for the reaction

$$2 \text{ H}_2(g) + \text{O}_2(g) \rightarrow 2 \text{ H}_2\text{O}(l)$$

This reaction does not occur reversibly under ordinary laboratory conditions; however, a galvanic cell can be constructed for which this reaction is the cell reaction. The appropriate half reactions are combined as follows:

$$
\begin{array}{llllll}
\text{O}_2 & + & 4 \text{ H}^+ & + & 4 \text{ e}^- & \rightarrow & 2 \text{ H}_2\text{O} & \epsilon^{\circ} = 1.23 \text{ V} \\
& & 2 \text{ H}_2 & \rightarrow & 4 \text{ H}^+ & + & 4 \text{ e}^- & \epsilon^{\circ} = 0.00 \text{ V} \\
\hline
& \text{O}_2 & + & 2 \text{ H}_2 & \rightarrow & 2 \text{ H}_2\text{O} & \epsilon^{\circ}_{cell} = 1.23 \text{ V}
\end{array}
$$

$$\Delta G^{\circ}_{298} = -nF\epsilon^{\circ} = -(4 \text{ moles e}^-)(96,500 \text{ C/mole e}^-)(1.23 \text{ V})$$
$$= -475,000 \text{ J} = -114 \text{ kcal}$$

From Table 3–2, for the formation of water from hydrogen and oxygen,

$$\Delta H^{\circ}_f = -68.32 \text{ kcal/mole}$$

Therefore, ΔS° for the above reaction can be calculated as follows:

$$\Delta S^{\circ}_{298} = \frac{\Delta H^{\circ} - \Delta G^{\circ}}{T} = \frac{2(-68.32) - (-114)}{298 \text{ K}} = -0.076 \text{ kcal/K} = -76 \text{ cal/K}$$

5-9 Combination of Two Half Reactions to Obtain a Third

Potential (like temperature) is intensive and is independent of the quantities of materials involved in a given reaction. When two redox half reactions are combined in a manner which yields a third half reaction, the standard potential of the third half reaction is *not* merely a combination of the standard potentials of the original two half reactions. Every half reaction is a relationship involving moles of reactants and moles of electrons and thus denotes quantities of matter undergoing change. The free energy change in each process does depend on the quantities of matter involved, and so the free energy changes for a half reaction which is a combination of two others can be obtained by combining the free energy changes of the original half reactions. Once its free energy change is obtained, the potential of the new half reaction may be calculated from the relationship

$$\Delta G = -n\text{F}\epsilon$$

Example

Using data from Table 5–2, calculate the standard reduction potential for the half reaction

$$2\,OCl^- \;+\; 2\,H_2O \;+\; 2\,e^- \;\rightarrow\; Cl_2(g) \;+\; 4\,OH^-$$

Since the free energy changes for half reactions are additive, the function $-\Delta G°/\text{F} = n\epsilon°$ is also additive. Appropriate half reactions from the table must involve OCl^- and Cl_2 along with a reduction product common to them. The couples OCl^-/Cl^- and Cl_2/Cl^- are selected. Doubling the first and reversing the second yields

	$\epsilon°$	n	$-\dfrac{\Delta G°}{\text{F}} = n\epsilon°$
$2\,OCl^- \;+\; 2\,H_2O \;+\; 4\,e^- \;\rightarrow\; 2\,Cl^- \;+\; 4\,OH^-$	0.94 V	4	3.76
$2\,Cl^- \;\rightarrow\; Cl_2(g) \;+\; 2\,e^-$	-1.36 V	2	-2.72

Adding the chemical equations and the $(-\Delta G°/\text{F})$ values yields

$2\,OCl^- \;+\; 2\,H_2O \;+\; 2\,e^- \;\rightarrow\; Cl_2(g) \;+\; 4\,OH^-$	2	1.04

$$\epsilon° = \frac{-\Delta G°/\text{F}}{n} = \frac{1.04}{2} = 0.52 \text{ V}$$

5-10 The Nernst Equation

In defining the standard cell potential, it was specified that all substances must be at unit activity. However, in practice it is unlikely that all of the substances making up a cell will be at unit activity, and the observed cell potential in general will not be $\epsilon°_{cell}$. The relationship between the observed cell potential and the activities of the reacting species is given by the **Nernst equation.** (This equation can be derived from theoretical principles, but the derivation is beyond the scope of this textbook.) For a general half reaction, written as a reduction,

$$a\,A \;+\; b\,B \;+\; n\,e^- \;\rightarrow\; c\,C \;+\; d\,D$$

the Nernst equation is written as

$$\epsilon = \epsilon° - \frac{2.303\,RT}{n\text{F}}\log\frac{[C]^c[D]^d}{[A]^a[B]^b}$$

where R is the gas constant (expressed as 8.314 J/mole · K), T is the absolute temperature, F is the faraday, and n is the number of moles of electrons transferred in the half reaction as written. The square brackets enclosing the symbols of the reagents denote their concentrations in moles per liter and the exponents are the coefficients from the equation for the half reaction. For example, $[A]^a$ represents the molar concentration of reagent A raised to the ath power. At 298 K, the factor $2.303\ RT/F$ has the value 0.0592 V. Hence

$$\epsilon = \epsilon^\circ - \frac{0.0592}{n}\log\frac{[C]^c[D]^d}{[A]^a[B]^b}$$

Example

What is the potential of an electrode consisting of zinc metal in a solution in which the zinc ion concentration is 0.0100 M?

$$Zn^{2+}\ +\ 2\,e^-\ \rightarrow\ Zn\qquad \epsilon^\circ = -0.763\ V$$

$$\epsilon = \epsilon^\circ - \frac{0.0592}{n}\log\frac{a_{Zn}}{[Zn^{2+}]}$$

The activity of pure zinc metal, a_{Zn}, is 1.00, and the Zn^{2+} concentration is 0.0100 M.

$$\epsilon = -0.763 - \frac{0.0592}{2}\log\frac{1.00}{0.0100} = -0.763 - \frac{0.0592}{2}\log(100)$$

$$= -0.763 - 0.0592 = -0.822\ V$$

It can be shown that for an overall cell reaction, such as

$$w\,W\ +\ m\,M\ \rightarrow\ p\,P\ +\ q\,Q$$

the Nernst equation has the form

$$\epsilon = \epsilon^\circ - \frac{0.0592}{n}\log\frac{[P]^p[Q]^q}{[W]^w[M]^m}$$

where in this case n represents the number of moles of electrons transferred between w moles of W and m moles of M.

Example

Calculate the potential of the cell

$$Pt\,|\,H_2(0.50\ atm)\,|\,H^+(0.10\ M)\,\|\,MnO_4^-(0.10\ M),\,Mn^{2+}(1.0\ M),\,H^+(0.10\ M)\,|\,Pt$$

The half reactions are combined as follows:

$$
\begin{array}{llllll}
& & 5\,H_2 & \rightarrow\ 10\,H^+\ +\ 10\,e^- & \epsilon^\circ = & 0.00\ V \\
16\,H^+\ +\ 2\,MnO_4^-\ + & 10\,e^- & \rightarrow\ 2\,Mn^{2+}\ +\ 8\,H_2O & \epsilon^\circ = & 1.51\ V \\
\hline
6\,H^+\ +\ 2\,MnO_4^-\ + & 5\,H_2 & \rightarrow\ 2\,Mn^{2+}\ +\ 8\,H_2O & \epsilon^\circ_{cell} = & +1.51\ V
\end{array}
$$

$$\epsilon_{cell} = \epsilon^{\circ}_{cell} - \frac{0.0592}{10} \log \frac{[Mn^{2+}]^2(a_{H_2O})^8}{[H^+]^6[MnO_4^-]^2(P_{H_2})^5}$$

In dilute aqueous solutions, water is present in large excess, and its activity is assumed to be 1.00. The activity of the hydrogen gas is assumed to be equal to its pressure in atmospheres, and the activities of the aqueous ions are assumed to be equal to their molar concentrations:

$$\epsilon_{cell} = +1.51 - \frac{0.0592}{10} \log \frac{(1.0)^2(1.00)^8}{(0.10)^6(0.10)^2(0.50)^5}$$

$$= +1.51 - (5.92 \times 10^{-3}) \log (3.2 \times 10^9)$$

$$= +1.51 - (5.92 \times 10^{-3})(+9.505) = +1.45 \text{ V}$$

5-11 Concentration Cells

When two aqueous solutions of a given solute having different concentrations are placed in contact with each other, they will mix spontaneously until the entire mixture has a uniform concentration. The free energy change associated with such a process can be measured by constructing a galvanic cell in which the two half cells contain the same sets of species but at different activities (or concentrations). A cell having a potential resulting only from concentration differences is called a **concentration cell.** For example, consider the following cell:

$$Zn \mid Zn^{2+}(0.0100 \ M) \parallel Zn^{2+}(0.500 \ M) \mid Zn$$

The cell reaction and potential are determined as follows:

$$Zn^{2+}(0.500 \ M) \ + \ 2 e^- \ \rightarrow \ Zn \qquad \epsilon = -0.763 - \frac{0.0592}{2} \log \frac{1}{0.500}$$

$$Zn \ \rightarrow \ Zn^{2+}(0.0100 \ M) \ + \ 2 e^- \qquad \epsilon = +0.763 - \frac{0.0592}{2} \log 0.0100$$

$$Zn^{2+}(0.500 \ M) \ \rightarrow \ Zn^{2+}(0.0100 \ M) \qquad \epsilon_{cell} = -\frac{0.0592}{2} \log \frac{0.0100}{0.500}$$

$$= -0.0296(-1.70) = +0.0503 \text{ V}$$

The free energy change for the cell reaction is calculated as follows:

$$\Delta G = -nF\epsilon_{cell} = -(2 \text{ moles } e^-)\left(\frac{96,500 \text{ C}}{\text{mole } e^-}\right)(0.0503 \text{ V}) = -9710 \text{ J}$$

$$= -9710 \text{ J}\left(\frac{1.00 \text{ kcal}}{4184 \text{ J}}\right) = -2.32 \text{ kcal}$$

The negative value of the free energy change indicates that the cell reaction proceeds spontaneously as written. The more concentrated solution spontaneously tends to become dilute, and the dilute solution tends to become more concentrated until the concentrations become equal. The same result would be expected if the two solutions were mixed directly.

Concentration cells have a number of important practical applications, especially in analytical chemistry. Certain types of electrodes are particularly sensitive to concentration changes. One example is the glass electrode, which is used to measure hydrogen ion activity. This electrode will be described further in Section 7–12.

5-12   Potentiometric Titrations

Since the potential of a cell changes with the concentrations of its reactants, it is possible to follow the course of a reaction potentiometrically by observing how the voltage of a cell containing one reagent is affected as the other reagent is added to it. The titration of iron(II) ion with cerium(IV) ion will be used as an example to illustrate the principles involved. The equation for the titration reaction is

$$Ce^{4+} + Fe^{2+} \rightarrow Ce^{3+} + Fe^{3+}$$

A typical experimental setup is shown in Figure 5–10. The titration cell consists of a reference calomel electrode and an indicator electrode—in this case a platinum electrode dipping into the iron(II) solution. The potential of the reference electrode remains constant throughout the experiment because it is not involved in the titration reaction, and the concentrations of its constituents do not change. On the other hand, as the titration proceeds, the potential of the indicator electrode, ϵ_{ind}, changes because the ratios $[Fe^{2+}]/[Fe^{3+}]$ and $[Ce^{3+}]/[Ce^{4+}]$ change as the reagent, Ce^{4+} ion, is added. The change in ϵ_{ind} may be predicted by use of the Nernst equation. Consider that 100.0 ml of a solution of iron(II) of unknown concentration is titrated with a 2.500 M solution of cerium(IV) ion. When sufficient Ce^{4+} has been added so that 10% of the

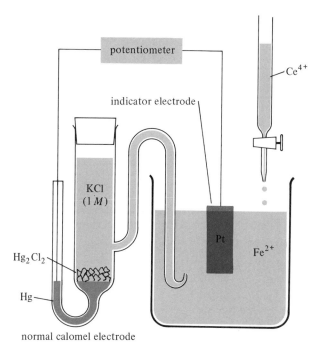

Figure 5-10. Apparatus for Potentiometric Titration

The Fe^{2+} is directly oxidized by the Ce^{4+} to Fe^{3+}. At each point in the titration, the measured potential corresponds to the reduction of Fe^{3+} to Fe^{2+} or of excess Ce^{4+} to Ce^{3+}.

Fe^{2+} is oxidized to Fe^{3+}, the ratio $[Fe^{2+}]/[Fe^{3+}] = 9$. (It must be noted that the indicator electrode is the cathode and the *electrode* reaction is the *reduction* of Fe^{3+} to Fe^{2+}.)

$$\epsilon_{ind} = \epsilon^{\circ}_{Fe^{2+}/Fe^{3+}} - 0.0592 \log \frac{[Fe^{2+}]}{[Fe^{3+}]} = 0.771 - 0.0592 \log 9 = 0.715 \text{ V}$$

When 50% of the iron(II) has been oxidized, the ratio of iron(II) concentration to iron(III) concentration will be equal to 1:

$$\epsilon_{ind} = 0.771 - 0.0592 \log 1 = 0.771 \text{ V} = \epsilon^{\circ}_{Fe^{2+}/Fe^{3+}}$$

When 90% of the Fe^{2+} has been oxidized, the ratio equals $\frac{1}{9}$. Hence

$$\epsilon_{ind} = 0.771 - 0.0592 \log \tfrac{1}{9} = 0.827 \text{ V}$$

At the equivalence point, the number of moles of added Ce^{4+} equals the number of moles of Fe^{2+} originally present. Hence, $[Ce^{4+}] = [Fe^{2+}]$ and $[Ce^{3+}] = [Fe^{3+}]$. Thus

$$\frac{[Ce^{3+}]}{[Ce^{4+}]} = \frac{[Fe^{3+}]}{[Fe^{2+}]}$$

and therefore

$$\frac{[Ce^{3+}][Fe^{2+}]}{[Ce^{4+}][Fe^{3+}]} = 1$$

At this point, the indicator electrode is best considered as both an iron(II)/ iron(III) electrode and a cerium(III)/cerium(IV) electrode. Addition of the Nernst equations for the two indicator electrodes produces the following:

$$2\epsilon_{ind} = \epsilon^{\circ}_{Ce^{3+}/Ce^{4+}} + \epsilon^{\circ}_{Fe^{2+}/Fe^{3+}} - 0.0592 \log \frac{[Ce^{3+}][Fe^{2+}]}{[Ce^{4+}][Fe^{3+}]}$$

Since $[Ce^{3+}][Fe^{2+}]/[Ce^{4+}][Fe^{3+}] = 1$, this expression reduces to

$$\epsilon_{ind} = \frac{\epsilon^{\circ}_{Fe^{2+}/Fe^{3+}} + \epsilon^{\circ}_{Ce^{3+}/Ce^{4+}}}{2} = \frac{1.61 + 0.771}{2} = 1.19 \text{ V}$$

Upon addition of cerium(IV) ion beyond the equivalence point, the potential of the indicator electrode is more conveniently calculated as a cerium ion electrode. Thus when 10% excess Ce^{4+} is added, the mole ratio

$$\frac{[Ce^{3+}]}{[Ce^{4+}]} = 10$$

$$\epsilon_{ind} = 1.61 - 0.0592 \log \frac{[Ce^{3+}]}{[Ce^{4+}]} = 1.55 \text{ V}$$

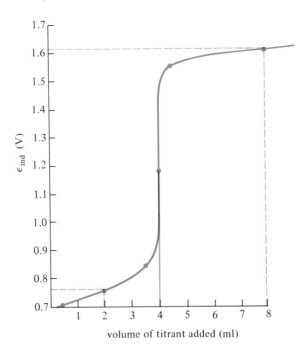

Figure 5-11. $Ce^{4+}-Fe^{2+}$ Titration Curve

Since the Fe^{3+}/Fe^{2+} and the Ce^{4+}/Ce^{3+} couples both exist in the same solution, their potentials at every point in the titration must be equal. It is more convenient to calculate the potential using the Fe^{3+}/Fe^{2+} couple before the equivalence point, and the Ce^{4+}/Ce^{3+} couple after that point. The values of ϵ_{ind} when the concentration ratios $[Fe^{2+}]/[Fe^{3+}]$ and $[Ce^{3+}]/[Ce^{4+}]$ are equal to unity, 0.77 V and 1.61 V, respectively, are shown by horizontal dashed lines. The equivalence point is shown by the vertical gray line.

When 100% excess cerium((IV) ion has been added, the concentrations of cerium in the two oxidation states are equal, and the indicator electrode potential equals $\epsilon°$ for the Ce^{3+}/Ce^{4+} electrode:

$$\epsilon_{ind} = 1.61 - 0.0592 \log 1 = 1.61 \text{ V} = \epsilon°_{Ce^{3+}/Ce^{4+}}$$

All the values of ϵ_{ind} calculated above, along with the corresponding volumes of titrant added, are plotted in Figure 5-11. It should be noted that at the equivalence point, a small volume of added reagent produces a large change in ϵ_{ind}. A vertical line drawn through the steepest point of the curve will intersect the axis at the equivalence point, in this case 4.000 ml. Hence, the concentration of the iron(II) in the original solution may be calculated as follows:

$$\text{mmole } Fe^{2+} = \text{mmole } Ce^{4+}$$
$$100.0 \text{ ml } [Fe^{2+}] = 4.000 \text{ ml} \times 2.500 \text{ } M$$
$$[Fe^{2+}] = 0.1000 \text{ } M$$

5-13   Practical Cells

Galvanic cells are convenient and efficient devices for storing energy as chemical energy and later transforming it into other forms of energy. For this reason, much research and considerable ingenuity goes into developing new types of cells. In principle, any oxidation-reduction reaction can be adapted to make a cell, but in designing a cell suitable for practical use several factors are important. The cell reaction must be reasonably rapid. Ideally, the cell must have a reasonably large

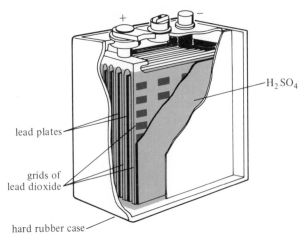

Figure 5-12. **Lead Storage Cell**
The basic features of the lead storage cell are electrodes of lead and of lead dioxide, immersed in concentrated aqueous sulfuric acid. The cell reaction is as follows:

anode: $\qquad Pb(s) + HSO_4^- \rightarrow PbSO_4(s) + H^+ + 2e^-$

cathode: $PbO_2(s) + HSO_4^- + 3H^+ + 2e^- \rightarrow PbSO_4(s) + 2H_2O$

cell: $\quad Pb(s) + PbO_2(s) + 2HSO_4^- + 2H^+ \rightarrow 2PbSO_4(s) + 2H_2O$

Thus both electrode reactions produce insoluble $PbSO_4$, which adheres to the electrodes. When the cell discharges, H_2SO_4 is used up and H_2O is produced. Since the density of water is about 70% that of the H_2SO_4 solution, the state of charge of the cell can be determined by measuring the density of the electrolyte solution. When the cell is recharged, the electrode reactions are reversed.

potential, which should remain relatively constant over the lifetime of the cell. It must have a long life and yet be small enough to be handled conveniently. It should be rugged and portable. It is an added advantage if the cell is rechargeable. Despite all these requirements, the cell should be relatively inexpensive.

The design of commercial cells involves compromises of these features. To obtain a high potential, several cells may be connected in series.[6] Such an arrangement is popularly called a **battery.** Constant voltage is obtained by use of redox couples in the form of solids or solids and saturated solutions. Long life with a correspondingly small size can be obtained by using redox systems in which the reactants have low formula weights and hence large numbers of moles per unit mass. Aluminum or magnesium or carbon are good for this purpose, but materials having larger formula weights, such as iron, nickel, cadmium, zinc, or lead and their compounds, turn out to be more suitable chiefly because cells using these materials can be recharged. Some practical cells are diagrammed in Figures 5-12, 5-13, and 5-14.

[6] When cells are connected in series, as in the accompanying diagram, the same current passes through each cell. The total potential is the sum of the potentials of the individual cells.

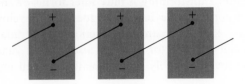

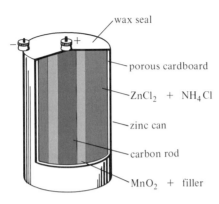

wax seal

porous cardboard

$ZnCl_2$ + NH_4Cl

zinc can

carbon rod

MnO_2 + filler

Figure 5-13. **Dry Cell**

The dry cell is relatively inexpensive and quite portable. It has many uses, such as in flashlights, portable radios, and automatic cameras. The anode consists of a zinc can in contact with a moist paste of $ZnCl_2$ and NH_4Cl. A carbon rod surrounded by MnO_2 and filler is the cathode. The cell reaction appears to vary with the rate of discharge, but at low power the probable reactions are

anode: $\qquad\qquad\qquad\qquad\qquad Zn(s) \rightarrow Zn^{2+} + 2e^-$

cathode: $2MnO_2(s) + Zn^{2+} + 2e^- \rightarrow ZnMn_2O_4(s)$

cell: $\qquad\qquad Zn(s) + 2MnO_2(s) \rightarrow ZnMn_2O_4(s)$

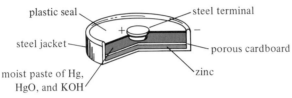

plastic seal

steel terminal

steel jacket

porous cardboard

moist paste of Hg, HgO, and KOH

zinc

Figure 5-14. **Mercury Cell**

Because of its compact size and constant voltage, the so-called mercury battery is used in such devices as hearing aids and "electric eye" cameras. The cell operates in a basic medium, as follows:

anode: $\qquad\qquad Zn + 2OH^- \rightarrow ZnO + H_2O + 2e^-$

cathode: $HgO + H_2O + 2e^- \rightarrow Hg + 2OH^-$

cell: $\qquad\qquad Zn + HgO \rightarrow ZnO + Hg$

5-14   Fuel Cells

It would be desirable to use ordinary fuels and oxygen as raw materials for a galvanic cell. Electricity generated commercially using fossil fuels, such as coal, oil, and natural gas, is produced indirectly at present. Heat liberated in the combustion of the fuels is used to produce steam, which in turn drives turbine generators. By this means, central power stations convert chemical energy into electrical energy, but with a maximum efficiency of about 40%. On the other hand, if electricity were produced directly from the oxidation of the fuel in the galvanic cell, efficiencies greater than 75% would be possible.

An ideal fuel cell reaction would involve oxygen (from the air) and a fossil fuel. The cell should have inert electrodes, and as energy is drawn from the cell, the reactants could be replenished. Unfortunately, for this purpose, conventional fuels are quite unreactive except at high temperatures. Hydrogen and hydrazine, N_2H_4, are two fuels which can be used in a cell operating at low temperature, but both are still too expensive for use on a large scale.

The hydrogen/oxygen cell operates in basic solution. The cell reaction does not proceed directly to water but produces the anion of hydrogen peroxide, HO_2^-. This ion subsequently decomposes to oxygen gas and hydroxide ion. The relevant reactions are anode: $\qquad$ $H_2(g) + 2\,OH^-(aq) \rightarrow 2\,H_2O(l) + 2\,e^-$

cathode: $\qquad$ $O_2(g) + H_2O(l) + 2\,e^- \rightarrow HO_2^-(aq) + OH^-(aq)$

cell: $\qquad$ $O_2(g) + H_2(g) + OH^-(aq) \rightarrow HO_2^-(aq) + H_2O(l)$

decomposition: $\qquad$ $HO_2^-(aq) \rightarrow \tfrac{1}{2}O_2(g) + OH^-(aq)$

net: $\qquad$ $\tfrac{1}{2}O_2(g) + H_2(g) \rightarrow H_2O(l)$

A schematic diagram of the hydrogen/oxygen fuel cell is shown in Figure 5-15. The electrodes are hollow tubes of porous compressed carbon. Finely divided platinum or palladium is incorporated into the anode, while silver(I) oxide or cobalt(II) oxide is incorporated into the cathode. These substances act as catalysts which speed up the electrode reactions.

A cell which uses natural gas as the fuel operates at temperatures about $500\,°C$ and involves the following reactions in molten KOH (using methane, CH_4, as a typical constituent of natural gas).

anode: $\qquad$ $CH_4(g) + 10\,OH^-(l) \rightarrow CO_3^{2-}(l) + 7\,H_2O(g) + 8\,e^-$

cathode: $\qquad$ $O_2(g) + 2\,H_2O(g) + 4\,e^- \rightarrow 4\,OH^-(l)$

cell: $CH_4(g) + 2\,O_2(g) + 2\,OH^-(l) \rightarrow CO_3^{2-}(l) + 3\,H_2O(g)$

A high temperature cell is diagrammed in Figure 5-16. A number of problems are encountered in the operation of such cells. At the high temperatures involved, (1) corrosion reactions are accelerated, (2) undesirable changes occur in the electrodes and the catalysts, and (3) leaks develop, allowing direct reaction between the fuel and the oxidant. The search for suitable construction materials to overcome these problems continues.

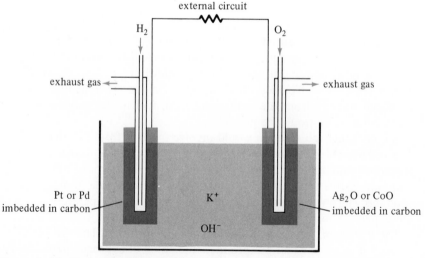

Figure 5-15. **Hydrogen/Oxygen Fuel Cell**

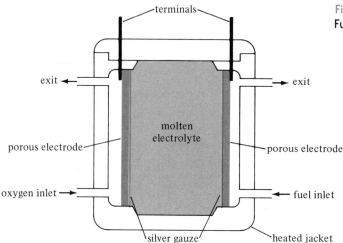

Figure 5-16. **High Temperature Fuel Cell (Schematic)**

5-15 Other Applications of Electrochemistry

In some cases, the products of an electrolysis experiment can be predicted from a consideration of the electrode potentials of the possible reactants. For example, when a copper anode is used in the electrolysis of 1 M $CuCl_2$ solution, copper metal is oxidized ($\epsilon^\circ_{red} = 0.34$ V) rather than chloride ion ($\epsilon^\circ_{red} = 1.36$ V). If different cations are present which would be deposited at very different potentials, the applied potential may be adjusted so that only the cation having the highest reduction potential deposits on a previously weighed platinum cathode. Then the cathode is removed, dried, and weighed. In this manner mixtures of metal ions may be analyzed by selective electrolysis.

Example

If a mixture of copper(II) ion and zinc(II) ion is electrolyzed, what will be the ratio of copper(II) ion concentration to zinc(II) ion concentration when the half-cell potentials are equal?

$$Zn^{2+} \quad + \quad 2\,e^- \quad \rightarrow \quad Zn \qquad \epsilon_{Zn} = -0.76 - \frac{0.0592}{2} \log \frac{1}{[Zn^{2+}]}$$

$$Cu^{2+} \quad + \quad 2\,e^- \quad \rightarrow \quad Cu \qquad \epsilon_{Cu} = +0.34 - \frac{0.0592}{2} \log \frac{1}{[Cu^{2+}]}$$

Equating the half cell potentials yields

$$\epsilon_{Zn} = \epsilon_{Cu}$$

$$-0.76 - \frac{0.0592}{2} \log \frac{1}{[Zn^{2+}]} = +0.34 - \frac{0.0592}{2} \log \frac{1}{[Cu^{2+}]}$$

Rearranging gives

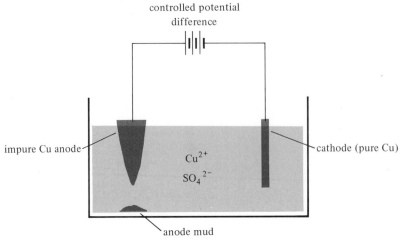

Figure 5-17. **Electrorefining of Copper**

$$-1.10 = \frac{0.0592}{2} \log \frac{[Cu^{2+}]}{[Zn^{2+}]}$$

$$\frac{[Cu^{2+}]}{[Zn^{2+}]} = 10^{-37}$$

It is apparent that practically all the copper is deposited before the zinc starts to be deposited.

Another example of the application of this principle is the metallurgical refining of copper metal. A large anode of impure copper and a small cathode of pure copper are immersed in a copper(II) ion solution (Figure 5–17). The potential is adjusted to that just required for the dissolving of copper from the anode. Any metal impurities having a higher reduction potential (lower oxidation potential) than copper, such as silver and gold, are not oxidized and merely fall to the bottom of the vessel, forming what is known as "anode mud." Any metal having a lower reduction potential than copper, such as iron, is dissolved from the anode and transported to the cathode along with the copper(II) ion. However, the potential at the cathode is not sufficient to reduce the iron, and only pure copper is plated out. Thus 99.96% pure copper is obtained. The silver and gold impurities are recovered from the anode mud, and the value of these metals is more than sufficient to cover the cost of the electrorefining process.

Electroplating of metals is discussed briefly in Section 14–5.

5-16 Overvoltage and Corrosion

Many electrode processes require a greater electrode potential than the thermodynamic electrode potential. Such extra potential is called **overvoltage** and represents the activation required to make the electrode reaction proceed at a reasonable rate. For example, the reduction of hydrogen ion to hydrogen gas has an appreciable overvoltage; consequently, when pure zinc is placed in acid, the expected evolution of hydrogen proceeds at a slow rate. However, if a trace of copper(II) ion is added to the mixture, the reaction proceeds rapidly. Apparently the copper(II) ion first reacts with the zinc, and the resulting activated surface is more readily attacked by H^+.

The corrosion of metals by moist air may be viewed as an electrochemical process. When iron rusts, the oxide $Fe_2O_3 \cdot xH_2O$ is formed. The initial reactions which take place are

$$Fe \rightarrow Fe^{2+} + 2e^-$$
$$O_2 + 2H_2O + 4e^- \rightarrow 4OH^-$$

Apparently because of the overvoltage effects, the second half reaction above takes place only at impurity specks or at strain points on the surface of the iron. Very pure iron is resistant to rusting. The iron(II) ion is further oxidized by oxygen as follows:

$$4Fe^{2+} + O_2 + (4 + 2x)H_2O \rightarrow 2Fe_2O_3 \cdot xH_2O + 8H^+$$

The H^+ reacts with the OH^-, and the net reaction is

$$4Fe + 3O_2 + 2xH_2O \rightarrow 2Fe_2O_3 \cdot xH_2O$$

One way to prevent the corrosion of iron is by **cathodic protection.** For example, stakes of zinc or magnesium are driven into the ground and connected electrically to an iron pipe which is to be protected. The iron acts as a cathode, and the metal having the lower reduction potential is preferentially oxidized. Similarly, the zinc coating of galvanized iron acts as an anode, but in addition the zinc reacts with oxygen to form a film of zinc oxide which protects the surface from further corrosion. Tin is also used as a protective coating on iron. ("Tin cans" are actually made of steel having a very thin coating of tin on its surface.) The tin has a higher resistance to corrosion than iron because the reduction potential of tin is higher, but for that reason tin does not provide cathodic protection. Once the tin coating is scratched or punctured, corrosion of the steel is apt to be faster than if the tin were not present.

5-17 Exercises

Basic Exercises

1. Express each of the following combinations of electrical units as a single unit: (a) volt · ampere, (b) ampere · second, (c) volt/ampere, (d) joule/volt, (e) watt/ampere · ohm, (f) joule/second, (g) joule/ampere · second, (h) joule/ampere² · second.

2. Two cells containing silver electrodes in aqueous silver nitrate and copper electrodes in aqueous copper(II) sulfate, respectively, are connected in series so that the same current must pass through each of the cells. In a given experiment, 2.000 grams of silver metal is deposited on the cathode of the first cell. Calculate the mass of silver dissolved from the anode of the first cell, and the mass of copper deposited at the cathode of the second cell.

3. Make a schematic diagram of a cell for each of the following cell reactions. Indicate on the diagram the anode, the cathode, the directions in which the anions and cations migrate through the cell and the salt bridge, if any, and the direction in which the electrons migrate through the external circuit. Calculate $\epsilon°$ for each cell.

(a) $Sn + 2Ag^+ \rightarrow Sn^{2+} + 2Ag$

(b) $2Cr + 3H_2O + 3OCl^- \rightarrow 2Cr^{3+} + 3Cl^- + 6OH^-$

(c) $H_2 + I_2 \rightarrow 2H^+ + 2I^-$

4. Calculate the free energy change per mole of copper(II) ion formed in a cell consisting of a copper/copper(II) ion half cell suitably connected to a silver/silver ion half cell of sufficient size that the concentration of the ions is not changed from 1.00 M.

5. From the definition of a faraday, calculate the charge on one electron.

6. Calculate the mass of copper which can be deposited by the passage of 10.0 A for 20.0 min through a solution of copper(II) sulfate.

7. Predict whether 1.0 M Mn^{2+} and 1.0 M MnO_4^- will react in acid solution to produce MnO_2.

8. **(a)** Using the Nernst equation for the cell reaction:

$$Pb + Sn^{2+} \rightarrow Pb^{2+} + Sn$$

calculate the ratio of cation concentrations for which $\epsilon = 0$. **(b)** Distinguish clearly between the meaning of $\epsilon = 0$ and $\epsilon° = 0$. Give an example of a reaction in which $\epsilon° = 0$.

9. Calculate the reduction potential of each of the following half cells with each metal immersed in 1.00×10^{-6} M solution of the cation. Arrange the half cells in order of decreasing reduction potential, and compare this list with the relative positions of these half cells in Table 5-2.   **(a)** Cu^{2+}/Cu,   **(b)** Cd^{2+}/Cd,   **(c)** Cr^{3+}/Cr,   **(d)** H^+/H_2(1 atm)(Pt),   **(e)** Pb^{2+}/Pb,   **(f)** Al^{3+}/Al.

10. How long will it take for a uniform current of 6.00 A to deposit 78.0 grams of gold from a solution of $AuCl_4^-$? What mass of chlorine gas will be formed simultaneously at the anode of the electrolytic cell?

11. What is the average current (in amperes) if 100 grams of nickel is deposited from a nickel(II) ion solution in 3 hours 20 minutes?

12. Calculate the reduction potential of a half cell consisting of a platinum electrode immersed in 2.0 M Fe^{2+} and 0.020 M Fe^{3+} solution.

13. Write the equation for the half reaction in which HNO_3 is reduced to NO. Under which of the following sets of conditions does the half-cell potential equal the standard reduction potential?

    (a) 1 M NO_3^-, 1 M H^+, 1 atm NO

    (b) 1 M NO_3^-, 4 M H^+, 1 atm NO

    (c) 1 M NO_3^-, 1 M H^+, 1 atm air

    (d) 1 M NO_3^-, 4 M H^+, 1 atm air

14. What desirable features, such as are enumerated in the text, are characteristic of a lead storage battery? What undesirable features, if any, are characteristic of the lead storage battery?

15. Explain why blocks of magnesium are often strapped to the steel hulls of ocean-going ships.

16. What purpose(s) does chrome plating of steel serve? nickel plating?

17. Explain why a homeowner should avoid attaching aluminum downspouts to galvanized steel gutters.

18. Explain concisely why a porous plate or a salt bridge is not required in a lead storage cell.

General Exercises

19. Calculate the number of kilowatt hours of electricity necessary to produce 1.00 ton (1000 kg) of aluminum by the Hall process in a cell operating at 15.0 V. Where in the United States are aluminum plants likely to be found?

20. A current of 2.00 A passing for 5.00 hours through a molten tin salt deposits 22.2 grams of tin. What is the oxidation state of the tin in the salt?

21. A current of 0.200 A is passed for 600 sec through 50.00 ml of 0.1000 M NaCl. If only chlorine gas is produced at the anode and if water is reduced to hydrogen gas at the cathode, what will be the hydroxide ion concentration in the solution after the electrolysis?

22. What reaction, if any, would zinc(II) ion undergo in the copper(II) half of a Daniell cell? What reaction, if any, would copper(II) ion undergo in the zinc half of a Daniell cell? Which of these ions might actually get into the other half cell during discharge of the cell? during recharge? Explain why a Daniell cell cannot be fully recharged.

23. For each of the following cell reactions, write each half cell reaction, and calculate ϵ and $\epsilon°$ for the cell.

    (a) $Hg_2Cl_2(s) \rightarrow 2\,Hg(l) + Cl_2(0.80\text{ atm})$

    (b) $6\,Fe^{2+}(1.0\ M) + Cr_2O_7^{2-}(0.50\ M) + 14\,H^+(3.0\ M) \rightarrow$
    $2\,Cr^{3+}(0.71\ M) + 6\,Fe^{3+}(2.0\ M) + 7\,H_2O(l)$

24. Calculate the potential of a cell consisting of an anode of silver in 0.10 M silver nitrate solution and a cathode of platinum immersed in a solution containing 1.5 $M\,Cr_2O_7^{2-}$, 0.75 $M\,Cr^{3+}$, and 0.25 $M\,H^+$.

25. Calculate the concentration of Sn^{4+} ion in solution with 1.00 $M\,Sn^{2+}$ ion in a half cell which would have a zero potential when suitable connected to a standard hydrogen/hydrogen ion half cell. Would Sn^{2+} ion tend to be oxidized or would Sn^{4+} ion tend to be reduced under these conditions?

26. What reaction, if any, would be expected in the following experiments? (a) Hg metal is shaken with 1.0 M AgNO$_3$ solution. (b) Solid AgCl is shaken with 1.0 M FeCl$_2$ solution. (c) 1.0 $M\,Cr_2O_7^{2-}$ solution is added to 1.0 M HBr solution.

27. Explain why aluminum metal cannot be produced by electrolysis of aqueous solutions of aluminum salts. Explain why aluminum is produced by the electrolysis of a molten mixture of Al_2O_3 and Na_3AlF_6 rather than by electrolysis of molten Al_2O_3 alone.

28. Using the information in Table 5-2, explain why copper(I) sulfate does not exist in aqueous solution.

29. Calculate the mass of Hg_2Cl_2 which can be prepared by the reduction of mercury(II) ion in the presence of chloride ion by the passage of a 5.00 A current for 3.00 hours.

30. A solution containing Na^+, Sn^{2+}, NO_3^-, Cl^-, and SO_4^{2-} ions, all at unit activity, is electrolyzed between a silver anode and a platinum cathode. What changes occur at the electrodes when current is passed through the cell?

31. Calculate the standard potential for each of the cells in which the following reactions occur. State which reactions will proceed spontaneously as written.

    (a) $2\,Fe^{2+} + Zn^{2+} \rightarrow Zn + 2\,Fe^{3+}$

    (b) $Fe + 2\,Fe^{3+} \rightarrow 3\,Fe^{2+}$

    (c) $Cd + 2\,CuI \rightarrow Cd^{2+} + 2\,Cu + 2\,I^-$

    (d) $Ag + Cu^{2+} + 2\,Cl^- \rightarrow AgCl + CuCl$

32. A certain electrode has a reduction potential of 0.140 V when measured against a saturated calomel electrode. Calculate its potential versus a standard hydrogen electrode.

33. From the data of Table 5-2, compute the standard reduction potential for each of the following:

    (a) $Cu^+ + e^- \rightarrow Cu$

    (b) $2\,Cu^+ + 2\,e^- \rightarrow 2\,Cu$

34. Prove that for two half reactions having potentials ϵ_1 and ϵ_2 which are combined to yield a third half reaction, having a potential ϵ_3,

$$\epsilon_3 = \frac{n_1\epsilon_1 + n_2\epsilon_2}{n_3}$$

35. Standard reduction potentials for two elements, X and Y, in various oxidation states are as follows:

$$
\begin{array}{llll}
X^{4+} & + & e^- & \rightarrow & X^{3+} & \epsilon^\circ = +0.6 \text{ V} \\
X^{3+} & + & e^- & \rightarrow & X^{2+} & \epsilon^\circ = -0.1 \text{ V} \\
X^{2+} & + & 2\,e^- & \rightarrow & X & \epsilon^\circ = -1.0 \text{ V} \\
Y^{3+} & + & e^- & \rightarrow & Y^{2+} & \epsilon^\circ = +0.6 \text{ V} \\
Y^{2+} & + & e^- & \rightarrow & Y^+ & \epsilon^\circ = +0.1 \text{ V} \\
Y^+ & + & e^- & \rightarrow & Y & \epsilon^\circ = +1.0 \text{ V}
\end{array}
$$

Predict the results of each of the following experiments:
(a) X^{2+} is added to $1\ M\ H^+$ solution.
(b) Y^+ is added to water.
(c) $1\ M\ Y^{2+}$ in $1\ M\ H^+$ solution is treated with O_2.
(d) Y is added in excess to $1\ M\ X^{3+}$.
(e) 25 ml of $0.14\ M\ X^{4+}$ solution is added to 75 ml of $0.14\ M\ Y^{2+}$ solution.

36. Assuming that a constant current is delivered, how many kilowatt hours of electricity can be produced by the reaction of 1.00 mole of zinc with copper(II) ion in a Daniell cell in which all concentrations remain $1.00\ M$?

37. For the cell

$$\text{Pt} \mid H_2(0.75 \text{ atm}) \mid \text{HCl}(0.25\ M) \parallel Sn^{2+}(1.50\ M),\ Sn^{4+}(0.60\ M) \mid \text{Pt}$$

(a) write the half-cell reactions, (b) write the cell reaction, (c) calculate the cell potential, and (d) calculate the ratio of concentrations of tin(II) to tin(IV) which would cause the potential to be zero.

38. Write the Nernst equation in terms of free energy change instead of potential.

39. Given the concentration cell

$$\text{Zn} \mid Zn^{2+}(1.0\ M) \parallel Zn^{2+}(0.15\ M) \mid \text{Zn}$$

write equations for each half reaction. Calculate ϵ. As the cell discharges, does the difference in the concentrations of the two solutions become smaller or larger?

40. In the reduction of gold(III) from $AuCl_4^-$ ion solution, the solution must be stirred rapidly during electrolysis. Explain concisely why stirring is so important in this process.

41. Describe the products formed when an aqueous solution of aluminum sulfate is electrolyzed between aluminum electrodes.

42. Explain why iron electrodes in a solution of $Fe(NO_3)_3$ would not be appropriate for an electrolysis experiment.

43. From the data of Table 5–2, show that neither Cu^+ nor Co^{3+} is stable in aqueous solution, whereas Fe^{2+} is stable.

44. The reversible reduction potential of pure water is -0.414 V under 1.00 atm H_2 pressure. If the reduction is considered to be $2\ H^+\ +\ 2\,e^-\ \rightarrow\ H_2$, calculate the hydrogen ion concentration of pure water.

45. $Ni(OH)_2$ and NiO_2 are insoluble in NaOH solution. Design a practical, rechargeable cell using these materials. Include in your description all chemical equations for each electrode reaction during charging and discharging. Could the state of charge of your cell be determined easily?

46. From the data for the two calomel, Hg_2Cl_2, half reactions in Table 5–2, calculate the concentration of saturated KCl at 25°C.

47. At what potential should a solution containing $1\ M\ CuSO_4$, $1\ M\ NiSO_4$, and $2\ M\ H_2SO_4$ be electrolyzed so as to deposit essentially none of the nickel and all of the copper, leaving $1.0 \times 10^{-9}\ M\ Cu^{2+}$?

48. To perform an analysis of a mixture of metal ions by electrodeposition, the second metal to be deposited must not begin plating out until the concentration ratio of the second to

the first is about 10^6. What must be the minimum difference in standard potential of two metals which form dipositive ions in order for such an analysis to be feasible?

49. Given the reaction

$$2 M \ + \ 6 H^+ \ \rightarrow \ 2 M^{3+} \ + \ 3 H_2$$

for which $\Delta H^\circ_{298} = -3.00$ kcal. the entropies are 6.5 for M, -22.2 for M^{3+}, 31.2 for H_2, and -10.0 cal/K for H^+. ΔG°_f for H^+ is 0.00. Calculate **(a)** the standard free energy of formation of M^{3+} and **(b)** ϵ° for the half reaction $M^{3+} \ + \ 3 e^- \ \rightarrow \ M$.

50. From the data of Table 5–2, calculate the standard free energy change for the reaction $Cu^+ \ + \ I^- \ \rightarrow \ CuI$.

Advanced Exercises

51. Determine a value for Avogadro's number, using the charge on the electron, 1.60×10^{-19} C (see Section 9–3), and the fact that 96,500 C deposits 107.9 grams of silver from its solution.

52. Current efficiency is defined as the extent of a desired electrochemical reaction divided by the theoretical extent of the reaction, times 100 to convert to percent. What is the current efficiency of an electrodeposition of copper metal in which 9.80 grams of copper is deposited by passage of a 3.00 A current for 10,000 sec?

53. When half reactions are added, the change in free energy of the total is merely the sum of the changes in free energy of the two halves. Show that this statement implies that the potentials are additive for the process in which half reactions are added to yield an overall reaction, but that they are not additive when added to yield a third half reaction.

54. Given 1 mole of copper atoms and 2 moles of iodine atoms, calculate, from the data of Table 5–2, which of the following systems is lowest in free energy: **(a)** $Cu \ + \ I_2$, **(b)** $CuI \ + \ \frac{1}{2} I_2$, **(c)** CuI_2.

55. Calculate the energy obtainable from a lead storage battery in which 0.100 mole of lead is consumed. Assume a constant concentration of 10.0 M H_2SO_4.

56. Calculate the potential of an indicator electrode, versus the standard hydrogen electrode, which originally contained 0.100 M MnO_4^- and 0.800 M H^+ and which has been treated with 90% of the Fe^{2+} necessary to reduce all the MnO_4^- to Mn^{2+}.

57. A galvanic cell consists of three compartments separated by porous barriers. The first contains a cobalt electrode in 5.00 liters of 0.100 M cobalt(II) nitrate; the second contains 5.00 liters of 0.100 M KNO_3; the third contains a silver electrode in 5.00 liters of 0.100 M $AgNO_3$. Assuming that the current within the cell is carried equally by the positive and negative ions, tabulate the concentrations of ions of each type in each compartment of the cell after the passage of 0.100 mole of electrons.

58. For a cell consisting of an inert electrode in a solution containing 0.10 M $KMnO_4$, 0.20 M $MnCl_2$, and 1.0 M HCl suitably connected to another inert electrode in a solution containing 0.10 M $K_2Cr_2O_7$, 0.20 M $CrCl_3$, and 0.70 M HCl, calculate ϵ **(a)** by combining ϵ values calculated separately for each half cell and **(b)** by combining ϵ° values from the half cells and using the Nernst equation for the overall cell. (Note: when performed correctly, procedures **(a)** and **(b)** must give the same result.)

59. MnO_4^- is reduced to Mn^{2+} in solutions of low pH, and to MnO_2 in solutions of intermediate to high pH. Explain why a potentiometric titration of 0.100 M MnO_4^- should be performed in 2.00 M H^+ rather than in 0.800 M H^+.

60. Assume that impure copper contains only iron, silver, and gold as impurities. After passage of 140 A for 482.5 sec, the mass of the anode decreased by 22.260 grams and the cathode increased in mass by 22.011 grams. Estimate the percent iron and the percent copper originally present. The apparatus is that shown in Figure 5–17.

61. By means of integral calculus, calculate the total energy theoretically obtainable from a Daniell cell which has a zinc electrode weighing 65.37 grams immersed in 1.000 liter of 1.000 M zinc ion and a copper electrode in 1.000 liter of 1.100 M copper(II) ion.

62. Calculate the standard potential for the reaction

$$Hg_2Cl_2 \;+\; Cl_2 \;\rightarrow\; 2\,Hg^{2+} \;+\; 4\,Cl^-$$

63. Calculate the number of coulombs delivered by a Daniell cell, initially containing 1.00 liter each of 1.00 M copper(II) ion and 1.00 M zinc(II) ion, which is operated until its potential drops to 1.00 V.

64. Determine the potential of a Daniell cell, initially containing 1.00 liter each of 1.00 M copper(II) ion and 1.00 M zinc(II) ion, after passage of 100,000 C of charge.

6

Chemical Equilibrium

A major objective of chemists is to understand chemical reactions—to know whether under a given set of conditions two substances will react when mixed, to determine whether a given reaction will be exothermic or endothermic, and to predict the extent to which a given reaction will proceed before equilibrium is established. An **equilibrium state,** produced as a consequence of two opposing reactions occurring simultaneously, is a state in which there is no net change as long as there is no change in conditions. In this chapter it will be shown how one can predict the equilibrium states of chemical systems from thermodynamic data, and conversely how the experimental measurements on equilibrium states provide useful thermodynamic data. Thermodynamics alone cannot explain the rate at which equilibrium is established, nor does it provide details of the mechanism by which equilibrium is established. Such explanations can be developed from considerations of the quantum theory of molecular structure and from statistical mechanics, aspects of which will be discussed later in this text. It will be shown in this chapter that a substantial number of chemical phenomena can be organized and understood in terms of enthalpies of formation, free energies of formation, and equilibrium constants.

To appreciate fully the nature of the chemical equilibrium state, it is necessary first to have some acquaintance with the factors which influence reaction rates. (The subject of rates of reaction will be developed fully in Chapter 17.) The factors which influence the rates of a chemical reaction are temperature, concentrations of reactants (or partial pressures of gaseous reactants), and presence of a catalyst. In general, for a given reaction, the higher the temperature, the faster the reaction will occur. The concentrations of reactants or partial pressure of gaseous reactants will affect the rate of reaction; an increase in concentration or partial pressure increases the rate of most reactions. Substances which accelerate a chemical reaction but which themselves are not used up in the reaction are called **catalysts.**

6-1 Dynamic Equilibrium

In many cases, direct reactions between two substances appear to cease before all of either starting material is exhausted. Moreover, the products of chemical reactions themselves often react to produce the starting materials. For example, nitrogen and hydrogen combine at 500°C in the presence of a catalyst to produce ammonia:

$$N_2 + 3 H_2 \rightarrow 2 NH_3$$

At the same temperature and in the presence of the same catalyst, pure ammonia decomposes into nitrogen and hydrogen:

$$2\,NH_3 \;\rightarrow\; 3\,H_2 \;+\; N_2$$

For convenience, these two opposing reactions are denoted in one equation by use of a double arrow:

$$N_2 \;+\; 3\,H_2 \;\rightleftharpoons\; 2\,NH_3$$

The reaction proceeding toward the right is called the **forward reaction;** the other is called the **reverse reaction.**

If either ammonia or a mixture of nitrogen and hydrogen is subjected to the above conditions, a mixture of all three gases will result. The rate of reaction between the materials which were introduced into the reaction vessel will decrease after the reaction starts, because their concentrations are decreasing. Conversely, after the start of the reaction the material being produced will react faster, since there will be more of it. Thus the faster forward reaction becomes slower, and the slower reverse reaction speeds up. Ultimately the time comes when the rates of the forward and reverse reactions become equal, and there will be no further *net* change. This situation is called **equilibrium.** Equilibrium is a dynamic state because both reactions are still proceeding; but since the two opposing reactions are proceeding at equal rates, no *net* change is observed.

All chemical reactions ultimately proceed toward equilibrium. In a practical sense, however, some reactions go so far in one direction that the reverse reaction cannot be detected, and they are said to go to **completion.** The principles of chemical equilibrium apply even to these, and it will be seen that for many of them, the extent of reaction can be expressed quantitatively.

6-2   Le Châtelier's Principle

Once equilibrium is established in a system, no further change is apparent as long as the external conditions remain unchanged. If the external conditions on the system are altered, however, the system will shift to a new state of equilibrium. Whether the equilibrium will shift toward products or reactants can be predicted by the application of **Le Châtelier's principle,** which is stated as follows:

> *If a stress is applied to a system at equilibrium, the equilibrium will shift to reduce the stress.*

Stresses are merely those external factors which influence the rate of a reaction. Included among stresses are changes of concentration, changes of pressure (if gases are involved), and changes of temperature. However, addition of a catalyst to a system at equilibrium does not shift the equilibrium point because catalysts alter the rates of the forward and reverse reactions equally.

Example

What effect will the addition of more nitrogen have on the following equilibrium, observed in a vessel at constant volume?

$$N_2(g) \; + \; 3\,H_2(g) \; \rightleftharpoons \; 2\,NH_3(g)$$

The equilibrium must shift to the right to use up some of the added nitrogen, thus lowering its concentration and thereby reducing the stress. A new equilibrium is established. Note that the equilibrium does not shift back to the left because of the additional ammonia formed, since that was produced by the reaction itself. Note also that Le Châtelier's principle alone does not tell *how much* the equilibrium will be shifted. No matter how large the quantity of nitrogen added, some hydrogen will be present at the new equilibrium state.

In the application of Le Châtelier's principle, heat energy may be treated as one of the reactants or products of the reaction. A more complete description of the reaction between nitrogen and hydrogen includes the heat produced:

$$N_2(g) \; + \; 3\,H_2(g) \; \rightleftharpoons \; 2\,NH_3(g) \; + \; 22\,kcal$$

Example

What is the effect of an increase in temperature on the above equilibrium at constant pressure?

Heat must be added to raise the temperature of the system. Therefore in order to use up some of the added heat, the equilibrium must shift to the left.

A change in total pressure on a gaseous system can cause a shift in the equilibrium if the partial pressures of the gaseous reactants are affected. An increase in pressure due to a reduction in volume will cause the equilibrium to shift in the direction which produces fewer moles of gas.

Example

$$N_2 \; + \; 3\,H_2 \; \rightleftharpoons \; 2\,NH_3 \; + \; 22\,kcal$$

What is the effect on the above equilibrium of halving the volume, thus initially doubling the total pressure?

The reaction of a total of 4 moles of hydrogen and nitrogen would produce 2 moles of ammonia. In a given volume, fewer moles of gas exert less pressure; thus, as required by Le Châtelier's principle, if additional pressure is applied, the equilibrium will shift to the right, in the direction of fewer moles of gas.

Example

What is the effect of halving the pressure by doubling the volume on the following system at 500°C?

$$H_2(g) \; + \; I_2(g) \; \rightleftharpoons \; 2\,HI(g)$$

Since the equation states that 2 moles of gaseous reactant(s) would produce 2 moles of gaseous product(s), changing the total pressure will not shift the equilibrium at all.

Changing the pressure affects equilibria involving gases to a much greater extent than it affects equilibria involving solids and liquids only. If the system involves solids (or liquids) in equilibrium with gases, the effect of changing the pressure is predicted by considering the gases only.

Example

What is the effect of reducing the volume on the system described below?

$$2\,C(s) \;+\; O_2(g) \;\rightleftharpoons\; 2\,CO(g)$$

The equation states that the reaction of 1 mole of gaseous reactant (and 2 moles of solid reactant) produces 2 moles of gaseous product. Reducing the volume affects the gases much more than it does the solid, and so this equilibrium will be shifted to the left by the increased pressure caused by the reduction in volume.

6-3   Equilibrium Constants

Equilibrium is a state of dynamic balance between two opposing processes. For a general reaction at a given temperature,

$$A \;+\; B \;\rightleftharpoons\; C \;+\; D$$

at the point of equilibrium, the following ratio must be a constant:

$$K = \frac{[C][D]}{[A][B]}$$

The constant, K, is called the **equilibrium constant** of the reaction. It has a specific value at a given temperature. If the concentration of any of the components in the system at equilibrium is changed, the concentrations of the other components will change in such a manner that the defined ratio remains equal to K as long as the temperature does not change. The equilibrium constant expression quantitatively defines the equilibrium state.

More generally, for the reversible reaction

$$a\,A \;+\; b\,B \;\rightleftharpoons\; c\,C \;+\; d\,D$$

the equilibrium constant expression is written as follows:

$$K = \frac{[C]^c[D]^d}{[A]^a[B]^b}$$

By convention, the concentration terms of the reaction products are always placed in the numerator of the equilibrium constant expression. It should be noted that the exponents of the concentration terms in the equilibrium constant expression are the coefficients of the respective species in the balanced chemical equation.

Example

Write an equilibrium constant expression for each of the following reactions. What relationship do the constants have to one another?

a. $H_2(g) + I_2(g) \underset{500°C}{\rightleftharpoons} 2 HI(g)$ **b.** $2 HI(g) \underset{500°C}{\rightleftharpoons} H_2(g) + I_2(g)$

The equilibrium constant expressions for the two reactions are reciprocals of each other:

a. $K = \dfrac{[HI]^2}{[H_2][I_2]}$ **b.** $K = \dfrac{[H_2][I_2]}{[HI]^2}$

The equilibrium state for a system involving gases may be defined in terms of a slightly different equilibrium constant, K_p. In the equation defining K_p pressures of the gases are used instead of concentrations.

Example

For the reaction at 800 K

$$3 H_2 + N_2 \rightleftharpoons 2 NH_3$$

determine the relationship between the values of K and K_p.

$$K_p = \frac{P_{NH_3}^2}{P_{N_2} P_{H_2}^3}$$

By use of the ideal gas law,

$$K_p = \frac{(n_{NH_3} RT/V)^2}{(n_{N_2} RT/V)(n_{H_2} RT/V)^3}$$

Since concentration is equal to n/V,

$$K_p = \frac{[NH_3]^2 (RT)^2}{[N_2](RT)[H_2]^3 (RT)^3} = \left(\frac{[NH_3]^2}{[N_2][H_2]^3}\right)\left(\frac{1}{(RT)^2}\right)$$

$$= K/(RT)^2 = K/(0.0821)^2(800)^2 = 2.32 \times 10^{-4} K$$

To be strictly correct, equilibrium constant expressions should be written in terms of activities rather than concentrations or pressures. However, concentrations and/or partial pressures can be used in equilibrium constant expressions with little error. For convenience, in this book concentrations and pressures will be used despite the small inaccuracy introduced.

The value of the equilibrium constant for a given reaction at a given temperature must be determined experimentally, either directly from concentration ratios or indirectly from thermodynamic data. The value of K at a given temperature is a fundamental property of the system, and once determined it is not subject to change other than as a result of improved measurement. The experimental determination of equilibrium constants is an important activity, and compilations of equilibrium constants are important sources of chemical data.

Example

Calculate the value of the equilibrium constant for the reaction

$$A + B \rightleftharpoons C + D$$

if at equilibrium there are 1.0 mole of A, 2.0 moles of B, 6.0 moles of C, and 20 moles of D in a 1.0 liter vessel.

$$K = \frac{[C][D]}{[A][B]} = \frac{(6.0 \text{ moles/liter})(20 \text{ moles/liter})}{(1.0 \text{ mole/liter})(2.0 \text{ moles/liter})} = 60$$

It is not necessary to measure the equilibrium concentrations of all the species involved in the reaction in order to determine the equilibrium constant. If one or two concentrations are measured at equilibrium, the others may be calculated from the initial concentrations and a knowledge of the stoichiometry of the system.

Example

For the reaction

$$E + F \rightleftharpoons G + H$$

one starts with 6.0 moles of E and 7.0 moles of F in a 1.0 liter vessel. When equilibrium is attained, 4.5 moles of G is formed. Calculate the value of the equilibrium constant for the reaction.

As shown by the balanced chemical equation, if 4.5 moles of G has been produced, then 4.5 moles of H has also been produced and 4.5 moles each of E and F have been consumed. The quantities of reactants present at equilibrium are equal to those initially present minus what has been used up. The quantities of products present at equilibrium are equal to those initially present plus what has been produced. It is often helpful to tabulate these quantities as follows:

	Initially Present	Consumed by the Reaction	Produced by the Reaction	Present at Equilibrium
E	6.0	4.5		1.5
F	7.0	4.5		2.5
G	0.0		4.5	4.5
H	0.0		4.5	4.5

The concentrations in the two middle columns of such a table always have values related by the balanced chemical equation.

$$K = \frac{[G][H]}{[E][F]} = \frac{(4.5 \text{ moles/liter})(4.5 \text{ moles/liter})}{(1.5 \text{ moles/liter})(2.5 \text{ moles/liter})} = 5.4$$

Note on Solving Equilibrium Constant Problems

There are several different kinds of equilibrium constants, but most equilibrium problems can be solved systematically by using the following scheme. Each step is illustrated below using the following problem as an example: The equilibrium constant for the dissociation of HY into H_2 and Y is 4.0×10^{-15}. One mole of HY is placed in a 1.0 liter flask and allowed to come to equilibrium. What is the concentration of Y at equilibrium?

	Steps		*Example*

1. Write the balanced chemical equation for the equilibrium reaction.

$$2\,HY \;\rightleftharpoons\; H_2 \;+\; 2\,Y$$

2. Write the equilibrium constant expression corresponding to the chemical equation.

$$K = \frac{[H_2][Y]^2}{[HY]^2}$$

3. Determine the equilibrium concentration of each of the species involved in the expression for K. In this example, it is necessary to write an expression for each of the concentrations in terms of an algebraic unknown, e.g., x. (This happens when the value of the constant is given and the equilibrium concentrations are to be determined.) Any algebraic substitutions should be explicitly noted. In other examples, equilibrium concentrations might be given in the statement of the problem or might be directly obtained by chemical reasoning from the initial concentrations of reagents. In any case, the concentration of each species should be explicitly noted.

Often it is easiest to denote the concentration of the substance lowest in concentration as x, and express the other concentrations in terms of x. Therefore, let $[H_2] = x$.

Then $[Y] = 2x$
 $[HY] = (1.0 - 2x)$

Note that $2x$ is THE equilibrium concentration of Y and not twice the equilibrium concentration of Y. (In this case [Y] happens to be twice the concentration of H_2.)

4. Substitute the equilibrium concentrations from step 3 into the equilibrium constant expression in step 2 and solve.

$$K = \frac{(x)(2x)^2}{(1.0 - 2x)^2} = 4.0 \times 10^{-15}$$

5. Sometimes it is possible to obtain approximate solutions by neglecting insignificant terms. Note that while a small term, x, may be neglected when added to or subtracted from a much larger quantity; the same term may *not* be neglected when multiplied or divided.

Let $1.0 - 2x \cong 1.0$.

$$K \cong \frac{(x)(2x)^2}{(1.0)^2} = 4x^3$$

$$x = 1.0 \times 10^{-5}$$
$$[Y] = 2x = 2.0 \times 10^{-5}$$

Notes: $2x$ is negligible when subtracted from 1 but not when divided by 1. [Y] was required, not x.

6. Check quantitatively if possible, or use Le Châtelier's principle to check qualitatively, the results obtained in step 4 or 5 to see that they are reasonable.

Check:

$$\frac{(1.0 \times 10^{-5})(2.0 \times 10^{-5})^2}{[1.0 - (2 \times 10^{-5})]^2} = 4.0 \times 10^{-15}$$

6-4   Equilibria Between Solids and Gases

As discussed in Section 6–2, the effect of pressure on equilibria between solids and gases depends on the number of moles of gaseous reactants used up or of gaseous products produced in the reaction but not on the number of moles of solid reacting. The reason for this behavior is apparent when one considers that the equilibrium constant depends on activities (effective concentrations). The activity of a solid is unaffected by a change in pressure, but at constant temperature the number of moles per liter of gas varies proportionally with pressure.

For example, the equilibrium constant expression for the reaction

$$CaCO_3(s) \;\rightleftharpoons\; CaO(s) \;+\; CO_2(g)$$

could be written

$$K = \frac{a_{CaO} P_{CO_2}}{a_{CaCO_3}}$$

However, the activities of the solids are unity, and in this case the equilibrium constant expression becomes

$$K_P = P_{CO_2}$$

This expression states that the pressure (or concentration) of CO_2 in equilibrium with $CaCO_3$ and CaO must be constant at any given temperature. Above a mixture of $CaCO_3$ and CaO, no matter how much of each is present, there can be only one equilibrium pressure of CO_2. If the CO_2 pressure were greater than the equilibrium pressure, the excess would be absorbed by the CaO; if the CO_2 pressure were lower, then more $CaCO_3$ would decompose until the equilibrium pressure was again established.

Example

For the reaction

$$CaCO_3(s) \;\rightleftharpoons\; CaO(s) \;+\; CO_2(g)$$

$K_P = 1.16$ atm at 800°C. If 20.0 grams of $CaCO_3$ was put into a 10.0 liter container and heated to 800°C, what percent of the $CaCO_3$ would remain unreacted at equilibrium?

$$K_P = P_{CO_2} = 1.16 \text{ atm}$$

$$n_{CO_2} = \frac{PV}{RT} = \frac{(1.16 \text{ atm})(10.0 \text{ liter})}{(0.0821 \text{ liter} \cdot \text{atm/mole} \cdot \text{K})(1073 \text{ K})} = 0.132 \text{ mole}$$

$$\text{moles } CaCO_3 \text{ initially present} = \frac{20.0 \text{ grams}}{100 \text{ grams/mole}} = 0.200 \text{ mole}$$

$$\frac{0.132 \text{ mole } CO_2}{0.200 \text{ mole } CaCO_3} \times 100 = 66.0\% \text{ decomposed}$$

Hence 34.0% remains undecomposed.

6-5 Equilibrium Constants and Free Energy

It is a rule of nature that all systems tend to approach equilibrium spontaneously. Once a system is in the equilibrium state, if left undisturbed, it remains there indefinitely. This phenomenon is perfectly general and applies to chemical reactions as well as to other types of systems. In the discussion which follows, oxidation-reduction reactions will be used to illustrate this concept, but the principles developed apply equally well to all types of chemical systems.

In the case of an oxidation-reduction reaction, the system can be constructed in the form of a galvanic cell, and as was shown in Section 5-6, the tendency toward reaction will be indicated by the magnitude of the cell potential. For the reaction

$$a\,\mathrm{A} \;+\; b\,\mathrm{B} \;\rightleftharpoons\; c\,\mathrm{C} \;+\; d\,\mathrm{D}$$

the cell potential, ϵ, can be determined by use of the Nernst equation:

$$\epsilon = \epsilon^{\circ} - \frac{2.30\,RT}{n\mathrm{F}} \log \frac{[\mathrm{C}]^{c}[\mathrm{D}]^{d}}{[\mathrm{A}]^{a}[\mathrm{B}]^{b}}$$

where ϵ° represents the standard cell potential; R, T, and F have the meanings given in Chapter 5; and n is the number of moles of electrons transferred between a moles of A and b moles of B.

If the concentrations of the substances involved in the reaction are such that $\epsilon = 0$, then the reaction will not proceed spontaneously either forward or backward; therefore the concentrations correspond to a set of equilibrium concentrations. Since the concentrations correspond to equilibrium concentrations, and the ratio $[\mathrm{C}]^{c}[\mathrm{D}]^{d}/[\mathrm{A}]^{a}[\mathrm{B}]^{b}$ is exactly that defined for the equilibrium constant,

$$\epsilon^{\circ} = \frac{2.30\,RT}{n\mathrm{F}} \log \frac{[\mathrm{C}]^{c}[\mathrm{D}]^{d}}{[\mathrm{A}]^{a}[\mathrm{B}]^{b}} = \frac{2.30\,RT}{n\mathrm{F}} \log K$$

$$n\mathrm{F}\epsilon^{\circ} = 2.30\,RT \log K$$

It was shown in Section 5-8 that

$$\Delta G^{\circ} = -n\mathrm{F}\epsilon^{\circ}$$

Therefore

$$\Delta G^{\circ} = -2.30\,RT \log K$$

This equation shows that a given temperature the equilibrium constant and hence *the equilibrium state is determined entirely by the standard free energy change, ΔG°.* Although an oxidation-reduction system was used as a basis for this derivation, the result is perfectly valid for all systems at chemical equilibrium.

Quantitative information on the equilibrium state is given by either the value of ΔG° or the value of K. A value of K greater than 1 corresponds to a *negative* value of ΔG°, which means that when the reactants and products are present in their standard states, the reaction will proceed spontaneously toward equilibrium as written. When the equilibrium constant is extremely large or extremely small, its value is best determined by calculation from a measured value of $\Delta G°$. On the other hand, when K is conveniently measured, $\Delta G°$ for the reaction can be calculated from it.

Example

At 490°C, the value of the equilibrium constant, K_p, is 45.9 for the reaction

$$H_2(g) + I_2(g) \rightleftharpoons 2 HI(g)$$

Calculate the value of $\Delta G°$ for the reaction at that temperature.

$$\Delta G° = -2.30 \, RT \log K_p$$
$$= -2.30(1.987 \text{ cal/mole} \cdot \text{K})(763 \text{ K})(\log 45.9) = -5.79 \times 10^3 \text{ cal/mole}$$
$$= -5.79 \text{ kcal/mole}$$

6-6 Absolute Entropies—The Third Law of Thermodynamics

If data for enthalpy change and entropy change are available for a given reaction, the relationship

$$\Delta G = \Delta H - T\Delta S$$

permits calculation of the free energy change for the reaction. Unfortunately, for most chemical reactions there is no convenient way to measure ΔS directly. However there is a way of calculating ΔS. A postulate known as the **third law of thermodynamics** states that the *entropies of all pure crystalline solids are zero at 0 K.* (The specifications *pure* and *crystalline* are necessary because mixtures and noncrystalline solids have some degree of randomness even at 0 K.) By the methods of calculus it is possible to calculate the difference between the entropies of a given substance at two temperatures from heat capacity data. The heat capacities as a function of temperature have been determined for a wide variety of substances, and extensive compilations are now available. Entropy changes for the process of raising the temperature of various substances from 0 K to any higher temperature have been calculated. Since the initial value is 0, the calculated entropy values are the **absolute entropies,** $S°$, of the respective substances. Typical data are given in Table 6–1. Note that consistent with the nature of entropy, gases have higher entropies than solids, while liquids have intermediate values. Solid elements which are hard, such as boron, have lower entropies than solid elements which are soft, such as sodium. Among compounds, the entropy increases with the increasing number of atoms per molecule or formula unit.

Note that both elements and compounds have nonzero entropies at 25°C. As a corollary, the absolute entropy of a compound is *not* the entropy of formation from its elements. The entropy change for any process is given by

$$\Delta S = \sum S_{\text{products}} - \sum S_{\text{reactants}}$$

Hence for the formation of a compound,

$$\Delta S_f = S_{\text{compound}} - \sum S_{\text{elements}}$$

TABLE 6-1. Absolute Entropies at 298 K

Solid Elements	$S°$ cal/mole · K	$S°$ J/mole · K	Solid Compounds	$S°$ cal/mole · K	$S°$ J/mole · K
Ag	10.2	42.7	$AlCl_3$	40	167
Al	6.77	28.3	Al_2O_3	12.19	51.00
B	1.6	6.7	AgCl	23.0	96.2
C[a]	1.36	5.69	AgBr	25.6	107
Ca	9.95	41.6	BaO	16.8	70.3
Cu	7.96	33.3	$BaSO_4$	31.6	132
Fe	6.49	27.2	CaO	9.5	40
I_2	27.8	116	$Ca(OH)_2$	18.2	76.1
Mg	7.77	32.5	CuO	10.4	43.5
Na	12.2	51.0	Cu_2O	24.1	101
S	7.62	31.9	FeO	12.9	54.0
Si	4.51	18.9	Fe_2O_3	21.5	90.0
Zn	9.95	41.6	$MgCO_3$	15.7	65.7
			NH_4Cl	22.6	94.6
			NaCl	17.30	72.38
			SiO_2	10	42
			ZnO	10.5	43.9

Liquids	$S°$ cal/mole · K	$S°$ J/mole · K	Liquids	$S°$ cal/mole · K	$S°$ J/mole · K
Br_2	36.4	152	CCl_4	51.2	214
CH_3OH	30.3	127	H_2O	16.73	70.00
C_6H_6	41.3	173	Hg	18.17	76.02

Gases	$S°$ cal/mole · K	$S°$ J/mole · K	Gases	$S°$ cal/mole · K	$S°$ J/mole · K
He	30.13	126.1	CO	47.3	198
Ne	34.95	146.2	CO_2	51	210
Ar	36.98	154.7	H_2O	45.1	189
Kr	39.19	164.0	H_2S	49.1	205
Xe	40.53	170.0	NH_3	46.01	192.5
HF	41.5	174	NO	50.3	210
HCl	44.6	187	NO_2	57.5	241
HBr	47.4	198	SO_2	59.4	249
HI	49.3	206	SO_3	61.24	256.2
H_2	31.21	130.6	CH_4	44.5	186
F_2	48.6	203	C_2H_6	54.8	229
Cl_2	53.3	223	C_3H_8	64.5	270
Br_2	58.6	245	C_4H_{10}	74.10	310.0
N_2	45.7	191	C_2H_4	52.45	219.5
O_2	49.0	205	C_2H_2	49.99	209.2

[a] Graphite.

Example

Calculate the standard entropy change and the free energy change when 1.00 mole of water is formed from its elements at 25°C.

$$H_2(g) \ + \ \tfrac{1}{2}O_2(g) \ \rightarrow \ H_2O(l)$$

$$\Delta S_f^\circ = S_{(H_2O)}^\circ - S_{(H_2)}^\circ - \tfrac{1}{2}S_{(O_2)}^\circ = 16.73 - 31.21 - 24.50 = -38.98 \text{ cal/mole} \cdot K$$

$$\Delta G^\circ = \Delta H^\circ - T\Delta S^\circ$$

Using ΔH_f° from Table 3–2 yields

$$\Delta G_f^\circ = -68,320 - 298(-38.98) = -56,700 \text{ cal/mole} = -56.7 \text{ kcal/mole}$$

Note that the formation of a mole of liquid water from a total of 1.5 moles of gases results in a large *decrease* in entropy. The reaction is spontaneous because of the highly exothermic nature of the reaction. The highly negative value of ΔH yields a negative value of ΔG despite the decrease in randomness of the system.

Example

Predict whether sulfur dioxide will reduce copper(II) oxide at 298 K.

$$SO_2(g) \ + \ CuO(s) \ \rightarrow \ Cu(s) \ + \ SO_3(g)$$

The enthalpy change for the above reaction, calculated from the values of enthalpies of formation from Table 3–2, is 14.1 kcal. Similarly, from the data of Table 6–1, the entropy change is calculated to be -0.60 cal/K.

$$\Delta G^\circ = \Delta H^\circ - T\Delta S^\circ$$
$$= 14.1 \text{ kcal} - (298 \text{ K})(-0.60 \times 10^{-3} \text{ kcal/K}) = 14.3 \text{ kcal}$$

The change in free energy is calculated to be positive, and the reaction will not occur.

6-7   Exercises

Basic Exercises

1. What is the effect on the following equilibrium if each of the indicated stresses is applied?

$$\tfrac{1}{2}N_2 \ + \ O_2 \ \rightleftharpoons \ NO_2 \ + \ heat$$

   (a) increase in N_2 concentration, (b) decrease in temperature, (c) increase in volume, (d) decrease in O_2 concentration, (e) addition of a catalyst.

2. Write equilibrium constant expressions for the following reactions:
   (a) $NO_2(g) \ + \ SO_2(g) \ \rightleftharpoons \ NO(g) \ + \ SO_3(g)$
   (b) $2\,SO_2(g) \ + \ O_2(g) \ \rightleftharpoons \ 2\,SO_3(g)$
   (c) $2\,SO_2(g) \ + \ O_2(g) \ \rightleftharpoons \ 2\,SO_3(g) \ + \ heat$
   (d) $Ca(HCO_3)_2(s) \ \rightleftharpoons \ CaO(s) \ + \ 2\,CO_2(g) \ + \ H_2O(g)$
   (e) $CO(g) \ + \ Cl_2(g) \ \rightleftharpoons \ COCl_2(g)$

3. A system at equilibrium is described by the equation

$$heat + SO_2Cl_2 \rightleftharpoons SO_2 + Cl_2$$

Why does the temperature of the system increase when Cl_2 is added to the equilibrium mixture at constant volume?

4. Predict the effect on each of the following equilibria of decreasing the volume of the system:

(a) $heat + MgCO_3(s) \rightleftharpoons MgO(s) + CO_2(g)$

(b) $2 C(s) + O_2(g) \rightleftharpoons 2 CO(g) + heat$

5. For the reaction

$$CO(g) + H_2O(g) \rightleftharpoons CO_2(g) + H_2(g)$$

the equilibrium constant at 1250 K is 0.63. (a) Calculate $\Delta G°$ for the reaction at 1250 K. (b) At 1250 K, does CO(g) react with $H_2O(g)$ to form $CO_2(g)$ and $H_2(g)$ in a system in which all four are present at 1.0 atm, or does the reverse reaction occur?

6. For the reaction of XO with O_2 to form XO_2, the equilibrium constant at 398 K is 1.0×10^{-4} liter/mole. If 1.0 mole of XO and 2.0 moles of O_2 are placed in a 1.0 liter vessel and allowed to come to equilibrium, what will be the equilibrium concentration of each of the species?

7. Determine the value of the equilibrium constant for the reaction

$$A + 2 B \rightleftharpoons 2 C$$

if 1.0 mole of A and 1.5 moles of B are placed in a 2.0 liter vessel and allowed to come to equilibrium. The equilibrium concentration of C is 0.35 mole/liter.

8. Determine the value of the equilibrium constant for the reaction

$$A + B \rightleftharpoons 2 C$$

if 1.0 mole of A, 1.4 moles of B, and 0.50 mole of C are placed in a 1.0 liter vessel and allowed to come to equilibrium. The final concentration of C is 0.75 mole/liter.

9. Determine the equilibrium concentration of each of the species which react according to the equation

$$A + B \rightleftharpoons C + 2 D$$

if the value of the equilibrium constant is 1.8×10^{-6} mole/liter, after 1.0 mole of C and 1.0 mole of D are placed in a 1.0 liter vessel and allowed to come to equilibrium.

10. Determine if possible what shift each of the following combinations of stresses would cause in the following equilibrium system:

$$2 CO + O_2 \rightleftharpoons 2 CO_2 + heat$$

(a) addition of CO and removal of CO_2 at constant volume.

(b) increase in temperature and decrease in volume.

(c) addition of O_2 and decrease in volume.

(d) addition of a catalyst and decrease in temperature.

(e) addition of CO and increase in temperature at constant volume.

11. In the Haber process for the production of ammonia from N_2 and H_2, a total pressure of 200 atm is used. Explain why such a high pressure is desirable.

12. Using data from Tables 3–2 and 6–1, calculate $\Delta G°_{298}$ for the reaction of 50.0 grams of nitrogen with oxygen according to the equation

$$N_2(g) \ + \ O_2(g) \ \rightarrow \ 2\,NO(g)$$

Can NO decompose into its elements at 298 K?

13. The balanced chemical equation

$$2\,CO \ + \ O_2 \ \rightleftharpoons \ 2\,CO_2$$

signifies which one(s) of the following? **(a)** One can add to a vessel only 2 moles of CO for each mole of O_2 added. **(b)** No matter how much of these two reagents are added to a vessel, only 1 mole of O_2 will react, and it will react with 2 moles of CO. **(c)** When they react, CO reacts with O_2 in a $2:1$ mole ratio. **(d)** When 2 moles of CO and 1 mole of O_2 are placed in a vessel, they will react to give 2 moles of CO_2.

14. From the data of Tables 3–2 and 6–1, calculate ΔG_f° of 1.00 mole of $NO_2(g)$ at 298 K. Is NO_2 stable at 298 K with respect to decomposition into its elements?

General Exercises

15. **(a)** What effect would the introduction of He gas have on the partial pressure of each gas in a system containing N_2, H_2, and NH_3 at equilibrium? **(b)** What effect, if any, will the added He have on the position of the equilibrium?

16. Derive the relationship between K and K_p for a reaction in which gases are involved

$$K_p = K(RT)^{\Delta n}$$

where Δn is the difference in the number of moles of gases between products and reactants.

17. Determine the value of the equilibrium constant for the reaction

$$A \ + \ 2\,B \ \rightleftharpoons \ 2\,C$$

if 1.0 mole of A, 2.0 moles of B, and 3.0 moles of C are placed in a 1.0 liter vessel and allowed to come to equilibrium. The final concentration of C is 1.4 moles/liter.

18. Ammonium hydrogen sulfide dissociates as follows:

$$NH_4HS(s) \ \rightleftharpoons \ H_2S(g) \ + \ NH_3(g)$$

If solid NH_4HS is placed in an evacuated flask at a certain temperature, it will dissociate until the total gas pressure is 500 torr. **(a)** Calculate the value of the equilibrium constant for the dissociation reaction. **(b)** Additional NH_3 is introduced into the equilibrium mixture without change in temperature until the partial pressure of ammonia is 700 torr. What is the partial pressure of H_2S under these conditions? What is the total pressure in the flask?

19. Calculate the value of the equilibrium constant for the reaction, at 25°C,

$$2\,H_2 \ + \ O_2 \ \rightleftharpoons \ 2\,H_2O(l)$$

from the data of Tables 3–2 and 6–1.

20. At a certain temperature 1.00 mole of $PCl_3(g)$ and 2.00 moles of $Cl_2(g)$ were placed in a 3.00 liter container. When equilibrium was established, only 0.700 mole of PCl_3 remained. Calculate the value of the equilibrium constant for the reaction

$$PCl_3 \ + \ Cl_2 \ \rightleftharpoons \ PCl_5(g)$$

21. At a certain temperature, the equilibrium constant for the reaction of CO with O_2 to produce CO_2 is 5.0×10^3 liter/mole. Calculate [CO] at equilibrium if 1.0 mole each of CO and O_2 are placed in a 2.0 liter vessel and allowed to come to equilibrium.

22. For the reaction $A + 2B \rightleftharpoons C + D$, the equilibrium constant is 1.0×10^8. **(a)** Calculate the equilibrium concentration of A if 1.0 mole of A and 3.0 moles of B are placed in a 1.0 liter vessel and allowed to come to equilibrium. **(b)** If 1.0 mole of C and 3.0 moles of D were placed in a 1.0 liter vessel, calculate the equilibrium concentration of B.

23. At a certain temperature, the value of the equilibrium constant corresponding to the equation

$$N_2 + 2O_2 \rightleftharpoons 2NO_2$$

is 100. Write the equilibrium constant expressions for each of the following reactions, and calculate the value of the equilibrium constant for each:
(a) $2NO_2 \rightleftharpoons N_2 + 2O_2$
(b) $NO_2 \rightleftharpoons \frac{1}{2}N_2 + O_2$

24. Calculate the value of the equilibrium constant for the reaction below if there are present at equilibrium 5.0 moles of N_2, 7.0 moles of O_2, and 0.10 mole of NO_2 in a 1.5 liter vessel at a certain temperature.

$$N_2 + 2O_2 \rightleftharpoons 2NO_2 + \text{heat}$$

If the temperature is increased, would the value of the equilibrium constant for the reaction increase, decrease, or remain unchanged?

25. Using appropriate data from Chapter 3, calculate the equilibrium constant for the reaction at 25°C

$$CH_4 + Cl_2 \rightleftharpoons CH_3Cl + HCl$$

Advanced Exercises

26. Calculate the standard free energy change of the following reaction. Determine the value of its equilibrium constant at 298 K.

$$CH_4 + 2O_2 \rightleftharpoons CO_2 + 2H_2O(g)$$

27. **(a)** Show that

$$2.30 \log K = -\frac{\Delta H°}{RT} + \frac{\Delta S°}{R}$$

(b) Given the data below for the system $H_2(g) + \frac{1}{2}S_2(g) \rightleftharpoons H_2S(g)$, and assuming that $\Delta H°$ and $\Delta S°$ are constant, calculate the value of $\Delta H°$ in the temperature range 1000 K to 1700 K.

T (K)	$\log K_p$ (atm$^{-1/2}$)	T (K)	$\log K_p$ (atm$^{-1/2}$)
1023	2.025	1473	0.643
1218	1.305	1667	0.257
1362	0.902		

28. At 500°C, the equilibrium constant of 3.9×10^{-3}/atm is found for the reaction

$$\tfrac{1}{2} N_2 \;+\; \tfrac{3}{2} H_2 \;\rightleftharpoons\; NH_3$$

If sufficient ammonia were introduced into an evacuated container at 500°C to give a pressure of 1.00 atm before any decomposition occurred, what would be the partial pressures of N_2, H_2, and NH_3 at equilibrium?

29. At 90°C, the following equilibrium is established:

$$H_2(g) \;+\; S(s) \;\rightleftharpoons\; H_2S(g) \qquad K = 6.8 \times 10^{-2}$$

If 0.20 mole of hydrogen and 1.0 mole of sulfur are heated to 90°C in a 1.0 liter vessel, what will be the partial pressure of H_2S at equilibrium?

30. NO and Br_2 at initial partial pressures of 98.4 and 41.3 torr, respectively, were allowed to react at 300 K. At equilibrium the total pressure was 110.5 torr. Calculate the value of the equilibrium constant and the standard free energy change at 300 K for the reaction $2 NO(g) \;+\; Br_2(g) \;\rightleftharpoons\; 2 NOBr(g)$.

31. Using data from Tables 3–3 and 6–1, ascertain which of the following reactions is feasible at 298 K with both reactants in their standard states:

$$C_6H_6 \;\rightarrow\; 3\,C_2H_2$$
$$3\,C_2H_2 \;\rightarrow\; C_6H_6$$

32. Given the reaction

$$CaCO_3(s) \;\rightleftharpoons\; CaO(s) \;+\; CO_2(g)$$

estimate the pressure of CO_2 in equilibrium with a mixture of $CaCO_3$ and CaO at 298 K.

33. For the reaction

$$C_2H_6(g) \;\rightleftharpoons\; C_2H_4(g) \;+\; H_2(g)$$

$\Delta G°$ is 5.35 kcal/mole at 900 K. Calculate the mole percent of hydrogen present at equilibrium if pure C_2H_6 is passed over a suitable dehydrogenation catalyst at 900 K and 1.00 atm pressure.

34. Assuming that the heat capacities of $H_2(g)$, $N_2(g)$, and $NH_3(g)$ do not vary with temperature and further assuming that ΔS is independent of temperature for the reaction

$$3 H_2(g) \;+\; N_2(g) \;\rightarrow\; 2 NH_3(g)$$

estimate the minimum temperature at which this reaction will occur spontaneously with all reactants at unit activity.

7
Acid-Base Equilibria in Aqueous Solution

Such a large number of chemical reactions are carried out in water solution that it is worthwhile to consider equilibrium in aqueous solution as a separate topic. Indeed, in many cases the water functions both as a reactant and as a solvent. Of three important types of equilibrium not yet considered, acid-base equilibria in aqueous solution will be taken up in this chapter; the dissociation of complex ions and equilibria between dissolved ions in solution and solid ionic materials will be considered in Chapter 8.

7-1 Brønsted Theory of Acids and Bases

The role of water as a reactant in acid-base equilibria is best understood in terms of a theory of acids and bases developed independently by J. N. Brønsted of Denmark and T. M. Lowry of England. According to this theory, an **acid** is defined as a substance which donates a proton[1] to another substance; a **base** is a substance which accepts a proton from another substance. Some examples are shown in the following equations:

$$NH_3 + HCl \rightleftharpoons NH_4^+ + Cl^-$$
$$H_2O + HCl \rightleftharpoons H_3O^+ + Cl^-$$
$$NH_3 + H_2O \rightleftharpoons NH_4^+ + OH^-$$

The first example shows the reaction of the base NH_3 with the acid HCl to form a new acid-base pair. In the second equation, HCl is the acid which donates a proton to water, which therefore acts as a base. Note that the products—**hydronium ion,** H_3O^+, and chloride ion—are an acid and a base, respectively. Although in earlier chapters the symbol H^+ has been used to denote the hydrogen ion, free H^+ is not a stable chemical species. In aqueous solution the proton is always solvated and exists as the hydronium ion.

In the third equation, NH_3 is the base which accepts a proton from the acid, H_2O, to form the acid NH_4^+ and the base OH^-. Water can act as either an acid or a base, depending on whether it reacts with a base or an acid. Water is said to be **amphoteric;** it is capable of either donating or accepting a proton.

[1] In this sense a proton is a hydrogen atom which has lost its electron, i.e., a hydrogen ion. The protons located in the nuclei of other atoms are not involved in acid-base equilibria.

Three important features of the Brønsted-Lowry theory are illustrated in the above examples. First, when an acid molecule or ion donates its proton, the remainder of the molecule, now capable of accepting the proton, is a base. The original base, which has received the proton, is converted by the process into an acid. The acid-base relationship involves the exchange of a proton, analogous to the exchange of electrons between the members of a redox couple:

$$\text{acid} \rightleftharpoons \text{base} + \text{proton} \qquad \text{(an acid-base couple)}$$

In this case the members of the couple are called the **conjugate acid and base.** The second feature of the theory is that a complete acid-base reaction is merely the exchange of a proton between two sets of conjugates. The first of the reactions described above is a combination of two of these "couples":

$$\begin{array}{rcl} NH_3 + H^+ & \rightleftharpoons & NH_4^+ \\ HCl & \rightleftharpoons & H^+ + Cl^- \\ \hline \text{net:} \quad NH_3 + HCl & \rightleftharpoons & NH_4^+ + Cl^- \end{array}$$

The third aspect of this theory is that the extent of exchange depends on the relative base "strength" and acid "strength" of the species involved.

7-2 Relative Strengths of Acids and Bases

Acids and bases may be classified as strong, weak, or extremely weak, according to their tendencies to donate protons to a given reference base. In aqueous solution, water is the reference base. Water also serves as the reference acid for comparison of the strengths of aqueous bases. A **strong acid** is one which donates its proton completely to the reference base. For example, in water, the reaction

$$\underset{\text{strong acid}}{HCl} + \underset{\substack{\text{very weak base} \\ }}{H_2O} \rightarrow \underset{\substack{\text{strong acid} \\ }}{H_3O^+} + \underset{\substack{\text{extremely weak base} \\ }}{Cl^-}$$

goes to completion; therefore HCl is a strong acid. Its water solution is called hydrochloric acid. The chloride ion is an **extremely weak base;** it cannot compete successfully with H_3O^+ for the proton. In contrast, when pure acetic acid is placed in water, the ionization reaction proceeds only slightly to the right. Accordingly, acetic acid is classified as **weak.**

$$\underset{\text{weak acid}}{HC_2H_3O_2} + \underset{\substack{\text{very weak base} \\ }}{H_2O} \rightleftharpoons \underset{\substack{\text{strong acid} \\ }}{H_3O^+} + \underset{\substack{\text{stronger base} \\ }}{C_2H_3O_2^-}$$

The acetate ion is a stronger base than chloride ion, and is able to remove a proton from H_3O^+ to a large degree.

A strong acid necessarily implies an extremely weak conjugate base, and the conjugate of a strong base is necessarily a very poor proton donor. In Table 7–1, some

TABLE 7-1. Relative Acid and Base Strengths

	Conjugate Acid		Conjugate Base		
Strong	perchloric acid	$HClO_4$	ClO_4^-	chlorate ion	**Extremely Weak**
	hydroiodic acid	HI	I^-	iodide ion	
	sulfuric acid	H_2SO_4	HSO_4^-	hydrogen sulfate ion	
	hydrobromic acid	HBr	Br^-	bromide ion	
	nitric acid	HNO_3	NO_3^-	nitrate ion	
	chloric acid	$HClO_3$	ClO_3^-	chlorate ion	
	hydrochloric acid	HCl	Cl^-	chloride ion	
	hydronium ion	H_3O^+	H_2O	water	
Weak	hydrogen sulfate ion	HSO_4^-	SO_4^{2-}	sulfate ion	**Weak**
	sulfurous acid	H_2SO_3	HSO_3^-	hydrogen sulfite ion	
	hydrofluoric acid	HF	F^-	fluoride ion	
	acetic acid	$HC_2H_3O_2$	$C_2H_3O_2^-$	acetate ion	
	pyridinium ion	$C_5H_5NH^+$	C_5H_5N	pyridine	
	hydrogen sulfite ion	HSO_3^-	SO_3^{2-}	sulfite ion	
	hydrosulfuric acid	H_2S	HS^-	hydrogen sulfide ion	
	ammonium ion	NH_4^+	NH_3	ammonia	
	methylammonium ion	$CH_3NH_3^+$	CH_3NH_2	methylamine	
	hydrogen carbonate ion	HCO_3^-	CO_3^{2-}	carbonate ion	
Extremely Weak	water	H_2O	OH^-	hydroxide ion	**Strong**
	hydrogen sulfide ion	HS^-	S^{2-}	sulfide ion	
	hydroxide ion	OH^-	O^{2-}	oxide ion	
	hydrogen	H_2	H^-	hydride ion	

acids and bases are listed according to their strengths in water solution. The acids located above the H_3O^+/H_2O conjugate pair are strong acids in aqueous solution. That is, these acids react completely with water to form H_3O^+ and the corresponding conjugate bases, which are extremely weak. The acids located below the H_2O/OH^- conjugate pair in the table are extremely weak acids in aqueous solution. Their conjugates are strong bases. That is, the reaction

$$A^- + H_2O \rightleftharpoons HA + OH^-$$

where HA represents an extremely weak acid, goes completely to the right.

On the other hand, the acids located between the H_3O^+/H_2O conjugate pair and the H_2O/OH^- conjugate pair are weak acids. The reaction between any weak acid, HA, and water

$$HA + H_2O \rightleftharpoons H_3O^+ + A^-$$

proceeds only until a few percent of the molecules are ionized.

With few exceptions, all salts are strong electrolytes—that is, in aqueous solution they exist only as their constituent ions. The cations and anions in solution may act as acids and bases, respectively, depending on their abilities to exchange protons with water or other species in solution. For example, in water, the ionic ammonium chloride, NH_4Cl, is completely dissociated into NH_4^+ and Cl^- ions. However, in aqueous solution some NH_4^+ ions donate protons to the solvent:

$$NH_4^+ + H_2O \rightleftharpoons NH_3 + H_3O^+$$

Thus the solution is acidic.

The hydronium ion is the strongest acid which can exist in aqueous solution; the hydroxide ion is the strongest base. Any stronger acid or base will react completely with the water to produce H_3O^+ or OH^- ions, respectively.

7-3 Dissociation Constants for Weak Acids and Bases

An equilibrium constant expression for the reaction of any weak acid, HA, with water

$$HA + H_2O \rightleftharpoons H_3O^+ + A^-$$

is of the form

$$K = \frac{[H_3O^+][A^-]}{[HA][H_2O]}$$

In dilute solutions of weak acids, the concentration of water is not changed appreciably by the ionization reaction which occurs. The molarity of pure water can be calculated as follows:

$$\left(\frac{1000 \text{ grams}}{\text{liter}}\right)\left(\frac{1 \text{ mole}}{18.0 \text{ grams}}\right) = 55.5 \text{ moles/liter} = 55.5 \ M$$

The concentration of water in a solution containing a 1 M solute is only about 2% lower. Hence, for practical purposes the concentration of water in dilute aqueous solutions may be regarded as constant. Both sides of the equilibrium constant expression may be multiplied by the concentration of water, $[H_2O]$, to obtain a new constant, K_a:

$$K_a = K[H_2O] = \frac{[H_3O^+][A^-]}{[HA]}$$

K_a is called the **acid dissociation constant** or the **ionization constant** of the acid.

In aqueous solutions of bases having strengths between those of H_2O and OH^-, the following type of equilibrium is established:

$$B + H_2O \rightleftharpoons HB^+ + OH^-$$

Using reasoning analogous to that used for weak acids,

$$K_b = \frac{[HB^+][OH^-]}{[B]}$$

TABLE 7-2

TABLE 7-2. Some Ionization Constants of Acids and Bases at 25°C

Acid	Formula	Ionization Constant, K_a
acetic	$HC_2H_3O_2$	1.8×10^{-5}
benzoic	$HC_6H_5CO_2$	6.5×10^{-5}
boric	H_3BO_3 (1st proton)	7.3×10^{-10}
carbonic	H_2CO_3 (1st proton)	3.5×10^{-7}
hydrogen carbonate ion (bicarbonate ion)	HCO_3^- (2nd proton)	5×10^{-11}
chloroacetic	$HC_2H_2ClO_2$	1.4×10^{-3}
formic	$HCHO_2$	1.7×10^{-4}
hydrocyanic	HCN	7.2×10^{-10}
hydrofluoric	HF	6.7×10^{-4}
hydrogen sulfate ion	HSO_4^- (2nd proton)	1.2×10^{-2}
hydrosulfuric	H_2S (1st proton)	1.0×10^{-7}
hydrogen sulfide ion	HS^- (2nd proton)	1×10^{-14}
nitrous	HNO_2	4.5×10^{-4}
oxalic	$H_2C_2O_4$ (1st proton)	5.9×10^{-2}
	$HC_2O_4^-$ (2nd proton)	6.4×10^{-5}
phenol	HOC_6H_5	1.3×10^{-10}
phosphoric	H_3PO_4 (1st proton)	7.5×10^{-3}
	$H_2PO_4^-$ (2nd proton)	6.2×10^{-8}
	HPO_4^{2-} (3rd proton)	4.8×10^{-13}

Base	Formula	Ionization Constant, K_b
ammonia	NH_3	1.8×10^{-5}
aniline	$C_6H_5NH_2$	4.3×10^{-10}
ethylenediamine	$NH_2CH_2CH_2NH_2$ (1st proton)	5.2×10^{-4}
	$NH_2CH_2CH_2NH_3^+$ (2nd proton)	3.7×10^{-7}
methylamine	CH_3NH_2	4.4×10^{-4}
pyridine	C_5H_5N	2.3×10^{-9}

where K_b is known as the **base dissociation constant** or the **ionization constant** of the base.

The numerical magnitudes of K_a and K_b are characteristic of the particular acid or base, and are quantitative measures of their strengths. Like all equilibrium constants, they vary with temperature. Some representative constants are listed in Table 7–2. In contrast to weak acids, such as those listed in the table, strong acids have very high values of K_a. For example, HCl has a K_a about 1×10^5: a 0.10 M HCl solution is 99.9999% ionized. Acid-base equilibria are the same as any other type of chemical equilibrium. Therefore the procedures for solving equilibrium problems suggested in Chapter 6 can be applied, as shown in the following examples.

Example

When 0.100 mole of ammonia, NH_3, is dissolved in sufficient water to make 1.00 liter of solution, the solution is found to have a hydroxide ion concentration of $1.3 \times 10^{-3} M$. Calculate K_b for ammonia.

$$NH_3 \ + \ H_2O \ \rightleftharpoons \ NH_4^+ \ + \ OH^-$$

$$K_b = \frac{[NH_4^+][OH^-]}{[NH_3]}$$

$$[OH^-] = 1.34 \times 10^{-3} \ M$$

Ionization of the base produces equal concentrations of hydroxide and ammonium ions:

$$[NH_4^+] = [OH^-] = 1.34 \times 10^{-3} \ M$$

Also

$$[NH_3] = 0.100 - (1.34 \times 10^{-3}) = 0.099 \ M$$

$$K_b = \frac{(1.34 \times 10^{-3})^2}{(0.099)} = 1.81 \times 10^{-5}$$

Example

Calculate the hydronium ion concentration of a solution containing 0.200 mole of $HC_2H_3O_2$ in 1.00 liter of solution. $K_a = 1.80 \times 10^{-5}$. What is the percent ionization of the acid?

$$HC_2H_3O_2 \ + \ H_2O \ \rightleftharpoons \ H_3O^+ \ + \ C_2H_3O_2^-$$

$$K_a = \frac{[H_3O^+][C_2H_3O_2^-]}{[HC_2H_3O_2]} = 1.80 \times 10^{-5}$$

At equilibrium, let $[H_3O^+] = x$

then $[C_2H_3O_2^-] = x$

and $[HC_2H_3O_2] = 0.200 - x$

$$\frac{x^2}{0.200 - x} = 1.80 \times 10^{-5}$$

$$x^2 + (1.80 \times 10^{-5})x - (3.60 \times 10^{-6}) = 0$$

Application of the quadratic formula[2] yields

$$x = \frac{-1.80 \times 10^{-5} + \sqrt{(3.24 \times 10^{-10}) + (1.44 \times 10^{-5})}}{2}$$

$$= 1.89 \times 10^{-3} \ M = [H_3O^+]$$

The percent ionization of the acid is the number of moles which ionized, divided by the number of moles of acid originally present, times 100. The moles which ionized are represented by the acetate ion, $C_2H_3O_2^-$. Since the ion and the nonionized acid are both in the same volume of solution, their concentrations can be used instead of moles.

$$\% \ \text{ionized} = \frac{[C_2H_3O_2^-]}{[HC_2H_3O_2]_{\text{original}}} \times 100 = \frac{1.89 \times 10^{-3}}{0.200} \times 100 = 0.945\%$$

The acid is less than 1% ionized.

[2] The quadratic formula $x = (-b \pm \sqrt{b^2 - 4ac})/2a$ may be used to solve an equation that is in the form $ax^2 + bx + c = 0$.

In the above example it is seen that the concentrations of $C_2H_3O_2^-$ ions and of H_3O^+ ions are small compared to the concentration of unreacted $HC_2H_3O_2$. This situation is expected because the small value of K_a indicates that the reaction of $HC_2H_3O_2$ with water goes to a limited extent. Whenever the ratio of the equilibrium constant to the concentration of acid (or base) is very small (about 10^{-4} or less), an approximate solution may be obtained by neglecting small quantities when added to or subtracted from larger quantities. In the case of 0.200 M $HC_2H_3O_2$, $K_a/[HC_2H_3O_2] = 9.0 \times 10^{-5}$; hence in the expression

$$\frac{x^2}{(0.200 - x)} = 1.80 \times 10^{-5}$$

it can be assumed that x is small with respect to 0.200; thus $0.200 - x \cong 0.200$.

$$\frac{x^2}{0.200} = 1.80 \times 10^{-5}$$

$$x = \sqrt{(0.200)(1.80 \times 10^{-5})} = 1.90 \times 10^{-3} \, M = [H_3O^+]$$

The approximate result obtained in this case agrees well with that obtained by the more precise calculation done using the quadratic formula.

7-4 Autoionization of Water

Since water is amphoteric, it is not surprising that it undergoes reaction with itself:

$$H_2O \ + \ H_2O \ \rightleftharpoons \ H_3O^+ \ + \ OH^-$$

This equilibrium exists in pure water and in all dilute aqueous solutions, and the following equilibrium constant expression applies to it:

$$K = \frac{[H_3O^+][OH^-]}{[H_2O]^2}$$

This expression may be rearranged as follows:

$$K[H_2O]^2 = [H_3O^+][OH^-]$$

However, the concentration of water, $[H_2O]$, may be regarded as a constant, and the term K_w is defined as $K[H_2O]^2$. Hence

$$K_w = [H_3O^+][OH^-]$$

At 25°C, K_w has the value 1.0×10^{-14}. The equilibrium between water and its ions exists in all aqueous solutions regardless of the presence of acid or base. If either acid or base is present in the water, the equilibrium will shift in such a manner that K_w remains satisfied.

Example

Calculate the hydronium ion concentration and the hydroxide ion concentration in pure water at 25°C.

$$2 H_2O \rightleftharpoons H_3O^+ + OH^-$$

$$K_w = [H_3O^+][OH^-] = 1.0 \times 10^{-14}$$

Let $x = [H_3O^+] = [OH^-]$

Hence $x^2 = 1.0 \times 10^{-14}$

$$x = 1.0 \times 10^{-7} M = [H_3O^+] = [OH^-]$$

Example

Calculate the hydronium ion concentration of a 0.100 M NaOH solution.

The hydroxide ion concentration from the autoionization of water is negligible compared to that provided from the NaOH, and the solution consists of 0.100 M Na$^+$ and 0.100 M OH$^-$. Hence

$$[OH^-] = 0.100 \, M$$

In any dilute aqueous solution

$$K_w = [H_3O^+][OH^-] = 1.0 \times 10^{-14}$$
$$[H_3O^+](0.100) = 1.0 \times 10^{-14}$$
$$[H_3O^+] = 1.0 \times 10^{-13}$$

As predicted by Le Châtelier's principle, the OH$^-$ from the sodium hydroxide represses the ionization of water, so that in the NaOH solution the H$_3$O$^+$ concentration is 1×10^{-13} M, compared to 1×10^{-7} M in pure water.

7-5   Common Ion Effect

In accordance with Le Châtelier's principle, addition of the products of the reaction to a system at equilibrium will cause the equilibrium to shift in the direction of the reactants. Therefore, addition of a strong acid to a solution of a weak acid will suppress the ionization of the latter, because the H$_3$O$^+$ ion is a *common* dissociation product of both acids.

Example

Calculate the percent dissociation of acetic acid in a solution 0.200 M in HC$_2$H$_3$O$_2$ and 0.100 M in HCl.

Since the HCl is completely dissociated in water, the equilibrium of interest is the ionization of the acetic acid in the presence of 0.100 M H$_3$O$^+$:

$$HC_2H_3O_2 + H_2O \rightleftharpoons H_3O^+ + C_2H_3O_2^-$$

$$K_a = \frac{[H_3O^+][C_2H_3O_2^-]}{[HC_2H_3O_2]}$$

At equilibrium, let $[C_2H_3O_2^-] = x$

$$[HC_2H_3O_2] = 0.200 - x \cong 0.200$$
$$[H_3O^+] = 0.100 + x \cong 0.100$$

Then

$$K_a = \frac{(0.100)x}{0.200} = 1.8 \times 10^{-5}$$

$$x = 3.6 \times 10^{-5} \, M = [C_2H_3O_2^-]$$

Since all the acetate ion resulted from the dissociation of acetic acid, the percent dissociation is given by

$$\% \text{ dissociation} = \frac{x}{0.200} \times 100 = \frac{3.6 \times 10^{-5}}{0.200} \times 100 = 0.018\%$$

amt. of acetic acid that dissociated

If there were no HCl in this solution, the acetic acid would have been 0.945% dissociated, as shown in the example on page 174. Thus the added strong acid represses the ionization of the acetic acid.

As could be shown by an analogous calculation, a salt containing the anion of a weak acid will suppress the ionization of that acid. A useful result is obtained from measuring the hydronium ion concentration in a solution of a weak acid to which a salt of the acid (the conjugate base) has been added.

Consider the general reaction

$$HA \; + \; H_2O \; \rightleftharpoons \; H_3O^+ \; + \; A^-$$

for which

$$K_a = \frac{[H_3O^+][A^-]}{[HA]}$$

which rearranges to

$$[H_3O^+] = K_a \frac{[HA]}{[A^-]}$$

This equation shows that for a given acid at a given temperature, the hydronium ion concentration depends only on the ratio $[HA]/[A^-]$ and not on the absolute magnitudes of the concentrations. It is apparent that when the concentrations of the acid and its conjugate base are equal, the hydronium ion concentration is equal to K_a. This fact provides a convenient way of obtaining numerical values of K_a for weak acids by a direct measurement of $[H_3O^+]$.

Relationships analogous to those derived for weak acids can be obtained for weak bases. In the latter case, the common ions would be OH^- if a strong base is added, or the conjugate acid of the weak base if a salt is added. Thus for the reaction

$$B \; + \; H_2O \; \rightleftharpoons \; BH^+ \; + \; OH^-$$

$$[OH^-] = K_b \frac{[B]}{[BH^+]}$$

7-6   The pH Scale

If 0.10 mole of strong acid is added to pure water to give 1.0 liter of solution, the hydronium ion concentration is changed from 1.0×10^{-7} M in water to 1.0×10^{-1} M in the solution, a millionfold change. If 0.20 mole of OH^- ion in the form of a strong base is added to this solution, the H_3O^+ concentration changes by a factor of a trillion (10^{12}). Such changes in hydronium ion concentration are common in water solution. Moreover, the actual magnitudes of the hydronium ion concentration in dilute solutions may be extremely small. The **pH scale** was devised to represent widely different hydronium ion concentrations conveniently on a graph and to express the hydronium ion concentrations of various solutions without the necessity of using numbers in exponential form. The pH scale is defined by the equation

$$pH = -\log [H_3O^+] = \log \frac{1}{[H_3O^+]}$$

A tenfold change in hydronium ion concentration is thus represented by a change of one pH unit.

Example

In pure water at 25°C, $[H_3O^+] = 1.0 \times 10^{-7}$. What is the pH of pure water at 25°C?

$$pH = -\log (1.0 \times 10^{-7}) = 7.00$$

The pOH scale is similarly defined:

$$pOH = -\log [OH^-]$$

At 25°C, the pH of pure water and of any aqueous solution having equal concentrations of hydronium ion and hydroxide ion is 7. Solutions having a pH value less than 7 are acidic, while those with a pH value greater than 7 are basic. Values of pH and pOH for various solutions may be tabulated as follows:

pH	$[H_3O^+]$		$[OH^-]$	pOH
−1	10	acid	1×10^{-15}	15
0	1		1×10^{-14}	14
1	1×10^{-1}		1×10^{-13}	13
⋮	⋮		⋮	⋮
7	1×10^{-7}	neutral	1×10^{-7}	7
⋮	⋮		⋮	⋮
14	1×10^{-14}		1	0
15	1×10^{-15}	base	10	−1

This type of notation has been extended to include the term pK, referring to equilibrium constants. From the value of K_w it is readily seen that in aqueous solution at 25°C

$$pH + pOH = 14.00$$

Thus one can define

$$pK_w = -\log K_w = 14.00$$

Also pK_a and pK_b refer to the negative logarithms of acid and base dissociation constants, respectively. It should be emphasized that the pK is merely a way of expressing the value of the dissociation constant. Hence for a given acid or base, pK is a constant at a given temperature. In contrast, pH and pOH are ways of expressing hydronium and hydroxide ion concentrations; therefore pH and pOH values will vary, depending on the concentrations of H_3O^+ and OH^-.

Example

Calculate the pH of a solution which has a hydronium ion concentration of 6.0×10^{-8} M.

$$pH = -\log(6.0 \times 10^{-8}) = -\log 6.0 - \log 10^{-8} = -0.78 + 8 = 7.22$$

Example

Calculate the hydronium ion concentration of a solution which has a pH of 11.73.

$$pH = -\log[H_3O^+] = 11.73$$

In order that it may be located in the logarithm table, the part of the logarithm to the right of the decimal point (the mantissa) must be positive. Hence

$$\log[H_3O^+] = -11.73$$
$$[H_3O^+] = 10^{-11.73} = 10^{0.27} \times 10^{-12}$$

From the definition of logarithms (Appendix A–7), $10^{0.27} = 1.9$. Hence

$$[H_3O^+] = 1.9 \times 10^{-12}$$

7-7   Cationic Acids and Anionic Bases—Hydrolysis

When a salt of a weak acid and a strong base, such as sodium acetate, is dissolved in water, the anion reacts to some extent with the water to form its conjugate acid and OH^- ion:

$$A^- + H_2O \rightleftharpoons HA + OH^-$$

This type of reaction is often called **hydrolysis;** essentially it is an acid-base reaction. The equilibrium constant expression for the reaction is formulated from the chemical equation in the usual manner, with a subscript h to denote hydrolysis; the equilibrium constant is

$$K_h = \frac{[HA][OH^-]}{[A^-]}$$

Similarly, salts of a weak base and a strong acid, such as ammonium chloride, dissolve in water and react to form solutions with a pH less than 7, owing to the formation of excess H_3O^+ ions:

$$BH^+ \;+\; H_2O \;\rightleftharpoons\; B \;+\; H_3O^+$$

$$K_h = \frac{[B][H_3O^+]}{[BH^+]}$$

A table of hydrolysis constants is not necessary, because K_h values are easily calculated from the values of the ionization constants of the conjugate acids or bases. For example, multiplying the equilibrium constant expressions of the two conjugates HA and A^- one obtains

$$K_h K_a = \left(\frac{[HA][OH^-]}{[A^-]}\right)\left(\frac{[H_3O^+][A^-]}{[HA]}\right) = [OH^-][H_3O^+] = K_w$$

Therefore, for a conjugate base of a weak acid,

$$K_h = \frac{K_w}{K_a}$$

Similarly, for a weak base and its conjugate acid it can be shown that $K_w = K_h K_b$ and $K_h = K_w/K_b$.

Example

Calculate the pH of a 0.200 M solution of NH_4Cl.
Cl^-, the conjugate base of a strong acid, does not react with water, but NH_4^+, the conjugate acid of NH_3 ($K_b = 1.8 \times 10^{-5}$) does react:

$$NH_4^+ \;+\; H_2O \;\rightleftharpoons\; H_3O^+ \;+\; NH_3$$

$$K_h = \frac{[H_3O^+][NH_3]}{[NH_4^+]} = \frac{K_w}{K_b}$$

Let $[H_3O^+] = x$

$\qquad [NH_3] = x$

$\qquad [NH_4^+] = 0.200 - x \cong 0.200$

$$K_h = \frac{x^2}{0.200} = \frac{1.0 \times 10^{-14}}{1.8 \times 10^{-5}} = 5.6 \times 10^{-10}$$

$$x = 1.1 \times 10^{-5} = [H_3O^+]$$

$$pH = 4.96$$

The solution of NH_4Cl in water is acidic, as expected.

There need be no set way to approach the solution of a problem involving a weak acid and its conjugate. Any equilibrium constant expression which includes all of the species present at equilibrium may be chosen; however, the choice of a particular expression may permit the calculation to be made more easily.

Example

Calculate the pH of a solution which results from the mixing of 50.0 ml of 0.300 M HCl with 50.0 ml of 0.400 M NH_3.

The simplest approach to this problem is to consider that the reaction

$$NH_3 \;+\; HCl \;\rightarrow\; NH_4^+ \;+\; Cl^-$$

goes to completion (see exercise 57, Chapter 2). The concentrations of all the species which would be present after that reaction are then calculated and used as the initial concentrations for the equilibrium reaction. In this case, on mixing, 15.0 mmole of NH_4^+ is produced and 5.0 mmole of NH_3 remains unreacted, in a total of 100.0 ml of solution. This solution is the same as that of a solution containing 0.050 M NH_3 and 0.150 M NH_4^+. The equilibrium for the ionization of the base is now considered:

$$NH_3 \;+\; H_2O \;\rightleftharpoons\; NH_4^+ \;+\; OH^-$$

$$K_b = \frac{[NH_4^+][OH^-]}{[NH_3]} = 1.8 \times 10^{-5}$$

Let $[OH^-] = x$

$$[NH_4^+] = x + 0.150 \cong 0.150$$
$$[NH_3] = 0.050 - x \cong 0.050$$

$$K_b = \frac{(0.150)(x)}{0.050} = 1.8 \times 10^{-5}$$

$$x = 6.0 \times 10^{-6} = [OH^-]$$

$$pOH = -(0.78 - 6) = 5.22$$

$$pH = 14.00 - pOH = 8.78$$

7-8 Buffer Solutions

A solution containing a mixture of a weak acid and its salt (conjugate base) or a weak base and its salt (conjugate acid) is called a **buffer solution.**

A buffer solution scarcely changes its pH even when relatively large quantities of strong acid or base are added to it. The pH changes only slightly because the weak acid present reacts with any added base or the weak base present reacts with any added acid. For example, consider a solution which contains a mixture of acetic acid and sodium acetate. Because of the common ion effect, the hydronium ion concentration of the solution is lower than that expected for acetic acid alone in solution. The condition at equilibrium is as follows:

$$HC_2H_3O_2 \;+\; H_2O \;\rightleftharpoons\; H_3O^+ \;+\; C_2H_3O_2^-$$

| relatively large quantity | huge excess | relatively small quantity | relatively large quantity |

If a strong acid is added to the buffer solution, the added hydronium ion reacts with the excess acetate ion present, and undissociated acetic acid is formed. If a strong base is added, the hydroxide ions of the base react with the hydronium ions present to form water. But then more acetic acid dissociates to replace some of the hydronium ion.

Example

Assuming that the final volume in each case remains the same, compare the effect of adding 0.010 mole of solid sodium hydroxide to **(a)** 1.0 liter of a solution 1.8×10^{-5} M in HCl and **(b)** 1.0 liter of a solution containing 0.10 mole of $NaC_2H_3O_2$ and 0.10 mole of $HC_2H_3O_2$.

The initial hydronium ion concentration of both solutions is 1.8×10^{-5} M. (The reader should verify this statement.)

(a) The 1.8×10^{-5} mole of HCl in the solution will react completely with 1.8×10^{-5} mole of the added NaOH, leaving the excess base in solution.

$$\text{excess base} = 0.010 \text{ mole} - (1.8 \times 10^{-5}) \text{ mole} \cong 0.010 \text{ mole}$$

$$[OH^-] = \frac{0.010 \text{ mole}}{1.0 \text{ liter}} = 0.010 \text{ M}$$

The hydroxide ion concentration would be virtually 0.010 M, and the hydronium ion concentration is calculated to be 1.0×10^{-12} M.

(b) The hydroxide ion of the NaOH would react with the 1.8×10^{-5} mole of H_3O^+ present. But as that reaction takes place, more acetic acid ionizes, since removal of the hydronium ion is a stress that shifts the weak acid ionization equilibrium to the right. More and more weak acid dissociates until all the added hydroxide has been used up. Acetate ion is produced in the process. Thus the solution will have a lower acetic acid concentration and a higher acetate ion concentration than it had originally. The solution will be the same as if it had been prepared originally from 0.11 mole of sodium acetate and 0.09 mole of acetic acid in enough water to make 1.0 liter of solution.

	Initial	Assuming Complete Chemical Reaction with NaOH	At Equilibrium
$[H_3O^+]$			x
$[HC_2H_3O_2]$	0.10	0.09	$0.09 - x \cong 0.09$
$[C_2H_3O_2^-]$	0.10	0.11	$0.11 + x \cong 0.11$

$$K_a = \frac{x(0.11)}{(0.09)} = 1.8 \times 10^{-5}$$

$$x = 1.5 \times 10^{-5} = [H_3O^+]$$

The hydronium ion concentration has been reduced only to five-sixths of its original value by the addition of about 500 times as much OH^- as H_3O^+ originally present. In the buffer solution, the hydronium ion concentration remains relatively constant. Added to the unbuffered solution of HCl, the same quantity of base caused an 18 million-fold change in hydronium ion concentration.

7-9 Preparation of Buffer Solutions

Buffer solutions have many important applications. In biological systems buffer action controls the hydronium ion concentration within the limits necessary for life. For example, buffers in human blood maintain the pH of the system in the range 7.35

to 7.45. Variations of only 0.2 pH unit beyond this range can cause serious illness and even death.

In chemical analysis, cations may be separated from each other by selective precipitation under controlled pH. Several examples of this procedure will be described in Chapter 8. Many industrial processes, such as wine making and paper manufacture, are monitored at various stages by a measurement of pH. The solutions being processed are usually compared to standard buffers.

To prepare a buffer with a chosen pH, an appropriate weak acid or base and its salt are mixed. In addition to the ratio of concentrations, several other factors must be considered. Among these are

1. The **buffer range,** that is, the range of pH over which the buffer is effective.
2. The **buffer capacity,** which determines the quantity of strong acid or strong base which may be added without producing a significant change in pH.
3. The chemical behavior of the system, which must be such that unwanted reactions are minimized.

Each of these factors will be further considered below.

It is sometimes convenient to express the hydronium ion concentration of a buffer solution directly in terms of pH as follows. For the system

$$HA + H_2O \rightleftharpoons H_3O^+ + A^-$$

$$[H_3O^+] = K_a \frac{[HA]}{[A^-]}$$

Taking the negative of the logarithms of the two sides of this expression yields

$$pH = pK_a + \log \frac{[A^-]}{[HA]}$$

An analogous treatment yields an expression for the pOH of a buffer system composed of a weak base and its conjugate:

$$B + H_2O \rightleftharpoons BH^+ + OH^-$$

$$pOH = pK_b + \log \frac{[BH^+]}{[B]}$$

For buffer solutions, the optimum ratio of salt to weak acid for maximum resistance to change in pH and maximum buffer capacity is $1:1$. Under these conditions, $pH = pK_a$. Therefore, in the selection of a buffer system, an acid is chosen which has a pK_a close to the desired pH. In all cases, it is desirable to keep the ratio $[A^-]/[HA]$ within the limits 10^{-1} to 10^1. Within these concentration limits, buffers have a practical range of about 2 pH units.

The buffer capacity of a solution is determined by the actual concentrations of the weak acid and salt, not by their ratio. The more moles of conjugates originally present, the more added acid or base can be absorbed without significant change in pH. In many commercial buffer solutions, the actual concentrations of weak acid and conjugate base are of the order of 1 M or 0.1 M.

Example

Using the data of Table 7-2, select a buffer system having a pH of 4.00 and state how you would prepare exactly 1 liter of the solution using 0.10 M solutions of the respective conjugates. What practical considerations influence your choice?

For acetic acid: $K_a = 1.8 \times 10^{-5}$ and $pK_a = 4.75$

For nitrous acid: $K_a = 4.5 \times 10^{-4}$ and $pK_a = 3.35$

Either acid would be suitable to make the required buffer having pH = 4.00, except that nitrous acid readily undergoes oxidation-reduction reactions and its usefulness is restricted. For acetic acid,

$$pH = 4.00 = 4.75 + \log \frac{[C_2H_3O_2^-]}{[HC_2H_3O_2]}$$

$$\log \frac{[C_2H_3O_2^-]}{[HC_2H_3O_2]} = 4.00 - 4.75 = -0.75$$

$$\frac{[C_2H_3O_2^-]}{[HC_2H_3O_2]} = 0.18$$

Since 1.00 liter of the buffer is to be made up from 0.10 M solutions of salt and acid and the volumes of the respective solutions must be in the ratio 0.18 : 1.00,

let x = the volume of sodium acetate solution in liters

then   1.00 − x = the volume of acetic acid solution in liters

$$\frac{x}{1.00 - x} = 0.18$$

$$x = 0.15 \text{ liter } NaC_2H_3O_2 \text{ solution}$$

$$1.00 - x = 0.85 \text{ liter } HC_2H_3O_2 \text{ solution}$$

Example

How many moles of sodium hydroxide can be added to 1.00 liter of a solution 0.100 M in NH_3 and 0.100 M in NH_4Cl without changing the pOH by more than 1.00 unit? Assume no change in volume.

$$K_b = 1.8 \times 10^{-5} \qquad pK_b = 4.75 \qquad \frac{[NH_4^+]}{[NH_3]} = 1.00$$

$$pOH = pK_b + \log \frac{[NH_4^+]}{[NH_3]} = 4.75 + 0.00 = 4.75$$

The original pOH = 4.75. The pOH after addition of NaOH cannot be less than 3.75.

$$pOH = 3.75 = 4.75 + \log \frac{[NH_4^+]}{[NH_3]}$$

$$\log \frac{[NH_4^+]}{[NH_3]} = -1.00$$

$$\frac{[NH_4^+]}{[NH_3]} = 0.10$$

Hence NaOH can be added until the ratio of $[NH_4^+]$ to $[NH_3]$ is 0.10. Initially $[NH_4^+]$ + $[NH_3]$ = 0.200. Although the reaction with OH^- converts NH_4^+ into NH_3, the sum of these two concentrations remains 0.200.

$$[NH_4^+] + [NH_3] = 0.200$$
$$[NH_4^+] = 0.10\,[NH_3] \quad \text{(from above)}$$
$$0.10\,[NH_3] + [NH_3] = 0.200$$
$$1.10\,[NH_3] = 0.200$$
$$[NH_3] = 0.182\ M$$

Hence $\quad [NH_4^+] = 0.018\ M$

Assuming no change in volume, $0.100 - 0.018 = 0.082$ mole of NaOH can be added without changing the pOH by more than 1.00 pOH unit.

7-10 Polyprotic Acids

Polyprotic acids are acids such as H_2S and H_3PO_4 which ionize in two or more steps. For H_2S, the steps are as follows:

$$H_2S + H_2O \rightleftharpoons H_3O^+ + HS^-$$
$$HS^- + H_2O \rightleftharpoons H_3O^+ + S^{2-}$$

An equilibrium constant expression can be written for each of these steps, and the concentration of each species present in solution at equilibrium must satisfy both of the equilibrium constant expressions simultaneously:

$$K_1 = \frac{[H_3O^+][HS^-]}{[H_2S]} \qquad K_2 = \frac{[H_3O^+][S^{2-}]}{[HS^-]}$$

H_2S is a weak acid ($K_1 = 1.0 \times 10^{-7}$), but HS^- is considerably weaker as an acid ($K_2 = 1 \times 10^{-14}$). In a solution of H_2S, the ionization of HS^- is further suppressed by the hydronium ion produced by the first ionization (common ion effect).

Example

Calculate the hydronium ion concentration and the sulfide ion concentration of a 0.100 M solution of H_2S.

The second step in the ionization process is the only source of sulfide ions, but neither the hydronium ion concentration nor the HS^- concentration is significantly changed by this step. Thus $[H_3O^+]$ and $[HS^-]$ can be calculated from K_1 only.

$$K_1 = \frac{[H_3O^+][HS^-]}{[H_2S]} = 1.0 \times 10^{-7}$$

Let $\quad [H_3O^+] = x$

then $\quad [HS^-] = x$

and $\quad [H_2S] = 0.100 - x \cong 0.100$

$$\frac{x^2}{0.100} = 1.0 \times 10^{-7}$$

$$x = 1.0 \times 10^{-4} = [H_3O^+] = [HS^-]$$

The sulfide ion concentration is calculated by use of K_2 and the values of $[H_3O^+]$ and $[HS^-]$ just determined.

$$K_2 = \frac{[H_3O^+][S^{2-}]}{[HS^-]} = 1.0 \times 10^{-14}$$

$$[H_3O^+] = [HS^-] = 1.0 \times 10^{-4}$$

$$K_2 = \frac{(1.0 \times 10^{-4})[S^{2-}]}{(1.0 \times 10^{-4})} = 1.0 \times 10^{-14}$$

$$[S^{2-}] = 1.0 \times 10^{-14}$$

The quantity of additional hydronium ion produced and the quantity of hydrogen sulfide ion used up in the second ionization step are each equal to the quantity of sulfide ion produced and are small compared to the quantities of H_3O^+ and HS^- produced in the first step. As was previously stated, the second ionization does not change the hydronium ion concentration or the hydrogen sulfide ion concentration significantly.

7-11 Indicators

An acid-base indicator is a weak acid or weak base with conjugate forms which have different colors. The color of at least one conjugate must be very intense so that only a small quantity of indicator need be used. For an indicator which is a weak acid, HInd, the following equation may be written.

$$\text{HInd} \quad + \quad H_2O \quad \rightleftharpoons \quad H_3O^+ \quad + \quad \text{Ind}^-$$
one color $\qquad\qquad\qquad\qquad\qquad\qquad$ another color

$$K_{\text{Ind}} = \frac{[\text{Ind}^-][H_3O^+]}{[\text{HInd}]}$$

At high hydronium ion concentrations, the solution will have the color characteristic of HInd, while at low hydronium ion concentrations, the color of the conjugate base will predominate. Rearranging the above equation and taking logarithms of both sides yields the following equation:

$$\text{pH} = pK_{\text{Ind}} + \log \frac{[\text{Ind}^-]}{[\text{HInd}]}$$

Thus the pH range of color change for an indicator is determined by its dissociation constant. It takes about a tenfold excess of one of the conjugates for its color to predominate. The color change then corresponds to a 100-fold change in the ratio of concentrations of the conjugates (from 1/10 to 10/1). Therefore the color change occurs over a range of about 2 pH units. For a titration procedure, it is desirable that one drop of titrant change the concentration ratio of the indicator conjugates by a

TABLE 7-3. pH Ranges of
Common Indicators

Indicator	pH Range
methyl violet	−0.3– 1.8
methyl orange	2.8– 3.8
Congo red	2.8– 4.8
methyl red	3.8– 6.1
bromothymol blue	6.0– 7.9
litmus	5.0– 8.1
phenol red	6.8– 8.6
cresol red	7.0– 9.1
thymol blue	0.5– 1.8
	and 7.6– 9.2
phenolphthalein	8.0– 9.6
thymolphthalein	10.2–11.7
tropeolin O	11.1–12.6

factor of 100, and so very little indicator must be present. However, the color change must be observable; therefore at least one form must have an intense color. Many substances are available which can be used as indicators, and one can usually be found which changes color within a desired pH range. The pH ranges of some indicators are depicted in Figure 7-1 and are listed in Table 7-3.

7-12 Measurement of pH

The measurement of the pH of a solution may be made either **colorimetrically** or **potentiometrically.** In the colorimetric measurement, the color of a test solution to which a few drops of indicator are added is compared with standard buffer solutions containing the same indicator. First, a rough determination of acidity or basicity is made using a special pH paper impregnated with several indicators chosen so as to give a characteristic color at any given range of pH. Then more careful comparisons are made using an indicator which is suitable for the specific pH range. In this manner measurements of pH accurate to 0.5 pH unit can be made.

The potentiometric method permits measurements which are accurate to 0.005 pH unit or better. As seen earlier (Chapter 5), the electrode potentials of many redox couples depend on the hydronium ion concentration. For example, the Nernst equation for the hydrogen electrode at 25°C and 1 atm hydrogen pressure is as follows:

$$\epsilon = \epsilon° - 0.0592 \log \frac{1}{[H_3O^+]} = -0.0592(pH)$$

The pH of a solution can be determined by measuring the potential of a cell consisting of a hydrogen electrode combined with a suitable reference electrode, such as the saturated calomel electrode. Such a cell could be diagrammed as follows:

$$Pt \mid H_2(g, 1 \text{ atm}) \mid H_3O^+(x \, M) \| KCl(sat) \mid Hg_2Cl_2 \mid Hg$$

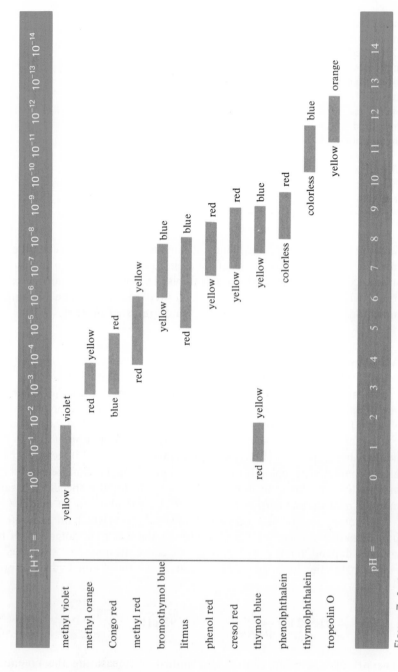

Figure 7-1. Some Acid–Base Indicators

The figure gives the approximate hydrogen ion concentrations and pH values required to produce the indicated change of color in dilute solutions (0.001%) of some of the more common indicators. (From L. F. Hamilton and S. G. Simpson, *Quantitative Analysis*, 12th ed., Macmillan, New York, 1964.)

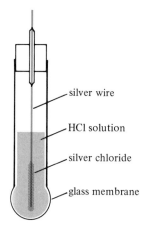

silver wire

HCl solution

silver chloride

glass membrane

Figure 7–2. **Glass Electrode**

A glass electrode consists of a silver/silver chloride electrode immersed in a dilute HCl solution contained in a thin-walled glass bulb. The glass electrode and a calomel electrode are placed in the solution whose pH is to be measured, and a cell of the following type is produced:

| Ag | AgCl(s) | HCl(aq) | glass | unknown solution | KCl(sat) | Hg$_2$Cl$_2$(s) | Hg |

glass electrode   saturated calomel electrode

The observed potential arises from the following sources: (1) the potential of the Ag/AgCl couple, (2) the potential of the calomel electrode, (3) the potential between the glass and the HCl solution within the glass electrode, (4) the "junction potential" between the calomel electrode and the unknown solution, and (5) the potential between the glass membrane and the solution of unknown pH. For a given cell, the first three sources of potential are fixed and constant. The fourth source is compensated for by adjusting the potentiometer when the electrodes are immersed in a buffer solution of known pH. Thereafter, the observed potential will depend only on the pH of an unknown solution. A special glass which has a high affinity for water is used to construct the glass membrane. The mechanism of its interaction with H$_3$O$^+$ is not fully understood.

where H$_3$O$^+$(x M) represents the solution of unknown hydronium ion concentration. The cell potential is given by

$$\epsilon_{cell} = \epsilon_{cal} - \epsilon = \epsilon_{cal} + 0.0592(pH)$$

$$pH = \frac{\epsilon_{cell} - \epsilon_{cal}}{0.0592} \qquad (\text{at } 25°C)$$

The hydrogen electrode is not convenient for routine pH measurement because it requires gaseous hydrogen. Moreover, the solution being tested might contain substances which would react with the hydrogen gas or which might destroy the catalytic activity of the platinized surface of the electrode. Glass electrodes (Figure 7–2) are often used to measure pH. The operation of this device depends on the difference in potential across a glass membrane separating two solutions of different pH.

7-13   Titration Curves

A description of how the concentration of acid or base in a solution may be determined by a titration procedure was given in Section 2–11. In this procedure, the equivalence point is signaled by the change in color of an indicator as the titrating solution is added dropwise to the test solution. With the progressive addition of acid to base (or vice versa) there is a corresponding change in the pH of the reaction mixture. The selection of a proper indicator is therefore an important consideration because the indicator chosen must change color at the pH corresponding to the equivalence point. To illustrate this concept, the calculated changes in pH which

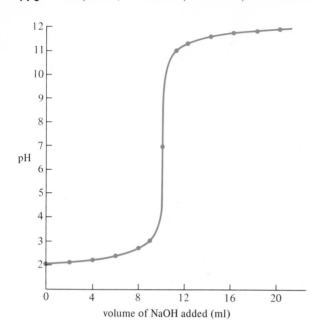

Figure 7-3. **Titration of HCl with NaOH**
1.000 liter of 0.01000 M HCl is titrated with 1.000 M NaOH.

occur when a strong acid, HCl, is titrated with a strong base, NaOH, are shown in Figure 7-3. (In this example, to simplify the calculations, the concentration of the base solution is 100 times greater than that of the acid; hence the change in volume of the solution upon addition of base is considered to be negligible.) It is seen that the pH rises slowly at first, but at the equivalence point, the addition of a very small quantity of base causes a very large change in pH. Beyond the equivalence point, the pH again increases very gradually. The curve obtained in Figure 7-3 is characteristic of the titration of any strong acid with a strong base, where the essential reaction is

$$H_3O^+ \; + \; OH^- \; \rightleftharpoons \; 2\,H_2O$$

Any indicator which changes color between pH 4 and pH 10 could be used to determine the end point. On the other hand, if the test solution itself were highly colored or if it were turbid, it would be difficult if not impossible to detect the end point by observing a change in the color of an indicator. In such cases, the titrations may be done potentiometrically (Section 5–12) using a standard reference electrode and an electrode which is sensitive to hydrogen ion concentration. A plot of measured pH versus added titrant would give a curve similar to that of Figure 7-3, and the equivalence point would be read from the graph.

When a weak acid is titrated with a strong base, or when a weak base is titrated with a strong acid, the titration curve obtained is quite different from that of a strong acid with a strong base. In the first place, before any titrant is added the weak acid does not have a pH as low as that of the strong acid of the same molarity. Then, as strong base is added, the solution being titrated is converted into a buffer solution, containing both weak acid and the salt of that acid. At the equivalence point, the pH of the solution is not 7, because when all of the original acid is just converted to salt, hydrolysis of the conjugate ion causes the solution to be basic. The converse situation prevails upon titration of a weak base with a strong acid. These facts are illustrated in

TABLE 7-4. Titration of 1.000 Liter of 0.01000 M NH$_3$ with 1.000 M HCl

HCl Added (ml)	HCl Added (mmole)	NH$_4^+$ Formed (mmole)	$\dfrac{[\text{NH}_4^+]}{[\text{NH}_3]}$	pOH	pH
2.0	2.0	2.0	0.25	4.14	9.86
4.0	4.0	4.0	0.67	4.57	9.43
5.0	5.0	5.0	1.00	4.74	9.26
6.0	6.0	6.0	1.50	4.92	9.08
8.0	8.0	8.0	4.00	5.32	8.68
9.0	9.0	9.0	9.00	5.69	8.31
10.0	10.0	10.0[a]	4.2×10^3	8.38	5.62
11.0	11.0	10.0[a]	1.8×10^6	11.0	3.0
12.0	12.0	10.0[a]	3.6×10^6	11.3	2.7
14.0	14.0	10.0[a]	7.2×10^6	11.6	2.4
16.0	16.0	10.0[a]	1.1×10^7	11.8	2.2
18.0	18.0	10.0[a]	1.4×10^7	11.9	2.1
20.0	20.0	10.0[a]	1.8×10^7	12.0	2.0

[a] The quantity present as NH$_3$ is negligible.

Table 7–4 and Figure 7–4, where calculated pH values corresponding to the points in a titration of 0.01000 M NH$_3$ solution with 1.000 M HCl are given. From the titration curve it is seen that an indicator which changes color in the pH range 4 to 7 should be selected for this titration.

Example

Explain the pH values in Table 7–4 corresponding to **(a)** 5.0 ml of HCl added, **(b)** 10.0 ml of HCl added, **(c)** 11.0 ml of HCl added.

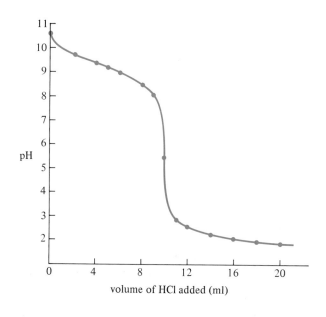

Figure 7-4. **Titration of NH$_3$ with HCl**
1.000 liter of 0.01000 M NH$_3$ is titrated with 1.000 M HCl.

(a) At 5.0 ml of HCl added, half the NH_3 will have been converted to NH_4^+, yielding a buffer solution corresponding to the equation

$$NH_3 \ + \ H_2O \ \rightleftharpoons \ NH_4^+ \ + \ OH^-$$

for which

$$K_b = \frac{[NH_4^+][OH^-]}{[NH_3]}$$

A negligible quantity of NH_3 dissociates in the presence of the NH_4^+ produced by the HCl reaction, and the concentrations of NH_3 and NH_4^+ are approximately equal. Thus

$$[OH^-] = K_b = 1.8 \times 10^{-5}$$
$$pOH = 4.74$$
$$pH = 9.26$$

(b) When 10.0 ml of the HCl has been added, exactly the same number of moles of HCl and NH_3 are present. They will react to yield the same solution as would be produced by 0.0100 mole of NH_4Cl in 1.010 liters of solution. A hydrolysis calculation (Section 7–7) yields a pH value of 5.62.

(c) When 11.0 ml of HCl has been added, all the NH_3 has been neutralized and 1.0 mmole of HCl is present in excess. The excess H_3O^+ represses the hydrolysis of the NH_4^+ present, and the pH is determined solely by the 1.0 mmole of H_3O^+ in 1.010 liter; that is, $pH = 3.0$.

Example

Select an indicator from Table 7–3 which would be suitable to indicate the equivalence point when 50.00 ml of 0.1000 M HCN is titrated with 0.1000 M NaOH.

At the equivalence point there will be equal numbers of millimoles of NaOH and HCN, which, if they could react completely, would give 5.000 mmole of NaCN. The solution is the equivalent of a 0.05000 M solution of NaCN, and the following equilibrium will be established:

$$CN^- \ + \ H_2O \ \rightleftharpoons \ HCN \ + \ OH^-$$

$$K_h = \frac{K_w}{K_a} = \frac{[HCN][OH^-]}{[CN^-]} = 1.4 \times 10^{-5} \qquad (K_a \text{ from Table 7–2})$$

Let $[OH^-] = [HCN] = x$

$$[CN^-] = 0.05000 - x \cong 0.05000$$

$$\frac{x^2}{0.05000} = 1.4 \times 10^{-5}$$

$$x = 8.4 \times 10^{-4} = [OH^-]$$
$$pOH = 3.08$$
$$pH = 10.92$$

Thymolphthalein would be a suitable indicator.

Numerical values of K_a (or K_b) for each acid (or base) can be determined by direct measurement of pH. When equimolar concentrations of a weak acid and its conjugate are present, the hydronium ion concentration of the solution is numerically equal to K_a (Section 7-9). The point at which equal numbers of moles of the two conjugates is present is easily determined from a titration curve.

7-14   Exercises

Basic Exercises

1. (a) Distinguish between acid strength and acid concentration. (b) Distinguish between a weak base and an insoluble base.

2. Draw an electron dot diagram to illustrate that "ammonium hydroxide" could not exist as a *weak* base. Explain the weakly basic properties of a solution of ammonia in water.

3. Which one(s) of the following reagents is (are) strong electrolytes? (a) NH_3, (b) NH_4Cl, (c) $HC_2H_3O_2$, (d) $NaC_2H_3O_2$, (e) HCl, (f) NaCl.

4. Write equilibrium constant expressions for the following reactions:
   (a) $HC_2H_3O_2 + H_2O \rightleftharpoons H_3O^+ + C_2H_3O_2^-$
   (b) $HC_2H_3O_2 + CH_3OH \xrightleftharpoons[\text{no solvent}]{} H_2O + C_2H_3O_2CH_3$
   (c) In which of these expressions may $[H_2O]$ be omitted?

5. Calculate the pH of each of the following solutions from the given hydronium ion molarity: (a) 1.0×10^{-3},   (b) 3.0×10^{-11},   (c) 3.9×10^{-8},   (d) 0.15, (e) 1.23×10^{-15}, (f) 1.0, (g) 2.0.

6. Calculate the hydronium ion concentration of each of the following solutions from the given pH value: (a) 5.00, (b) 7.572, (c) 12.12, (d) 0.00, (e) 13.85.

7. Calculate the pH of $1.0 \times 10^{-3} M$ solutions of each of the following: (a) HCl, (b) NaOH, (c) $Ba(OH)_2$, (d) NaCl.

8. Which one(s) of the reagents listed below could be added to water to make $0.10 M$ solutions of each of the following ions? (a) NH_4^+, (b) $C_2H_3O_2^-$, (c) Cl^-.

   (i) NH_3 (ii) NH_4Cl (iii) $HC_2H_3O_2$
   (iv) $NaC_2H_3O_2$ (v) HCl   (vi) NaCl

9. Consider the reaction $A^- + H_3O^+ \rightleftharpoons HA + H_2O$. The K_a value for the acid HA is 1.0×10^{-6}. What is the value of the K for this reaction?

10. Explain why a solution of NH_4Cl is acidic.

11. Calculate the pH of a solution of $0.10 M$ HA and $0.20 M$ NaA. $K_a = 1.0 \times 10^{-7}$.

12. It is desired to prepare a buffer solution consisting of $0.10 M HC_2H_3O_2$ and $0.10 M NaC_2H_3O_2$. Assuming no volume change upon the addition of the pure compounds, state what reagents, and in what quantities, should be added to 1.00 liter of each of the following solutions to prepare the desired buffer solution: (a) $0.10 M HC_2H_3O_2$, (b) $0.20 M HC_2H_3O_2$,   (c) $0.20 M NaC_2H_3O_2$,   (d) $0.10 M NaC_2H_3O_2$,   (e) $0.10 M$ NaOH.

13. Sodium chloride, hydrogen chloride gas, and water are put into a vessel. Sodium acetate, acetic acid, and water are put into a second vessel. (a) Write a balanced chemical equation for the reaction which occurs in the first vessel. (b) Does the added NaCl influence the reaction in any way? (c) Is it necessary for every substance which is placed in a vessel to be a reactant in any reaction which might occur? (d) Write a balanced equation for the equilibrium reaction which occurs in the second vessel. (e) Does the added sodium acetate influence the reaction in any way? (f) Are all the substances in the second vessel reactants?

14. Calculate the pH of a solution containing $0.10\ M\ H_3BO_3$ and $0.18\ M\ NaH_2BO_3$.

15. (a) Calculate the concentration of each ion, Mg^{2+}, H_3O^+, and Br^-, in a solution prepared by dissolving 0.20 mole of $MgBr_2$ and 0.10 mole of HBr in enough water to make 1.0 liter of solution. Both substances are completely soluble and ionic in solution. (b) Calculate the concentration of each ion, H_3O^+, Mg^{2+}, and $C_2H_3O_2^-$, in a solution prepared by dissolving 0.20 mole of $HC_2H_3O_2$ and 0.050 mole of $Mg(C_2H_3O_2)_2$ in enough water to make 1.0 liter of solution.

16. (a) Calculate K_a for an acid whose 0.10 M solution has a pH of 4.50. (b) Calculate K_b for a base whose 0.10 M solution has a pH of 10.50.

17. Calculate the pH of a 0.10 $M\ NH_3$ solution.

18. What is the value of K_w (a) in 0.10 M NaOH solution? (b) in 0.10 M NaCl solution?

19. Calculate the hydronium ion concentration and the percent dissociation of an acid with $K_a = 1.0 \times 10^{-6}$ in (a) 0.10 M solution, (b) 0.0010 M solution. (c) Explain precisely the difference between the terms *higher hydronium ion concentration* and *greater percent dissociation,* as applied to weak acid solutions.

20. Calculate the pH of 0.0030 $M\ Ba(OH)_2$ solution.

21. Explain why a solution containing a strong base and its salt does not act as a buffer solution.

22. (a) Which of the following 0.10 M solutions are acidic, which are basic, and which are neutral? (b) Arrange them in order of increasing pH. (i) NH_4Cl, (ii) NaOH, (iii) $HC_2H_3O_2$, (iv) NaCl, (v) $NH_3\ +\ NH_4Cl$, (vi) NH_3, (vii) HCl.

23. For each of the following salts, state which ion (if either) hydrolyzes more in water at 25°C: (a) $NaC_2H_3O_2$, (b) NH_4Cl, (c) MgS, (d) NaCl, (e) $NH_4C_2H_3O_2$.

24. Which of the following combinations of solutes would result in the formation of a buffer solution? (a) $NaC_2H_3O_2\ +\ HC_2H_3O_2$, (b) $NH_4Cl\ +\ NH_3$, (c) HCl + NaCl, (d) HCl + $HC_2H_3O_2$, (e) NaOH + HCl, (f) NaOH + $HC_2H_3O_2$ in a 1:1 mole ratio, (g) $NH_3\ +\ $ HCl in a 2:1 mole ratio, (h) $HC_2H_3O_2\ +\ $ NaOH in a 2:1 mole ratio.

25. Which equilibrium constant(s) or ratio of equilibrium constants should be used to calculate the pH of 1.00 liter of each of the following solutions? (a) KOH, (b) NH_3, (c) $HC_2H_3O_2$, (d) $HC_2H_3O_2\ +\ NaC_2H_3O_2$, (e) $KC_2H_3O_2$, (f) 0.10 mole $HC_2H_3O_2\ +\ $ 0.050 mole NaOH, (g) H_2S, (h) 0.10 mole $NH_4Cl\ +\ $ 0.050 mole NaOH, (i) 0.10 mole $HC_2H_3O_2\ +\ $ 0.10 mole NaOH.

26. Calculate the pH of a solution of 0.10 M acetic acid. Calculate the pH after 50.0 ml of this solution is treated with 25.0 ml of 0.10 M NaOH.

27. (a) Determine the pH of a 0.10 M solution of pyridine, C_5H_5N. (b) Predict the effect of addition of pyridinium ion, $C_5H_5NH^+$, on the position of the equilibrium. Will the pH be raised or lowered? (c) Calculate the pH of 1.00 liter of 0.10 M pyridine solution to which 0.15 mole of pyridinium chloride, $C_5H_5NH^+Cl^-$, has been added, assuming no change in volume.

28. In an electrochemical cell, what value does the cell potential have when the reaction is at equilibrium? What is the value at equilibrium of the ratio of concentrations which is part of the Nernst equation?

29. Write complete net ionic equations for the following processes. Which combinations of reactants will react less than 2% of the theoretically possible extent? Which one(s) will react until more than 98% of the limiting quantity is used up. (a) $HC_2H_3O_2\ +\ H_2O$, (b) $C_2H_3O_2^-\ +\ H_2O$, (c) $C_2H_3O_2^-\ +\ H_3O^+$, (d) NaOH + $HC_2H_3O_2$, (e) $NaC_2H_3O_2\ +\ $ HCl(aq), (f) HCl(g) + H_2O, (g) $Cl^-\ +\ H_3O^+$, (h) $Cl^-\ +\ H_2O$, (i) $NH_4^+\ +\ $ NaOH, (j) $NH_4^+\ +\ OH^-$, (k) $NH_3\ +\ H_2O$, (l) $NH_3\ +\ H_3O^+$, (m) $NH_3\ +\ $ HCl(aq), (n) $Na^+\ +\ OH^-$.

30. At what pH will a 1.0×10^{-3} M solution of an indicator with $K_b = 1.0 \times 10^{-10}$ change color?

General Exercises

31. Calculate the pH of 0.10 M NH_4NO_3 solution.

32. From the data of Table 7–2, choose an acid and a base, neither of which is strong, which when mixed in equimolar quantities gives a solution with pH nearest 7.

33. Distinguish between the *end point* (where a titration with indicator is stopped) and the *equivalence point* of a titration. Explain, using these two terms, the importance of choosing the proper indicator for a given titration.

34. Calculate the sulfide ion concentration of a solution prepared by adding 0.100 mole of H_2S and 0.300 mole of HCl to enough water to make 1.00 liter of solution.

35. Calculate the pH of a solution prepared by mixing 50.00 ml of 0.0200 M NaOH and 50.00 ml of 0.0400 M $HC_2H_3O_2$.

36. Calculate the pH of a solution resulting from the addition of 50.0 ml of 0.100 M HCl to 50.0 ml of a solution containing 0.150 M $HC_2H_3O_2$ and 0.200 M $NaC_2H_3O_2$.

37. Calculate the pH of a solution made by mixing 50.0 ml of 0.200 M NH_4Cl and 75.0 ml of 0.100 M NaOH.

38. Calculate the pH of a 0.100 M solution of $NH_2CH_2CH_2NH_2$, ethylenediamine (en). Determine the enH_2^{2+} concentration of the solution.

39. Calculate the OH^- concentration and the H_3PO_4 concentration of a solution prepared by dissolving 0.100 mole of Na_3PO_4 in sufficient water to make 1.00 liter of solution.

40. Calculate the sulfate ion concentration in 0.15 M H_2SO_4.

41. Calculate the mass of $NaC_2H_3O_2$(s) and the volume of a 5.00 M $HC_2H_3O_2$ solution, plus water as necessary, which are required to prepare 1.00 liter of a buffer solution containing 0.200 M $HC_2H_3O_2$ with a pH = 4.00.

42. What is the pH of a 0.100 M $NH_4C_2H_3O_2$ solution?

43. Calculate the acetic acid to acetate ion concentration ratio in a buffer solution whose pH is 7.00. Explain how it is possible to have *any* acid in a neutral solution.

44. Calculate the values of the equilibrium constants for the reactions with water of $H_2PO_4^-$, HPO_4^{2-}, and PO_4^{3-} as bases. Comparing the relative values of the two equilibrium constants of $H_2PO_4^-$ with water, deduce whether solutions of this ion in water are acidic or basic. Deduce whether solutions of HPO_4^{2-} are acidic or basic.

45. Calculate the pH of a 0.100 M Na_2HPO_4 solution. State any approximations which are necessary for this calculation.

46. What indicator should be used for the titration of 0.10 M KH_2BO_3 with 0.10 M HCl?

47. Calculate the pH of 50.0 ml of a 0.100 M acetic acid solution to which each of the following quantities of 0.100 M NaOH has been added: **(a)** 15.0 ml, **(b)** 25.0 ml, **(c)** 40.0 ml, **(d)** 49.0 ml, **(e)** 50.0 ml, **(f)** 51.0 ml, **(g)** 60.0 ml, **(h)** 70.0 ml. Draw a smooth curve through the points which result when pH (on the vertical axis) is plotted against volume (on the horizontal axis).

48. Calculate the pH of a solution containing 0.100 M HCO_3^- and 0.150 M CO_3^{2-}.

Advanced Exercises

49. The number 10.92 has how many significant figures? The number 0.92 has how many? If these were pH values, how many significant figures should be reported for the corresponding hydronium ion concentrations?

50. Construct the titration curve which would be expected for the titration of 100 ml of 0.010 M H_2S with 1.0 M NaOH.

51. What type of titration curve would be expected for the titration of a weak acid with a weak base? Is such a titration feasible?

52. The pH of blood is 7.4. Assuming that the buffer in blood is carbon dioxide, hydrogen carbonate ion, calculate the ratio of conjugate base to acid necessary to maintain blood at its proper pH. What would be the effect of rapid, forced breathing (panting) on the pH of blood?

53. Calculate the ammonia concentration of a solution prepared by dissolving 0.150 mole of $NH_4C_2H_3O_2$ in sufficient water to make 1.0 liter of solution.

54. Calculate the pH of a $1.0 \times 10^{-8}\ M$ solution of HCl.

55. Calculate the percent error in the hydronium ion concentration made by neglecting the ionization of water in a $1.0 \times 10^{-6}\ M$ NaOH solution.

56. Calculate the pH of a solution prepared by addition of sufficient water to make 1.00 liter of solution to 0.100 mole of $HC_2H_3O_2$, 0.130 mole of $NaC_2H_3O_2$, and **(a)** 0.090 mole of NaOH, **(b)** 0.090 mole of HCl.

57. Write equilibrium constant expressions for the following equations. Show how they are related.

(a) $Na_2CO_3\ +\ HCl\ \rightleftharpoons\ NaHCO_3\ +\ NaCl$
(b) $2\,Na^+\ +\ CO_3^{2-}\ +\ H^+\ +\ Cl^-\ \rightleftharpoons\ 2\,Na^+\ +\ HCO_3^-\ +\ Cl^-$
(c) $CO_3^{2-}\ +\ H^+\ \rightleftharpoons\ HCO_3^-$

58. Determine the equilibrium carbonate ion concentration after equal volumes of 1.0 M sodium carbonate and 1.0 M HCl are mixed.

Further Concepts of Equilibrium

In this chapter, the concepts of equilibrium will be extended to include (1) acid-base reactions of a more general type, as, for example, the reactions of metal ions with electron pair donor molecules; (2) equilibria between ions in solution and undissolved ionic solids; and (3) systems involving more than one equilibrium reaction occurring simultaneously. Finally, it will be shown how all the concepts of equilibrium in solution can be applied in the systematic study of the chemistry of aqueous ions. It will be seen that thermodynamic data are very useful for estimating equilibrium constants which cannot be determined directly. Conversely, the equilibrium state can often be used to deduce values for free energy changes and entropy changes of processes which occur in aqueous solution.

This chapter completes the survey of the various types of reactions which occur between ions in aqueous solution. Inasmuch as all such reactions proceed to a state of equilibrium, it is important to know which equilibrium constant(s) is(are) applicable to a given system. Furthermore, the principles studied thus far can be applied to chemical analysis as a means of separating and identifying various ions in a mixture. Also, the same principles can be applied fruitfully in the preparation of pure elements and compounds.

8-1 Lewis Theory of Acids and Bases

When the reactions for the protonation of water and of ammonia are written using electron dot formulas, it is apparent that the proton becomes attached to an unshared pair of electrons on the oxygen and nitrogen atoms, respectively. Such unshared pairs of electrons are called **lone pairs.**

$$\text{H}^+ + \overset{\cdot\cdot}{\underset{\text{H}}{\text{:O:H}}} \rightleftharpoons \left[\text{H}\overset{\cdot\cdot}{\underset{\text{H}}{\text{:O:H}}}\right]^+$$

$$\text{H}^+ + \overset{\text{H}}{\underset{\text{H}}{\text{:N:H}}} \rightleftharpoons \left[\text{H}\overset{\text{H}}{\underset{\text{H}}{\text{:N:H}}}\right]^+$$

When copper(II) ion solution is treated with excess aqueous ammonia, the color changes from light blue to an intensely dark blue. The intense color is due to the presence of a species which has the formula $Cu(NH_3)_4^{2+}$. The reaction between copper(II) ion and ammonia molecules in aqueous solution may be represented in a manner analogous to the reaction of ammonia with the proton:

$$Cu^{2+} \;+\; 4 \; :NH_3 \;\rightleftharpoons\; \left[\begin{array}{c} \overset{..}{N}H_3 \\ H_3N : \overset{..}{C}u : NH_3 \\ \underset{..}{N}H_3 \end{array} \right]^{2+}$$

In all three equations given above, the formation of covalent bonds is represented as the sharing of a pair of electrons donated by only one of the atoms of the bonded pair. Bonds formed in this manner are called **coordinate covalent bonds.**

   In 1923, G. N. Lewis suggested that these types of reaction may be used for a new definition of acids and bases. According to his definition, a **Lewis acid** is any substance which can accept a share of a lone pair of electrons to form a covalent bond. A **base** is any substance which can donate a pair of electrons to be shared. These definitions are much broader than those of the Brønsted-Lowry system. All of the substances considered to be bases in the Brønsted-Lowry theory are capable of providing an electron pair and are therefore bases in the Lewis system. In the Lewis system, the acids include the proton itself, as well as metal ions and even some neutral molecules. For example, because the boron atom in boron trifluoride lacks a complete octet of electrons, the molecule can act as a Lewis acid by forming a coordinate covalent bond with an electron pair donor, such as ammonia:

$$\begin{array}{ccccc} & :\overset{..}{F}: & & H & \\ :\overset{..}{F}:\overset{..}{B} & + & :\overset{..}{N}:H & \rightleftharpoons & :\overset{..}{F}:\overset{..}{B}:\overset{..}{N}:H \\ & :\overset{..}{F}: & & H & \end{array}$$

$$BF_3(g) \;+\; NH_3(g) \;\rightleftharpoons\; F_3BNH_3(s)$$

As this example shows, acid-base reactions in the Lewis system are not limited to reactions involving the transfer of protons or to reactions occurring in aqueous solution.

8-2   Complex Ions

Complex compounds, often referred to as **coordination compounds,** are substances consisting of combinations of species which are capable of independent existence, but which have properties, even in solution, which are not merely those of the simpler substances. For example, anhydrous $CuSO_4$, a white solid, reacts with water to form the familiar blue hydrated copper(II) sulfate, $CuSO_4 \cdot 5H_2O$, in which the copper ion is bonded to four water molecules as $Cu(H_2O)_4^{2+}$ ion. Ions such as this one are called **complex ions,** and may be explained as the products of the formation of coordinate covalent bonds in Lewis acid-base reactions.

   In aqueous solution, the metal ions, which are Lewis acids, form coordinate covalent bonds with water, a Lewis base. Relatively large ions of low charge, such as the metal ions of periodic groups I A and II A, form loosely bonded hydrates of uncertain composition. Ions having relatively small size and/or high charge, especially the transition metal ions, form hydrates of considerable stability. (It should be emphasized that in aqueous solution, some water is always associated with any dissolved species, and even loosely bonded water may influence the magnitudes of equilibrium constants.) However, for convenience, just as the symbol H^+ is often used to represent H_3O^+, symbols such as Cu^{2+} or $Cu^{2+}(aq)$ are often used instead of formulas showing water explicitly, e.g., $Cu(H_2O)_4^{2+}$.

In an aqueous solution, other molecules or ions which are capable of donating electron pairs may replace some or all of the coordinately bonded water molecules. The bonded water molecules and other such electron pair donors are called **ligands.** The number of atoms which are covalently bonded to a given metal ion is called the **coordination number** of the metal ion. The coordination number is characteristic of a given ion in a given oxidation state; for example, silver(I) ion always has a coordination number of 2. The most commonly observed coordination numbers are 6, 4, and 2. Theoretical explanations of the coordination numbers and further descriptions of the structures of coordination compounds will be given in Chapter 14.

The charge on a complex ion is the net of the charge of the metal ion plus the charges of the ligands. These concepts are illustrated in the following examples:

Ligand Type	M^I	M^{II}	M^{III}
neutral	$[Ag(NH_3)_2]^+$	$[Cu(H_2O)_4]^{2+}$	$[Cr(NH_3)_6]^{3+}$
mononegative	$[Ag(CN)_2]^-$	$[CuCl_4]^{2-}$	$[FeF_6]^{3-}$

In naming complex ions, the ligands are named first. The number of like ligands is denoted by a Greek prefix, such as *di-, tri-, tetra-, penta-,* and *hexa-,* for 2, 3, 4, 5, and 6, respectively. If the ligand is a neutral molecule, its name is used unchanged, except for common ligands such as water, called **aquo;** ammonia, called **ammine;** and carbon monoxide, called **carbonyl.** For ligands which are originally negative ions, the ordinary ending is changed to *o,* as, for example, **chloro** for Cl^- and **sulfato** for SO_4^{2-}. To designate that a complex ion is an anion, the ending *ate* is used as part of its name. The oxidation number of the central metal ion is appended as a Roman numeral in parentheses to the name of the complex ion. The following examples illustrate these rules:

$[Cu(NH_3)_4]^{2+}$	is	tetraamminecopper(II) ion
$[Fe(CN)_6]^{3-}$	is	hexacyanoferrate(III) ion
$[Zn(H_2O)_2(OH)_2]$	is	dihydroxodiaquozinc(II)

More detailed rules for naming complex ions will be given in Section 14-3. The rules given here are sufficient for naming complex ions involved in the equilibria described in this chapter.

8-3   Formation Constants

The equation for the formation (or dissociation) of a complex ion is written in a manner analogous to any other type of acid-base reaction. When the complex ion is regarded as the product of the reaction, the corresponding equilibrium constant is called the **formation constant** of that ion. Thus for the formation of tetraamminecopper(II) ion by the stepwise replacement of water from tetraaquocopper(II) ion, the equilibria are written as follows:

$$[Cu(H_2O)_4]^{2+} + NH_3 \overset{1}{\rightleftharpoons} [Cu(H_2O)_3(NH_3)]^{2+} + H_2O$$

$$[Cu(H_2O)_3(NH_3)]^{2+} + NH_3 \overset{2}{\rightleftharpoons} [Cu(H_2O)_2(NH_3)_2]^{2+} + H_2O$$

$$[Cu(H_2O)_2(NH_3)_2]^{2+} + NH_3 \overset{3}{\rightleftharpoons} [Cu(H_2O)(NH_3)_3]^{2+} + H_2O$$

$$[Cu(H_2O)(NH_3)_3]^{2+} + NH_3 \overset{4}{\rightleftharpoons} [Cu(NH_3)_4]^{2+} + H_2O$$

The equilibrium constant expression for each of these equilibria is written with the concentration of water, $[H_2O]$, incorporated into the equilibrium constant:

$$K_1 = \frac{[Cu(H_2O)_3(NH_3)^{2+}]}{[Cu(H_2O)_4{}^{2+}][NH_3]}$$

$$K_2 = \frac{[Cu(H_2O)_2(NH_3)_2{}^{2+}]}{[Cu(H_2O)_3(NH_3)^{2+}][NH_3]}$$

$$K_3 = \frac{[Cu(H_2O)(NH_3)_3{}^{2+}]}{[Cu(H_2O)_2(NH_3)_2{}^{2+}][NH_3]}$$

$$K_4 = \frac{[Cu(NH_3)_4{}^{2+}]}{[Cu(H_2O)(NH_3)_3{}^{2+}][NH_3]}$$

The overall formation constant, which represents the reaction

$$[Cu(H_2O)_4]^{2+} + 4\,NH_3 \rightleftharpoons [Cu(NH_3)_4]^{2+} + 4\,H_2O$$

is the product of the stepwise formation constants:

$$K = \frac{[Cu(NH_3)_4{}^{2+}]}{[Cu(H_2O)_4{}^{2+}][NH_3]^4} = 1 \times 10^{12} = K_1 \cdot K_2 \cdot K_3 \cdot K_4$$

Overall formation constants at 25°C of several complex ions are given in Table 8–1.
The large value of the formation constant of $[Cu(NH_3)_4]^{2+}$ indicates that in the presence of excess ligand, the reaction is virtually complete, or conversely that the

TABLE 8-1. Overall Formation Constants, K_f, for Selected Complex Ions in Water at 25°C

Ammine Complexes		Cyano Complexes		Halo Complexes	
$[Ag(NH_3)_2]^+$	1×10^8	$[Ag(CN)_2]^-$	1×10^{21}	$[AgCl_2]^-$	3×10^5
$[Cu(NH_3)_4]^{2+}$	1×10^{12}	$[Fe(CN)_6]^{3-}$	1×10^{31}	$[AlF_6]^{3-}$	7×10^{19}
$[Zn(NH_3)_4]^{2+}$	5×10^8	$[Fe(CN)_6]^{4-}$	1×10^{24}	$[CuCl]^+$	1.0
		$[Ni(CN)_4]^{2-}$	1×10^{30}	$[CuCl_4]^{2-}$	4×10^5
Hydroxo Complexes		$[Zn(CN)_4]^{2-}$	5×10^{16}	$[CuCl_2]^-$	5×10^4
				$[PbCl_4)]^{2-}$	4×10^2
$[Al(OH)_4]^-$	2×10^{28}			$[SnF_6]^{2-}$	1×10^{25}
$[Zn(OH)_4]^{2-}$	5×10^{14}				

complex ion is stable toward dissociation. In the following discussions it will generally be assumed that excess ligand is present, and the intermediate steps in the formation or dissociation of complex ions will not be considered.

Example

Calculate the concentration of $[Cu(H_2O)_4]^{2+}$ (represented by $[Cu^{2+}]$) in 1.0 liter of a solution made by dissolving 0.10 mole of Cu^{2+} in 1.00 M aqueous ammonia.

$$Cu^{2+} + 4 NH_3 \rightleftharpoons [Cu(NH_3)_4]^{2+}$$

$$K = \frac{[Cu(NH_3)_4{}^{2+}]}{[Cu^{2+}][NH_3]^4} = 1 \times 10^{12}$$

The reaction between Cu^{2+} and NH_3 goes almost to completion, as is suggested by the large magnitude of the formation constant. Thus the equilibrium concentration of the tetraamine complex is approximately 0.10 M. (If x moles per liter of Cu^{2+} ion still remains uncoordinated, the concentration of the complex will be 0.10 − x.) With the formation of 0.10 M $Cu(NH_3)_4{}^{2+}$, which requires 4 moles of NH_3 per mole of complex, the ammonia concentration becomes 1.00 − 4(0.10) = 0.60 M. (Dissociation of the complex provides $4x$ moles of NH_3 per x moles of Cu^{2+}.) In summary:

	Initial Concentration	Used Up	Produced	Equilibrium Concentration
$[Cu^{2+}]$	0.10	(0.10 − x)		x
$[NH_3]$	1.00	4(0.10 − x)		0.60 + 4x ≅ 0.60
$[Cu(NH_3)_4{}^{2+}]$	0.00		0.10 − x	0.10 − x ≅ 0.10

$$K = \frac{0.10}{x(0.60)^4} = 1 \times 10^{12}$$

$$x = 8 \times 10^{-13} M$$

8-4 Amphoterism

A substance which can react either as an acid or a base is said to be **amphoteric.** Amphoterism can be exhibited by complex ions as well as simple ions. In a complex ion containing water ligands, when the bonds between the metal ion and the oxygen atoms of the water molecules are strong, the hydrated ion may behave as a Brønsted-Lowry acid. For example, owing to the reaction

$$[Al(H_2O)_6]^{3+} + H_2O \rightleftharpoons H_3O^+ + [Al(H_2O)_5(OH)]^{2+}$$

solutions of aluminum salts in water are acidic. If the equilibrium hydronium ion concentration of the solution is lowered by the addition of base, further loss of protons occurs stepwise until a precipitate of hydrated aluminum hydroxide is formed:

$$[Al(H_2O)_6]^{3+} + 3 OH^- \rightleftharpoons [Al(H_2O)_3(OH)_3](s) + 3 H_2O$$

Aluminum hydroxide is amphoteric, and further addition of base forms a soluble complex anion:

$$[Al(H_2O)_3(OH)_3](s) \; + \; OH^- \; \rightleftharpoons \; [Al(H_2O)_2(OH)_4]^- \; + \; H_2O$$

Addition of acid to the solid aluminum hydroxide yields the soluble hexaaquo-aluminum ion:

$$[Al(H_2O)_3(OH)_3](s) \; + \; 3\,H_3O^+ \; \rightleftharpoons \; [Al(H_2O)_6]^{3+} \; + \; 3\,H_2O$$

Other amphoteric hydroxides include $[Zn(H_2O)_2(OH)_2]$, $[Sn(H_2O)_2(OH)_2]$, and $[Cr(H_2O)_3(OH)_3]$. Often, for convenience, the water of hydration of these compounds is not shown, and the reactions are represented by the following type of equations:

$$Zn(OH)_2 \; + \; 2\,H_3O^+ \; \rightleftharpoons \; Zn^{2+} \; + \; 4\,H_2O$$
$$Zn(OH)_2 \; + \; 2\,OH^- \; \rightleftharpoons \; [Zn(OH)_4]^{2-}$$

If an ion forms an amphoteric hydroxide, this property can be used to separate it from a mixture containing ions which do not form amphoteric hydroxides. For example, an excess of base may be added to a mixture of Al^{3+} and Fe^{3+}, causing the iron to precipitate as $Fe(OH)_3$ and also forming the soluble $Al(OH)_4(H_2O)_2{}^-$ ion. After the insoluble $Fe(OH)_3$ is removed by filtration, careful addition of acid to the filtrate will cause precipitation of $Al(OH)_3(H_2O)_3$.

8-5   Solubility

Various types of solutions and their properties will be discussed in Chapter 16. For the present, discussion will be limited to aqueous solutions of electrolytes. In many cases, the quantity of solute held by a given quantity of solvent at a given temperature is limited. For example, if a solid is added gradually to a given quantity of water, it dissolves until a point is reached where there is no net increase in the concentration of dissolved materal no matter how much undissolved substance may be present. The solution is then said to be **saturated.** This state can be shown to be an equilibrium state in one of the following ways: (1) A solution may be prepared which contains no undissolved solute, and the solvent may be allowed to evaporate at constant temperature until some solute precipitates from the solution. The concentration of the solute in solution at this point will be identical to that of the solution to which excess solute was added. If more solvent is removed, no further change in concentration will occur. (2) A mixture which contains an excess of undissolved solute is heated until all of the excess dissolves. Then the solution is allowed to cool to the original temperature. The excess solute will precipitate and the concentration of the solution will again be that of the original solution containing an excess of undissolved solute.[1]

The concentration of a substance in a saturated solution is called the **solubility** of the substance at that temperature.

---

[1] In some cases, after cooling to the original temperature the excess solute does not precipitate. The solution is said to be **supersaturated.** Supersaturated solutions are inherently unstable, and shaking the solution or scratching the container with a glass rod or adding a minute quantity of solid solute will cause precipitation of the excess solute.

8-6   Solubility Product Constants

When an ionic solid is dissolved in water, its ions are the actual solutes. For example, if solid silver chloride is shaken with water, the following equilibrium is established:

$$AgCl(s) \rightleftharpoons Ag^+(aq) + Cl^-(aq)$$

Similarly, when solutions of silver ion and chloride ion are mixed, equilibrium is approached from the reverse direction:

$$Ag^+(aq) + Cl^-(aq) \rightleftharpoons AgCl(s)$$

The extent to which either reaction occurs is determined by the equilibrium constant expression for the solution process, which can be written as follows:

$$K = \frac{[Ag^+][Cl^-]}{[AgCl]}$$

However, since the concentration (activity) of solid silver chloride is constant, it can be incorporated into a new constant, defined by the equation

$$K_{sp} = [Ag^+][Cl^-]$$

where K_{sp} is called the **solubility product constant.** By convention, this equilibrium constant applies to the process in which the forward reaction is the dissolving of the ionic solid. Some K_{sp} values are listed in Table 8-2.

For slightly soluble ionic substances having unequal numbers of cations and anions, the equilibrium constant expression is written with exponents equal to the coefficients in the balanced equation for the solution process. This procedure is shown in the following example:

TABLE 8-2

TABLE 8-2. Solubility Product Constants at 25°C

AgBr	5×10^{-13}	CdS	8×10^{-27}	Mn(OH)$_2$	4×10^{-14}
AgCl	1×10^{-10}	CoS	3×10^{-26}	MnS	2.5×10^{-10}
Ag$_2$CrO$_4$	9×10^{-12}	CuI	5×10^{-12}	Ni(OH)$_2$	2×10^{-16}
AgI	1.5×10^{-16}	Cu(OH)$_2$	1×10^{-19}	NiS	2×10^{-21}
AgOH	1.5×10^{-8}	CuS	8.5×10^{-36}	PbCO$_3$	3.3×10^{-14}
Ag$_2$S	1.6×10^{-49}	Fe(OH)$_2$	1.6×10^{-14}	PbCl$_2$	1.7×10^{-4}
Al(OH)$_3$	2×10^{-33}	Fe(OH)$_3$	1.1×10^{-36}	PbF$_2$	3.6×10^{-8}
BaCO$_3$	5×10^{-9}	FeS	3.7×10^{-19}	PbI$_2$	1.4×10^{-8}
BaCrO$_4$	2.4×10^{-10}	Hg$_2$Cl$_2$	2×10^{-18}	PbSO$_4$	2×10^{-8}
BaF$_2$	1.7×10^{-6}	Hg$_2$I$_2$	1.2×10^{-28}	SrSO$_4$	2.8×10^{-7}
BaSO$_4$	1×10^{-10}	MgCO$_3$	2.6×10^{-5}	Zn(OH)$_2$	1.8×10^{-14}
CaCO$_3$	1×10^{-8}	MgF$_2$	6.5×10^{-9}	ZnS	1.2×10^{-22}
CaF$_2$	3.4×10^{-11}	Mg(OH)$_2$	1.2×10^{-11}		
CaSO$_4$	2×10^{-4}				

$$Mg(OH)_2 \rightleftharpoons Mg^{2+} + 2 OH^-$$

$$K_{sp} = [Mg^{2+}][OH^-]^2$$

If the molar concentrations of the ions of a slightly soluble substance are known, the numerical value of K_{sp} may be obtained. Conversely, if K_{sp} is known, the concentrations of the ions in equilibrium with the undissolved solid may be calculated.

Example

When a sample of solid AgCl is shaken with water at 25°C, a solution containing $1.0 \times 10^{-5} M$ silver ions is produced. Calculate K_{sp}.

$$AgCl(s) \rightleftharpoons Ag^+ + Cl^-$$

$$K_{sp} = [Ag^+][Cl^-]$$
$$[Ag^+] = [Cl^-] = 1.0 \times 10^{-5}$$
$$K_{sp} = (1.0 \times 10^{-5})(1.0 \times 10^{-5}) = 1.0 \times 10^{-10}$$

Example

Calculate the solubility of $Mg(OH)_2$ in water. $K_{sp} = 1.2 \times 10^{-11}$

$$Mg(OH)_2 \rightleftharpoons Mg^{2+} + 2 OH^-$$

$$K_{sp} = [Mg^{2+}][OH^-]^2$$

Let $[Mg^{2+}] = x$
Then $[OH^-] = 2x$

$$K_{sp} = (x)(2x)^2 = 4x^3 = 1.2 \times 10^{-11}$$

Note that $2x$ is THE concentration of OH^-, and in this example $[OH^-]$ is twice the concentration of magnesium ion. According to the expression for K_{sp}, it is THE hydroxide ion concentration which must be squared, no matter what its value.

$$x = 1.4 \times 10^{-4} = [Mg^{2+}]$$
$$2x = 2.8 \times 10^{-4} = [OH^-]$$

To check,

$$K_{sp} = (1.4 \times 10^{-4})(2.8 \times 10^{-4})^2 = 1.1 \times 10^{-11}$$

8-7 Applications of K_{sp}

One important use of the solubility product constant is the prediction of the concentrations of ions remaining in solution after a precipitation reaction.

Example

If 50.0 ml of a solution containing 0.0010 mole of silver ion is mixed with 50.0 ml of 0.100 M HCl solution, how much silver ion remains in solution?

The initial quantity of Cl^- ion is given by

$$(50.0 \text{ ml})(0.100 \ M) = 5.00 \text{ mmole} = 0.00500 \text{ mole } Cl^-$$

At equilibrium, let $x =$ number of moles of silver ion remaining in solution. Therefore $0.0010 - x \cong 0.0010$ mole of silver ion and an equal fraction of a mole of chloride ion have precipitated, leaving 0.0040 mole of chloride ion remaining in solution. The total volume of the solution is 100.0 ml.

$$[Cl^-] = \frac{0.0040 \text{ mole}}{0.100 \text{ liter}} = 0.040 \ M$$

$$K_{sp} = [Ag^+][Cl^-] = [Ag^+](0.040) = 1 \times 10^{-10}$$

$$[Ag^+] = \frac{1 \times 10^{-10}}{0.040} = 2.5 \times 10^{-9} \ M$$

$$x = (2.5 \times 10^{-9} \ M)(0.100 \text{ liter}) = 2.5 \times 10^{-10} \text{ mole}$$

K_{sp} is also used to determine whether a precipitate will form when two solutions containing ions which could form a slightly soluble ionic solid are mixed. For precipitation to occur, the product of the concentrations of the ions involved raised to the appropriate powers must exceed the K_{sp} of the insoluble substance. If the concentrations of ions in solution exceed those required for equilibrium, the system will shift in the direction to achieve equilibrium, and precipitation should occur.

Example

Determine whether a precipitate will form when 100 ml of 0.100 M Pb^{2+} solution is mixed with 100 ml of 0.30 M Cl^- solution.

Assuming that no precipitate forms and the final volume is 200 ml, the lead and chloride ion concentrations would be 0.050 and 0.15 M, respectively.

$$PbCl_2 \rightleftharpoons Pb^{2+} + 2 Cl^-$$

$$K_{sp} = [Pb^{2+}][Cl^-]^2 = 1.7 \times 10^{-4}$$

But if there were no precipitation,

$$[Pb^{2+}][Cl^-]^2 = (0.050)(0.15)^2 = 1.1 \times 10^{-3}$$

Since this product is greater than the value of K_{sp}, the concentrations of ions exceed that required for equilibrium, and precipitation will occur.

Example

What is the maximum pH of a solution 0.10 M in Mg^{2+} from which $Mg(OH)_2$ will not precipitate?

$$Mg(OH)_2 \rightleftharpoons Mg^{2+} + 2\,OH^-$$

$$K_{sp} = [Mg^{2+}][OH^-]^2 = 1.2 \times 10^{-11}$$

Any OH^- concentration higher than that contained in a saturated solution would cause precipitation. Hence, the solution must be at the point of attaining equilibrium, and the concentrations of ions in solution must be no greater than those required to satisfy the solubility product constant. In this solution $[Mg^{2+}] = 0.10\ M$.

$$[Mg^{2+}][OH^-]^2 = 1.2 \times 10^{-11}$$

$$[OH^-]^2 = \frac{1.2 \times 10^{-11}}{0.10} = 1.2 \times 10^{-10}$$

$$[OH^-] = 1.1 \times 10^{-5}$$

$$pOH = 4.96$$

$$pH = 9.04$$

A very important application of the solubility product concept is in devising experimental procedures for separating mixtures of ions in solution. If the K_{sp} values for various salts of these ions are known, precipitating reagents can be added in a sequence such that the least soluble substance will precipitate first. After the solution is filtered, addition of a second reagent will cause precipitation of a second, etc. To effect good separations, the precipitating agents must be chosen so that there are large differences in the solubilities of the precipitates. A variation of this technique is the separation of a mixture of ions by means of a single precipitating agent under conditions of controlled concentration. For example, suppose it is desired to separate Fe^{2+} from Mg^{2+} in a solution in which each is $0.10\ M$. According to the example worked above, Mg^{2+} ion will not precipitate from a $0.10\ M$ solution as the hydroxide unless the pH exceeds 9.04. At pH 9.00 the hydroxide ion concentration is 1.0×10^{-5}. From the K_{sp} of $Fe(OH)_2$, it can be shown that almost all the iron(II) will precipitate from a solution buffered at pH 9.00, while all the magnesium ion will remain in solution:

$$Fe(OH)_2 \rightleftharpoons Fe^{2+} + 2\,OH^-$$

$$K_{sp} = [Fe^{2+}][OH^-]^2 = 1.6 \times 10^{-14}$$
$$[Fe^{2+}](1.0 \times 10^{-5})^2 = 1.6 \times 10^{-14}$$
$$[Fe^{2+}] = 1.6 \times 10^{-4}$$

8-8   Determination of K_{sp} Using Galvanic Cells

Many substances are so slightly soluble that the concentrations of the ions in equilibrium with the solid cannot be determined by the usual type of chemical analysis. For example, the solubility product constant for mercury(II) sulfide is 1×10^{-50}. This value of K_{sp} corresponds to one mercury(II) ion and one sulfide ion in 17 liters of solution! In cases such as this, K_{sp} must be determined indirectly by application of the principles of thermodynamics. In Section 6–5 it was shown that the equilibrium constant is related to the standard free energy change and the standard cell potential as follows:

$$-\Delta G^\circ = nF\epsilon^\circ = 2.30\, RT \log K = 0.0592F \log K$$

If a galvanic cell is designed such that the cell reaction corresponds to the desired equilibrium, a numerical value for K can be calculated from the measured value of ϵ°:

$$\log K = \frac{n\epsilon^\circ}{0.0592}$$

This equation applies to precipitation equilibria as well as to equilibria between substances in solution or in the gas state.

Example

Design a cell by means of which K_{sp} for AgI may be determined. The desired equilibrium reaction is

$$AgI(s) \;\rightleftharpoons\; Ag^+(aq) \;+\; I^-(aq)$$

This reaction can be made to occur in the following cell:

$$Ag \mid Ag^+(1\,M) \parallel I^-(1\,M) \mid AgI \mid Ag$$

At 25°C the electrode reactions and ϵ° values are as follows:

$$
\begin{array}{lll}
AgI + e^- \rightleftharpoons Ag + I^- & \epsilon^\circ = -0.151\ V \\
Ag \rightleftharpoons Ag^+ + e^- & \epsilon^\circ = -0.799\ V \\
\hline
\text{cell:}\quad AgI \rightleftharpoons Ag^+ + I^- & \epsilon^\circ_{cell} = -0.950\ V
\end{array}
$$

$$\log K = \frac{n\epsilon^\circ}{0.0592} = \frac{-0.950}{0.0592} = -16.05$$

$$K = 8.9 \times 10^{-17} = K_{sp}$$

The relationship between the equilibrium constant and the standard reduction potential may also be used to obtain ϵ° values from K_{sp}.

Example

Estimate the standard reduction potential for the copper/copper sulfide electrode. For CuS, $K_{sp} = 8.5 \times 10^{-36}$.

One can design a cell in which the half reaction $CuS + 2\,e^- \rightleftharpoons Cu + S^{2-}$ occurs. The following cell is suitable:

$$Cu \mid Cu^{2+} \parallel S^{2-} \mid CuS \mid Cu$$

For this cell

$$
\begin{array}{lll}
CuS + 2\,e^- \rightleftharpoons Cu + S^{2-} & \epsilon^\circ_{Cu/CuS} = ? \\
Cu \rightleftharpoons Cu^{2+} + 2\,e^- & \epsilon^\circ_{Cu/Cu^{2+}} = -0.34\ V \\
\hline
\text{cell:}\quad CuS \rightleftharpoons Cu^{2+} + S^{2-} & \epsilon^\circ_{cell} = \dfrac{0.0592}{2}\log K
\end{array}
$$

$$\epsilon^\circ_{cell} = \epsilon^\circ_{Cu/Cu^{2+}} + \epsilon^\circ_{Cu/CuS} = \frac{0.0592}{2} \log(8.5 \times 10^{-36}) = -1.04 \text{ V}$$

$$\epsilon^\circ_{Cu/CuS} = -1.04 \text{ V} - (-0.34 \text{ V}) = -0.70 \text{ V}$$

8-9   Limitations on the Use of K_{sp}

There are cases where the solubility product principle does not seem to apply:

1. Although in principle a precipitate should form whenever the product of the concentrations of the ions raised to the appropriate powers exceeds K_{sp}, it takes about 10^{-5} gram of solid per milliliter to give a visible turbidity to a water solution. If there is only a minute excess of reagent present, the quantity of precipitate formed may be too small to be seen with the naked eye.
2. In some cases, even when K_{sp} is exceeded, the precipitation reaction occurs so slowly that supersaturated solutions are formed. Unless deliberate measures are taken to encourage the formation of precipitate, such as vigorous stirring or heating followed by cooling, the onset of precipitation may be delayed indefinitely.
3. Competitive reactions may occur to such an extent that the K_{sp} of the material is not exceeded even though a large excess of precipitating reagent is added to the solution. For example, if carbonate ion is used as a precipitating agent, it may react with water as follows:

$$CO_3^{2-} + H_2O \rightleftharpoons HCO_3^- + OH^-$$

Therefore, the actual carbonate ion concentration in solution may not be sufficient to exceed K_{sp}. Similarly in water the reaction

$$S^{2-} + H_2O \rightleftharpoons HS^- + OH^-$$

occurs extensively in neutral or acid solution. Since this reaction reduces the sulfide ion concentration, the equilibrium between an insoluble sulfide and its ions will be shifted toward the formation of more of the ions in solution. Hence, a number of metal sulfides dissolve to a greater extent than would be predicted from their K_{sp} values by assuming no other reaction takes place.
4. Upon addition of a large excess of precipitating agent, further reaction leading to the formation of soluble complex ions may occur. In such cases the original precipitate may dissolve. For example, if too great an excess of concentrated HCl is used to precipitate silver chloride or lead chloride, the following reactions will occur:

$$AgCl(s) + Cl^- \rightleftharpoons AgCl_2^-(aq)$$
$$PbCl_2(s) + 2 Cl^- \rightleftharpoons PbCl_4^{2-}(aq)$$

5. The K_{sp} principle applies rigorously only for slightly soluble salts, where the concentrations of ions in solution are small. In relatively dilute solutions the concentrations of the ions approach their activities. In the case of moderately soluble salts, where the activities differ from the concentrations, the equilibria

between ions and undissolved salt cannot be described in terms of their concentrations. Even in the cases of very slightly soluble salts, if an appreciable concentration of other ions is also present in solution, interionic attractions and hydration effects change the activities of all the solutes to such an extent that simple K_{sp} considerations no longer apply exactly.

8-10   Simultaneous Equilibria

In practically all aqueous solutions, several equilibria involving a given species occur simultaneously. The distribution of the solutes as reactants and products of the various equilibria is such that every equilibrium constant for the solution must be satisfied. For example, for 1 mole of weak acid dissolved in 1 liter of water, the concentrations of HA, H_3O^+, and A^- are governed by the magnitude of K_a. Simultaneously, the magnitudes of the H_3O^+ and OH^- concentrations are governed by K_w. The hydronium ion concentration must be such that at equilibrium both of these equilibrium constants are satisfied.

Examples of simultaneous equilibria which have already been discussed include the successive ionizations of polyprotic acids and the stepwise formation of complex ions. In both of these cases the stepwise equilibria could be combined to give an overall reaction for which the overall constant is the product of the individual constants:

$$K_{overall} = K_1 \cdot K_2 \cdot K_3 \cdot K_4 \cdots$$

A useful application of the above principles is encountered in the precipitation of insoluble metal sulfides by saturation of a solution with gaseous H_2S. The overall dissociation constant for H_2S, corresponding to the loss of both protons, is

$$K_{1,2} = K_1 \cdot K_2 = \frac{[H_3O^+]^2[S^{2-}]}{[H_2S]} = 1 \times 10^{-21}$$

When an aqueous solution is saturated with H_2S at 1.0 atm pressure and 25°C, the concentration of dissolved H_2S is 0.10 M. Hence

$$\frac{[H_3O^+]^2[S^{2-}]}{0.10} = 1 \times 10^{-21}$$

$$[H_3O^+]^2[S^{2-}] = 1 \times 10^{-22} \qquad \text{(for saturated H_2S solutions)}$$

$$[S^{2-}] = \frac{1 \times 10^{-22}}{[H_3O^+]^2}$$

The last equation states that the sulfide ion concentration of a solution saturated with H_2S is inversely proportional to the square of the hydronium ion concentration. By adjusting the hydronium ion concentration, it is possible to obtain a wide range of sulfide ion concentrations; hence various metal sulfides can be selectively precipitated.

Example

Calculate what hydronium ion concentration must be maintained in order to just prevent precipitation of ZnS from a solution which contains 0.010 M Zn^{2+} and 0.010 M Cd^{2+} when the solution is saturated with H_2S. What concentration of Cd^{2+} will remain in the solution under these conditions?

$$K_{sp} = [Zn^{2+}][S^{2-}] = 1.2 \times 10^{-22}$$
$$[Zn^{2+}] = 0.010 \ M$$
$$[S^{2-}] = 1.2 \times 10^{-20}$$

This sulfide ion concentration is the maximum possible without precipitation of ZnS. For that sulfide ion concentration, the hydronium ion concentration can be calculated from the equation presented above for saturated H_2S solutions:

$$[H_3O^+]^2[S^{2-}] = 1 \times 10^{-22}$$
$$[H_3O^+]^2 \cong 8 \times 10^{-3}$$
$$[H_3O^+] \cong 0.09 \ M$$

For cadmium sulfide, $K_{sp} = 8 \times 10^{-27}$. The product of the sulfide ion concentration and the original cadmium ion concentration is 1.2×10^{-22}, which exceeds the value of K_{sp}, and therefore CdS is expected to precipitate. The cadmium ion left in solution will be determined by K_{sp} for CdS.

$$[Cd^{2+}][S^{2-}] = [Cd^{2+}](1.2 \times 10^{-20}) = 8 \times 10^{-27}$$
$$[Cd^{2+}] = 7 \times 10^{-7}$$

Thus in 0.09 M H_3O^+ solution saturated with H_2S, the cadmium ion concentration will have been reduced from 0.010 M to $7 \times 10^{-7} \ M$, a factor of 7×10^{-5}, while the zinc ion concentration will be unchanged. It should be noted that during the precipitation reaction, H_2S is used up and H_3O^+ is generated:

$$Cd^{2+} \ + \ H_2S \ + \ 2 \ H_2O \ \rightleftharpoons \ CdS(s) \ + \ 2 \ H_3O^+$$

The hydronium ion concentration may be kept constant by means of a buffer, and the hydrogen sulfide concentration must be kept constant by continuously saturating the solution with H_2S gas.

8-11   Dissolving Precipitates

A precipitated ionic solid can be dissolved by removing one or more of its ions from solution, thus causing the equilibrium to shift in the direction to form more ions in solution. This procedure is simply an application of Le Châtelier's principle.

Example

Tenth molar solutions of Cu^{2+} and of Zn^{2+} are each saturated with H_2S to produce CuS and ZnS, respectively. What concentrations of H_3O^+ will cause the precipitates to redissolve completely in the saturated H_2S solution?

If the solids are completely dissolved, assuming no change in volume, the concentrations of the metal ions will be 0.10 M. Thus for ZnS

$$K_{sp} = [Zn^{2+}][S^{2-}] = 1.2 \times 10^{-22}$$
$$(0.10)[S^{2-}] = 1.2 \times 10^{-22}$$
$$[S^{2-}] = 1.2 \times 10^{-21}$$

$$\frac{[H_3O^+]^2[S^{2-}]}{[H_2S]} = 1 \times 10^{-21}$$

$$[H_3O^+]^2 = \frac{(1 \times 10^{-21})(0.10)}{1.2 \times 10^{-21}} = 8 \times 10^{-2}$$

$$[H_3O^+] = 0.3 \ M \quad \text{(to one significant figure)}$$

But for CuS,

$$[Cu^{2+}][S^{2-}] = 8.5 \times 10^{-36}$$
$$(0.10)[S^{2-}] = 8.5 \times 10^{-36}$$
$$[S^{2-}] = 8.5 \times 10^{-35}$$

$$\frac{[H_3O^+]^2[S^{2-}]}{[H_2S]} = 1 \times 10^{-21}$$

$$\frac{[H_3O^+]^2(8.5 \times 10^{-35})}{(0.10)} = 1 \times 10^{-21}$$

$$[H_3O^+] = 1 \times 10^6 \ M = 1 \text{ million molar}$$

Since it is impossible to achieve such a large hydronium ion concentration, CuS cannot be dissolved by addition of strong acid alone. To dissolve CuS, the equilibrium sulfide ion concentration must be reduced by some method other than the formation of H_2S. A suitable method is the oxidation of the sulfide ion to sulfur by means of HNO_3:

$$CuS \ \rightleftharpoons \ Cu^{2+} \ + \ S^{2-}$$
$$8 \ H^+ \ + \ 2 \ NO_3^- \ + \ 3 \ S^{2-} \ \rightarrow \ 3 \ S \ + \ 2 \ NO \ + \ 4 \ H_2O$$

Precipitates can also be dissolved by removal of the cation from solution as a complex ion.

Example

What is the minimum quantity of ammonia which must be added to 1.0 liter of solution in order to dissolve 0.10 mole of silver chloride by forming $[Ag(NH_3)_2]^+$?

$$AgCl(s) \ + \ 2 \ NH_3 \ \rightleftharpoons \ [Ag(NH_3)_2]^+ \ + \ Cl^-$$

When 0.10 mole of AgCl is converted to a complex ion in 1.0 liter of solution, 0.10 M chloride ion is also formed. The maximum concentration of silver ion which can coexist with that concentration of chloride ion is given by the K_{sp} expression:

$$K_{sp} = [Ag^+][Cl^-] = 1 \times 10^{-10}$$
$$[Ag^+] = 1 \times 10^{-9}$$

Therefore the concentration of ammonia in the solution must be sufficient to prevent the silver ion concentration from exceeding 1×10^{-9} M.

$$Ag^+ \quad + \quad 2\,NH_3 \quad \rightleftharpoons \quad Ag(NH_3)_2^+$$

$$K_f = \frac{[Ag(NH_3)_2^+]}{[Ag^+][NH_3]^2} = 1 \times 10^8$$

Since the complex ion is practically 0.10 M and the silver ion is 1×10^{-9} M, the ammonia concentration can be determined to be

$$[NH_3]^2 = \frac{0.10}{(1 \times 10^{-9})(1 \times 10^8)} = 1.0$$

$$[NH_3] = 1.0\ M$$

Therefore the minimum quantity of ammonia which must be added to 1.0 liter of solution is at least 1.2 moles; 0.20 mole for the formation of the 0.10 mole of $Ag(NH_3)_2^+$ and 1.0 mole in excess to maintain an ammonia concentration such that the equilibrium silver ion concentration is held at 1×10^{-9} M.

8-12   Entropies of Aqueous Ions

In Chapters 3 and 6 it was shown how tables of enthalpy data and entropy data are useful in calculating enthalpy changes and free energy changes for various processes. Tables of equilibrium constants, in addition to their usefulness in calculations of equilibrium processes, are useful to complement these sets of data, since

$$-\Delta G^\circ = 2.30\ RT \log K$$

Conversely, if appropriate free energy change data are available, it is possible to calculate values for equilibrium constants. Thus data on the entropies of aqueous ions could be of value in estimating the properties of solutions of these ions. As in the case of enthalpies of aqueous ions, a set of relative entropies for ions may be obtained by *defining* the entropy of the hydrogen ion, at 298 K and unit activity, as zero.

Example

Using appropriate data, determine the entropy of the OH^- ion at 298 K and unit activity. The standard enthalpy change (Tables 3–2 and 3–5) for the reaction

$$H_2O \quad \rightleftharpoons \quad H^+ \quad + \quad OH^-$$

is 13.4 kcal. The equilibrium constant for this reaction is $K_w = 1.0 \times 10^{-14}$. Hence ΔG° and ΔS° can be calculated:

$$\Delta G^\circ = -2.30\ RT \log K = 19.1\ \text{kcal/mole}$$

$$\Delta S^\circ = \frac{\Delta H^\circ - \Delta G^\circ}{T} = \frac{(13.4 - 19.1)(10^3)}{298} = -19.2\ \text{cal/mole} \cdot \text{K}$$

But

$$\Delta S° = S°_{products} - S°_{reactants}$$
$$\Delta S° = S°_{(H^+)} + S°_{(OH^-)} - S°_{(H_2O)}$$

From Table 6-1, $S°_{(H_2O)} = 16.7$ cal/mole · K

$$-19.2 = 0.0 + S°_{(OH^-)} - 16.7$$
$$S°_{(OH^-)} = -2.5 \text{ cal/mole} \cdot \text{K}$$

Some representative standard entropy values for aqueous ions are listed in Table 8–3.

Example

Using data from Tables 3–2, 3–5, 6–1, and 8–3, estimate K_{sp} for AgCl and compare the value obtained with the tabulated value, 1×10^{-10}.

The process is represented by

$$AgCl(s) \rightleftharpoons Ag^+ + Cl^-$$

$$\Delta H° = \Delta H°_{f(Ag^+)} + \Delta H°_{f(Cl^-)} - \Delta H°_{f(AgCl)}$$
$$= 25.3 + (-40.0) - (-30.36) = 15.7 \text{ kcal/mole}$$
$$\Delta S° = S°_{(Ag^+)} + S°_{(Cl^-)} - S°_{(AgCl)}$$
$$= 17.7 + 13.2 - 23.0 = 7.9 \text{ cal/mole} \cdot \text{K}$$
$$\Delta G° = \Delta H° - T\Delta S° = 15,700 - (298)(7.9) = 13,350 \text{ cal}$$

$$-\log K = \frac{\Delta G°}{2.30 \, RT} = \frac{13,350}{(2.30)(1.99)(298)} = 9.79$$
$$K = 10^{-9.79} = 1.6 \times 10^{-10}$$

For this type of data, the agreement between the values determined by two different methods is acceptable. The order of magnitude of the number is established, but its precision is poor. It is apparent from this result why many equilibrium constants are given to only one significant figure.

TABLE 8-3. Standard Entropies of Aqueous Ions at 298 K

	$S°$			$S°$	
	cal/mole · K	J/mole · K		cal/mole · K	J/mole · K
H^+	0.0	0.0	OH^-	−2.5	−10.5
Na^+	14.4	60.2	F^-	3.0	12.6
K^+	24.5	102.5	Cl^-	13.2	55.2
Ag^+	17.7	74.1	Br^-	19.3	80.8
Ba^{2+}	3.0	12.6	I^-	26.1	109.2
Ca^{2+}	−13.2	−55.2	HS^-	14.6	61.1
Cu^{2+}	−23.6	−98.7	S^{2-}	5.3	22.2
Zn^{2+}	−25.4	−106.3	SO_4^{2-}	4.1	17.2
			CO_3^{2-}	−12.7	−53.1

8-13   Systematic Chemistry of Aqueous Ions

By use of appropriate data on standard electrode potentials and the various types of equilibrium constants, many chemical properties of a given ion in aqueous solution may be predicted in a systematic way. Even in cases where concentration data are not available, equilibrium constants can be calculated, or at least estimated, from thermodynamic data. Also, using these data it is possible to devise schemes for separating a given ion from other ions which may be in the same solution. The aqueous chemistry of the zinc(II) ion will be used to illustrate how the many types of data can be used to predict chemical properties.

The value of the standard reduction potential of Zn^{2+} indicates that this ion can be prepared from the element by treatment with a strong acid. The relatively low reduction potential (-0.76 V) indicates that Zn^{2+} is not apt to be reduced to the metal by reducing agents such as H_2O_2, Sn^{2+}, HSO_3^-, or I^-. The solubility product constants indicate that the ion can be quantitatively removed from solution as $Zn(OH)_2$ or ZnS by addition of hydroxide or sulfide ions, respectively. However, $Zn(OH)_2$ is amphoteric and dissolves in excess base to form $[Zn(OH)_4]^{2-}$.

It may be calculated that $Zn(OH)_2$ is readily soluble in ammonia solution owing to the formation of the complex ion, $[Zn(NH_3)_4]^{2+}$, which is soluble. On the other hand, ZnS will not dissolve in the presence of ammonia because the K_{sp} value is such that $ZnS(s)$ provides a smaller zinc ion concentration that does $[Zn(NH_3)_4]^{2+}$. Zinc sulfide does dissolve in the presence of strong acids because the concentration of the sulfide ion is lowered by the H_3O^+ below that necessary to maintain equilibrium with solid ZnS. In contrast, weak acids such as acetic acid do not dissolve ZnS because they do not furnish sufficient hydronium ion concentrations in solution.

The principles discussed above may be applied to the preparation of pure reagents. For example, to prepare pure zinc acetate from zinc chloride in aqueous solution, sodium hydroxide is added until zinc hydroxide precipitates. The solid is separated from the resulting NaCl solution and washed with distilled water. Thus the chloride ion is completely removed. Then the zinc hydroxide is treated with acetic acid:

$$Zn(OH)_2 \;+\; 2\,HC_2H_3O_2 \;\rightleftharpoons\; Zn^{2+} \;+\; 2\,C_2H_3O_2^- \;+\; 2\,H_2O$$

The pure salt is obtained by careful evaporation of most of the water, whereupon crystallization occurs.

The same principle can be used to prepare soluble salts of other metals. In cases where K_{sp} is favorable, a metal ion, M^{2+}, can be precipitated as the carbonate rather than as the hydroxide. The carbonate anion can also be completely removed by treatment with an acid:

$$MCO_3 \;+\; 2\,H_3O^+ \;\rightleftharpoons\; M^{2+} \;+\; 3\,H_2O \;+\; CO_2$$

8-14   Qualitative Analysis

Qualitative analysis involves the separation and identification of the various ions in a mixture. The study of the subject provides many examples of the application of principles of chemical equilibrium in solving chemical problems. The laboratory procedures demand careful control of experimental conditions and they develop skill in manipulation and in visual observation of chemical reactions.

A number of successful qualitative analysis schemes have been devised. A typical scheme may be broken down into the following steps:

1. Separation of the ions into broad groups according to some property, such as solubility.
2. Separation of the individual ions within each group.
3. Identification of the individual ions by means of specific tests.

A scheme in common use, which divides the common cations into five groups, is shown in chart form as Figure 8-1. Addition of dilute HCl to a solution containing any or all of the listed cations causes precipitation of the cations of group I. These

Figure 8-1. **A Qualitative Analysis Scheme**

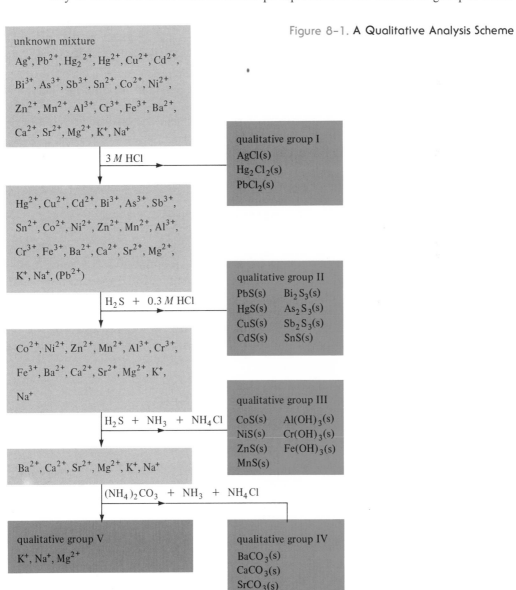

chloride precipitates are separated from the original solution containing the cations of the remaining four groups, and appropriate steps are taken to separate and identify the individual cations present.

Group II includes those cations which have sulfides with such small values of K_{sp} that the sulfide ion concentration in a solution containing 0.3 M HCl and saturated with H_2S is sufficient to precipitate them quantitatively. The cations of group III form sulfides which have larger values of K_{sp}; they do not precipitate from an acidic solution which is saturated with H_2S, but they form insoluble sulfides or hydroxides when their solutions buffered at pH 8 are saturated with H_2S. The group IV cations are precipitated as carbonates; the precipitation is done at pH values greater than 10 so that the carbonate ion concentration is maintained high enough to exceed the K_{sp} values of the metal carbonates. Group V includes those cations which have soluble chlorides, sulfides, hydroxides, and carbonates.

The separation and identification of the individual ions within each of these groups will not be described here. Details may be obtained from books on qualitative analysis. However, the principles of selective precipitation under controlled conditions, as was used to separate the groups, can be applied to separate the individual cations within each group.

For the purposes of qualitative analysis, anions can be classified as follows:

1. Conjugate bases of weak acids.
2. Oxidizing agents.
3. Reducing agents.
4. Those which form insoluble barium salts.
5. Those which form insoluble silver salts.
6. Those which form only soluble salts.

In Table 8–4 some common anions are classified according to this scheme.

By appropriate combinations of tests it is possible to identify the common anions without performing group separations. For example, carbonate ion gives an insoluble barium salt which reacts with acid to yield a gas, CO_2.

One area in which the techniques of qualitative analysis are used extensively is the field of radiochemistry (see Section 20–4). For example, if copper metal is subjected

TABLE 8–4. Classification of Some Common Anions for Qualitative Analysis

Conjugate Bases of Weak Acids	Oxidizing Agents	Reducing Agents	Insoluble Ba Salts	Insoluble Ag Salts	Generally Soluble Salts
$C_2H_3O_2^-$	CrO_4^{2-}	Br^-	AsO_4^{3-}	Br^-	$C_2H_3O_2^-$
CO_3^{2-}	MnO_4^-	I^-	$H_2BO_3^-$	Cl^-	NO_3^-
F^-	SO_4^{2-}	S^{2-}	CO_3^{2-}	I^-	NO_2^-
S^{2-}	NO_3^-	$C_2O_4^{2-}$	CrO_4^{2-}	S^{2-}	ClO_3^-
$C_2O_4^{2-}$	NO_2^-	NO_2^-	F^-	NCS^-	ClO_4^-
NO_2^-	ClO_4^-	SO_3^{2-}	$C_2O_4^{2-}$		MnO_4^-
AsO_4^{3-}	ClO_3^-		PO_4^{3-}		
$H_2BO_3^-$			SO_4^{2-}		
PO_4^{3-}			SO_3^{2-}		
NCS^-					

to a nuclear reaction, radioactive copper, radioactive zinc, and radioactive nickel might be formed. It would be desirable to identify the respective radioactive species. Since only a few thousand radioactive atoms are produced, it is impossible to isolate them from the comparatively large quantity of unreacted copper. Therefore the entire sample is dissolved, and small quantities of nonradioactive zinc and nonradioactive nickel are added to the solution. Using the techniques of qualitative analysis, the copper, zinc, and nickel are then separated. The chemical behavior of the radioactive atoms is exactly the same as that of their nonradioactive isotopes; hence the radioactivity found in the separated zinc is due to the zinc atoms produced by the nuclear reaction. Similarly, the separated nickel and copper samples will have the radioactivity corresponding to the nickel and copper isotopes produced by the nuclear reaction.

8-15   Limitations of Thermodynamics

In this and preceding chapters, models for predicting the direction of spontaneous change and for defining the equilibrium state of various reactions have been developed in terms of the concepts of thermodynamics. The power of thermodynamics lies in the fact that useful information is obtained by considering only the initial state and the final state of a system; intermediate states are not described. The description of the mode or mechanism of a chemical reaction is beyond the scope of thermodynamics. Similarly, thermodynamic data provide no direct information about the structures of molecules. Therefore much of what interests modern chemists must be described in terms of other models.

The next several chapters of the textbook will be concerned with experimental evidence and theories which are the basis for present concepts of atomic and molecular structure. There is a qualitative relationship between thermodynamics and molecular structure which is implied in the idea that the stabilities of molecules are related to the strengths of the bonds between their constituent atoms. The energy changes observed in a chemical reaction reflect the relative bond energies of the reactant and product molecules. In the strictly thermodynamic approach only large assemblies of molecules are considered. In the next several chapters models will be developed in terms of the characteristics of individual atoms and/or molecules.

8-16   Exercises

Basic Exercises

1. What is the distinction between a covalent bond and a coordinate covalent bond? How many covalent bonds and how many coordinate covalent bonds are there in the NH_4^+ ion? Is it possible to distinguish between the coordinate covalent bond(s) and the other covalent bond(s) in this ion? Explain.

2. Name the following complex ions: (a) $[PdBr_4]^{2-}$, (b) $[CuCl_2]^-$, (c) $[Au(CN)_4]^-$, (d) $[AlF_6]^{3-}$, (e) $[Cr(NH_3)_6]^{3+}$, (f) $[Zn(NH_3)_4]^{2+}$, (g) $[Fe(CN)_6]^{3-}$.

3. Determine the oxidation number of the central metal ion in each of the following: (a) $[Co(NH_3)_6]^{3+}$, (b) $Ni(CO)_4$, (c) $[CuCl_4]^{2-}$, (d) $[Ag(CN)_2]^-$, (e) $[Co(NH_3)_4(NO_2)_2]^+$.

4. Write the formula for **(a)** dichlorotetraamminerhodium(III) ion, **(b)** tetrahydroxodi-aquoaluminate(III) ion, **(c)** tetrachlorozincate(II) ion, **(d)** aluminum nitrate, **(e)** hexaamminecobalt(III) tetrachlorodiamminechromate(III).

5. Calculate the silver ion concentration in a solution of $[Ag(NH_3)_2]^+$ prepared by adding 1.0×10^{-3} mole of $AgNO_3$ to 1.0 liter of 0.100 M NH_3 solution.

6. Calculate the Fe^{2+} concentration in a solution containing 0.200 M $[Fe(CN)_6]^{4-}$ and 0.100 M CN^-.

7. Calculate the concentration of silver ion which is in equilibrium with 0.15 M $[Ag(NH_3)_2]^+$ and 1.5 M NH_3.

8. Write stepwise equations for the formation of tetrahydroxodiaquochromate(III) ion, $[Cr(H_2O)_2(OH)_4]^-$, by addition of OH^- to $[Cr(H_2O)_6]^{3+}$ in aqueous solution.

9. Calculate the nickel ion concentration of a solution prepared by shaking $Ni(OH)_2$ with water until equilibrium is established.

10. Calculate the solubility of AgCl in 0.20 M $AgNO_3$ solution.

11. Calculate the silver ion concentration in a solution prepared by shaking solid Ag_2S with saturated H_2S (0.10 M) in 0.15 M H_3O^+ until equilibrium is established.

12. Complete the following equations:
    (a) $Al(OH)_4^- \ + \ H_3O^+ \ \rightarrow$
    (b) $Al(OH)_4^- \ + \ 4 H_3O^+ \ \rightarrow$
    (c) $Ag(NH_3)_2^+ \ + \ Cl^- \ + \ 2 H_3O^+ \ \rightarrow$

13. In the quantitative determination of silver ion as AgCl, a solution of NaCl is used as the precipitating agent. Why should a large excess of that reagent be avoided? (*Hint:* See Table 8–1.)

14. Calculate the solubility of CoS in 0.10 M H_2S and 0.15 M H_3O^+.

15. Which has a greater molarity in water, AgCl or $Mg(OH)_2$? Can relative solubilities be predicted on the basis of the relative magnitudes of the K_{sp} values alone? Explain.

16. The values of K_{sp} for the slightly soluble salts MX and QX_2 are each equal to 4.0×10^{-18}. Which salt is more soluble? Explain your answer fully.

17. Using data from Tables 5–2 and 8–2, estimate the standard electrode potential for a cobalt/cobalt sulfide electrode.

18. Should a precipitate of barium fluoride be obtained when 100 ml of 0.25 M NaF and 100 ml of 0.015 M $Ba(NO_3)_2$ are mixed?

19. What mass of Pb^{2+} ion is left in solution when 50.0 ml of 0.20 M $Pb(NO_3)_2$ is added to 50.0 ml of 1.5 M NaCl?

20. Determine the mass of PbI_2 that will dissolve in **(a)** 500 ml of water, **(b)** 500 ml of 0.10 M KI solution, **(c)** 500 ml of a solution containing 1.33 grams of $Pb(NO_3)_2$.

21. Calculate the pH at which $Mg(OH)_2$ will just begin to precipitate from a 0.100 M $Mg(NO_3)_2$ solution by addition of NaOH.

22. The number of moles of CoS which dissolves per liter of solution exceeds the solubility predicted by the value of K_{sp} for CoS. Explain by means of appropriate chemical equations why this behavior is observed.

23. A certain insoluble compound of M^{2+}, when shaken with water, provides an M^{2+} concentration of 1.0×10^{-4} M. A ligand is added to the system in a quantity which forms a soluble complex with M^{2+} and leaves 1.0×10^{-6} M M^{2+} in solution. Will the insoluble compound tend to dissolve? Explain.

24. What mass of $BaSO_4$ will dissolve in 450 ml of aqueous solution?

25. The solubility of ML_2 (formula weight, 60) in water is 2.4×10^{-5} gram/100 ml of solution. Calculate the solubility product constant for ML_2.

26. Using thermodynamic data from Tables 3–2, 3–5, 6–1, and 8–3, estimate the K_{sp} of AgBr.

27. Given the reagents NH_3, NaOH, HCl, and H_2S, which one could be used to separate the ions in each of the following mixtures? **(a)** Cu^{2+} and Zn^{2+}, **(b)** Cu^{2+} and Al^{3+}, **(c)** Zn^{2+} and Al^{3+}.

General Exercises

28. If $0.10\ M\ Ag^+$ and $0.10\ M\ H_3O^+$ are present in 1.0 liter of solution to which 0.010 mole of gaseous NH_3 is added, with which cation would the ammonia react to a greater extent? Support your answer by calculation.

29. Calculate the copper ion concentration in a solution obtained by shaking CuS with saturated H_2S $(0.10\ M)$ in which $0.15\ M$ HCl is present.

30. Calculate the hydroxide ion concentration of a solution after 100 ml of $0.100\ M\ MgCl_2$ is added to 100 ml of $0.200\ M$ NaOH.

31. H_2S is bubbled into a solution containing $0.15\ M\ Cu^{2+}$ until no further change takes place. Calculate the concentrations of H_3O^+ produced and of Cu^{2+} remaining in solution.

32. Determine the copper ion concentration resulting from the addition to 500 ml of $0.050\ M\ Cu^{2+}$ solution of **(a)** gaseous H_2S until no further change takes place, **(b)** 500 ml of saturated H_2S solution.

33. Calculate the concentration of Fe^{2+} in a solution prepared by mixing 75.0 ml of $0.030\ M\ FeSO_4$ with 125.0 ml of $0.20\ M$ KCN.

34. When a solution of Zn^{2+} was added to a solution of NaOH, a clear solution was obtained. When NH_4Cl was added to the clear solution, $Zn(OH)_2$ precipitated. Using balanced chemical equations, explain these observations.

35. How many moles of NH_3 must be added to 1.0 liter of $0.750\ M\ AgNO_3$ in order to reduce the silver ion concentration to $5.0 \times 10^{-8}\ M$?

36. Write equations showing all of the equilibrium reactions occurring in aqueous solutions containing each of the following sets of reagents: **(a)** NaCl, **(b)** NaOH, **(c)** $NaC_2H_3O_2$ + $HC_2H_3O_2$, **(d)** Na_2S + CuS, **(e)** NH_4Cl + NH_3 + $Mg(OH)_2(s)$.

37. Using the value of K_{sp} for Hg_2I_2 from Table 8–2, calculate the concentrations of cation and anion in a saturated solution of Hg_2I_2 in water.

38. A mixture of water and AgCl is shaken together until a saturated solution is obtained. Then the solid is filtered, and to 100 ml of the filtrate is added 100 ml of $0.030\ M$ NaBr. Should a precipitate be formed? If so, will it be visible to the naked eye?

39. Calculate the solubility of A_2X_3 in pure water, assuming that neither kind of ion reacts with water. For A_2X_3, $K_{sp} = 1.1 \times 10^{-23}$.

40. Arrange the following solutions in order of decreasing silver ion concentration: **(a)** $1\ M\ [Ag(CN)_2]^-$, **(b)** saturated AgCl, **(c)** $1\ M\ [Ag(NH_3)_2]^+$ in $0.10\ M\ NH_3$, **(d)** saturated AgI.

41. Explain the following observations: **(a)** When a sample of H_2SO_4 is treated with an equal number of moles of a base such as NaOH, the entire sample of acid is half neutralized. **(b)** When a sample of $Ba(OH)_2$ is treated with an equal number of moles of an acid such as HCl, half of the sample of base is completely neutralized. **(c)** When a given number of moles of H_2SO_4 is treated with half the number of moles of $Ba(OH)_2$, half of the acid is completely neutralized (and the other half does not react at all, in contrast to the situation in part **(a)**).

42. Explain why $0.10\ M\ NH_3$ solution **(a)** will precipitate $Fe(OH)_2$ from a $0.10\ M$ solution of Fe^{2+}, **(b)** will not precipitate $Mg(OH)_2$ from a solution which is $0.20\ M$ in NH_4^+ and $0.10\ M$ in Mg^{2+}, **(c)** will not precipitate AgOH from a solution which is $0.010\ M$ in Ag^+.

43. What is the maximum possible concentration of Ni^{2+} ion in a solution which is also $0.15\ M$ in HCl and $0.10\ M$ in H_2S?

44. Calculate the hydronium ion concentration necessary to just prevent precipitation of ZnS from a solution $0.20\ M$ in Zn^{2+} which is saturated with H_2S.

45. Calculate the equilibrium concentrations of each of the indicated species necessary to reduce an initial $0.20\ M\ Zn^{2+}$ solution to $1.0 \times 10^{-4}\ M\ Zn^{2+}$. In each case, repeat the calculation to determine the concentrations necessary to reduce the $0.20\ M\ Zn^{2+}$ to

$1.0 \times 10^{-10} \, M \, Zn^{2+}$.   **(a)** NH_3 and $Zn(NH_3)_4^{2+}$ (assume no partial complexation), **(b)** OH^- in equilibrium with $Zn(OH)_2(s)$, **(c)** OH^- and $Zn(OH)_4^{2-}$.   **(d)** Calculate the OH^- concentration which would be produced by each equilibrium concentration of NH_3 in part **(a)**. Using these values, predict whether $Zn(OH)_2$ or $Zn(OH)_4^{2-}$ would form in preference to $Zn(NH_3)_4^{2+}$ upon addition of sufficient ammonia to produce the equilibrium concentrations calculated in part **(a)**.   **(e)** Describe what would be observed if concentrated ammonia solution were added slowly to a $0.20 \, M$ solution of Zn^{2+}.

46. The solubility of CuS in pure water is 3.3×10^{-4} gram/liter at $25°C$. Using this number, calculate the apparent value of K_{sp} for CuS. By precise measurement, the value of K_{sp} for CuS at $25°C$ is found to be 8.5×10^{-36}. Explain why CuS is more soluble than predicted by the K_{sp}.

47. Aluminum sulfide reacts with water to form hydrogen sulfide and a white precipitate. Identify the latter substance and write equations accounting for its formation.

48. Assuming that the only source of periodic group IIA metals is an equimolar mixture of NaCl, $BaCl_2$, and $MgCl_2$, use the principles described in this and earlier chapters to suggest ways of preparing pure samples of **(a)** $MgSO_4$,   **(b)** Ba metal, **(c)** $Ba(C_2H_3O_2)_2$.

49. Calculate $\Delta H°$ for the reaction

$$Ag^+(aq) \;+\; Br^-(aq) \;\rightleftharpoons\; AgBr(s)$$

(a) directly from the data of Tables 3–2 and 3–5 and   **(b)** using data of Table 6–1 and other sources.   **(c)** Compare the results.

50. Calculate the enthalpy of solution, $\Delta H_{298}°$, of AB, which dissolves as follows:

$$AB(s) \;\rightleftharpoons\; A^+(aq) \;+\; B^-(aq)$$

given that $K_{sp} = 8 \times 10^{-12}$ and that $\Delta S° = 25.3$ cal/mole $\cdot$ K.

Advanced Exercises

51. Estimate the cell potential of a Daniell cell (Figure 5–6) having $1.00 \, M \, Zn^{2+}$ and originally having $1.00 \, M \, Cu^{2+}$ after sufficient ammonia has been added to the cathode compartment to make the NH_3 concentration $2.00 \, M$. Does that half cell continue to function as the cathode?

52. A saturated solution of silver benzoate, $AgOCOC_6H_5$, has a pH of 8.63. K_a for benzoic acid is 6.5×10^{-5}. Estimate the value of K_{sp} for silver benzoate.

53. Assuming no change in volume, calculate the minimum mass of NaCl necessary to dissolve 0.010 mole of AgCl in 100 liters of solution.

54. From the data of Table 8–1, calculate the concentration of each of the ions produced when 1.5 moles of $CuCl_2 \cdot 2H_2O$ is dissolved in enough water to make 1.0 liter of solution.

55. Calculate the concentrations of silver ion, bromide ion, chloride ion, $[Ag(NH_3)_2]^+$ ion, ammonium ion, and hydroxide ion in a solution which results from shaking excess AgCl and AgBr with $0.0200 \, M$ ammonia solution. Assume that no monoammine complex is formed.

56. One liter of a certain solution contained 0.15 mole of Cu^{2+} and 0.15 mole of Fe^{2+}. The solution was treated with H_2S until no further change occurred and the solution was saturated. Calculate the concentrations of Cu^{2+} and Fe^{2+} in the resulting solution.

57. In the qualitative analysis scheme described in this chapter, after removal of the group I cations, the group II cations are separated from the remaining groups by precipitation as sulfides from acid solution. Would these cations precipitate as sulfides from neutral solution? from basic solution? Explain why acid solution is used.

58. Using CO_2, NH_3, NH_4NO_3, and K_2CrO_4 as the only reagents, devise a qualitative analysis scheme for separating and identifying the following ions, which might all be present in the same mixture: Ba^{2+}, Ca^{2+}, Mg^{2+}, Na^+, Pb^{2+}. Assume that each cation present is 0.10 M. State the conditions of pH and the reagent concentrations which are required in each step.

59. Using appropriate thermodynamic data, compare $\Delta H°$, $\Delta S°$, and $\Delta G°$ for the precipitations of $CaCO_3$ and $BaCO_3$ from aqueous solution at 25°C. Show the contributions of $\Delta H°$ and $T\Delta S°$ to $\Delta G°$ by plotting the results on an energy diagram analogous to the enthalpy diagrams used in Chapter 3. Discuss the factors which make the solubilities of $CaCO_3$ and $BaCO_3$ so similar. Suggest an explanation of why the enthalpy change is greater for the precipitation of $CaCO_3$.

60. The overall formation constant for the reaction of 6 moles of CN^- with cobalt(II) is 1×10^{19}. The standard reduction potential for the reaction

$$Co(CN)_6{}^{3-} \quad + \quad e^- \quad \rightarrow \quad Co(CN)_6{}^{4-}$$

is -0.83 V. Using the data of Table 5-2, calculate the overall formation constant of $Co(CN)_6{}^{3-}$.

Experimental Basis
for the Atomic Theory

A major goal of chemistry is to explain the observed properties of matter in terms of **structure**—the particular arrangements of the constituent atoms in given substances and the forces which hold such atoms together. It is also recognized that the structures of the atoms themselves must influence the structures of the molecules which they form. Modern theories of atomic structure are based on the interpretation of three types of evidence: (1) chemical experiments, (2) experiments in electrostatics and magnetism, and (3) experiments which can be interpreted only by means of the quantum theory.

In 1804, John Dalton proposed that each element was made up of atoms having characteristic atomic weights. This concept accounted for the observed compositions of compounds and permitted the writing of chemical formulas. The experiments of Faraday (1825) demonstrated that there were quantitative relationships between electrical energy and atomic weights. The periodic law of Mendeleev and Meyer (1864) stated that the properties of the elements were related in some way to their atomic weights. Other research led to the discovery that the atoms in molecules are arranged in definite geometric configurations. Nevertheless, there was no successful explanation of the observed atomic weights, nor was there any satisfactory way of accounting for the observed "valences" of the elements. There was no explanation of why some substances were electrolytes, and no way of predicting the geometries of molecules on the basis of their constituent atoms. In short, there was no way of correlating the properties of atoms with their structures. Chemical theory of the mid-nineteenth century was limited to making correlations based on a few empirical rules.

In this chapter some experiments and theories which provided insight into the structure of the atom will be described. These experiments, performed in the latter part of the nineteenth century and the early twentieth century, led to some revolutionary ideas about the nature of matter. The brief account given here can present only the essential concepts and ideas, any one of which might well be expanded to fill an entire chapter. It is worth noting that several of the principles and/or concepts discussed are the bases for the operation of modern research instruments. For example, the laws of interaction of charged particles with each other and with magnetic fields are applied in the mass spectrometer (see Figure 9–7).

Just 100 years after Dalton, Max Planck proposed that the interactions of energy with individual atoms and molecules are quantized; that is, energy is emitted and absorbed in discrete units called quanta. When the energy is in the form of light, the magnitude of a quantum, E, is given by

$$E = h\nu$$

where h is Planck's constant and ν is the frequency of the light. The quantum theory is in effect an "atomic theory" applied to energy. Thus light has a dual nature—that of waves and that of particles. Light quanta are also known as photons.

The quantum theory provided the impetus for a new approach to the structure of matter. The absorption and emission of discrete quantities of energy suggested the existence of definite energy levels in atoms and molecules. As a corollary to the dual nature of light, it was proposed and demonstrated that matter too has wave properties. By means of the quantum theory, it has been possible to deduce the electronic structures of atoms, to explain the nature of bonding between atoms, and to account for the geometries of molecules. Before discussing these theories, it is necessary to review briefly some concepts of electricity and magnetism.

9-1 A Review of Some Concepts of Electricity and Magnetism

Static electricity is an accumulation of excess electric charge on a body. It can be generated by friction, as when a hard rubber rod is rubbed with a piece of fur or when a glass rod is rubbed with a piece of silk. The accumulation of charge can be demonstrated by experiments with electroscopes, as shown in Figure 9-1. There are two opposite kinds of charge, called positive and negative, and it can easily be proved that opposite charges attract while like charges repel each other. (The attraction of opposite charges holds electrons to nuclei and also holds positive and negative ions together in an ionic crystal such as sodium chloride.) A quantitative expression of the force of attraction or repulsion between charged bodies in a vacuum is **Coulomb's law of electrostatics:**

$$f = k\,\frac{q_1 q_2}{d^2}$$

where f is the force, q_1 and q_2 are the magnitudes of the charges, and d is the distance

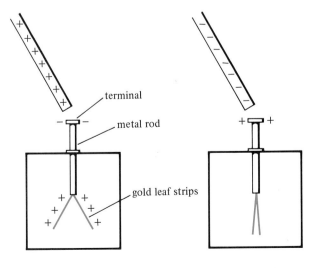

Figure 9-1. Electroscope

When a positively charged rod is brought near the terminal (drawing on the left), the electroscope is charged by induction. The terminal becomes negatively charged, and the leaves become positive. Since the leaves are both charged alike, they diverge. If the terminal is grounded momentarily (by touching it with a large neutral object) and then the charged rod is removed, the electroscope remains charged, and the leaves remained diverged. If now a negatively charged rod is brought near the terminal (drawing on the right), the negative charges on the terminal are repelled toward the leaves, which collapse. Approach of a rod with a greater positive charge would have caused further separation of the leaves.

terminal

metal rod

gold leaf strips

between the charges. The numerical value of the constant, k, depends on the units[1] used for all the other quantities. Possible values for k include

9.0×10^{18} dyne $\cdot$ cm²/C²	9.0×10^9 Nt $\cdot$ meter²/C²
9.0×10^{18} erg $\cdot$ cm/C²	9.0×10^9 J $\cdot$ meter/C²
1.0 dyne $\cdot$ cm²/(esu)²	1.0 erg $\cdot$ cm/(esu)²

Example

An electron and a body with a $+1.0$ C charge on it are 2.0 meters apart. Calculate the force of attraction between them. The charge on the electron is -1.6×10^{-19} C.

$$f = k \frac{q_1 q_2}{d^2} = (9.0 \times 10^9 \text{ Nt} \cdot \text{meter}^2/\text{C}^2) \left(\frac{(-1.6 \times 10^{-19} \text{ C})(1.0 \text{ C})}{(2.0 \text{ meter})^2} \right)$$

$$= -3.6 \times 10^{-10} \text{ Nt}$$

The negative sign implies an attractive force.

The **electric field** of a charged object is the region of space about it in which a second charged object would experience an appreciable force. It follows from Coulomb's law that the strength of a field about a point charge, q, varies inversely as the square of the distance from the charge. A uniform field may be created by use of charged parallel plates. In such a field, the force on a point charge would be independent of the position of the point charge in the field (Figure 9–2). The force exerted on a unit charge determines the **field strength, $\mathcal{E}$**. The force on a given particle of charge q is merely the field strength times the magnitude of the charge:

$$f = \mathcal{E}q$$

When two bodies with electric charges of the same sign are moved closer to each other, work must be done against the force of repulsion. Conversely, when two oppositely charged bodies are moved closer to each other, energy is released. The energy required or released in moving two charged bodies from a distance where there is no significant attractive or repulsive force (called infinite distance) to a distance of separation d is given by the following equation, which may be derived from Coulomb's law by the methods of calculus:

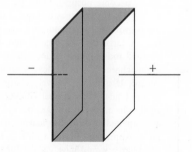

Figure 9-2. **Charged Parallel Plates Generate a Uniform Electric Field**

[1] See Appendix A-3 for a discussion of units.

$$E = k \frac{q_1 q_2}{d}$$

Example

How much energy will be released when a sodium ion and a chloride ion, originally at infinite distance, are brought together to a distance of 2.76 Å (the shortest distance of approach in a sodium chloride crystal)? $1 \text{ Å} = 1 \times 10^{-10}$ meter. Assume that the ions act as point charges, each with a magnitude of 1.60×10^{-19} C (the electronic charge).

$$E = k \frac{q_1 q_2}{d} = \left(\frac{9.0 \times 10^9 \text{ J} \cdot \text{meter}}{\text{C}^2} \right) \left(\frac{(1.60 \times 10^{-19} \text{ C})^2}{2.76 \times 10^{-10} \text{ meter}} \right) = 8.3 \times 10^{-19} \text{ J}$$

This energy corresponds to 119 kcal per mole of Na^+, Cl^- ion pairs. Of course in a crystal of NaCl, there are attractions between a given ion and several ions of opposite charge as well as repulsions between ions of like charge; hence for a mole of ion pairs in a crystal the energy of attraction is 185 kcal/mole, approximately 1.5 times as great (see Section 18–9).

A **magnetic field** is the region of space about a magnetic pole in which a second pole would experience an appreciable force of attraction or repulsion. The strength of a magnetic field, H, expressed in gauss, is determined by the force exerted by the field on a unit pole. A magnetic field moving relative to an electrical conductor will induce a potential along the conductor. If a wire which is part of a complete circuit is passed through a magnetic field or if a magnet is passed over the wire, a current is induced in the circuit. This principle is the basis for such practical devices as the generator and the dynamo.

Conversely, an electric charge in motion generates a magnetic field which has a direction perpendicular to the direction of motion of the charge. Electromagnets and electric motors are practical devices based on this principle. If a magnet is placed near the path of moving charged particles, the magnetic field generated by the moving charges will interact with the field of the magnet. The magnetic force on the charged particles is a function of the magnitudes of the applied field, H, the charges, q, and the velocity of the particles, v:

$$f = Hqv$$

If the charged particle is moving freely through space, the interaction of its magnetic field with the applied field will cause the particle to be deflected in a direction perpendicular to the applied magnetic field. The use of an applied magnetic field is an important technique for modifying the path (trajectory) of a moving charged particle.

9-2   Cathode Rays

When a glass tube containing two electrodes, such as is shown in Figure 9–3, is evacuated to a pressure about 0.01 atm, and a sufficiently high voltage is applied, electricity is conducted across the tube and the residual gas becomes luminous. Neon lighting is a familiar example of this phenomenon. If the tube is further evacuated to

Figure 9-3. **Discharge (Cathode-Ray) Tube**

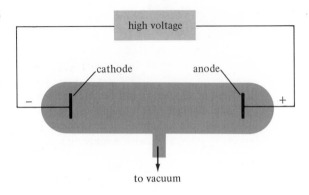

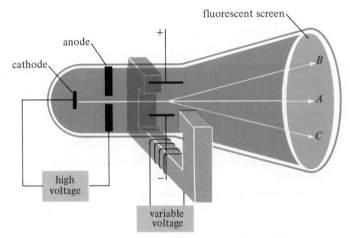

Figure 9-4. **Apparatus for the Determination of the Charge to Mass Ratio for the Electron**
Electrons having a mass m and a charge e are emitted from the cathode. After passing through a hole in the anode with velocity v, they strike the screen at point A. Electric plates located above and below the electron beam generating a field $\mathcal{E}$ can exert an electric force, $\mathcal{E}e$, which deflects the stream of electrons upward so that they finally arrive at point B, as shown in the figure. Similarly, a magnetic field of field strength H applied in the absence of the electric field deflects the beam perpendicularly to the field to point C. The force due to the magnetic field is equal to Hev. This force is equal to the mass of the electron times its acceleration, which for a particle traveling in a circular arc of radius r is equal to mv^2/r:

$$Hev = \frac{mv^2}{r} \qquad \text{hence} \qquad \frac{e}{m} = \frac{v}{Hr}$$

To determine the velocity of the electrons, the electric and magnetic fields are applied simultaneously such that the two forces are exactly equal in magnitude and no deflection occurs. In such a case,

$$Hev = \mathcal{E}e \qquad \text{and} \qquad v = \mathcal{E}/H$$

Substituting for v above yields e/m in terms of the measurable quantities $\mathcal{E}$, H, and r:

$$\frac{e}{m} = \frac{\mathcal{E}}{H^2 r}$$

The magnitude of e/m so obtained is 1.7588×10^8 C/gram.

extremely low pressures, the luminosity disappears; however, the electrical conductance persists, and the walls of the glass tube begin to glow with a faint fluorescence.
The fluorescence is due to the bombardment by rays, which originate at the cathode.
These **cathode rays** travel in straight lines until they are stopped by a solid object in
their path or until they are deflected by an applied electrostatic or magnetic field.
When a uniform electric field is applied by means of charged plates, the rays are
deflected away from the negative plate and toward the positive plate. When a
magnetic field is applied, the rays are bent in a direction perpendicular to the applied
field. The results indicate that the cathode rays are negatively charged carriers of
electricity.

In 1897, J. J. Thomson, assuming that the cathode rays consisted of particles,
observed their behavior in the presence of electric and magnetic fields. A schematic
representation of his apparatus is shown in Figure 9–4. The design of his experiments
was such that only the ratio of charge, e, to mass, m, could be obtained. However,
even this information was important because the value of e/m obtained for cathode
rays was the largest observed for any charged particle. Thomson obtained a value of
1.7588×10^8 C/gram for e/m of cathode rays regardless of the nature of the metal
used as a cathode. The ratio was also independent of the nature of the residual gas in
the tube and of the materials used in the construction of the tube. Thus the cathode
rays were considered to be composed of fundamental constituents of all materials. In
1874, G. J. Stoney had suggested that the fundamental particles of electricity be
called **electrons.** Cathode rays are therefore streams of electrons.

9-3 Charge on the Electron

R. A. Millikan, in 1909, demonstrated that all electrons bear the same charge. He
surmised that static electricity was due either to an accumulation of extra electrons in
an object or to a deficiency of electrons in the object. He designed a simple apparatus
to measure the static charges on droplets of oil. Details of the experiment are given in
Figure 9–5. Although various charges were found on the droplets, the charge on each
could be expressed as an integral multiple of 1.60×10^{-19} C. Hence Millikan assumed that this value corresponded to the charge on a single electron.

Millikan's result can be combined with the measured charge to mass ratio for
electrons to obtain the mass of an electron:

$$m = \frac{e}{e/m} = \frac{1.60 \times 10^{-19} \text{ C}}{1.7588 \times 10^8 \text{ C/gram}} = 9.10 \times 10^{-28} \text{ gram}$$

By comparison, a single hydrogen atom has a mass 1.67×10^{-24} gram.

9-4 Positive Rays

Matter as it is usually encountered is electrically neutral. Hence if negatively
charged electrons are constituents of atoms, there must also be constituents having
positive charges. Electrons possess relatively little mass compared to the mass of an
atom; most of the mass of the atom is associated with its positively charged constituents.

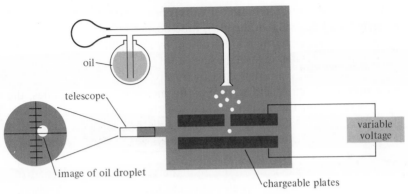

Figure 9-5. Oil Drop Experiment

A fine mist of oil droplets is sprayed into the apparatus. The act of spraying produces static electric charges on the droplets. Some of the droplets fall through a hole in the upper plate, and an individual droplet between the plates is observed by means of a telescope. In the absence of any field, the droplets fall through the air with increasing speed until they achieve a constant **terminal velocity,** which is measured by measuring the time required for the droplet to pass across calibrated lines on the telescope lens. The velocity, v, is constant because the resistance of the air in the apparatus offsets the acceleration due to gravity. According to Stokes' law:

$$v = \frac{mg}{6\pi\eta r}$$

where m is the mass of the droplet, r is its radius, η is the viscosity of air, and g is the gravitational constant. Another relationship between the mass and the radius of the droplet (which is nearly spherical) is obtained from the density d, of the oil:

$$d = \frac{3m}{4\pi r^3}$$

Hence for any droplet, m and r are determined from the terminal velocity and the density. If the electric field strength between the plates, $\mathcal{E}$, is adjusted until the drop is accelerated neither upward nor downward, the coulombic force on the charged drop is equal in magnitude to the gravitational force:

$$\mathcal{E}q = mg$$
$$q = mg/\mathcal{E}$$

Millikan found that the charge on each oil droplet could be expressed as

$$q = ne$$

where n is a positive integral number and e is the charge on the electron. The magnitude of e is 1.60×10^{-19} C.

Using a modified cathode-ray tube (Figure 9-6), E. Goldstein, in 1886, demonstrated that positive rays were produced simultaneously with cathode rays. Further experiments showed that the charge to mass ratio for these positive rays was not constant but that it varied with the residual gas in the tube. The largest value of e/m for positive rays was obtained when hydrogen was the residual gas. This value was 9.574×10^4 C/gram, approximately 1/1837 of the e/m value for electrons. If one

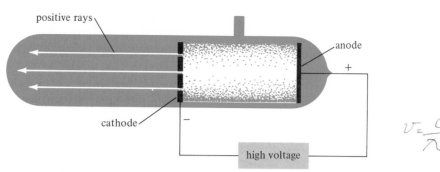

$v = \dfrac{c}{\lambda}$

Figure 9-6. Production of Positive Rays

A discharge tube is constructed with the cathode near its center. The cathode contains a number of holes, or **canals.** When sufficient potential is applied across the electrodes, electrons migrating from the cathode to the anode knock other electrons from molecules of residual gas in the discharge tube, creating positive ions. The positive ions migrate toward the cathode. Some positive ions pass through the canals and may be detected by means of a fluorescent screen at the end of the tube. Further analysis of these rays shows that the charge to mass ratios for positive rays are less than one thousandth the charge to mass ratio of the electron. The magnitude of the charge to mass ratio of the positive rays depends on the residual gas in the discharge tube, the largest value being obtained when hydrogen is the residual gas.

assumes that the bonds of some H_2 molecules are broken in the high-energy discharge and that the magnitude of the charge on the positive particle is identical with that of the negative charge on the electron, it is possible to calculate the mass of the positive particle obtained from hydrogen:

$$m = \frac{e}{e/m} = \frac{1.60 \times 10^{-19}\,\text{C}}{9.574 \times 10^{4}\,\text{C/gram}} = 1.67 \times 10^{-24}\,\text{gram}$$

This mass is approximately that of a hydrogen atom.

The fact that various positive particles have different e/m values makes possible the separation of particles according to their mass by means of the **mass spectrometer** (Figure 9-7). The operation of the mass spectrometer is based on the principle that a moving charged particle is deflected into a curved trajectory by a magnetic field perpendicular to its direction of motion. Atoms or molecules are ionized and accel-

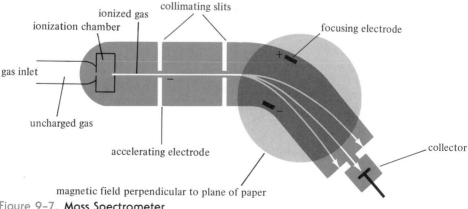

Figure 9-7. Mass Spectrometer

erated by a high voltage in the ionization chamber. They pass through the slits, which are provided to insure that all the particles which pass are traveling in the same direction. The charged particles pass through the electric field created by the focusing electrodes and at the same time through the magnetic field created by a magnet with poles (not shown here) perpendicular to the path of the particle. Each field tends to cause some deflection, depending on the velocity of the particle, its mass, its charge, and the strength of each field. The two fields are adjusted until only the mass of the particle makes a difference in the curvature of the path of the particle. Once the particles are separated on the basis of their different masses, they are collected and the number with each mass is measured in the collector. The magnetic field strength can be varied to collect particles of different mass.

9-5   Radioactivity

The phenomenon of radioactivity provided a means for investigating the internal structure of atoms and led to the nuclear model of the atom. It was only later shown that radioactivity itself is a nuclear phenomenon.

In 1897, while investigating the phosphorescence[2] produced in various materials by light, H. Becquerel discovered radioactivity. One of his samples was a compound of uranium. Quite accidentally, he observed that this sample emitted "rays" without prior excitation by any kind of light. These highly energetic rays were capable of passing through several layers of materials, such as paper, which were opaque to ordinary light. It is now known that **radioactivity** is the spontaneous emission of radiation by an atom due to changes occurring in its nucleus. Three types of radiations are emitted from radioactive elements. These are called **alpha, beta,** and **gamma rays,** respectively. A summary of the properties of these rays is given in Table 9–1 and Figure 9–8.

Marie Curie, a student of Becquerel, found that the radioactivity of uranium compounds was independent of the specific compound. The quantity of radioactivity depended only on the quantity of uranium present. Similarly, thorium compounds exhibited radioactivity which was determined only by the thorium content of the sample. She concluded that radioactivity is characteristic of certain elements rather

TABLE 9-1. Properties of Radioactive Emissions

Name	Mass (D)	Charge (e)	Identity
alpha	4.0026	+2	helium nucleus
beta	1/1837	−1	electron
gamma	0	0	high-energy electromagnetic radiation

[2] **Phosphorescence** is the emission of light of a certain wavelength by an object during and after excitation by light of another wavelength. The time lag distinguishes this effect from **fluorescence,** which is the emission of a given wavelength only while the object is being excited.

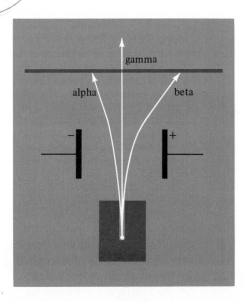

Figure 9-8. **Radioactive Emissions**

If radiations from a radioactive sample are passed through an electric field, the alpha rays are deflected toward the negative plate and the beta rays are deflected toward the positive plate. Alpha and beta rays are actually streams of charged particles. However, because of their relatively large mass and momentum, alpha particles are deflected to a smaller extent than beta particles of equal energy. If these rays are passed through a magnetic field, alpha and beta rays will be deflected perpendicularly to the field, but again in opposite directions. Gamma rays are undeflected by either electric or magnetic fields. Gamma rays are high-energy light waves which are capable of penetrating layers of matter opaque to ordinary light.

than their compounds. In testing the activity of various uranium-containing materials, Mme. Curie observed in one of them, pitchblende, an activity which was several times greater than that expected from its uranium content. After painstaking work, she isolated about 0.2 gram of the bromide of a new element from about 1 ton of the original pitchblende. She named the highly radioactive element **radium.**

The alpha, beta, and gamma rays emitted from a radioactive atom possess tremendously large energies. Detailed descriptions of radiations from radioactive substances will be given in Chapter 20. For the present it is useful to know that since alpha particles have been identified as doubly charged helium ions, a substance which emits alpha particles can be used as a source of nuclear-sized projectiles.

9-6 The Nucleus

After it was demonstrated that matter (atoms) contains negatively charged particles (electrons) and positive particles, the questions remained as to how these constituents were arranged to make up the structures of atoms and in what ways structures determine the properties of the various elements. At the turn of the twentieth century, the prevailing view was that the negative and positive particles in an atom were intimately mixed so that no region of the atom would have a net charge. In light of this view, the results of some experiments carried out in 1911 in Lord Rutherford's laboratory were quite unexpected. Thin foils of various metals were bombarded with alpha particles. A schematic diagram of Rutherford's apparatus is shown in Figure 9-9. Most of the alpha particles penetrated the foils with no appreciable deviation from their original trajectory. However, a number of them were deflected through rather large angles, and some were even scattered backward.

If the atoms making up the foils were a homogeneous mixture of positive and negative particles, then the energetic alpha particles should pass through the foil with practically no deviation. To explain why some alpha particles suffered large deflec-

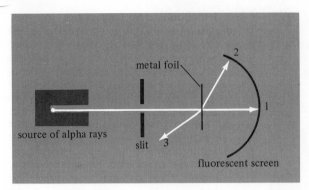

Figure 9-9. **Schematic Diagram of the Alpha Particle Scattering Experiment**
Alpha particles from radium (or polonium) are directed through a narrow slit onto a thin metal foil. Most of the particles pass through the foil and are detected on a fluorescent screen at point 1. Some small number of particles is deflected upon passing through the foil; they are detected in directions such as that of point 2. An even smaller number of particles is deflected backward, as represented by point 3.

tions, Rutherford postulated that the atoms making up the foils contained massive centers, as is represented in Figure 9-10. Since the atom is electrically neutral and its negative part consists of electrons, which have very little mass, the greater portion of the mass of an atom must also be located in its positive center, or **nucleus.**

Whenever the positively charged alpha particles passing through the foil came near a nucleus, they were repelled and deflected from their original path. Since most of the alpha particles penetrated the foil without deflection, the positive centers must be very small and spaced relatively far apart. Rutherford showed mathematically that the number of alpha particles of a given energy that will be deflected through a given angle depends on the mass and charge of the nucleus and on the distance of closest approach. If the charge on the nucleus of the metal atom is taken to be the same as its atomic number, the calculated distance of closest approach can be regarded as an

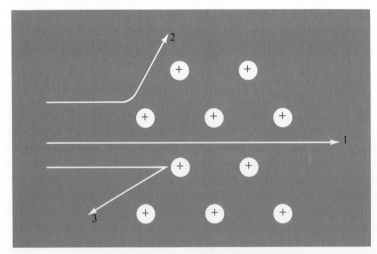

Figure 9-10. **Schematic Representation of Alpha Particle Scattering by Nuclei**
The nuclei of the atoms making up the metal foil are located relatively far away from each other compared to their own dimensions. (In the figure, the nuclei are shown out of proportion to the distance between them.) They are centers of positive charge and contain most of the mass of the atoms. Most of the alpha particles pass through the foil without coming close to any nuclei (path 1). When an alpha particle does come close to a nucleus, the positive charges repel each other, and the alpha particle deviates from its original direction and emerges from the foil along a path such as 2. Path 3 represents an almost "head-on" collision of an alpha particle with a nucleus.

upper limit to the radius of the nucleus. His calculations indicated that the radius of a nucleus is on the order of 10^{-13} cm. The calculated value of the radius of an entire atom is about 10^{-8} cm. The tiny size of the nucleus as compared to the apparent size of the atom was indeed an amazing result. Thus the greatest part of the volume of an atom is empty space, in which the electrons are found. Hence, in any interactions of atoms with each other or with external sources of energy, the electrons rather than the nuclei are likely to be involved.

9-7 Light

Of the investigations of the structure of matter, those dealing with the interactions of light energy and matter have been among the most useful. Therefore, the nature of light will be considered next. Light is a form of energy which has properties associated with wave motion. However, in some of its interactions with matter, light also displays properties of a stream of particles.

Some characteristics of waves are illustrated in Figure 9–11. Waves are variations within some medium such that there is a displacement or disturbance of the medium which increases and decreases in a periodic manner. The **wavelength** is the distance from crest to crest (or from trough to trough) along a wave. The **frequency** of a wave is defined as the number of waves passing a given point per second. A property analogous to frequency is **wavenumber,** defined as the number of waves per unit length. Frequency is expressed in units of reciprocal seconds—1/sec or sec^{-1}—or hertz, Hz (1/sec = 1 Hz). The units of wavenumber are reciprocal centimeters, 1/cm, or cm^{-1}. The frequency, ν, is related to wavelength, λ, by the relationship

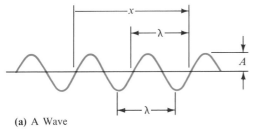

(a) A Wave

Figure 9–11. **Characteristics of Waves**
Wavelength is represented by λ; amplitude by A. If distance x is equal to 1 cm, the wavenumber, $\bar{\nu}$, is 2 cm^{-1}.

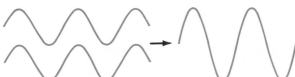

(b) Constructive Interference

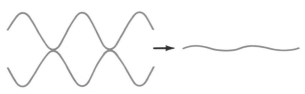

(c) Destructive Interference

$$v = \frac{\text{velocity}}{\lambda}$$

whereas the wavenumber, $\bar{\nu}$, is the reciprocal of wavelength:

$$\bar{\nu} = \frac{1}{\lambda}$$

Another property of waves is **amplitude.** In the case of waves in material objects, such as water waves or vibrations of a violin string, the amplitude is the maximum displacement of the individual particles of the medium from their equilibrium positions.

Waves also exhibit **interference.** When two waves in the same medium come together so that their crests coincide and their troughs coincide, they are said to be **in phase,** and they combine by constructive interference into a wave having a greater amplitude but the same wavelength and frequency as the original waves. When two waves combine so that the crest of one coincides with the trough of the other, they interfere with one another and are said to be **out of phase.** The result may be a wave with a decreased amplitude, or both waves may be completely canceled. These phenomena are shown in Figure 9–11.

The wave character of light is demonstrated by its ability to undergo diffraction, as is illustrated in Figure 9–12. If light of a given wavelength **(monochromatic light)** falls on an opaque barrier containing a narrow slit, some light will pass through the slit and be observed as a bright line on a screen. If a second slit is made parallel to the first at a suitable, very small distance away, a series of light and dark lines will appear on the screen (Figure 9–12a). In this case, the points B and B' will be dark, while point A midway between B and B' will be bright. These observations can be explained as follows. Each slit acts as a source of light waves which leave the slits in phase. The distances from each slit to point A on the screen are the same, and the waves arriving at this point will arrive in phase, causing reinforcement and brightness. However, the distances from the slits to point B differ by one-half wavelength, so the waves which left the slits in phase will arrive out of phase. Therefore the waves interfere with each other and darkness results at point B. Similarly, point B' will be dark. The distance from slit 1 to point C is exactly one wavelength less than the distance from slit 2 to point C. At this point, the waves arrive crest to crest again, and a bright line is observed. Applying similar reasoning to points up and down the screen allows explanation of the observed pattern of alternating bright and dark lines.

A **diffraction grating** can be made of a translucent material on which many parallel, closely and equally spaced lines are scratched to make these portions scatter light. The areas between the scratched portions transmit light. The grating functions in the same manner as the two parallel slits, but because of the multiple reinforcements and interference, the lines obtained are much sharper. Light of different wavelengths reinforces at different places on the screen. When **white light** (light containing all wavelengths in the visible region of the spectrum, from about 4000 to 7000 Å) is passed through the grating, continuous spectra can be seen (Figure 9–12b). Reflective diffraction gratings are used in many modern instruments, rather than transmission gratings.

Unlike matter waves, light waves require no medium for propagation. Light can travel in a vacuum. The waves are actually oscillating electric and magnetic fields,

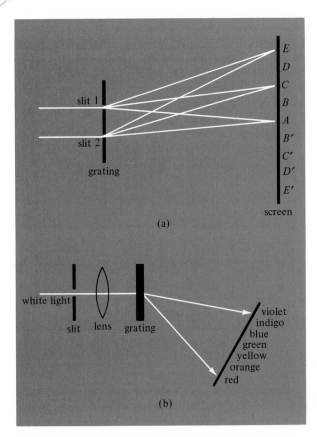

Figure 9-12. **Diffraction of Light**

and light is thus a form of **electromagnetic radiation.** In a vacuum, the velocity of light (and all other forms of electromagnetic radiation), customarily designated by the letter c, is 2.9979×10^{10} cm/sec. The frequency, ν, of light is related to its wavelength, λ, by

$$\nu = \frac{c}{\lambda}$$

The intensity (brightness) of the light is a measure of the amplitude of the light waves.

The wavelengths of electromagnetic radiations vary from a few billionths of a centimeter up to many kilometers. The parts of the electromagnetic spectrum are illustrated in Figure 9-13. Visible light is but a small portion of the electromagnetic spectrum. By passing white light through a diffraction grating and then through a suitable arrangement of slits in order to screen out all other wavelengths, nearly monochromatic light can be obtained. Systematic studies of the effects of light on matter are best made using monochromatic light.

9-8 X Rays and Atomic Numbers

In 1896, W. K. Roentgen discovered that when cathode rays were directed against a metal target, highly penetrating rays were emitted. He called them X rays, because

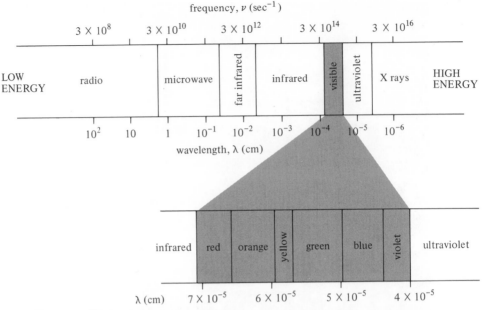

Figure 9–13. **Electromagnetic Spectrum**

their origin was not understood, nor was it known how to measure their wavelengths exactly. A tube for the production of X rays is shown in Figure 9–14. In 1912, Max von Laue found that crystals of salts could be used as three-dimensional diffraction gratings for X rays. Hence, by using a crystal which has known internal dimensions (exercise 45) the wavelengths of X rays can be measured.

In 1913, H. G. J. Moseley used Laue's technique to measure the wavelengths of X rays emitted when various elements were used as targets in an X-ray tube. He was able to show that the atomic number of an element has a more fundamental significance than the somewhat arbitrary numbering of the element in the periodic table. Moseley found that a given element emitted X rays of several different wavelengths but that these could be divided into series. For example, for each element the X rays having the shortest wavelength belong to one series. Moseley

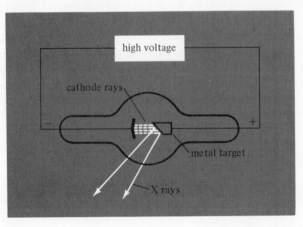

Figure 9–14. **Production of X Rays**

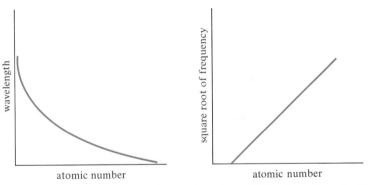

Figure 9-15. **Variation with Atomic Number of Wavelength and of Square Root of Frequency of X Rays.**

observed that the wavelengths of X rays in a given series decreased as the atomic number of the elements increased. The frequencies could be predicted using the equation

$$\sqrt{v} = a(Z - b)$$ *linear relationship*

where a and b are constants characteristic of the series and Z is the atomic number of the element. Plots of this relationship are shown in Figure 9-15. When such a plot was first made, there were places missing corresponding to the atomic numbers of elements which had not yet been discovered. When the missing elements were discovered, their atomic numbers were confirmed by showing that the square roots of the frequencies of their characteristic X rays fell at the expected points on Moseley's plot.

9-9   The Quantum Theory of Light

When a substance is heated to a sufficiently high temperature, it glows "red hot." If its temperature is raised further, it becomes "white hot," as more light of shorter wavelength is emitted. As shown in Figure 9-16, the distribution of energy as a function of the wavelength of the light from an incandescent object depends on the temperature of the object. Prior to 1900, all attempts to explain such distributions, in terms of a wave theory of light, were unsuccessful. In 1904, Max Planck proposed a theory which considered the molecules of a heated solid to be acting as minute vibrating oscillators which could absorb and emit energy, E, only in quantities equal to the frequency of oscillation, v, times a constant, h. The light emitted had a frequency equal to the frequency of the oscillator, v, and

$$E = hv$$

The proportionality constant, h, is known as Planck's constant. It is numerically equal to 6.63×10^{-27} erg · sec. Thus on a molecular scale, energy is absorbed or emitted in discrete quantities called **quanta,** equal in magnitude to hv. This important theory

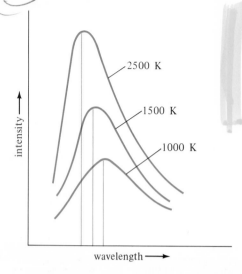

Figure 9-16. **Energy Distribution in Radiations from Black Bodies**

An ideal **black body** is a perfect absorber and a perfect emitter of radiation. When such a body is at a given temperature, the emitted radiation has a characteristic distribution of wavelengths. The higher the temperature, the greater is the quantity of energy emitted, but also the greater is the percentage which is emitted at shorter wavelengths.

advances the concept that light energy is corpuscular in nature, consisting of **photons** which possess energy and momentum analogous to particles of matter.

Example

How many photons of light having a wavelength of 4000 Å are necessary to provide 1.00 erg of energy?

$$E_{photon} = h\nu = h\frac{c}{\lambda}$$

$$= \frac{(6.63 \times 10^{-27} \text{ erg} \cdot \text{sec})(3.00 \times 10^{10} \text{ cm/sec})}{4000 \times 10^{-8} \text{ cm}} = 4.97 \times 10^{-12} \text{ erg}$$

For light having a wavelength of 4000 Å, the number of photons per erg is

$$\frac{1.00}{4.97 \times 10^{-12}} = 2.01 \times 10^{11} \text{ photons/erg}$$

The quantum theory has allowed solutions to several problems which could not be solved in terms of Newtonian physics and the wave theory of light. For example, in 1905, Albert Einstein showed that the quantum theory could explain a puzzling phenomenon known as the **photoelectric effect.** When a clean metallic surface in a vacuum is irradiated with monochromatic light of sufficiently high frequency, electrons are emitted from the surface. To produce photoelectrons, the incident light must have a frequency greater than a threshold frequency, ν_0. Different metals have different threshold frequencies. Light having a frequency below ν_0 for a given metal will produce no electrons regardless of the intensity of the light. If the frequency of the incident light is greater than the ν_0 of the metal, electrons will be emitted. The kinetic energies of the electrons increase linearly with the frequency of the incident light, as shown in Figure 9–17. For a given frequency above the threshold frequency, increasing the intensity of the light increases the rate of emission of electrons but does not increase their maximum kinetic energy.

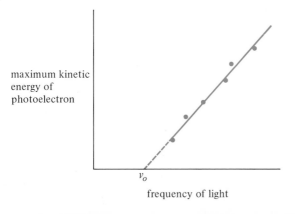

Figure 9–17. **Kinetic Energy of Photoelectrons as a Function of Frequency of Irradiating Light**

maximum kinetic energy of photoelectron

frequency of light

These facts can be explained if it assumed that light consists of photons having energy $E = h\nu$ and that a given electron interacts with *only one* photon. When a photon strikes the surface of the metal, it gives up its energy to the electron. A certain portion of the energy is used to overcome the attractive force between the electron and the metal; the remainder imparts kinetic energy to the ejected photoelectron. Therefore, the kinetic energy of the photoelectron is given by

$$KE = \tfrac{1}{2}mv^2 = h\nu - h\nu_0$$

As shown in Figure 9–17, a plot of the maximum kinetic energy versus the frequency of the incident light gives a straight line. Its slope is equal to h. Extrapolation of the line to zero kinetic energy yields the threshold frequency, ν_0.

Example

The minimum energy necessary to overcome the attractive force between the electron and the surface for silver metal is 7.52×10^{-12} erg. What will be the maximum kinetic energy of the electrons ejected from silver which is being irradiated with ultraviolet light having a wavelength 360 Å?

$$h\nu_0 = 7.52 \times 10^{-12} \text{ erg}$$

$$KE = h\nu - h\nu_0 = \frac{hc}{\lambda} - h\nu_0$$

$$= \frac{(6.63 \times 10^{-27} \text{ erg} \cdot \text{sec})(3.00 \times 10^{10} \text{ cm/sec})}{3.60 \times 10^{-6} \text{ cm}} - 7.52 \times 10^{-12} \text{ erg}$$

$$= 4.77 \times 10^{-11} \text{ erg}$$

9-10   Atomic Spectra

After an atom absorbs a quantum of energy, it is said to be in an **excited** state relative to its normal (**ground**) state. When an excited atom returns to the ground state or to a lower excited state, it emits light. For example, the yellow light observed when glass is heated in a flame is due to excited sodium atoms in the glass returning to their ground state. Similarly, the familiar red light of neon signs is due to neon atoms

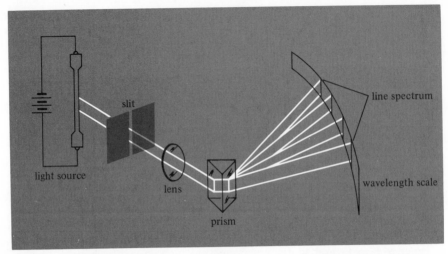

Figure 9-18. Schematic Diagram of a Spectroscope

which have been excited by an electrical discharge returning to a lower excited state. When light from excited atoms is viewed through a spectroscope (Figure 9–18), images of the slit appear along the scale of the instrument as a series of colored lines. The various colors correspond to light of definite wavelengths, and the series of lines is called a **line spectrum.** The line spectrum of each element is so characteristic of that element that its spectrum may be used to identify it.

The simplest spectrum is that of hydrogen, the simplest element. The part of the spectrum which appears as visible light is shown in Figure 9–19. It should be noted that the lines at shorter wavelengths have progressively lower intensities. The wavelengths of successive lines are closer and closer together until they finally become a continuum, a region of continuous faint light. In 1885, J. J. Balmer suggested that the wavelengths of the lines in the visible spectrum of hydrogen could be represented by the equation

$$\lambda = 3646 \left(\frac{n^2}{n^2 - 4} \right)$$

where n is an integer greater than 2, and λ is given in Ångström units.

J. R. Rydberg later suggested the following related equation, which represents the infrared and ultraviolet spectra of hydrogen as well as its visible spectrum:

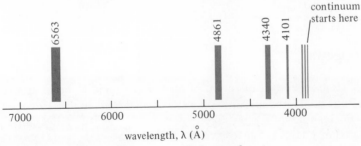

Figure 9-19. Balmer Series in the Hydrogen Spectrum

$$\bar{\nu} = \frac{1}{\lambda} = R\left(\frac{1}{n_1{}^2} - \frac{1}{n_2{}^2}\right)$$

where $\bar{\nu}$ is the wavenumber, and n_1 and n_2 are integers. For the visible spectrum of hydrogen, $n_1 = 2$ and n_2 has values 3, 4, 5, ..., ∞. The value of the constant R is 109,677.581 cm^{-1}. It is known as the Rydberg constant.

9-11   The Bohr Theory

Although the Rydberg equation successfully represented the hydrogen spectrum, there was no theoretical justification for it. In 1914, Niels Bohr proposed a theory of the hydrogen atom which explained the origin of its spectrum and which also led to an entirely new concept of atomic structure. The Bohr model of the hydrogen atom was based on four postulates. Stated in modern terms, these are

Postulate 1. The hydrogen atom consists of a nucleus containing a proton (and therefore having a charge $+e$), and an electron (with a charge of $-e$) moving about the nucleus in a circular orbit of radius r (Figure 9-20). According to (classical) Coulomb's law, the force of attraction[3] between the nucleus and the electron is

$$f = k\frac{e^2}{r^2} = \frac{e^2}{r^2}$$

This force is equal to the centripetal force on the electron, mv^2/r, where m is the mass of the electron and v its velocity:

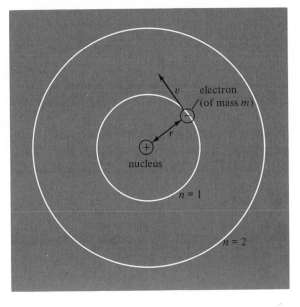

Figure 9-20. **Bohr Model of the Hydrogen Atom**

---

[3] When force is measured in dynes, distance in centimeters, and charge in esu, the magnitude of k is 1.

$$\frac{mv^2}{r} = \frac{e^2}{r^2}$$

$$v^2 = \frac{e^2}{rm}$$

The kinetic energy of the electron may be determined from this equation also:

$$KE = \tfrac{1}{2}mv^2 = \frac{\tfrac{1}{2}e^2}{r}$$

Postulate 2. Not all circular orbits are permitted for the electron. Only the orbits which have angular momentum of the electron, mvr, equal to integral multiples of $h/2\pi$ are allowed:

$$mvr = \frac{nh}{2\pi} \qquad n = 1, 2, 3, 4, 5, \ldots$$

$$v = \frac{nh}{2\pi rm}$$

$$v^2 = \frac{n^2h^2}{4\pi^2 r^2 m^2}$$

Equating the two expressions for v^2 yields an expression for r, the radius of an allowed orbit:

$$r = \frac{n^2h^2}{4\pi^2 me^2}$$

Therefore

$$r \propto n^2$$

The value $n = 1$ defines the first (smallest) orbit. Larger orbits have higher values for n. It is customary to refer to the integer n as a **quantum number.**

Postulate 3. As a consequence of the restrictions on the angular momentum of an orbit, the energy of an electron in a given orbit is fixed. As long as the electron stays in that orbit, it neither absorbs nor radiates energy. The total energy, E, of an electron is the sum of its potential energy, $-e^2/r$, and its kinetic energy, $\tfrac{1}{2}mv^2 = \tfrac{1}{2}e^2/r$ (from postulate 1). Therefore

$$E = -\frac{e^2}{r} + \frac{e^2}{2r} = -\frac{e^2}{2r}$$

Substituting the expression for r from above yields

$$E = -\frac{2\pi^2 me^4}{n^2h^2}$$

All of the quantities on the right-hand side of this equation are known constants except for the arbitrary integer n. Hence the possible energies of the electron are determined by the values of n.

$$-E \propto \frac{1}{n^2}$$

It should be noted that the minus sign in the energy expression means that the larger the numerical value of n, the higher (the less negative) will be the energy of the electron. The energy scale is chosen such that when the electron is completely separated from the atom, $E = 0$.

Postulate 4. To change from one orbit to another, the electron must absorb or emit a quantity of energy exactly equal to the difference in energy between the two orbits. When light energy is involved, the photon has a frequency given by

$$h\nu = E_2 - E_1$$

Substitution of the corresponding energy expression yields

$$h\nu = \frac{hc}{\lambda} = \frac{2\pi^2 m e^4}{h^2} \left(\frac{1}{n_1^2} - \frac{1}{n_2^2} \right)$$

Hence

$$\bar{\nu} = \frac{1}{\lambda} = \frac{2\pi^2 m e^4}{ch^3} \left(\frac{1}{n_1^2} - \frac{1}{n_2^2} \right)$$

$$\frac{1}{\lambda} = R \left(\frac{1}{n_1^2} - \frac{1}{n_2^2} \right)$$

The last equation is identical to the Rydberg equation. The constant R may be evaluated by using the numercial values of π, m, e, c, and h, and the result agrees closely with the experimentally determined value for the Rydberg constant.

Example

Calculate the radius of the first allowed Bohr orbit for hydrogen.

For the first orbit, $n = 1$. Substituting the values of the other constants in the equation of postulate 2 yields

$$r = \frac{n^2 h^2}{4\pi^2 m e^2}$$

$$= \frac{1^2 (6.63 \times 10^{-27} \text{ erg} \cdot \text{sec})^2}{4(3.14)^2 (9.109 \times 10^{-28} \text{ gram})(4.80 \times 10^{-10} \text{ esu})^2} = 0.529 \times 10^{-8} \text{ cm} = 0.529 \text{ Å}$$

Experimental methods of determining the effective radius of the hydrogen atom yield the value 0.53 Å.

Example

Calculate the energy of an electron in the first Bohr orbit of hydrogen.

$$E = -\frac{e^2}{2r} = -\frac{(4.80 \times 10^{-10} \text{ esu})^2}{2(0.529 \times 10^{-8} \text{ cm})} = -2.178 \times 10^{-11} \text{ erg}$$

This value agrees closely with the experimentally determined energy required to remove an electron from a gaseous hydrogen atom.

It follows from the Bohr theory, since n_1 can have values 1, 2, 3, 4, 5, etc., that several other series of lines should exist in the hydrogen spectrum besides the Balmer series. Other series have been discovered in the ultraviolet, infrared, and far infrared regions of the spectrum. All of these series have lines with wavelengths which are given by the Rydberg equation when appropriate values of n_1 and n_2 are used. See Table 9-2.

Example

Calculate the wavelengths of the first line and the series limit for the Lyman series for hydrogen.

First line: $\dfrac{1}{\lambda} = 109{,}678 \left(\dfrac{1}{1^2} - \dfrac{1}{2^2} \right) = 82{,}259 \text{ cm}^{-1}$

$\lambda = 1.2157 \times 10^{-5} \text{ cm}$

Series limit: $\dfrac{1}{\lambda} = 109{,}678 \left(\dfrac{1}{1^2} - \dfrac{1}{\infty} \right) = 109{,}687 \left(\dfrac{1}{1^2} - 0 \right) = 109{,}687 \text{ cm}^{-1}$

$\lambda = 9.1176 \times 10^{-6} \text{ cm}$

The Bohr theory accounts for the hydrogen spectrum in the following manner. When hydrogen atoms are excited, their electrons occupy orbits having higher energies. When an electron returns to a more stable orbit, it emits a photon of energy corresponding to the energy difference between the orbits. The transition back to the ground state can occur directly or stepwise, yielding one photon or several. The various spectral series correspond to transitions between higher orbits and those having a given value of n (Figure 9-21). For example, the transitions from higher orbits to the second orbit correspond to the Balmer series.

The Bohr theory can be applied with equal success to "hydrogen-like atoms": ions containing only one electron (He^+, Li^{2+}, etc.). The expression for the energy of the electron in the nth orbit for these ions is

TABLE 9-2. The Hydrogen Spectrum

$$\bar{\nu} = \frac{1}{\lambda} = R \left(\frac{1}{n_1{}^2} - \frac{1}{n_2{}^2} \right)$$

Series Name	n_1	n_2 Values	Spectral Region
Lyman	1	2 to ∞	ultraviolet
Balmer	2	3 to ∞	visible
Paschen	3	4 to ∞	infrared
Brackett	4	5 to ∞	far infrared
Pfund	5	6 to ∞	far infrared

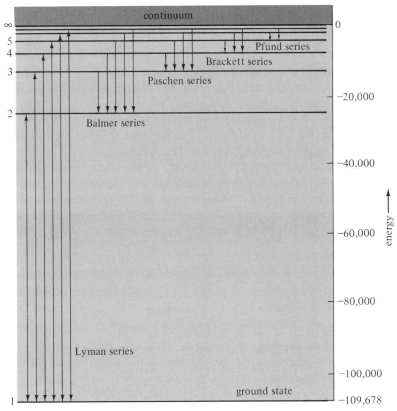

Figure 9-21. Electronic Transitions Corresponding to the Spectral Series of Hydrogen

$$E_n = -\frac{2\pi^2 m Z^2 e^4}{n^2 h^2}$$

where Z represents the atomic number of the element.

On the other hand, the Bohr theory cannot explain the spectra of atoms containing more than one electron. Even for atoms with one electron, the theory does not predict the relative intensities of the lines or the splitting of the lines observed when the atoms are excited in a magnetic field (the Zeeman effect). Even in the absence of external fields, the spectral lines were found to be more complex when examined with high resolution instruments. Some "improvements" in the Bohr theory were made by arbitrarily introducing additional quantum numbers, but significant improvement was not made.

Although the Bohr theory gave quantitative results only for hydrogen-like atoms, the notion of energy levels or "shells of electrons" was seized upon to explain the properties of more complex atoms. Most notable were the proposals of W. Kossel and of G. N. Lewis, which later led to the concepts of the octet rule and the electron pair bond described in Section 1–8. The notion of energy levels in atoms also provides an explanation of the production of X rays and accounts for the relationship between the frequency of X rays and atomic number discovered by Moseley (Section 9–8).

In the X-ray tube, a high energy electron from the cathode interacts with an atom of the metal target and ejects an electron from its innermost shell. Then an electron from the next higher energy level of the atom fills the vacancy, simultaneously emitting light having a frequency corresponding to the energy difference between the two energy levels. The energy of the X ray is high because the electron is strongly attracted by the relatively high charge of the nucleus. The vacancy left by the first electron which moved is filled in turn by an electron from an even higher orbit. Thus several photons are emitted, and their frequencies correspond to the frequencies of the series of X rays characteristic of the particular target metal. According to the Bohr theory, the energies of the levels, and therefore the energy difference between two levels of the same atom, are proportional to Z^2. Since the energy of the photon produced is equal to the energy difference between two levels in an atom, the energy and also the frequency of the photon are proportional to Z^2. Hence the square root of the frequency is proportional to the atomic number. This relationship is similar to that discovered by Moseley, $\sqrt{\nu} = a(Z - b)$, for atoms with many electrons.

9-12 Exercises

Basic Exercises

1. The magnitude of the charge on the electron is 4.8×10^{-10} esu. What is the magnitude of the charge on the proton? on the nucleus of a helium atom?
2. Calculate the force between two bodies 2.00 cm apart, each having a 1.0×10^{-5} C charge.
3. Calculate the energy required to move a 1.0×10^{-10} C negatively charged body from infinite distance to a point (a) 1.0 cm from a 1.0 C negatively charged body, (b) 0.10 cm from a 1.0 C negatively charged body, (c) from position (a) to position (b). (d) How would the answer to (a) be changed if the first body were positively charged?
4. What, if any, is the effect of a magnetic field on (a) a static electric charge in the field? (b) an electric charge moving through the field?
5. Why does the charge to mass ratio of positive rays depend on the residual gas in the discharge tube? Why is the charge to mass ratio of all cathode rays the same?
6. Distinguish between (a) a proton and a photon, (b) a photon and a quantum.
7. State at least four ways in which positive (canal) rays differ from cathode rays.
8. What is the equivalent of the unit 1.00 cm^{-1} in (a) ergs per photon? (b) kilocalories per mole of photons? (c) kilojoules per mole of photons?
9. Tabulate for the longest and the shortest wavelengths of visible light the following: (a) wavelength, (b) wavenumber, (c) joules per mole, (d) ergs per photon, (e) frequency, (f) kilocalories per mole.
10. How many photons of light of 7000 Å wavelength are equivalent to 1.00 J of energy?
11. Calculate the ratio of protons in the atomic nuclei to sodium atoms in a sample of NaCl. Is it necessary to know the size of the sample, the total number of atoms in the sample, or the total number of protons in the sample? To determine the charge to mass ratio of the electron, is it necessary to know either the charge or the mass of the electron? Under what conditions do the strengths of the magnetic and electric fields applied in a cathode ray tube permit the determination of the charge to mass ratio of the electron?
12. Which of the following relate to light as wave motion, to light as a stream of particles, or to both? (a) diffraction, (b) interference, (c) photoelectric effect, (d) $E = mc^2$, (e) $E = h\nu$.
13. Which has the greater energy—a photon of violet light or a photon of green light?

14. In a photoelectric effect experiment, irradiation of a metal with light of frequency 2.00×10^{16} Hz yields electrons with maximum kinetic energy 7.5×10^{-11} erg. Calculate ν_0 for the metal.

15. The unit of frequency is $\sec^{-1}$. Explain why the unit is not cycles/sec. If the second hand on a watch rotates 60 cycles/hour, what is the period (the time required for 1 cycle)? What is the relationship between frequency and period? between their units?

16. One photon of ultraviolet light can excite an electron from the surface of a certain metal. When the same metal surface is irradiated with two photons of red light having a total energy equal to that of the ultraviolet photon, no photoelectrons are produced. Explain these facts in terms of Einstein's theory of the photoelectric effect.

17. Calculate the frequency of light emitted for an electron transition from the sixth to the second orbit of the hydrogen atom. In what region of the spectrum does this light occur?

18. Calculate the energy of an electron in the second Bohr orbit of a hydrogen atom.

19. The third line in the Balmer series corresponds to an electronic transition between which Bohr orbits in hydrogen?

20. An electron volt (eV) is the energy necessary to move an electronic charge (e) through a potential of exactly 1 V. Express this energy in **(a)** ergs, **(b)** kilocalories per mole of electrons, **(c)** kilojoules per mole of electrons.

21. Evaluate the quotient from the Bohr theory, $2\pi^2 me^4/ch^3$. Compare the result to the Rydberg constant, R. Use $e = 4.80 \times 10^{-10}$ esu.

22. Express the Rydberg constant, $R = 109{,}678$ cm^{-1}, in **(a)** joules per mole, **(b)** ergs per atom.

23. Using conversion factors from Appendix A–3, show that

$$9.0 \times 10^{18} \text{ dyne} \cdot \text{cm}^2/\text{C}^2 = 9.0 \times 10^{18} \text{ erg} \cdot \text{cm}/\text{C}^2 = 9.0 \times 10^9 \text{ Nt} \cdot \text{meter}^2/\text{C}^2$$
$$= 9.0 \times 10^9 \text{ J} \cdot \text{meter}/\text{C}^2 = 1.0 \text{ dyne} \cdot \text{cm}^2/(\text{esu})^2$$

General Exercises

24. Two 1.0 gram carbon disks 1.00 cm apart have opposite charges of equal magnitude such that there is a 1.00 dyne force between them. Calculate the ratio of excess electrons to total atoms on the negatively charged disk.

25. Would it take more energy to move a negatively charged body from 3.0 cm to 2.0 cm from a second negatively charged body, or from 2.0 cm to 1.0 cm from the second body? Explain. Would it take more energy to move a charged body from midway between two charged parallel plates 1.0 cm toward the positive plate or from this second (off center) position to a position 1.0 cm closer to the positive plate. Explain.

26. In an oil drop experiment, the following charges (in arbitrary units) were found on a series of oil droplets: 2.30×10^{-15}, 6.90×10^{-15}, 1.38×10^{-14}, 5.75×10^{-15}, 3.45×10^{-15}, 1.96×10^{-14}. Calculate the magnitude of the charge on the electron (in the same units).

27. When white light which is passed through sodium vapor is viewed through a spectroscope, the observed spectrum has a dark line at 5890 Å. Explain this observation.

28. When a certain metal was irradiated with light of frequency 3.2×10^{16} sec^{-1}, the photoelectrons emitted had twice the kinetic energy as did photoelectrons emitted when the same metal was irradiated with light of frequency 2.0×10^{16} sec^{-1}. Calculate ν_0 for the metal.

29. For silver metal, ν_0 is 1.13×10^{17}/sec. What is the maximum energy of the photoelectrons produced by shining ultraviolet light of 15.0 Å wavelength on the metal?

30. In Figure 9–8 it may be seen that, although alpha particles have a larger charge, beta particles are deflected more than alpha particles in a given electric field. Explain this observation.

31. The radius of a nucleus, in centimeters, can be estimated by the equation

$$R \cong 1.4 \times 10^{-13} A^{1/3}$$

where A is the mass number of the atom. Calculate the approximate density of a polonium nucleus.

32. (a) Calculate the radii of the first two Bohr orbits of Li^{2+}. (b) Calculate the difference in potential energy between these two orbits. (c) Calculate the difference in total energy between these orbits.

33. The characteristic X-ray wavelengths for the lines of the K_α series in magnesium and chromium are 9.87 and 2.29 Å, respectively. Using these values, determine the constants a and b in Moseley's equation, and predict the wavelengths of X rays in this series for strontium and for chlorine.

34. Using Figure 9–13, estimate the minimum difference in energy between two Bohr orbits such that an electronic transition would correspond to the emission of an X ray. Assuming that the electrons in other shells exert no influence, at what Z (minimum) would a transition from the second energy level to the first result in the emission of an X ray?

35. The energy of an electron in the first Bohr orbit for hydrogen is -13.6 eV (see exercise 20). Which one(s) of the following is(are) possible excited state(s) for electrons in Bohr orbits of hydrogen? (a) -3.4 eV, (b) -6.8 eV, (c) -1.7 eV, (d) $+13.6$ eV.

36. Assuming that it were possible for such an atom to exist, calculate the energy of a positron (a positive electron) in the first Bohr orbit of a hydrogen atom.

37. Transitions between which orbits in He^+ ions would result in the emission of visible light?

Advanced Exercises

38. Assuming that the oil was sufficiently nonvolatile, would it be possible to perform Millikan's oil drop experiment in an evacuated apparatus? Explain.

39. Using a sound wave as an example, describe in familiar terms the disturbance of the medium, the velocity, the amplitude, and the frequency. Explain why the pitch seems higher when the source of sound is traveling toward you, but lower when the source is traveling away.

40. Analogous to the accomplishment of Balmer, fit the following three series of numbers into an equation involving integers:

Series 1 Wavelength (Å)	Series 2 Wavelength (Å)	Series 3 Wavelength (Å)
68.26	22.76	11.38
91.02	34.13	18.20
102.40	40.96	22.76
109.22	45.51	

41. A photon was absorbed by a hydrogen atom in its ground state, and the electron was promoted to the fifth orbit. When the excited atom returned to its ground state, visible and other quanta were emitted. In this process, radiation of what wavelength *must* have been emitted? Explain.

42. Derive an expression for the velocity of an electron in any Bohr orbit of a hydrogen-like atom. Calculate the velocity of an electron in the first orbit of a hydrogen atom. What is the ratio of this velocity to the velocity of light in a vacuum?

43. In an oil drop experiment, the terminal velocity of an oil droplet was observed to be 1.00 mm/sec. The density of the oil is 0.850 gram/cm^3, and the viscosity of air is 1.83×10^{-4} dyne · sec/cm^2. Calculate the mass and the radius of the oil droplet.

44. The frequency (in wavenumbers) of the first line in the Balmer series of hydrogen is 15,200 cm^{-1}. What is the frequency of the first line in the Balmer series of Be^{3+}?

45. The density of crystalline CsCl is 3.988 grams/cm^3. **(a)** Calculate the volume effectively occupied by a single CsCl ion pair in the crystal. **(b)** Calculate the smallest Cs to Cs internuclear distance, equal to the length of the side of a cube corresponding to the volume of one CsCl ion pair. **(c)** Calculate the smallest Cs to Cl internuclear distance in the crystal, assuming each Cs ion to be located in the center of a cube with Cl ions at each corner of the cube (see Figure 18-6). **(d)** Light of what portion of the electromagnetic spectrum has a wavelength corresponding to this distance?

10

Electronic Structure of the Atom

As noted in Chapter 9, the Bohr theory had fundamental defects. Early attempts to improve the theory included the proposal that the electrons revolved about the nucleus in elliptical orbits as well as in circular orbits. For this approach it was necessary arbitrarily to introduce additional quantum numbers because the mechanical properties of particles moving in elliptical orbits are different from those of particles moving in circular orbits. Nevertheless, the theory still had defects.

In this chapter a more sophisticated theory of the electronic structure of the atom will be described. This theory, called **wave mechanics,** not only gives results as good as the Bohr theory for the case of the hydrogen atom, but also can be used to explain the details of the spectra of atoms with more than one electron and in addition gives a better picture of chemical bonding. Fortunately, it is possible to describe many important results of wave mechanics without recourse to the detailed mathematics involved, and in this chapter it will be shown how the treatment of the hydrogen atom can be applied qualitatively to describe more complex atoms.

10-1   The Wave Equation

In 1924, a young French scientist, Louis de Broglie, proposed that, as light does, a stream of electrons might have wave properties in addition to particle properties. He suggested that a particle traveling at a velocity v possessed a wavelength λ which is given by

$$\lambda = \frac{h}{mv}$$

where h is Planck's constant and m is the mass of the particle. In 1927, C. J. Davisson and L. H. Germer, two Americans, and independently G. P. Thomson, an Englishman, measured the wavelength associated with a beam of electrons. By passing beams of electrons through thin gold foils, Thomson obtained diffraction patterns analogous to those observed when X rays are passed through a crystal. Thus, de Broglie's hypothesis was confirmed.

In 1926, Erwin Schrödinger postulated an equation similar to those which describe wave motion which could be applied to describe the behavior of an electron in an atom. The solutions of the Schrödinger equation are exact for the hydrogen atom. By use of approximation methods and electronic computers, satisfactory solutions of the Schrödinger equation can be obtained for atoms with more than one electron. Moreover, the Schrödinger equation can be generalized to provide interpretations of the behavior of electrons in atoms, vibrations of chemical bonds, and many other phenomena. In contrast to the Bohr theory, which postulated that the electrons occupied definite orbits within an atom, solutions to the Schrödinger equation describe the probabilities of finding the electrons in given regions of space about the nucleus or within a molecule.

The solutions of the Schrödinger equation involve integers which determine the energies and momenta of the electrons. The integers correspond to the quantum numbers of the Bohr theory, but in this case they are required by the mathematical form of the wave equation, whereas the quantum numbers of the Bohr theory were assumed arbitrarily. At this point, it is not necessary to learn how to solve the Schrödinger equation. It is only necessary to know that the solutions are used to describe the arrangement of electrons in atoms and that the quantum numbers are the most characteristic feature of the solutions. The arrangements of electrons in atoms are accounted for in terms of the quantum numbers alone. The properties of electrons which are determined by this arrangement include their energies, their orientations in space, and their interactions with other electrons within the same atom and with electrons of other atoms. The chemical and physical properties of an atom depend on the arrangement of its electrons; therefore the quantum numbers are immediately useful in providing chemical information.

10-2   Quantum Numbers

Each electron in an atom may be assigned a set of four **quantum numbers,** which define its energy, its orientation in space, and its possible interaction with other electrons. The **principal** quantum number is designated by the letter n. In the *hydrogen atom,* it defines the total energy of the electron in the same manner as the quantum number n of the Bohr theory.

A second quantum number, designated by the letter l, is called the **orbital angular momentum** quantum number. As its name implies, it specifies the angular momentum of the electron. In any atom other than the hydrogen atom, n and l together define the energy of the electron.

The **magnetic** quantum number, designated by m_l, defines the possible orientations of the angular momentum in space with respect to some arbitrarily defined axis. The magnetic quantum number becomes important in situations in which the electron interacts with external electric or magnetic fields, including the fields generated by other electrons.

As previously indicated, the principal, orbital angular momentum, and magnetic quantum numbers are direct results of the solutions of the Schrödinger equation. To account for the fine structure observed in atomic spectra, it is necessary to introduce a fourth quantum number, called the **spin magnetic** quantum number, m_s. One may think of this quantum number as describing the spinning of the electron about its axis as it moves above the nucleus. It can have values of $+\frac{1}{2}$ or $-\frac{1}{2}$ only.

For a given electron, only certain values of the four quantum numbers are permitted. These are listed in Table 10-1.

TABLE 10-1. Permitted Values for the Quantum Numbers

Symbol	Name	Permitted Values	Examples
n	principal quantum number	any positive integer	$1, 2, 3, \ldots$
l	orbital angular momentum quantum number	any integer from 0 up to $(n-1)$	$0 \cdots (n-1)$
m_l	magnetic quantum number	any integer from $-l$ through 0 to $+l$	$-l \cdots 0 \cdots +l$
m_s	spin magnetic quantum number	minus or plus $\frac{1}{2}$	$-\frac{1}{2}, +\frac{1}{2}$

Example

What are the possible values of l for an electron with $n = 3$?
l can be 0, 1, or 2.

Example

What are the possible values of m_l for an electron with $l = 2$.
The value of m_l can vary from $-l$ to $+l$; in this case m_l can have values $-2, -1, 0, +1$, and $+2$.

Example

What are the possible values of m_s for an electron with $m_l = 0$?
m_s must be $-\frac{1}{2}$ or $+\frac{1}{2}$. These are the only permitted values, regardless of the values of the other quantum numbers.

10-3 Orbitals

A solution of the Schrödinger equation, expressed in terms of a set of permitted values of n, l, and m_l, describes an **orbital.** The term orbital is a carryover from the original Bohr theory, but there is a fundamental difference between the concept of an orbital and the discrete orbits of the earlier model. The n, l, and m_l quantum numbers ascribed to each electron are actually the quantum numbers of the orbital which it occupies.

The most characteristic property of an orbital is its energy. Since in the absence of an external field the value of m_l does not influence the energy, orbitals are grouped into sets called subshells, which are denoted by the values of n and l only. Customarily, orbitals with l values of 0, 1, 2, and 3 are denoted as s, p, d, and f, respectively. Thus $1s$ denotes an orbital having $n = 1$ and $l = 0$; $4d$ denotes orbitals with $n = 4$ and $l = 2$. *Why can there be no 3f orbital?*

According to the **Heisenberg uncertainty principle,** the more exactly one attempts to measure the position of a small particle such as an electron, the less exactly one can simultaneously determine its momentum. Conversely, if its momentum is known very accurately, only probability statements can be made about the position of the electron. For example, an electron having a known momentum and kinetic energy may be directed toward a fluorescent screen. When the electron hits the screen, a flash of light is emitted and its position at that instant is established. However, as a result of the collision with the screen, energy is lost in the form of light and the momentum of the electron is changed. As a result of establishing the position of the electron, its momentum is no longer known.

The energy of an electron in a given orbital can be specified precisely from the values of its quantum numbers. However, the position of the electron is quite uncertain. Therefore the physical significance of an orbital is that while the solutions to the Schrödinger equation give the energies of electrons precisely, they can give only the probability of finding the electron in any particular region around the nucleus.

For the hydrogen atom, for which exact solutions of the Schrödinger equation are possible, the probabilities may be expressed diagrammatically in several ways. For example, Figure 10–1(a) illustrates the probability of finding the electron in a 1s orbital of hydrogen at certain positions near the nucleus. In the figure the density of the dots in any region is proportional to the probability of finding the electron in that region. It should be noted that this figure represents a two-dimensional cross section of the three-dimensional distribution of probability about the nucleus. Another useful diagram is a plot of the probability of finding the electron in the 1s orbital at a certain distance, r, from the nucleus, as shown in Figure 10–1(b). For the hydrogen atom it

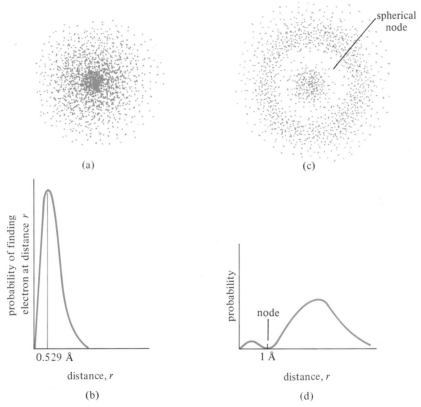

Figure 10–1. **Representations of Probability Distributions for an Electron in a 1s [(a) and (b)] or 2s [(c) and (d)] Orbital of Hydrogen.**

In the 2s orbital, there is a spherical surface at a certain distance from the nucleus at which the probability of finding the electron is zero. Such a surface is called a **node.**

is seen that there is maximum probability of finding the electron at a distance of 0.529 Å from the nucleus. This distance is the same as the accepted radius of the hydrogen atom and also agrees with the radius derived by means of the Bohr theory.

Similar representations for the 2s orbital of hydrogen are shown in Figures 10–1(c) and (d). It is seen that for the 2s orbital there is a greater probability of finding the electron farther away from the nucleus than in the case of the 1s orbital, but there is still some chance of finding the electron very close to the nucleus. Also significant is the fact that at an intermediate distance there is a surface at which the probability of finding the electron is zero.

Surfaces at which the probability goes to zero are called **nodes.** For any orbital having principal quantum number n, there are always $(n - 1)$ nodes (besides the node at infinity). Of these, there are l planar nodes and $(n - l - 1)$ spherical nodes. This nodal character of the probability distributions is consistent with the assumption that the motion of the electron has the character of a wave.

Diagrams such as Figures 10–1(a) and 10–1(c) are sometimes described as "electron clouds." The number of dots do *not* represent a number of electrons; rather they represent a number of probable instantaneous positions of a single electron.

Another way to look at the diagrams is to imagine taking photographs of a vast number of hydrogen atoms under similar conditions and then printing them as a single picture. If the actual positions of the electrons could be thus photographed, which is impossible, the resulting composite print might resemble Figure 10–1(a) or 10–1(c).

10-4 Plots of Orbitals

The probabilities of finding the electron in a given element of volume with respect to the atomic nucleus are customarily plotted using the Cartesian coordinate system, with the nucleus located at the origin. Because of the dependence of the solutions of the Schrödinger equation on the quantum numbers, plots corresponding to the various types of orbitals have characteristic shapes.

The plots may be depicted in the form of three-dimensional angular dependence diagrams, which give the relative magnitude of the probability of finding the electron in any direction from the nucleus. Alternatively, the plots may be depicted in the form of boundary surface diagrams, which show the smallest volume of space which encloses 90% of the probability distribution of the electron. Boundary surface diagrams for s, p, and d orbitals of the hydrogen atom are illustrated in Figure 10–2. It should be noted that while the s orbital is depicted as a sphere, the probability distribution inside the surface is not uniform but varies with r, for example, as shown in Figure 10–1(a) for a 1s orbital and in Figure 10–1(c) for a 2s orbital.

In the case of the p orbitals, there are three possible values of the quantum number m_l. Three different orbitals are depicted. These are oriented along the Cartesian axes. It is customary to label these orbitals p_x, p_y, and p_z. The labeling of the five d orbitals can also be associated with the Cartesian axes, as follows: the d_{z^2} orbital is symmetrical about the z axis, the $d_{x^2-y^2}$ orbital is distributed along the x and y axes, and the d_{xy}, d_{xz}, and d_{yz} orbitals lie in the xy, xz, and yz planes of the coordinate system,

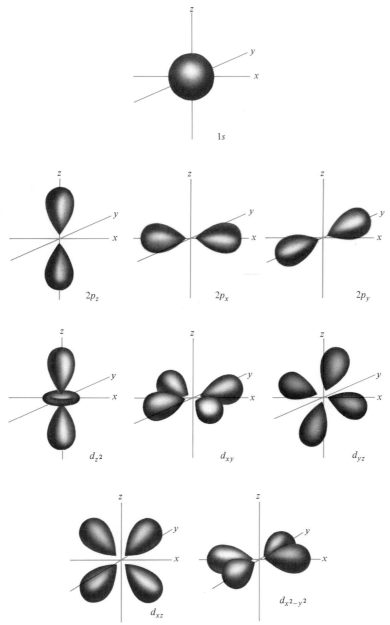

Figure 10-2. **Boundary Surfaces for Various Orbitals**

respectively. It is important to become thoroughly familiar with these representations of orbitals because they will be referred to again and again in the discussions which follow. They are referred to as "hydrogen-like" orbitals because they have been precisely derived only for the hydrogen atom. Similar orbitals are assumed to exist for atoms of the other elements.

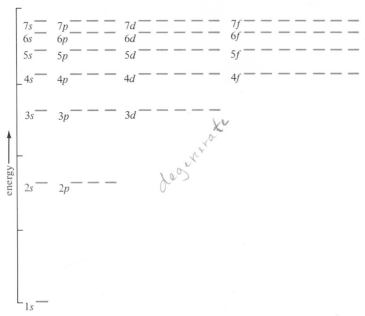

Figure 10-3. **Energy Level Diagram for Hydrogen (not drawn to scale)**

10-5 Energy Level Diagrams

A diagram showing the relative energies of the various orbitals of the hydrogen atom is given in Figure 10-3. The actual energies may be determined experimentally from the spectrum of hydrogen or values may be calculated using the Schrödinger equation. In the diagram, each orbital is depicted as a line, and energy increases toward the top. The displacement of some of the orbitals toward the right of the diagram is done merely for clarity and has no physical significance. In the case of hydrogen, orbitals having the same principal quantum number have the same energy. Orbitals with the same energies are said to be **degenerate.**

In terminology taken from the Bohr theory, energy levels with a common value of n are said to belong to the same **shell.** Spectroscopists use the following capital letter designations for the shells:

n value:	1	2	3	4	5	6	7
Shell designation:	K	L	M	N	O	P	Q

Orbitals in a given shell and also having a common value of the quantum number l are grouped into subshells. Thus the L shell has two subshells, the M shell has three subshells, and so on.

Energy level diagrams for atoms other than hydrogen are more complicated. When two or more electrons are present in the atom, they tend to repel each other because of their like charges, and the effect of the nuclear charge on each one is modified. The effect is different for different types of orbitals. This fact can be deduced from Figure 10-1, where it is seen that its probability distribution places an electron in the $2s$

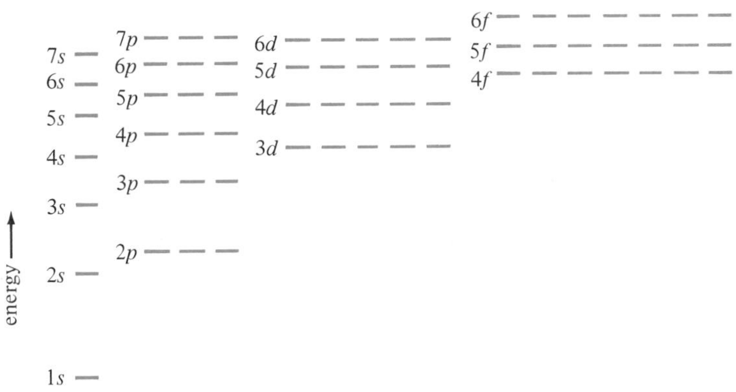

Figure 10-4. **Energy Level Diagram for Atoms Containing More Than One Electron (not drawn to scale)**

orbital somewhat farther from the nucleus than an electron in the $1s$ orbital. There-fore electrons in the $1s$ orbital would act as a screen of negative charge between the nucleus and electrons in the second shell, so these latter electrons should be subject to a diminished attractive force and have higher energies. However, Figure 10–1(c) shows that there is some probability that an electron in the $2s$ orbital will be found very close to the nucleus, even closer than the most probable location of the electron in the $1s$ orbital. While in such close proximity, the $2s$ electron would be subject to the full attractive force of the nucleus. An electron in the $2p$ orbital, on the other hand, does not penetrate the innermost shell, and its energy is always modified by the screening effect. When an electron in the $2s$ orbital penetrates the innermost shell, its energy is lowered somewhat compared to an electron in the $2p$ orbital. Consequently, in atoms containing more than one electron, the $2s$ and $2p$ subshells are no longer degenerate.

This argument can be extended to include shells having higher values of n, and as a result of the screening and penetration effects, an energy level diagram for many-electron atoms is that shown in Figure 10–4. It should be noted in the diagram that although the various subshells of a given shell are no longer degenerate with respect to each other, the orbitals within each subshell remain degenerate. (In the presence of a magnetic or an electrostatic field, however, the orbitals in a given subshell lose their degeneracy.)

The relative order of the energies of the various subshells depicted in Figure 10–4 may be approximated by the so-called $(n + l)$ rule. This mnemonic states that the order of increasing energy of the subshells in neutral atoms is determined by the order of increasing value of the sum $(n + l)$. As a corollary, when two subshells have the same value of the sum $(n + l)$, the one with the lower n value will usually be lower in energy. This rule applies to the outermost occupied energy levels. As the atomic numbers increase from one atom to the next in the periodic table, there are changes in the relative energies of the subshells, as shown in Figure 10–5. For example, in potassium, calcium, and scandium, the $4s$ subshell is slightly lower in energy than the $3d$ subshell. As atomic numbers increase, the energy of the $3d$ subshell decreases faster than that of the $4s$ subshell. After zinc, the $3d$ subshell is lower in energy. The energies are said to have "crossed."

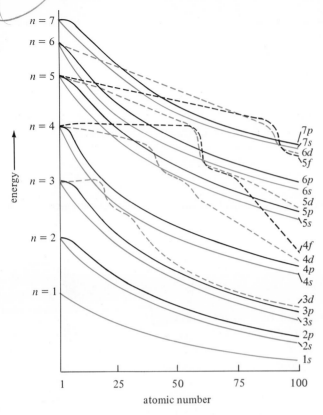

———— s
———— p
- - - - d
----- f

Example

Arrange the following orbitals, described only by their n and l quantum numbers, in order of increasing energy for atoms having fewer than 20 electrons.

Orbital	n	l
i	3	1
ii	3	2
iii	4	0
iv	2	1

The orbital with the lowest sum $(n + l)$ is iv, where the sum is 3. The orbitals i and iii have equal values of $(n + l)$, so i, the one with the lower value of n, is lower in energy. Highest of the four is ii, with a value of $(n + l) = 5$. Thus the order of increasing energy is iv, i, iii, ii. Verify that these orbitals correspond to the $2p$, $3p$, $4s$, and $3d$ subshells, respectively.

Outside an electric or magnetic field the values of the principal and orbital angular momentum quantum numbers are the only ones which affect the energy of an orbital. Orbitals differing only in m_l value will not differ in energy. Thus the order of increasing energy of the subshells is $1s$, $2s$, $2p$, $3s$, $3p$, $4s$, $3d$, $4p$, $5s$, $4d$, $5p$, $6s$, $4f$, $5d$,

$6p$, $7s$, $5f$, $6d$, $7p$, etc. Verify that this sequence is in accord with the $(n + l)$ rule and with Figure 10-4.

10-6   Electronic Configurations of Atoms

Electrons in atoms occupy orbitals in the order of increasing energy, and the atoms of each element have an **electronic configuration** characteristic of that element. Using Figure 10-4, it is possible to deduce the configuration of electrons in any atom merely by adding a number of electrons equal to the atomic number of the element, starting with the lowest energy level and working upward. In so doing, it is necessary to observe the **Pauli exclusion principle,** which states that no two electrons in the same atom can have the same set of four quantum numbers.

For example, the quantum numbers for a $1s$ orbital are $n = 1$, $l = 0$, $m_l = 0$. If two electrons are put into this orbital, the values of their spin magnetic quantum numbers must be $m_s = -\frac{1}{2}$ and $m_s = +\frac{1}{2}$, respectively. The Pauli principle is not violated, and the electrons are said to have **paired** spins. If a third electron were to be put into the same $1s$ orbital, its spin magnetic quantum number would have to be $-\frac{1}{2}$ or $+\frac{1}{2}$, since there is no other choice. But then it would have the same set of four quantum numbers as one of the first two electrons. This situation is not allowed. Indeed, for any specified orbital the values of n, l, and m_l are fixed; hence the orbital can accommodate only two electrons with spins paired. Of course, two different orbitals in the same atom can each contain an electron having the identical value for the spin magnetic quantum number because at least one of the other quantum numbers will be different.

When its electrons are occupying orbitals having the lowest available energies, an atom is in its most stable state. This arrangement of electrons is called the **ground state.** Any other allowed arrangement will correspond to an **excited state** of the atom. Transitions between the ground state and the excited states and vice versa will be accompanied by the absorption or emission of discrete amounts of energy.

In its ground state the hydrogen atom has a single electron in its $1s$ orbital. Helium (atomic number 2) has two electrons in the $1s$ orbital with paired spins. Lithium (atomic number 3) has two electrons in the $1s$ orbital and one electron in the $2s$ orbital. Thus the ground state electronic configuration of the elements may be built up either by starting with the energy level diagram and feeding in all the electrons, or more simply by adding one "last electron" to the electronic configuration of the atom having the next lower atomic number. The electronic configuration of each of the elements is given in Table 10-2.

It is customary to write the electronic configuration of any element in terms of the numbers of electrons which occupy each of the subshells. The subshells are listed in order of increasing energy, and superscripts are added to denote the number of electrons occupying each. For example, the electronic configuration of potassium (atomic number 19) is

$$1s^2\ 2s^2\ 2p^6\ 3s^2\ 3p^6\ 4s^1$$

It should be obvious that the sum of the superscripts is merely the number of electrons in the atom.

Z	Element	1 s	2 s	2 p	3 s	3 p	3 d	4 s	4 p	4 d	4 f	5 s	5 p	5 d	5 f	6 s	6 p	6 d	7 s	First Ionization Potential (eV)
1	H	1																		13.595
2	He	2																		24.481
3	Li	2	1																	5.39
4	Be	2	2																	9.32
5	B	2	2	1																8.296
6	C	2	2	2																11.256
7	N	2	2	3																14.53
8	O	2	2	4																13.614
9	F	2	2	5																17.41
10	Ne	2	2	6																21.559
11	Na	2	2	6	1															5.138
12	Mg	2	2	6	2															7.644
13	Al	2	2	6	2	1														5.984
14	Si	2	2	6	2	2														8.149
15	P	2	2	6	2	3														10.484
16	S	2	2	6	2	4														10.357
17	Cl	2	2	6	2	5														13.01
18	Ar	2	2	6	2	6														15.755
19	K	2	2	6	2	6		1												4.339
20	Ca	2	2	6	2	6		2												6.111
21	Sc	2	2	6	2	6	1	2												6.54
22	Ti	2	2	6	2	6	2	2												6.82
23	V	2	2	6	2	6	3	2												6.74
24	Cr	2	2	6	2	6	5	1												6.764
25	Mn	2	2	6	2	6	5	2												7.432
26	Fe	2	2	6	2	6	6	2												7.87
27	Co	2	2	6	2	6	7	2												7.86
28	Ni	2	2	6	2	6	8	2												7.633
29	Cu	2	2	6	2	6	10	1												7.724
30	Zn	2	2	6	2	6	10	2												9.391
31	Ga	2	2	6	2	6	10	2	1											6
32	Ge	2	2	6	2	6	10	2	2											7.88
33	As	2	2	6	2	6	10	2	3											9.81
34	Se	2	2	6	2	6	10	2	4											9.75
35	Br	2	2	6	2	6	10	2	5											11.84
36	Kr	2	2	6	2	6	10	2	6											13.996
37	Rb	2	2	6	2	6	10	2	6			1								4.176
38	Sr	2	2	6	2	6	10	2	6			2								5.692
39	Y	2	2	6	2	6	10	2	6	1		2								6.38
40	Zr	2	2	6	2	6	10	2	6	2		2								6.84
41	Nb	2	2	6	2	6	10	2	6	4		1								6.88
42	Mo	2	2	6	2	6	10	2	6	5		1								7.10
43	Tc	2	2	6	2	6	10	2	6	6		1								7.28
44	Ru	2	2	6	2	6	10	2	6	7		1								7.36
45	Rh	2	2	6	2	6	10	2	6	8		1								7.46
46	Pd	2	2	6	2	6	10	2	6	10										8.33
47	Ag	2	2	6	2	6	10	2	6	10		1								7.574
48	Cd	2	2	6	2	6	10	2	6	10		2								8.991
49	In	2	2	6	2	6	10	2	6	10		2	1							5.785
50	Sn	2	2	6	2	6	10	2	6	10		2	2							7.342
51	Sb	2	2	6	2	6	10	2	6	10		2	3							8.639
52	Te	2	2	6	2	6	10	2	6	10		2	4							9.01

TABLE 10-2.

TABLE 10-2. Electronic Configurations and First Ionization Potentials (continued)

Z	Element	1s	2s	2p	3s	3p	3d	4s	4p	4d	4f	5s	5p	5d	5f	6s	6p	6d	7s	First Ionization Potential (eV)
53	I	2	2	6	2	6	10	2	6	10		2	5							10.454
54	Xe	2	2	6	2	6	10	2	6	10		2	6							12.127
55	Cs	2	2	6	2	6	10	2	6	10		2	6			1				3.893
56	Ba	2	2	6	2	6	10	2	6	10		2	6			2				5.21
57	La	2	2	6	2	6	10	2	6	10		2	6	1		2				5.61
58	Ce	2	2	6	2	6	10	2	6	10	2	2	6			2				5.6
59	Pr	2	2	6	2	6	10	2	6	10	3	2	6			2				5.46
60	Nd	2	2	6	2	6	10	2	6	10	4	2	6			2				5.51
61	Pm	2	2	6	2	6	10	2	6	10	5	2	6			2				
62	Sm	2	2	6	2	6	10	2	6	10	6	2	6			2				5.6
63	Eu	2	2	6	2	6	10	2	6	10	7	2	6			2				5.67
64	Gd	2	2	6	2	6	10	2	6	10	7	2	6	1		2				6.16
65	Tb	2	2	6	2	6	10	2	6	10	9	2	6			2				5.98
66	Dy	2	2	6	2	6	10	2	6	10	10	2	6			2				6.8
67	Ho	2	2	6	2	6	10	2	6	10	11	2	6			2				
68	Er	2	2	6	2	6	10	2	6	10	12	2	6			2				6.08
69	Tm	2	2	6	2	6	10	2	6	10	13	2	6			2				5.81
70	Yb	2	2	6	2	6	10	2	6	10	14	2	6			2				6.2
71	Lu	2	2	6	2	6	10	2	6	10	14	2	6	1		2				
72	Hf	2	2	6	2	6	10	2	6	10	14	2	6	2		2				7
73	Ta	2	2	6	2	6	10	2	6	10	14	2	6	3		2				7.88
74	W	2	2	6	2	6	10	2	6	10	14	2	6	4		2				7.98
75	Re	2	2	6	2	6	10	2	6	10	14	2	6	5		2				7.87
76	Os	2	2	6	2	6	10	2	6	10	14	2	6	6		2				8.5
77	Ir	2	2	6	2	6	10	2	6	10	14	2	6	7		2				9
78	Pt	2	2	6	2	6	10	2	6	10	14	2	6	9		1				9.0
79	Au	2	2	6	2	6	10	2	6	10	14	2	6	10		1				9.22
80	Hg	2	2	6	2	6	10	2	6	10	14	2	6	10		2				10.43
81	Tl	2	2	6	2	6	10	2	6	10	14	2	6	10		2	1			6.106
82	Pb	2	2	6	2	6	10	2	6	10	14	2	6	10		2	2			7.415
83	Bi	2	2	6	2	6	10	2	6	10	14	2	6	10		2	3			7.287
84	Po	2	2	6	2	6	10	2	6	10	14	2	6	10		2	4			8.43
85	At	2	2	6	2	6	10	2	6	10	14	2	6	10		2	5			9.5
86	Rn	2	2	6	2	6	10	2	6	10	14	2	6	10		2	6			10.746
87	Fr	2	2	6	2	6	10	2	6	10	14	2	6	10		2	6		1	4
88	Ra	2	2	6	2	6	10	2	6	10	14	2	6	10		2	6		2	5.277
89	Ac	2	2	6	2	6	10	2	6	10	14	2	6	10		2	6	1	2	6.9
90	Th	2	2	6	2	6	10	2	6	10	14	2	6	10		2	6	2	2	6.95
91	Pa	2	2	6	2	6	10	2	6	10	14	2	6	10	2	2	6	1	2	
92	U	2	2	6	2	6	10	2	6	10	14	2	6	10	3	2	6	1	2	6.08
93	Np	2	2	6	2	6	10	2	6	10	14	2	6	10	4	2	6	1	2	
94	Pu	2	2	6	2	6	10	2	6	10	14	2	6	10	6	2	6		2	5.1
95	Am	2	2	6	2	6	10	2	6	10	14	2	6	10	7	2	6		2	
96	Cm	2	2	6	2	6	10	2	6	10	14	2	6	10	7	2	6	1	2	
97	Bk	2	2	6	2	6	10	2	6	10	14	2	6	10	8	2	6	1	2	
98	Cf	2	2	6	2	6	10	2	6	10	14	2	6	10	10	2	6		2	
99	Es	2	2	6	2	6	10	2	6	10	14	2	6	10	11	2	6		2	
100	Fm	2	2	6	2	6	10	2	6	10	14	2	6	10	12	2	6		2	
101	Md	2	2	6	2	6	10	2	6	10	14	2	6	10	13	2	6		2	
102	No	2	2	6	2	6	10	2	6	10	14	2	6	10	14	2	6		2	
103	Lr	2	2	6	2	6	10	2	6	10	14	2	6	10	14	2	6	1	2	

10-7   Electronic Configurations of Ions

The electronic configurations of ions may be written similarly to those of atoms. If the ion has a negative charge, a sufficient number of electrons in excess of the atomic number is added to the appropriate orbitals. The electronic structure of a positive ion is obtained from that of the neutral atom by assuming that the required number of electrons is removed. However, for positive ions, the electrons with the highest values of $(n + l)$ are *not* necessarily removed first. In the ionization process the electrons which are removed first are the **valence shell** electrons—those with the highest values of n. If more than one subshell of the valence shell is occupied, the one with the highest l value loses the electrons first. It is particularly important to remember these rules in deriving the electronic configurations of ions of the transition and inner transition metals.

Example

Write the electronic configurations of S^{2-} and Ni^{2+}.

$$(S \qquad 1s^2\, 2s^2\, 2p^6\, 3s^2\, 3p^4)$$
$$S^{2-} \qquad 1s^2\, 2s^2\, 2p^6\, 3s^2\, 3p^6$$
$$(Ni \qquad 1s^2\, 2s^2\, 2p^6\, 3s^2\, 3p^6\, 4s^2\, 3d^8)$$
$$Ni^{2+} \qquad 1s^2\, 2s^2\, 2p^6\, 3s^2\, 3p^6\, 4s^0\, 3d^8$$

10-8   Hund's Rule

When a more detailed description of the electronic structure of an atom is desired, the arrangement of electrons within a partially filled subshell becomes important. Consider boron, with the electronic configuration $1s^2\, 2s^2\, 2p^1$. It makes no difference which of the three $2p$ orbitals contains the electron. On the other hand, for carbon, $1s^2\, 2s^2\, 2p^2$, there are several possibilities: One of the $2p$ orbitals could contain both electrons with their spins paired, or the electrons could occupy two different orbitals, for example, the $2p_x$ and $2p_y$, with spins paired or with spins unpaired. In the ground state of carbon, the electrons actually occupy separate orbitals with their spins unpaired. This configuration is an application of **Hund's rule of maximum multiplicity,** which states that electrons will occupy the degenerate orbitals of a subshell singly with spins parallel (unpaired) until the subshell is half filled, and then successive electrons will begin to pair up in the already singly occupied orbitals.

Example

Write the detailed electronic configuration for the nitrogen atom.

Nitrogen has seven electrons to be placed in the $1s$, $2s$, and $2p$ orbitals. Each of the s orbitals will contain a pair of electrons. Since these orbitals are completely filled and since they are not degenerate, Hund's rule does not apply to them (for both reasons). The remaining electrons will occupy the $2p$ orbitals, all unpaired, one in each orbital. The detailed electronic configuration for the nitrogen atom therefore is $1s^2\, 2s^2\, 2p_x^{\,1}\, 2p_y^{\,1}\, 2p_z^{\,1}$.

A corollary to Hund's rule is that half-filled or completely filled subshells have enhanced stabilities. This effect, along with the possibility that the energies of two subshells will cross as the atomic numbers of the atoms increase (Figure 10–5), helps to explain several apparent discrepancies in the electronic configurations given in Table 10–2. For example, the electronic configuration[1] of chromium is $[Ar]\,4s^1\,3d^5$ rather than $[Ar]\,4s^2\,3d^4$. The former arrangement is the more stable because of the half-filled subshells. In copper, the configuration is $[Ar]\,4s^1\,3d^{10}$ rather than $[Ar]\,4s^2\,3d^9$. Here the increased stability of a completely filled ($3d$) and a half filled ($4s$) subshell is a contributing factor.

10-9   Magnetic Properties

The magnetic properties of substances, whether composed of atoms, ions, or molecules, are direct consequences of their electronic structures. All stable chemical species have at least one pair of electrons. Many substances have some unpaired electrons as well. There are two different types of interactions of substances with magnetic fields—diamagnetism and paramagnetism—described below.

Diamagnetic substances are repelled by magnetic fields. These substances have electronic structures in which all the electrons are paired. In fact, some diamagnetism exists in all stable chemical species, even those which also have unpaired electrons. However, if even one unpaired electron is present per atom, ion, or molecule, the diamagnetism will be swamped by the much larger effect of paramagnetism.

Paramagnetic substances are attracted into magnetic fields. They have one or more unpaired electrons. The interactions of the unpaired electrons with the magnetic field are much greater than those of paired electrons, so every species with unpaired electrons will tend to be drawn into a magnetic field despite the presence of paired electrons. The latter only diminish the attraction somewhat. The approximate magnetic moment due to the spin of the unpaired electrons is expressed by the "spin-only formula":

$$\mu = \sqrt{n(n + 2)}$$

where μ is the magnetic moment in Bohr magnetons[2] and n is the number of unpaired electrons. It must be pointed out that the orbital motions of the electrons also contribute to the magnetic moments of substances; consequently the spin-only formula does not always agree with the observed magnetic moment. The magnetic moments of several ions are listed in Table 10–3.

Another kind of magnetic behavior is ferromagnetism, of which a familiar example is the ordinary magnet. **Ferromagnetic** materials are strongly attracted into magnetic fields. This attraction results from the cooperative interaction of the unpaired electrons in one atom of the material with those of the other atoms. Although ferromagnetism is of great practical importance, it is not directly related to the electronic structures of individual atoms or ions and is not of immediate interest. Other types of magnetic behavior also exist, but they will not be discussed in this text.

---

[1] [Ar] signifies the electronic configuration of Ar. The other subshells are added to this core.

[2] One Bohr magneton (1 B.M.) $= \dfrac{eh}{4\pi mc} = 9.273$ ergs/gauss.

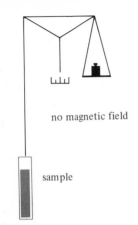

no magnetic field

sample

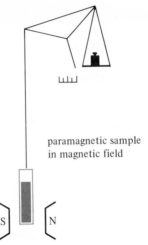

paramagnetic sample
in magnetic field

S   N

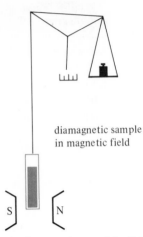

diamagnetic sample
in magnetic field

S   N

(a) A sample in a sealed glass tube is suspended from one arm of a balance, and its weight is determined in the absence of a magnetic field.

(b) Then a strong magnetic field is placed around the sample. A paramagnetic material will be attracted into the field and will appear to weigh more. If the strength of the field is about 10,000 gauss, the change in apparent weight of a paramagnetic sample may amount to about 1%.

(c) A diamagnetic material will be repelled by the magnetic field and appear to weigh less. The effect of diamagnetism is about $10^{-3}\%$.

Figure 10-6. Experimental Determination of Magnetic Susceptibility

Some details regarding the experimental determination of magnetic moments are given in Figure 10-6.

Example

Predict the magnetic moment of the Cu^{2+} ion.

The electronic configuration of Cu^{2+} is [Ar] $3d^9$. The $3d$ orbitals are occupied as represented in the following diagram, where each electron is represented by an arrow pointing either up or down to denote its spin.

TABLE 10-3. Some Calculated and Observed Magnetic Moments

Ion	Number of Unpaired Electrons	Calculated Moment (B.M.)	Observed Moments (B.M.)
V^{4+}	1	1.73	1.7–1.8
Cu^{2+}	1	1.73	1.7–2.2
V^{3+}	2	2.83	2.6–2.8
Ni^{2+}	2	2.83	2.8–4.0
Cr^{3+}	3	3.87	~3.8
Co^{2+}	3	3.87	4.1–5.2
Fe^{2+}	4	4.90	5.1–5.5
Co^{3+}	4	4.90	~5.4
Mn^{2+}	5	5.92	~5.9
Fe^{3+}	5	5.92	~5.9

$$Cu^{2+} \quad \underline{\uparrow\downarrow} \; \underline{\uparrow\downarrow} \; \underline{\uparrow\downarrow} \; \underline{\uparrow\downarrow} \; \underline{\uparrow} \qquad \underline{}$$
$$\quad\qquad\qquad 3d \qquad\qquad\qquad 4s$$

Hence Cu^{2+} has one unpaired electron, and its moment is given by

$$\mu = \sqrt{1(1+2)} = 1.73 \text{ B.M.}$$

Consequences of Electronic Structure

10-10   Electronic Configurations and the Periodic Table

The electronic energy levels, diagrammed in Figure 10–4, are the underlying basis of the periodic table. Consider the questions "How many electrons can be added to an atom in its ground state before the first electron is introduced into the $n = 2$ shell? into the $n = 3$ shell? into the $n = 4, 5, 6, 7$ shells?" The answers to these questions are as follows:

before the $n = 2$ shell $\cdots$ 2
$n = 3$ shell $\cdots$ 10
$n = 4$ shell $\cdots$ 18
$n = 5$ shell $\cdots$ 32
$n = 6$ shell $\cdots$ 54
$n = 7$ shell $\cdots$ 86

These numbers are the atomic numbers of the noble gases. The differences between succeeding numbers, (2), 8, 8, 18, 18, 32, are the numbers of elements in the successive rows of the periodic table.

The electronic structures of the elements also account for the group relationships in the periodic table. The elements lithium, sodium, potassium, rubidium, and cesium, for example, all in the same group in the periodic table, have very similar chemical properties. Their electronic configurations provide a clue as to why this is so:

Li	$1s^2\, 2s^1$
Na	$1s^2\, 2s^2\, 2p^6\, 3s^1$
K	$1s^2\, 2s^2\, 2p^6\, 3s^2\, 3p^6\, 4s^1$
Rb	$1s^2\, 2s^2\, 2p^6\, 3s^2\, 3p^6\, 4s^2\, 3d^{10}\, 4p^6\, 5s^1$
Cs	$1s^2\, 2s^2\, 2p^6\, 3s^2\, 3p^6\, 4s^2\, 3d^{10}\, 4p^6\, 5s^2\, 4d^{10}\, 5p^6\, 6s^1$

The outermost shell of each element contains a single electron in an s orbital. The underlying shells have the electronic configuration of a noble gas element. The members of each of the other groups of the periodic table likewise have electronic configurations in their outer shells which are characteristic of the respective groups. Since the members of a particular periodic group all have similar chemical properties, the conclusion is that these properties are determined by their electronic configurations.

As can be seen in Table 10–2, the elements which are members of transition metal groups in the periodic table are those in which the differentiating electrons are found in d orbitals. These are called the transition elements, or, more aptly, the "d-block elements." Similarly, the elements having atomic numbers 58 to 71, the **lanthanide series,** and 90 to 103, the **actinide series,** can be referred to as "f-block elements." In these cases the differentiating electrons occupy f orbitals two shells below the outermost shell. It is convenient to use the periodic table as an aid in deducing the electronic structures of the atoms, as shown in Figure 10–7. Since the electronic configuration of the atoms is the basis of the periodic table, the electronic configuration of any element can be deduced from its position in the table. It is especially convenient to use the periodic table to ascertain the configuration of the outermost electrons by starting at the beginning of the period in which the element is found.

Example

What is the electronic configuration of lead?

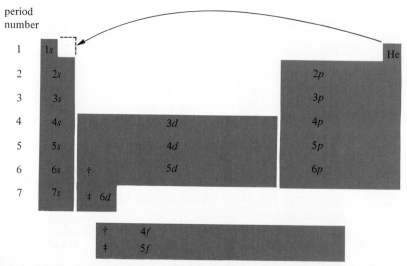

Figure 10–7. Use of the Periodic Table to Determine the Electronic Configurations of the Elements

To deduce the electronic configuration of an element using the periodic table as a mnemonic device, regard the regular table as if it had the form shown in this figure. Note that for this purpose, He is placed in the 1s block. Also note that the elements in the s and p blocks correspond to electrons having principal quantum numbers equal to the period number; the d-block elements correspond to electrons having principal quantum numbers one less than the period number; and the f-block elements correspond to electrons having principal quantum numbers two less than the period number.

One starts with hydrogen and merely counts, in the order of increasing atomic number, the elements in the various blocks as if they were electrons in the corresponding orbitals, up to and including the element in question. To ascertain the electronic configuration of sodium, for example, one counts the two s-block elements H and He as 1s^2, the two s-block elements Li and Be as 2s^2, the six p-block elements B through Ne as 2p^6, and Na as 3s^1. Thus sodium has the configuration 1s^2 2s^2 2p^6 3s^1.

After gaining some experience, one can readily find an element's *outermost* electronic configuration merely by starting the process at the beginning of that element's period. Thus the outermost electronic configuration of Sc is 4s^2 3d^1, and the underlying electrons are those of Ar.

Although this procedure cannot be applied in cases of exceptions to the general rules of electronic configurations, such as Cr or Cu, it is useful in the majority of cases.

Lead is in period 6. Its inner structure corresponds to that of xenon; the 55th through 82nd electrons occupy subshells as follows:

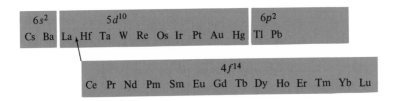

The configuration is $[Xe] 6s^2 5d^{10} 4f^{14} 6p^2$.

The lower stability of an electron in a given valence shell orbital due to the presence of electrons in inner orbitals is called the **screening effect.** Electrons in s orbitals screen other electrons from the nucleus better than electrons in orbitals with higher values of l. The enhanced stability of an electron occupying an orbital which has a probability of being closer to the nucleus at times than orbitals of lower principal quantum number is called the **penetration effect.** Within a given shell, d orbital electrons penetrate less than p orbital electrons, which in turn penetrate less than s orbital electrons. If the only occupied orbitals below the valence shell are s and p orbitals, the shielding effect will predominate. If the shell(s) beneath the valence shell contain(s) occupied d and/or f orbitals, the penetration effect will be quite large.

The differences between the main group elements and the corresponding d-block elements can be explained in terms of their electronic structures. Consider periodic group I A, the alkali metals, and group I B, the coinage metals. The former are very reactive. They form stable compounds; hence they are never found in nature in the free state. Copper, silver, and gold are rather unreactive and often occur in nature as the free metals. The free metals are easily obtained from their compounds. Their only similarities to the group I A elements are that they are metals and they form compounds having analogous formulas; for example, all form monochlorides. Comparing the electronic structures of potassium and copper, it is seen that each contains a single electron in the $4s$ orbital. However, in the case of potassium, this electron is outside a shell having the configuration $3s^2 3p^6$. The inner s and p electrons are rather effective in screening the $4s$ electron from the positive nuclear charge, and although the $4s$ electron does penetrate within the inner shell electrons, for potassium the screening effect is more important than the penetration. Consequently, the $4s$ electron of potassium is not strongly bonded, and, for example, the metal can be readily oxidized by water. In the case of copper, the $4s$ electron is outside a shell having the configuration $3s^2 3p^6 3d^{10}$. The electrons in the $3d$ subshell are not very effective at screening the $4s$ electron from the nuclear charge (Figure 10–8), and for copper the penetration effect is more important than the screening effect. Since the nuclear charge of the

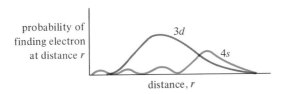

Figure 10–8. **Graph Showing Penetration of $4s$ Electron Probability Density Within That of the $3d$ Electrons**

copper atom is ten units greater than that of the potassium atom and the extra electrons do not screen well, the 4s electron is more difficult to remove. As one result, copper is not oxidized by water or by hydronium ions even in strongly acid solutions.

10–11 Atomic Sizes

The size of an individual atom is difficult to define. The maxima in the electron probability distribution of atoms occur at definite distances from the nuclei, but there is some probability of finding the electron farther from the nucleus, as far out as infinity [see Figure 10–1(b)]. From a practical viewpoint, one can tell the size of the atom from data which show how closely two atoms, bonded or unbonded, can approach each other. These experimentally determined atomic sizes are useful for comparisons of the sizes of different atoms, but they depend on the method of measurement and the physical state of the sample. Atomic radii deduced from X-ray and spectroscopic studies are given in Figure 10–9. Details of how these values were obtained will be presented in Chapter 13.

The apparent size of an atom is a consequence of the nuclear charge of the atom and of its electronic configuration. It is apparent that for a given number of electrons, the larger the nuclear charge, the greater will be the attraction and the smaller will be the volume occupied by the electrons. With a given nuclear charge, the greater the number of electrons in an atom or ion, the greater will be the interelectronic repulsion and therefore the larger will be the volume. Thus positive ions will be smaller than neutral atoms containing the same number of electrons, while negative ions will be larger than neutral atoms having the same nuclear charge.

Comparisons between species which have neither equal nuclear charges nor equal numbers of electrons are more complex. In going from one atom to the next in the periodic table, the number of protons and the number of electrons each increase by 1. Therefore the difference in size between two adjoining elements depends mainly on the orbital into which the added electron goes and also on which orbitals are already occupied. The important considerations are whether (1) a new electronic shell is started, (2) the electron enters a more or less penetrating orbital, and (3) the added electron effectively shields electrons already present.

There are definite correlations between the sizes of the atoms and their positions in the periodic table. In going from left to right across a period, the outermost electrons of consecutive elements all have the same value for n, the principal quantum number. There is a general decrease in size as the nuclear charge increases. A large decrease is found when the added electron is placed in a penetrating orbital, as illustrated by the large reduction in size between sodium and magnesium. The added 3s electron penetrates the inner electron shells and in addition does not screen the other 3s electron from the increased nuclear charge.

With the d-block elements, the added electron goes into an underlying shell, which provides some screening for the outermost s electrons, and a smaller reduction in size is expected. However, the d electrons do not penetrate, and when sufficient numbers are present, interelectronic repulsion leads to gradual increase in size. As a result of these opposing tendencies, the sizes of the d-block elements decrease to a minimum and then increase gradually.

Electrons in f orbitals are not very effective in screening other electrons from the nuclear charge, and there is a gradual decrease in size of the lanthanide elements

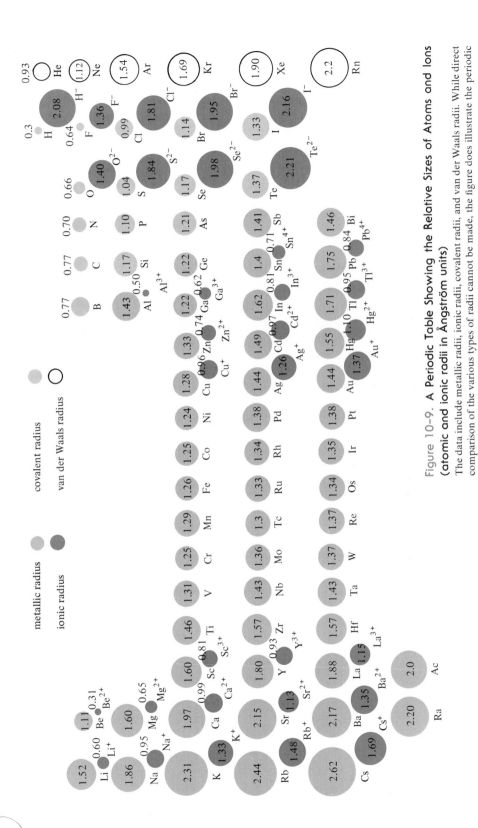

Figure 10-9. A Periodic Table Showing the Relative Sizes of Atoms and Ions (atomic and ionic radii in Ångström units)

The data include metallic radii, ionic radii, covalent radii, and van der Waals radii. While direct comparison of the various types of radii cannot be made, the figure does illustrate the periodic trends in sizes.

metallic radius

ionic radius

covalent radius

van der Waals radius

which is termed the **lanthanide contraction.** Although the effect is given this name, it is not an unusual effect. The name is useful to describe the reason that the sizes of the transition elements following the lanthanide series are smaller than might be expected. Because of the lanthanide contraction, atoms of the elements starting with hafnium are very close in size to atoms of the elements immediately above them in the periodic table. As a consequence, the properties of these sets of elements are quite similar; for example, hafnium is so similar to zirconium that the two are very difficult to separate.

Generally, in any periodic group, size increases with increasing atomic number. This effect is expected since more shells are being added as the atomic number is increased, and the type(s) of orbital(s) occupied in the outermost shell(s) is(are) kept constant. The effect of adding new shells of electrons generally overpowers the effect of adding a number of protons equal to the number of electrons.

10–12 Ionization Potentials

The energy necessary to remove an electron from a gaseous atom to produce a gaseous ion is called the **ionization potential,** IP, of the element. For example, the energy required for the following reaction is the ionization potential of sodium:

$$Na(g) \rightarrow Na^+(g) + e^-$$

This process is not the one which takes place in most ordinary chemical reactions, because here both the reactant and the product are in the gas phase. Note that energy is always required to remove an electron from an atom. The energy required to

TABLE 10-4. Successive Ionization
Potentials of Several Elements (eV)

	First	Second	Third	Fourth
He	24.481	54.14		
Ne	21.559	40.9	63.2	
Na	5.138	47.06	70.72	
Mg	7.644	14.96	79.72	108.9
Al	5.984	18.74	28.31	119.37
K	4.339	31.66	46.5	
Ca	6.111	11.82	50.96	69.7
Sc	6.54	12.8	24.61	
Ni	7.633	18.2		
Cu	7.724	20.34	29.5	
Zn	9.391	17.89	40.0	
Ag	7.574	21.4	35.9	
La	5.61	11.4		
Ce	5.6	14.8		
Pr	5.46			
Au	9.22	19.95		
O_2	12.08			

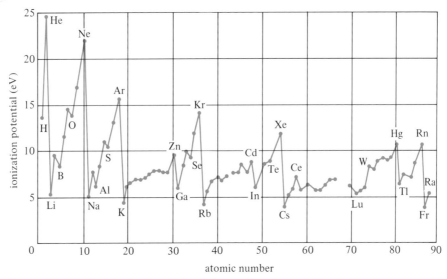

Figure 10–10. **First Ionization Potential as a Function of Atomic Number**

remove a second electron is called the second ionization potential. Since the second electron is removed from a positive ion, considerably more energy is required; second ionization potentials are always much larger than the corresponding first ionization potentials. Successive ionization potentials for several elements are listed in Table 10–4.

Like other properties depending on electronic configuration, the ionization potentials of the elements vary in a periodic manner. The first ionization potentials of the elements are listed in Table 10–2 and are plotted against atomic number in Figure 10–10. The lowest values of ionization potential are observed for the alkali metals. From lithium to cesium, the ionization potentials of the metals become progressively smaller. This effect is consistent with the trend in atomic sizes of the alkali metals. Owing to the enhanced stability of the $ns^2\,np^6$ electronic configuration, the noble gas elements have the highest ionization potentials.

With respect to the periodic table, the ionization potentials are highest in the upper right-hand corner and lowest in the lower left. In general, as one goes from left to right across a period, ionization potential increases. However, the difference between electrons in penetrating and nonpenetrating orbitals and the increased stability of half-filled or fully filled subshells make for an irregular increase; hence the jagged appearance of the plot in Figure 10–8.

10–13 Electron Affinity

The **electron affinity,** EA, of an element is the energy *released* when an electron is added to a gaseous atom to form a gaseous ion. Like ionization potential, electron affinity is a periodic property. Except for the noble gases, electron affinity like ionization potential increases toward the upper right-hand corner of the periodic table. Note that both the energy of addition of an electron (electron affinity) and the energy of removal of an electron (ionization potential) increase toward the upper

TABLE 10-5. Electron Affinities of Some Elements

	Energy (eV)[a]		Energy (eV)[a]		Energy (eV)[a]
H	0.754	K	0.501	Te	1.971
He	−0.22	Ca	−1.62	I	3.061
Li	0.620	Cr	0.66	Xe	−0.42
Be	−2.5	Ni	1.15	Cs	0.472
B	0.24	Cu	1.276	Ba	−0.54
C	1.270	Ga	0.37	Ta	0.8
N	0.0	Ge	1.20	W	0.5
O	1.465	As	0.80	Re	0.15
F	3.339	Se	2.020	Pt	2.128
Ne	−0.30	Br	3.363	Au	2.309
Na	0.548	Kr	−0.40	Tl	0.5
Mg	−2.4	Rb	0.486	Pb	1.05
Al	0.46	Sr	−1.74	Bi	1.05
Si	1.24	Mo	1.0	Po	1.8
P	0.77	Ag	1.303	At	2.8
S	2.077	In	0.35	Rn	−0.42
Cl	3.614	Sn	1.25	Fr	0.456
Ar	−0.36	Sb	1.05		

[a] The energy, in electron volts, required to add an electron to a gaseous atom to form a gaseous ion at 0 K.

right-hand corner of the periodic table because one is defined in terms of energy needed and the other in terms of energy released.

The electron affinities of some elements are listed in Table 10–5. The halogen atoms have relatively large electron affinities because the resulting negative ions have the stable electronic configuration of the noble gases. Group II A elements have the lowest values of electron affinity.

10-14 Oxidation States of the Elements

In Figure 10–11 the oxidation states of the main group and transition elements are plotted against atomic number. Despite the fact that the oxidation states are assigned according to rather arbitrary rules (Section 4–2), the periodic trends (Figure 10–11) suggest that electronic configurations influence their values.

The metallic elements may be classified into several types in terms of their electronic structures. The first type includes all the elements whose positive ions have noble gas configurations. These elements generally have only one positive oxidation state, as, for example, the alkali and alkaline earth metals. A second type includes elements which form positive ions which have incompletely filled d subshells. These elements characteristically have several positive oxidation states above $+1$, varying in steps of 1 or more, as, for example, Fe^{II} and Fe^{III}. After the overlying s orbital electrons are removed, a variable number of d electrons may be lost. A third type of electronic structure includes elements with positive ions which do not have a noble gas configuration but which do have completely filled d subshells. In this type of element, the s orbitals of the valence shell penetrate the d subshell. The completely filled d subshell is very stable, and as a result these elements have either a single positive oxidation state, as, for example, Zn^{II} and Cd^{II}, or two positive oxidation states differing by two units, as, for example, Tl^I and Tl^{III}. In Tl^I, only the $6p^1$ electron is lost upon oxidation, while in Tl^{III} the $6s^2$ electron pair is removed as well. Since the p electrons are more easily removed than the pair of s electrons, the latter are sometimes referred to as the "inert pair." Additional examples of this type of configuration are Pb^{II} and Bi^{III}.

In the case of nonmetallic elements, positive oxidation states are observed only when the atom is in combination with fluorine, oxygen, or another element located in the periodic table above or to the right of the element in question. For an element which can form compounds with different numbers of oxygen atoms, the oxidation state of the element varies in steps of 2. For example, the most important positive

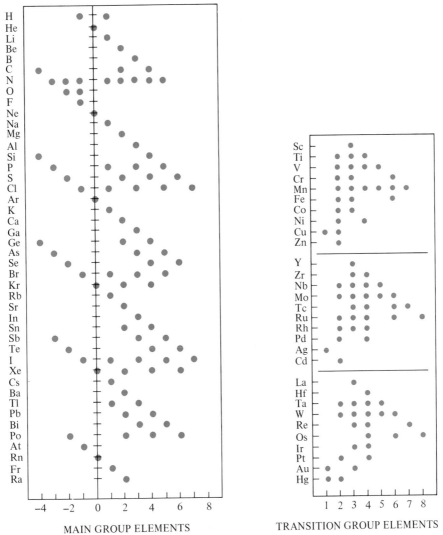

MAIN GROUP ELEMENTS   TRANSITION GROUP ELEMENTS

Figure 10-11. **Common Oxidation States of the Main Group and Transition Elements**
Oxidation states of the inner transition elements are shown in Figure 22-12.

oxidation states of chlorine are $+1$, $+3$, $+5$, and $+7$, corresponding to ClO^-, ClO_2^-, ClO_3^-, and ClO_4^-.

In their negative oxidation states, the nonmetallic elements usually assume the electronic configuration of a noble gas.

10-15 Exercises

Basic Exercises

1. Describe briefly the contribution to the present theory of atomic structure made by each of the following: **(a)** Bohr, **(b)** de Broglie, **(c)** Heisenberg, **(d)** Hund, **(e)** Pauli, **(f)** Schrödinger.

2. **(a)** What values are permitted for the orbital angular momentum quantum number l for an electron with principal quantum number $n = 4$? **(b)** How many different values for the magnetic quantum number are possible for an electron with orbital angular momentum quantum number $l = 3$? **(c)** How many electrons can be placed in each of the following subshells: s, p, d, f? **(d)** What is the maximum number of electrons which can be placed in an atomic orbital which has an orbital angular momentum quantum number $l = 3$? **(e)** What is the lowest shell which has an f subshell?

3. In what ways do the spatial distributions of the orbitals in each pair differ from each other? **(a)** $1s$ and $2s$, **(b)** $2s$ and $2p_x$, **(c)** $2p_x$ and $2p_z$.

4. What is the maximum number of electrons which can be accommodated **(a)** in the shell with $n = 4$? **(b)** in the $4f$ subshell? **(c)** in an *atom* in which the highest principal quantum number value is 4?

5. Match each of the following quantum number values with the proper letter designation (K, L, M, N, s, p, d, f): **(a)** $n = 1$, **(b)** $l = 2$, **(c)** $l = 0$, **(d)** $n = 3$.

6. How many electrons can be placed **(a)** in the shell with $n = 2$? **(b)** in the shell with $n = 3$? **(c)** in the shell with $n = 3$ before the first electron enters the shell with $n = 4$?

7. Give the symbol for an atom (if any) and also for an ion (if any) whose ground state corresponds to each of the following electronic configurations:
   (a) $1s^2 2s^2 2p^6 3s^2 3p^5$
   (b) $1s^2 2s^2 2p^6 3s^2 3p^6 4s^2 3d^{10} 4p^6 4d^8$
   (c) $1s^2 2s^2 2p^6 3s^2 3p^6 4s^2 3d^{10} 4p^6 5s^2 4d^{10} 5p^6 4f^7 6s^2$
   (d) $1s^2 2s^2 2p^6 3s^2 3p^6 4s^2 3d^{10} 4p^6 5s^2 4d^{10}$
   (e) $1s^2 2s^2 2p^6$

8. Explain why chromium has only one electron in its $4s$ subshell.

9. Predict the magnetic moment for Co^{3+}.

10. Which properties of the elements depend on the electronic configuration of the atoms and which do not?

11. Explain in terms of electronic configuration why the halogens have similar chemical properties. Why do they not have identical properties?

12. Write the electronic configuration and identify the periodic group for each of the elements whose atomic numbers are listed: 6, 16, 26, 36, 56.

13. **(a)** List all the elements whose atoms have only one electron in a p subshell. **(b)** List all the elements whose atoms have only one electron in an s subshell. **(c)** Which of these two sets of elements contains elements in only one periodic group? **(d)** How many periodic groups are represented in the other set?

14. Write the electronic configuration for each of the following atoms or ions. State the number of unpaired electrons contained in each. **(a)** C, **(b)** Cu^{2+}, **(c)** Zn, **(d)** Cl^-, **(e)** Eu, **(f)** Gd^{3+}, **(g)** Tl^+, **(h)** Fe^{3+}, **(i)** Ni^{2+}, **(j)** U, **(k)** Au.

15. Account for the great chemical similarity of the lanthanide elements (atomic numbers 57 to 71).

16. Select from each of the following groups the one which has the largest radius: (a) Co, Co^{2+}, Co^{3+}; (b) S^{2-}, Ar, K^+; (c) Li, Na, Rb; (d) C, N, O; (e) Ne, Na, Mg; (f) La, Lu; (g) Cu, Ag, Au; (h) Ba, Hf.

17. Write the electronic structure of each of the following ions: (a) Pb^{2+}, (b) Tl^+, (c) Sn^{2+}. (d) Explain on the basis of these ions what is meant by the term *inert pair*. Is the pair truly inert, or merely low in reactivity?

18. For each of the following species, tabulate the electronic configuration, the number of unpaired electrons, and the type of magnetic behavior: (a) Co^{3+}, (b) Se^{2-}, (c) Gd, (d) Ni, (e) Gd^{3+}.

19. Determine the *total* ionization potential for the first three electrons of aluminum and for the first two electrons of sodium. Prove on the basis of these data that ionization potential is not the sole factor governing the chemical stability of aluminum(III).

20. Explain why for sodium the second ionization potential is so much larger than the first, whereas for magnesium the magnitude of the difference between the first and second ionization potentials is much less.

21. Write electronic structures for (a) Ar and S^{2-}; (b) Fe and Ni^{2+}. (c) Which pair is isoelectronic? Explain. (d) Consult Table 10–2 to ascertain whether any ions of a transition element are isoelectronic with free elements.

22. Without consulting Table 10–2, select from each of the following groups the element which has the largest ionization potential: (a) Na, P, Cl; (b) Na, K, Rb; (c) He, Ne, Ar; (d) O, F, Ne.

23. (a) Explain in words the meaning of a negative value of electron affinity. Does any element have a negative ionization potential? (b) Write equations to demonstrate the difference between electron affinity and the *reverse* of ionization potential.

General Exercises

24. Given Einstein's relationship, $E = mc^2$, and Planck's hypothesis, $E = h\nu$, derive the relationship between the wavelength of a photon and its mass and velocity. Compare this relationship to that derived by de Broglie for the wavelength of an electron.

25. Explain, using copper atoms and copper(II) ions as examples, why the electronic configurations of the ions are rather easily predictable despite the fact that the electronic configurations of the corresponding atoms do not obey the $(n + l)$ rule.

26. From the data for the F atom and the trend shown in the radii for F^-, Ne, and Na^+, estimate the radius of Mg^{2+}. Estimate a reasonable radius for Na^{2+}.

27. (a) What differences are there between the ionization potential of an element and the oxidation potential of the element? (b) Determine the value (including the sign) of the electron affinity of Na^+.

28. Arrange the species in each group in order of increasing ionization potential, and in each case, explain the reason for the sequence: (a) K^+, Ar, Cl^-; (b) Fe, Fe^{2+}, Fe^{3+}; (c) Na, Mg, Al; (d) K, Ca, Sc; (e) N, O, F; (f) C, N, O; (g) Cu, Ag, Au; (h) Be, B, C; (i) K, Rb, Cs.

29. Describe the term *penetration* as it applies to electronic configuration. The properties of which one of the following elements are most modified by penetration, and the properties of which one are least modified: Zn, Ca, Br, H?

30. The nucleus of an atom is located at $x = y = z = 0$. (a) If the probability of finding an s-orbital electron in a tiny volume around $x = a$, $y = z = 0$, is 1.0×10^{-5}, what is the probability of finding the electron in the same sized volume around $x = z = 0$, $y = a$? (b) What would be the probability at the second site if the electron were in a p_z orbital? Explain.

31. Explain each of the following observations:
    (a) The radius of Cd^{2+} is less than that of Sr^{2+}.
    (b) It is easy to separate V from Nb in a mixture, but difficult to separate Nb from Ta in a mixture.
    (c) Eu can be separated from the other lanthanide elements much more easily than Gd can.
    (d) Sc and Y have higher ionization potentials than do Ga and Tl.
    (e) The ionization potentials increase in the series Ru, Rh, Pd, but decrease in the series Fe, Co, Ni.
    (f) The members of the second period of the periodic table are not typical representatives of their respective groups.
    (g) The difference in size between Hf and Ba is much greater than that between Zr and Sr.
    (h) The electronic configurations of the positive ions of transition elements are more regular than those of the neutral atoms.
32. Explain why so few transition elements have +1 oxidation states. List those which do (Figure 10–11), and explain why they do.
33. Explain why the +2 oxidation states of tin and lead are more stable than those of carbon and silicon. Explain why that of lead is more stable than that of tin.
34. Compare the first ionization potentials of sodium, magnesium, and aluminum, with those of potassium, calcium, and scandium. Explain the difference in trend of ionization potential in the two neighboring series.
35. Explain why the first ionization potential for copper is higher than that for potassium, whereas the second ionization potentials have the reverse trend.

Advanced Exercises

36. Draw a representation of a $3p$ orbital, including in your sketch the information that it has the planar node characteristic of p orbitals but in addition has a spherical node.
37. PtF_6 is a powerful oxidizing agent, capable of effecting the following reaction:

$$O_2(g) + PtF_6(s) \rightarrow O_2PtF_6(s)$$

Compare the first ionization potential of O_2 (Table 10–4) with those of the noble gases (Table 10–2). Which, if any, of the noble gases might undergo a comparable reaction with PtF_6? Are there factors other than oxidation potential which might determine whether a given noble gas would react with PtF_6? What are these? Confirm your conclusions by consulting the article on the discovery of the noble gas compounds by H. H. Hyman, "The Chemistry of the Noble Gases," *Journal of Chemical Education,* **41,** 174 (1964).
38. An electron in a hydrogen atom in its ground state absorbs 1.50 times as much energy as the minimum required for it to escape from the atom. What is the wavelength of the emitted electron?
39. Show that de Broglie's hypothesis applied to an electron moving in a circular orbit leads to Bohr's postulate of quantized angular momentum.
40. (a) Suppose a particle has four quantum numbers such that the permitted values are those given below. How many particles could be fitted into the $n = 1$ shell? into the $n = 2$ shell? into the $n = 3$ shell?

n: $1, 2, 3, \ldots$
l: $(n - 1), (n - 3), (n - 5), \ldots$ but no negative number
j: $(l + \frac{1}{2})$ or $(l - \frac{1}{2})$ if the latter is not negative
m: $-j$ in integral steps to $+j$

(b) These quantum numbers apply to protons and neutrons in atomic nuclei. Explain the stability of $_2^4\text{He}$, $_8^{16}\text{O}$, and $_{20}^{40}\text{Ca}$.

41. Explain how the magnetic properties of a compound might be used to determine whether it contained gold(II) or an equal number of moles of gold(I) and gold(III).

42. If there were three possible values for the spin magnetic quantum number m_s, how many elements would there be in the second period of the periodic table? (All other quantum numbers are as previously described.) Construct a periodic table showing the first 54 elements in such a hypothetical situation.

43. The solution to the Schrödinger equation for an electron in the ground state of the hydrogen atom is

$$\psi_{1s} = \frac{1}{\sqrt{\pi a_0^3}} e^{-r/a_0}$$

where r is the distance from the nucleus and a_0 is 0.529×10^{-8} cm. The probability of finding an electron ot any *point* in space is proportional to $|\psi|^2$. Using calculus, show that the maximum probability of finding the electron in the $1s$ orbital of hydrogen occurs at $r = a_0$.

44. Account for the difference in ionization potential **(a)** between Cu and K, **(b)** between K^+ and Ca^+, **(c)** between Cu^+ and Zn^+.

45. The uncertainty principle may be stated mathematically

$$\Delta(mv)\Delta x \cong h/4\pi$$

where $\Delta(mv)$ represents the uncertainty in the momentum of a particle and Δx represents the uncertainty in its position. **(a)** If a 1.00 gram body is travelling along the x axis at 100 cm/sec within 1 cm/sec, what is the theoretical uncertainty in its position? **(b)** If an electron is travelling at 10,000 cm/sec within 100 cm/sec, what is the theoretical uncertainty in its position? Explain why the uncertainty principle is not important for macroscopic bodies.

11

Chemical Bonding

A chemical bond is an attraction between atoms (or groups of atoms) which is sufficiently strong to allow the aggregate to be recognized as a distinct entity. The properties of a substance can often be explained in terms of the nature of the bonds holding the constituent atoms of the substance together. It is convenient to recognize three extreme types of bonds—ionic bonds, covalent bonds, and metallic bonds—even through there are gradations between these extremes. In this chapter some aspects of covalent bonding will be discussed. Ionic bonding will be treated in Chapter 18, and details of metallic bonding will be taken up in Chapter 19. Intermolecular forces are also involved in chemical systems and affect the properties of substances substantially, even though such forces are too weak to be classified as chemical bonds. Examples of such intermolecular forces will also be presented in this chapter.

A covalent bond is the attractive force resulting from pairs of electrons shared between atoms. Theories of covalent bonding help answer such questions as why some elements occur as polyatomic molecules, such as H_2, O_2, and S_8. These theories account for the relatively high boiling point of water, and they explain why carbon dioxide is so volatile compared to silicon dioxide. Experimental verifications of the theories are sought through studies of molecular structure, that is, the shapes of molecules (the distances and angles between the constituent atoms of the molecule) and the distribution of electric charge within molecules.

It can be assumed as a general principle that the number of bonds formed by a given atom and the geometry of the resulting molecule are consequences of a trend toward achieving the most stable configuration. In this chapter it will be shown that in some cases consideration of the repulsion between the electron pairs about a covalently bonded atom is one way to predict the geometry of the molecule of which the atom is a part. Stability is enhanced when there is a minimum repulsion between electron pairs. However, the concept of electron pair repulsion does not allow prediction of such important properties as bond energies or the possible existence of centers of positive and negative charge within a molecule.

Since in an atom, two electrons having paired spins can occupy a given orbital, it is plausible to assume that the electron pairs in molecules also occupy some sort of orbitals. Orbitals are solutions to the Schrödinger equation (Chapter 10), and a successful theory of bonding should provide a means of obtaining solutions to the Schrödinger equation which are appropriate for the description of electrons in molecules. In this chapter two methods of obtaining approximate solutions are described: the valence bond method and the molecular orbital method. Neither method is complete or fully satisfactory, but both are useful in accounting for the properties of molecules.

11-1   Energy of the Bonded State

In an isolated atom each electron is under the influence of only the nucleus and the other electrons in that atom. When two atoms come together, the electrons of each atom come under the influence of the electrons and the nucleus of the other. If the interaction results in an attraction between the two atoms, the new electronic arrangement corresponds to the formation of a chemical bond. The formation of a more stable state suggests that the molecule represents a system having a lower energy than the separated atoms. Figure 11-1 shows schematically the change in potential energy which occurs when two hydrogen atoms are brought together to form the H_2 molecule. When the two atoms are far apart, there is no appreciable interaction between them. As they come closer together, the nucleus of each atom exerts an attractive force on the electron of the other, and the nuclei repel each other because of their like charges. The electrons also tend to repel each other because of their like charges. If the spins of the electrons are parallel, both cannot occupy the same space between the nuclei (the Pauli principle, Section 10-6), and a net repulsion occurs. If the spins of the two electrons are opposed, both can occupy the same region of space between the nuclei. In this case, the total attraction is greater than the total repulsion. Therefore, if the two approaching atoms have electrons with opposing spins, the overall potential energy of the system (see Appendix A-2) decreases as the atoms get closer together. At a distance corresponding to the distance between the two atoms in the stable H_2 molecule, R_{H_2}, the potential energy has a minimum value, corresponding to maximum attraction. If the hydrogen atoms are pushed even closer together, the effects of repulsion of the two nuclei and of the two electrons increase drastically, as does the potential energy. The difference between the minimum energy and the energy of the isolated atoms is called the **bond energy** (Figure 11-1). At the internuclear distance corresponding to the minimum energy, the two electrons belong to both atoms simultaneously; therefore the bond is covalent.

If the approaching atoms have electrons with parallel spins, the potential energy increases continuously as the atoms are brought together. The interaction of two hydrogen atoms containing electrons with parallel spins is shown by the dashed line in Figure 11-1. Only repulsion and a corresponding increase in potential energy is observed no matter what the distance between the atoms.

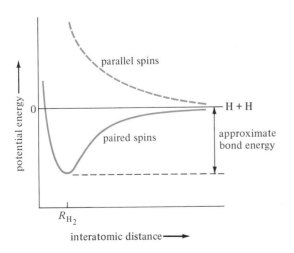

Figure 11-1. **Bond Energy and Interatomic Distance in H_2**

parallel spins

0 ⊢ H + H

paired spins

approximate bond energy

potential energy ⟶

R_{H_2}

interatomic distance ⟶

The formation of a covalent bond between any two atoms may be represented in a manner somewhat similar to the formation of H_2. The **bond length**—that is, the average distance between the respective nuclei in the molecule—corresponds to the distance at which the potential energy of the system is at a minimum. Whenever the familiar electron dot formulas for the valence electrons (Section 1–8) are written to denote molecules, it is implied that the pairs of dots represent electrons having opposed spins. Whenever the electron dot formula of a molecule is written so that the electrons are not paired, it is implied that the molecule is paramagnetic (Section 10–9). Conversely, if a substance is known experimentally to be paramagnetic, its formula should be written to show an appropriate number of unpaired electrons.

Example

Draw electron dot structures for (a) NO_2 (paramagnetic), (b) PF_3 (diamagnetic), and (c) CO_2 (diamagnetic).

(a) $:\ddot{O}:N::\ddot{O}:$ (b) $:\ddot{F}:\ddot{P}:\ddot{F}:$ (c) $:\ddot{O}::C::\ddot{O}:$
$\qquad\qquad\qquad\qquad\qquad\quad :\ddot{F}:$

11-2   Bond Energies

For diatomic molecules, the **bond energy,** D, is the energy required to break the bond. It is equal in magnitude but opposite in sign to the energy of formation of the molecule from its gaseous atoms.

$$A(g)\ +\ B(g)\ \rightarrow\ AB(g)$$

For example, enthalpy changes have been measured for the reactions

$$H_2(g)\ \rightarrow\ 2\,H(g)\qquad D_{H_2} = \Delta H^\circ_{298} = 104.2\ \text{kcal/mole } H_2$$
$$Cl_2(g)\ \rightarrow\ 2\,Cl(g)\qquad D_{Cl_2} = \Delta H^\circ_{298} = 58.0\ \text{kcal/mole } Cl_2$$

Each of these reactions consists of breaking 1 mole of bonds of the respective diatomic molecules. Accordingly, it is considered that the H—H bond energy is 104.2 kcal/mole and that the Cl—Cl bond energy is 58.0 kcal/mole.

Example

Using the bond energies of H_2 and Cl_2 and the enthalpy of formation of $HCl(g)$ from Table 3–2, calculate the energy of the HCl bond, D_{HCl}.

$$HCl(g)\ \rightarrow\ \tfrac{1}{2}H_2(g)\ +\ \tfrac{1}{2}Cl_2(g)\qquad -\Delta H^\circ_{f(HCl)} = 22.1\ \text{kcal}$$
$$\tfrac{1}{2}H_2(g)\ \rightarrow\ H(g)\qquad\qquad\qquad\ \tfrac{1}{2}D_{H_2} = 52.1\ \text{kcal}$$
$$\tfrac{1}{2}Cl_2(g)\ \rightarrow\ Cl(g)\qquad\qquad\qquad\ \tfrac{1}{2}D_{Cl_2} = 29.0\ \text{kcal}$$

The bond energy of HCl, D_{HCl}, is obtained by combining these equations and the respective enthalpy changes.

$$HCl(g) \rightarrow H(g) + Cl(g) \qquad D_{HCl} = -\Delta H^{\circ}_{f(HCl)} + \tfrac{1}{2}D_{H_2} + \tfrac{1}{2}D_{Cl_2}$$
$$= 22.1 \text{ kcal} + 52.1 \text{ kcal} + 29.0 \text{ kcal} = 103.2 \text{ kcal}$$

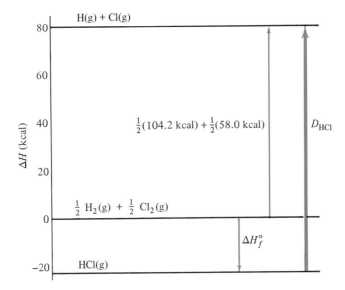

For more complex molecules the assignment of bond energy values for the individual bonds is somewhat arbitrary. It has been found that the energy necessary to break a particular type of bond between given atoms is approximately the same regardless of the molecule in which the bond occurs. Thus the C—H bond in methane has about the same strength as the C—H bond in ethane. A list of typical bond energies is given in Table 11–1. For polyatomic molecules, these are average bond

TABLE 11-1. Selected Bond Energies, D^a

Single Bond	kcal/mole	kJ/mole	Single Bond	kcal/mole	kJ/mole	Multiple Bond	kcal/mole	kJ/mole
H—H	104.2	436.0	N—N	32	134	C=C	146	611
H—C	100	418	N—F	56	234	C≡C	200	837
H—N	93	389	N—Cl	37	155	C=N	147	615
H—O	111	464	O—O	33	138	C≡N	213	891
H—F	135	565	O—F	45	188	C=O	177	741
H—S	81	339	O—Si	106	444	C≡O	256	1070
H—Cl	103.2	431.8	O—Cl	50	209	N=N	100	418
H—Br	88	368	O—S	124	519	N≡N	226	946
H—I	71	297	F—F	37.8	158	bond in O_2	119.2	498.7
C—C	82	343	P—Cl	78	326			
C—N	64	268	S—S	49	205			
C—O	83	347	Cl—Cl	58	243			
C—F	102	427	Cl—I	50	209			
C—S	61	255	Br—Br	46	192			
C—Cl	79	330	I—I	36	151			
C—Br	66	276						
C—I	52	218						

$^a D = -\Delta H_f$ from gaseous atoms, at 298 K, per mole of bonds.

energies because in the calculation of these bond energies it has been assumed that all like bonds have the same strength. In the actual breakup of a polyatomic molecule, the bonds may be broken in sequence, and the dissociation energy of the first bond may differ from the dissociation energies of the remaining bonds.

Example

Estimate the C—H bond energy in methane, CH_4. The required enthalpy data are the enthalpy of formation of methane, -17.9 kcal/mole; the bond energy of hydrogen 104.2 kcal/mole; and the enthalpy of sublimation of carbon, 171.7 kcal/mole.

The C—H bond energy in methane is *defined* as one fourth of the energy required to break the four C—H bonds.

$$CH_4(g) \rightarrow C(g) + 4 H(g)$$

To determine the enthalpy change of this reaction the data are combined in the following manner:

$C(s) \rightarrow C(g)$		$\Delta H_{sub} =$	171.7 kcal
$2 H_2(g) \rightarrow 4 H(g)$		$2D_{H_2} =$	2(104.2) kcal
$CH_4(g) \rightarrow C(s) + 2 H_2(g)$		$-\Delta H_f =$	17.9 kcal
$CH_4(g) \rightarrow C(g) + 4 H(g)$		$\Delta H =$	398.0 kcal

The C—H bond energy is one fourth of the energy required to break all four bonds, 99.5 kcal/mole.

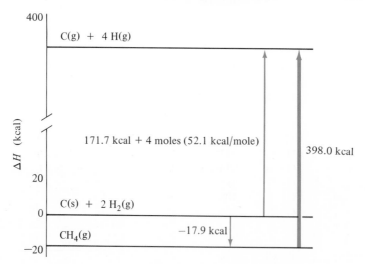

Bond energies are the negative of the enthalpies of formation of the bonds from gaseous atoms. Some enthalpies of formation of gaseous atoms are listed in Table 11–2. For those substances which normally exist as gaseous diatomic molecules, the enthalpy of formation of the atoms is simply the dissociation energy of the molecules. For elements which normally exist as solids, the enthalpy of formation of the gaseous atoms is determined by using the enthalpy of sublimation of the element, or the enthalpies of fusion and vaporization of the element.

TABLE 11-2. Enthalpies of Formation of Gaseous Atoms at 298 K

	kcal/mole	kJ/mole		kcal/mole	kJ/mole
H	52.1	218	N	113.0	472.8
D	53.0	222	P	79.8	334
Na	25.8	108	Sb	60.8	254
K	21.3	89.1	O	59.6	249
Be	78.3	328	S	65.7	275
Mg	35.3	148	F	18.9	79.1
Ca	46.0	192	Cl	28.9	121
B	133	556	Br	26.7	112
Al	78.0	326	I	25.5	107
C	171.7	718.4	Cu	81.5	341
Si	106	444	Ag	69.1	289
Ge	78.4	328	Zn	31.2	131
Sn	72	300	Hg	14.7	61.5
Pb	46.7	195			

A set of bond energies is useful for finding enthalpy changes of reactions which involve a compound for which ΔH_f° is not known. Such a situation might occur if the particular compound has never been synthesized or if it is so reactive that a suitable calorimetric determination cannot be performed.

The breaking and formation of bonds during a chemical reaction can be shown schematically for the reaction

$$A_2 \;+\; B_2 \;\rightarrow\; 2\,AB$$

as

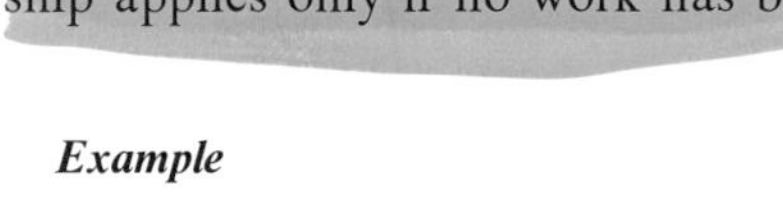

A—A		A A		A A
	$\rightarrow$		$\rightarrow$	$\mid\quad\mid$
B—B		B B		B B

If the A—B bonds are stronger than the average of the A—A and the B—B bonds, that is, if the A—B bonds are lower in energy, then the reaction should be exothermic, and the excess energy should correspond to the enthalpy of reaction. This relationship applies only if no work has been done on or by the surroundings.

Example

Estimate the enthalpy of formation of ammonia from bond energy data.

$$N_2 \;+\; 3\,H_2 \;\rightarrow\; 2\,NH_3$$

Noting that there are three N—H bonds in each NH_3 molecule; one obtains

$3\,H(g)$	$+\;N(g)$	$\rightarrow\;NH_3(g)$	$-3D_{N-H} = -279\ \text{kcal}$
	$\tfrac{1}{2}N_2(g)$	$\rightarrow\;N(g)$	$\tfrac{1}{2}D_{N_2} = \quad 113\ \text{kcal}$
	$\tfrac{3}{2}H_2(g)$	$\rightarrow\;3\,H(g)$	$\tfrac{3}{2}D_{H_2} = \quad 156\ \text{kcal}$
$\tfrac{1}{2}N_2(g)$	$+\;\tfrac{3}{2}H_2(g)$	$\rightarrow\;NH_3(g)$	$\Delta H_f = \quad -10\ \text{kcal}$

The estimated enthalpy of formation is of the order of magnitude of the experimental value, -11 kcal/mole.

11-3   Bond Lengths and Covalent Radii

It is convenient to define the terms (covalent) **bond length** or **bond distance** as the distance between the nuclei of covalently bonded atoms. It has been found experimentally that bond lengths are characteristic of the atoms involved. When the bond distance of a **homonuclear** diatomic molecule (a molecule with two atoms of the same element, such as F_2 or Cl_2) is determined, one half of the observed distance is arbitrarily considered to be the covalent bond radius of the individual atoms (Figure 11–2). Covalent radii for atoms which do not form gaseous diatomic molecules have been estimated by other methods. For example, the C—C bond distance in diamond and in a large number of organic compounds is found to be 1.54 ± 0.01 Å; hence the covalent radius of carbon is taken as half that distance, 0.77 Å. Gaseous N_2 and O_2 are diatomic, but they have multiple bonds. Their single bond radii must be determined from compounds such as hydrazine, H_2N—NH_2, and hydrogen peroxide, HO—OH, respectively. Double- and triple-bonded carbon atom radii are obtained from molecules such as ethylene, $H_2C{=}CH_2$, and acetylene, $HC{\equiv}CH$. It is found experimentally that the lengths of multiple bonds between a given pair of atoms are less than the single bond distance between the same atoms. Accordingly, the double bond radius of a given atom is smaller than its single bond radius, and the triple bond radius is even smaller. A compilation of covalent radii is given in Table 11–3. These radii refer to the atoms covalently bonded in compounds, as opposed to the free atoms and ions.

The covalent radii of different atoms can be added to obtain reasonable values of the bond lengths of bonds between unlike atoms. For example, the covalent radii of carbon and chlorine add up to 1.76 Å, in agreement with the C—Cl bond distance in CCl_4.

The data of Table 11–3 give good agreement between calculated and experimental values only when the respective bonds are the same type as those used to establish the table in the first place. When considerable discrepancies are observed, it is indicative that additional factors must be involved in the formation of the particular bond. One of these factors, resonance, will be discussed in Section 11–10.

bond length

Figure 11–2. **Definition of Covalent Radius**

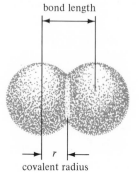

r

covalent radius

TABLE 11-3. Covalent Bond Radii (Å)

Single Bond Radii						Multiple Bond Radii	
H	0.28	P	1.10	Te	1.37	C=	0.67
C	0.77	As	1.21	F	0.64	C≡	0.61
Si	1.17	Sb	1.41	Cl	0.99	N=	0.63
Ge	1.22	O	0.66	Br	1.14	N≡	0.55
Sn	1.40	S	1.04	I	1.33		
N	0.70	Se	1.17				

Example

Using Table 11-3, calculate the bond lengths in the molecules H_2O and H_2. Compare the results with the observed values, 0.94 Å and 0.75 Å, respectively, and explain any discrepancy.

The sum of the H— and O— radii is 0.94 Å. The actual bond distance in H_2O is 0.94 Å. The sum of the radii for two hydrogen atoms is 0.56 Å, but the observed bond distance is 0.75 Å. The discrepancy in the latter case is due to the fact that the experimental covalent radius of hydrogen was determined using molecules other than H_2. H_2 is unique; there is no inner electron cloud. In other covalent molecules the proton "buries" itself in the surrounding electron cloud, causing a somewhat smaller effective radius.

11-4   Bond Angles and the Shapes of Molecules

In 1957, R. J. Gillespie and Ronald S. Nyholm showed that it is possible to predict the shapes of many simple molecules from their electron dot formulas. In the simplest case, in which all the outermost electrons of a central atom are bonded to atoms of one other element with single bonds, such as in SF_6, CCl_4, NH_4^+, BCl_3, and BeH_2, the shapes can be predicted by using the principle that two electrons having opposed spins (spins paired) can be located close together, but electrons having parallel spins tend to be as far away from each other as possible. Shared and unshared electron pairs and groups of electrons in multiple bonds also tend to be as far away from other such combinations as possible. The distributions of various numbers of electron pairs about a spherical atom are shown schematically in Figure 11-3.

In many molecules, such as H_2O and NH_3, the electron pairs on the central atom are not all used in bonding. The approximate geometry of this type of molecule can be deduced by using the principle that unshared electron pairs and electron pairs in multiple bonds repel the other electrons slightly more than do electron pairs in single bonds. The shape of a molecule is described as the relative positions of the molecule's *atoms* rather than as the locations of its electron pairs. Of course, the atoms are located in the directions of the *bonding* electron pairs. The observed bond angles in molecules will differ somewhat from the idealized positions shown in Figure 11-3 according to the types of electron pair repulsions which exist in each particular case.

Example

Predict the geometry of each of the following molecules: BeH_2, BF_3, CH_4, PF_5, SF_6, NH_3, H_2O, XeF_4, and CO_2.

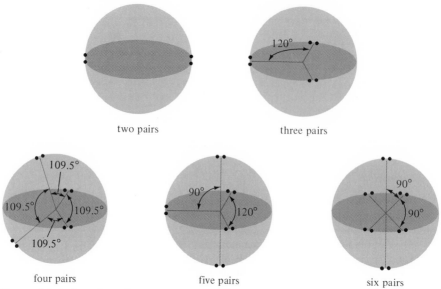

Figure 11-3. Repulsion Between Single Electron Pairs

The predicted geometries are shown in Figure 11-4. For BeH_2, the two electron pairs will be located on opposite sides of the beryllium atom, yielding a linear molecule. For BF_3, the three electron pairs will be located at angles of 120° with respect to each other, giving the molecule a planar, triangular shape.

For CH_4, so that the four electron pair bonds will be located as far as possible from each other, they must assume a regular tetrahedral orientation, with angles of 109.5° with respect to each other. By reference to Figure 11-3, it can be predicted that the structures of PF_5 and SF_6 will be trigonal bipyramidal and octahedral, respectively.

The electron dot formula for ammonia suggests a tetrahedral arrangement of the four electron pairs. However, the unshared pair of electrons exerts a comparatively larger repulsion, and therefore the angles between the bonded pairs are somewhat smaller than 109.5° (actually 107°). The molecule has the shape of a triangular pyramid, with the nitrogen at the apex. The electron pair which is not involved in bonding is not included in the description of the geometry of the molecule.

In the H_2O molecule there are two sets of unshared electron pairs. The arrangement of bonding electron pairs is more distorted from the tetrahedral angle than in NH_3, yielding an angular molecule (with a bond angle of 105°). The electron dot formula for XeF_4 contains two pairs of unshared electrons and four pairs of shared electrons on the xenon atom. The former repel other electron pairs most and therefore lie on opposite sides of the xenon atom. The other four pairs are arranged at 90° angles about the xenon atom, all in one plane. A square planar molecule results.

The electron dot formula of carbon dioxide is written with two double bonds—two groups of electrons. Since there are no other electron pairs in the outermost shell of carbon, these double bonds will be located on opposite sides of the carbon atom, and the molecule is linear.

Example

The observed H—C—H angle in the molecule $H_2C{=}O$ is 111°, rather than 120°. Explain this observation.

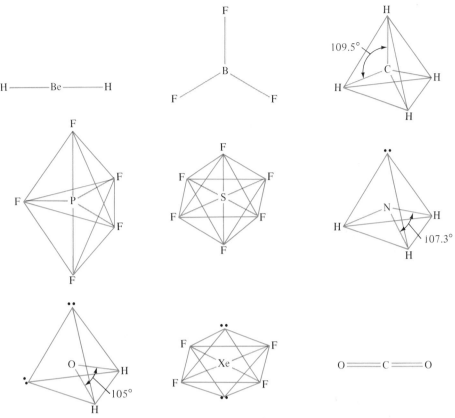

Figure 11-4. **Geometries of Some Simple Molecules**

The electron dot formula, with three groups of electrons, suggests a planar molecule:

:O:
::
.C.
H· ·H

The electron pairs in the double bond repel the others more than the electron pairs of the single bonds repel each other. The angle between the single bonds in expected to be somewhat less than 120°.

Examples of molecular shapes are summarized in Table 11-4.

11-5   Electronegativity

When a covalent bond exists between atoms of two different elements, it is very likely that one of the atoms will attract the electron pair more strongly than the other. The atom (or group of atoms) which exerts the stronger attraction is said to be the more electronegative. In other words, **electronegativity** is defined as that quality of an atom which determines the relative attraction of the atom for the electrons *in a*

TABLE 11-4. Shapes of Simple Molecules

Number of Groups of Electrons	Orientation of Electron Groups	Number of Atoms Bonded	Geometry of Atoms	Examples
2	180°	2	linear	CO_2, BeH_2, HCN
3	120°	2	angular	SO_2, $CH_2{=}NH$ (N atom)
		3	trigonal	SO_3, BF_3, $CH_2{=}NH$ (C atom)
4	109.5°	2	angular	H_2O, H_2S, SCl_2
		3	pyramidal	NH_3, H_3O^+, XeO_3
		4	tetrahedral	CH_4, NH_4^+, XeO_4
5	90° and 120°	2	linear	I_3^- (using axial groups)
		3	T-shaped	ICl_3 (axial plus one other group)
		4	nonplanar	SF_4
		5	trigonal bipyramidal	$PCl_5(g)$
6	90°	4	square planar	XeF_4
		5	square pyramidal	$XeOF_4$, ICl_5
		6	octahedral	SF_6

covalent bond. It is not possible to determine absolute values of the electronegativities of atoms; however, by assigning an arbitrary value to one atom, a scale of relative electronegativities can be established. A number of electronegativity scales have been proposed, corresponding to several ways of estimating the relative attraction for electrons by bonded atoms. The most widely used scale is that proposed by L. Pauling and subsequently modified by others. On this scale the most electronegative element, fluorine, is assigned an electronegativity value of 4.0. The basis for Pauling's scale is discussed in Section 11–14. Electronegativities of the elements are listed in chart form in Figure 11–5. It is useful to note that in the second period of the periodic table,

H 2.1																	He
Li 1.0	Be 1.5											B 2.0	C 2.5	N 3.0	O 3.5	F 4.0	Ne
Na 0.9	Mg 1.2											Al 1.5	Si 1.8	P 2.1	S 2.5	Cl 3.0	Ar
K 0.8	Ca 1.0	Sc 1.3	Ti 1.5	V 1.6	Cr 1.6	Mn 1.5	Fe 1.8	Co 1.8	Ni 1.8	Cu 1.9	Zn 1.6	Ga 1.6	Ge 1.8	As 2.0	Se 2.4	Br 2.8	Kr
Rb 0.8	Sr 1.0	Y 1.2	Zr 1.4	Nb 1.6	Mo 1.8	Tc 1.9	Ru 2.2	Rh 2.2	Pd 2.2	Ag 1.9	Cd 1.7	In 1.7	Sn 1.8	Sb 1.9	Te 2.1	I 2.5	Xe
Cs 0.7	Ba 0.9	La 1.1	Hf 1.3	Ta 1.5	W 1.7	Re 1.9	Os 2.2	Ir 2.2	Pt 2.2	Au 2.4	Hg 1.9	Tl 1.8	Pb 1.9	Bi 1.9	Po 2.0	At 2.2	Rn
Fr 0.7	Ra 0.9	Ac 1.1															
			Ce 1.1	Pr 1.1	Nd 1.1	Pm 1.1	Sm 1.1	Eu 1.1	Gd 1.1	Tb 1.2	Dy 1.2	Ho 1.2	Er 1.2	Tm 1.2	Yb 1.2	Lu 1.2	
			Th 1.3	Pa 1.5	U 1.7	Np 1.3	Pu 1.3	Am 1.3	Cm 1.3	Bk 1.3	Cf 1.3	Es 1.3	Fm 1.3	Md 1.3	No 1.3	Lr 1.3	

Figure 11–5. **Electronegativities of the Elements**

successive elements differ in electronegativity by 0.5 units. Within any period the least electronegative element is the alkali metal and the most electronegative is the halogen.

Electronegativity differences between the constituent atoms of a substance strongly influence the properties of that substance. As will be shown, both physical properties and chemical behavior can be explained in terms of electronegativity.

11–6   Polar Bonds and Dipolar Molecules

When atoms of two different elements are joined by a covalent bond, the more electronegative atom exerts a relatively stronger attraction for the shared electrons. Thus one end of the bond is relatively negative and the other relatively positive. Such a bond is said to be **polar.** For example, in the hydrogen halide molecules the electronegativity differences are as follows: HF, 1.9; HCl, 0.9; HBr, 0.7; HI, 0.4; therefore the bond between hydrogen and fluorine is the most polar.

Some atoms in a molecule with polar bonds possess a partial negative charge, and others possess a partial positive charge. (Nevertheless, they still are not ions; neither has as much as a single electronic charge, either positive or negative.) For molecules with more than two atoms, the cumulative effect of the respective charges, either positive or negative, can be regarded as being centered at one position in the molecule. If, as a result of its containing polar bonds, the molecule has a very small degree of positive charge, $\delta+$, centered at one place and an equal negative charge, $\delta-$, centered at a *different place*, the molecule is said to possess a **dipole moment** or to be **dipolar.** The dipole moment, μ, of such a molecule is defined as the product of the charge, δ, at one center times the distance between the positive and negative centers, d:

$$\mu = \delta d$$

When dipolar molecules are placed in an electric field, for example, between two charged plates, they tend to become oriented with the positive centers toward the negative plate and the negative centers toward the positive plate, as shown in Figure 11–6. If there is a vacuum between a given set of plates, the quantity of charge which the plates can hold will depend on the distance of their separation. The orientation of dipolar molecules between such plates has the same effect as pushing the plates closer together, thus increasing their capacity to hold a charge under a constant potential. This phenomenon is the basis for an experimental method of determining the magnitudes of dipole moments.

For a molecule to have a dipole moment it must (1) possess polar bonds and (2) have those bonds arranged so that an unsymmetrical distribution of charge results. Some examples of some simple molecules, both with and without dipole moments, are given in Table 11–5.

Experimentally determined dipole moments provide an insight into the geometry of the molecule, especially the angles between the bonds. For example, Table 11–5 shows that the carbon dioxide molecule has no dipole moment. However, there is a large difference in electronegativity between carbon and oxygen, and the carbon-

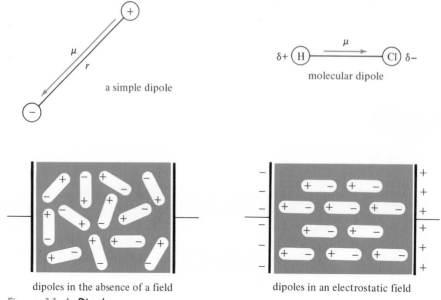

a simple dipole

molecular dipole

dipoles in the absence of a field   dipoles in an electrostatic field

Figure 11-6. **Dipoles**

oxygen bonds must be polar. Therefore, the CO_2 molecule must be linear because only then would the centers of positive and negative charge coincide:

$$O—C—O$$
$$\delta-\ \ \delta+\ \ \delta-$$

On the other hand, the large dipole moment of water molecules indicates both that water molecules have polar bonds and that they are not linear (Figure 11-7).

Example

Using data from Tables 11–3 and 11–5, compare the magnitude of the partial charge on the hydrogen atom in HCl **(a)** with that on the hydrogen atom of HI and **(b)** with that of a singly charged positive ion, which is 4.80×10^{-10} esu.

TABLE 11–5. Dipole Moments of Some Simple Molecules

Molecule AX_n	Electronegativity Difference Between A and X	Dipole Moment (D^a)	Molecular Geometry
HF	1.9	1.91	linear
HCl	0.9	1.03	linear
HI	0.4	0.38	linear
NH_3	0.9	1.5	pyramidal
CH_4	0.4	0.0	tetrahedral
H_2O	1.4	1.85	angular
CO_2	1.0	0.0	linear
BF_3	2.0	0.0	trigonal (planar)

[a] Dipole moments are measured in Debyes; $1\ D = 1 \times 10^{-18}$ esu $\cdot$ cm.

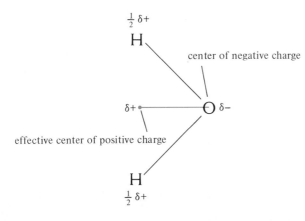

Figure 11-7. **Dipole of Water Molecule**

The effect of the partial positive charges of the hydrogen atoms is the same as the effect of a partial positive charge located midway between them.

(a) If it is assumed that the partial positive and negative charges are centered on the hydrogen and halogen atoms, respectively, the distances between charge centers are calculated to be 1.27 and 1.61 Å for HCl and HI, respectively, from data in Table 11-3. The charge, δ, equals the dipole moment divided by the separation, d.

For HCl: $\delta = \dfrac{\text{dipole moment}}{d} = \dfrac{1.03 \times 10^{-18} \text{ esu} \cdot \text{cm}}{1.27 \times 10^{-8} \text{ cm}} = 8.11 \times 10^{-11} \text{ esu}$

For HI: $\delta = \dfrac{0.38 \times 10^{-18} \text{ esu} \cdot \text{cm}}{1.61 \times 10^{-8} \text{ cm}} = 2.4 \times 10^{-11} \text{ esu}$

The charge on the hydrogen atom of HCl is over three times that on the hydrogen atom of HI, in accordance with the greater electronegativity difference between hydrogen and chlorine compared to the difference between hydrogen and iodine.

(b) The charge on the hydrogen atom of HCl is about one sixth the charge on a mono-positive ion.

In molecules containing more than two atoms, the numerical magnitude of the dipole moment depends on several factors. These include

1. The electronegativity differences between the bonded atoms.
2. The directions of the bond dipole moments.
3. The angles between the various bonds.
4. The influence of unshared electron pairs.

11-7   Intermolecular Forces

The fact that substances composed of independent molecules exist in the form of liquids and solids indicates that forces of attraction exist between the molecules to hold them in close proximity. The magnitude of intermolecular forces is reflected in such properties as the enthalpies of fusion or of vaporization (Table 3-4), and less directly in the melting point or the boiling point of the condensed phase. Some data for a series of isoelectronic molecules are given in Table 11-6. As shown in the extreme cases of H_2O and Ne, the intermolecular forces vary considerably.

TABLE 11-6. Effects of Intermolecular Forces

Compound	Molecular Weight	Melting Point (°C)	Enthalpy of Fusion		Normal Boiling Point (°C)	Enthalpy of Vaporization	
			cal/gram	J/gram		cal/gram	J/gram
CH_4	16	−183	14.5	60.7	−59	138	577
NH_3	17	−75	108.1	452.3	−33	327	1368
H_2O	18	0	79.7	333	100	540	2259
Ne	20	−249			−246	22	92

In the case of molecules which have no dipole moment, such as CH_4, the chief intermolecular force is known as the **van der Waals force.** This type of force results from the instantaneous polarization of the molecules due to oscillations of the electron cloud around the atomic nuclei (Figure 11-8). When the electron cloud is distributed for an instant more to one side of the molecule than toward the other, a small electric moment (separation of charge) is set up, which in turn produces a similar moment in adjacent molecules, and a weak attractive force results. Other things being equal, the greater the number of electrons per molecule, the more diffuse and easily polarized is the electron cloud; hence the strength of the van der Waals force is dependent on the number of electrons in the molecule. When molecules associate through van der Waals forces alone, the system is lowered in energy by up to about 1 kcal/mole.

Example

Predict the order of increasing boiling points of the noble gases.

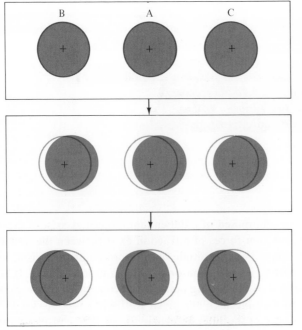

Figure 11-8. Origin of the van der Waals Force

The vibration of the cloud of electrons about the nucleus of atom A polarizes the electron clouds of adjacent atoms, B and C, creating a weak attractive force. The gray outlines represent the undistorted positions of the electron cloud.

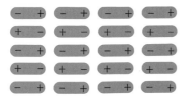

Figure 11-9. **Idealized Orientation of Polar Molecules in a Crystalline Solid**

The greater the atomic number, the greater the number of electrons in each atom, and the greater is the van der Waals force. The greater the forces between the molecules, the higher should be the boiling point. The actual boiling points of the noble gases increase with increasing atomic number, as expected: He, 4 K; Ne, 27 K; Ar, 87 K; Kr, 120 K; Xe, 166 K; Rn, 211 K.

Intermolecular forces between molecules with dipole moments are so much larger than the van der Waals forces that the latter are often negligible by comparison. The (permanent) dipoles tend to align themselves so that attractions between adjacent molecules are enhanced. Thus molecules with dipole moments form solids with higher melting points and liquids which are less volatile than molecules with the same number of electrons but which have no dipole.

Example

Would Br_2 or ICl be expected to have the higher boiling point?
Both of these molecules have the same number of electrons, but ICl has a dipole moment. ICl should have the higher boiling point. At 1 atm pressure the measured boiling points are ICl, 97.4°C, and Br_2, 58.78°C.

In the solid state, molecules having dipoles tend to be oriented so that oppositely charged centers are adjacent, giving the solid an ordered structure (Figure 11-9). When such crystals are melted, energy must be provided to overcome the attraction between the molecular dipoles. Consequently the solid has a large enthalpy of fusion. In the liquid state the molecules are less well ordered, but there is still some attraction between the molecular dipoles, giving rise to higher boiling points and enthalpies of vaporization than otherwise expected.

11-8   Hydrogen Bonds

When a hydrogen atom is covalently bonded to a very small, highly electronegative atom, such as F, O, or N, the resulting bond is highly polar. The hydrogen atom has such a large positive partial charge that its side which lies away from the other atom acts somewhat like a bare proton. Therefore, it will be attracted to the negative center of an adjacent molecule with an appreciable intermolecular force. This attraction is called **hydrogen bonding.** The strengths of hydrogen bonds range from about 2 to 10 kcal/mole, making them about one tenth as strong as ordinary covalent bonds. A schematic representation of hydrogen bonding in liquid water is shown in Figure 11-10.

Hydrogen bonds may be formed to fluorine, oxygen, and nitrogen, and to chlorine in some compounds. Larger atoms have more diffuse electron clouds in which the

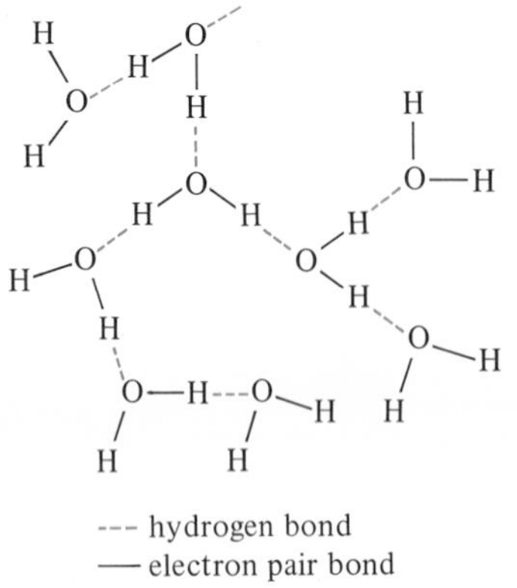

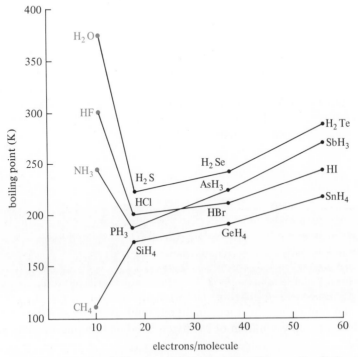

Figure 11-10. Schematic Diagram of Hydrogen Bonding in Liquid Water

--- hydrogen bond
— electron pair bond

hydrogen atom becomes buried, and with them hydrogen bonding is not possible. As would be expected from considerations of electronegativity and size, hydrogen bonds involving fluorine atoms are the strongest. Nitrogen and chlorine have about the same electronegativities, but nitrogen, being smaller, forms much stronger hydrogen bonds.

The influence of hydrogen bonding on boiling points is shown in Figure 11-11, where the boiling points of sets of hydrogen compounds are plotted against the number of electrons in each. The boiling points of the molecules without dipoles, CH_4 to SnH_4, increase regularly with increasing numbers of electrons, as is expected for

Figure 11-11. The Influence of Hydrogen Bonding on the Boiling Points of NH_3, H_2O, and HF

situations where the chief intermolecular force is the van der Waals force. In contrast, owing to the existence of hydrogen bonding between molecules, the hydrides of nitrogen, oxygen, and fluorine show "abnormally high" boiling points.

The importance of hydrogen bonding in biochemistry will be discussed in Chapter 23.

11-9   Solubility of Polar Covalent Molecules in Water

Some covalent molecules are sufficiently polar to dissolve in water quite readily. For example, formaldehyde, CH_2O, and acetone, CH_3COCH_3, mix with water in all proportions. In such cases solubility is enhanced by the attraction of the dipolar water molecules for the dipoles of the dissolved substance. Molecules which contain covalently bonded OH groups are particularly soluble in water because of hydrogen bonding. Thus methyl alcohol, CH_3OH, ethyl alcohol, C_2H_5OH, and propyl alcohol, C_3H_7OH, all dissolve in water in all proportions. However, alcohols with long chains of carbon atoms are only slightly soluble. Molecules containing several OH groups, such as glycerol, $CH_2OHCHOHCH_2OH$, and table sugar, $C_{12}H_{22}O_{11}$, are very soluble in water. It should be noted that aqueous solutions of these polar covalent compounds do not conduct electricity. Despite their solubility, these substances are nonelectrolytes.

11-10   Resonance

In a number of cases the method of depicting covalent bonds by pairs of dots or by lines (Section 1-8) fails to give an accurate representation of the structure of a molecule. For example, the structure of the SO_3 molecule can be written to show all the atoms obeying the octet rule:

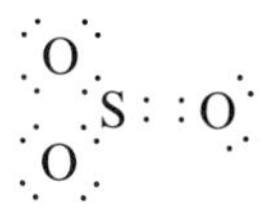

The structure, containing two S—O single bonds and one S=O double bond, suggests that there should be two different S-to-O bond lengths in the molecule. Moreover, such a molecule would have an unsymmetrical electronic arrangement, and it should have a measurable dipole moment. Actually, however, the molecule has a planar structure in which all of the S-to-O bond lengths are identical, and the dipole moment is zero. In addition, each bond length is shorter than that predicted for an S—O single bond but longer than that predicted for an S=O double bond. Since it is impossible to write a single structure in the usual manner which shows that each bond is a blend of both single and double bonds, it is customary to draw three structures:

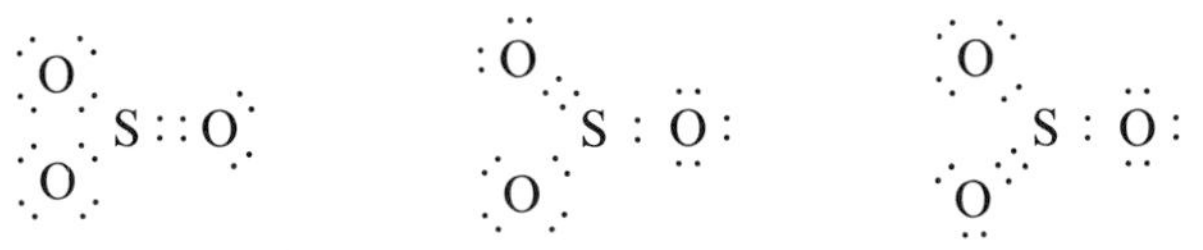

The true structure of SO_3 is none of these structures but is a sort of average of all three. The actual molecule is said to be a **resonance hybrid** of the three structures. Each of the S-to-O bonds is intermediate between a single and a double bond, and can be said to have about one-third double bond character. SO_3 is sometimes represented as

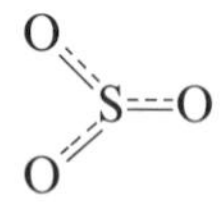

Other species which have structures similar to SO_3, and which can be represented by analogous resonance forms, are the CO_3^{2-} ion and the NO_3^- ion.

The resonance concept is particularly useful in accounting for the properties of some organic molecules. For example, it is found experimentally that the carbon atoms in benzene, C_6H_6, are arranged in a plane in the shape of a regular hexagon. Two possible structures are shown in the accompanying figure as (a) and (b). But neither one of these structures, having alternate single and double bonds, would have a regular hexagonal shape. Moreover, as was mentioned in Section 1–11, the chemical properties of benzene are not similar to those of aliphatic molecules having double bonds. Hence benzene is considered to be a resonance hybrid of structures (a) and (b). It is sometimes denoted as structure (c).

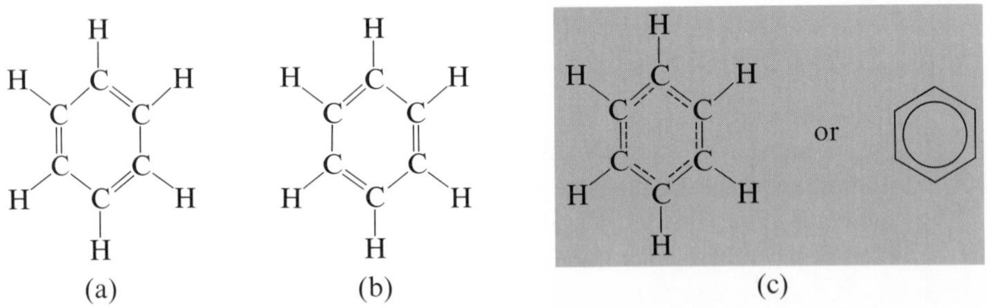

(a) (b) (c)

The enthalpy of formation of either structure (a) or (b) can be calculated from the sum of the bond energies, 6 C—H $+$ 3 C—C $+$ 3 C=C, to give -1284 kcal/mole. The actual enthalpy of formation of gaseous benzene, as determined from its enthalpy of combustion, the enthalpy of sublimation of carbon, the enthalpy of dissociation of H_2, and the enthalpies of formation of water and carbon dioxide (see Chapter 3), is -1323 kcal/mole. Thus the actual benzene molecule is more stable by about 39 kcal/mole than would be predicted for a molecule having a structure like (a) or (b). A greater stability than would be expected on the basis of individual structures suggests the existence of a resonance hybrid structure. In the case of benzene, this **resonance stabilization energy** is about -39 kcal/mole.

When one writes resonance structures, certain rules must be observed. The positions of the atoms and the magnetic susceptibility of the real molecule can be determined experimentally. Thus all of the proposed resonance structures must have (1) the corresponding atoms located at the same sites and (2) the same number of unpaired electrons in each resonance form as found in the actual molecule.

Example

Which one(s) of the following structures *cannot* represent resonance forms for (diamagnetic) NNO?

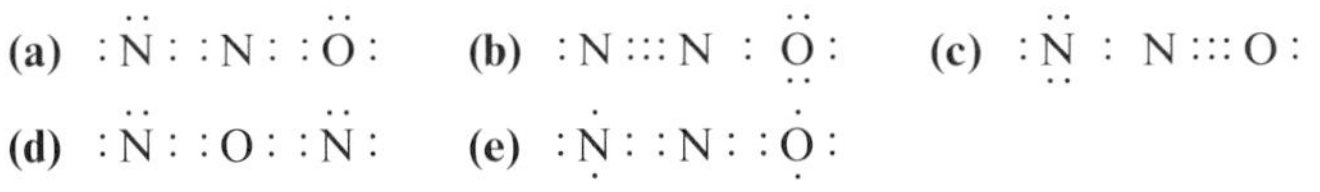

(a) $:\!\ddot{N}:\!:\!N:\!:\!\ddot{O}:$ (b) $:\!N:::\!N:\!\ddot{O}:$ (c) $:\!\dot{N}:\!N:::\!O:$

(d) $:\!\ddot{N}:\!:\!\ddot{O}:\!:\!\dot{N}:$ (e) $:\!\dot{N}:\!:\!\dot{N}:\!:\!\ddot{O}:$

The forms shown as **(d)** and **(e)** cannot be used. In **(d)** the atoms are arranged with the oxygen atom between the nitrogen atoms. This arrangement is not the same as in the other forms or as the positions found experimentally. The form shown as **(e)** has four unpaired electrons, and this is not possible, since the molecule is diamagnetic.

A great many resonance structures for a given molecule might be written according to these rules. However, only those which have relatively low energies will actually be important resonance forms. The more nearly equal in energy the resonance hybrids are, the greater the stabilizing effect of the resonance. If a particular structure is very much higher in energy than another, the high energy form will not be important and may be ignored in writing resonance structures.

Example

Would resonance stabilization be greater in   **(a)** $CO_3{}^{2-}$ or   **(b)** H_2CO_3?
The possible resonance hybrids are

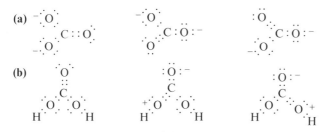

higher energy resonance forms

In $CO_3{}^{2-}$, three hybrids of equal energy contribute equally to the resonance structure. In H_2CO_3, the last two forms are of higher energy than the first and therefore are less important. Because it has three forms of equal energy, $CO_3{}^{2-}$ has a much greater resonance stabilization than does H_2CO_3.

Theoretical Approaches to Covalent Bonding

Although the use of electron dot notation and the concept of electronegativity are sufficient for qualitative descriptions of covalent bonding, more sophisticated methods provide for mathematical predictions of molecular properties. The results of two of these methods—the valence bond method and the molecular orbital method—are discussed in the rest of this chapter. The valence bond method allows prediction of molecular geometries. The molecular orbital method is more suitable for calculating energies and for predicting the magnetic properties of molecules. Complete details of either method are beyond the scope of this text. Therefore, only the essential features of each will be given as sets of postulates or rules which can be applied in certain limited situations.

11-11 Valence Bond Method

In the **valence bond method** it is assumed that when two atoms are brought close together, their atomic orbitals are in a suitable position to overlap. A pair of electrons (with opposed spins) can occupy these overlapping orbitals to form a covalent bond. Several examples illustrate these principles. In the formation of the hydrogen molecule, H_2, the $1s$ orbitals on the two atoms overlap, and the electrons in these orbitals pair up. The pair is then shared between the two atoms. If two helium atoms were brought together until the orbitals overlapped, no reaction would occur because the electrons in the $1s$ atomic orbitals are already paired. A very large energy would be required to produce unpaired electrons on each atom. Hence He_2 is not a stable molecule. In the forming of an Li_2 molecule, the respective $2s$ electron can pair up to give a stable bond. In the gaseous state lithium does indeed form Li_2 molecules. From reasoning analogous to that concerning He_2, one would not expect the molecule Be_2 to exist; nor is it found to exist.

Postulates of the valence bond theory are as follows:

1. The greater the overlap of the atomic orbitals, the more stable is the resulting bond.
2. In covalently bonded molecules, the electrons may occupy hybrid orbitals rather than orbitals like those on isolated atoms. The **hybrid orbitals,** new sets of orbitals, are mathematical combinations of atomic orbitals on a central atom which have greater ability to overlap with orbitals of other atoms.
3. When a given atom forms more than one covalent bond to other atoms, the bond angles approximate the angles between the orbitals which are used for bonding.
4. A single bond results from the sharing of a pair of electrons between two atoms in one set of overlapping orbitals, and a double or triple bond results from the sharing of two or three pairs in two or three sets of overlapping orbitals, respectively.

11-12 Hybrid Orbitals

From what has been stated, perhaps it is unexpected to find that the molecule BeH_2 does exist. The two Be—H bonds are found experimentally to be identical to each other, and the molecule is found to be linear. If the electronic configuration of the beryllium atom were changed from the atom's ground state, $1s^2 2s^2$, to the higher energy configuration $1s^2 2s^1 2p^1$, the atom would be able to form two bonds to hydrogen atoms. These bonds would be expected to have different properties because the s and p orbitals would overlap with the hydrogen orbitals by different amounts. If the beryllium atom assumed the configuration $1s^2\,2s^0\,2p_x{}^1\,2p_y{}^1$, identical bonds would be expected, but the angle between them would be expected to be $90°$, instead of the experimentally determined $180°$, because the p orbitals on the central atom are oriented at $90°$ angles from one another. None of these possibilities is in agreement with the observed facts. There is a description, however, which is consistent with the observed facts. The $2s$ orbital and a $2p$ orbital are combined into two new sp hybrid orbitals. The two hybrid orbitals are identical except that one has its greatest electron probability density in a direction $180°$ from that of the other. The result is shown schematically in the boundary surface diagrams of Figure 11–12. It is easy to imagine two equivalent bonds of a linear molecule being formed from these hybrid orbitals.

The concept of hybrid orbitals can be extended to explain the bonding about any central atom. In such explanations, the following principles apply:

1. The number of hybrid orbitals must be equal to the number of ground state orbitals from which the hybrid orbitals are constructed.
2. In contrast to the atomic orbitals from which they are constructed, hybrid orbitals of the same type all have the same energy.
3. Hybrid orbitals always have a greater potential for overlapping than do the ground state orbitals from which they are constructed.
4. The bond angles of actual molecules are very close to the directions of the hybrid orbitals.

Without using mathematics beyond the scope of this book, it cannot be demonstrated why each combination of atomic orbitals leads to particular angles between the hybrid orbitals; however the resulting geometries may be learned through association with some typical examples (see Table 11–7).

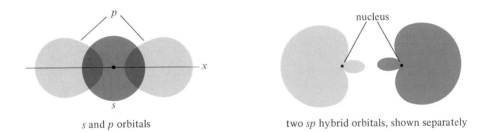

s and p orbitals   two sp hybrid orbitals, shown separately

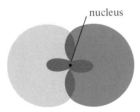

two sp hybrid orbitals
in position about atomic nucleus

Figure 11–12. **Construction of sp Hybrid Orbitals from Atomic s and p Orbitals**

TABLE 11-7. Hybrid Orbitals

Hybrid Type[a]	Atomic Orbitals	Number of Orbitals	Bond Angles	Example	Geometry
sp	$s + p$	2	$180°$	BeH_2	linear
sp^2	$s + $ two p	3	$120°$	BF_3	planar
sp^3	$s + $ three p	4	$109°$	CH_4	tetrahedral
dsp^2	$d + s + $ two p	4	$90°$	$PtCl_4^{2-}$	square planar
sp^3d	$s + $ three $p + d$	5	$\begin{Bmatrix} 90° \\ 120° \end{Bmatrix}$	$PCl_5(g)$	trigonal bipyramidal
d^2sp^3	two $d + s + $ three p	6	$90°$	CrF_6^{3-}	octahedral
sp^3d^2	$s + $ three $p + $ two d	6	$90°$	SF_6	octahedral

[a] The d orbitals are listed first if their principal quantum number is lower than that of the s orbital.

The combination of an s orbital with two p orbitals to form three equivalent sp^2 hybrid orbitals is shown in Figure 11-13. In sp^2 orbitals the directions of maximum probability density lie in the same plane, at angles $120°$ from each other. This type of hybridization is observed in compounds like BF_3.

Four equivalent sp^3 hybrid orbitals can be formed from one s orbital and three p orbitals. The geometry of a set of these orbitals is shown in Figure 11-14, where the directions of maximum probability point to the corners of a regular tetrahedron, providing for bond angles of $109.5°$. Actual substances having this geometry include CH_4, NH_4^+, SnF_4, and BH_4^-, as well as less symmetrical molecules such as CH_3Cl and $SnBrCl_3$.

Hybrid orbitals involving atomic d orbitals are shown in Figure 11-15. The combination of one d orbital, one s orbital, and two p orbitals produces four equivalent dsp^2 hybrid orbitals which are in a plane and directed toward the corners of a square and which provide for bond angles of $90°$. The combination of a d orbital with an s orbital and three p orbitals produces five hybrid orbitals which are not all equivalent. Three of these orbitals are directed at angles of $120°$ from each other in a plane, and the other two orbitals are perpendicualr to the plane. The combination of two d orbitals, an s orbital, and three p orbitals gives a set of six d^2sp^3 orbitals, which are directed toward the corners of a regular octahedron. Descriptions of all these hybrid orbitals are summarized in Table 11-7.

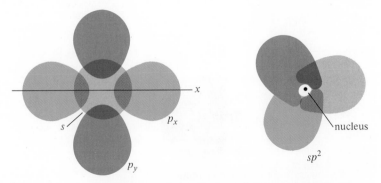

Figure 11-13. **Construction of the Three sp^2 Hybrid Orbitals from Atomic s and p Orbitals. (Only the principal lobes of the sp^2 orbitals are shown.)**

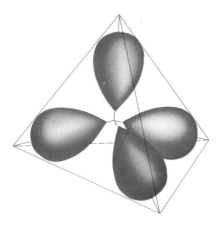

Figure 11-14. Major Lobes of the Four sp^3 Hybrid Orbitals

It should be noted specifically that the calculated energies of electrons in hybrid orbitals are higher than the energies of electrons in orbitals of atoms in their ground states. Energy must be provided to unpair electrons, if necessary, in addition to promoting them to energy states corresponding to the hybrid orbitals. However, the gain in stability resulting from the formation of bonds with the hybrid orbitals more than compensates for this additional energy. The relative energies involved are illustrated in Figure 11-16 for the case of a carbon atom forming a molecule such as methane, CH_4. It should be noted that the states illustrated do not necessarily have any real existence, nor is this description a mechanism for the reaction. The states are merely proposed to allow better understanding of the energy terms involved.

Even in cases where the number of bonds to be formed in a molecule is fewer than the number of pairs of electrons available, it is possible to explain the observed bond angles in terms of hybridization. For example, it can be assumed that in the water molecule the oxygen atom uses four sp^3 hybrid orbitals to accommodate the four electron pairs. Two of these orbitals overlap with $1s$ orbitals from the two hydrogen atoms, forming the bonds, and the other two electron pairs occupy the remaining two orbitals. As a result, the observed H—O—H bond angle is 105°, quite close to the 109° angle expected for regular tetrahedral molecules. Greater repulsion between the unshared pairs of electrons might explain the small difference. The nitrogen atom in ammonia can be considered in the same manner, with three of the sp^3 hybrid orbitals overlapping the $1s$ orbitals of the three hydrogen atoms. In this case, since there are

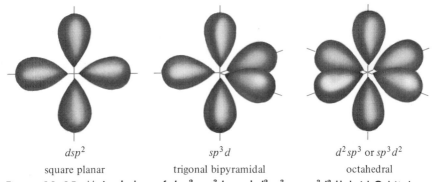

| dsp^2 | sp^3d | d^2sp^3 or sp^3d^2 |
| square planar | trigonal bipyramidal | octahedral |

Figure 11-15. Major Lobes of dsp^2, sp^3d, and d^2sp^3 or sp^3d^2 Hybrid Orbitals

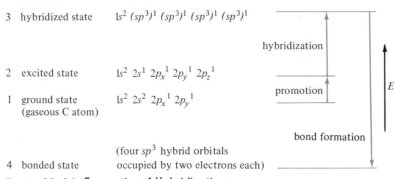

3	hybridized state	$1s^2 \ (sp^3)^1 \ (sp^3)^1 \ (sp^3)^1 \ (sp^3)^1$
2	excited state	$1s^2 \ 2s^1 \ 2p_x{}^1 \ 2p_y{}^1 \ 2p_z{}^1$
1	ground state (gaseous C atom)	$1s^2 \ 2s^2 \ 2p_x{}^1 \ 2p_y{}^1$
4	bonded state	(four sp^3 hybrid orbitals occupied by two electrons each)

Figure 11-16. **Energetics of Hybridization**

fewer unshared electron pairs, the observed bond angle of 107° is even closer to the tetrahedral bond angle.

11-13   Valence Bond Concept of Resonance

Calculations of the bond energy of the hydrogen molecule have been made using the valence bond method of approximating the solutions of the Schrödinger equation. The results of such calculations give values of the bond energy close to the experimentally observed value, provided that it is assumed that the hydrogen molecule is a resonance hybrid of the following structures:

H : H   H:$^-$ H$^+$ H$^+$:H$^-$

In other words, the best solution to the Schrödinger equation for the hydrogen molecule is a combination of individual solutions corresponding to pure covalent bonding and to pure ionic bonding. The contribution of the ionic forms is about 5% of the total. Thus the bond in the hydrogen molecule is said to have 5% ionic character. The calculated energy using resonance is lower and much nearer the experimental energy than the calculated energy using the covalent form alone and very much closer than that using either ionic form alone. The difference between the calculated energy using only the covalent structure and that using the combination of structures is called the **ionic resonance energy.**

It must be emphasized that the valence bond concept of resonance is a mathematical construction. There is no independent ionic structure or independent covalent structure for hydrogen, but one can calculate energies for such structures. Like the individual structures themselves, these energies are hypothetical. However, there is an actual bond energy which can be measured experimentally. The fact that the solution based on a combination of ionic and covalent energies yields a calculated bond energy which agrees with experiment is justification for the use of the resonance concept.

11-14   Basis for the Electronegativity Scale

The valence bond concept of resonance is the basis for the electronegativity scale formulated by Pauling (Section 11-10). In bonds between unlike atoms, A—B, one atom is likely to attract the electron pair more than the other. Hence the contribution of ionic resonance structures to the bond energy should be considerable. On the other hand, if A and B are of equal electronegativity, the contribution of ionic terms to the bond energy should be no greater than the ionic terms in the molecules A_2 and B_2. Therefore, the extra ionic resonance energy of the A—B bond is assumed to be the difference between the observed bond energy and the geometric mean of the bond energies of A_2 and B_2:

$$\Delta = D_{AB} - \sqrt{(D_{AA})(D_{BB})}$$

where Δ is the extra ionic resonance energy and D_{AA}, D_{BB}, and D_{AB} are the experimentally determined bond energies of the respective molecules.

Example

From the bond energies given in Table 11-1, calculate the extra ionic resonance energy in the HF molecule.

$$\Delta_{HF} = 135 - \sqrt{(104)(38)} = 72 \text{ kcal/mole}$$

Pauling proposed that the electronegativity difference between two atoms, A and B, is given by the equation

$$EN_A - EN_B = 0.208 \sqrt{\Delta}$$

where EN_A is the electronegativity of the more electronegative element, and EN_B is that of the less electronegative element. Since only differences in electronegativity can be obtained by this method, Pauling assigned the most electronegative element, fluorine, the electronegativity value 4.0 and calculated the other values from that. Different values are obtained for the electronegativity of a given element when its electronegativity is determined from data for different compounds. In the tabulation of electronegativities, a weighted average of such values is taken.

Example

Calculate the electronegativity of chlorine from the bond energy of ClF (61 kcal/mole) and the data of Table 11-1.

$$\Delta = D_{ClF} - \sqrt{D_{F_2}D_{Cl_2}} = 61 - \sqrt{(38)(58)} = 14 \text{ kcal/mole}$$
$$EN_F - EN_{Cl} = 0.208 \sqrt{\Delta}$$
$$EN_{Cl} = EN_F - 0.208 \sqrt{\Delta} = 4.0 - 0.78 = 3.2$$

The electronegativity of 3.0 tabulated for chlorine is a rounded-off average of values from many compounds.

11-15   Molecular Orbital Method

The **molecular orbital method** begins with the assumption that the electrons in molecules are in orbitals which belong to the molecule as a whole. In principle, such molecular orbitals can be defined mathematically. However, the procedure is complex, and a number of simplifying assumptions are made. The most widely used is a linear combination of atomic orbitals, where functions used to define atomic orbitals are appropriately modified and combined to give approximate molecular orbitals. Although this technique is known to involve a number of approximations, the results obtained are still quite useful, and such molecular orbitals can be visualized by use of boundary surface diagrams analogous to the boundary surface diagrams of atomic orbitals (Section 10–4).

If two atomic orbitals, one from each of the two different atoms, are combined, two molecular orbitals will be produced. One of these will have a lower energy than either of the atomic orbitals, and the other will have a higher energy than either of the atomic orbitals. As in the case of atomic orbitals, each molecular orbital can contain two electrons with opposed spins, or one electron, or no electrons. Also analogous to atomic orbitals, in cases where the molecular orbitals are degenerate, the electrons will occupy each orbital singly until the degenerate orbitals are half filled; then successive electrons will pair up in the orbitals, in accordance with Hund's rule (see Section 10–8). These principles are illustrated in the several examples which follow.

The construction of molecular orbitals for the hydrogen molecule is shown in Figure 11–17. Two hydrogen nuclei are located at the known internuclear distance of the H_2 molecule, and two molecular orbitals are constructed by combinations of their $1s$ orbitals. One molecular orbital represents an increased probability of finding the electrons between the nuclei. This location would produce attraction, and the orbital is thus a **bonding molecular orbital.** The second molecular orbital constructed from the two $1s$ orbitals represents the probability density as having a **nodal plane** (a plane where there is no electron density) between the nuclei. There is a very small probability of finding the electrons between the nuclei in this orbital, and as a result there is appreciable internuclear repulsion. This type of molecular orbital is **antibonding.** Both the bonding and antibonding orbitals are symmetrical about an imaginary line, called the bond axis, running through the two nuclei. Orbitals with such symmetry are called σ (sigma) orbitals. Antibonding orbitals are conventionally denoted by an asterisk, so in the cases cited the two orbitals are labeled σ_{1s} and σ_{1s}^*, where the subscripts denote that they were formed from $1s$ atomic orbitals.

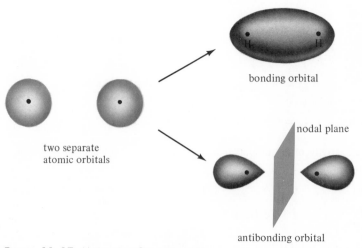

two separate
atomic orbitals

bonding orbital

nodal plane

antibonding orbital

Figure 11–17. **Molecular Orbitals of H_2**

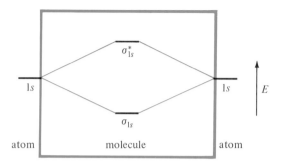

Figure 11-18. **Energy Levels in the Hydrogen Molecule**

The energy of the σ_{1s} molecular orbital is lower than the energy of the 1s orbital of the hydrogen atom. Conversely, the energy of the σ_{1s}^* orbital is higher than that of the 1s atomic orbital of hydrogen. These energy relationships are shown in Figure 11-18. In the hydrogen molecule two electrons having opposite spins enter the σ_{1s} orbital. Since this orbital is at a lower energy level than the 1s orbitals in the separated atoms, the H_2 molecule is in a lower energy state than two unbonded H atoms, and so it is stable.

The same set of orbitals and energy level diagram can be used to discuss the possible existence of He_2. Since each helium atom has two 1s electrons, a total of four electrons must be accommodated in the molecular orbitals. Two electrons cound enter the σ_{1s} orbital, and two could enter the σ_{1s}^* orbital. Since the former orbital is bonding and the latter is more antibonding, equal occupation of these orbitals by electrons would result in no net bonding. Hence the He_2 molecule is not stable with respect to two separate He atoms.

This procedure can be extended to include atoms with valence electrons having higher principal quantum numbers. In homonuclear diatomic molecules (molecules with two identical atoms) the combination of two 2s atomic orbitals will result in a σ_{2s} orbital and a σ_{2s}^* orbital. These molecular orbitals are analogous to the σ_{1s} and σ_{1s}^* orbitals.

The combination of two atomic p orbitals into molecular orbitals can be achieved in two ways, as shown in Figure 11-19. In these diagrams it is assumed that the two atomic nuclei approach each other along a mutual z axis. At a distance appropriate to the bond length, the two p_z orbitals combine to give two molecular orbitals as shown in Figure 11-19(a). One of these has a maximum probability density between the two

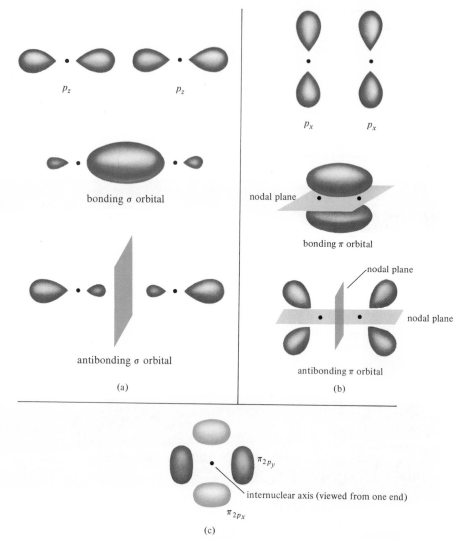

Figure 11-19. **Molecular Orbitals from Atomic p Orbitals**
(a) Combinations of p orbitals to form σ orbitals.
(b) Combinations of p orbitals to form π orbitals.
(c) Two bonding π orbitals viewed down the internuclear axis.

nuclei and is therefore at a lower energy; it is a bonding orbital. The other has a nodal plane between the nuclei, has a higher energy, and is antibonding. However, both of these molecular orbitals are symmetrical about the internuclear axis, and therefore they are σ orbitals. They are denoted σ_{2p} and σ_{2p}^*, respectively.

The p_x atomic orbitals are perpendicular to the p_z orbitals, but at the internuclear distance in a molecule they can still be combined to form bonding and antibonding molecular orbitals. There is a maximum probability density for the bonding molecular orbital between the nuclei despite the fact that this orbital has a nodal plane which includes the line joining the nuclei (Figure 11-19b). The antibonding molecular orbital has a nodal plane between the nuclei as well as one which includes the internuclear axis. It must be emphasized that the somewhat complex probability distribution of this antibonding orbital represents only one molecular orbital. Orbitals having one and only one nodal plane which includes the internuclear axis are called π (pi) orbitals. Hence the bonding and antibonding orbitals constructed from $2p_x$ orbitals are known as π_{2p_x} and $\pi^*_{2p_x}$ orbitals, respectively. Combination of the $2p_y$ atomic orbitals similarly yields π_{2p_y} and $\pi^*_{2p_y}$ molecular orbitals. These are situated $90°$ around the internuclear axis from the π_{2p_x} orbital, as shown in Figure 11–19(c).

Because of interaction with the σ_{2s} and σ^*_{2s} orbitals, the energy of the σ_{2p} orbital varies with the atomic number of the elements. In boron, carbon, and nitrogen, the interaction is relatively great. The σ_{2s} and σ^*_{2s} orbitals are lowered in energy by the interaction, and the σ_{2p} orbital is raised in energy. Its energy turns out to be higher than that of the π_{2p} orbitals, as shown in Figure 11–20(a). As the $2s$ and $2p$ orbitals of the atoms get further apart in energy with increasing atomic number, as in oxygen, fluorine, and neon, the interaction between the σ_{2s}, the σ^*_{2s}, and the σ_{2p} orbitals is less. The σ_{2p} orbital is not raised as much, and is lower in energy than the π_{2p} orbital, as shown in Figure 11–20(b). The σ^*_{2p} orbital always has a higher energy than the π^*_{2p} orbitals.

11-16 Electronic Configurations for Homonuclear Diatomic Molecules

The energy level diagrams of Figure 11–20 can be used as filling diagrams from which the electronic configurations of diatomic molecules formed from second period elements can be deduced. The procedure has already been illustrated for the cases of H_2 and He_2. In the case of Li_2 both the bonding and antibonding molecular orbitals from the combination of the $1s$ atomic orbitals are filled. The electrons in the σ_{1s} bonding orbital add to the stability of the molecule, and those in the σ^*_{1s} antibonding orbital destabilize it, so that these orbitals together do not contribute to the bonding. However, two electrons also fill the σ_{2s} bonding orbital. Hence the molecule should be stable. Spectroscopic studies of lithium vapor reveal that Li_2 molecules do exist.

The Be_2 molecule would have two more electrons than Li_2, and these two electrons would occupy the σ^*_{2s} antibonding orbital. Since the number of bonding and antibonding electrons in this molecule would be equal, no net bonding would result, and the molecule is not stable. This species has not been found experimentally. These results are qualitatively the same as were obtained using the valence bond method.

Continuing this procedure for homonuclear diatomic molecules of the elements of the second period gives the results listed in Table 11–8. It should be noted that the greater the net number of bonding electrons over antibonding electrons, the more stable is the bond. The number of net bonding electrons divided by 2 is called the **bond order**. The higher the bond order, the more stable is the bond. "Molecules" having no *net* bonding electrons are not expected to be stable.

The case of the oxygen molecule deserves special mention. The electrons in filled σ_{1s} and σ^*_{1s} orbitals do not contribute to the bonding. The filling of the molecular orbitals with the 12 valence electrons can be represented by an energy level diagram (Figure 11–21). It is seen that in accordance with Hund's rule each of the π^* orbitals must contain an unpaired electron. The bond order is 2. Thus the molecular orbital theory accounts directly for both the paramagnetism and the bond strength of O_2, whereas the valence bond theory cannot explain these facts without additional assumptions. This logical explanation of the experimental facts concerning the oxygen molecule is one of the early successes of the molecular orbital theory.

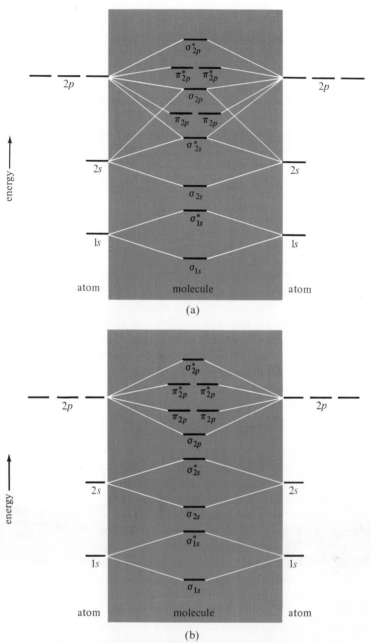

Figure 11-20. **Energy Level Diagrams for Homonuclear Diatomic Molecules of Second Period Elements**

TABLE 11-8. Outer Electron Distributions and Other Properties of Diatomic Molecules[a]

	Li_2	Be_2	B_2	C_2	N_2	O_2	F_2	Ne_2
bonding electrons	2	2	4	6	8	8	8	8
antibonding electrons	0	2	2	2	2	4	6	8
net bonding electrons	2	0	2	4	6	4	2	0
dissociation energy (eV)	1.14	unk[b]	3.6	6.35	9.75	5.08	2.8	unk[b]
bond distance (Å)	2.67		[c]	1.31	1.09	1.20	1.42	

[a] The $2s$ and $2p$ electrons only are considered.
[b] Unknown species.
[c] Uncertain value.

11-17 Heteronuclear Diatomic Molecules

Molecular orbitals for diatomic molecules containing atoms of different elements can be constructed. The basic principles are the same as for homonuclear diatomic molecules. If the same types of atomic orbitals are available in each atom, such as the $2p$ orbitals of nitrogen and oxygen in NO, the situation is very close to that for homonuclear molecules. Indeed, if the energies of the combining orbitals are not too different, the energy level diagrams given in Figure 11-20 can be used. (Since the atomic orbitals available for combination from two different atoms are not necessarily the same, the subscripts of the homonuclear molecular orbitals are not used for heteronuclear molecules.)

Example

What is the bond order in NO?
The electronic configuration of NO is shown in Figure 11-22. The six bonding and one antibonding electrons yield a net of five bonding electrons, for a bond order of $2\frac{1}{2}$.

A somewhat different energy level diagram results from the combination of two atomic orbitals, one from each atom, which have significantly different energies. The

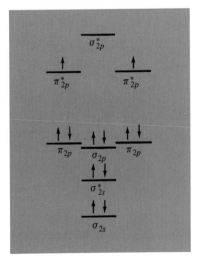

Figure 11-21. Electronic Configuration of O_2
The order of energies for the σ_{2p} and the π_{2p} orbitals results from the crossing of energies of these two sets in atoms containing about 15 electrons.

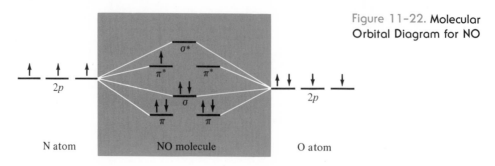

Figure 11-22. Molecular Orbital Diagram for NO

bonding orbital which results is more like the atomic orbital of lower energy than like that of higher energy. The electrons which fill the bonding molecular orbital can be thought of as "belonging" more to the former atom, and conversely the electrons (if any) which occupy the antibonding orbital are more the property of the other atom (Figure 11-23). Moreover, the two different atoms may differ in the number of atomic orbitals available to form molecular orbitals. Electrons in the orbitals which are not involved in bonding remain characteristic of the separated atoms. The HF molecule illustrates both of these points. A σ bond, formed from the $2p_z$ orbital of fluorine and the $1s$ orbital of hydrogen, connects the two atoms. The other two p orbitals of fluorine do not interact with the hydrogen $1s$ orbital because their orientations are not suitable. Their energies are not affected, and in the molecule they are termed **nonbonding** orbitals (not to be confused with antibonding orbitals). Electron pairs in nonbonding orbitals are often referred to as **lone pairs.** These lone pairs are localized about the fluorine atom.

It must be noted explicitly that although the $1s$ and $2p$ orbitals of a given atom have widely different energies, the case in point refers to $1s$ and $2p$ orbitals of different atoms. The $1s$ orbital of hydrogen and the $2p$ orbital of fluorine have energies of the same order of magnitude.

11-18 Multiple Bonding

The data of Table 11-8 show that the multiple bonding in N_2 and O_2 is correctly predicted by the molecular orbital theory. Specifically, it can be shown that the

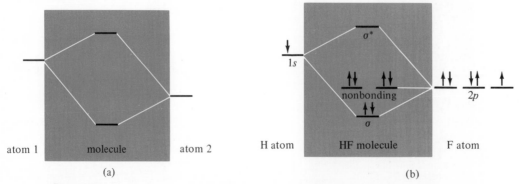

Figure 11-23. Molecular Orbital Energy Levels Arising from Atomic Orbitals of Different Energies

bonding electrons in N_2 occupy the σ orbital and two π orbitals, and in O_2 these orbitals are filled and the two antibonding π orbitals are half filled. In multiple bonding, more than one pair of electrons is shared; hence more than one molecular orbital must be utilized. One molecular orbital can be constructed from hybridized atomic orbitals, as shown in Figure 11–24(a), and the other molecular orbital(s) can be constructed from simple atomic orbitals. For example, it is assumed that in the ethylene molecule, C_2H_4, there are three sp^2 hybrid orbitals formed on each carbon atom. Two of these, one from each atom, combine to form a σ molecular orbital between the carbon atoms. The two other sp^2 hybrid orbitals on each carbon atom combine with $1s$ orbitals of hydrogen to form σ orbitals between the carbon and hydrogen atoms. One atomic p orbital of each carbon has not been used in forming the sp^2 hybrids and is available to form a bonding π orbital, which will be able to accommodate two electrons. There are electrons enough to fill all of the bonding orbitals described, but none in excess, so that the antibonding orbitals will be empty. Thus the double bond in ethylene consists of four electrons accommodated in one σ orbital and one π orbital. The H—C—H bond angle in C_2H_4 is $116°$, slightly less than the angle expected between sp^2 hybrid orbitals. The π bond keeps the six atoms all coplanar. The structure of the double bond in ethylene is shown in Figure 11–24(b).

In the formation of the N_2 molecule, ten electrons have to be accommodated in molecular orbitals. It is assumed that two sp hybrid orbitals are formed from the $2s$ and $2p_z$ orbitals of each of the two nitrogen atoms. Two of these hybrid orbitals then combine to form a σ molecular orbital between the nuclei, as shown in Figure 11–24(a). The $2p_x$ and $2p_y$ orbitals on the respective nitrogen atoms are then able to form two sets of bonding π orbitals. A shared pair of electrons is held in the σ orbital, and a shared pair is found in each of the π orbitals. The remaining two pairs of electrons are unshared and occupy the remaining nonbonding sp hybrid orbital on each nitrogen atom. The resulting boundary surface diagram is shown in Figure 11–24(c).

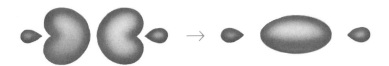

(a) Combination of two sp hybrid orbitals to form a bonding σ orbital.

(b) Bonding π orbital in ethylene. The plane of the atoms is the nodal plane.

(c) Boundary surface diagram of electron density in N_2. The bond directly between the nuclei is a σ bond. The combination of the two π bonds surrounds the internuclear axis symmetrically. Extending from the nitrogen atoms are the lone pairs.

Figure 11–24. **Molecular Orbitals from Hybrid Orbitals**

Double bonds generally consist of two pairs of electrons, one pair accommodated in a σ orbital and one pair in a π orbital. Triple bonds always consist of three electron pairs, one of which occupies a σ orbital and two of which are in π orbitals.

Example

Describe the HCN molecule in terms of molecular orbitals derived from hybrid orbitals.

The structure will be analogous to that of N_2 in that one of the two sp hybrid orbitals of the carbon atom combines with one of the sp hybrid orbitals of the nitrogen atom to form a σ bond. The second sp hybrid orbital of the carbon atom combines with the $1s$ orbital of hydrogen, and the other sp hybrid orbital of the nitrogen atom is occupied by a lone pair. The unhybridized p orbitals of carbon and nitrogen combine to form two bonding π orbitals (as well as two antibonding ones). The ten electrons are then distributed in the C—N and C—H σ orbitals, in the unshared sp orbital of nitrogen, and in the two bonding π orbitals between the carbon and nitrogen. Owing to the sp hybridization of the carbon atom, the molecule is linear.

11-19 Note on Deducing Hybridization and Multiple Bonding from Chemical Formulas

Sometimes, when given the formula of a molecule or ion, one is asked to deduce the hybridization of orbitals of a constituent atom and to predict the geometry of the species. The number of hybrid orbitals required on a central atom is not necessarily equal to the number of bonds which the atom forms. How does one determine merely from the formula how many electrons are promoted and how many hybrid orbitals are involved? The number of electrons to be promoted on a central atom to which atoms with unpaired electrons are to be bonded will be one of the following:

1. If the number of atoms bonded to the central atom plus lone pairs on the central atom is four or fewer, there must be a maximum number of unpaired electrons in the s and p subshells together. Any of these unpaired electrons which are not used in σ bonds or in lone pairs will be used in π bonds.
2. If the number of atoms bonded to the central atom plus lone pairs on the central atom exceeds four, there must be one unpaired electron for each bond to be formed. In this case, d orbitals will be used.

The number of hybrid orbitals required is merely the sum of the number of atoms bonded to the central atom plus the number of lone pairs (or unpaired electrons).

Example

Describe the promotion and the hybrid orbitals in **(a)** each carbon atom in acetylene, HC≡CH, **(b)** SF_6, **(c)** ICl_3.

(a) The ground state carbon atom configuration is

$$\underset{2s}{\underline{\uparrow\downarrow}} \qquad \underset{2p}{\underline{\uparrow}\;\underline{\uparrow}\;\underline{}}$$

One electron is promoted to increase the number of unpaired electrons in s and p subshells to the maximum, in this case four.

$$\underset{2s}{\uparrow} \qquad \underset{2p}{\uparrow \; \uparrow \; \uparrow}$$

Since there are two atoms bonded to each carbon atom, and no lone pairs, two hybrid orbitals are required. The hybrid orbitals formed are sp, which are used to form σ bonds to the bonded atoms. The other two unpaired electrons on each atom pair up in π bonds with the similar electrons on the other carbon atom. A linear molecule results.

  (b) The sulfur atom has a ground state

$$\underset{3s}{\uparrow\downarrow} \qquad \underset{3p}{\uparrow\downarrow \; \uparrow \; \uparrow} \qquad \underset{3d}{- \; - \; - \; - \; -}$$

Since six uncharged fluorine atoms are to be added, six unpaired electrons are required in six hybrid orbitals. The sp^3d^2 hybrids which result cause formation of an octahedral molecule.

  (c) The iodine atom has a ground state

$$\underset{5s}{\uparrow\downarrow} \qquad \underset{5p}{\uparrow\downarrow \; \uparrow\downarrow \; \uparrow} \qquad \underset{5d}{- \; - \; - \; - \; -}$$

Since there are to be three atoms bonded, together with two lone pairs, there must be three unpaired electrons. After promotion, the configuration is

$$\underset{5s}{\uparrow\downarrow} \qquad \underset{5p}{\uparrow\downarrow \; \uparrow \; \uparrow} \qquad \underset{5d}{\uparrow \; - \; - \; - \; -}$$

The sp^3d hybrid orbitals orient the electron pairs toward the corners of a trigonal bipyramid; the atoms will be attached to three of these, resulting in a "T-shaped" molecule (Figure 11–25).

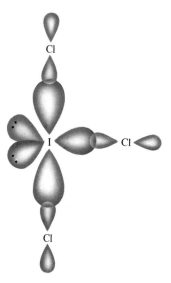

Figure 11–25. **Structure of ICl$_3$**

The lone pairs repel other electrons more than shared pairs do. The lone pairs therefore occupy the equatorial positions, where they are 90° removed from only two shared pairs. If they were to occupy the axial positions, they would be 90° away from three shared pairs.

In Chapter 14 the hybridization of empty orbitals to accommodate electron pairs from ligands to form coordinate covalent bonds will be described.

11-20   Macromolecular Solids

Macromolecular solids are substances in which the atoms in the entire crystal are linked together by covalent bonds. There are no individual molecules; in a sense the entire solid may be regarded as one giant molecule. Because covalent bonding extends throughout the solid, these substances are usually high melting and insoluble. Typical examples of such solids are diamond, graphite, and silicon dioxide, whose structures are diagrammed in Figure 11–26.

The carbon-to-carbon bonds in diamond are sp^3 hybrid bonds. The bond lengths are all 1.54 Å, the bond angles are all tetrahedral, and the bond energy for each bond is 85 kcal/mole, typical of carbon-to-carbon single bonds. The diamond structure is not easily deformed because of the strength of the bonds, and the crystal is tough and very high melting, as well as extremely hard. In fact, diamond is one of the hardest substances known.

Graphite is another crystalline modification of carbon. The graphite crystal consists of plane layers of carbon atoms connected by sp^2 hybrid σ bonds to give a regular hexagonal pattern. The carbon atoms are further connected by delocalized π bonds. Within each layer the carbon-to-carbon bond distance is 1.41 Å, close to the bond length in benzene. However, the distance between layers is 3.4 Å. This distance suggests that the layers are not held to each other by covalent bonds, but rather by van der Waals forces. The forces between the layers are comparatively weak, and the layers separate easily under applied stress. A "lead" pencil contains graphite. As one writes, layers of carbon are deposited on the paper by the force of friction. On the other hand, as might be expected because of the strong bonds within each layer, graphite has a high melting point (higher than that of diamond). The strength of the bonds within the layers together with the weakness between layers makes graphite a

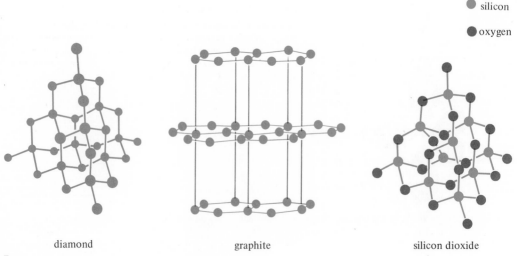

● silicon

● oxygen

diamond   graphite   silicon dioxide

Figure 11–26. **Structures of Macromolecular Solids**

Figure 11-27. **Naturally Occurring Quartz Crystals**

good lubricant. The layers slip past one another quite easily but are not themselves readily broken apart; thus the lubricating properties are retained over long periods of use. In contrast to diamond, which has no electrical conductivity, graphite has an electrical conductivity which is large in directions parallel to the layers of atoms but small in directions perpendicular to the layers. The conductivity is apparently due to the mobility of the electrons in the delocalized molecular orbitals (see Section 12-1).

In silicon dioxide the silicon and oxygen atoms are bonded in σ orbitals made from the combination of sp^3 hybrid orbitals from each type of atom. However, the oxygen atoms form only two covalent bonds. The mineral quartz (Figure 11-27), which is almost pure SiO_2, is an example of the gross shape of a crystal resulting from an infinite network of Si—O—Si bonds (Figure 11-26).

11-21   Exercises

Basic Exercises

1. Calculate the bond lengths in   (a) NH_3,   (b) SCl_2,   (c) CH_2Cl_2,   (d) $HOCl$, (e) H_3PO_4,   (f) HCN,   (g) CH_3NH_2.
2. Arrange C—C, C=C, and C≡C in order of   (a) increasing bond energy,   (b) increasing bond length.

3. In what ways does the periodic variation of the electronegativities of the elements differ from the periodic variation of ionization potentials?

4. (a) Can a molecule have a dipole moment if it has no polar covalent bonds? (b) How is it possible for a molecule to have polar bonds but no dipole moment?

5. Distinguish between electronegativity and electron affinity.

6. Arrange in order of decreasing polarity of the bonds: SbH_3, AsH_3, PH_3, NH_3.

7. Arrange in order of increasing dipole moment: BF_3, H_2S, H_2O.

8. Write electron dot structures for and predict the geometry of (a) H_3O^+, (b) $CH_2=NH$, (c) ClO_2^-, (d) NH_4^+, (e) N_2H_4.

9. In which of the following molecules is the van der Waals force likely to be the most important in determining the melting point and boiling point? (a) ICl, (b) Br_2, (c) HCl, (d) H_2S, (e) CO.

10. Deduce the shape of (a) SO_3, (b) SO_3^{2-}, (c) BF_3, (d) BF_4^-, (e) NF_3.

11. Which one of the following is expected to have the highest melting point? PH_3, NH_3, $(CH_3)_3N$. Explain why.

12. Which one of each of the following pairs is expected to exhibit hydrogen bonding? (a) CH_3CH_2OH and CH_3OCH_3; (b) CH_3NH_2 and CH_3SH; (c) CH_3OH and $(CH_3)_3N$.

13. From Figure 11-11 estimate the boiling point which water would have if there were no hydrogen bonding.

14. Arrange the following types of interactions in order of increasing stability: covalent bond, van der Waals force, hydrogen bonding, dipole attraction.

15. Compare the shapes of a p orbital and an sp hybrid orbital. Which one has a greater directional orientation? Explain.

16. In each of the following pairs, select the species having the greater resonance stabilization: (a) HNO_3 and NO_3^-, (b) $H_2C=O$ and $HC-O^-$.
$$\overset{\|}{O}$$

17. Which of the sets of hybridized orbitals (Table 11-7) do(does) not maximize the angles between electron pairs?

18. What hybridization is expected on the central atom of each of the following molecules? (a) BeH_2, (b) CH_2Br_2, (c) PF_6^-, (d) BF_3.

19. (a) Which molecule, AX_3, AX_4, AX_5, AX_6, is most likely to have a trigonal bipyramidal structure? (b) If the central atom, A, has no lone pairs, what type of hybridization will it have?

20. Determine the geometry of each of the following molecules and the hybridization about the central atom in each: (a) $BeF_2(g)$, (b) AlH_3, (c) NH_3, (d) $HC\equiv CH$.

21. Predict whether the He_2^+ ion in its electronic ground state is stable toward dissociation into He and He^+.

22. Compare and contrast the concepts of hybrid orbitals and molecular orbitals with respect to (a) the number of atoms involved, (b) the number of orbitals produced from a given number of ground state orbitals, (c) the energies of the resulting orbitals with respect to one another.

23. Distinguish between nonbonding orbitals and antibonding orbitals.

24. Draw a molecular orbital energy level diagram for each of the species in Table 11-8. Verify that each has the number of bonding and antibonding electrons listed.

25. Write the molecular orbital electronic configuration, state the bond order, and indicate whether the species is paramagnetic: (a) B_2, (b) C_2, (c) N_2, (d) O_2, (e) CN^-, (f) Br_2, (g) Cl_2^+, (h) NO, (i) NO^+, (j) CO.

26. Which of the following molecules has the highest bond order? (a) BN, (b) CO, (c) NO, (d) Ne_2, (e) F_2.

27. Which of the following is paramagnetic? (a) CO, (b) O_2^{2-}, (c) BN, (d) NO^+, (e) B_2.

28. Explain why N_2 has a greater dissociation energy than N_2^+, whereas O_2 has a lower dissociation energy than O_2^+.

29. Make a table giving (i) number of orbitals with a given energy, (ii) maximum number of electrons per orbital, and (iii) maximum number of electrons at a given energy for the following types of orbitals: **(a)** s, **(b)** p, **(c)** sp^2, **(d)** sp^3, **(e)** σ, **(f)** σ^*, **(g)** π^*.

30. **(a)** Using data from Table 11-3, calculate the molecular lengths of C_2H_2 and HCN. **(b)** From the observed oxygen-to-oxygen distance in the CO_2 molecule of 2.323 Å and data from Table 11-3, estimate the covalent radius of double-bonded oxygen atoms.

General Exercises

31. Which one of each of the following pairs is expected to have the larger bond angle? **(a)** H_2O and NH_3, **(b)** SF_2 and BeF_2, **(c)** BF_3 and BF_4^-, **(d)** PH_3 and NH_3, **(e)** NH_3 and NF_3.

32. Draw an electron dot structure for Br_3^-. Deduce an approximate value for the bond angle, and explain your deduction.

33. Would the peroxide ion, O_2^{2-}, have a longer or shorter bond length than O_2? Explain.

34. Write electron dot diagrams for each of the following molecules. Then write an electron dot diagram for an ion which is isostructural with each, that is, which has the same geometry and the same numbers of shared and unshared electrons: **(a)** CH_4, **(b)** F_2, **(c)** HNO_3, **(d)** CF_4, **(e)** He, **(f)** NH_3, **(g)** SO_2, **(h)** N_2O_4, **(i)** N_2, **(j)** CS_2.

35. NO_2 gas is paramagnetic at room temperature. When a sample of the gas is cooled below $0°C$, its molecular weight increases and it loses its paramagnetism. When it is reheated, the behavior is reversed. **(a)** Using electron dot structures, write an equation which accounts for these observations. **(b)** How does this phenomenon differ from resonance?

36. Carborundum, SiC, and corundum, Al_2O_3, are important industrial abrasives. Diagram structures for these compounds which show why they have such hardness.

37. A diatomic molecule has a dipole moment of 1.2 D. If its bond distance is 1.0 Å, what fraction of an electronic charge, e, exists on each atom?

38. At 300 K and 1.00 atm pressure, the density of gaseous HF is 3.17 grams/liter. Explain this observation, and support your explanation by calculations.

39. Calculate the electronegativity of nitrogen from the bond energies given in Table 11-1 and the electronegativity of fluorine. Compare the result to the tabulated electronegativity of nitrogen, and suggest a reason for any difference.

40. Using bond energy data from Table 11-1 and taking the electronegativity of hydrogen as 2.1, estimate the electronegativities of sulfur and chlorine.

41. Predict the shapes of the following species and describe the type of hybrid orbitals on the central atom: **(a)** $PbCl_4$, **(b)** SbF_6^-, **(c)** BH_4^-, **(d)** PCl_3, **(e)** N_2Cl_4.

42. What hybrid orbitals are ascribed to carbon in the (short-lived) CH_2^{2+} ion? What is the geometry of this ion?

43. Show schematically, according to valence bond theory, the orbital occupancy of electrons in the chlorine atom in **(a)** ClO_3^-. **(b)** ClO_4^-. **(c)** Describe the geometries of these two species.

44. Describe and compare the geometries of the molecules and the hybridization of the carbon and boron atoms in $F_2C=C=CF_2$ and $F_2B-C\equiv C-BF_2$. Compare the relative orientations of the sets of fluorine atoms in the two cases. In which case is it impossible for all four fluorine atoms to lie in the same plane?

45. Using Figures 11-20(a) and then (b), write the electronic configuration of O_2 and F_2. Repeat the process for B_2 and C_2. In which case(s) does(do) the choice of energy level diagram make a difference in the number of unpaired electrons? Explain.

46. The bonding σ_{2s} orbital has a higher energy than the antibonding σ_{1s}^* orbital, as shown in Figure 11-20. Why is the former a bonding orbital while the latter is antibonding?

47. What would be the expected electronic arrangement and magnetic moment in **(a)** the superoxide ion, O_2^-? **(b)** the peroxide ion, O_2^{2-}?

48. Represent by electron dot structures and by molecular orbital boundary surface diagrams the electronic structure of **(a)** CO, **(b)** CO_2, **(c)** NO_2, **(d)** NO_2^-, **(e)** NO_2^+, **(f)** SO_3.

49. Is the HHe molecule apt to be stable toward dissociation into atoms? Given two hydrogen atoms and two helium atoms, which one of the following combinations (if any) has the lowest energy? **(a)** 2 HHe, **(b)** H_2 + He_2, **(c)** He_2 + 2 H, **(d)** H_2 + 2 He, **(e)** 2 H + 2 He.

Advanced Exercises

50. Explain on the basis of hydrogen bonding the existence of the compound $NaHF_2$. Write equations for a two-step dissociation of HF. The pH of 0.10 M HF is 2.23. Calculate a value for K_1 on the basis of this two-step dissociation.

51. Show that if two atoms bond along their z axes, the various d orbitals can combine to form σ, π, or δ (delta) type molecular orbitals.

52. Analogous to the formation of π_{2p} orbitals from separate p orbitals, diagram the approach of the d_{xy} orbitals of two atoms along their z axes, to form bonding δ and antibonding δ^* orbitals. What are the distinguishing characteristics of δ orbitals?

53. Deduce the geometry of each of the following molecules: **(a)** XeF_6, **(b)** XeO_3, **(c)** XeF_4, **(d)** BrF_5.

54. Deduce the geometry and diagram the electronic structure of **(a)** I_3^-, **(b)** ClO_3^-, **(c)** ClO_3^+, **(d)** F_2SeO, **(e)** $IO_2F_2^-$.

55. Boric acid, $B(OH)_3$, forms hexagonal crystals that cleave easily into thin layers, indicating weak interplanar forces. Diagram a possible structure for a layer of crystalline $B(OH)_3$.

56. Two different bond lengths are observed in the PF_5 molecule, but only one bond length is observed in SF_6. Explain the difference.

57. In water, the H—O—H bond angle is 105°. Using the dipole moment of water and the covalent radii of the atoms, determine the magnitude of the charge on the oxygen atom in the water molecule.

58. At a given potential, a certain pair of parallel plates can hold charges in vacuum, and with HCl between them, as tabulated. The concentration of HCl at the two temperatures is the same. Explain these data.

	Charge at 0°C	Charge at 100°C
with HCl	5×10^{-8} C	3×10^{-8} C
vacuum	1×10^{-8} C	1×10^{-8} C

59. The two molecules indicated below are capable of intramolecular hydrogen bonding. Which is likely to form more stable hydrogen bonds? Suggest a reason for your choice.

60. Al_2Cl_6(g) has a bridged structure in which the two aluminum atoms share two chlorine atoms, and the effect is that of two tetrahedra sharing a common edge. In terms of Lewis notation, how is this possible? In what ways is this structure analogous to and in what ways is it different from the case of a double-bonded molecule such as C_2Cl_4?

61. From the enthalpies of formation of PH_3 and $P(g)$ (Tables 3-2 and 11-2), calculate the average P—H bond energy. Then, using appropriate electronegativity values, estimate the bond energy of a P_2 molecule.

62. The percent ionic character of a single bond may be estimated from the ratio of the observed dipole moment to the calculated moment, assuming oppositely charged ions located at a distance equal to the bond length. The observed dipole moments of HCl, HBr, and HI are 1.03, 0.79, and 0.38 D, respectively. Calculate the percent ionic character of the bond in each of these compounds. Do the results obtained parallel the magnitudes of the extra ionic resonance energies, Δ, for these molecules?

63. In 1934, R. S. Mulliken proposed that a value of electronegativity for an element could be defined as the average of its ionization potential and its electron affinity. Using units of electron volts for the ionization potential and electron affinity, determine the electronegativity of each halogen using this system. Compare the relative electronegativities of these elements on this scale with that given by the Pauling scale.

64. An electronic excited state of He_2 has been observed, but no ground state has been found. Explain why He_2 molecules in an excited state could be stable toward dissociation into atoms.

65. In terms of molecular orbital theory, using the F_2 molecule as an example, criticize the postulate that a single bond consists of one σ bond only.

66. Either of the following hybridizations of a central atom can lead to a square planar molecule. Give one example of each. **(a)** dsp^2, **(b)** sp^3d^2.

12

Structures of
Organic Compounds

Prior to modern concepts of bonding and structure, a "vital force" was believed necessary for the preparation of organic molecules. That is, it was thought that only living plants and animals could synthesize them. It was known that simple organic compounds could be converted to other simple organic compounds, since the starting materials were thought to already contain the vital force. It was not thought possible to synthesize complex organic compounds from simpler ones outside living organisms. However, in 1828 Friedrich Wöhler synthesized the organic compound urea, a product of animal metabolism, from potassium cyanate and ammonium chloride. Both NH_4Cl and $KCNO$ were considered to be inorganic compounds because they are typical salt-like, ionic materials not derived from living systems. The net ionic reaction for Wöhler's synthesis is

$$NH_4^+(aq) \ + \ CNO^-(aq) \ \xrightarrow[\text{heat}]{} \ NH_2\!\!-\!\!\underset{\underset{\text{O}}{\|}}{C}\!\!-\!\!NH_2$$

urea

Hence, this reaction was the first evidence that a vital force was not necessary to prepare organic compounds. It is now recognized that the same principles of bonding and structure apply to all molecules, and today the term *organic chemistry* is used to refer to the chemistry of the compounds of carbon, regardless of their source.

In this chapter, the principles of bonding and structure presented in Chapter 11 will be applied to some typical classes of organic compounds. The functional groups commonly encountered in organic molecules will be described, because the presence of a functional group in a given molecule has a bearing on the structure of the molecule as a whole. This chapter also provides a background for the discussion of experimental methods of structure determination given in Chapter 13. Further details of the properties and reactions of organic compounds are presented in Chapter 23.

12-1 Aliphatic Hydrocarbons

In molecules of saturated hydrocarbons, electrons in sp^3 hybrid orbitals of the carbon atoms form bonds to other carbon atoms and to hydrogen atoms (Section 11–12). Accordingly, the methane molecule, CH_4, has a tetrahedral configuration, and the ethane molecule, C_2H_6, has the structure of two tetrahedra joined at one

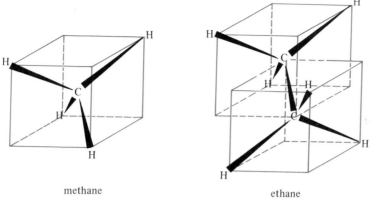

methane   ethane

Figure 12-1. Tetrahedral Shapes of Methane and Ethane
Lines drawn from the center of a cube to alternate corners are directed toward the apexes of a tetrahedron. The structure of methane is represented by a carbon atom at the center and hydrogen atoms at alternate corners of the cube. The structure of ethane can be viewed as being derived from two interlocking cubes, as shown.

corner. These structures are shown in Figure 12-1. In the latter case, rotation about the carbon-carbon single bond allows an infinite number of conformations, the two extremes of which are called eclipsed and staggered. The staggered conformation is

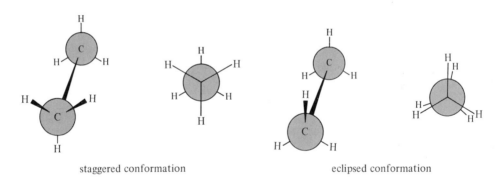

staggered conformation   eclipsed conformation

the most stable because that form has the least repulsion between the hydrogen atoms on neighboring carbon atoms. However, the energy difference between the different conformations is small, and at ordinary temperatures the energy required for rotation about the carbon–carbon bond is available from the thermal motion of the molecules (Section 15–1). The eclipsed and staggered conformations and all intermediate ones are readily interconvertible, and there is said to be **free rotation** about the single bond. The bond angles of longer chain hydrocarbons are also tetrahedral, and free rotation about their carbon–carbon single bonds permits a large number of different conformations.

The cyclic alkanes (Section 1–11) provide interesting examples of the effects of geometry on both bonding and conformation. The closed ring structures of such compounds limits the number of possible conformations. For example, in cyclopropane, H_2C———CH_2, there can be only one conformation, having the three carbon CH_2

atoms in a plane. The C—C—C bond angles are 60°, considerably less than the angle of 109.5° normally formed by sp^3 hybrid orbitals (Section 11–12). Similarly, in cyclobutane the C—C—C bond angles are 90°. In molecules having such small bond angles, there is less overlap of the sp^3 orbitals of a given carbon atom with the orbitals of the adjacent carbon atoms.

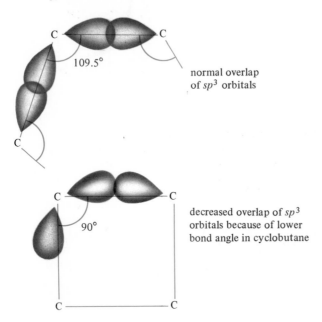

109.5°

normal overlap
of sp^3 orbitals

90°

decreased overlap of sp^3
orbitals because of lower
bond angle in cyclobutane

Therefore the carbon–carbon bonds in cyclopropane and cyclobutane are considerably weaker than the carbon–carbon bonds in noncyclic molecules. If cyclopentane

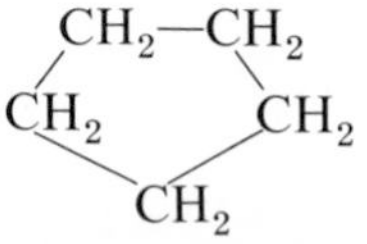

$$\begin{array}{c} CH_2\!-\!CH_2 \\ CH_2 \qquad CH_2 \\ CH_2 \end{array}$$

had a planar five-membered ring structure, it would have bond angles of 108°, very close to the tetrahedral angle of 109.5°. In a planar conformation, however, all the hydrogen atoms would be eclipsed. The molecule attains a somewhat lower energy by a slight puckering of the ring.

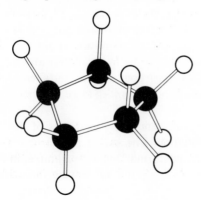

chair form   boat form

Figure 12-2. **The Chair and Boat Conformations of Cyclohexane**

For cyclohexane and larger cycloalkanes, bond angles of 109.5° are easily attained in several possible nonplanar conformations. The *chair* and *boat* forms of cyclohexane are shown in Figure 12–2. Because of ring strain, cyclopropane and cyclobutane are more reactive than are the other cycloalkanes.

Example

The enthalpy of combustion of gaseous cyclohexane is −944.4 kcal/mole, while that of gaseous cyclopropane is −496.8 kcal/mole. In each case calculate the enthalpy change per mole of O_2 reacted and explain the difference.

For cyclohexane, the enthalpy change is −104.9 kcal/mole of O_2; for cyclopropane, the enthalpy change is −110.4 kcal/mole of O_2. The higher energy state of the strained three-membered ring of cyclopropane is responsible for the relatively greater release of energy when this substance reacts with oxygen.

Bonding in unsaturated hydrocarbons can be described in terms of molecular orbital theory (Section 11–18). The procedure for developing molecular orbital models consists of two steps: (1) construction of a σ bond framework using localized bond orbitals and (2) construction of π bonds from those *p* orbitals not used in σ bonding. For example, the structure of ethylene, $H_2C{=}CH_2$, is depicted as follows: Three sp^2 hybrid orbitals on a given carbon atom overlap with *s* orbitals of two hydrogen atoms and an sp^2 orbital of the other carbon atom to form σ bonds.

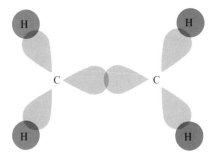

One atomic *p* orbital still remains on each carbon atom, and these two are available to form a bonding π orbital between the two carbon atoms (Figure 11–24). Thus the ethylene molecule contains five bonding σ orbitals and one bonding π orbital. There are exactly enough valence electrons in the carbon and hydrogen atoms to fill these

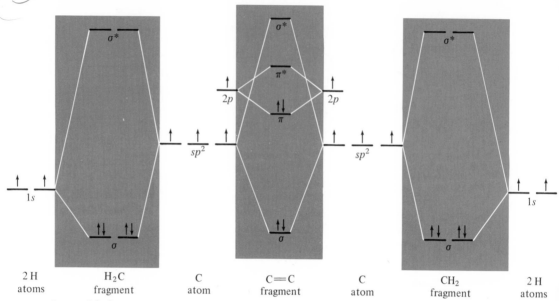

Figure 12-3. **Energy Level Diagram for Ethylene**

molecular orbitals. The four electrons occupying the σ and π orbitals located between the two carbon atoms constitute a double bond. The H—C—H bond angle in ethylene is 116°, only slightly less than that expected between sp^2 hybrid orbitals. The π bond keeps all six atoms coplanar. An approximate energy level diagram for the ethylene molecule is given in Figure 12-3.

In contrast to saturated hydrocarbons, molecules containing double bonds have a limited ability to form different conformations, because free rotation about the double bond does not occur. Rotation would require that the atomic p orbitals be twisted with respect to each other, rupturing a π bond.

Acetylene, H—C≡C—H, is the simplest hydrocarbon containing a triple bond. A molecular orbital description of the acetylene molecule is analogous to that given for the nitrogen molecule (see Section 11–18).

$$:N:::N: \qquad H:C:::C:H$$

In molecules having double bonds alternating with single bonds, such as butadiene, CH_2=CH—CH=CH_2, the π molecular orbitals have a more or less uniform probability density about all the atoms involved in the bonding, sometimes extending over the entire molecule. Such molecular orbitals are said to be **delocalized.** In butadiene, s and p orbitals on the four carbon atoms combine to form a σ-bonded skeleton of sp^2 hybrid bonds, leaving four p orbitals (one on each carbon atom) to form four π type molecular orbitals, two of which are bonding and two of which are antibonding. The π molecular orbitals, along with a diagram of their relative energies, are shown in Figure 12–4.

In the molecule allene, CH_2=C=CH_2, two σ bonds are formed by the central carbon atom by means of sp hybrid orbitals. As diagrammed in Figure 12–5, two perpendicular atomic p orbitals on the central carbon atom are available for π

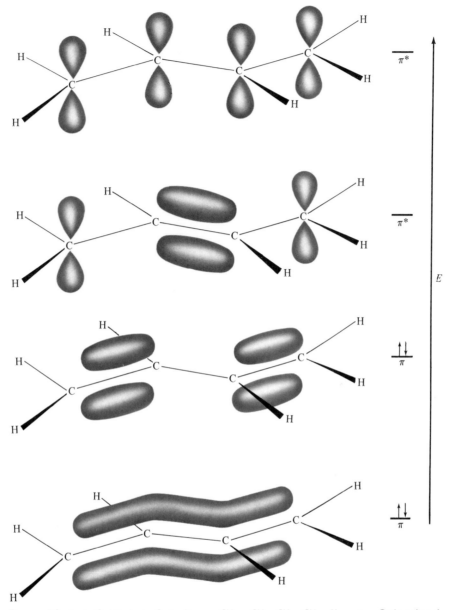

Figure 12-4. π Orbitals in Butadiene, $CH_2{=}CH{-}CH{=}CH_2$, Showing Delocalized Molecular Orbitals with Their Relative Energies

bonding. In this case, the molecular π orbitals are oriented $90°$ apart around the internuclear axis, and there is no delocalization.

12-2   Delocalized Molecular Orbitals in Benzene

In Section 11-10 it was pointed out that the benzene molecule, C_6H_6, cannot be represented by a single valence bond structure. To account for the bond angles, bond

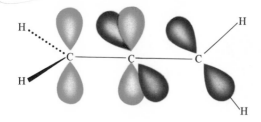

atomic *p* orbitals

Figure 12–5. **Localized π Bonding in Allene**
The center carbon atom has p_x (in color) and p_y (in gray) orbitals available for π bonding. The p_x orbital overlaps with the p_x orbital of the left carbon atom, and the p_y orbital overlaps with the p_y orbital of the right carbon atom. Two π molecular orbitals are formed which are located 90° away from each other about the line joining the carbon atom centers. No delocalization can occur because of the difference in positioning of the π bonds.

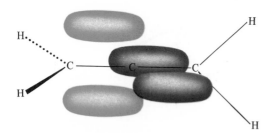

molecular π orbitals

distances, and relatively high stability of benzene, the concept of resonance was invoked. Using the molecular orbital approach, the structure of benzene can be explained without resort to resonance structures.

The benzene molecule has a planar, cyclic, hexagonal structure. Each carbon in the ring is attached to two adjacent carbon atoms and to a hydrogen atom by *sp²* hybrid orbitals. The σ-bonded planar ring which results is diagrammed in Figure 12–6(a). There remains on each carbon atom one *p* orbital perpendicular to the plane of the ring. From these six *p* orbitals, six π molecular orbitals are formed, three of which are bonding and three of which are antibonding. The benzene molecule contains just enough electrons to fill all the bonding σ orbitals and the bonding π orbitals. No antibonding π orbitals are occupied. The three pairs of electrons occupy the three bonding π orbitals, which are shown in Figure 12–6(b). The empty antibonding π orbitals are shown in Figure 12–6(c). These π orbitals, extending over the entire molecule, are completely delocalized. In effect, each carbon to carbon bond consists of a σ bond and a share of the π bonds equivalent to half a π bond. The observed six equal bond lengths in benzene lie between the usual carbon to carbon single bond length and the usual carbon to carbon double bond length. It is conventional to portray the benzene molecule as follows:

The circle signifies the delocalized molecular orbitals, while the hydrogen atoms are not shown.

Calculations show that the energy of the delocalized π-bonding orbitals is lower

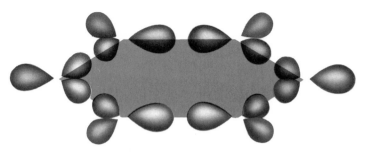

(a) In benzene, the major lobes of sp^2 hybridized orbitals of six carbon atoms combine to form six σ molecular orbitals in one plane.

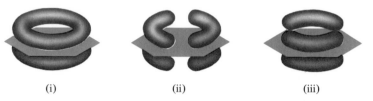

(i)   (ii)   (iii)

(b) The three bonding π orbitals of benzene.

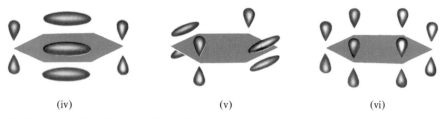

(iv)   (v)   (vi)

(c) The three antibonding π^* orbitals of benzene.

Figure 12-6. **Molecular Orbitals in Benzene**

than that of localized π-bonding orbitals (i.e., π orbitals between two specific carbon atoms). The added stability of the delocalized orbitals is comparable to the resonance energy of the valence bond theory.

A representation of the relative energies of the π orbitals in benzene is given in Figure 12–7. Only the bonding π orbitals contain electron pairs.

Benzene and compounds related to it are called *aromatic*. Historically, these compounds were given that name because of their distinctive, pungent odors. (The odor of one kind of mothballs, made from naphthalene, is a familiar example.)

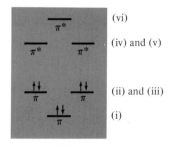

(vi)

(iv) and (v)

(ii) and (iii)

(i)

Figure 12-7. **Energies of π Orbitals in Benzene**
The Roman numerals identify the orbitals pictured in Figure 12-6.

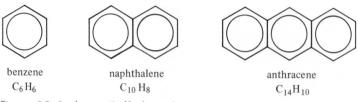

benzene   naphthalene   anthracene
C_6H_6 $C_{10}H_8$ $C_{14}H_{10}$

Figure 12-8. **Aromatic Hydrocarbons**

However, many compounds with the same ring structure, and which are still classified as aromatic, have little or no odor. In the terminology of molecular orbital theory, **aromatic** describes cyclic compounds having a special stability because of bonding electrons in delocalized π orbitals.

The characteristic structure of aromatic hydrocarbons includes at least one six-membered ring of carbon atoms (Section 1–11). Conventional representations for benzene, naphthalene, and anthracene are shown in Figure 12–8.

12-3   Nomenclature

In the first few members of the alkane series (Section 1–11), there are few structural isomers; to name them one can use prefixes such as *iso* as in *iso*pentane (the first *iso*mer of pentane, $CH_3CH_2CHCH_3$) or *neo*, as in *neo*pentane (the "new" isomer of
$\qquad\qquad\qquad\qquad\qquad\qquad\qquad\quad | $
$\qquad\qquad\qquad\qquad\qquad\qquad\qquad\quad CH_3$

pentane, $CH_3-\overset{\overset{\textstyle CH_3}{|}}{\underset{\underset{\textstyle CH_3}{|}}{C}}-CH_3$. However, with increasing numbers of carbon atoms, the

number of structural isomers increases drastically. Hence the IUPAC system of nomenclature has been adopted for naming organic compounds.

In the IUPAC system, the continuous chain hydrocarbons are named as listed in Table 1–7. **Hydrocarbon radicals** are portions of alkanes having one hydrogen atom removed from a carbon atom. A radical having a hydrogen removed from an end carbon is named from its parent hydrocarbon by replacing the *ane* ending of the name of the alkane with a *yl* ending. Thus, CH_3-, the radical derived from methane, CH_4, is called methyl. Similarly, $CH_3CH_2CH_2-$ is propyl, and $CH_3CH_2CH_2CH_2-$ is butyl. The radical derived from benzene, C_6H_5-, has the special name *phenyl*. Branched-chain hydrocarbons and molecules with functional groups (Section 12–4) are denoted systematically using these radical names. For example, CH_3CHCH_3 is
$\qquad\qquad\qquad\qquad\qquad\qquad\qquad\qquad\qquad\qquad\qquad\quad |$
$\qquad\qquad\qquad\qquad\qquad\qquad\qquad\qquad\qquad\qquad\quad CH_3$

denoted methylpropane. More than one branch of the same kind is denoted by a
$\qquad\qquad\qquad\qquad\qquad\qquad\qquad\qquad\qquad\qquad\qquad\qquad\quad CH_3$
$\qquad\qquad\qquad\qquad\qquad\qquad\qquad\qquad\qquad\qquad\qquad\qquad\quad |$
prefix—*di*, *tri*, *tetra*, etc. The systematic name for neopentane, CH_3-C-CH_3, is
$\qquad\qquad\qquad\qquad\qquad\qquad\qquad\qquad\qquad\qquad\qquad\qquad\qquad\quad |$
$\qquad\qquad\qquad\qquad\qquad\qquad\qquad\qquad\qquad\qquad\qquad\qquad\quad CH_3$

therefore dimethylpropane.

In hydrocarbon molecules with long carbon chains, the name of the compound is derived from the longest continuous chain. The position number of the carbon atom in the longest continuous chain to which a branch is attached is used to denote the position of the branch, for example

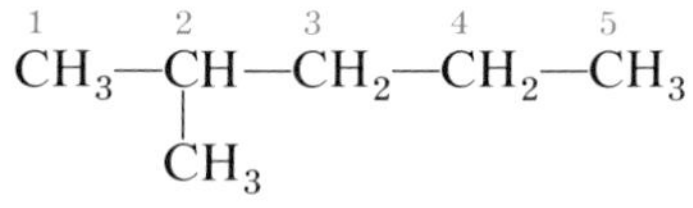

$$\overset{1}{CH_3}-\overset{2}{CH}-\overset{3}{CH_2}-\overset{4}{CH_2}-\overset{5}{CH_3}$$
$$\qquad\quad |$$
$$\qquad\quad CH_3$$

$$\overset{1}{CH_3}-\overset{2}{CH_2}-\overset{3}{CH}-\overset{4}{CH_2}-\overset{5}{CH_3}$$
$$\qquad\qquad\qquad |$$
$$\qquad\qquad\qquad CH_2$$
$$\qquad\qquad\qquad |$$
$$\qquad\qquad\qquad CH_3$$

2-methylpentane 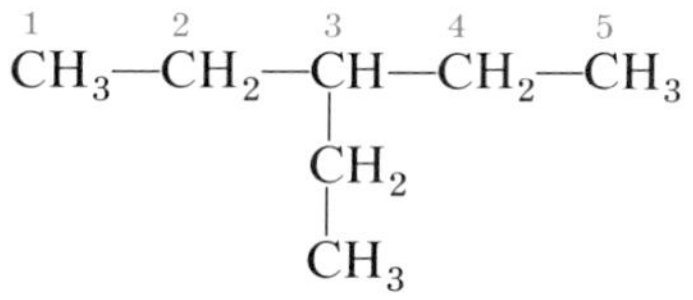   3-ethylpentane

To avoid having two equally appropriate names for the same compound, the smaller number (or set of numbers) is conventionally used to locate the branch or branches. Thus, for

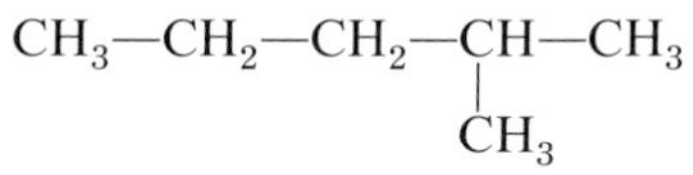

$$CH_3-CH_2-CH_2-CH-CH_3$$
$$\qquad\qquad\qquad\quad |$$
$$\qquad\qquad\qquad\quad CH_3$$

the name 2-methylpentane is used rather than 4-methylpentane.

Example

Name the following compounds:

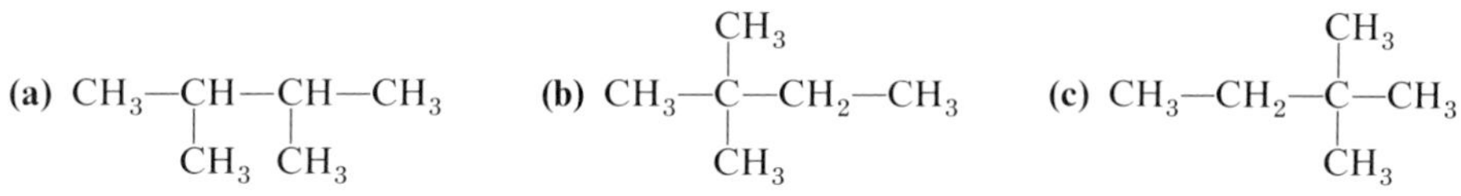

(a) $CH_3-CH-CH-CH_3$
   $\quad\quad | \quad\, |$
   $\quad\quad CH_3 \; CH_3$

(b) $CH_3-\underset{\underset{CH_3}{|}}{\overset{\overset{CH_3}{|}}{C}}-CH_2-CH_3$

(c) $CH_3-CH_2-\underset{\underset{CH_3}{|}}{\overset{\overset{CH_3}{|}}{C}}-CH_3$

(a) 2,3-dimethylbutane,   (b) and (c) 2,2-dimethylbutane.

Example

Write formulas for   (a) 2-methyl-3-ethylhexane,   (b) 2,2-dimethylhexane,   (c) 2,3,6-trimethylheptane.

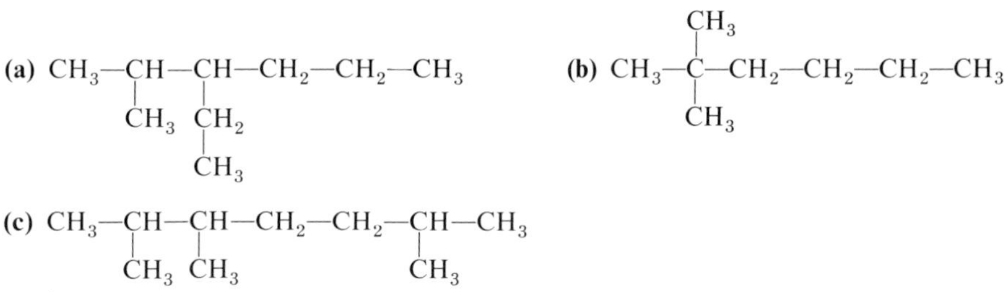

(a) $CH_3-CH-CH-CH_2-CH_2-CH_3$
   $\qquad\quad |\quad\,\, |$
   $\qquad\quad CH_3 \; CH_2$
   $\qquad\qquad\qquad |$
   $\qquad\qquad\qquad CH_3$

(b) $CH_3-\underset{\underset{CH_3}{|}}{\overset{\overset{CH_3}{|}}{C}}-CH_2-CH_2-CH_2-CH_3$

(c) $CH_3-CH-CH-CH_2-CH_2-CH-CH_3$
   $\qquad\quad |\quad\,\, | \qquad\qquad\quad |$
   $\qquad\quad CH_3 \; CH_3 \qquad\qquad CH_3$

Example

Explain why 2-ethylheptane is not a standard IUPAC name.

The longest continuous chain would include the ethyl group, and the compound would have the standard name 3-methyloctane.

$$\underset{3}{CH_3}\underset{}{\overset{}{\underset{|}{C}}}\underset{4}{H}\underset{}{CH_2}\underset{5}{CH_2}\underset{6}{CH_2}\underset{7}{CH_2}\underset{8}{CH_3}$$

$$_2CH_2$$

$$_1CH_3$$

12-4   Functional Groups

Every organic molecule containing other elements besides carbon and hydrogen may be regarded as consisting of a part which is derived from a hydrocarbon and another part (or parts) which is called a **functional group.** Functional groups are atoms or groups of atoms which impart specific chemical properties to a molecule, regardless of the nature of the hydrocarbon portion of the molecule. Thus double and triple bonds, even in hydrocarbons, are also included as functional groups. Compounds are classified according to their functional groups because all molecules containing a specific functional group will have properties characteristic of that group. Even in a single molecule having two or more functional groups, each group will tend to react in its characteristic manner more or less independently of the others. Some important functional groups are shown in Table 12-1. Molecules containing functional groups are denoted in the IUPAC system with names having characteristic endings, which are also shown in the table. The chemistry of the hydrocarbon portion of a molecule is predictable from the chemistry of the parent hydrocarbon, and it is common practice to denote this portion of the molecule as a radical, R. If the radical, R, is derived from an alkane, it is called an **alkyl** radical; if it is derived from an aromatic hydrocarbon, it is called an **aryl** radical.

Example

Classify the following molecules according to functional group:

(a) $CH_3CH_2CH_2OCH_3$

(b)
$$\begin{array}{cc} CH_2CH_2 & \\ CH_2 & CHOH \\ CH_2CH_2 & \end{array}$$

(c) $CH_3 - \underset{\underset{O}{\|}}{C} - CH_2CH_3$

(d) $CH_2{=}CHCH_2C_6H_5$

(e)
$$\begin{array}{cc} CH_2CH_2 & \\ CH_2 & O \\ CH_2CH_2 & \end{array}$$

(f)
$$\begin{array}{cc} CH_2CH_2 & \\ CH_2 & C{=}O \\ CH_2CH_2 & \end{array}$$

(g) $CH_2{=}CH\underset{\underset{CH_2}{\|}}{C}H$

Molecules **(a)** and **(e)** have the ROR′ type formula and are ethers; **(b)**, having the form ROH, is an alcohol; **(c)** and **(f)** are ketones because they contain the $C{-}\underset{\underset{O}{\|}}{C}{-}C$ group; **(d)** is an alkene which also contains an aromatic group; and **(g)** is a diene, with two double bonds connecting carbon atoms.

The first few members in each series of compounds are generally the most familiar. Many are known by common names which are older than the present nomenclature

TABLE 12-1. Important Functional Groups

Class of Compound	Functional Group	Formula[a] of Compound Type	Ending	Example Formula	Example Name
halide	X = F, Cl, Br, I	RX		CH_3Cl	chloro-methane
alcohol	—OH	ROH	-ol	CH_3OH	methanol
ether	—O—	ROR′	ether	CH_3OCH_3	dimethyl ether
aldehyde	—C(=O)—H	R—C(=O)—H[b]	-al	CH_3CHO	ethanal
ketone	—C(=O)—	R—C(=O)—R′	-one	CH_3COCH_3	propanone
acid	—C(=O)—OH	R—C(=O)—OH[b]	-oic acid	$CH_3CH_2CO_2H$	propanoic acid
ester	—C(=O)—O—	R—C(=O)—O—R′[c]	-ate	$CH_3CH_2CO_2CH_3$	methyl propanoate
amine	—NH₂, ＼NH, ＼N／	RNH_2, R_2NH, R_3N	amine	CH_3NH_2	methyl-amine
amide	—C(=O)—NH—	R—C(=O)—NH—R′[d]	-amide	$CH_3CH_2CH_2CONH_2$	butanamide
alkene	C=C	$R_2C=CR_2'$[d]	-ene	$CH_3CH=CHCH_3$	2-butene
alkyne	C≡C	RC≡CR′[d]	-yne	$CH_3C≡CCH_3$	2-butyne
aromatic	⬡			CH_3-⬡	toluene (methyl-benzene)

[a] R and R′ refer to hydrocarbon radicals except where otherwise indicated.
[b] In this case R can represent a hydrogen atom or a hydrocarbon radical.
[c] In this case R but not R′ can represent a hydrogen atom.
[d] In this case either R, R′, or both can represent a hydrogen atom or a hydrocarbon radical.

system. Chemists tend to retain and use the familiar names for these compounds, just as they use water instead of dihydrogen oxide for H_2O. Table 12-2 shows the first member(s) of several series of compounds with their common names.

Example

Write formulas for the following compounds: (a) butyl alcohol, (b) propyl phenyl ether, (c) ethyl acetate, and (d) diphenylamine.

(a) $CH_3CH_2CH_2CH_2OH$ (b) $CH_3CH_2CH_2OC_6H_5$

(c) $CH_3COC_2H_5$ (often written $CH_3CO_2C_2H_5$)   (d) $(C_6H_5)_2NH$
    $\overset{\|}{O}$

TABLE 12-2. Common Names of Some Small Molecules

Class	Common Name	Systematic Name	Formula
alkene	ethylene	ethene	$H_2C{=}CH_2$
alkyne	acetylene	ethyne	$HC{\equiv}CH$
alcohol	methyl alcohol	methanol	CH_3OH
	ethyl alcohol	ethanol	CH_3CH_2OH
aldehyde	formaldehyde	methanal	$HCHO$
	acetaldehyde	ethanal	CH_3CHO
acid	formic acid	methanoic acid	HCO_2H
	acetic acid	ethanoic acid	CH_3CO_2H
ketone	acetone	propanone	CH_3COCH_3
aromatic compounds	toluene	methylbenzene	$C_6H_5CH_3$
	phenol	hydroxybenzene	C_6H_5OH
	aniline	aminobenzene	$C_6H_5NH_2$
	benzaldehyde	benzaldehyde	C_6H_5CHO
	benzoic acid	benzoic acid	$C_6H_5CO_2H$
halide	methyl chloride	chloromethane	CH_3Cl

The systematic names of compounds containing functional groups are not always based on the longest continuous chain of carbon atoms. In molecules in which the functional group is indicated in the name by a suffix (see Table 12-1)—alkenes, alkynes, alcohols, and so forth—the name is based on the longest continuous chain which includes the functional group, even if this chain is not the longest continuous chain in the molecule. The numbering of the chain begins at the end which gives the position of the functional group or multiple bond the lower number. Further, the position of a multiple bond is denoted by the lower-numbered of the two carbon atoms joined by that bond. For example, the compound

$$CH_3CH_2\underset{\overset{|}{\underset{CH_2}{\|}}}{C}CH_2CH_3$$

has five carbon atoms in a continuous chain, but its name, 2-ethyl-1-butene, is based on the four-carbon chain containing the double bond. Molecules which have functional groups denoted as prefixes, such as the chloro hydrocarbons, have names based on the longest continuous chain in the molecule.

Example

Name the following compounds:

(a) $CH_3{-}CH_2{-}\underset{\overset{|}{\underset{\overset{|}{OH}}{CH_2}}}{CH}{-}CH_2{-}CH_3$

(b) $HO{-}CH_2{-}CH_2{-}\underset{\overset{|}{CH_3}}{CH}{-}CH_3$

(c) $HO{-}\underset{\overset{|}{\underset{\overset{|}{CH_3}}{CH_2}}}{CH}{-}CH_2{-}CH_2{-}CH_3$

(d) $CH_2{=}CH{-}CH_2{-}\underset{\overset{|}{CH_3}}{CH}{-}CH_3$

(e) $CH_3-CH=CH-CH_2-CH_3$

(f) $CH_3CH_2CHCH_2CH_3$
$$\underset{\underset{Cl}{|}}{\overset{|}{CH_2}}$$

(a) 2-Ethyl-1-butanol, **(b)** 3-methyl-1-butanol,
(c) 3-hexanol, **(d)** 4-methyl-1-pentene, **(e)** 2-pentene,
(f) 3-(chloromethyl)pentane.

It is customary to designate a given carbon atom in a molecule as primary, secondary, tertiary, or quaternary if it is attached to one, two, three, or four other carbon atoms, respectively.

Example

In the following representation of a hydrocarbon molecule, designate the primary, secondary, tertiary, and quaternary carbon atoms:

$$CH_3-\overset{\overset{H}{|}}{\underset{\underset{H-\overset{\overset{H}{|}}{\underset{\underset{H}{|}}{C}}-\overset{\overset{H}{|}}{\underset{\underset{H}{|}}{C}}-H}{C}}}{\underset{e}{C}}-\overset{\overset{\overset{a}{CH_3}}{|}}{\underset{d}{C}}-\overset{c}{C}\overset{H}{\underset{H}{\diagup}}{CH_3}$$

Atoms a, b, and f are primary; c, g, and h are secondary; e is tertiary; and d is quaternary.

It is worthwhile noting whether a functional group in a given molecule is attached to a primary, secondary, or tertiary carbon atom because its position may have a bearing on the chemical reactivity of the compound. Several examples of this effect will be given in Section 23–3.

12–5 Alcohols

Alcohols may be primary, secondary, or tertiary, depending on the type of carbon atom to which the OH group is attached. Alcohols may be considered as derivatives of water in which one hydrogen atom has been replaced by an organic group, R. The oxygen is bonded to R and to hydrogen by sp^3 type hybrid orbitals, and the R—O—H bond angle is about $105°$. In the liquid state, alcohol molecules, like water, tend to associate by hydrogen bonding, but the interaction is weaker and less extensive than in water. Also like water, alcohols have some tendency to act as acids and bases. However, except for aromatic hydroxy compounds, alcohols are so weak that in aqueous solution they react neither as acids nor as bases. As acids, they are about as strong as water. Alcohols react with very active metals like sodium to form alcoholates, $Na^+\,OR^-$, and with concentrated acids like H_2SO_4 to form ROH_2^+. The conjugate bases, RO^-, are somewhat stronger than OH^-, and in aqueous solution they react to form an equilibrium with OH^- ion and the alcohol:

$$RO^- + H_2O \rightleftharpoons ROH + OH^-$$

In contrast, the hydroxy derivatives of aromatic hydrocarbons are stronger acids than water. For example, phenol, C_6H_5OH, the simplest aromatic hydroxy derivative, has

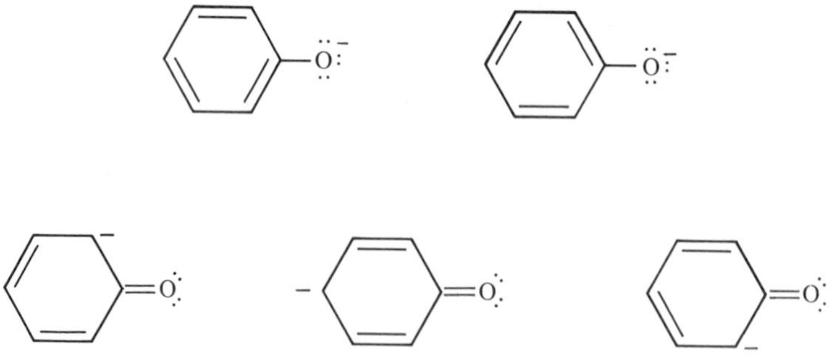

Figure 12-9. **Resonance of the Phenolate Ion**
The top two resonance forms are also found in phenol, C_6H_5OH, itself. The bottom three resonance forms represent extra stability of the phenolate ion, which makes it a weaker base than it would be otherwise.

a pK_a of 9.89. As shown in Figure 12-9, the conjugate base of phenol is stabilized by resonance. Its added stability makes its reaction with water less likely, and conversely makes the ionization of phenol as an acid proceed to a greater extent than the ionization of aliphatic alcohols.

12-6   Ethers

Ethers may be considered to be derived from water by replacement of both hydrogen atoms by alkyl or aryl groups. As in water, the bond angle about the oxygen is 105°. The two pairs of unshared electrons and the two bonding pairs on the oxygen atom are in sp^3 hybrid orbitals. Ethers can act as Brønsted bases or Lewis bases, but not as Brønsted acids or Lewis acids. Cyclic ethers are fairly good Lewis bases because the ring structures of their molecules makes the electron pairs on the oxygen atoms more readily available.

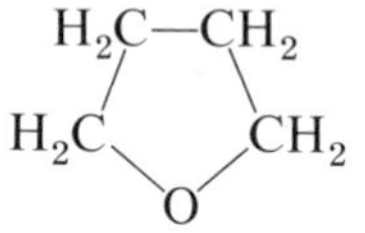

tetrahydrofuran,
a typical cyclic ether

12-7   Simple Aldehydes and Ketones

Both aldehydes and ketones are characterized by the carbonyl group, $\ce{>C=O}$. The simplest aldehyde is formaldehyde, $H_2C=O$, which is isoelectronic with ethylene and therefore has an analogous electronic structure. The molecular orbital representation of formaldehyde includes a σ-bonded framework of sp^2 hybrid orbitals on the carbon and oxygen atoms (see Section 11–19). The three sp^2 orbitals on the carbon atom accommodate the electron pairs shared between this atom and the two hydrogen atoms and the oxygen atom. The sp^2 orbitals on the oxygen atom accommodate the electron pair shared with the carbon atom and the two unshared pairs of electrons.

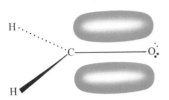

Figure 12-10. π Bonding in Formaldehyde

The unhybridized p orbitals, one on the carbon atom and one on the oxygen atom, can be combined to produce a π bonding orbital, as shown in Figure 12-10. The corresponding energy level diagram is shown in Figure 12-11.

Other aldehydes may be considered to be derived from formaldehyde by the replacement of one, but not both, of the hydrogen atoms of formaldehyde by an organic group, R. The general formula of an aldehyde is thus R—C=O.
                                                             H

Ketones may be considered as being derived from formaldehyde by replacement of both of the hydrogen atoms with radicals, to yield R—C—R. It is also possible to
                                                       O

have cyclic ketones, in which the carbon atom of the carbonyl group is a part of the ring.

Owing to the difference in electronegativity between carbon and oxygen, the

$$\underset{\diagup}{\overset{\diagdown}{}}\overset{\delta+}{C}=\overset{\delta-}{O}$$

carbonyl group is highly polar. Accordingly, aldehydes and ketones have rather large dipole moments, in the range of 2 to 3 D.

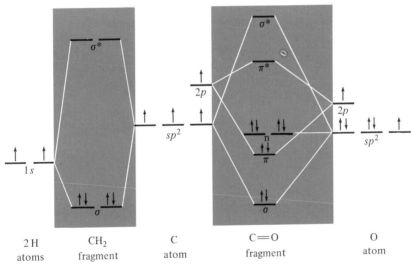

Figure 12-11. **Energy Level Diagram for Formaldehyde**
The molecular orbital type energy levels for the CH_2 fragment of the molecule are shown between the diagrams for the two hydrogen atoms and the hybridized carbon atom. That for the C=O portion of the molecule is between the hybridized carbon and oxygen atoms. Note the two nonbonding orbitals in the carbonyl fragment, containing the lone pairs of the oxygen atom.

Ketones, in particular acetone and methyl ethyl ketone, are important industrial solvents. As will be discussed in Chapter 23, aldehydes and ketones are useful starting materials and intermediates in the synthesis of complex organic compounds.

12-8   Organic Acids and Esters

Organic acids contain the carboxyl group, $-\underset{\underset{O}{\|}}{C}-OH$. The simplest such acid is formic acid, HCO_2H. The fact that in the carboxyl group there are two oxygen atoms on the same carbon atom makes the O—H bond very polar. Carboxylic acids tend to associate by hydrogen bonding. Also, the acidic proton is readily donated to bases such as water (Chapter 7):

$$RCO_2H + H_2O \rightleftharpoons RCO_2^- + H_3O^+$$

The proton transfer reaction is facilitated because the resulting anion, RCO_2^-, is stabilized by resonance involving the two equivalent oxygen atoms of the anion:

On substitution of a halogen atom for a hydrogen on the carbon adjacent to the carboxylate group, there is an increase in the acid strength which roughly parallels the electronegativity of the halogen substituent. The acidities of some substituted acetic acids are compared in Table 12–3. Electronegative groups such as the halogens are said to have an electron-withdrawing **inductive effect.** The anion of such an acid

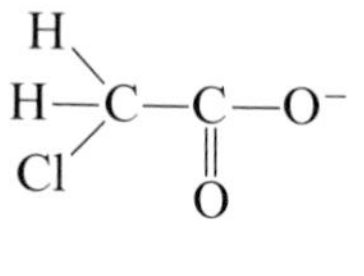

has a greater stability than the anion of the nonsubstituted acid owing to the

TABLE 12–3. Values of pK_a of Some Substituted Acetic Acids

Acid	pK_a
CH_3CO_2H	4.74
$ClCH_2CO_2H$	2.86
Cl_2CHCO_2H	1.30
$BrCH_2CO_2H$	2.87
FCH_2CO_2H	2.66
$(CH_3)_3SiCH_2CO_2H$	5.22

electrostatic attraction of the dipole created by the inductive effect of the electronegative group.

$$
\begin{array}{c}
\text{H} \\
\text{H} - \overset{|}{\underset{\underset{\delta-}{\text{Cl}}}{\text{C}}} - \overset{\underset{\|}{\text{O}}}{\text{C}} - \text{O}^- \\
\end{array}
$$

The reduction in negative charge on one carbon atom by the dipole stabilizes the full negative charge on the adjacent carboxylate group.

If the group attached to the carbon atom is less electronegative than carbon, it exerts an electron-releasing effect, which increases the electron density on the oxygen and results in a decrease in acid strength. The trimethylsilicon group, $(CH_3)_3Si—$, is an example of an electron-releasing group (see Table 12–3).

$$
\begin{array}{c}
\text{H} \\
\text{H} - \overset{|}{\underset{\underset{\delta+}{(CH_3)_3Si}}{\text{C}}} - \overset{\underset{\|}{\text{O}}}{\text{C}} - \text{O}^- \\
\end{array}
$$

Esters are the products of the reaction of an acid, RCO_2H, with an alcohol, $R'OH$, where the symbols R and R' indicate that the radicals *may* differ from each other.

$$RCO_2H \;+\; HOR' \;\rightleftharpoons\; RCO_2R' \;+\; H_2O$$

For example,

$$CH_3CO_2H \;+\; HOCH_2CH_3 \;\rightleftharpoons\; CH_3CO_2CH_2CH_3 \;+\; H_2O$$

The structure of an ester molecule has features of the structures of both its parent acid and its parent alcohol.

12-9 Amines

As a first approximation, amines may be considered as derivatives of ammonia. If one, two, or three of the hydrogen atoms of ammonia are replaced by R groups, primary, secondary, or tertiary amines, respectively, are produced. It should be noted that in amines the terms primary, secondary, and tertiary refer to the number of carbon atoms attached directly to the nitrogen atom. There are also quaternary ammonium ions, NR_4^+, analogous to ammonium ion, NH_4^+, in which all four hydrogen atoms are replaced by R groups. In all of these compounds, the nitrogen atom forms σ bonds by means of sp^3 type hybrid orbitals (Section 11–19).

There are two types of aromatic amines. The first type is simply the compound which results by the replacement of a hydrogen atom of ammonia by an aromatic radical. Aniline, $C_6H_5NH_2$, shown in Figure 12–12, is the simplest example of this type of compound. In the second type of aromatic amine, the nitrogen atom is an integral part of the ring structure. For example, pyridine, C_5H_5N, contains a nitrogen atom which completes the six member aromatic ring, as shown in Figure 12–12. A

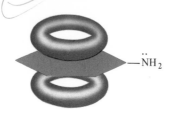

$-\ddot{N}H_2$

aniline

N:

pyridine

Figure 12–12. Two Types of Aromatic Amines
In aniline, the hydrogen of ammonia is replaced by a benzene ring. In pyridine, a CH group of the benzene ring is replaced by a nitrogen atom. In each case, the nitrogen atom possesses a lone pair of electrons.

nitrogen atom is isoelectronic with the CH group, and so delocalized π bonding persists when a nitrogen atom replaces a CH group in an aromatic ring. In this case, two of the three sp^2 hybrid orbitals on the nitrogen atom form bonds to its two neighboring carbon atoms, with the lone pair occupying the third sp^2 orbital. The unhybridized p orbital then becomes part of the delocalized π bonding of the planar ring.

Because of the lone pair of electrons on the nitrogen atoms, amines are basic toward proton donors and other Lewis acids. Amines are stronger bases than water, and in aqueous solution, amines react to give the conjugate acids and hydroxide ion:

$$RNH_2 \; + \; H_2O \; \rightleftharpoons \; RNH_3^+ \; + \; OH^-$$

In contrast to the amines, which are weakly basic, quaternary ammonium hydroxides, $NR_4^+OH^-$, are strong bases because of the presence of OH^- ion.

12–10 Geometric Isomerism

In Chapter 1, isomers were defined as compounds having identical compositions but different structures, as, for example, ethyl alcohol, C_2H_5OH, and dimethyl ether, CH_3OCH_3. **Geometric isomers** are substances which not only have identical compositions but also have each of the atoms bonded to the same set of other atoms. However, as the name implies, they have different geometries. Geometric isomerism is possible whenever the various possible conformations of a molecule cannot be interconverted without rupturing one or more bonds. Thus the eclipsed and staggered forms of ethane are not geometric isomers because they are readily interconverted by rotation about the C—C single bond. Similarly, the chair and boat forms of cyclohexane (Figure 12–2) are interconvertible without rupture of bonds, and these two conformations of the molecule are not geometric isomers.

However, consider the following possible structures for $C_2H_2Cl_2$:

$$\underset{cis}{\underset{H}{\overset{Cl}{>}}C=C\underset{H}{\overset{Cl}{<}}} \qquad \underset{trans}{\underset{H}{\overset{Cl}{>}}C=C\underset{Cl}{\overset{H}{<}}}$$

These correspond to two different compounds, which are geometric isomers of each other. They exist as isomers because rotation of the two halves of the molecule about the double bond is restricted (in contrast to free rotation about single bonds). These isomers are different compounds, with different properties, and can readily be separated and identified. For example, the *cis* isomer has a dipole moment and the *trans* isomer does not. A third isomer, called unsymmetrical dichloroethylene,

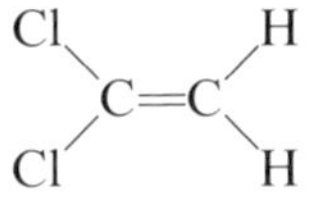

$$\begin{array}{c}\mathrm{Cl}\diagdown \\ \quad\mathrm{C}{=}\mathrm{C}\diagup \\ \mathrm{Cl}\diagup \quad\diagdown \mathrm{H}\end{array}\quad\begin{array}{c}\mathrm{H}\\ \\ \mathrm{H}\end{array}$$

is a structural isomer, not a geometric isomer, of the above pair, because the two chlorine atoms are both bonded to the same carbon atom.

Other examples of geometric isomerism are found in cyclic alkanes. For example, three of the isomers of dimethylcyclobutane are

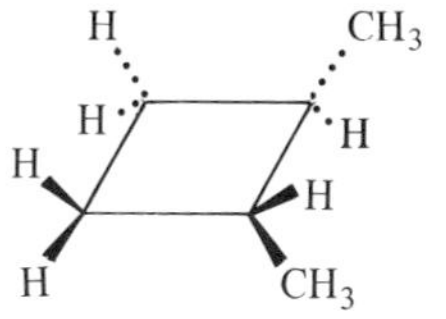

(a) 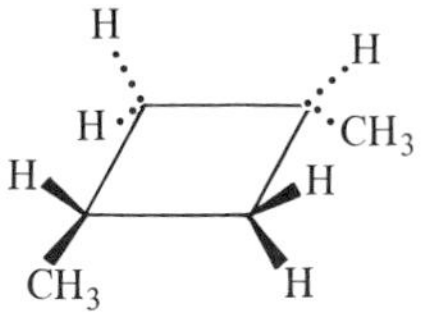   (b) 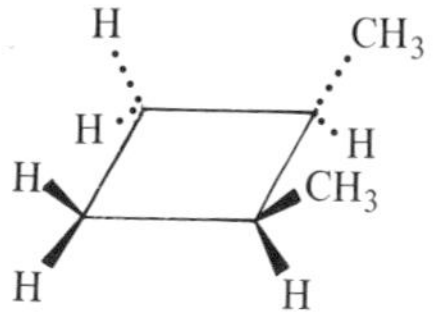   (c)

The three compounds are isomeric, but only (a) and (c) are geometric isomers. In both (a) and (c), the CH_3 groups are attached to adjacent carbon atoms; in one they are on the same side of the ring and in the other they are on opposite sides of the ring.

Example

Which of the following compounds can exist as geometric isomers? CH_2Cl_2, $CH_2Cl{-}CH_2Cl$, $CHBr{=}CHCl$, $CH_2Cl{-}CH_2Br$.
Only $CHBr{=}CHCl$ can exist as geometric isomers:

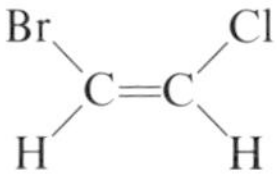

$$\begin{array}{c}\mathrm{Br}\diagdown \\ \quad\mathrm{C}{=}\mathrm{C}\\ \mathrm{H}\diagup\end{array}\begin{array}{c}\mathrm{Cl}\\ \\ \mathrm{H}\end{array}\quad\text{and}\quad\begin{array}{c}\mathrm{Br}\diagdown \\ \quad\mathrm{C}{=}\mathrm{C}\\ \mathrm{H}\diagup\end{array}\begin{array}{c}\mathrm{H}\\ \\ \mathrm{Cl}\end{array}$$

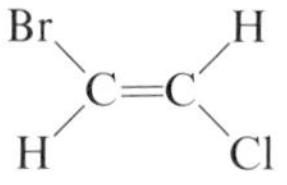

In $CH_2Cl{-}CH_2Cl$ and $CH_2Cl{-}CH_2Br$, the carbon atoms are connected by a single bond about which the groups can rotate relatively freely. Thus any conformation of the halogen atoms may be converted into any other simply by rotation about the single bond. In CH_2Cl_2, the configuration of the molecule is tetrahedral, and all interchanges of atoms yield exactly equivalent configurations.

12-11 Optical Isomerism

The existence of optical isomers provides a unique way of establishing the three-dimensional structure of molecules. Before discussion of this very important topic, it is necessary to describe **polarized light** and the phenomenon of optical activity. The production of plane polarized light is diagrammed in Figure 12–13. An **optically active** material is one which rotates the plane of polarized light. That is, when plane

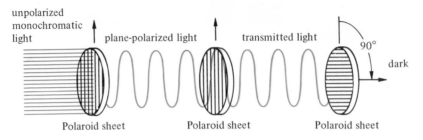

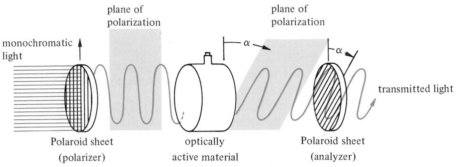

Figure 12-13. Plane-Polarized Light and the Measurement of Optical Activity

Normally, monochromatic light consists of rays vibrating in all possible planes perpendicular to the direction of propagation. If a beam of such light is passed through certain materials, such as a Polaroid sheet or a Nicol prism, only those rays vibrating in a single plane are transmitted. The transmitted beam consists of **plane-polarized light.** Another sheet of Polaroid, aligned exactly as the first, will transmit all of that light. A third sheet, aligned perpendicularly to the first two, will transmit none of the polarized light.

When an optically active substance is placed in a beam of plane-polarized light, it rotates the plane of polarization through an angle α either to the right or left. If a second sheet is to transmit completely all of the rotated beam of polarized light, it must be aligned at an angle α with respect to the first sheet. In a **polarimeter**, an instrument for measuring optical activity, the first Polaroid sheet is called a **polarizer**, and the second is called an **analyzer.**

polarized light is passed through the material, the plane of vibration of the emerging light will be changed. If, in a measurement of optical activity, the analyzer must be rotated to the right (viewed looking toward the light source) to obtain maximum transmission, the optically active material is said to be **dextrorotatory.** If the analyzer must be turned to the left, the material is said to be **levorotatory.**

During the nineteenth century, scientists had observed optical activity in certain liquids and solutions, such as turpentine and water solutions of sucrose (cane sugar). In 1844, Louis Pasteur observed that a sample of the sodium ammonium double salt of tartaric acid, $NaNH_4C_4H_4O_6$, which did not rotate the plane of polarized light, consisted of two kinds of crystals. These crystals were the same except that they were unsymmetrically shaped in such a manner that one kind was the mirror image of the other—there was a left-handed form and a right-handed form (Figure 12–14). Pasteur painstakingly separated the two kinds of crystals by hand, and he found that solutions of each were optically active. One form of crystal gave solutions that were dextrorotatory while solutions of the other form were levorotatory. It was apparent that there are at least two kinds of tartaric acid molecules which formed salts having different molecular structures. The shapes of the crystals of the salts reflected the difference in structures of the parent acid molecules.

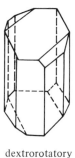 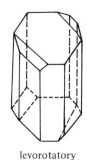

Figure 12-14. **Pasteur's Drawing of Hemihedral Crystals of Sodium Ammonium Tartrate**

dextrorotatory   levorotatory

About the time of Pasteur's work, knowledge of the structures of molecules was at the stage where it was recognized that certain elements combined with characteristic "valences." For example, the characteristic "valence" of carbon was known to be four. In 1874, a Dutch chemist, Jacobus H. van't Hoff, and a French chemist, Jules A. leBel, noted that in optically active compounds of carbon, a carbon atom was attached to four different atoms or groups of atoms. They reasoned that no structure in which the atoms lie in a single plane can lead to optical activity, since a planar molecule is its own mirror image. On the other hand, if the structures were nonplanar, arrangements of atoms giving rise to molecules which are mirror images but nonsuperimposable were possible (see Figure 12-15).

Of the possible nonplanar configurations, the one in which the groups are attached to the carbon atom in a tetrahedral arrangement is correct. Other nonplanar arrangements would allow the prediction of many more isomers of simple compounds than are observed (see exercise 28). The model of the tetrahedral arrangement of bonds around the carbon atom has been confirmed both theoretically and by many experiments.

A carbon atom to which four different atoms or groups of atoms are attached is said to be **asymmetric.** For example, the second carbon atom in 2-chlorobutane is asymmetric, being attached to a chlorine atom, a hydrogen atom, a methyl group, and an ethyl group. Two possible structures for this compound are diagrammed in Figure 12-16. These isomeric structures are seen to have a right-handed and left-handed character. (Indeed the right and left hands are good examples of objects which are mirror images but nonsuperimposable. With the hands placed palm to palm, the

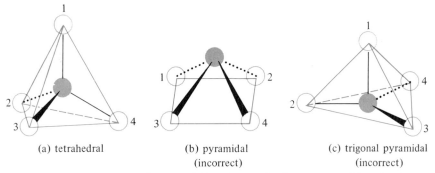

(a) tetrahedral   (b) pyramidal   (c) trigonal pyramidal
                         (incorrect)   (incorrect)

Figure 12-15. **Historically Proposed Nonplanar Configurations of Bonds about a Carbon Atom**

Figure 12–16. **Optically Active Forms of 2-Chlorobutane**

mirror

fingers reflect each other; but when both palms face in the same direction, the fingers no longer match.) In molecules, the property of optical activity is due to a lack of symmetry in structure such that two isomers can exist which are mirror images of each other, but which are not superimposable. The two isomers of 2-chlorobutane are optically active.

The criterion for optical activity is the existence of nonsuperimposable mirror images. The general property of an object having a nonsuperimposable mirror image is called **chirality.** Any object which has either a right-handed or a left-handed character contains a **chiral center.** For example, an asymmetric carbon atom is a chiral center. A chiral center is a necessary but not sufficient requirement for optical activity. If no chiral center exists, there can be no optical activity. If a chiral center exists in a molecule and if there is no second such center with exactly the same groups arranged in the opposite manner, then optical activity will be possible. Compounds having two asymmetric carbon atoms will have $2^2 = 4$ optical isomers, unless the activity caused by one of these chiral centers exactly cancels that caused by the other. In short, for any organic molecule, the maximum number of optical isomers possible is 2^n, where n is the number of chiral centers. For example, 2,3-dichlorobutane has two chiral carbon atoms; the second and third carbon atoms each have attached a hydrogen atom, a chlorine atom, a methyl group, and a chloroethyl group. This compound exists in only three forms, one of which is optically inactive because in that form each of the chiral centers exactly cancels the effect of the other (Figure 12–17).

Isomeric forms of a molecule which contain two chiral centers but which have a plane of symmetry within the molecule because the chiral centers exactly balance each other are called **meso** isomers. An equimolar mixture of two isomeric forms of a substance, one dextrorotatory and one levorotatory, does not rotate the plane of polarized light at all because the effect of one isomer cancels the effect of the other. The optically inactive *combination* of optically active forms is called a **racemic mixture.**

The dextro and levo forms of a substance are called **enantiomers.** Except for their ability to rotate plane-polarized light differently, enantiomers have almost identical properties. To separate the enantiomers in a racemic mixture, it is customary to allow the mixture to react with either the dextro form or the levo form of another optically active compound. The usual products of such a reaction, with a levo form, for example, are two compounds having a combination form such as dextro-levo and levo-levo. These compounds are not enantiomers, and often they can be separated by means of their different solubilities in a given solvent. The separated compounds are

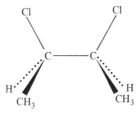

Figure 12–17. **Optical Isomers of 2,3-Dichlorobutane**

mirror

optically active forms

meso form (optically inactive)

then treated with an optically inactive material which restores the pair to their original compositions. The separation of enantiomers is called **resolution** of the mixture. The dextro isomer is conventionally denoted by (+); the levo isomer by (−).

Example

Describe a reaction sequence which might be used to resolve a racemic mixture of 3-chlorobutanoic acid.

A reaction sequence such as the following might be employed:

$$\left.\begin{array}{l} (+)\text{-}CH_3CHClCH_2CO_2H \\ (-)\text{-}CH_3CHClCH_2CO_2H \end{array}\right\} + 2((-)\text{-base}) \rightarrow \left\{\begin{array}{l} (+)\text{-}CH_3CHClCH_2CO_2((-)\text{-base}) \\ (-)\text{-}CH_3CHClCH_2CO_2((-)\text{-base}) \end{array}\right.$$

Since the pair of compounds formed in the reactions are not enantiomers, they may be separated by physical means. After separation, each is treated with excess HCl to restore the original acid.

$$(+)\text{-}CH_3CHClCH_2CO_2((-)\text{-base}) + HCl \rightarrow$$
$$(+)\text{-}CH_3CHClCH_2CO_2H + H((-)\text{-base})$$
$$(-)\text{-}CH_3CHClCH_2CO_2((-)\text{-base}) + HCl \rightarrow$$
$$(-)\text{-}CH_3CHClCH_2CO_2H + H((-)\text{-base})$$

It should be noted that in coordination compounds (Section 14–4), optical activity can exist without an asymmetric carbon atom, and indeed without an asymmetric atom of any kind. However, the existence of chirality in such compounds is evidence that they possess three-dimensional structures.

12-12 Exercises

Basic Exercises

1. Draw all isomeric forms of C_6H_{14} and name them according to the IUPAC system. Which one(s), if any, is(are) optically active?
2. What continuous chain hydrocarbon is isomeric with **(a)** 2-methyl-3-ethylhexane? **(b)** 2,2,4-trimethylpentane?
3. Which of the following is(are) called methyl chloride? **(a)** CH_3Cl, **(b)** CH_2Cl_2, **(c)** $CHCl_3$, **(d)** CCl_4. **(e)** Explain your choice.
4. Give the systematic name and formula for each of the following compounds: **(a)** phenyl ethyl ketone, **(b)** ethylene, **(c)** acetone, **(d)** acetaldehyde, **(e)** acetic acid, **(f)** form-aldehyde.
5. Name each of the following compounds: **(a)** CH_3CH_2OH, **(b)** $CH_3C_6H_5$, **(c)** CH_3COCH_3, **(d)** CH_3OCH_3, **(e)** CH_3CHO, **(f)** $CH_3CHOHCH_3$, **(g)** CH_3CO_2H, **(h)** $HOCH_2CH_3$, **(i)** C_6H_5OH, **(j)** $CH_3CO_2CH_3$, **(k)** $CH_3CHClCH_2CH_3$.
6. Molecules containing which of the functional groups act in aqueous solution **(a)** as Brønsted bases? **(b)** as Brønsted acids?
7. Write a structural formula for each of the following: **(a)** 4-ethylheptane, **(b)** 4-propyl-heptane, **(c)** 4-(1-methylethyl)heptane.
8. From the formulas listed below, select all examples of **(a)** primary amines, **(b)** sec-ondary amines, **(c)** secondary alcohols.
    (i) $CH_3CHNH_2CH_3$, **(ii)** $CH_3CHOHCH_2NHCH_3$, **(iii)** $(CH_3)_2COHCH_2NH_2$, **(iv)** $(CH_3)_2NCH(CH_3)_2$.
9. For each class of compounds listed in Table 12–1, **(a)** write a formula for the compound in the class which contains the fewest possible carbon atoms, **(b)** name each compound, **(c)** determine the oxidation number of the carbon atoms in each compound.
10. Select from the following the pairs of **(a)** geometric isomers, **(b)** optical isomers, **(c)** structural isomers.

    (i)
 Cl, Br / C=C / H, CH₃
    (ii)
 Cl, H / C=C / Br, CH₃
    (iii)
 Cl, CH₃ / C=C / H, Br

    (iv)
 Cl, Cl / C / H, H
    (v)
 H, Cl / C / Cl, H
    (vi)
 H, H / C / Cl, Cl

    (vii)
 Cl, Br / C / H, CH₃
    (viii)
 Cl, Br / C / CH₃, H
    (ix)
 Br, Cl / C / CH₃, H

11. Name and classify the following compounds: **(a)** $CH_2{=}CH{-}CH{=}CH{-}CH{=}CH_2$, **(b)** $(C_6H_5)_2O$, **(c)** $(C_2H_5)_2NH$, **(d)** $CH_3CH_2CONHCH_3$, **(e)** $(C_2H_5)_2NH_2{}^+Cl^-$.
12. Write formulas for **(a)** ethylamine, **(b)** propionaldehyde, **(c)** butanone, **(d)** ethyl propionate, **(e)** butyl formate, **(f)** bromobenzene, **(g)** acetylene, **(h)** phenylacety-lene.
13. Write formulas for **(a)** 2-butene, **(b)** methyl ethyl ether, **(c)** propanal, **(d)** 2-propa-nol, **(e)** 2,4-pentandione (acetylacetone).
14. Draw structures representing the resonance in each of the following: **(a)** benzene, **(b)** naphthalene, **(c)** anthracene.

15. Are all optical isomers necessarily optically active? Explain.

16. Explain the difference between a meso isomer and a racemic mixture. What characteristic(s) do they have in common?

17. Draw structures of all isomers of C_3H_5Cl.

18. (a) Explain why the name *butanol* is not specific, whereas the name *butanone* represents one specific compound. (b) Is the name *pentanone* specific?

19. Using Table 11–1, calculate $\Delta H°$ for the reaction at 25°C

$$C_2H_4(g) + 3 O_2(g) \rightarrow 2 CO_2(g) + 2 H_2O(g)$$

20. Using Table 11–1, estimate whether CH_4 or C_8H_{18} would yield more energy per gram upon complete combustion.

21. Using Table 11–1, and assuming a Kekule (localized double bond) structure, calculate the enthalpy of combustion of benzene to CO_2 and water. Compare this value with that experimentally determined for benzene, given in Table 3–3, and account for the difference.

General Exercises

22. In aqueous solution, tetramethylammonium hydroxide, $(CH_3)_4NOH$, is a strong base; aqueous solutions of trimethylamine are weakly basic. Explain the difference in terms of the structures of these substances.

23. (a) Calculate the oxidation number of the carbon atom in formaldehyde and in methandiol, $HOCH_2OH$. (b) Using bond energy data, calculate the enthalpies of formation of these two substances. (c) Determine $\Delta H°$ for the following reaction, and predict if one of the compounds would be stable with respect to reaction to produce the other.

$$HCHO + H_2O \rightarrow HOCH_2OH$$

24. Choose from the following list those molecules which are (a) Brønsted acids, (b) Brønsted bases, (c) Lewis bases.
(i) C_2H_5OH, (ii) C_5H_5N, (iii) $C_2H_5OCH_3$, (iv) C_6H_5OH, (v) $(C_2H_5)_2NH_2^+$, (vi) $C_6H_5CO_2H$, (vii) $(C_2H_5)_2NH$.

25. What kinds of hybrid orbitals are formed on the carbon atoms in cyclohexane? in cyclopropane?

26. Draw resonance structures for

27. Draw a structure showing intramolecular hydrogen bonding in . How many atoms are in the additional ring formed?

28. Demonstrate that CH_2Cl_2 could exist in more than one isomeric form if the bonds about the carbon atom in saturated compounds were square pyramidal, suggested as incorrect in Figure 12–15(b), or trigonal pyramidal, as in Figure 12–15(c).

29. Write electron dot structures and describe the geometry of the following molecules: (a) hydrazine, NH_2NH_2, (b) hydroxylamine, NH_2OH, (c) methylenimine, $CH_2{=}NH$, (d) acetyl chloride, $CH_3\underset{\underset{O}{\|}}{C}{-}Cl$.

30. In terms of electron delocalization, explain why amides are much weaker bases than amines.
31. Write a structural formula for an alcohol which has the formula $C_4H_{10}O$ and which can exist as optically active enantiomers.
32. How many optical isomers can exist for 2,3-butanediol? Would all of these be optically active?
33. If the double bonds in dichlorobenzene, $C_6H_4Cl_2$, were localized between specific carbon atoms, how many isomers of this compound would exist? How many isomers actually exist?
34. Tartaric acid is a dihydroxydicarboxylic acid, $HOCOCHOHCHOHCO_2H$. Identify any chiral centers, draw all optical isomers, and explain the total number of such isomers of tartaric acid in terms of the 2^n rule.
35. Write the formula for 1-phenyl-1-butanone. Explain why the designation "1-butanone" is somewhat unusual.
36. **(a)** Identify the asymmetric carbon atom(s) in each of the following substances. **(b)** Identify each of the molecules which has a meso isomer.

(i) $CH_3CHNH_2CONH_2$, (ii) $HOCH_2CHOHCH_2CH_3$, (iii) $HOCH_2CHOHCH_2OH$,

(iv)
$$\begin{array}{c} H_3C \\ \diagdown \\ CH{-}CH_2 \\ H_3CCHCH_2 \\ CH_2{-}CH_2 \end{array}$$

(v)
$$\begin{array}{c} CH_2{-}CH_2 \\ H_3CCHCH_2 \\ CH{=}CH \end{array}$$

37. Draw structures for all isomers of C_3H_4ClBr. State which of them, if any, will be optically active.
38. Name the following compounds:

(a) $CH_3CCH_2CH_2Cl$, (b) $CH_3(CH_2)_6CO_2H$, (c) $CH_3CH_2CHCH_2CH_2CH_3$.
$\parallel$ $\mid$
CH_2 CHO

39. The carbon atoms of the benzene ring may be numbered for identification of substituent groups, just as continuous chains of carbon atoms are numbered. Again, the smallest set of numbers designating the substituents is the preferred set. Draw structures for each of the following: **(a)** 1,3-dichlorobenzene, **(b)** 2,4,6-trinitrotoluene, **(c)** 1,4-diethylbenzene.

Advanced Exercises

40. Give the systematic name of the following compound:

$$CH_3CH_2CHCH_2CH_2CH_2CH_3$$
$$\mid$$
$$CH{-}CH_3$$
$$\mid$$
$$CH_3$$

41. Show why the phenolate ion, $C_6H_5O^-$, has a greater resonance stabilization than phenol, C_6H_5OH.
42. Construct molecular models of 1,2-dichlorocyclobutane. **(a)** How many asymmetric carbon atoms does this molecule contain? **(b)** Can this substance exist as a meso isomer? **(c)** What is the total number of isomers containing a four-membered ring corresponding to the formula $C_4H_6Cl_2$?

43. A perspective drawing of the ethylene molecule is shown in Figure 11–24(b). Draw a diagram of ethylene, including all atoms and the π bond, from the top view, the front view, and the side view. Repeat this process for the allene molecule (Figure 12–5).

44. Construct a molecular model of a noncyclic organic compound containing only carbon–carbon single bonds in which it is impossible to have complete rotation about a carbon–carbon bond.

45. Construct molecular models of the chair and the boat forms of cyclohexane. **(a)** Predict which form predominates in cyclohexane at room temperature, and justify your prediction. **(b)** Identify in the chair form the sets of hydrogen atoms which are referred to as axial and equatorial, respectively. **(c)** Change the conformation of the chair form into the boat form and then back into a chair form which is not the same as the original conformation. What positions do the hydrogen atoms which were originally axial now occupy? **(d)** How many isomers are there of chlorocyclohexane?

46. Construct a molecular model of the following derivative of cyclohexane:

(a) How many asymmetric carbon atoms does this molecule contain? **(b)** Deduce the number of optically active isomers possible for this structural isomer. **(c)** Are there any meso forms?

47. Construct a molecular model of 2,3-pentadiene. **(a)** Does this molecule contain any asymmetric carbon atoms? **(b)** Does it contain a chiral center? **(c)** Can it exist in optically active isomeric forms?

48. The dicarboxylic acids maleic acid and fumaric acid are *cis* and *trans* isomers, respectively, of $HOCOCH=CHCO_2H$. One of them has a K_1 value 10 times that of the other. For each isomer draw an appropriate structure of the anion which results from the loss of one proton. Deduce which isomer has the larger K_1 value, and suggest a reason for the difference. On this basis, which K_2 value should be larger?

49. It has been suggested that aromatic behavior stems from the occurrence of certain numbers of electrons in delocalized π orbital systems. The effective numbers are given by the formula $4n + 2$, where n is an integer. Based on this fact, explain the aromatic behavior of benzene, anthracene, and azulene. What determines the value of n? Draw resonance structures for azulene.

azulene

13

Experimental Determination of Structure

In previous chapters, the shapes of molecules were described with such confidence that it could have been thought that they had been examined visually. But it is not possible to handle individual molecules or to measure their dimensions directly. How, then, are the dimensions of molecules measured? The methods used are deductive rather than direct; that is, they depend on the measurements of some property or properties of a substance which might be related to the structural characteristics of the individual molecules.

The design of experiments by means of which molecular structures can be deduced is one of the most challenging and fascinating aspects of chemical research. In some cases, quantitative results are obtained, as in the measurement of bond lengths. In other cases, only the shapes of molecules are inferred. For example, from the knowledge that CO_2 molecules have zero dipole moment, it can be inferred that they are linear and symmetrical. On the other hand, the fact that sulfur dioxide molecules have a finite dipole moment means either that they are nonlinear,

S/O O   or   O/O S

or that they are linear and nonsymmetrical,

O—O—S

To determine which of these possibilities is correct, additional experiments beyond the measurement of dipole moment must be performed.

There is no single procedure for establishing the structure of molecules. Every substance presents its own challenge, and the skillful chemist may employ a combination of techniques to arrive at a satisfactory model of the structure. These techniques include the measurement of physical properties, such as solubilities, and the determination of molecular weights by boiling-point-elevation and/or freezing-point-lowering methods. Details of these types of determinations will be given in Sections 16–12, 16–14, and 16–18.

In this chapter, some of the more elaborate techniques of structure determination will be discussed. These include the use of sophisticated equipment such as infrared spectrometers, nuclear magnetic resonance spectrometers, and X-ray diffraction apparatus. The discussion will point out the types of measurements made, the kinds of data obtained, and the conclusions which can be drawn from the data. The main purpose of this chapter is to develop an appreciation for the power and limitations of these methods.

13-1   Mass Spectrometry

Mass spectrometry is used to determine the masses of molecules (molecular weights) as well as the masses of fragments of the molecules. Any proof of structure should begin with an accurate chemical analysis of the substance in question. Using the percent composition data obtained by chemical analysis, the empirical formula of the substance is calculated (see Section 2–5). Then if the molecular weight is determined, a molecular formula can be established. However, the molecular formula does not always make apparent the structure of the molecule. For example, it is not possible to distinguish between isomeric compounds by molecular formula alone. But if the substance is examined by mass spectrometry, it is often possible not only to obtain its molecular weight but also to distinguish between possible isomers and even to obtain some information on relative bond energies in the molecule.

Principles of the operation of the mass spectrometer are presented in Figure 9–7. A liquid or gaseous sample is injected into the ionization chamber of the instrument, where it is bombarded with high energy electrons. A number of different kinds of (highly unstable) positive ions are formed, which may include that of the parent molecule from which a single electron has been removed. Numerous fragment ions, formed by the rupture of chemical bonds and the removal of electrons from the fragments, are also produced. Before they have had a chance to decompose, the ions are accelerated into the magnetic field of the spectrometer. The magnitudes of their deflections in the field are determined by their mass/charge ratios. The intensity of the detector response at various mass/charge ratios, recorded as a series of peaks, constitutes the **mass spectrum.**

The peak of the positively charged parent molecule, **the molecular ion,** is usually the peak of highest mass. Small peaks due to ions containing isotopic atoms may be observed as masses one or two units higher. The exact molecular weight of the compound, in daltons, is determined from all these peaks. In this respect, mass spectrometry is superior to other methods of molecular weight determination, where values within 1 to 2% of the true molecular weight are considered acceptable (see Section 16–18).

As an example, the mass spectrum of the substance $HOCH_2CH_2SH$ is shown in Figure 13–1. The peak at the mass/charge ratio of 78 corresponds to the molecular weight of the compound. Two small peaks at 79 and 80 are recognized as being due to the presence of a small percentage of molecules containing the naturally occurring isotopes ^{13}C, ^{18}O, ^{33}S, and ^{34}S. The compositions of several fragment ions are noted on the figure. The absence of fragments at mass/charge ratio of 77 suggests that the S—H, C—H, and O—H bonds are strong under electron bombardment compared to the bonds between the heavier atoms.

Suppose that only the empirical formula of this compound were known, but not its structure. The mass spectrum allows immediate deductions about its structure. The molecular weight is established, yielding the molecular formula. The absence of a strong peak at a mass/charge ratio of $78 - 15 = 63$ rules out the possibility that the compound has a structure containing a CH_3 group, such as CH_3OCH_2SH, CH_3SCH_2OH, or $CH_3CH(SH)(OH)$. Consequently, the most probable structural formula would be $HOCH_2CH_2SH$.

As the above example suggests, mass spectrometry is particularly useful in the field of organic chemistry where there are many possibilities for isomerism and where

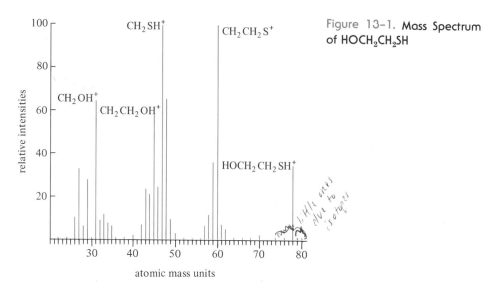

Figure 13-1. **Mass Spectrum of HOCH$_2$CH$_2$SH**

many substances are sufficiently volatile to be injected into the instrument in the gaseous state.

Example

What major differences would be expected between the mass spectra of the isomeric compounds CH$_3$CH$_2$OH and CH$_3$OCH$_3$?

The mass spectrum of CH$_3$CH$_2$OH would contain a peak at the mass/charge ratio of 29, CH$_3$CH$_2^+$, while the spectrum of the ether should not contain this peak. Both spectra should contain a peak at $m/e = 31$.

To summarize the procedure of identifying a compound by mass spectrometry, first the molecular weight is determined. Then all the known information about the compound (for example, the empirical formula) is used to write possible structural formulas for it. The experimental mass spectrum is matched with each type of structure until the one most likely to have a fragmentation pattern corresponding to the observed mass/charge peaks is identified. For example, if the mass spectrum of a sample shows no peak corresponding to an ethyl group, there clearly is no need to consider isomeric compounds containing this group. Sometimes, all the possible isomers of a given molecule must be considered. If one isomer can be matched with the spectrum, a final identification can be made.

The technique of writing possible structural formulas and matching them with experimental data is widely used in all types of structural determinations. Therefore it is important to understand how to write such structures and also how to predict the shapes of the proposed molecules. In writing possible structures, it is useful to note that hydrogen is always bonded to one other atom; oxygen is bonded with two single bonds or with one double bond; nitrogen usually has three single bonds; a double bond and a single bond; or a triple bond; and carbon has some combination of bonds which give a total bond order of four (Table 1–7).

Example

Write formulas for all the possible structural isomers of $C_2H_8N_2$. Which isomer would NOT have a peak at a mass/charge ratio of 45?

(a) $H_2NCH_2CH_2NH_2$ (b) $CH_3NHCH_2NH_2$ (c) $CH_3NHNHCH_3$ (d) $CH_3-\underset{\underset{NH_2}{|}}{N}-CH_3$

(e) $CH_3CH_2NHNH_2$

Isomer (a) has no CH_3 group and could not yield a significant peak at a mass/charge ratio of 45.

13-2   Absorption Spectroscopy

Molecular spectroscopy is based on the concept that the interactions of individual atoms and molecules with energy are quantized. The difference between two energy levels is expressed in terms of frequency, ν, by

$$E_2 - E_1 = h\nu$$

where h is Planck's constant. This relationship applies to both emission spectroscopy and absorption spectroscopy. In **emission spectroscopy** the molecule is excited by some means; and when it returns to its ground state, light is emitted of frequencies corresponding to allowed transitions within the system. In **absorption spectroscopy**, the molecules originally in their ground states absorb light of certain frequencies which correspond to allowed transitions to excited states.

When energy is absorbed by molecules, transitions may occur between electronic energy levels, vibrational energy levels, and/or rotational energy levels. As shown in the energy level diagram (Figure 13–2), all of these transitions are quantized. Transitions between electronic levels involve the greatest energies; transitions between vibrational states are intermediate in energy; and transitions between rotational levels require comparatively low energies. These energy requirements are summarized in Figure 13–3, where it is seen that electronic transitions are produced by the absorption of light mainly in the ultraviolet and visible regions of the spectrum, while molecular vibrations are excited by infrared light, and the rotational states of molecules have energies corresponding to light in the far infrared and microwave regions.

A schematic diagram of the experimental arrangement for obtaining absorption spectra is shown in Figure 13–4. Light of a range of wavelengths is passed through the sample, and those wavelengths corresponding to allowed molecular transitions are absorbed. The transmitted light is then passed through a resolving device such as a prism or a grating and then onto a detector. The intensity of transmitted light of various wavelengths is thus determined. The record of the detector response compared to its response when no sample is present constitutes the **absorption spectrum.**[1]

---

[1] Many commercial instruments operate on the "double beam" principle. Light from the source is divided into two beams, one of which passes through the sample while the other simultaneously passes through air (or if a solution is being studied, through pure solvent). The two beams follow similar paths through the instrument and the difference in response at the detectors is recorded directly.

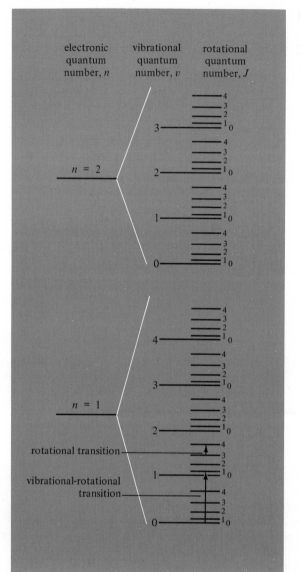

Figure 13-2. **Molecular Energy Levels**

A rotational transition and a vibrational-rotational transition are shown (not drawn to scale).

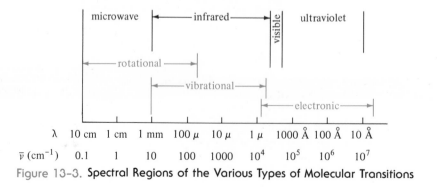

Figure 13-3. **Spectral Regions of the Various Types of Molecular Transitions**

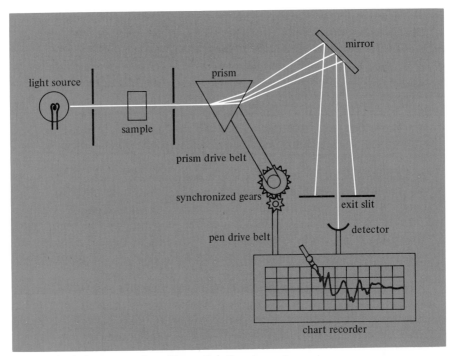

Figure 13-4. **Diagram of an Absorption Spectrometer**
Light of a range of wavelengths is emitted by the source and is allowed to pass through the sample, where certain wavelengths are absorbed. The light transmitted by the sample is resolved into its various wavelengths by means of a prism and is then reflected from a mirror through an exit slit and onto a suitable detector. By rotating the prism, it is possible to focus light of a given wavelength upon the detector. The response of the detector is recorded on a chart by means of a motor-driven pen. The horizontal pen drive is synchronized with the prism mechanism, and when the prism is rotated so that light of each given wavelength is focused on the detector, the pen will be located at the corresponding value on the wavelength scale of the chart. The instrument is adjusted so that the pen responds vertically either as percent of light transmitted or as absorbance. These quantities are defined by the following equations, where I is the intensity of the light transmitted and I_0 is the intensity of the light before it enters the sample:

$$\% \text{ transmittance} = \frac{I}{I_0} \times 100 \qquad \text{absorbance} = \log \frac{I_0}{I}$$

13-3   Vibrational Spectra

At any temperature, the atoms making up a molecule are continuously in vibrational motion. These motions include the stretching and contracting of bonds, the changing of bond angles through bending and twisting, rocking motions of one part of the molecule with respect to the other, and torsion or twisting about a given bond. However, a genuine vibration produces no net translation (movement of the center of gravity) or rotation of the molecule as a whole. All of the vibrations of a molecule can be described as one or a combination of a certain number of **fundamental modes** of vibration. A linear molecule has $3n - 5$ fundamental modes of vibration, where n is the number of atoms in the molecule. A nonlinear molecule has $3n - 6$ such modes. (Three directions of motion for each atom minus three translational and three rotational motions for the molecule as a whole.)

When a photon of infrared radiation is absorbed by a molecule, a vibration corresponding to one of its fundamental modes becomes excited. Figure 13-2 shows that a molecule in any given vibrational energy state can exist in any one of several rotational energy states. Hence transitions between vibrational states may be accompanied by changes in rotational state, and a given transition occurs over a narrow range of energies corresponding to an **absorption band.** The $3n - 6$ vibrations might be expected to absorb photons having $3n - 6$ different wavelengths corresponding to transitions between vibrational states of each fundamental mode, and the infrared spectrum of a molecule should consist of a number of bands centered at several different wavelengths.[2] However, of several possible fundamental vibrations of a given molecule, only those vibrations accompanied by a change in dipole moment will be infrared active. Hence not all possible vibrations will result in an infrared band. Moreover, some vibrational transitions are **degenerate;** that is, two or more different changes in vibration occur at the same energy. In such cases only one absorption band is observed for the two transitions. In these cases, the observed spectrum will contain fewer bands than the $3n - 6$ (or $3n - 5$) fundamental modes of vibration.

On the other hand, certain combinations of fundamental vibrations may absorb radiation, and the actual vibrational spectrum may contain more than $3n - 6$ absorption bands. However, the bands due to fundamental vibrations are usually more intense than those due to combinations of vibrations.

Example

Predict the number of fundamental vibrations and the number of infrared absorption bands in the following molecules: **(a)** H_2, **(b)** CO_2, **(c)** SO_2.

(a) H_2 is diatomic and hence linear; there should be $3n - 5 = 1$ vibration in the molecule—the H—H stretch. However, for this vibration there is no change in dipole moment (the dipole moment remains zero no matter what the bond distance) and no infrared absorption band is expected.

(b) CO_2 is a linear triatomic molecule; there should be $3n - 5 = 4$ fundamental vibrational modes.

←O···C···O→	←O···C→←O	degenerate bending modes ($\oplus$ in $\odot$ out)
symmetrical stretch	asymmetrical stretch	

Actually, only two bands are observed for CO_2. The symmetrical stretch does not result in a change in dipole moment, and there is no infrared band due to this mode. A band due to the asymmetrical stretch is observed. The second band corresponds to the two degenerate bending modes.

(c) SO_2 is nonlinear; hence there should be $3n - 6 = 3$ modes.

symmetrical stretch	asymmetrical stretch	deformation

[2] If the spectra are recorded graphically, the bands consist of a series of maxima and minima corresponding to the response of the detector to light which has passed through the sample.

All of these modes produce a change in dipole moment, and thus all three are infrared active.

If the stretching vibration of the bond in a molecule AB is regarded as analogous to the stretching of a spring, the frequency of the oscillation, ν, will be given by

$$\nu = \frac{1}{2\pi}\sqrt{\frac{k}{\mu}}$$

where

$$\mu = \frac{m_A m_B}{m_A + m_B}$$

μ is called the **reduced mass** and k is the **stretching force constant.** The force constant is a measure of the "stiffness of the spring," that is, the strength of the bond. Although this analogy cannot be expected to hold exactly for more complicated molecules, it is reasonable to expect that the stretching of strong bonds will give rise to absorptions at relatively high frequencies. Also, the greater the value of μ, the lower will be the stretching frequency.

Example

The O—H stretching frequency in the infrared spectrum of CH_3OH in a dilute solution in CCl_4 is 3680 cm^{-1}. Estimate the O—D stretching frequency in CH_3OD, the compound containing the ^2H (D) isotope of hydrogen.

Since H and D are isotopes, their chemical properties are very similar. Consequently, the O—H and O—D stretching force constants should be about equal, and the observed O—H and O—D frequencies should vary inversely as the square roots of the respective reduced masses. Since the vibration is largely localized in the O—H bond, the reduced masses will be calculated without considering the mass of the CH_3 group.

$$\frac{\nu_{OD}}{\nu_{OH}} = \frac{\sqrt{\dfrac{f}{\mu_{OD}}}}{\sqrt{\dfrac{f}{\mu_{OH}}}} = \frac{\sqrt{\mu_{OH}}}{\sqrt{\mu_{OD}}} = \frac{\sqrt{\dfrac{m_0 m_H}{17.0}}}{\sqrt{\dfrac{m_0 m_D}{18.0}}} = \sqrt{\frac{m_H}{m_D}\frac{18.0}{17.0}} = \sqrt{\frac{18.0}{34.0}} = 0.728$$

$$\nu_{OD} = \nu_{OH}(0.728) = (3680 \text{ cm}^{-1})(0.728) = 2680 \text{ cm}^{-1}$$

The actual value of the O—D stretching frequency in CH_3OD is 2720 cm^{-1}.

Consistent with the fact that the energies of bonds between given pairs of atoms are reasonably constant, a number of groups of atoms have characteristic group absorption frequencies. Thus the stretching of the carbonyl group, $\diagup C{=}O$, causes absorption about 1700 cm^{-1}, and the stretching of C—H bonds causes absorption in the range 2800 to 3300 cm^{-1}. Some commonly encountered group absorption frequencies are listed in Table 13–1. Such tabulations of frequencies are useful in the identification of these groups in molecules. Moreover, shifts of observed group frequencies can often be used by the spectroscopist to infer the nature of the bonding of adjacent groups in the molecule. Several examples of the latter concept will be described in Sections 14–2 and 14–4.

Every chemical compound has a characteristic infrared spectrum. Obtaining its infrared spectrum is often referred to as "fingerprinting" the compound, and an unknown material can be identified by showing that its infrared spectrum is identical with that of a known compound. Some representative spectra are shown and interpreted in Figure 13–5. These spectra should be studied with reference to Table 13–1 in order that the shapes of the bands can be associated with the group frequencies.

Interpretation of an infrared spectrum of a complex molecule is as much an art as a science, and only after considerable practice with a variety of types of compounds do the spectra become relatively easy to interpret. Very often, other types of information, such as mass spectra, are used in conjunction with infrared spectra to identify compounds. In attempting to interpret infrared spectra, the beginning student should proceed in a stepwise manner as shown on pages 358–59.

Figure 13–5. Some Interpreted Infrared Spectra

Note that a number of the weaker bands are not assigned to any type of vibration. These are overtone and/or combination bands. Usually they are not characteristic of a particular type of vibration. However, in the fingerprint method of identification, in which the spectrum of an unknown compound is matched with that of an authentic sample of a compound, even these relatively weak bands become important points of comparison.

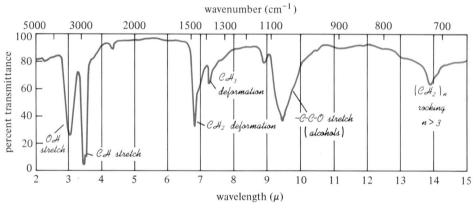

(a) Infrared spectrum of dodecanol, $C_{11}H_{23}CH_2OH$. Note the bands at 3300 and 1075 cm^{-1}, characteristic of alcohols. Also note the band near 725 cm^{-1}, which is due to a chain of eleven CH_2 units.

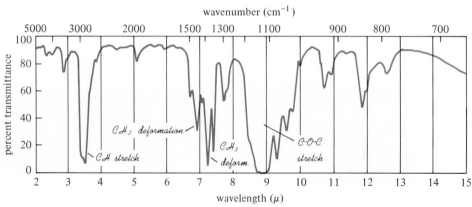

(b) Infrared spectrum of diethyl ether, $CH_3CH_2OCH_2CH_3$. The broad complex band in the frequency range 1200 to 1000 cm^{-1} is characteristic of a C—O—C linkage. Note also the *absence* of a band near 725 cm^{-1}.

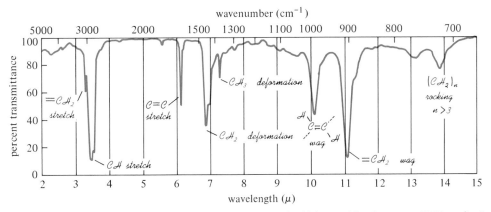

(c) Infrared spectrum of 1-octene, $C_6H_{13}CH{=}CH_2$. Note the high stretching frequency (3010 cm^{-1}) of hydrogen atoms attached to a doubly bonded carbon atom.

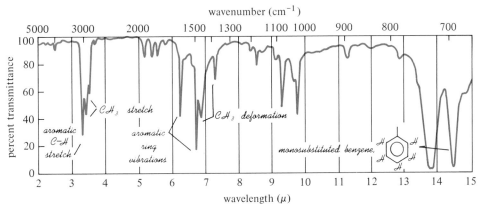

(d) Infrared spectrum of methylbenzene (toluene), $C_6H_5CH_3$. The sharp bands at 1600 and 1500 cm^{-1} are characteristic of aromatic compounds. Also, *two* very strong bands around 730 and 690 cm^{-1} are indicative of the $C_6H_5{-}$ group.

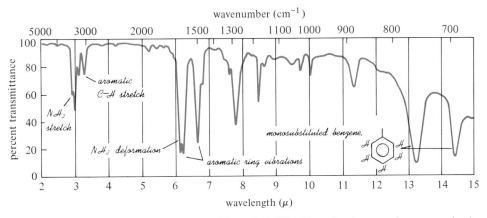

(e) Infrared spectrum of aminobenzene (aniline), $C_6H_5NH_2$. Note that in aromatic compounds, the frequencies of the C—H stretching vibrations occur above 3000 cm^{-1}.

TABLE 13-1. Characteristic Group Absorption Frequencies

Band	Group	Description[a]	Approximate Wavenumbers (cm^{-1})
C—H	CH$_3$, alkane	stretching	2960, 2870
C—H	CH$_2$, alkane	stretching	2920, 2850
C—H	—CH=CH—	stretching	3095–3010
C—H	≡C—H	stretching	3300
C—H	—C—H, aromatic	stretching	3100–3030
C—H	—CH$_2$—	bending	1470–1400
C—H	—C—CH$_3$	bending	1380–1360
C—H	—CH=CH—, *trans*	bending	1300, 970
C—H	—CH=CH—, *cis*	bending	690
C—C	(CH$_3$)$_3$C—	skeletal	1250
C—C	—(CH$_2$)$_n$— ($n > 3$)	skeletal	725
C—C	—C=C—	stretching	1680–1620
C—C	—C≡C—	stretching	2230
C—H	C$_6$H$_5$— (monosubstituted aromatic)	CH wag	1600, 1500
O—C	>C=O	stretching	1760–1690
O—C	=C—O—	stretching	1250–1000
C—N	>C=N—	stretching	1690–1630
O—H	O—H (free)	stretching	3620
O—H	O—H (hydrogen bonded, alcohols)	stretching	3600–3300
N—H	N—H$_x$ ($x = 1, 2, 3$)	stretching	3500–3300
S—H	S—H	stretching	2600
Cl—C	Cl—C	stretching	800–700

[a] The terms bending, wag, etc., are technical jargon describing the various kinds of vibrational motions.

Steps	*Example*
	A compound has the molecular formula C$_2$H$_6$OS. Its infrared spectrum is given in Figure 13–6. Identify the compound.
1. If the empirical formula is available, write down structures of possible compounds or types of compounds.	Possible structures: CH$_3$OCH$_2$SH CH$_3$SCH$_2$OH HSCH$_2$CH$_2$OH CH$_3$CH(OH)(SH)
2. Scan the entire spectrum for characteristic group frequencies, i.e., OH stretching, carbon–carbon double bonds, carbonyl (>C=O) vibration, aromatic ring modes, etc. Consult Table 13–1 or similar tables or "correlation charts," which may be found in reference books on spectroscopy.	Overall examination reveals a strong band at 3360 cm^{-1}, indicative of OH stretching, but no bands due to aromatic rings or unsaturation.

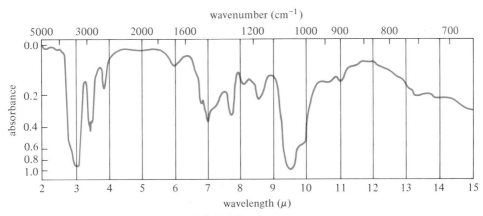

Figure 13-6. Infrared Spectrum of C_2H_6OS

3. Examine the C—H stretching region (2700 to 3300 cm^{-1}) for possible unsaturation or aromatic rings. Then check the 1360 to 1380 cm^{-1} and 1400 to 1470 cm^{-1} regions for CH$_3$— and for —CH$_2$— deformation modes.

C—H stretching vibrations occur below 3000 cm^{-1}, confirming the lack of aromatic rings or unsaturation. Absence of a band at 1370 to 1380 cm^{-1} indicates that no CH$_3$ groups are present, but the bands at 1400 and 1450 cm^{-1} suggest that CH$_2$ groups are present.

4. If the compound contains atoms (X) other than carbon, hydrogen, and oxygen, look for bands attributable to X—H or X—C types of vibrations.

The compound contains sulfur, and the band at 2560 cm^{-1} may be attributed to S—H.

5. If the compound is not aromatic, a broad band at 725 cm^{-1} indicates a chain of four or more CH$_2$ groups.

The compound does not contain a long carbon chain.

6. Look for special features, such as two bands of equal intensity at 1380 cm^{-1}, indicating the group (CH$_3$)$_2$C, or bands at 1500 and 1600 cm^{-1}, indicating aromatic rings.

The very broad band at 1050 cm^{-1} is characteristic of the group —CH$_2$OH (a primary alcohol).

The spectrum fits very closely the structure HSCH$_2$CH$_2$OH. The alternative structures have been eliminated on the basis of the absence of bands due to a CH$_3$ group.

13-4   The Raman Effect

The interpretation of infrared spectra is often complemented by Raman spectroscopy. Applied together, these two techniques often allow the determination of the overall shape of a molecule. When a beam of monochromatic visible or ultraviolet

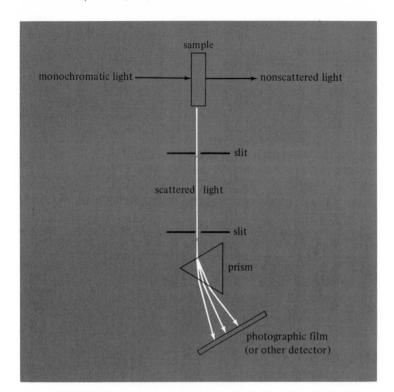

Figure 13-7. **Schematic Diagram of Raman Spectroscopy**

light is passed through a homogeneous medium, some of the light may be absorbed, some will be transmitted, and some of it is scattered, e.g., deflected perpendicular to its original direction. The wavelength of most of the scattered light is unchanged, but a small portion of it will have interacted with the material to produce both longer and shorter wavelength light than that of the incident beam. Figure 13–7 shows schematically how a Raman spectrum is obtained.

The differences between the frequencies of the scattered light and the frequency of the incident beam correspond to transitions between vibrational and rotational energy levels in the molecules of the sample; thus Raman spectroscopy should provide the same kinds of information as infrared spectroscopy. However, there is an important difference. A given molecular vibration will be Raman active only if there is a change in the polarizability of the molecule during the vibration. The concept of

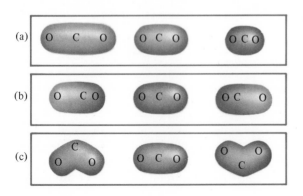

Figure 13–8. **Deformation of Electron Probability Density During Vibrations of the Carbon Dioxide Molecule**

(a) Symmetrical stretching vibration, ν_1. (b) Asymmetrical stretching vibration, ν_2. (c) Bending vibration (one of two degenerate bending modes), ν_3.

polarizability was discussed in Section 11-7 in connection with van der Waals forces. Qualitatively it is associated with the deformation of the electron cloud surrounding a molecule. If the movements of the atoms cause the electron cloud to become more compact or more diffuse in going from one extreme vibrational configuration to the other, the polarizability changes during the vibration. For example, in the symmetrical stretching mode of the carbon dioxide molecule, the electron cloud goes from diffuse to compact, as shown in Figure 13-8. On the other hand, the compactness of the electron cloud does not change during the asymmetrical stretching vibration or during the degenerate bending modes. Hence only the symmetrical stretching vibration of CO_2 leads to a change in polarizability. Therefore in the Raman spectrum of CO_2, the difference in frequency between the scattered light and that of the incident beam should correspond to the frequency of the symmetrical stretching mode. It should be noted that this mode produces no change in the dipole moment of the molecule and hence is not infrared active. The expected four fundamental vibrations of carbon dioxide thus result in two infrared bands (one of which corresponds to two degenerate vibrations) and one Raman active band. These results are consistent with the symmetrical linear structure proposed for the carbon dioxide molecule solely on the basis of its zero dipole moment.

On the other hand, the infrared spectrum of the nonlinear triatomic molecule, SO_2, has three absorption bands corresponding to its three fundamental vibrations. These three vibrations also give three Raman shifts, as would be predicted from considerations of changes in polarizability. Infrared and Raman data for carbon dioxide and sulfur dioxide are presented in Table 13-2.

The numbers of fundamental molecular vibrations which are infrared active and/or Raman active can be predicted for any molecular geometry. The proposed geometry of a molecule is confirmed by the agreement of its observed spectrum with the predicted spectrum. The methods of predicting the activities of vibrational modes will not be discussed here; however, the results of such predictions for several types of molecules are listed in Table 13-3.

Example

In the infrared spectrum of BF_3 there are three absorption bands which correspond to the fundamental vibrations, appearing at 480, 691, and 1446 cm^{-1}. Raman shifts of 481 and 880 cm^{-1} are observed for this compound. Using Table 13-3 deduce the geometry of the BF_3 molecule.

A planar AB_3 molecule would have three infrared active and two Raman active fundamentals, only one of which is observed in the infrared spectrum. The actual data agree exactly with this prediction for a planar molecule.

TABLE 13-2. Observed Frequencies, ν (cm^{-1}), of the Fundamental Vibrations of CO_2 and SO_2

		Infrared	Raman			Infrared	Raman
CO_2	ν_1	—	1351	SO_2	ν_1	1151	1151
	ν_2	667	—		ν_2	519	524
	$\nu_3{}^a$	2349	—		ν_3	1361	1336

^{a} Doubly degenerate vibration.

TABLE 13-3. Theoretical Infrared and Raman Activities of Representative Types of Compounds, AB_x

Geometry	Formula Type	Total Number of Vibrations	Infrared Active[a]	Raman Active[a]	Inactive[a]
linear	BAB	4	$\nu_2\nu_3^*$	ν_1	
linear	ABB	4	$\nu_1\nu_2\nu_3^*$	$\nu_1\nu_2\nu_3^*$	
angular	AB_2	3	$\nu_1\nu_2\nu_3$	$\nu_1\nu_2\nu_3$	
planar	AB_3	6	$\nu_2\nu_3^*\nu_4^*$	$\nu_1\nu_4^*$	
pyramidal	AB_3	6	$\nu_1\nu_2\nu_3^*\nu_4^*$	$\nu_1\nu_2\nu_3^*\nu_4^*$	
square planar	AB_4	9	$\nu_2\nu_5^*\nu_6^*$	$\nu_1\nu_3\nu_4^*$	
tetrahedral	AB_4	9	$\nu_3^\dagger\nu_4^\dagger$	$\nu_1\nu_2^*\nu_3^\dagger\nu_4^\dagger$	
octahedral	AB_6	15	$\nu_3^\dagger\nu_4^\dagger$	$\nu_1\nu_2^*\nu_5^\dagger$	$\nu_6^\dagger$

[a]Each fundamental frequency is denoted by a numerical subscript; doubly degenerate vibrations are denoted by an asterisk; triply degenerate vibrations are denoted by a dagger. For a given geometry, if ν_n ($n = 1, 2, 3, \ldots$) appears in both the infrared active and the Raman active columns, the same frequencies will be observed (within experimental error) in the spectra obtained by the two methods.

13-5   Electronic Spectra

It is seen in Figure 13-2 that compared to transitions between rotational and vibrational energy states, transitions between electronic energy levels in molecules involve relatively large energies. Accordingly, absorption spectra due to transitions between two different electronic states are observed mainly in the visible and ultraviolet regions of the spectrum. As can be deduced from Figure 13-2, when an electron is excited from the ground state to a higher electronic state, the transition may occur with a change in rotational quantum number and/or vibrational quantum number. Thus a given change in electronic energy levels will occur over a range of energies. Consequently the electronic spectra of simple molecules consist of many, many closely spaced "lines." For more complex molecules, absorption spectra in the visible and ultraviolet regions are recorded as more or less broad bands, characterized by maxima at particular wavelengths (Figure 13-9).

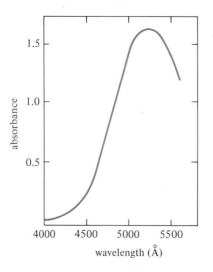

Figure 13-9. **Visible Absorption Spectrum of I_2 in CCl_4 Solution**

In contrast to infrared spectra, electronic spectra are rarely used to indicate the presence of individual functional groups but rather to provide information about the excited states of molecules as well as their ground states. Stable compounds of low molecular weight are excited only by quanta in the ultraviolet region of the spectrum. Thus most compounds of this type, for example, H_2O, CO_2, and C_2H_5OH, are colorless. Color in a compound indicates the existence of a relatively low energy excited state, from 30 to 70 kcal/mole above the ground state. This energy range corresponds to quanta in the visible region of the spectrum, having wavelengths from 4000 Å (violet) to 7600 Å (red). In organic molecules, certain groups of atoms, such as $-NO_2$, $\mathord{>}C{=}O$, and $-N{=}N-$, are called **chromophores.** Chromophores contain electrons in π orbitals (Section 12–1) which are readily excited to π^* orbitals only a few kilocalories per mole higher in energy. For example, in aldehydes and ketones—molecules containing the $\mathord{>}C{=}O$ group—absorption of light around 2800 Å is due to a transition from a nonbonding orbital to a π^* orbital, involving one of the four nonbonding electrons of the oxygen atom (see Figure 12–11). Many compounds whose molecules contain chromophores are colored.

In contrast to compounds of main group elements, compounds of the transition elements are often colored. In such cases, an electron in one d orbital is excited to a higher-lying d orbital as the atom absorbs a quantum of energy corresponding to the visible region of the spectrum. Several examples of the application of electronic spectroscopy to transition elements will be given in Section 14–7.

Example

Explain the trend of colors in the halogen family, from the colorless F_2, to the yellow-green Cl_2, to red Br_2, to violet I_2.

The outer electronic configurations of the halogen molecules are as follows:

$$F_2:\ \sigma_{2s}^{\,2}\,\sigma_{2s}^{*\,2}\,\sigma_{2p}^{\,2}\,\pi_{2p}^{\,4}\,\pi_{2p}^{*\,4}$$
$$Cl_2:\ \sigma_{3s}^{\,2}\,\sigma_{3s}^{*\,2}\,\sigma_{3p}^{\,2}\,\pi_{3p}^{\,4}\,\pi_{3p}^{*\,4}$$
$$Br_2:\ \sigma_{4s}^{\,2}\,\sigma_{4s}^{*\,2}\,\sigma_{4p}^{\,2}\,\pi_{4p}^{\,4}\,\pi_{4p}^{*\,4}$$
$$I_2:\ \sigma_{5s}^{\,2}\,\sigma_{5s}^{*\,2}\,\sigma_{5p}^{\,2}\,\pi_{5p}^{\,4}\,\pi_{5p}^{*\,4}$$

Each of these molecules has a bond order of 1, equivalent to a single σ bond. However, in Cl_2, the energy of the π_{3p}^* orbitals is close to the energy of a set of nonbonding orbitals originating from the $3d$ orbitals of the chlorine atoms. Electrons in the π_{3p}^* orbitals can be excited to these latter orbitals by the absorption of visible light. Similarly the electrons in the π^* orbitals of Br_2 and I_2 can be excited to orbitals originating from the $4d$ and $5d$ atomic orbitals of Br and I, respectively. The change in color from chlorine to iodine is due in general to a progressive shift of the exciting light to longer wavelengths, since the nonbonding orbitals get closer in energy to the originally occupied π^* orbitals. On the other hand, fluorine atoms have no d orbitals from which molecular orbitals of suitable energy can be formed. Thus there are no low-lying empty molecular orbitals into which electrons can be excited by visible light, and F_2 is colorless.

13–6   Nuclear Magnetic Resonance

Nuclei possess intrinsic, quantized spins which are the resultant of the spins of the individual protons and neutrons which they contain. The net spin of the nucleus

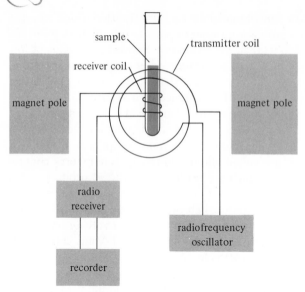

Figure 13-10. Schematic Diagram of a Nuclear Magnetic Resonance Spectrometer

creates a tiny magnetic field, which is capable of interaction with external fields. This interaction yields clues to the structure of the molecule containing the nucleus.

Nuclei are characterized by nuclear spin numbers, I, which have the values $0, \frac{1}{2}, 1, \frac{3}{2}, 2, \ldots$ (expressed in units of $h/2\pi$). For a given nucleus with spin number I, there are $2I + 1$ possible orientations of the nucleus in a magnetic field. Thus for $I = \frac{1}{2}$, there are two orientations, represented as $\uparrow$ and $\downarrow$. The spin states are quantized, and a discrete quantum of energy is required to cause a transition from one allowed orientation to another. To cause transitions between nuclear spin states in a sample in the magnetic field of a typical commercial instrument requires energies in the radiofrequency range of the electromagnetic spectrum. Therefore, an oscillator which generates waves in the radiofrequency range is used as a power source. When the oscillator is tuned to a frequency exactly corresponding to the energy required to change the nucleus from one spin state to another, the nuclei lined up in the magnetic field absorb the energy by "resonance"[3] and change their orientations in the field. This phenomenon is **nuclear magnetic resonance** (nmr).

Nuclei with even atomic numbers and even mass numbers, e.g., $^{12}_{6}C$, $^{16}_{8}O$, and $^{32}_{16}S$, have nuclear spins of zero. These nuclei have only one nuclear spin state and cannot be excited by radiofrequency waves. They do not exhibit nuclear magnetic resonance. For nuclei with spin states $I = \frac{1}{2}$, there are two $(2(\frac{1}{2}) + 1)$ possible orientations in the field. Nuclei with spin states of $\frac{1}{2}$ include $^{1}_{1}H$, $^{13}_{6}C$, $^{15}_{7}N$, $^{19}_{9}F$, and $^{31}_{15}P$. These nuclei are of greatest current interest for nuclear magnetic resonance studies.

A simplified diagram of a nuclear magnetic resonance spectrometer is presented in Figure 13-10. The resonance frequency of the nucleus, ν_0, depends on the strength of the field, H_0, the magnetic moment of the nucleus, μ, and its spin number, I:

$$h\nu_0 = \frac{\mu H_0}{I}$$

---

[3] The exchange of energy by resonance is analogous to the transfer of energy between two tuning forks of the same pitch. When they are held near each other and one of them is struck, both forks vibrate.

When the frequency of the oscillating field satisfies this equation, transitions between spin states occur in the sample. The transitions are detected as a voltage oscillation induced in the receiver coil. In most commercial instruments, the field of the large magnet is fixed, as is the transmitted radiofrequency. The resonance condition is achieved by superimposing a small variable magnetic field onto the field of the large magnet. Changing the small variable field varies the total field until the resonance condition occurs.

13-7   Chemical Shift

Since the nucleus undergoing resonance is that of an atom in a molecule, it is surrounded by electrons. The nucleus will be shielded somewhat from the magnetic field by the diamagnetism of the electron pairs.[4] This shielding effect amounts to only 10^{-2} gauss, compared to the applied field strength, which is of the order of magnitude of 10,000 gauss. Thus the shielding effect is only a few *parts per million* of the external field. Nuclear magnetic resonance measurements can be made so precisely that this effect can be determined to within 1% accuracy.

The shielding effect is the phenomenon that makes nuclear magnetic resonance useful for the determination of molecular structure. Differences in the distributions of the electrons about a given nmr active nucleus changes the extent of shielding. Such changes are due to changes in the nature of the chemical bonding, as, for example, when an electronegative atom is replaced from its position adjacent to an nmr active nucleus by a less electronegative atom. As a result, the resonance frequency of each type of nmr active nucleus will be different in different chemical environments. The difference between the resonance frequency of the given atom in two chemical environments is known as the **chemical shift**. The chemical shift, δ, is expressed in parts per million (ppm) and is measured relative to some standard substance. In the case of proton resonance, some standards are $CHCl_3$, C_6H_{12}, and tetramethylsilane, $(CH_3)_4Si$, called TMS. Chemical shifts are expressed by the relationship

$$\delta = \frac{H_{reference} - H_{sample}}{H_{reference}}$$

where $H_{reference}$ and H_{sample} are the field strengths at which the standard substance and the sample, respectively, are in resonance for a given frequency, ν_0. If the sample nucleus requires a higher field strength than that of the reference, it is more highly shielded. Its resonance is said to be "upfield." Thus the direction of the chemical shift is an indication of the extent of shielding of the nucleus in the molecule.

Most commercial nuclear magnetic resonance (nmr) spectrometers are applied to the study of protons. Instruments of this type operate at frequencies of 60×10^6 or 100×10^6 hertz (Hz; 1 Hz = 1 cycle/sec). The approximate chemical shifts of protons in some typical organic compounds, relative to tetramethylsilane as the standard, are shown in Figure 13-11. The structural information provided by proton magnetic resonance spectra includes

[4] If unpaired electrons are present, the resulting paramagnetism complicates the interpretation of the effect of the nuclear spins.

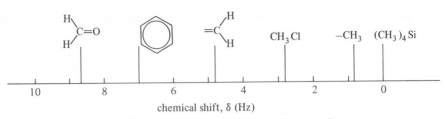

chemical shift, δ (Hz)

Figure 13–11. **Typical Chemical Shifts of Protons in Organic Environments**

1. The number of different proton environments in the molecule.
2. The number of protons in each of these environments.
3. The relative extent of shielding produced by each environment.
4. The influence of neighboring active nuclei upon each other.

The spectrum of $CH_3CH_2CCl_2CHCl_2$, shown in Figure 13–12, illustrates how these types of information are obtained. In the low resolution spectrum (Figure 13–12a) three signals are observed, indicating that there are three different proton environments in the molecule. The signal at the lowest magnetic field strength is attributed to the hydrogen atom bonded to the terminal CCl_2 group. The electrons in the $CHCl_2$ group are shifted toward the very electronegative chlorine atoms, and the proton of this group is less shielded than the other protons in the molecule. The protons of the CH_2 group are somewhat more shielded, while those of the CH_3 group, farthest from any chlorine atoms, are the most highly shielded. Their resonance frequencies occur successively upfield. The areas under the three peaks are in the ratio 1/2/3, corresponding to the relative numbers of protons in the three environments.

The high resolution spectrum shown in Figure 13–12(b) requires further explanation. Magnetic nuclei on adjacent or relatively close atoms interact with each other by the coupling of their magnetic spins. The result is that even in a given *chemical* environment, the nuclei may experience a different magnetic field, depending on the spin states of adjacent nuclei. In the case of the $CH_3CH_2CCl_2CHCl_2$ molecule, the

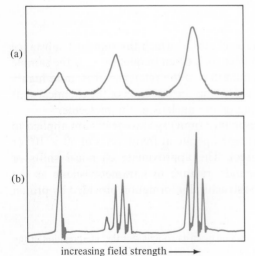

(a)

(b)

increasing field strength ⟶

Figure 13–12. **Nuclear Magnetic Resonance**

Spectrum of

$$H-\overset{\overset{\displaystyle H}{|}}{\underset{\underset{\displaystyle H}{|}}{C}}-\overset{\overset{\displaystyle H}{|}}{\underset{\underset{\displaystyle H}{|}}{C}}-\overset{\overset{\displaystyle Cl}{|}}{\underset{\underset{\displaystyle Cl}{|}}{C}}-\overset{\overset{\displaystyle Cl}{}}{\underset{\underset{\displaystyle Cl}{}}{C}}-H$$

(a) Low resolution. (b) High resolution, showing spin-spin splitting.

TABLE 13-4. Spin Arrangements for Protons in $CH_3CH_2CCl_2CHCl_2$

CH$_2$ Protons	Total Spin	CH$_3$ Protons	Total Spin
↑ ↑	1	↑ ↑ ↑	$\frac{3}{2}$
↑ ↓ ↓ ↑	0	↓ ↑ ↑ ↑ ↑ ↓ ↑ ↓ ↑	$\frac{1}{2}$
↓ ↓	−1	↓ ↓ ↑ ↓ ↑ ↓ ↑ ↓ ↓	$-\frac{1}{2}$
		↓ ↓ ↓	$-\frac{3}{2}$

CH_2 protons and the CH_3 protons can have the spin arrangements shown in Table 13-4. There are four possible ways in which the spin of a given proton of the CH_2 group can interact with the total spin of the protons in the CH_3 group, since there are effectively four different total spins possible in the latter. In the many molecules in a sample, each of the possibilities will occur many times, and four different peaks will be produced. Moreover, there are three ways in which the total spin of $+\frac{1}{2}$ (or $-\frac{1}{2}$) can be achieved, compared to only one way to obtain a total spin of $+\frac{3}{2}$ (or $-\frac{3}{2}$). Thus there is three times the probability of a CH_2 proton being adjacent to a CH_3 group in the former state(s) than in the latter, and the peak corresponding to this interaction is three times as large. That is, a **quartet** is obtained, with the area under the two middle peaks three times the area under the exterior peaks. Therefore in the high resolution spectrum of $CH_3CH_2CCl_2CHCl_2$, the peak attributed to CH_2 is seen to be a quartet, with the two inner peaks having an intensity three times that of the two outer peaks. A given proton in the CH_3 group can interact with the total spin in the CH_2 group in one of three ways. Of these three ways, the interaction with the total spin of zero is twice as likely to occur as either of the other two. The peak of the CH_3 protons is split into a **triplet,** with the inner peak twice as intense as either of the outer peaks. It should be noted that the centers of the spin-spin patterns occur at the same positions as the peaks of the low resolution spectrum. Thus the chemical shifts are the same in both cases.

To summarize, three factors affect the magnitude of the magnetic field at the site of each group of hydrogen atoms, and therefore the energy at which a reversal of spin direction will occur:

1. The external field strength, which is a function of the instrument.
2. Shielding by the electrons of the compound.
3. Small magnetic fields created by the protons on adjacent atoms.

The shielding of the electrons causes the chemical shift, which is dependent on the external field strength. The interaction with adjacent groups of protons, which causes spin-spin splitting, is independent of the external field strength. This independence of the external field sometimes allows use of different external field strengths to unscramble overlapping peaks.

In the analysis of a high resolution spectrum, the following points must be kept in mind:

1. Protons on the same atom, and some other special groups of protons, such as those of the benzene molecule, are equivalent members of a set all of which experience the same chemical shift. The signal of a given proton in a set is not split by spin-spin interactions with other members of the set.

2. The number of lines into which the signal of a given set of protons is split by a set of protons on an adjacent atom is $n + 1$, where n is the number of protons on the adjacent atom. This second fact can be used to determine which sets of protons are on adjacent atoms.

3. Protons on atoms adjacent to highly electronegative atoms are strongly deshielded, and their chemical shifts are downfield. The chemical shifts of protons bonded directly to very electronegative atoms are even more deshielded and are farther downfield.

4. The magnitudes of the spin-spin splittings for protons on adjacent carbon atoms range from 2 to 10 Hz; that is, the symmetrical patterns of lines may be as much as 10 Hz wide. Protons on other than adjacent atoms exert a much weaker effect and will not be discussed here.

Example

A nuclear magnetic resonance spectrum of a compound having the molecular formula C_2H_6OS has the following relative chemical shifts:

Chemical Shift (Hz)	Relative Intensity	Structure
+1.6	3	singlet
+2.6	2	narrow doublet
+3.3	1	narrow triplet

What is the structure of the compound?
Some possible structures may be represented as follows:

I. $HOCH_2CH_2SH$
II. CH_3OCH_2SH
III. CH_3SCH_2OH

Structure I can be eliminated at once since it has no groups of protons having three times as many protons as another group. Moreover, the observed spectrum is too simple for this molecule, for which at least four peaks would be expected. Structures II and III contain the CH_3 group, which should appear as a singlet with the observed intensity. However, the resonance of the singlet appears farthest downfield, which makes it more likely that the CH_3 group is adjacent to a strongly deshielding atom such as oxygen. Thus structure II is favored. Further support for structure II comes from the fact that the peak corresponding to the single proton occurs the farthest upfield, suggesting that it is attached to S. (A proton attached to O would be expected to resonate farthest downfield.) The middle resonance peak corresponds to the CH_2 protons. The spin-spin patterns are in agreement with those expected for structure II.

Example

Predict the spectrum expected for the molecule CH_3CHO.
Two peaks corresponding to protons in two different chemical environments are expected for this molecule. The one located downfield would represent the proton of the CHO group; this peak would be a quartet. The peak of the CH_3 protons is upfield and should appear as a doublet. The ratio of intensities of the quartet and doublet is $1:3$.

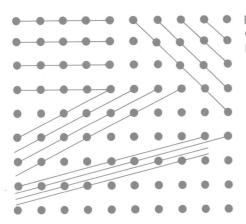

Figure 13-13. **Two-Dimensional Model of a Crystal Lattice Showing Various Sets of Planes of Atoms**

13-8   Diffraction Methods

The most powerful tools available for the study of the structures of materials are X-ray diffraction, neutron diffraction, and electron diffraction. These techniques provide numerical values for bond distances and bond angles. However, the calculations required are quite involved and are performed with the aid of high speed electronic computers. At this point it is not necessary to describe the actual handling of the raw data, and only a general description of each technique will be given.

X-ray diffraction is the most widely used of these techniques. It was mentioned in Section 9-8 that crystalline solids act as three-dimensional diffraction gratings for X rays. In any crystal the constituent atoms are arranged in an ordered pattern and there are many series of equally spaced planes of atoms throughout the crystal. Some examples of the series of planes are diagrammed (in two dimensions) in Figure 13-13.

The experimental apparatus for determination of the structure of a crystalline sample is diagrammed in Figure 13-14. A narrow beam of X rays of a known

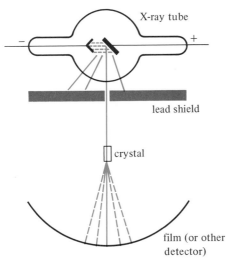

Figure 13-14. **Diagram of an X-Ray Diffraction Experiment**

wavelength, λ, impinging on a crystal will be diffracted through various angles, θ, according to the Bragg diffraction equation:

$$n\lambda = 2d \sin \theta$$

where n is an integer and d is the distance between the planes of a given set of atoms. (For a derivation of the Bragg equation, see Figure 13–15.) After passing through the crystal, the X rays may be detected by a photographic film. A representation of a developed film is shown as Figure 13–16. It shows a central spot through which the undeflected X rays passed, and also a series of spots on concentric rings produced by the X rays which were diffracted through various angles from respective planes of atoms in the crystal.

When the crystal consists of molecules, these will be oriented in the crystal in a pattern which depends on the molecular geometry. Accordingly, the various crystal planes will reflect the arrangement of the atoms in the molecules. In certain instances the distances between planes will correspond to bond lengths in the molecule.

The mathematical analysis of the angles through which the X rays are scattered

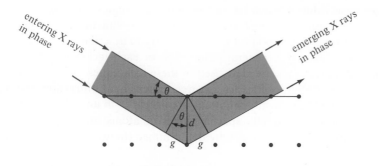

Figure 13–15. **The Bragg Equation**

The diffracted X rays must be in phase as they leave the crystal just as they were as they entered the crystal. That means that the extra length traveled to the second layer of atoms, 2g, must be equal to some integral multiple of the wavelength of the X rays used.

$$2g = n\lambda$$

If the angle of incidence of the X rays is θ, then the angle of the small triangle opposite the side of length g must also be θ, since both of these angles are complementary to the angle between them. Then

$$\sin \theta = \frac{g}{d}$$

$$2g = 2d \sin \theta$$

$$n\lambda = 2d \sin \theta$$

This equation is known as the Bragg equation, named for its discoverer. From the measured angles at which diffraction occurs with X rays of known wavelength, the value of the interatomic distance, d, can be determined.

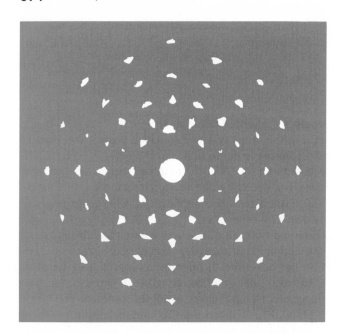

Figure 13-16. **Representation of an X-Ray Diffraction Pattern**

and of the intensities of the scattered X rays as a function of the angles of diffraction provides the data from which numerical values for bond distances and bond angles may be calculated. The process of calculation is not direct. In one method, an "educated guess" is first made as to the structure of the crystal. On the basis of this structure, calculations are made to determine the angles and intensities of diffracted X rays which the model would produce. The model is then amended, and another calculation is made. Successive changes are made until the calculated angles and intensities for the model give reasonable agreement with the experimental values. Such calculations are made by high speed computer. The model structure for which the calculation gives closest agreement with the experimentally determined values of angle and intensity is considered to be the structure of the crystal.

Fortunately, the establishment of the original model is not a completely random guess. Information about bond lengths and angles in similar compounds may be used to establish a reasonable idea of a preliminary structure. Any information about the structure from such techniques as infrared spectroscopy, nuclear magnetic resonance spectroscopy, and/or dipole moment studies as well as the observed chemical behavior of the compound give further clues for establishing a reasonable model.

X-ray diffraction as a technique for determining structure has limitations owing to the fact that the X rays are diffracted by the electrons of atoms. Light atoms, having few electrons, especially hydrogen, are not readily detected by this method. Similarly, other light atoms which may be in the same crystal with much heavier atoms cannot be easily detected.

Determination of structures of substances containing light atoms is better done by neutron diffraction, because the neutrons are scattered by atomic nuclei rather than by electrons, and the neutron-scattering power of light and heavy atoms is of the same order of magnitude. Beams of neutrons have wave characteristics, just as do beams of electrons and beams of X rays. The wavelength associated with a beam of neutrons is given by de Broglie's equation:

$$\lambda = \frac{h}{mv} = \frac{h}{\sqrt{2mE}}$$

where E is the energy of the neutrons. Thermal neutrons from a nuclear reactor (see Section 20–12) have wavelengths in the range of 1 to 2 Å, which is the same order of magnitude as the internuclear distances in molecules.

Since a nuclear reactor is required as a source of the neutron beam, and somewhat elaborate equipment is required for detecting the scattered neutrons, the use of neutron diffraction is generally limited to those applications where the required information cannot be obtained in any other way.

In contrast to X-ray diffraction and neutron diffraction, where the substances are examined as solids, electron diffraction experiments are usually performed on gaseous samples. Electron beams do not penetrate through many layers of matter, and their use on solids is limited to the examination of the structure of surfaces. Electron beams used for diffraction experiments have energies of the order of 40,000 eV, corresponding to wavelengths about 0.1 Å. Electron diffraction techniques are most useful for the evaluation of bond lengths and angles of simple molecules; as the number of atoms per molecule increases, the difficulty of assigning the diffraction pattern to any of several possible alternative structures increases drastically.

13-9   Use of Structural Data

The results of structural determinations are used in many ways. They are used to test theoretical predictions about molecular geometry, of course. They are also important aids in interpreting the chemical behavior of substances, as is shown in Section 12–8, for example, where the strengths of acids are related to their structures. Moreover, the physical properties such as color, hardness, and melting point are influenced by the molecular and electronic structures of the particular substance.

Structural determinations often reveal that the structure of a given substance depends on its physical state. For example, X-ray diffraction experiments show that in the solid state PCl_5 is ionic; the cation is PCl_4^+ and the anion is PCl_6^-. However, upon sublimation, the entire structure changes. Infrared spectroscopy and electron diffraction studies show that in the gaseous state PCl_5 exists as trigonal bipyramidal molecules.

Structural changes may also occur when a substance is put into solution, particularly if the solvent is capable of interacting with the dissolved molecules. Infrared studies of BCl_3 dissolved in a solvent such as CCl_4, which has no dipole moment, show that the BCl_3 has a planar structure. But when BCl_3 is dissolved in an ether, such as $C_2H_5OC_2H_5$, its infrared spectrum changes owing to the formation of a coordinate covalent bond between the boron atom and the oxygen atom of the ether:

$$BCl_3 + :\ddot{O}\!\!\begin{array}{c} {}^{C_2H_5} \\ {}_{C_2H_5} \end{array} \longrightarrow \begin{array}{c} Cl \\ \\ Cl \end{array}\!\!\!B\!:\!\ddot{O}\!\!\begin{array}{c} {}^{C_2H_5} \\ {}_{C_2H_5} \end{array}$$

13-10   Exercises

Basic Exercises

1. Explain how to distinguish the isomers of $C_2H_4Cl_2$ by mass spectroscopy and nmr spectroscopy.

2. Write structural formulas for all isomers corresponding to the molecular formula C_2H_7N. Explain how to distinguish each of these using (a) mass spectrometry, (b) infrared spectroscopy, (c) nmr spectroscopy.

3. How many fundamental modes of vibration would be expected in each of the following molecules? (a) C_2H_2, (b) C_2H_4.

4. The presence of a mass/charge ratio for the parent peak at an odd value in the mass spectrum of an organic compound is indicative of an amine or amide. Explain why amines and amides exhibit this type of behavior, whereas alcohols, hydrocarbons, ethers, etc., do not.

5. The nmr spectrum of a compound containing 64.9% carbon, 13.5% hydrogen, and 21.6% oxygen consists of a quartet and a triplet, the latter 1.5 times an intense as the former. What is the structure of the compound?

6. Which one(s) of the following has(have) only a single peak in the nmr spectrum?

(a) CH_3Cl, (b) CH_3CH_3, (c) CH_2COCH_2, (d) C_6H_6, (e) $Cl-\!\!\bigcirc\!\!-Cl$.
      $\overset{|}{C}H_2$

7. The infrared spectrum of a compound having the formula C_3H_9N has no peak at 3300 cm^{-1}. Predict the nmr spectrum of this compound in terms of the number of peaks expected and their relative intensities.

8. The nmr spectrum of a compound of molecular formula C_4H_9NO has two singlets with an intensity ratio of 2:1. Identify the compound.

9. Select from compounds (i)–(vi) below the one(s) which has(have) (a) a quartet and a triplet as a part of its nmr spectrum, (b) a major peak of 30 in its mass spectrum, (c) a major mass spectral peak at 71, (d) infrared peaks at 1660 and 3300 cm^{-1}, (e) the lowest probability of a major peak 1 less than the molecular ion peak in the mass spectrum, (f) a mass spectrum which is changed significantly by repeated contact of the sample with D_2O, (g) major mass spectral peaks of 29 and 30 and an infrared band at 1660 cm^{-1}, (h) a parent peak in the mass spectrum having an odd value of mass/charge.

(i) $CH_3CH_2-\!\!\bigcirc$ (ii) $\bigcirc\!\!-\overset{\displaystyle H}{\underset{\displaystyle O}{C}}$ (iii) $NH_2CH_2CH_2NH_2$,

(iv) $CH_3CH_2NH_2$,   (v) $NH_2CH_2CH\!=\!NCH_2CH_3$,   (vi) $\bigcirc\!\!\overset{CHO}{\underset{OH}{}}$

10. The nmr spectrum of benzene shows all six hydrogen atoms to be in a single set—they do not split each other. Which one(s) of the dichlorobenzenes behave similarly? Explain why.

11. Predict the nmr spectrum of CH_2ClCH_2Br.

12. The wavelength of a beam of light is 24.0 microns. What is (a) its wavelength in centimeters? (b) its frequency? (c) its wavenumber? (d) the energy of one of its photons?

13. In an absorption spectrum, the absorbance of the sample at a certain wavelength was 0.440. What percent of the incident light was transmitted?

14. A molecule absorbs electromagnetic radiation at the following wavenumbers: 2.5, 250, 250,000 cm^{-1}. **(a)** Calculate the wavelength for each absorption. **(b)** Identify the portion of the electromagnetic spectrum of each absorption. **(c)** Calculate the energy of a photon in each.

15. The asymmetric stretching mode of CO_2 may be diagrammed *in part* as follows:

$$O{\to}C\cdots O{\to} \quad \text{or} \quad {\leftarrow}O\cdots C{\leftarrow}O$$

(a) Complete these diagrams by indicating the motion of the carbon atom. **(b)** Explain why the carbon atom must move during the asymmetric stretching vibration.

16. Tell how to distinguish between the two molecules CH_3COCH_3 and CH_3CH_2CHO on the basis of nmr spectroscopy alone.

17. Arrange the following elements in order of increasing ability to scatter X rays: zinc, uranium, carbon, hydrogen, antimony.

18. X rays having a wavelength of 1.56 Å are diffracted from a certain crystal at 13.2°. Calculate the distance between the planes of atoms in the crystal which are responsible for this diffraction.

19. Which one(s) of the following methods can be used to obtain bond distance data? X-ray diffraction, nuclear magnetic resonance spectroscopy, dipole moment measurement, electron diffraction.

20. The infrared spectrum of 1-aminobutane is given. Identify the bands indicated on the spectrum by the letters A through F.

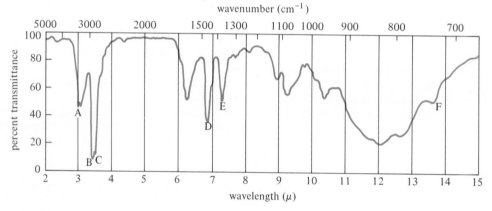

21. The infrared spectrum of methyl alcohol is given. Identify the bands indicated on the spectrum by the letters A through C.

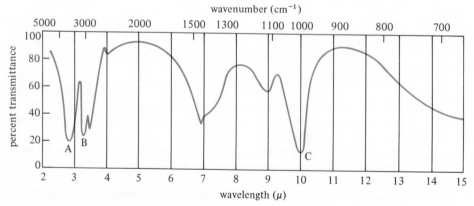

General Exercises

22. Explain why, on the basis of the characteristic group vibrational motions in the molecules, 1,6-dichlorohexane has absorption bands in the infrared spectrum at 730, 1470, and 2800 cm^{-1}, but not above 2900 cm^{-1}.

23. Predict the nmr spectrum of diethyl ether, $CH_3CH_2OCH_2CH_3$.

24. What peaks are to be expected in the mass spectrum of CH_3Cl?

25. A certain organic compound has the molecular formula C_4H_9NO. Its infrared spectrum shows a band at 3300 cm^{-1}; its mass spectrum has peaks at 15, 29, and 30, among others; and its low resolution nmr spectrum has 4 peaks with relative intensities 3:2:3:1. Deduce the structure of the compound.

26. Select at least two instrumental methods that you would use to differentiate between the two isomers in each of the following sets. Explain how each isomer would be identified.

    (a) $(CH_3)_2CHCONH_2$ and $CH_3CH_2CONHCH_3$

    (b) Cl—⟨○⟩—Cl   and   Cl—⟨○⟩
                                              Cl

    (c) CH_3CH_2OH and CH_3OCH_3

    (d) $CH_3CH_2CH_2CH_3$ and $(CH_3)_2CHCH_3$

    (e) $CH_2{=}ClCH_2CH_3$ and $CH_3Cl{=}CHCH_3$

27. The three isomers of dichlorobenzene are experimentally distinguishable by nmr spectroscopy, but not by mass spectrometry. Explain. State what features in the nmr spectrum should be expected for each of the isomers.

28. Identify the peak occurring at a mass/charge ratio of 46 in the mass spectrum of $HSCH_2CH_2OH$ (see Figure 13–1).

29. Two isomeric compounds, A and B, have the molecular formula C_3H_9N. The important peaks in their mass spectra are tabulated. Identify the compounds.

Compound A		Compound B	
Mass Peak	Relative Abundance	Mass Peak	Relative Abundance
59	100	59	100
58	10	58	5
44	23	44	40
43	30	30	30
16	38	29	30
15	20	15	30

30. The infrared spectra described in this chapter (cf. Figure 13–6) were plotted on a scale which is linear in wavelength; that is, equal intervals represent equal differences in wavelength. In what ways would the infrared spectra differ if they were plotted on a scale which is linear in wavenumber? What advantages are there in using the latter scale?

31. From its infrared spectrum and molecular formula, identify unknown compound A.

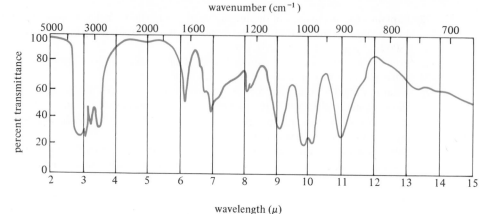

Unknown Compound A. The molecular formula for this compound is C_3H_6O.

32. From its infrared spectrum and molecular formula, identify unknown compound B.

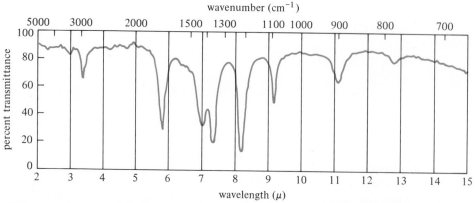

Unknown Compound B. The molecular formula for this compound is C_3H_6O.

33. From its infrared spectrum and composition data, identify unknown compound C.

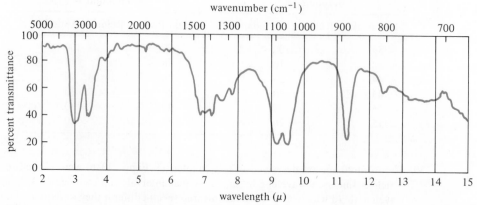

Unknown Compound C. This compound contains only C, H, and O. Its molecular weight is 46.

34. From its infrared spectrum and composition data, identify unknown compound D.

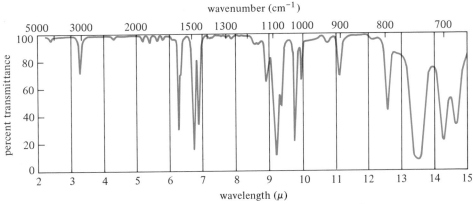

Unknown Compound D. This compound contains only C, H, and Cl.

35. A compound contains 54.55% carbon, 9.09% hydrogen, and 36.36% oxygen. Its nmr spectrum consists of two peaks. The downfield peak is a quartet and is one third as intense as the second peak, which is a doublet. Propose a structure for the compound.

36. The high resolution nmr spectrum of a compound with the molecular formula C_4H_8 has two peaks. The smaller peak is downfield and is one third the intensity of the other. Write detailed structures for two possible isomers corresponding to these data.

37. Sketch the nmr spectrum you would expect for each of the following. Indicate the order of magnitude of the chemical shifts, the relative intensities of the peaks, and the fine structure due to spin-spin splitting, if any. **(a)** $CH_3CO_2CH_3$, **(b)** $(CH_3)_3N$, **(c)** benzene, C_6H_6.

Advanced Exercises

38. What instrumental technique described in this chapter is best suited to distinguish $(+)$-2-chlorobutane from its $(-)$ isomer?

39. Explain why carbon–carbon multiple bonds are reactive centers in chemical reactions, but are relatively stable in mass spectrometry.

40. Explain why chloroethene, $H_2C{=}CHCl$, has three distinct peaks in its nmr spectrum, all of equal intensity. How is each of the three peaks split?

41. Every neighboring set of hydrogen atoms is expected to split a given set. If two groups adjacent to a given group are identical, they will interact the same, and a splitting pattern reflecting the total number of hydrogen atoms will occur. Explain why $CHCl_2CH_2CHCl_2$ contains two triplets, whereas $CHBr_2CH_2CHCl_2$ has a much more complicated pattern.

42. Predict the number of peaks and their relative intensities in the nmr spectrum of $CH_2ClCH_2CHBr_2$.

43. A compound has the composition 54.55% carbon, 9.09% hydrogen, and 36.36% oxygen (cf. exercise 35). Its nmr spectrum has three major peaks, from downfield to upfield in the ratio 2:3:3 in intensity. The first of these is a quartet, the second a singlet having a chemical shift close to that of the quartet, and the third peak is a triplet somewhat further upfield. What is the structure of the compound?

44. The nmr spectrum of which isomer of chloropropane would be expected to contain a septet? Explain.

45. The proton magnetic resonance spectrum of CH_3F consists solely of one doublet; that of CH_3Cl consists of one singlet. Explain these observations.

46. Copper(I) chloride was formerly written Cu_2Cl_2, on the basis of its vapor density. It is now characterized more commonly as CuCl, on the basis of its X-ray structure determination. Explain how these experimental results could both be accepted as correct.

47. (a) How would one prove that PCl_5 exists as discrete molecules in the gas phase but as $[PCl_4]^+[PCl_6]^-$ in the solid? **(b)** Explain why PBr_5 exists as $[PBr_4]^+Br^-$ in the solid phase.

48. A compound with empirical formula C_8H_8O has a molecular ion mass/charge peak of 120 and includes also a principal peak at 43. The nmr spectrum includes four peaks with relative intensities $3:2:2:1$. The infrared spectrum contains a band at 1700 cm^{-1} but none at 3300 cm^{-1}. Suggest a possible structure for the compound.

49. Explain how infrared spectroscopy, mass spectroscopy, and nmr spectroscopy can be used to distinguish among the esters having the molecular formula $C_4H_8O_2$.

50. The cation of an organic dye molecule has the structure

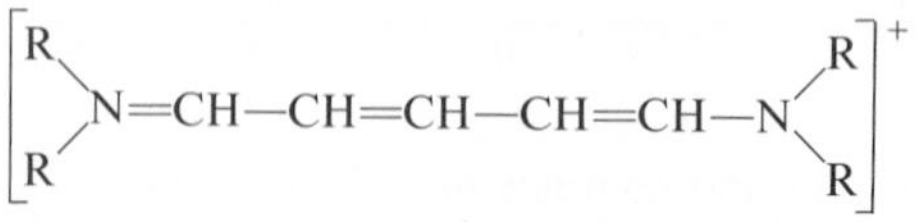

$$\left[\begin{array}{c} R \\ \diagdown \\ N=CH-CH=CH-CH=CH-N \\ \diagup \qquad\qquad\qquad\qquad\qquad \diagdown \\ R \qquad\qquad\qquad\qquad\qquad\qquad R \end{array} \begin{array}{c} R \\ \diagup \\ \\ \diagdown \\ R \end{array} \right]^+$$

which is stabilized by delocalized π bonding extending from one of the nitrogen atoms to the other. **(a)** Starting with de Broglie's relationship, $\lambda = h/mv$, and given that d is the distance between the nitrogen atoms, derive the equation for the energy levels of an electron in the delocalized π orbitals

$$E_n = \frac{n^2h^2}{8md^2}$$

(Note that the distance, d, must correspond to an integral number of half wavelengths.) **(b)** Draw an energy level diagram showing the five lowest energy states of the system, and indicate the respective quantum numbers (n) on the diagram. **(c)** According to the Pauli exclusion principle, only two electrons can occupy a given orbital. Show on your energy level diagram the occupancy of the orbitals by the six electrons in the delocalized π orbitals of the dye cation. **(d)** Using data from Table 11–3, calculate a value for d. Then compute the difference in energy between the highest energy level occupied by an electron and the next higher (empty) level on your diagram. **(e)** To what wavelength of absorbed light does this energy difference correspond?

14

Coordination Compounds

In the middle of the eighteenth century, an artist's pigment known as Prussian blue was first prepared when sodium carbonate and animal excrement were heated together in an iron pot. That a richly intense pigment could be produced from such an unlikely mixture was a source of wonder. It is now known that Prussian blue may be formed by the reaction of iron(III) ion with $K_4Fe(CN)_6$ to give $KFe^{III}[Fe^{II}(CN)_6]$. Early workers termed substances such as Prussian blue "complex compounds," and this name has persisted. Even today the term **complex compound** is used synonymously with the term **coordination compound.**

In Section 8–2, coordination compounds were defined as substances consisting of combinations of species which are capable of independent existence but having properties, even in solution, which are not merely those of the simpler species. For example, the formulas for the compounds listed in Table 14–1 are written to emphasize that they are combinations of $CoCl_3$, NH_3, and, in one case, H_2O. However, it is quite apparent that the properties are not simply those of the components. It is unlikely that the observed differences in color between the substances listed in Table 14–1 could be due solely to differences in the number of ammonia molecules per formula unit. Indeed, differences occur even when the number of ammonia molecules per formula unit is the same in two compounds, as between "roseo" and "purpureo" as well as between "praseo" and "violeo."

Most transition metal ions in water solution have characteristic colors. Those with partially filled d subshells have pale colors in water, but addition of other substances often causes a deepening of the color. For example, when ammonia is added to a solution of $[Cu(H_2O)_4]^{2+}$, the light blue color changes to the very deep blue characteristic of $[Cu(NH_3)_4]^{2+}$. Can the observed colors be related to the nature of the bonding in coordination compounds?

In this chapter the nature of coordination compounds will be explored in some detail. Experimental techniques used to characterize complex compounds and to

TABLE 14-1. Some Coordination Compounds of Cobalt(III)

Formula	Color	Historical Name
$CoCl_3 \cdot 6NH_3$	orange	luteo cobaltic chloride
$CoCl_3 \cdot 5NH_3 \cdot H_2O$	rose	roseo cobaltic chloride
$CoCl_3 \cdot 5NH_3$	purple	purpureo cobaltic chloride
$CoCl_3 \cdot 4NH_3$	green	praseo cobaltic chloride
$CoCl_3 \cdot 4NH_3$	violet	violeo cobaltic chloride
$CoCl_3 \cdot 3NH_3$	blue-green	—

379

distinguish between isomers will be described. Some practical uses and applications of coordination compounds will be discussed. Finally, it will be shown how the structures, colors, and magnetic properties of coordination compounds may be explained theoretically in terms of their electronic structures. Both the valence bond model and the molecular orbital model have been applied to explain the bonding in coordination compounds. In addition, a model of the coordination sphere known as the **crystal field theory** has been applied with considerable success toward the explanation of such phenomena as color and reactivity. A combination of the molecular orbital approach and the crystal field theory, known as the **ligand field theory,** provides quantitative results in some cases. Each approach provides some answers which the others do not, but none is completely successful. In this chapter the salient features of the valence bond, crystal field, and ligand field approaches will be discussed.

14-1   Werner Theory

The first successful theory explaining the properties of coordination compounds in terms of their structures was proposed by Alfred Werner around 1893, long before sophisticated techniques for determining molecular structures were developed and well before the concepts of electronic structure were applied to chemical systems. Werner's theory was so complete and his methods of confirming it were so ingenious that it is worthwhile to review his work in some detail. However, modern terminology will be used for this purpose.

Werner proposed that a complex ion contains a central metal atom to which some neutral molecules and/or some negative ions are attached. In modern terminology, the attached groups are called **ligands.** The aggregate of the central metal atom or ion and the ligands is called the **coordination sphere.** The number of donor atoms bonded to the central atom in the coordination sphere is called the **coordination number** of the metal. The coordination numbers 6, 4, and 2 are most often encountered. A particular metal may exhibit more than one coordination number, depending on its oxidation state. For example, copper(II) ion has a characteristic coordination number of 4, while copper(I) ion has a coordination number of 2.

When the ligand molecule or ion contains two donor atoms, it is possible for a single ligand to occupy two positions in the coordination sphere. For example, the four ammonia molecules in $[Cu(NH_3)_4]^{2+}$ may be replaced by two molecules of ethylenediamine, $NH_2CH_2CH_2NH_2$, to give $[Cu(NH_2CH_2CH_2NH_2)_2]^{2+}$. In this complex, both ends of each ethylenediamine molecule are attached to the copper ion, thus forming five-membered rings and maintaining the coordination number 4.

$$\left[\begin{array}{c} CH_2{-}NH_2 \\ | \\ CH_2{-}NH_2 \end{array} \!\! \diagdown \!\! \underset{\diagup}{\overset{\diagdown}{Cu}} \!\! \diagup \!\! \begin{array}{c} NH_2{-}CH_2 \\ | \\ NH_2{-}CH_2 \end{array} \right]^{2+}$$

This process of ring formation is called **chelation.** In that it is able to occupy two coordination positions, ethylenediamine is called a **bidentate** (two-toothed) ligand. For convenience, bidentate ligands are sometimes denoted in drawings simply by a curved line. Thus the complex $[Cu(NH_2CH_2CH_2NH_2)_2]^{2+}$ might be represented

$$\left[\bigcirc Cu \bigcirc \right]^{2+} \quad \text{or} \quad \left[\left(\begin{array}{c} N \quad N \\ \framebox{Cu} \\ N \quad N \end{array} \right) \right]^{2+}$$

Multidentate ligands are also known which have as many as six or more donor atoms; for example, ethylenediaminetetraacetate, EDTA, is a hexadentate ligand, with four oxygen atoms and two nitrogen atoms serving as the donor atoms:

$$\left[\begin{array}{c} \overset{O}{\overset{\|}{:OCCH_2}} \qquad\qquad \overset{O}{\overset{\|}{CH_2CO:}} \\ \diagdown NCH_2CH_2N \diagup \\ :\overset{}{OCCH_2} \qquad\qquad \overset{}{CH_2CO}: \\ \overset{\|}{O} \qquad\qquad\qquad \overset{\|}{O} \end{array} \right]^{4-}$$

Many organic ligands are multidentate. Some common ligands are presented in Table 14-2.

Werner further proposed that the ligands were arranged in the coordination sphere about the central metal ion in definite geometries, depending on the coordination number. For the coordination number 2, the ligands will be on opposite sides of the central metal, giving a linear configuration for the coordination sphere. For the coordination number 4, two different configurations are possible. The ligands might be arranged at the corners of a tetrahedron with the metal atom at the center, or they might be at the corners of a square with the central metal at the center. For the coordination number 6, the donor atoms occupy the corners of a regular octahedron, with the metal at the center (Figure 14-1). According to Werner's theory, the differences between the compounds listed in Table 14-1 are attributed to differences in their coordination spheres. Different arrangements of ligands within the coordination sphere are reflected in the different colors of the compounds as well as in differences in other properties. Before considering further examples, it is worthwhile to review some of the experimental techniques which have been used to determine the nature of the coordination sphere.

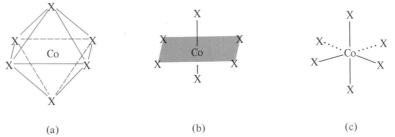

(a)   (b)   (c)

Figure 14-1. Representations of an Octahedral Coordination Sphere

Figure (a) best shows the geometry of the octahedron. Figure (c) shows the actual chemical bonds, but it does not represent the three-dimensional nature of the coordination sphere well. Figure (b) is a compromise three-dimensional representation and is most often used. It must be emphasized that the six positions in the octahedral coordination sphere are inherently equivalent.

TABLE 14-2. Some Common Ligands

Unidentate Ligands

Name	Formula	Number of Donor Atoms	Name as Ligand
ammonia	NH_3	1	ammine
halide ion	X^-	1	halo
cyanide ion	CN^-	1	cyano
hydroxide ion	OH^-	1	hydroxo
carbonate ion	CO_3^{2-}	1 or 2	carbonato
nitrogen oxide	NO	1	nitrosyl
carbon monoxide	CO	1	carbonyl
sulfate ion	SO_4^{2-}	1 or 2	sulfato
oxide ion	O^{2-}	1	oxo
nitrite ion	NO_2^-	1	nitro or nitrito[a]
thiocyanate ion	SCN^-	1	thiocyanato or isothiocyanato[a]

Multidentate Ligands

Name	Formula	Abbreviation[b]	Donor Atoms
ethylenediamine	$NH_2CH_2CH_2NH_2$	en	2
1,10-phenanthroline		o-phen	2
ethylenediaminetetraacetate	$(\overset{\text{O}}{\overset{\|}{\text{OC}}}CH_2)_2NCH_2CH_2N(CH_2\overset{\text{O}}{\underset{\|}{\text{CO}}})_2^{4-}$	EDTA^{4-}	6
oxalate ion		$C_2O_4^{2-}$	2
acetylacetonate ion	$CH_3\overset{\text{O}}{\overset{\|}{\text{CCH}}}{=}\overset{\text{O}^-}{\overset{\|}{\text{CCH}}}_3$	acac$^-$	2
dimethylglyoxime ion	$CH_3C{=}NO^-$ $CH_3C{=}NOH$	dmg$^-$	2
glycinate ion	$NH_2CH_2CO_2^-$	gly$^-$	2
cyclopentadienide ion		cp$^-$	5

[a] See Section 14-4.

[b] Because of the inconvenience of writing the formulas of such complex ligands in a coordination sphere, abbreviations are often used. Thus $[Cu(en)_2]^{2+}$ is used to represent $[Cu(NH_2CH_2CH_2NH_2)_2]^{2+}$.

14-2 Determination of the Nature of the Coordination Sphere

The charge on an ion can be inferred from conductance measurements on solutions of its salts. An apparatus for the measurement of the conductance of solutions is shown in Figure 14-2. In a conductance experiment, the resistance to the passage of electricity of a medium placed between two electrodes is measured. Conductance is expressed as the reciprocal of the resistance (Section 5–1). **Specific conductivity** is the conductance of 1 cm³ of a solution between electrodes exactly 1 cm apart.

Molar conductivity is defined as the conductance of a solution containing exactly 1 mole of dissolved electrolyte between electrodes which are placed exactly 1 cm apart and which are large enough to contain all of the solution between them. (Actually, it is not necessary to use a conductivity cell large enough to contain 1 mole of solute. Any size cell can be used if it is first calibrated by measuring the conductivity of a solution whose molar conductivity is known.) The molar conductivity of any salt varies with concentration, but as the solutions are made more and more dilute, the observed conductivities tend toward a limiting value. The observed molar conductivities at "infinite dilution" approach values whose magnitudes depend on the charge type of electrolyte being examined. For example, salts consisting of a uni-positive cation and a uninegative anion [represented $(1+, 1-)$], such as K^+Cl^-, have a lower molar conductivity at infinite dilution than salts containing a dipositive cation and two uninegative anions $(2+, 1-)$, such as $Ca^{2+}(Cl^-)_2$. The approximate ranges of molar conductivities at infinite dilution of salts of various charge types are listed in Table 14-3.

The molar conductivities of the cobalt complexes listed in Table 14-1 range from 432 ohm^{-1} cm^{-1} for $CoCl_3 \cdot 6NH_3$, characteristic of $(3+, 1-)$ salts, to a value for $CoCl_3 \cdot 3NH_3$, characteristic of nonelectrolytes (less than 10 ohm^{-1} cm^{-1}). The conductivities of the solutions of the other compounds listed in Table 14-1 change with time, so that precise measurements are not possible. However, the measured conductivities of these compounds are in the same general range as simple salts of the $(2+, 1-)$ and $(1+, 1-)$ types, respectively. The type of conductivity observed for each of the compounds is listed in Table 14-4. Since the compounds provide different numbers of ions in solution, it must be concluded that in some of them the chloride ions exist in the coordination sphere rather than as free ions. A chloride "ion" which is a part of the coordination sphere does not behave like a chloride ion of a simple salt; for example, it does not immediately react with silver ion to form silver chloride.

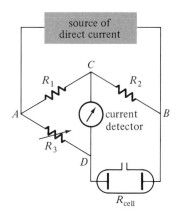

Figure 14-2. Apparatus for Measurement of Conductivity

The electrical resistance of a solution in the cell, R_{cell}, is measured. The conductivity is defined as the reciprocal of the resistance. The apparatus, called a Wheatstone bridge, consists of two resistors, R_1 and R_2, having known resistances connected in parallel to a variable resistor, R_3, and the conductivity cell with resistance R_{cell}. A potential is applied to points A and B, and R_3 is adjusted until points C and D are at the same potential. Under these conditions,

$$\frac{R_{cell}}{R_3} = \frac{R_2}{R_1} \quad \text{and} \quad R_{cell} = \frac{R_3 R_2}{R_1}$$

TABLE 14–3. Ranges of Molar Conductivities

Charge Type	Range ($ohm^{-1} \cdot cm^{-1}$)	Moles of Ions per Mole of Compound	Examples
$(4+, 1-)$	550	5	$Sn(NO_3)_4$
$(3+, 1-)(1+, 3-)$	430	4	$AlCl_3$, K_3PO_4
$(2+, 1-)(1+, 2-)$	250	3	$CaCl_2$, Na_2SO_4
$(1+, 1-)$	100	2	KCl

Hence, how much of the chloride is within the coordination sphere can sometimes be ascertained by determining the fraction of the total chloride which precipitates immediately upon addition of excess silver ion. These data are presented in Table 14–4 under the heading "Ratio." The ionic formulas listed in the table show the composition of the coordination sphere.

Example

The substance $CoBr_3 \cdot 4NH_3 \cdot 2H_2O$ has a molar conductivity of 420 $cm^{-1} \cdot ohm^{-1}$ at infinite dilution. Indicate the composition of the coordination sphere.

The conductivity corresponds to that of a $(3+, 1-)$ electrolyte; hence the substance should be represented as $[Co(NH_3)_4(H_2O)_2]^{3+}(Br^-)_3$, or, more simply, $[Co(NH_3)_4(H_2O)_2]Br_3$.

Information about the coordination sphere can often be provided by means of infrared spectroscopy (Section 13–3). For example, in the complex compound $[Co(NH_3)_6]_2(SO_4)_3 \cdot 5H_2O$, only two infrared active bands due to sulfur–oxygen vibrations are observed. These correspond to the expected two triply degenerate infrared active vibrations of a tetrahedral AB_4 species, SO_4^{2-} (see Table 13–3). Therefore in this compound, the sulfate groups exist as free ions outside the coordination sphere. On the other hand, in the compound $[Co(NH_3)_5SO_4]Br$, six separate absorption bands due to S—O vibrations are observed. This observation indicates that an oxygen atom of the sulfate group is covalently bonded to the metal.

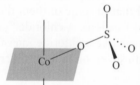

TABLE 14–4. Characteristics of Cobalt(III) Complexes

Compound	Conductivity Type	Ratio[a]	Ionic Formula
$CoCl_3 \cdot 6NH_3$	$AlCl_3$	3	$[Co(NH_3)_6]^{3+}(Cl^-)_3$
$CoCl_3 \cdot 5NH_3 \cdot H_2O$	$AlCl_3$	3	$[Co(NH_3)_5H_2O]^{3+}(Cl^-)_3$
$CoCl_3 \cdot 5NH_3$	$CaCl_2$	2	$[Co(NH_3)_5Cl]^{2+}(Cl^-)_2$
$CoCl_3 \cdot 4NH_3$	KCl	1	$[Co(NH_3)_4Cl_2]^+Cl^-$
$CoCl_3 \cdot 4NH_3$	KCl	1	$[Co(NH_3)_4Cl_2]^+Cl^-$
$CoCl_3 \cdot 3NH_3$	none	0	$[Co(NH_3)_3Cl_3]$

[a] *Ratio* signifies the ratio of the number of moles of silver chloride immediately precipitated per mole of complex by excess Ag^+ ion.

Since the sulfate group no longer vibrates as a perfectly tetrahedral molecule, the degeneracy (Section 10–5) is reduced, giving a greater number of infrared active bands. In the spectrum of $[Co(en)_2SO_4]Br$, eight bands assigned to S—O vibrations are observed. This finding indicates that all the possible degeneracies are removed; hence in this case the sulfate group must be acting as a bidentate ligand:

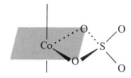

14-3   Nomenclature

The system for naming complex compounds was described briefly in Section 8–2. The following more detailed set of rules has been adopted by the International Union of Pure and Applied Chemistry:

1. If a compound is a salt, the cation is named first, whether it is a complex or simple ion; then the anion is named. Each ion is written as one word, with no spaces in its name.
2. The constituents of each coordination sphere are named in the following order: anions, neutral molecules, central metal ion.
3. Names of ligands which are anions before coordination end with the letter *o*; for example, hydroxo for OH^- and chloro for Cl^-.
4. Names of neutral ligands are used unchanged, except that H_2O is called aquo, NH_3 is called ammine; CO is called carbonyl, and NO is called nitrosyl.
5. The number of ligands of a particular type is denoted by a Greek prefix, di-, tri-, tetra-, penta-, hexa-, ..., for 2, 3, 4, 5, 6, ... In cases where the ligand itself is a complicated molecule, the prefixes bis-, tris-, tetrakis-, ... are used; for example, $[Cu(en)_2]^{2+}$ is called bis(ethylenediamine)copper(II) ion.
6. If the complex is a cation or a neutral molecule, the name of the central metal atom is used unchanged, followed immediately by a Roman numeral in parentheses to denote its oxidation state. If the complex is anionic, the ending -ate is appended to the name of the central metal, followed by the Roman numeral in parentheses to denote the oxidation state. Examples include that given above and hexacyanoferrate(III) ion for $[Fe(CN)_6]^{3-}$.
7. Geometric isomers are denoted with the words *cis* or *trans* placed before the name or formula (Section 14–4).

Example

Name the following compounds:

(a) $[Co(en)_2(CN)_2]ClO_3$ (b) $K_4[Co(CN)_6]$ (c) $[Ni(NH_3)_6]_3[Co(NO_2)_6]_2$

The names are, respectively, (a) dicyanobis(ethylenediamine)cobalt(III) chlorate, (b) potassium hexacyanocobaltate(II), and (c) hexaamminenickel(II) hexanitrocobaltate(III).

Example

Write formulas for the following compounds: **(a)** hexaamminecobalt(III) bromide, **(b)** dibromotetraamminecobalt(III)   tetrachlorozincate(II),   **(c)** dichlorodiammineplatinum(II).

   (a) $[Co(NH_3)_6]Br_3$,   **(b)** $[Co(NH_3)_4Br_2]_2[ZnCl_4]$,   **(c)** $[Pt(NH_3)_2Cl_2]$.

14-4   Isomerism in Coordination Compounds

Geometric Isomerism

Werner's theory provided for the existence of geometric isomers. If the coordination sphere has a definite geometry, some sets of ligands can occupy the sites in the coordination sphere in more than one way. In Table 14–1 the same formulas are used to represent both the "praseo" and violeo" compounds, although the two are different compounds, as evidenced, for example, by their different colors. Obviously, the two substances must be isomeric. For an octahedral complex represented by the formula $[Co(NH_3)_4Cl_2]^+$, there are only two possible arrangements. In one of these the two chloride ions are located at adjacent corners of the octahedron, forming an isomer called the *cis* isomer; in the other, the chloride ions are on opposite corners, giving the *trans* isomer.

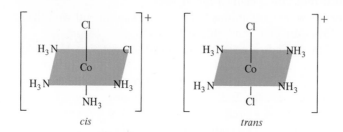

In complexes in which the coordination number of the central atom is 4, planar or tetrahedral geometry is possible, depending on the metal ion. For example, zinc(II) complexes have four donor atoms arranged tetrahedrally about the zinc. If two of the

ammonia molecules could be replaced by chloride ions, the resulting compound, $Zn(NH_3)_2Cl_2$, would exist in only one form, with no isomers. In contrast, if the geometry of a four-coordinate complex is square planar, then a disubstituted complex can exist in *cis* and *trans* isomeric forms. For example, two isomeric forms of

$Pt(NH_3)_2Cl_2$ can be prepared as follows:

$$K_2[PtCl_4] \; + \; 2\,NH_3 \; \rightarrow \; 2\,K^+ \; + \; 2\,Cl^- \; + \quad \underset{cis}{\overset{\displaystyle Cl \diagdown \quad \diagup Cl}{\underset{\displaystyle H_3N \diagup \quad \diagdown NH_3}{Pt}}}$$

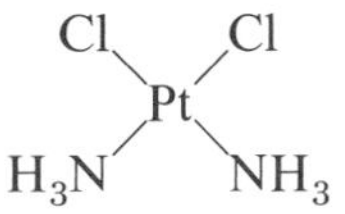

$$[Pt(NH_3)_4]Cl_2 \; + \; 2\,HCl \; \rightarrow \; 2\,NH_4^+ \; + \; 2\,Cl^- \; + \quad \underset{trans}{\overset{\displaystyle Cl \diagdown \quad \diagup NH_3}{\underset{\displaystyle H_3N \diagup \quad \diagdown Cl}{Pt}}}$$

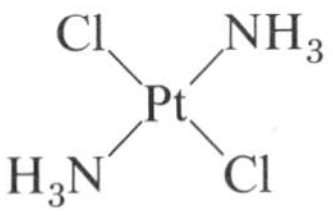

With the knowledge that most bidentate ligands cannot span *trans* positions, the configurations of the two isomers can be deduced from the following reaction sequences.

With the compound prepared from $K_2[PtCl_4]$:

$$Pt(NH_3)_2Cl_2 \xrightarrow{\;2\,AgNO_3\;} Pt(NH_3)_2(NO_3)_2$$

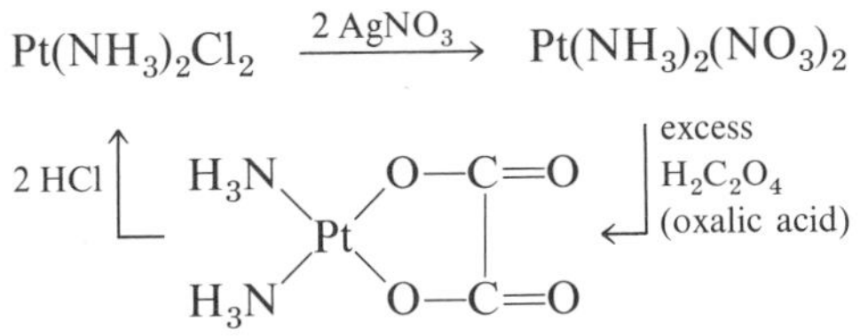

With the compound obtained from $[Pt(NH_3)_4]Cl_2$:

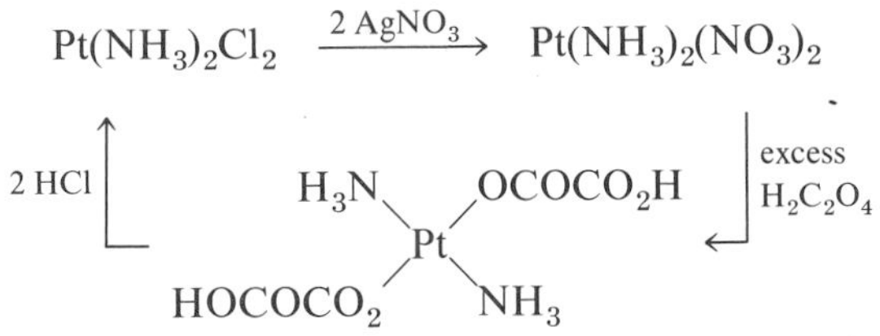

$$Pt(NH_3)_2Cl_2 \xrightarrow{\;2\,AgNO_3\;} Pt(NH_3)_2(NO_3)_2$$

If no changes in configuration occur during the above reactions, and the oxalate group can serve as a bidentate group only in the *cis* isomer, then the compound prepared from K_2PtCl_4 must be the *cis* isomer.

Dipole moment measurements are useful in distinguishing between *cis* and *trans* isomers, provided that the respective compounds are sufficiently soluble in nonpolar solvents to permit measurement of their dipole moments. Thus is has been shown that for compounds of the type $[Pt\{P(C_2H_5)_3\}_2Cl_2]$, the *trans* isomers have zero dipole moments, while the corresponding *cis* isomers have moments of 8–12 D.

Often infrared spectroscopy can be used to distinguish between *cis* and *trans* isomers. In a *trans* octahedral complex, such as $[Co(NH_3)_4Cl_2]^+$, or a *trans* square planar complex, such as $[Pt(NH_3)_2Cl_2]$, the Cl—metal—Cl symmetrical stretching vibration produces no change in the dipole moment of the molecule (Figure 14–3a and 14–3c), and thus no band corresponding to this vibration is observed in the infrared spectrum. However, in the *cis* form of each compound, the symmetrical stretching vibration (Figures 14–3b and 14–3d), as well as the asymmetrical stretching

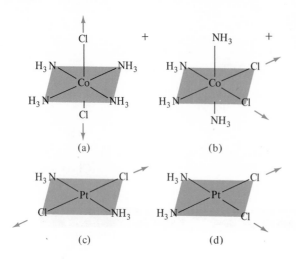

(a)

(b)

(c)

(d)

Figure 14-3. Symmetrical Cl—Metal—Cl Stretching Vibrations of Isomeric Coordination Compounds

vibration, produce appreciable changes in the dipole moment; hence, the infrared spectrum of the *cis* isomer will contain a larger number of bands because of the Cl—metal—Cl stretching.

Optical Isomerism in Coordination Compounds

One of the triumphs of Werner's theory was the prediction that coordination compounds could exist as optical isomers (Section 12–11). Since the ligands occupy specific positions in the coordination sphere, it is possible to prepare isomeric compounds having structures which are nonsuperimposable mirror images. Diagrammed below are two possible structures of the *cis* form of the complex [Co(en)$_2$XY]. These isomeric structures are seen to have a "right-handed" and a "left-handed" character. The two isomers of *cis*-[Co(en)$_2$XY] both have the ability to rotate the plane of polarized light. One form rotates the plane of light to the right and the other form rotates it to the left to an identical extent. The first compound is said to be dextrorotatory and the second, levorotatory.

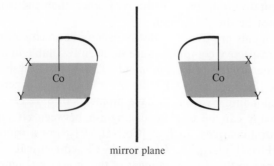

mirror plane

Werner supported his hypothesis of an octahedrally shaped coordination sphere by preparing and resolving a number of optically active six-coordinate complex salts. Werner's theory of the coordination sphere has also been confirmed by X-ray analysis of a variety of coordination compounds. Tris(ethylenediamine)cobalt(III) ion is another example of an optically active complex ion. This example also illustrates the point that optical activity does not depend on a central atom being bonded to different groups. In this case, chirality is achieved by the alternative sites of bonding of the ligands in the coordination sphere, such that "left-handed" and "right-handed" molecules are obtained.

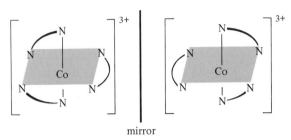

mirror

optically active forms of $[Co(en)_3]^{3+}$

Example

Describe a reaction sequence which might be used to resolve a racemic mixture of cis-$[Co(en)_2Cl_2]^+$.

A reaction sequence such as the following might be employed:

$$\left.\begin{array}{l}(+)\text{-}cis\text{-}[Co(en)_2Cl_2]Cl\\(-)\text{-}cis\text{-}[Co(en)_2Cl_2]Cl\end{array}\right\} + 2\,((-)\text{-base}^-) \rightarrow \begin{cases}(+)\text{-}cis\text{-}[Co(en)_2Cl_2]((-)\text{-base}) + Cl^-\\(-)\text{-}cis\text{-}[Co(en)_2Cl_2]((-)\text{-base}) + Cl^-\end{cases}$$

Since the last pair of compounds are not enantiomers, they may be separated by physical means. After separation, each is treated with excess HCl to restore its original composition.

$$(+)\text{-}cis\text{-}[Co(en)_2Cl_2]((-)\text{-base}) + HCl \rightarrow (+)\text{-}cis\text{-}[Co(en)_2Cl_2]Cl + H((-)\text{-base})$$
$$(-)\text{-}cis\text{-}[Co(en)_2Cl_2]((-)\text{-base}) + HCl \rightarrow (-)\text{-}cis\text{-}[Co(en)_2Cl_2]Cl + H((-)\text{-base})$$

It should be noted that *trans* isomers are so symmetrical that they are identical to their mirror images and do not exist in optically active forms. Thus if it is possible to prepare optically active forms of an octahedral complex, the original configuration may be assumed to have been *cis*. An example of the use of optical activity to determine the configurations of possible *cis-trans* isomers is shown in Figure 14-4.

Coordination Isomerism

In addition to geometric and optical isomerism, which depend on the spatial arrangement of ligands *within* the coordination sphere, other types of isomerism are possible in complex compounds. These usually involve both the cation and anion of a coordination compound. Some examples of these types of isomerism are described below.

When both the cation and the anion of a salt are complex ions, isomers exist in which the central ions of the respective coordination spheres are interchanged or in which some ligands are interchanged. This type of isomerism is called coordination isomerism and is illustrated by the following examples:

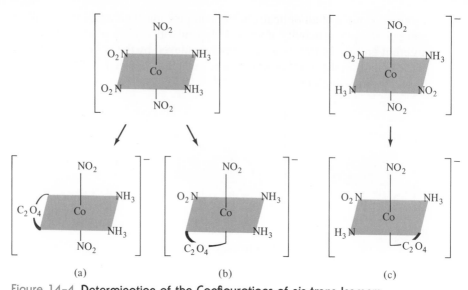

Figure 14-4. Determination of the Configurations of *cis-trans* Isomers

For the ion $[Co(NH_3)_2(NO_2)_4]^-$ there are two possible configurations, either *cis* or *trans* with respect to the NH_3 groups. The two can be distinguished by means of a reaction with a chelating ligand, such as oxalate ion. The *cis* isomer yields three monochelated derivatives, (a), (b), and the optical isomer of (b). The *trans* isomer yields only one monochelated derivative, (c), which is optically inactive.

$[Co(NH_3)_6][Cr(NO_2)_6]$ isomeric with $[Cr(NH_3)_6][Co(NO_2)_6]$

$[Cr(NH_3)_6][Cr(NO_2)_6]$ isomeric with $[Cr(NH_3)_4(NO_2)_2][Cr(NH_3)_2(NO_2)_4]$

Example

Other than by X-ray diffraction, how could the above pairs of isomers be distinguished from one another?

One way $[Co(NH_3)_6][Cr(NO_2)_6]$ can be distinguished from its coordination isomer is by electrolysis of an aqueous solution. In one case the cobalt(III) complex migrates toward the negative electrode, where cobalt would be deposited. In the other case, chromium would be deposited there. $[Cr(NH_3)_6][Cr(NO_2)_6]$ can be distinguished from $[Cr(NH_3)_4(NO_2)_2]$ $[Cr(NH_3)_2(NO_2)_4]$ by conductivity measurements. The former would conduct as a $(3+, 3-)$ electrolyte, while the latter would conduct as a $(1+, 1-)$ electrolyte.

Ionization Isomerism

Ionization isomers are salts, having identical compositions, which produce different ions when the complexes are dissolved in water. Examples are the compounds $[Co(NH_3)_5SO_4]Cl$ and $[Co(NH_3)_5Cl]SO_4$. The first compound gives an immediate precipitate of silver chloride when its solution is treated with silver ion but does not immediately precipitate $BaSO_4$ upon treatment with barium salts. The other isomer behaves in the opposite manner.

Linkage Isomerism

Linkage isomers are different compounds which contain a ligand capable of attachment to the central metal ion by either of two different donor atoms. For example, the nitrite ion, NO_2^-, may be attached either by an oxygen atom or by the nitrogen atom. Consequently, the isomeric ions $[Co(NH_3)_5NO_2]^{2+}$ and $[Co(NH_3)_5ONO]^{2+}$ can exist. Linkage isomers are often differentiated by using alternative names for the coordinated ligand. For example, NO_2^- bonded to the metal through the nitrogen atom is called the nitro group, whereas when bonded through an oxygen atom, it is called the nitrito group. Another ligand which exhibits linkage isomerism is the thiocyanate ion, SCN^-. Linkage isomers containing SCN^- may be distinguished by infrared spectroscopy. If the thiocyanate ion coordinates to the metal through the sulfur atom, the ligand is referred to as isothiocyanato, and the frequency of the C—S stretching vibration is lower than that observed for the free ion. On the other hand, if coordination is through the nitrogen atom, the ligand is named thiocyanato, and the frequency of the C—S stretching vibration is usually higher than that observed for the free ion.

Polymerization Isomerism

Isomers are different compounds which have the same molecular formula, hence the same molecular weight. Addition polymers have the same empirical formula as the simpler compound from which they are produced, but the actual formula weight of a polymer is an integral multiple of that of the simpler compound. For example, polyethylene has the same empirical formula, CH_2, as ethylene, C_2H_4, but the formula weight of polyethylene is many times that of ethylene. Some coordination compounds—for example, $[Co(NH_3)_3(NO_2)_3]$, $[Co(NH_3)_4(NO_2)_2][Co(NH_3)_2(NO_2)_4]$, $[Co(NH_3)_5(NO_2)][Co(NH_3)_2(NO_2)_4]_2$, and $[Co(NH_3)_6][Co(NH_3)_2(NO_2)_4]_3$—have the same empirical composition, $CoN_6H_9O_6$. In solution, the first is a nonelectrolyte while the others all conduct electricity. However, they cannot be distinguished by molecular weight determinations in solution (Section 16–18) because the average formula weight of each particle (molecule or ion) is the same in each compound. Hence, despite the fact that the actual formula weights are all different, it is not easy to tell that these substances are not isomers of each other. Compounds showing this type of relationship were called **polymerization isomers** by early investigators of coordination compounds. Another example of coordination isomerism is the series $[Pt(NH_3)_2Cl_2]$, $[Pt(NH_3)_3Cl][Pt(NH_3)Cl_3]$, and $[Pt(NH_3)_4][Pt(NH_3)Cl_3]_2$. Actually, no two of these compounds are isomers, nor are the larger members polymers.

Polymerization isomerism applies to coordination compounds but to no other types of compounds. For example, C_2H_2 and C_6H_6 are not classified as polymerization isomers because their molecular weights are easily determined to be different.

14-5   Practical Applications of Coordination Compounds

Coordination compounds have many practical applications, ranging from metallurgical to various biological processes. A few examples are described below.

Electrodeposition of many metals is done from solutions having very specific concentrations of metal ion. Therefore a controlled concentration of metal ion is maintained by complexing with a suitable ligand. (The formation constant of a complex ion determines the equilibrium concentration of free metal ion.) For example, copper is deposited from an ammonia solution in which the copper(II) forms $[Cu(NH_3)_4]^{2+}$ ion and in which the $Cu^{2+}(aq)$ concentration is very low. It has been determined empirically that under such conditions the electrodeposition process forms a bright layer of copper which does not discolor or flake off.

Cyanide complexes are used in the refining of silver and gold from their ores. In the presence of cyanide ion, these metals can be oxidized by air to their monopositive oxidation states:

$$H_2O + 2\,Ag + 4\,CN^- + \tfrac{1}{2}O_2 \rightarrow 2\,[Ag(CN)_2]^- + 2\,OH^-$$

The soluble complex is easily separated from insoluble impurities in the ore, and afterward the precious metal is reduced to the metallic state by zinc:

$$2\,[Ag(CN)_2]^- + Zn \rightarrow 2\,Ag + [Zn(CN)_4]^{2-}$$

The Mond process is a commercial method for the purification of nickel metal. Since nickel forms a gaseous carbonyl at $43\,^{\circ}C$, and it is easy to distill this compound from any impurities, the crude product of reduction of nickel ore by carbon is treated with carbon monoxide:

$$Ni(s) + 4\,CO(g) \rightarrow Ni(CO)_4(g)$$

The pure nickel carbonyl is then decomposed into pure nickel and carbon monoxide by heating above $50\,^{\circ}C$.

Many important biological materials are coordination compounds. The hemoglobin molecule contains a six-coordinate iron(II) ion surrounded in a plane by the four nitrogen donor atoms of the heme group (Figure 14–5). A fifth coordination position of the octahedron is occupied by the globin portion of the structure. The sixth coordination position is available to coordinate O_2 molecules, depending on its equilibrium concentration at a particular site. In the lungs, where the oxygen concentration is high, O_2 occupies the coordination site. In the muscle cells, where the O_2 has been converted to a great extent to CO_2, and its concentration is low, the O_2 leaves the coordination sphere. If air containing carbon monoxide is inhaled, the CO molecules become attached to the sixth coordination position of the hemoglobin; the CO complex is so stable that O_2 cannot displace it easily from the coordination sphere, and the hemoglobin molecules are effectively prevented from performing their oxygen-carrying function. If enough carbon monoxide is inhaled, carbon monoxide poisoning and suffocation soon occur. Other biologically important coordination compounds include vitamin B_{12}, chlorophyll (Figure 14–5), and enzymes that are coordination compounds of amino acids and metal ions.

There are many applications of coordination compounds in analytical chemistry. For example, in qualitative analysis, mixed precipitates of AgCl and Hg_2Cl_2 are separated by adding aqueous ammonia. The soluble $[Ag(NH_3)_2]^+$ ion is formed, which is then removed from the insoluble mixture of mercury metal and mercury amido chloride by filtration:

$$AgCl + 2\,NH_3 \rightleftharpoons [Ag(NH_3)_2]^+ + Cl^-$$
$$Hg_2Cl_2 + 2\,NH_3 \rightarrow Hg(s) + Hg(NH_2)Cl(s) + NH_4^+ + Cl^-$$

Other complexes serve as confirmatory tests for specific ions, as, for example,

$$Fe^{3+} \; + \; SCN^- \; \rightleftharpoons \; [FeNCS]^{2+}$$

deep red solution

$$Co^{2+} \; + \; 3\,K^+ \; + \; 7\,NO_2^- \; + \; 2\,HC_2H_3O_2 \; \rightarrow$$
$$K_3Co(NO_2)_6(s) \; + \; NO \; + \; 2\,C_2H_3O_2^- \; + \; H_2O$$

yellow

$$[Ni(NH_3)_4]^{2+} \; + \; 2\,dmgH \; \rightarrow \; Ni(dmg)_2 \; + \; 2\,NH_3 \; + \; 2\,NH_4^+$$

dimethylglyoxime   red precipitate

Some chelating agents, particularly EDTA, have proved especially valuable as reagents for the titration of metals. A solution of the disodium salt of EDTA, added to a solution of the metal ion, converts the hydrated metal ion into the EDTA complex. At the equivalence point, the ratio of metal ion to EDTA is usually 1/1, and there is a sharp decrease in the concentration of free metal ion. This change can be detected potentiometrically or by use of a suitable indicator. One type of indicator is an organic dye which forms a weaker complex with the metal than does the EDTA. As the solution containing the metal ion and indicator is titrated with EDTA reagent, the dye is displaced. At the end point, the color of the solution changes from that of the metal–indicator complex to that of the uncomplexed indicator ion.

Figure 14–5. **Structures of Hemin Cation and of Chlorophyll a**

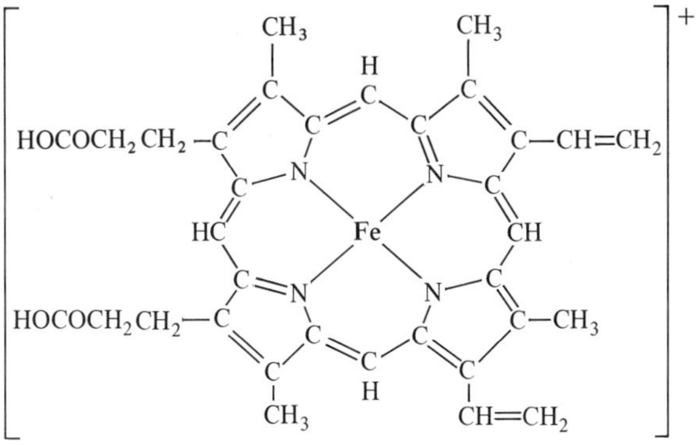

hemin cation

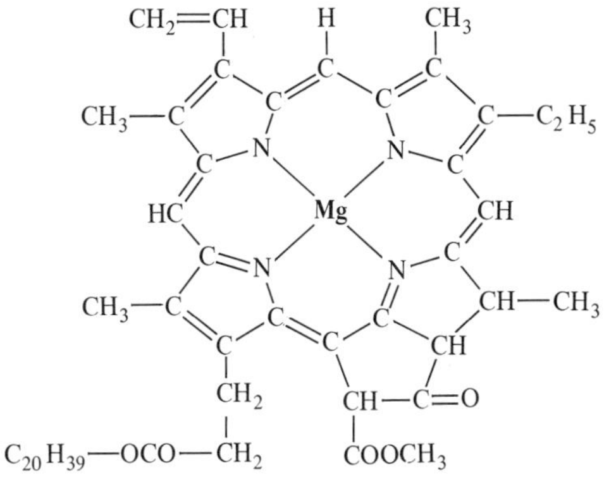

chlorophyll a

Metal–Ligand Bonding

The Werner theory describes the structures of coordination compounds and explains the existence of various types of isomers, but the nature of bonding within the coordination sphere remains to be discussed. Even with the recognition that the interaction between the central atom and the ligands is merely an acid-base interaction in the Lewis sense, it is not apparent why there are characteristic coordination numbers or varying geometries. For example, why are there so many complexes of transition metal ions which are octahedrally coordinated? Why are some four-coordinate complexes square planar while others are tetrahedral?

14-6   Valence Bond Method

The valence bond method, described in Section 11–11, readily allows prediction of the geometries of coordination spheres. Ligands are molecules or ions which are Lewis bases. They possess unshared pairs of electrons, which can react to form coordinate covalent bonds (Section 8–1). In coordination compounds, the central metal ion (atom) acts as a Lewis acid, accepting the electron pairs from the ligands into empty metal orbitals. Moreover, the interaction between metal and ligand is enhanced by hybridization of the metal ion orbitals. Since the hybrid orbitals have definite orientations in space, the coordination sphere will have a definite geometry.

For a given metal ion, the orbitals used to form hybrid orbitals may be determined by the following procedure: First, the outer electronic configuration of the "free" metal ion is written down in a manner which shows how the various orbitals capable of holding valence electrons are occupied. For example, the configuration of chromium is $1s^2 2s^2 2p^6 3s^2 3p^6 3d^5 4s^1$. Chromium(III) therefore has the outer electronic configuration $3d^3 4s^0$. The configuration of the ion may be represented as follows:

$$Cr^{3+} \quad \underset{3d}{\uparrow\ \uparrow\ \uparrow\ \underline{\quad}\ \underline{\quad}} \qquad \underset{4s}{\underline{\quad}} \qquad \underset{4p}{\underline{\quad}\ \underline{\quad}\ \underline{\quad}}$$

This diagram shows that in Cr^{3+} two $3d$ orbitals, the $4s$ orbital, and the three $4p$ orbitals are empty and available to form six d^2sp^3 hybrid orbitals which can accept six electron pairs from the ligands as follows (where the pairs of circles represent the pairs of electrons shared with the ligands, L):

$$[CrL_6]^{3+} \quad \underset{3d}{\uparrow\ \uparrow\ \uparrow\ \overset{\overset{\displaystyle d^2sp^3}{\overbrace{\qquad\qquad\qquad}}}{oo\ oo}} \quad \underset{4s}{oo} \quad \underset{4p}{oo\ oo\ oo}$$

Hence the geometry of the coordination sphere is octahedral.

It was pointed out in Section 10–9 that many transition metal ions are paramagnetic. This property is associated with species containing unpaired electrons. Often, the formation of complex ions is accompanied by striking changes in the magnetic susceptibility of the central metal ion.

In Co^{3+} there are six electrons which occupy the $3d$ subshell in accordance with Hund's rule:

Co^{3+} $\quad$ $\underset{3d}{\uparrow\downarrow\ \uparrow\ \uparrow\ \uparrow\ \uparrow}$ $\quad$ $\underset{4s}{—}$ $\quad$ $\underset{4p}{—\ —\ —}$

To form octahedral complexes of cobalt(III) using d^2sp^3 hybrid orbitals, such as $[Co(NH_3)_6]^{3+}$, there must be a rearrangement of the electrons so that two empty d orbitals are provided. This arrangement results in the pairing of all the electrons, as shown below:

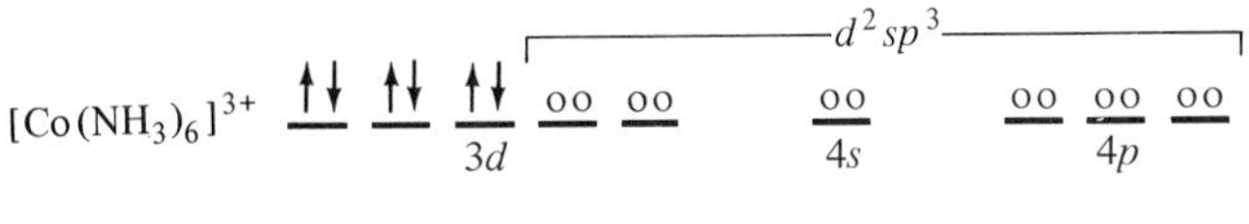

$[Co(NH_3)_6]^{3+}$ $\quad$ $\underset{3d}{\uparrow\downarrow\ \uparrow\downarrow\ \uparrow\downarrow}\ oo\ oo$ $\quad$ $\overset{\underline{\qquad\qquad d^2sp^3\qquad\qquad}}{\underset{4s}{oo}\quad\underset{4p}{oo\ oo\ oo}}$

It is assumed that d^2sp^3 hybrid orbitals form stronger bonds, providing the energy required to cause electron pairing. Since it contains no unpaired electrons, this complex is diamagnetic, in contrast to the free metal ion, which is paramagnetic.

Four-coordinate complexes may have either tetrahedral or square planar structures, depending on their electronic configurations. The configurations of free nickel(II) ion and two of its complexes illustrate this point:

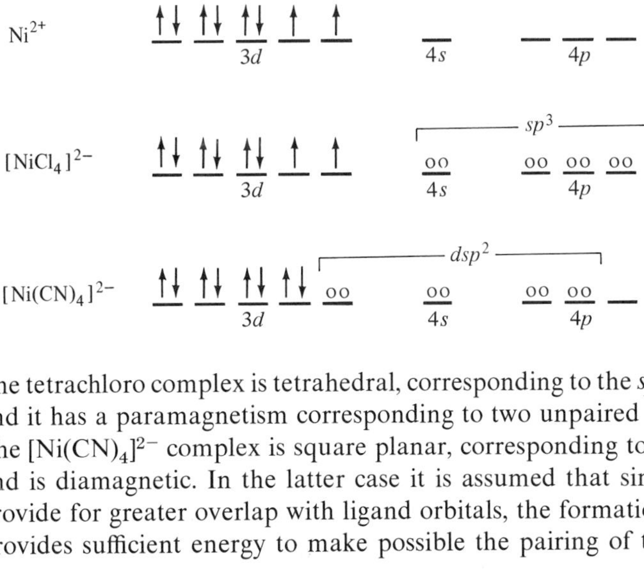

Ni^{2+} $\quad$ $\underset{3d}{\uparrow\downarrow\ \uparrow\downarrow\ \uparrow\downarrow\ \uparrow\ \uparrow}$ $\quad$ $\underset{4s}{—}$ $\quad$ $\underset{4p}{—\ —\ —}$

$[NiCl_4]^{2-}$ $\quad$ $\underset{3d}{\uparrow\downarrow\ \uparrow\downarrow\ \uparrow\downarrow\ \uparrow\ \uparrow}$ $\quad$ $\overset{\underline{\qquad sp^3\qquad}}{\underset{4s}{oo}\quad\underset{4p}{oo\ oo\ oo}}$

$[Ni(CN)_4]^{2-}$ $\quad$ $\underset{3d}{\uparrow\downarrow\ \uparrow\downarrow\ \uparrow\downarrow\ \uparrow\downarrow}\ oo$ $\quad$ $\overset{\underline{\qquad dsp^2\qquad}}{\underset{4s}{oo}\quad\underset{4p}{oo\ oo}\ —}$

The tetrachloro complex is tetrahedral, corresponding to the sp^3 hybridization shown, and it has a paramagnetism corresponding to two unpaired electrons, as does Ni^{2+}. The $[Ni(CN)_4]^{2-}$ complex is square planar, corresponding to the hybridization dsp^2, and is diamagnetic. In the latter case it is assumed that since dsp^2 hybrid orbitals provide for greater overlap with ligand orbitals, the formation of very strong bonds provides sufficient energy to make possible the pairing of the electrons.

Example

The magnetic moment of $[Mn(CN)_6]^{3-}$ is 2.8 B.M. The magnetic moment of $[MnBr_4]^{2-}$ is 5.9 B.M. What are the geometries of these complex ions?

From the spin-only formula (Section 10–9), it may be calculated that $[Mn(CN)_6]^{3-}$ has two unpaired electrons, while $[MnBr_4]^{2-}$ has five. (Note that the manganese is in different

396

oxidation states.) The electronic configurations corresponding to the simple ions and the complexes with the indicated numbers of unpaired electrons are as follows:

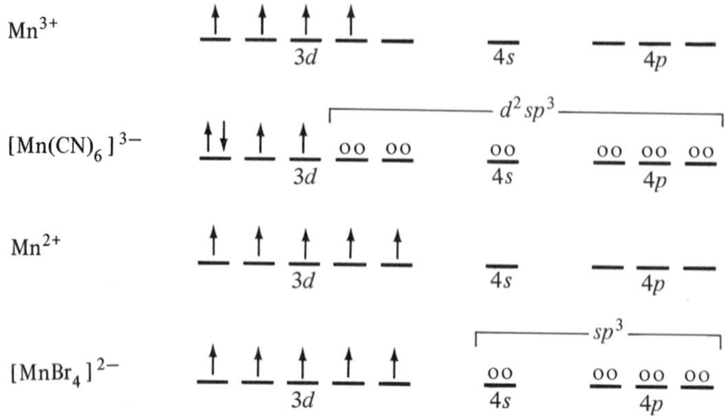

Mn^{3+}

$[Mn(CN)_6]^{3-}$ — $d^2 sp^3$

Mn^{2+}

$[MnBr_4]^{2-}$ — sp^3

The first complex is octahedral and the second is tetrahedral.

The valence bond theory has several shortcomings. For example, the theory offers no explanations of the colors observed for complex ions, nor can quantitative conclusions be drawn concerning such parameters as bond energies or formation constants. By means of the valence bond theory, it is usually possible to predict the geometry of a complex from a knowledge of its magnetic behavior. However, it is not always possible to make the reverse prediction. For example, it might be predicted that octahedral complexes of iron(III) would have a magnetic moment corresponding to a single unpaired electron, and that octahedral complexes of cobalt(III) would be diamagnetic. However, it is found experimentally that the (octahedral) complexes $[FeF_6]^{3-}$ and $[CoF_6]^{3-}$ have five and four unpaired electrons, respectively. To account for their octahedral geometries, the following type of configuration must be used:

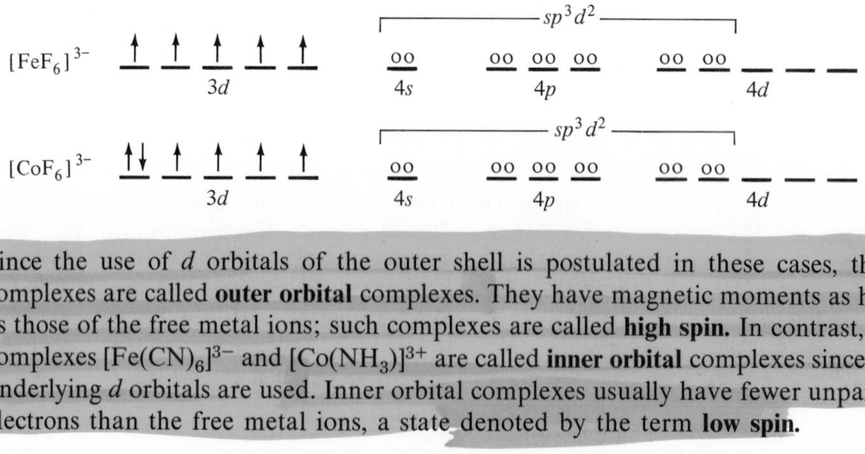

$[FeF_6]^{3-}$ — $sp^3 d^2$

$[CoF_6]^{3-}$ — $sp^3 d^2$

Since the use of d orbitals of the outer shell is postulated in these cases, these complexes are called **outer orbital** complexes. They have magnetic moments as high as those of the free metal ions; such complexes are called **high spin**. In contrast, the complexes $[Fe(CN)_6]^{3-}$ and $[Co(NH_3)]^{3+}$ are called **inner orbital** complexes since the underlying d orbitals are used. Inner orbital complexes usually have fewer unpaired electrons than the free metal ions, a state denoted by the term **low spin**.

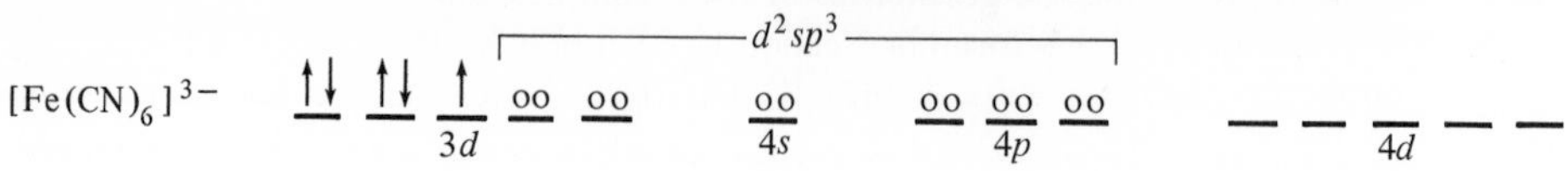

$[Fe(CN)_6]^{3-}$ — $d^2 sp^3$

Example

Show that all octahedral complexes of nickel(II) must be outer orbital complexes. The electronic configuration of Ni^{2+} is as follows:

Ni^{2+} ↑↓ ↑↓ ↑↓ ↑ ↑ — — — — — — — — — — —
             3d   4s   4p   4d

Since only one 3d orbital can be made available by pairing of the electrons, there cannot be inner orbital d^2sp^3 hybridization. The only octahedral hybridization possible is sp^3d^2, using outer d orbitals.

A further complication in the use of the valence bond theory arises when one attempts to write the electronic configuration of octahedral complexes of cobalt(II). For example, the only way a d^2sp^3 configuration which has only one unpaired electron can be written for $[Co(NO_2)_6]^{4-}$ is as follows:

$[Co(NO_2)_6]^{4-}$ ↑↓ ↑↓ ↑↓ oo oo   oo   oo oo oo   ↑ — — — — —
               3d   4s   4p   4d
(with d^2sp^3 bracket spanning 3d–4p)

Similarly, to explain the experimentally determined square planar geometry of copper(II) complexes, it is necessary either to use a 4d orbital for bonding when a lower energy 4p orbital is available or to promote one electron from the 3d subshell to make room for a pair of bonding electrons from the ligands. The resulting configurations would be written as follows:

$[Cu(acac)_2]$ ↑↓ ↑↓ ↑↓ ↑↓ ↑   oo   oo oo —   oo — — — —
               3d   4s   4p   4d
(with sp^2d bracket spanning 4s–4d)

$[Cu(acac)_2]$ ↑↓ ↑↓ ↑↓ ↑↓ oo   oo   oo oo ↑   — — — — —
               3d   4s   4p   4d
(with dsp^2 bracket spanning 3d–4p)

$[Cu(acac)_2]$ ↑↓ ↑↓ ↑↓ ↑↓ oo   oo   oo oo —   ↑ — — — —
               3d   4s   4p   4d
(with dsp^2 bracket spanning 3d–4p)

None of these possibilities is entirely plausible. The basic defect in the valence bond model is that attention is focused only on the orbitals of the central atom. Although it is recognized that the ligands are bonded to the central atom by covalent bonds, no further considerations are made of their influence on the properties of the complex.

14-7 Crystal Field Theory

In the crystal field theory, the interaction between the metal ion and the ligands is treated as a purely electrostatic problem in which the donor atoms act as point charges. From the outset it should be stated that the crystal field approach does not correspond to the actual bonding situation in coordination compounds. It has been shown experimentally by a variety of techniques that the bonding between ligands and the central metal ion is at least partly covalent. However, in the crystal field theory, covalent bonding is ignored. What is considered is the effect of the negative charges or dipoles of the ligands on the energies of the electrons in the d orbitals of the central metal ion. The interaction between the metal ion and the ligands is treated as analogous to the electrostatic theory of crystals, developed by physicists during the 1930s—hence the name **crystal field theory.** In this regard, the crystal field theory is merely a formalism; however, it provides a very simple way to treat numerically many aspects of the electronic structures of coordination compounds.

To apply the crystal field method, the spatial orientations of the d orbitals must be considered. These are shown again in Figure 14–6. In a given free metal ion, all five of these orbitals are degenerate; that is, they all have the same energy. Surrounded by

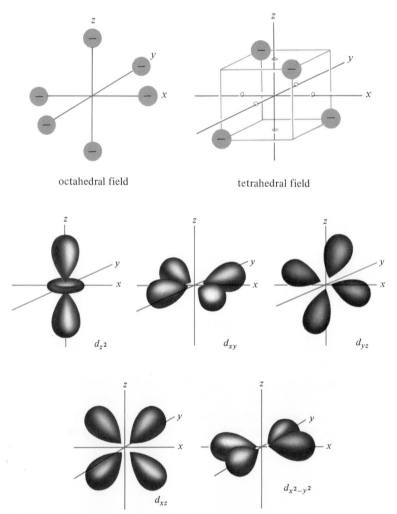

octahedral field   tetrahedral field

d_{z^2} d_{xy} d_{yz}

d_{xz} $d_{x^2-y^2}$

Figure 14–6. **Orientations of *d* Orbitals Relative to Octahedral and Tetrahedral Fields**

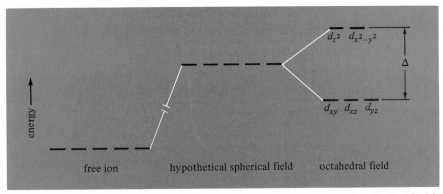

Figure 14-7. **Relative Energies of the d Orbitals in Spherical and Octahedral Fields**

a hypothetical spherical field of negative change, all the orbitals would be raised in energy equally. However, if negative charges are positioned near the central atom in a nonspherical distribution, such as at the corners of an octahedron or of a tetrahedron, the electrons in the d orbitals will be affected differently, depending on the orientation of the applied charges. For example, as shown in Figure 14-6, the electron probability density in the $d_{x^2-y^2}$ and the d_{z^2} orbitals is greatest in the directions which lie along the Cartesian coordinate axes. Therefore when six ligands are brought near the central ion along the axes, as in an octahedral complex, electrons in these orbitals are repelled by the ligand electron pairs more than electrons in the d_{xy}, d_{xz}, and d_{yz} orbitals. Thus in an octahedral field, the d subshell is split into a higher energy set of orbitals and a lower energy set. This splitting is diagrammed in Figure 14-7. The $d_{x^2-y^2}$ and d_{z^2} orbitals are often designated as the e_g set, while the d_{xy}, d_{yz}, and d_{xz} set of orbitals is designated as the t_{2g} set. (These terms are used in the mathematical description of these orbitals and may be regarded here merely as labels.)

The magnitude of the difference in energy between these two sets of orbitals in an octahedral field is usually designated Δ. The magnitude of Δ depends on the metal which is involved, the oxidation state of the metal, the charge or dipole moment of the ligands, and the metal-ligand bond length. In general, Δ is more easily obtained experimentally than theoretically. Values of Δ for an octahedral field for several transition metal ions with various ligands are listed in Table 14-5. The number of d electrons in each metal ion is also shown.

TABLE 14-5. Some Crystal Field Splittings, Δ (cm^{-1}), for Octahedral Complexes

Metal Ion	Number of d Electrons	Ligand			
		H_2O	Cl^-	NH_3	CN^-
Ti^{III}	1	20,400			
Cr^{III}	3	17,400	13,800	21,600	26,100
Mn^{III}	4	20,900			
Mn^{II}	5	15,900			
Co^{III}	6	18,200		22,900	33,800
Ni^{II}	8	8,500	7,200	10,800	

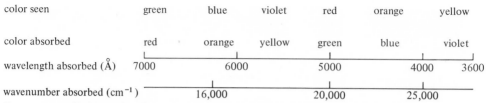

color seen		green	blue	violet	red	orange	yellow
color absorbed		red	orange	yellow	green	blue	violet

wavelength absorbed (Å)   7000   6000   5000   4000   3600

wavenumber absorbed (cm^{-1})   16,000   20,000   25,000

Figure 14-8. **Relationship Between Color and Wavelength of Absorbed Light**

One of the major successes of the crystal field theory is the explanation it provides for the observed colors of complex ions. The color of an object is due to the absorption of light of specific wavelengths from white light. As shown in Figure 14-8, the transmitted or reflected light will be the resultant of the colors not absorbed. In the simplest case, when light is absorbed by a complex ion, an electron in one of the lower energy (t_{2g}) orbitals is excited to one of the higher energy (e_g) orbitals. The energy corresponding to the frequency of the absorbed light is equal to Δ. For example, the $[Ti(H_2O)_6]^{3+}$ ion absorbs light of a wavelength of about 5000 Å (Figure 14-9). This absorption is due to the excitation of the single d electron in Ti^{III} from the t_{2g} to the e_g set of orbitals, as shown in Figure 14-10. In terms of wavenumbers,[1] this transition corresponds to a value of Δ of 20,400 cm^{-1}. In the case of $[Ti(H_2O)_6]^{3+}$, absorption in the 5000 to 6000 Å region corresponds to the removal from white light of the green and yellow components, and the resulting color is purple. However, in predicting the color in cases of ions having more than one electron in the d subshell, the chart must be used with caution because interelectronic interactions make possible other electronic transitions,[2] and the observed color might result from two or more absorption processes.

---

[1] These energies are customarily reported in terms of wavenumbers, which are directly proportional to energy: $\bar{\nu} = 1/\lambda = E/hc$. For example, 350 wavenumbers = 350 cm^{-1} = 1.0 kcal/mole.

[2] Other possible electronic transitions can be predicted, but by means which are beyond the scope of this text.

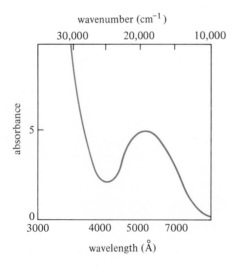

Figure 14-9. **Absorption Spectrum of $[Ti(H_2O)_6]^{3+}$**

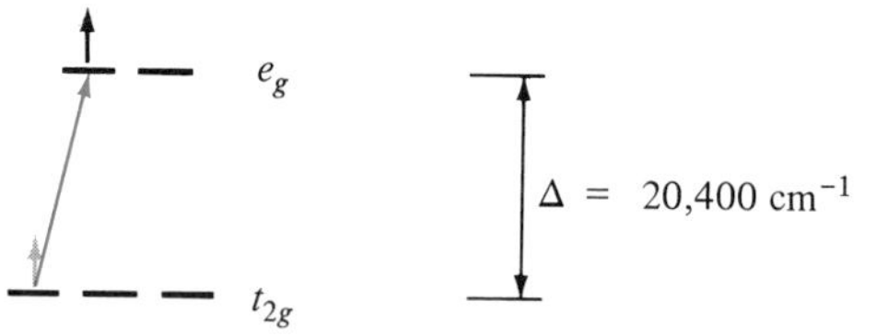

Figure 14-10. **The d Electron Transition in Ti^{III}**

$\Delta = 20,400$ cm^{-1}

Example

Predict the colors of [Co(NH$_3$)$_6$]$^{3+}$ and [Cr(H$_2$O)$_6$]$^{3+}$, using the data of Table 14-5. The energy difference between the d levels, Δ, in [Co(NH$_3$)$_6$]$^{3+}$ is 22,900 cm^{-1}:

$$\lambda = \frac{1}{\nu} = \frac{1}{22,900 \text{ cm}^{-1}} = 4.37 \times 10^{-5} \text{ cm} = 4370 \text{ Å}$$

An absorption band is expected at about 4400 Å, which corresponds to the absorption of blue light. The observed color of the complex is orange, as expected. In [Cr(H$_2$O)$_6$]$^{3+}$ the value of Δ is 17,400 cm^{-1}, corresponding to the absorption of light at about 5750 Å. The expected color is violet; however, the actual color is green. In this case an additional absorption occurs at 4050 Å; hence the color cannot be predicted from a consideration of Δ alone.

14-8   The Spectrochemical Series

When the absorption spectra of various complexes of a given metal ion in a given oxidation state are measured, different values of crystal field splitting are obtained. For example, in [Ni(NH$_3$)$_6$]$^{2+}$ the value of Δ is 10,600 cm^{-1}, while in [NiCl$_6$]$^{4-}$, the value of Δ is 7800 cm^{-1}. It is possible to arrange ligands according to the magnitudes of the values of Δ observed with a given metal ion. The sequence for common ligands with most metal ions is as follows:

$$I^- < Br^- < Cl^- < F^- < OH^- < C_2O_4^{2-} < H_2O < NH_3 < en < NO_2^- < CN^- < CO$$

This sequence is known as the **spectrochemical series.** This order is not invariant; for different metal ions, the positions of some adjacent members of the series may be interchanged.

14-9   Magnetic Effects of Crystal Field Splitting

The crystal field theory also offers a more rational explanation of the magnetic properties of transition metal complexes. The *average* energy of the five orbitals in a field of negative charge of any symmetry is the same as that in a spherical field. In an octahedral field, for example, the net increase in the stability of the three orbitals of lower energy must be exactly balanced by the decrease in stability of the two destabilized orbitals. The two e_g orbitals will be destabilized by an energy 0.6 Δ each, while the t_{2g} orbitals will each be stabilized by an energy 0.4Δ. Hence, each electron which occupies the t_{2g} set decreases in energy by 0.4Δ compared to its energy in the spherical field, while each electron occupying the e_g orbitals increases in energy by 0.6 Δ. The total gain in stability of the d orbital electrons over that of the spherical field case is known as the **crystal field stabilization energy** (CFSE).

Example

Confirm that the lowering of the stabilized orbitals is 0.4Δ while the raising of the destabilized orbitals is 0.6Δ.

The total energy separation is Δ.

Let x = energy of stabilization

and y = energy of destabilization

Then $y - x = \Delta$

Since the orbitals yield no net increase in energy when they are equally occupied,

$$3x + 2y = 0$$
$$3x + 2(\Delta + x) = 0$$
$$5x = -2\Delta$$
$$x = -0.4\Delta$$

The configuration of the electrons in the d orbitals of a central ion in a crystal field will be that which yields the largest CFSE. For d^1, d^2, and d^3 ions in an octahedral field, the electrons will occupy the t_{2g} orbitals separately. The crystal field stabilization energies will be -0.4Δ, -0.8Δ, and -1.2Δ, respectively. For a d^4 ion in an octahedral field there are two possibilities: a high spin state and a low spin state. In the low spin state, all four electrons occupy the t_{2g} orbitals, and one of these orbitals must contain an electron pair. Since in pairing electrons, energy is required to overcome electron–electron repulsion, the crystal field stabilization energy is $-1.6\Delta + P$, where P is the energy required to form each electron pair. In the high spin case, the electrons occupy singly the three t_{2g} orbitals and one of the e_g orbitals. The crystal field stabilization energy is -0.6Δ. These two cases are diagrammed in Figure 14–11. Whether a complex will be high spin or low spin depends on the relative magnitudes of Δ and P. If Δ is larger than P, the low spin complex is favored and vice versa. Crystal fields which result in low spin complexes are called **strong fields.** By contrast, a high spin complex is favored by a **weak field.** It follows that ligands high in the spectrochemical series, such as NO_2^- and CN^-, generally form strong field complexes, because these ligands usually generate a large crystal field splitting.

Example

For the $[Cr(H_2O)_6]^{2+}$ ion, the mean pairing energy, P, is found to be 23,500 cm^{-1}. The magnitude of Δ is 13,900 cm^{-1}. Calculate the crystal field stabilization energy for the complex in configurations corresponding to high spin and low spin states. Which is more stable?

For a d^4 ion in a high spin state,

$$\text{CFSE} = -0.6\Delta = -0.6(13{,}900 \text{ cm}^{-1}) = -8340 \text{ cm}^{-1}$$

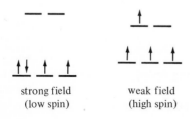

strong field
(low spin)

weak field
(high spin)

Figure 14-11. Configurations of a d^4 Ion in Strong and Weak Octahedral Fields

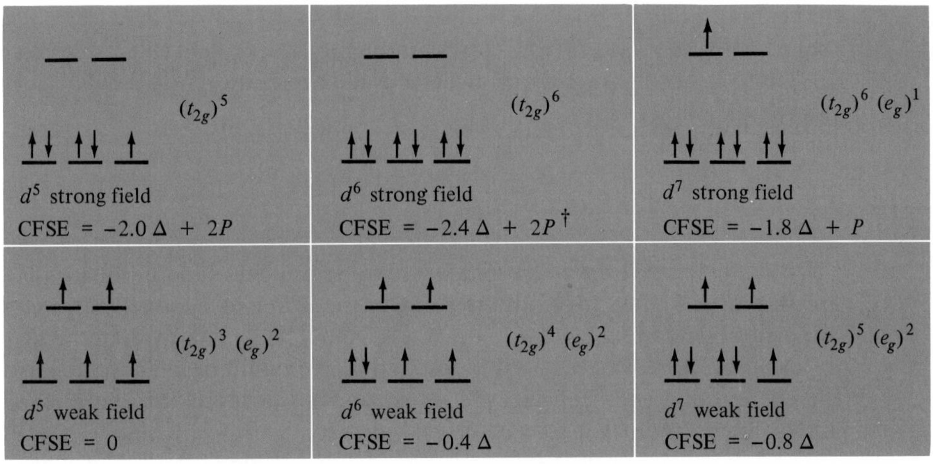

$(t_{2g})^5$

d^5 strong field

CFSE $= -2.0\,\Delta + 2P$

$(t_{2g})^6$

d^6 strong field

CFSE $= -2.4\,\Delta + 2P$ †

$(t_{2g})^6\,(e_g)^1$

d^7 strong field

CFSE $= -1.8\,\Delta + P$

$(t_{2g})^3\,(e_g)^2$

d^5 weak field

CFSE $= 0$

$(t_{2g})^4\,(e_g)^2$

d^6 weak field

CFSE $= -0.4\,\Delta$

$(t_{2g})^5\,(e_g)^2$

d^7 weak field

CFSE $= -0.8\,\Delta$

† An isolated d^6 ion already contains one pair of electrons. In forming the strong field complex of a d^6 ion, two additional electron pairs are formed; hence the CFSE is $-2.4\,\Delta + 2P$

Figure 14-12. **Electronic Distributions in d^5, d^6, and d^7 Ions in Strong and Weak Octahedral Fields**

For a d^4 ion in a low spin state, the net crystal field stabilization energy is

$$\text{CFSE} = -1.6\Delta + P = -1.6(13{,}900\ \text{cm}^{-1}) + 23{,}500\ \text{cm}^{-1} = +1260\ \text{cm}^{-1}$$

As is generally the case, the lower energy state will be the more stable. The ligand H_2O does not produce a sufficiently strong crystal field to yield a low spin cobalt(II) complex. Since $\Delta < P$, it should have been apparent from the outset that the high spin configuration would be more stable.

The weak and strong field electronic configurations of ions having five, six, and seven d electrons are shown in Figure 14-12, and the crystal field stabilization energies for all d^n configurations in an octahedral field are listed in Table 14-6. As is the case with d^1, d^2, and d^3 ions, d^8, d^9, and d^{10} ions have only one possible spin state, regardless of the magnitude of the crystal field. Thus magnetic data can be

TABLE 14-6. Crystal Field Stabilization Energies in an Octahedral Field

Number of d Electrons	Weak Field	Strong Field
1	-0.4Δ	-0.4Δ
2	-0.8Δ	-0.8Δ
3	-1.2Δ	-1.2Δ
4	-0.6Δ	$-1.6\Delta + P$
5	0.0Δ	$-2.0\Delta + 2P$
6	-0.4Δ	$-2.4\Delta + 2P$
7	-0.8Δ	$-1.8\Delta + P$
8	-1.2Δ	-1.2Δ
9	-0.6Δ	-0.6Δ
10	0.0Δ	0.0Δ

rationalized in terms of CFSE. For example, in $[Fe(H_2O)_6]^{3+}$ the electronic configuration of the Fe^{III} is $(t_{2g})^3(e_g)^2$, corresponding to a weak field (high spin) complex. In $[Fe(CN)_6]^{3-}$, in which Δ is large, the electronic configuration of the Fe^{III} is $(t_{2g})^5$, and the complex is low spin.

14-10   Four-Coordinate Complexes

Crystal field splittings of d orbitals in tetrahedral fields and in square planar fields are shown in Figure 14–13. In contrast to the order in the octahedral case, in the tetrahedral field the d_{z^2} and $d_{x^2-y^2}$ orbitals have a lower energy than the d_{xy}, d_{xz}, and d_{yz} orbitals. Moreover, for a given ion and a given kind of ligand, the magnitude of the energy difference between the upper and lower levels in a tetrahedral field is much smaller than that for an octahedral field.

The square planar field (with the ligands placed along the x and y axes) can be regarded as having been derived from the octahedral field by withdrawing the pairs of electrons located along the z axis of the complex. As a result, the two e_g orbitals no

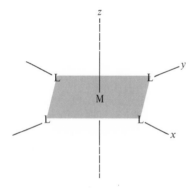

longer have the same energy. Electrons in the d_{z^2} orbital will become stabilized relative to the electrons in the $d_{x^2-y^2}$ orbital. (Since the average energy of all the

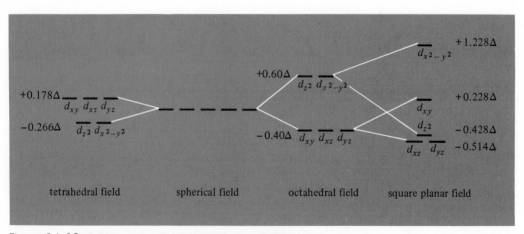

Figure 14–13. **Splitting of d Orbitals in Various Fields**

orbitals must remain the same, the $d_{x^2-y^2}$ orbital is destabilized.) Electrons in the d_{xy} orbital will be repelled by the square planar field of the ligands more than the d_{xz} and d_{yz} electrons; hence the t_{2g} set is also split in energy. In Figure 14–13, these trends are denoted by white lines.

The observed geometries of some complexes can be rationalized in terms of their crystal field stabilization energies. In the case of a d^8 ion, a large CFSE is obtained in a square planar field owing to the large crystal field splitting between the two highest orbitals. Since no other geometry will produce such a large CFSE for d^8 ions, complexes having square planar configurations are favored. In contrast, transition metal ions which have equally occupied d orbitals tend to form tetrahedral complexes; for example, $[Zn(NH_3)_4]^{2+}$ (d^{10}), $[MnCl_4]^{2-}$ (d^5), and $[FeCl_4]^-$ (d^5) are tetrahedral. A relatively small crystal field splitting is produced by a tetrahedral field and the electronic configuration of the metal ion is not altered by formation of the complex. Since in these cases the orbitals are equally occupied, the CFSE will be zero anyway.

14-11 Thermodynamic Properties and the Crystal Field Theory

One of the most satisfying applications of crystal field theory is in the explanation of the trends of certain thermodynamic properties of coordination compounds. For example, in Figure 14–14, the enthalpies of hydration of some dipositive transition metal ions are plotted against atomic number. The experimental points are the measured enthalpy changes for the following type of process:

$$M^{2+}(g) \quad + \quad n\,H_2O(l) \quad \rightarrow \quad [M(H_2O)_6]^{2+}(aq)$$

It is expected that for cations of a given charge, the enthalpy of hydration will increase regularly as the size of the cation decreases. For the series of consecutive cations from Ca^{2+} to Zn^{2+} a plot of enthalpies of hydration against atomic number would be expected to give a smooth curve. However, the experimental points give a curve containing two humps. The humps represent the added stability of the product ion due to its crystal field stabilization energy. Because the product, $[M(H_2O)_6]^{2+}$, is more stable than it would be without the crystal field stabilization energy, the reaction is more exothermic. If for each of the ions the crystal field stabilization energy is

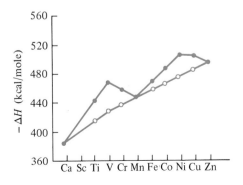

Figure 14–14. **Enthalpies of Hydration of Dipositive Transition Metal Ions**

The experimental values (•), when adjusted by subtracting the crystal field stabilization energies, give points (∘) which fall on the smoother curve. Thus crystal field stabilization energy quantitatively accounts for the two-humped curve.

subtracted from the enthalpy of reaction and the results are plotted against atomic number, the smooth curve of Figure 14–14 is obtained. Those ions having no crystal field stabilization energy, Ca^{2+}, Mn^{2+}, and Zn^{2+}, have experimental values which fail on the smooth curve. Thus, values of Δ can be estimated from thermochemical data as well as from electronic spectra.

Example

The enthalpy of hydration of Cr^{2+} is -460 kcal/mole. In the absence of crystal field stabilization energy, as estimated from Figure 14–14 the value for ΔH would be -435 kcal/mole. Estimate the value of Δ for $Cr(H_2O)_6^{2+}$.

The enthalpy of hydration of Cr^{2+} is 0.6Δ higher for this d^4 weak field ion than it would be in the absence of CFSE. Thus

$$-0.6\Delta = -460 \text{ kcal/mole} - (-435 \text{ kcal/mole}) = -25 \text{ kcal/mole}$$

$$\Delta = \left(\frac{-25 \text{ kcal/mole}}{-0.6}\right)\left(\frac{350 \text{ cm}^{-1}}{1 \text{ kcal/mole}}\right) = 14{,}600 \text{ cm}^{-1}$$

The actual value of Δ is $13{,}900$ cm^{-1}.

The chief limitations of the crystal field theory stem from its lack of correspondence with real molecules. First, since only electrostatic interactions of the ligands with the central atoms are considered, all properties resulting from covalent bonding of the ligands cannot be treated adequately. Second, although the spectrochemical series is defined in terms of crystal field theory, the theory does not explain the observed sequence of ligands in the series. Also, the important phenomenon of π bonding between ligands and central atoms cannot be explained by the crystal field theory. Another theoretical approach, the ligand field theory, more effectively treats these phenomena.

14-12   Ligand Field Theory

Ligand field theory uses the results of the crystal field theory to approximate the order of the molecular orbital energies in compounds of metals and ligands. Of all the theoretical approaches to bonding, the molecular orbital method is the most sophisticated and the most general. As explained in Section 11–15, molecular orbitals may be of a bonding, an antibonding, or a nonbonding nature. In the case of coordination compounds, the construction of molecular orbitals is complicated, and a number of approximations must be made. The total number and relative energies of molecular orbitals is determined from the number of metal ion orbitals and ligand orbitals of suitable energy and orientation. A reasonable representation of the energy levels of the molecular orbitals for an octahedral complex is shown in Figure 14–15.

In Figure 14–15, the six molecular orbitals of lowest energy are σ-bonding orbitals. They are constructed from suitable combinations of ligand orbitals and the ns, np, and two $(n - 1)d$ orbitals of the metal. The same sets of orbitals also combine to form σ^* antibonding orbitals. It is important to note that the d_{z^2} and $d_{x^2-y^2}$ orbitals specifically are used in forming the highest energy bonding orbitals and the lowest energy antibonding orbitals. The d_{xy}, d_{xz}, and d_{yz} orbitals do not combine with ligand orbitals to form σ molecular orbitals; instead they form three degenerate nonbonding

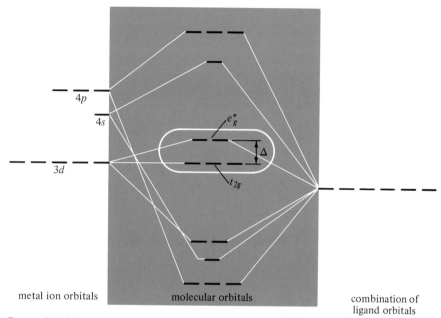

metal ion orbitals   molecular orbitals   combination of
                                                    ligand orbitals

Figure 14-15. **Molecular Orbital Diagram for an Octahedral Complex**
The molecular orbitals constructed from the metal ion orbitals and a combination of six ligand
orbitals show some of the characteristics of orbital splitting described by crystal field theory (see
encircled sets).

molecular orbitals composed of the t_{2g} orbitals of the metal ion only. The difference in
energy between this set of orbitals and the lowest antibonding set, which are derived
in part from the e_g set of metal orbitals, corresponds to the Δ of crystal field theory;
that is, the relative energies of the orbitals (encircled in Figure 14–15) above the six
lowest energy orbitals are analogous to the relative energies of the t_{2g} and e_g sets of
orbitals in an octahedral field as described using crystal field theory. The lowest six
orbitals are occupied by 12 electrons shared in pairs with the six ligands. Electrons
from the d subshell of the free metal ion then occupy the encircled orbitals in a
manner determined by the energy difference (Δ) and by the energy required for
electron pairing (P). Thus the concepts of the crystal field theory are all applicable in
the molecular orbital method. But, as will be demonstrated below, the molecular
orbital method is capable of explaining much more.

The origin of the spectrochemical series can be explained using the molecular
orbital theory. The lower the energy of a given bonding molecular orbital relative to
the separate orbitals which make it up, the higher will be the energy of the corresponding antibonding orbital. Thus if the bonding molecular orbital constructed from
the e_g set of the metal atom and suitable ligand orbitals has a very high stability
relative to the nonbonding t_{2g} set, the corresponding antibonding orbitals will be
highly destabilized. Consequently, the magnitude of Δ will be large. Conversely, if the
bonding orbitals are but slightly stabilized, the antibonding orbitals will not be
destabilized much, and the value of Δ will be small. These two cases are, of course,
analogous to the strong field and weak field effects described in terms of the crystal
field theory. Therefore ligands which form strong covalent bonds with the central
metal ion will have very stable bonding orbitals and highly destabilized antibonding

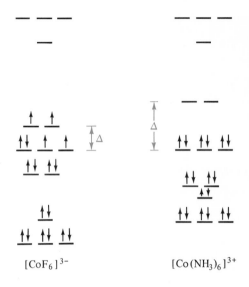

$[\mathrm{CoF_6}]^{3-}$

$[\mathrm{Co(NH_3)_6}]^{3+}$

Figure 14–16. Molecular Orbital Diagrams for Cobalt(III) Complexes

The eighteen electrons (twelve from the σ-bonding ligands and six from the metal ion) are situated in orbitals in which they achieve the lowest energy state. If the difference between the nonbonding t_{2g} and the antibonding e_g^* is small, as in $[\mathrm{CoF_6}]^{3-}$, the electrons occupy these levels according to Hund's rule, and the complex is high spin. If the energy difference is large, as in $[\mathrm{Co(NH_3)_6}]^{3+}$, the six electrons of highest energy pair up in the t_{2g} orbitals to give a low spin complex.

orbitals, and complexes of such ligands will have large values of Δ. Hence the order of the spectrochemical series reflects the relative strengths of the covalent bonds between the metal and the ligands.

The magnetic properties of complexes can also be explained in a manner which is analogous to the crystal field theory explanation (Figure 14–16).

14–13 π Bonding in Octahedral Complexes

In octahedral complexes, the d_{xy}, d_{xz}, and d_{yz} orbitals of the metal ion may be used for π bonding. In such cases for a given metal ion, the magnitude of Δ will be either larger or smaller than the value which would be expected if the metal formed no π bonds. Consider a complex containing six chloride ion ligands. As shown in Figure 14–17, each of the t_{2g}-type d orbitals overlaps with four ligand p orbitals to form π orbitals. In the resulting bonding molecular orbitals, some electronic charge is transferred toward the metal. This transfer of electronic charge is referred to as ligand to metal (L—M) π bonding. The t_{2g} orbitals are destabilized in the process and are raised in energy, while the e_g orbitals are not affected. As a consequence, the magnitude of Δ is less than in a purely σ-bonded complex (Figure 14–18a).

If, on the other hand, the ligand itself is one which has empty π^* molecular orbitals, such as is the case for CN^- or CO (Section 11–17), two types of π bonding are

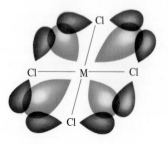

Figure 14–17. Overlap of a Metal d_{xy} Orbital with the Chloride Ion p Orbitals to Form Ligand-Metal π Bonds

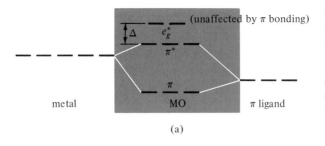

Figure 14–18. **Influence of π Bonding on the Magnitude of Δ**

(a) With ligand orbitals of lower energy than that of the d orbitals of the metal ion.

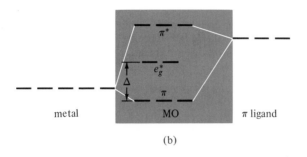

(b) With ligand orbitals of higher energy than that of the d orbitals of the metal ion.

possible. There can be L—M π bonding between the electrons in the π-bonding orbitals of the ligand and the t_{2g} orbitals of the metal ion. Also, electrons in the t_{2g} orbitals of the metal can be delocalized into the empty π^*-antibonding orbitals of the ligand. The latter type of bonding removes electron density from the metal and is called metal to ligand (M—L) π bonding. It is also known as **back donation** or **back bonding.** Back bonding lowers the energy of the t_{2g} orbitals relative to the non-π-bonded case, and as a result, the magnitude of Δ is greater than in a purely σ-bonded complex (Figure 14–18b).

The molecular orbital method provides a better explanation of the sequence of the spectrochemical series. Ligands which donate electrons to metal orbitals (strong L—M π bonding) are low in the series; ligands which can accept metal electrons into their π^*-antibonding orbitals (strong M—L π bonding) are high in the series. Ligands which have little or no π-bonding capacity are intermediate in the series. This trend is shown in Figure 14–19.

Among the π-bonded complexes of the L—M type are the cyclopentadienyl compounds derived from the ligand cyclopentadiene (Table 14–2). The best known of these is bis(cyclopentadienyl)iron(II), known as ferrocene. Ferrocene may be prepared by the reaction

$$2 \, C_5H_5MgBr \; + \; FeCl_2 \; \rightarrow \; Fe(C_5H_5)_2 \; + \; MgCl_2 \; + \; MgBr_2$$

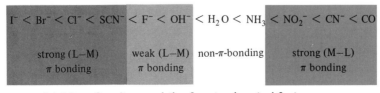

Figure 14–19. **π Bonding and the Spectrochemical Series**

Figure 14-20. **Ferrocene**

The product is an orange solid, melting at 174°C. It is soluble in alcohol and benzene but insoluble in water or sodium hydroxide solution. It is easily oxidized to the cation $[Fe(C_5H_5)_2]^+$, which has a blue color and a paramagnetism corresponding to one unpaired electron:

$$Ag^+ \;+\; Fe(C_5H_5)_2 \;\rightarrow\; Ag \;+\; [Fe(C_5H_5)_2]^+$$

X-ray diffraction studies indicate that the structure of ferrocene is the staggered "sandwich" configuration shown in Figure 14-20.

In addition to ferrocene, a number of cyclopentadienyl compounds of other transition metals have been prepared as well as some mixed cyclopentadienyl–carbonyl compounds. Some examples are listed in Table 14-7.

There are several important practical applications of π-bonded complexes. The refining of nickel metal by the formation of nickel tetracarbonyl has already been described. Another application is the purification of hydrocarbons. Mixtures of unsaturated and saturated hydrocarbons are bubbled through solutions containing copper(I), silver(I), mercury(II), or platinum(II) ions. The hydrocarbons having multiple bonds form π complexes, while the others are recovered unchanged.

Another application is the production of polyethylene from ethylene, an important commercial process:

$$n\,C_2H_4 \xrightarrow{\text{catalyst}} \;\; (\!\!-CH_2\!-\!CH_2\!-\!)_n$$

The catalyst used is obtained from the reaction of titanium(IV) chloride and tri-methylaluminum. Although the detailed mechanism of the catalytic action is not clear, it seems likely that π-bonded complexes of aluminum or titanium are formed as intermediates in the polymerization reaction.

TABLE 14-7. Some Sandwich Compounds

Compound[a]	Color	Unpaired Electrons
$(\pi\text{-}C_5H_5)_2Fe$	orange	0
$(\pi\text{-}C_5H_5)_2Cr$	scarlet	2
$(\pi\text{-}C_5H_5)_2Ni$	green	2
$(\pi\text{-}C_5H_5)_2Mn(CO)_3$		0

[a] $\pi\text{-}C_5H_5$ indicates that the organic ring is bonded to the metal by π bonding and not by localized bonding through any one carbon atom.

14-14   Effective Atomic Number Rule

The stoichiometric compositions of most, but not all, π-bonded complexes can be predicted by the effective atomic number rule. This rule states that the sum of the number of electrons originally possessed by the metal atom plus the number of electrons shared with it by the ligands (and by any other atoms bonded to it) equals the atomic number of the succeeding noble gas. This total number is called the **effective atomic number** of the complex. For example, in $[Fe(CN)_6]^{4-}$, the sum of the 24 electrons in the iron(II) ion plus the 12 electrons from the six ligands totals 36, the atomic number of krypton.

Example

What are the effective atomic numbers of the metal atoms in $Fe(CO)_5$ and in $Co_2(CO)_8$?
The effective atomic number of iron includes 26 electrons from iron(0) plus 10 electrons from five CO molecules, totaling 36, the atomic number of krypton. Each cobalt atom in $Co_2(CO)_8$ has 27 electrons of its own and shares one with the other cobalt atom in addition to the eight electrons from four CO molecules, again for a total of 36 electrons per Co atom.

Example

Deduce the value of x in the formula $Mo(CO)_x$.
Since the effective atomic number of molybdenum must be 54 and its atomic number is 42, 12 electrons ($54 - 42$) must be donated by the CO molecules. Hence, six CO molecules must each donate a pair of electrons, resulting in the formula $Mo(CO)_6$.

For the purpose of the effective atomic number rule, each molecule of NO is considered to donate three electrons to the metal atom on coordination. This is not an unreasonable procedure, since NO has one more electron than does CO, which donates two electrons. Moreover, complexes of nitrosyls are generally diamagnetic, and since the ligand has an odd number of electrons, an odd number must be shared. For example, $Co(NO)(CO)_3$ is diamagnetic despite a central metal atom and a ligand each of which has an odd number of unpaired electrons when uncombined. Donation of three electrons by NO to the metal allows all the electrons in the compound to become paired.

Similarly, the cyclopentadienide ion is considered for the purpose of this rule to donate five electrons to the central metal atom.

Example

Determine the effective atomic number of iron in $Fe(NO)_2(CO)_2$ and in $Fe(C_5H_5)_2$.
The effective atomic number of iron in $Fe(NO)_2(CO)_2$ includes 26 electrons from Fe, 6 electrons from the two NO molecules, and 4 electrons from the two CO molecules, for a total of 36—the atomic number of krypton. In $Fe(C_5H_5)_2$, the 26 electrons from the Fe are added to the 10 electrons from the two $C_5H_5^-$ ions, again for a total of 36.

Structures of several metal carbonyls are shown in Figure 14–21.

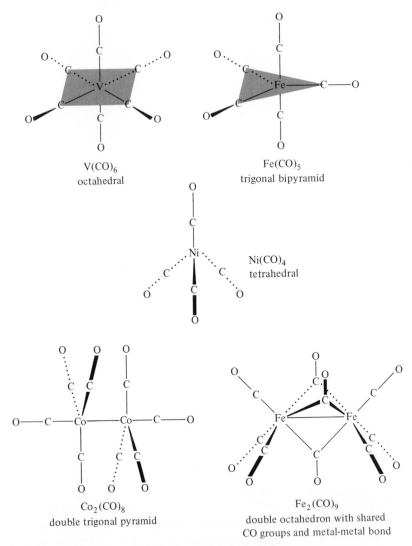

V(CO)$_6$
octahedral

Fe(CO)$_5$
trigonal bipyramid

Ni(CO)$_4$
tetrahedral

Co$_2$(CO)$_8$
double trigonal pyramid

Fe$_2$(CO)$_9$
double octahedron with shared
CO groups and metal-metal bond

Figure 14–21. **Structures of Some Metal Carbonyls**

14-15 Exercises

Basic Exercises

1. How many faces has an octahedron? How many corners? What is the coordination number of a central ion which has an octahedral coordination sphere?
2. Arrange the following compounds in order of increasing molar conductivity: **(a)** K[Co(NH$_3$)$_2$(NO$_2$)$_4$], **(b)** [Cr(NH$_3$)$_3$(NO$_2$)$_3$], **(c)** [Cr(NH$_3$)$_5$(NO$_2$)]$_3$[Co(NO$_2$)$_6$]$_2$, **(d)** Mg[Cr(NH$_3$)(NO$_2$)$_5$].
3. From your own knowledge and the information in this chapter, give the characteristic coordination number of each of the following central metal ions: **(a)** Cu^I, **(b)** Cu^{II}, **(c)** Co^{III}, **(d)** Al^{III}, **(e)** Zn^{II}, **(f)** Fe^{II}, **(g)** Fe^{III}, **(h)** Ag^I.

4. Indicate the oxidation state of the central metal ion in each of the following complex ions: **(a)** $[Cu(NH_3)_4]^{2+}$, **(b)** $[CuBr_4]^{2-}$, **(c)** $[Cu(CN)_2]^-$, **(d)** $[Cr(NH_3)_4(CO_3)]^+$, **(e)** $[PtCl_4]^{2-}$, **(f)** $[Co(NH_3)_2(NO_2)_4]^-$, **(g)** $Fe(CO)_5$, **(h)** $[ZnCl_4]^{2-}$, **(i)** $[Fe(en)_3]^{2+}$.

5. Complete the designation of the following coordination spheres by adding the charge: **(a)** $[Fe^{III}(CN)_6]$, **(b)** $[Pt^{IV}(NH_3)_3(H_2O)Cl_2]$, **(c)** $[Cr^{III}(NH_3)_2(H_2O)_2Cl_2]$, **(d)** $[Pd^{II}(en)Cl_2]$, **(e)** $[Al(H_2O)_2(OH)_4]$.

6. Name each of the following compounds or ions: **(a)** $[Pt(NH_3)_4Cl_2]^{2+}$, **(b)** $Cr(CO)_6$, **(c)** $[Co(en)Cl_3(H_2O)]$, **(d)** $[Co(NH_3)_5CO_3]_2[CuCl_4]$, **(e)** $Fe[PtCl_4]$.

7. Write formulas for the following compounds or ions: **(a)** dichlorotetraaquochromium(III) chloride, **(b)** bromochlorotetraamminecobalt(III) sulfate, **(c)** diamminesilver(I) hexacyanoferrate(II), **(d)** dichlorobis(ethylenediamine)chromium(III) tetrachloropalladate(II), **(e)** *cis*-dichlorotetraammineplatinum(IV) tetrachloroplatinate(II), **(f)** aluminum tetrachloroaurate(III), **(g)** bis(ethylenediamine)copper(II) ion.

8. Write formulas for the nine polymerization isomers of $Co(NH_3)_3(NO_2)_3$, including those listed on page 391.

9. Write structural formulas for all possible isomers corresponding to each of the following, and state the type(s) of isomerism: **(a)** $[Cr(NH_3)_4Br_2]^+$, **(b)** $[Cr(NH_3)_4Br_2]NO_2$, **(c)** $[Cr(en)_2Cl_2]^+$, **(d)** $[Cr(en)_3]^{3+}$, **(e)** $[Cr(NH_3)_4ClBr]Br$.

10. Ethylenediaminetetraacetic acid, EDTA, in the form of its calcium dihydrogen salt, is administered as an antidote for lead poisoning. Explain why this reagent might be an effective medicine. Why is the calcium salt administered rather than the free acid?

11. When $[Ni(NH_3)_4]^{2+}$ is treated with concentrated HCl, two compounds having the formula $Ni(NH_3)_2Cl_2$ (designated I and II) are formed. I can be converted into II by boiling in dilute HCl. A solution of I reacts with oxalic acid to form $Ni(NH_3)_2(C_2O_4)$. II does not react with oxalic acid. Deduce the configurations of I and II and the geometry of the nickel(II) complexes.

12. Select from the following the pairs of **(a)** geometric isomers, **(b)** optical isomers, **(c)** identical structures.

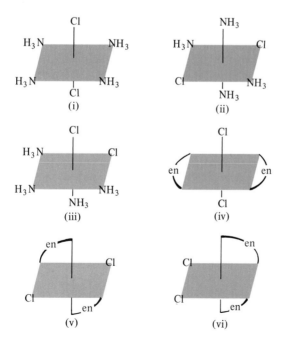

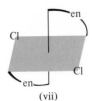

(vii)

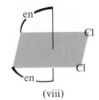

(viii)

13. Give the electronic configuration and determine the number of unpaired electrons in each of the following: **(a)** Fe, **(b)** Fe^{2+}, **(c)** Ni^{2+}, **(d)** Cu^{2+}, **(e)** Pt^{2+}, **(f)** Pt^{4+}.

14. Determine the number of unpaired electrons in $[Cr(NH_3)_6]^{3+}$. Why is it not necessary to designate the complex as inner orbital or outer orbital?

15. In $[MnBr_4]^{2-}$, electron pairs in sp^3 hybrid orbitals of the manganese atom form bonds to the bromine atoms. Determine the number of unpaired electrons in the complex.

16. Write valence bond electronic configurations for each of the following: **(a)** $[PtCl_6]^{2-}$, **(b)** $Cr(CO)_6$, **(c)** $[Ir(NH_3)_6]^{3+}$, **(d)** $[Pd(en)_2]^{2+}$.

17. **(a)** Explain why a knowledge of the magnetic susceptibility of a complex is often necessary for a correct assignment of the electronic configuration according to valence bond theory. **(b)** Draw valence bond representations of the electronic structures of (paramagnetic) $[CoF_6]^{4-}$ and (diamagnetic) $[Co(CN)_6]^{3-}$.

18. Using the valence bond method, **(a)** assign electronic configurations to the central metal atoms of the following complex ions, **(b)** predict their geometries, and **(c)** predict their magnetic moments. (i) $[Ag(CN)_2]^-$, (ii) $[Cu(CN)_4]^{2-}$, (iii) $[Fe(CN)_6]^{3-}$, (iv) $[Zn(CN)_4]^{2-}$.

19. The magnetic moment of $[CoI_4]^{2-}$ is above 3.5 B.M. Using the valence bond approach, diagram the electronic configuration of the central metal atom.

20. On the basis of valence bond theory, predict whether square planar complexes of palladium(II) are high spin or low spin.

21. **(a)** A complex of a certain metal ion has a magnetic moment of 4.90 B.M.; another complex of the same metal ion in the same oxidation state has a zero magnetic moment. The central metal ion could be which one of the following? Cr^{III}, Mn^{II}, Mn^{III}, Fe^{II}, Fe^{III}, Co^{II}. **(b)** If a metal ion had complexes with moments 4.90 and 2.83 B.M., which one of these central metal ions could it be?

22. What factor determines whether the crystal field in an octahedral complex is to be regarded as strong or weak? How many d electrons must be present in orbitals of the central atom for there to be an abrupt change in crystal field stabilization energy between strong and weak fields?

23. Diagram the electronic configuration of the central metal of each of the following ions from the viewpoint of the crystal field theory (compare exercises 16 and 17): **(a)** $[Pd(en)_2]^{2+}$, **(b)** $[Co(CN)_6]^{3-}$, **(c)** $[Ir(NH_3)_6]^{3+}$, **(d)** $[PtCl_6]^{2-}$.

24. $[Ti(H_2O)_6]^{3+}$ absorbs light of wavelength 5000 Å. Name one ligand which would form a titanium(III) complex absorbing light of lower wavelength than 5000 Å and one ligand which would form a complex absorbing light of wavelength higher than 5000 Å.

25. Determine the crystal field stabilization energy of a d^6 complex having $\Delta = 25,000$ cm^{-1} and $P = 15,000$ cm^{-1}.

26. For an ion located in a square planar field of negative charges situated in the x and y directions, which d orbital(s) will have the highest energy(ies)? the lowest?

27. Distinguish between the possibilities in complex ions of $\Delta = 0$ and CFSE $= 0$. Give an example of each.

28. What is the crystal field stabilization energy, in terms of Δ, for nickel(II) ion in a square planar field? in a tetrahedral field? in an octahedral field? for zinc(II) ion?

29. From the data of Table 14–5 and Figure 14–8, predict the colors of $[Cr(NH_3)_6]^{3+}$ and $[Cr(CN)_6]^{3-}$.

30. The existence of infrared spectral bands corresponding to metal–chloride vibrations in the spectrum of *cis*-[Cr(NH$_3$)$_4$Cl$_2$]$^+$ implies what type of metal–ligand bonding?

31. Compare the valence bond, crystal field, and ligand field approaches to bonding in coordination compounds by indicating with checks ($\checkmark$) in a table the features of the various approaches. Across the top of the table are the headings (1) predicts geometry, (2) predicts magnetic moments, (3) predicts strength of metal–ligand bond, (4) explains spectrochemical series, and (5) explains thermodynamic properties. Down the left side of the table include valence bond, crystal field, and ligand field.

General Exercises

32. Explain why a (2+, 1−) electrolyte is expected to have a higher *molar* conductivity than a (1+, 1−) electrolyte.

33. A freshly prepared aqueous solution of Pd(NH$_3$)$_2$Cl$_2$ does not conduct electricity. Is this compound to be regarded as a strong or weak electrolyte. Explain in terms of its structure.

34. Potassium alum, KAl(SO$_4$)$_2\cdot$12H$_2$O, is obtained in the form of octahedral-shaped crystals when a solution of K$_2$SO$_4$ and Al$_2$(SO$_4$)$_3$ is concentrated by evaporation. Suggest specific experiments which might be used to determine whether potassium alum is a coordination compound. Explain what results would be expected if it were and if it were not a coordination compound.

35. Which one(s) of the ions (i)–(v) would be optically active?

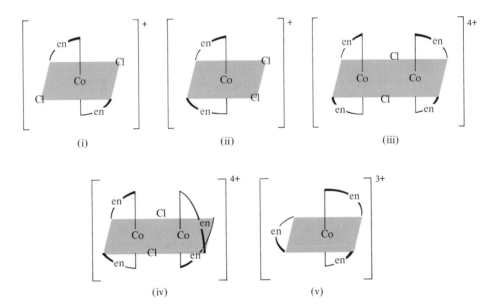

36. Neither optical isomers nor geometric isomers can be distinguished by mass spectroscopy. Suggest a reason for this fact.

37. The formula Co(NH$_3$)$_4$CO$_3$Br could be that of three isomers. Write their possible structures. Tell how you would distinguish the coordination isomers using **(a)** chemical means and **(b)** instrumental methods.

38. Describe the types of isomerism which can be associated with carbonatoaquotetraamminecobalt(III) chloride monohydrate.

39. Metal M forms a highly colored complex with ligand A, and a colorless complex with ligand B, which has a larger formation constant than the complex of M and A. Suggest a method to determine the concentration of M ions in a solution by titration, using A and B as reagents.

40. Two compounds have empirical formulas corresponding to $Cr(NH_3)_3(NO_2)_3$. In aqueous solution, one of these is a nonelectrolyte, while the other conducts electricity. What is the lowest possible formula weight of the conducting reagent? What is the highest possible formula weight for the nonconducting reagent?

41. The "ferroin" complex of iron(II), $[Fe(o\text{-phen})_3]^{2+}$, is highly colored. The corresponding complex of iron(III) is weakly colored. Suggest how ferroin can be used in an oxidation-reduction titration method.

42. Construct a table of crystal field stabilization energies for square planar complexes such as the one shown for octahedral complexes in Table 14–6. Assume that only the energy difference between the $d_{x^2-y^2}$ and d_{xy} orbitals is sufficiently large to cause electron pairing in some cases.

43. Explain why crystal field theory is not applied to complexes of main group metals.

44. Explain why d^8 complexes are more likely than other complexes to have a square planar geometry.

45. Diagram the electronic configuration of the central metal of each of the following ions from the viewpoint of the crystal field theory: **(a)** $[Pt(NH_3)_4]^{2+}$, **(b)** $[Cu(NH_3)_4]^{2+}$, **(c)** $[Cr(NH_3)_6]^{3+}$.

46. In terms of valence bond method, cite an example in which the term *inner orbital* does not imply low spin. In terms of crystal field theory, cite an example in which the term *strong field* does not imply low spin.

47. A solution of $[Ni(H_2O)_6]^{2+}$ is green, but a solution of $[Ni(CN)_4]^{2-}$ is colorless. Suggest an explanation for these observations.

48. An ion, M^{II}, forms the following complexes: $[M(H_2O)_6]^{2+}$, $[MBr_6]^{4-}$, and $[M(en)_3]^{2+}$. The colors of the complexes, though not necessarily in order, are green, red, and blue. Match the complex with the appropriate color, and explain.

49. The electronic configuration of a d^9 central ion is analogous to that of a hypothetical ion containing a single positron (e^+) in its d subshell (d^+). **(a)** Would the ground state of a d^+ ion be degenerate? **(b)** What would be the degeneracy of an *excited* state of the d^+ ion in an octahedral field? **(c)** What is the degeneracy of the ground state of a d^1 ion in an octahedral field? **(d)** Explain why in an octahedral field the d^9 configuration is regarded as the "inverse" of the d^1 configuration.

50. Given the following possible conditions with respect to the magnitudes of the crystal field splitting, Δ, and the electron pairing energy, P, for an octahedral complex, (1) $\Delta > P$, (2) $\Delta < P$, predict the spin states for the following types of ions: **(a)** d^9, **(b)** d^3, **(c)** d^4, **(d)** d^5.

51. Calculate the crystal field stabilization energy of a d^8 ion in a square planar field for both strong and weak field cases. Are square planar fields likely for complexes of ligands which generate weak fields?

52. What experimental evidence can be cited to prove that the magnitude of Δ in many octahedral complexes is less than the difference in energy between the $1s$ and $2s$ orbitals in the hydrogen atom? What orbitals in the hydrogen atom do have energy differences comparable to the magnitude of Δ in octahedral complexes?

53. Draw a crystal field splitting diagram, such as shown in Figure 14–13, for a complex in a linear field. Assume that the ligands lie on the z axis.

54. Explain why the difference in energy between the $d_{x^2-y^2}$ and d_{xy} orbitals in a square planar field is identical to the difference between the same orbitals in the octahedral field.

55. Predict the value of x in each of the following carbonyls: **(a)** $Co_2(CO)_x$, **(b)** $H_xCr(CO)_5$, **(c)** $H_xCo(CO)_4$.

56. The enthalpy of hydration of the Fe^{2+} ion is 11.4 kcal/mole higher than would be expected if there were no crystal field stabilization energy. Assuming the aquo complex to be high spin, estimate the magnitude of Δ for $[Fe(H_2O)_6]^{2+}$.

Advanced Exercises

57. Draw representations of the electronic structures of the following complex ions according to ligand field theory: **(a)** $[Co(NH_3)_6]^{2+}$,   **(b)** $[Pt(NH_3)_6]^{4+}$,   **(c)** $[CoF_6]^{4-}$.

58. The compound $Co(NH_3)_3(H_2O)(NO_2)(OCl)Br$ can exist in a variety of isomeric forms. Write structural formulas for at least 10 isomeric forms of this composition, and indicate the types of isomerism shown.

59. Write the formulas for all possible polymerization isomers of $[Pt(NH_3)_2(NO_2)_2]$.

60. Using the observed lattice energies tabulated below, estimate the magnitude of Δ in octahedrally coordinated crystals of VO, MnO, and FeO.

Oxide	Lattice Energy, U		Oxide	Lattice Energy, U	
	kcal/mole	kJ/mole		kcal/mole	kJ/mole
CaO	−828.2	−3465	FeO	−937.6	−3923
TiO	−927.8	−3882	CoO	−954.1	−3992
VO	−936.1	−3917	NiO	−974.2	−4076
MnO	−911.4	−3813	ZnO	−964.4	−4035

61. In aqueous solution, cobalt(III) ion is able to oxidize water.

$$Co^{3+} + e^- \rightarrow Co^{2+} \qquad \epsilon^\circ = 1.842 \text{ V}$$

The formation constant for $[Co(NH_3)_6]^{3+}$ is 5×10^{33} and for $[Co(NH_3)_6]^{2+}$, 1×10^5. Show that an aqueous solution of $[Co(NH_3)_6]^{3+}$ in $1\,M\,NH_3$ does not oxidize water.

62. The ion $[Co(en)_3]^{3+}$ exists in the form of two optical isomers. Show that this fact proves that the complex is not hexagonal or trigonal prismatic. Does it prove that the complex is octahedral?

63. The complex ion $[M(CN)(NO_2)(H_2O)(NH_3)]^+$ was found to be optically active. What does this finding signify about the configuration of the coordination sphere?

64. To demonstrate that optical activity was not related to the presence of carbon atoms in a molecule, Werner prepared $[Co\{(HO)_2Co(NH_3)_4\}_3]^{6+}$. Draw the structure of this complex ion, and show how Werner explained its optical activity.

65. $[Pt\{NH_2CH(C_6H_5)CH(C_6H_5)NH_2\}\{NH_2C(CH_3)_2CH_2NH_2\}]^{2+}$ was prepared from optically inactive materials and was found to be optically active. Show how this fact proves that the coordination sphere of the platinum is *not* tetrahedral. Does it *prove* that the arrangement is square planar?

66. The compound with the empirical formula $CsAuCl_3$ is diamagnetic. No metal-to-metal bonds are present in the compound.   **(a)** Is there gold(II) in this compound?   **(b)** Propose a structure for the compound.

67. Explain why platinum(II) and palladium(II) form square planar complexes almost exclusively but only a few nickel(II) complexes are square planar.

68. Explain each of the following phenomena, giving full details, such as pertinent equations and/or numerical justifications: **(a)** The complex $[CuCl_4]^{2-}$ exists, but $[CuI_4]^{2-}$ does not.   **(b)** Gold is not attacked by common acids, but "dissolves" in *aqua regia* ($HCl + HNO_3$).   **(c)** Copper metal will dissolve in aqueous KCN solution, with the evolution of hydrogen.   **(d)** Since chelate rings containing over six atoms are generally not stable, a single bidentate chelating ligand cannot replace two monodentate ligands which are located in the *trans* positions of a square planar complex.

69. The M—O bond distances in a series of six-coordinate crystal oxides are tabulated below. Plot these data against the atomic numbers of the metals, and explain the observed trends.

oxide:	CaO	TiO	VO	MnO	FeO	NiO	ZnO
d(Å):	2.4	2.1	2.05	2.2	2.18	2.08	2.1

70. Draw a crystal field splitting diagram for a cubic field (with the ligands at the corners of a cube which has the Cartesian axes going through the centers of the cube faces). Compare the magnitude of the splitting(s) with those of the octahedral, square planar, and tetrahedral crystal fields. Ions with how many d electrons would give the greatest crystal field stabilization energy in such a cubic field?

71. In terms of crystal field theory, explain why a d^9 octahedral complex with six identical ligands is *not* expected to have all six metal–donor atom distances identical.

72. Calculate the net coulombic energy of attraction of a dipositive ion in a square planar field of uninegative point charges at a distance of 2.5 Å. Calculate the equivalent energy for a tetrahedral field. Calculate the crystal field stabilization energy for a d^8 ion in each of these fields. Determine the minimum value of Δ which would favor a square planar geometry for this strictly ionic case.

15
The Kinetic
Molecular Theory

In science, the construction of intellectual "models" is a valuable technique which permits the prediction of new facts as well as the explanation of known observations. The Bohr theory (Section 9–11), the crystal field theory (Section 14–7), and the applications of the Schrödinger equation to problems of atomic and molecular structure (Chapters 10 and 11) are illustrations of this technique. In addition to these models of behavior on the molecular level, it is also possible to develop models to account for the behavior of the large assemblies of molecules which constitute matter in bulk. In contrast to the properties of individual atoms and molecules, bulk samples of matter have some properties which are determined by their physical states and which are largely independent of the chemical nature of the material. For example, all gases exert pressure and obey the gas laws approximately. The vapor pressures of liquids increase with rising temperature, and all solids absorb heat upon melting. Laws dealing with phenomena such as the critical point, phase equilibria, solubility, and lattice energy apply only to large assemblies of molecules. The laws of thermodynamics apply to large assemblies of molecules and are useful tools for predicting and explaining the behavior of matter in bulk. In this chapter, it will be shown how a model of the gaseous state, called the **kinetic molecular theory,** is useful in accounting for many of the characteristic properties of gases. The theory will be extended to explain some of the properties of matter in other states in Chapter 16, and to develop a mechanical model of chemical reactivity in Chapter 17.

15-1 Postulates of the Kinetic Molecular Theory

Within the limitations of low pressure and relatively high temperature, all gases obey the same general laws, and the chemical identity of the gas is of no immediate importance. Therefore, a model of matter in the gaseous state was proposed which accounts for the gas laws (Chapter 2) in terms of well-known laws of classical physics. The central idea is that molecules of gas are in continuous motion, hence the theory is called the **kinetic molecular theory.** The properties of the molecules of an ideal gas are postulated below. Molecules of real gases will behave similarly, especially at low pressures and relatively high temperatures.

The postulates of the kinetic molecular theory are as follows:

1. Gases consist of molecules in constant random motion.
2. The molecules exert no net force of attraction or repulsion upon one another.

3. Molecular collisions are **elastic;** that is, no kinetic energy is lost during a collision.
4. The molecules are negligibly small compared to the volume of their container.

A statement which may be derived from these four postulates and the ideal gas law is sometimes regarded as a fifth postulate.

5. The average translational kinetic energy of the molecules is directly proportional to the absolute temperature of the gas.

These postulates permit the development of an equation which expresses the product of the pressure and volume of a gas in terms of the mechanical properties of its molecules. Consider a cubic box (Figure 15–1) with each side having length l. Suppose that in the box there is only one molecule of mass m moving with a velocity v. Velocity is a vector quantity (having both magnitude and direction) and v can be resolved into the components v_x, v_y, and v_z, corresponding to the x, y, and z directions. The motion of the molecule in each direction may then be considered separately. When the molecule collides with wall A, which is perpendicular to the x direction, it will experience a component of momentum change in the x direction from mv_x to $-mv_x$. The magnitude of the momentum change $[mv_x - (-mv_x)]$ is $2mv_x$. Since the molecule must travel in the x direction the length of the box and back between collisions with wall A, the time between these collisions (equal to distance divided by velocity) is $2l/v_x$. The change in momentum in the x direction per unit time is

$$\frac{\text{change in momentum}}{\text{time}} = \frac{2mv_x}{2l/v_x} = \frac{mv_x^2}{l}$$

Since force may be defined as change in momentum per unit time, this quantity is the average force exerted by the repeated collisions of the molecule on wall A.

Suppose now that there are n' molecules in the box, each having mass m. Each of these molecules may have a different value of velocity in the x direction: $(v_x)_1$, $(v_x)_2$, $(v_x)_3, \ldots, (v_x)_{n'}$. The total force exerted on wall A by the n' molecules is thus

$$F_x = \frac{m}{l}[(v_x)_1^2 + (v_x)_2^2 + (v_x)_3^2 + \cdots + (v_x)_{n'}^2]$$

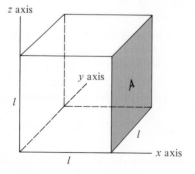

z axis

y axis

l

l

l

A

x axis

Figure 15–1. **Cubic Box**

If one defines u_x^2, the **mean square speed,** as

$$u_x^2 = \frac{(v_x)_1^2 + (v_x)_2^2 + (v_x)_3^2 + \cdots + (v_x)_{n'}^2}{n'}$$

the total force exerted by the molecules on wall A is then

$$F_x = \frac{m}{l}(n'u_x^2)$$

Since pressure is defined as force per unit area, the pressure on wall A, having area l^2, is given by

$$P_x = \frac{F_x}{l^2} = \frac{n'mu_x^2}{l^3} = \frac{n'mu_x^2}{V}$$

where $V = l^3$ is the volume of the box. By similar arguments to that given above,

$$P_y = \frac{n'mu_y^2}{V}; \qquad P_z = \frac{n'mu_z^2}{V}$$

Since the pressure, P, exerted by a confined gas is the same in all directions,

$$P = P_x = P_y = P_z$$

and therefore

$$u_x^2 = u_y^2 = u_z^2$$

As may be proved by the Pythagorean theorem, any vector such as u is related to its rectilinear components by the following type of equation:

$$u^2 = u_x^2 + u_y^2 + u_z^2$$

where u^2 is the mean square speed of the molecules, and $u = \sqrt{u^2}$ is called the **root mean square speed.** Since the molecules in the box have on the average the same distributions of velocities in the three directions,

$$u^2 = 3u_x^2$$

and therefore

$$P = \frac{n'mu^2}{3V}$$

$$PV = \tfrac{1}{3}n'mu^2$$

This equation is the **fundamental equation of the kinetic molecular theory of gases.**

To relate the average kinetic energy of its molecules with the absolute temperature of a gas, the fundamental equation may be restated as follows:

$$PV = \tfrac{2}{3}n'(\tfrac{1}{2}mu^2) = \tfrac{2}{3}n'(\overline{KE})$$

where $\overline{KE}$ is the average kinetic energy of the molecules due to translations (a translation is the movement of a molecule as a whole through space). Comparing this result to the equation of state for an ideal gas (Section 2–16),

$$PV = nRT$$

it is apparent that the average kinetic energy of the molecules of a gas is proportional to the absolute temperature of the gas, since both are directly proportional to PV:

$$nRT = \tfrac{2}{3}n'\overline{KE}$$

$$\overline{KE} = \frac{3n}{2n'}RT$$

Since n'/n is the number of molecules per mole of gas, Avogadro's number, N,

$$\overline{KE} = \frac{3}{2}\frac{R}{N}T = \tfrac{3}{2}kT$$

The constant

$$k = \frac{R}{N} = 1.381 \times 10^{-16}\ erg/molecule \cdot K = 1.381 \times 10^{-23}\ J/molecule \cdot K$$

is known as the **Boltzmann constant.** The equation above defines absolute temperature in terms of the translational kinetic energy of the molecules of an ideal gas. The absolute temperature is proportional to the average translational kinetic energy of the molecules; therefore absolute zero is the temperature at which the molecules have no translational motion.[1]

Example

Calculate the kinetic energy of gas molecules at 0°C.

$$\overline{KE} = \tfrac{3}{2}kT = (1.50)(1.381 \times 10^{-16}\ erg/molecule \cdot K)(273\ K) = 5.66 \times 10^{-14}\ erg/molecule$$

Note that in the calculation of average kinetic energy, the identity of the gas molecules is unimportant.

Example

Calculate the root mean square speed, u, of a hydrogen molecule at 0°C.

---

[1] It is found that at absolute zero all molecules possess a residual energy known as a "zero point energy." This energy is due to small oscillations of the atoms about their equilibrium positions in the crystal.

The root mean square speed is obtained by using the equation relating the kinetic energy of gas molecules with the absolute temperature of the gas:

$$\overline{KE} = \tfrac{1}{2}mu^2 = \tfrac{3}{2}kT$$

To obtain u in centimeters per second, it is necessary to use the mass of the molecule in grams, equal to the molecular weight, M, divided by Avogadro's number, N:

$$\frac{1}{2}\frac{M}{N}u^2 = \frac{3}{2}\frac{R}{N}T$$

$$u = \sqrt{\frac{3\,RT}{M}} = \sqrt{\frac{3(8.31 \times 10^7 \text{ erg/mole} \cdot \text{K})(273 \text{ K})}{2.01 \text{ grams/mole}}}$$

$$= \sqrt{3.39 \times 10^{10} \text{ cm}^2/\text{sec}^2} = 1.84 \times 10^5 \text{ cm/sec} = 1.84 \text{ km/sec} = 4140 \text{ miles/hr}$$

15-2   Experimental Confirmation of the Kinetic Molecular Theory

No specific gas was referred to in the development of the kinetic molecular theory; hence the theory is applicable to any gas. It is instructive to demonstrate how the theory accounts for the observed properties of gases as expressed by the various gas laws.

The facts that gases occupy the entire container in which they are found and that they are infinitely expandable are in agreement with the postulates that gases are composed of molecules which are very far apart compared to their own dimensions, that they exert negligible forces of attraction on one another, and that they are in constant random motion. They fill any container since they can always get farther apart. The existence of pressure is explained as the result of the bombardment of the gas molecules on the walls of the container.

Example

Predict from the kinetic molecular theory what the effect will be on the pressure of a gas inside a cubic box of sides l of reducing the size so that each side measures $l/2$. Assume no change in temperature.

The volume of the box will be one eighth the original volume: $(l/2)^3 = V/8$ (Figure 15-2). With no change in temperature, the molecules will retain the original average kinetic energy, and thus the original mean square speed:

$$\overline{KE} = \tfrac{3}{2}kT = \tfrac{1}{2}mu^2$$

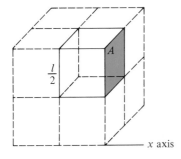

Figure 15-2. Smaller Cubic Box

The molecules will strike wall A (Figure 15–2) twice as often, since the distance in the x direction that they have to go is only half as great. Also, the area of wall A will be reduced to one fourth of the original area, and the pressure (force per unit area) will be increased on that account by a factor of 4. Thus the pressure due to collisions on wall A will be increased by a factor of 2 because the molecules hit the wall more often and by a factor of 4 because the wall is smaller, for a total increase of a factor of 8. This result is exactly that predicted by Boyle's law.

Example

Consider a sample of gas in a fixed volume container. From the arguments of the kinetic molecular theory, show that quadrupling the absolute temperature causes a quadrupling in pressure.

Quadrupling the absolute temperature causes a quadrupling in the average kinetic energy of the molecules. This increase in average kinetic energy causes an increase of the root mean square speed by a factor of 2:

$$\overline{KE}_1 = \tfrac{1}{2}mu_1^2$$
$$\overline{KE}_2 = \tfrac{1}{2}mu_2^2 = 4\,\overline{KE}_1 = \tfrac{1}{2}m(2u_1)^2$$

Thus the molecules are traveling twice as fast at the higher temperature, and therefore they hit the walls twice as often. In addition, because they are traveling twice as fast, their momentum is doubled, and their change in momentum at the wall also doubles. That is, they hit the wall twice as hard each time. The combined effect of hitting the wall twice as often with twice the change in momentum each time they hit causes a fourfold increase in pressure, in agreement with Charles' law.

15-3 Graham's Law

It may easily be demonstrated that gases diffuse through one another. Ammonia, a gas, when liberated in a room in which there are no air currents or breezes, soon can be detected throughout the room. The movement of the gas molecules through gases of another kind is called **diffusion.**

When a gas is allowed to escape from a container through a very small hole, the rate of escape, or **effusion,** is found to depend on the masses of the gas molecules. Both diffusion and effusion are described quantitatively by **Graham's law,** which states that at a given temperature the relative rates of effusion of two gases are inversely proportional to the square roots of their densities:

$$\frac{\text{rate}_1}{\text{rate}_2} = \frac{v_1}{v_2} = \sqrt{\frac{d_2}{d_1}}$$

Since at a given temperature and pressure, the density of a pure gas is proportional to its molecular weight, as may be determined from Avogadro's law, the rate of effusion of a gas is inversely proportional to the square root of its molecular weight, M. Thus two pure gases under the same conditions of temperature and pressure will have relative rates of effusion given by the equation

$$\frac{\text{rate}_1}{\text{rate}_2} = \sqrt{\frac{M_2}{M_1}} \qquad \text{(for pure gases)}$$

Example

Calculate the ratio of rates of effusion of H_2 and O_2, both at $0°C$ and 1 atm pressure.

$$\frac{\text{rate}_{H_2}}{\text{rate}_{O_2}} = \sqrt{\frac{32.0}{2.0}} = 4.0$$

H_2 effuses four times as fast as O_2 under these conditions.

Graham's law can be explained according to the kinetic molecular theory, as shown by the following example. Consider two different gases at the same temperature. The average kinetic energies of their molecules will be equal:

$$\tfrac{1}{2}m_1u_1{}^2 = \tfrac{1}{2}m_2u_2{}^2$$

Rearranging this equation yields

$$\frac{u_1}{u_2} = \sqrt{\frac{m_2}{m_1}} = \sqrt{\frac{M_2/N}{M_1/N}} = \sqrt{\frac{M_2}{M_1}}$$

where M represents the molecular weight of the gas and N is Avogadro's number. This equation is analogous to the equation expressing Graham's law.

A practical application of Graham's law is the gaseous diffusion method of separating isotopes. If a gaseous compound of the element whose isotopes are to be separated is allowed to effuse through a porous barrier, the molecules containing the lighter isotope will effuse faster, and the escaping gas will be enriched in it. A partial separation is effected. Recycling the enriched gas through the porous barrier further increases the concentration of the lighter isotope so that after many cycles, practically complete separation may be achieved. For example, the uranium isotopes ^{235}U and ^{238}U are separated in this manner at Oak Ridge, Tennessee. The volatile compound UF_6 is passed through thousands of porous towers. Although the ratio of molecular weights of $^{238}UF_6$ to $^{235}UF_6$ is only 1.0085, practically complete separation is achieved by repeated recycling.

15–4   Deviations from Ideal Behavior

According to Boyle's law, the product of the pressure and volume of a gas at a given temperature should remain constant at all pressures. However, all real gases show deviations from the predicted behavior; Figures 15–3 and 15–4 depict several examples of the nonideal behavior of real gases. The causes of the deviations from ideality can be understood by considering the simplifying assumptions of the postulates of the kinetic molecular theory. First, there is the assumption that the volume of the gas molecules themselves is negligible. Second, it is assumed that there are no forces of attraction exerted between molecules. When the distances between mole-

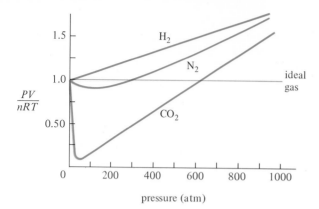

Figure 15-3. Nonideal Behavior of Real Gases at Very High Pressures

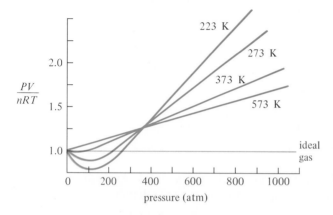

Figure 15-4. Nonideal Behavior of Nitrogen Gas at High Pressures at Various Temperatures

cules are small, as is the case of gases under high pressure, these assumptions lead to serious discrepancies between theory and experiment. Under this condition, the total volume of the gas and the volume occupied by the molecules themselves become closer in magnitude. Also, when the molecules are closer together, the van der Waals force of attraction (Section 11–7) between molecules comes more into play. For a given pressure, these attractive forces will have a more pronounced effect at lower temperatures, because the molecules have lower velocities and are within interacting distances of one another for longer periods of time. Thus in Figure 15–4 the curves at lower temperatures have more pronounced deviation from ideal behavior. Indeed it may be seen that each increase in temperature results in a smaller minimum.

15-5   The van der Waals Equation

Several equations of state relating pressure, volume, and temperature of real gases have been proposed. All of these equations use experimentally determined constants as corrections for the theoretical pressures and/or volumes. One of the best known is the van der Waals equation, which aims to "correct" the observed pressure and the observed volume of a real gas, as follows:

$$\left(P + \frac{n^2a}{V^2}\right)(V - nb) = nRT$$

where a and b are constants and P, V, T, n, and R have the same meanings they have in the equation of state for an ideal gas. The term n^2a/V^2 corrects the pressure for the effect of intermolecular attractions. The constant a is determined experimentally for each gas; typical values are reported in Table 15-1. The term nb is a correction for the volume of the molecules themselves. Thus $V - nb$ is the corrected volume. Values of b also are tabulated in Table 15-1. The form of the van der Waals equation of state using the corrected pressure and volume of the real gas has the same form as the equation of state for an ideal gas. Thus

$$P_{corr}V_{corr} = nRT$$

Example

Using the van der Waals equation, calculate the pressure exerted by 10.0 moles of carbon dioxide in a 2.00 liter vessel at 47°C. Repeat the calculation using the equation of state for an ideal gas. Compare these results with the experimentally observed pressure of 82 atm.

$$\left(P + \frac{n^2a}{V^2}\right)(V - nb) = nRT$$

From Table 15-1, $a = 3.59$ liter$^2 \cdot$ atm/mole and
$$b = 42.7 \text{ cm}^3/\text{mole} = 0.0427 \text{ liter/mole}$$

$$\left(P + \frac{100 \times 3.59}{4.00} \text{ atm}\right)\left[2.00 \text{ liters} - (10.0 \text{ moles})\left(0.0427 \frac{\text{liter}}{\text{mole}}\right)\right]$$

$$= 10.0(0.0821)(320) \text{ liter} \cdot \text{atm}$$
$$P = 80 \text{ atm}$$

From the ideal gas equation:

$$PV = nRT$$
$$(2.00 \text{ liters})P = 10.0(0.0821)(320) \text{ liter} \cdot \text{atm}$$
$$P = 131 \text{ atm}$$

Since the total pressure is so high, one would not expect the gas to behave ideally. The van der Waals equation gives a result much closer to the experimental value.

TABLE 15-1. Van der Waals Constants

Gas	a $\left(\dfrac{\text{liter}^2 \cdot \text{atm}}{\text{mole}^2}\right)$	b $\left(\dfrac{\text{cm}^3}{\text{mole}}\right)$	Gas	a $\left(\dfrac{\text{liter}^2 \cdot \text{atm}}{\text{mole}^2}\right)$	b $\left(\dfrac{\text{cm}^3}{\text{mole}}\right)$
He	0.0341	23.7	C_2H_4	4.47	57.1
Ne	0.2107	17.1	CO_2	3.59	42.7
H_2	0.244	26.6	NH_3	4.17	37.1
N_2	1.39	39.1	H_2O	5.46	30.5
CO	1.49	39.9	Hg	8.09	17.0
O_2	1.36	31.8			

TABLE 15–2. Critical Temperatures and Pressures

Gas	Critical Temperature (K)	Critical Pressure (atm)	Gas	Critical Temperature (K)	Critical Pressure (atm)
water	647.2	217.7	nitrogen	126.1	33.5
hydrogen chloride	224.4	81.6	ammonia	405.6	111.5
carbon dioxide	304.2	73	hydrogen	33.3	12.8
carbon disulfide	546	76	helium	5.3	2.26
oxygen	153.4	49.7	carbon tetrachloride	556.2	45.0
chlorine	417	76.1	Freon-12 (CCl_2F_2)	384.9	39.4

15–6   Critical Temperature and Pressure

If any real gas is subjected to high pressures and sufficiently low temperatures, it will liquefy. However, for each gas there is a temperature above which the gas will not liquefy no matter what pressure is applied. This temperature is called the **critical temperature** of the substance. The pressure necessary to liquefy a gas at its critical temperature is called its **critical pressure.** Above the critical temperature a substance can exist only as a gas, because the kinetic energies of the molecules are so great that intermolecular attractions are negligible by comparison, even at very high pressures. The value of the critical temperature of a substance thus depends on the magnitude of the attractive forces between the molecules. Substances having relatively strong intermolecular forces will have high critical temperatures, while nonpolar substances in which the intermolecular attractions are relatively small will have low critical temperatures. Critical temperatures and pressures for several substances are listed in Table 15–2.

15–7   Heat Capacity of Gases

The effect of adding energy to a system might be to increase its energy, to cause the system to do work, or a combination of these (Section 3–2). Gaseous systems can absorb energy at constant volume or at constant pressure. According to the kinetic molecular theory, the average kinetic energy of the molecules of a gas is proportional to its absolute temperature. Therefore, when a gaseous system absorbs the energy, its temperature must increase corresponding to that quantity of energy which is not converted into work. Consequently, the quantity of energy absorbed by the system may be metered by following its change in temperature.

Heat capacity is defined as the quantity of heat required to raise the temperature of a given quantity of substance by 1 degree Celsius (Section 3–3). When a given quantity of gas is heated at constant volume, its heat capacity will differ from that for the gas heated at constant pressure. If the volume is kept constant, there will be a rise in the average kinetic energy of the molecules and also an increase in the pressure of the gas. Consider 1 mole of a monatomic ideal gas:

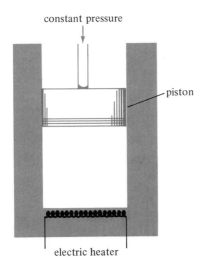

constant pressure

piston

electric heater

Figure 15-5. **Apparatus for Heating a Gas at Constant Pressure**

$$\overline{KE}_1 = \tfrac{3}{2}RT_1$$

At $T_2 = T_1 + 1°$,

$$\overline{KE}_2 = \tfrac{3}{2}RT_2 = \tfrac{3}{2}R(T_1 + 1)$$
$$\Delta\overline{KE} = \overline{KE}_2 - \overline{KE}_1 = (\tfrac{3}{2}RT_1 + \tfrac{3}{2}R) - \tfrac{3}{2}RT_1 = \tfrac{3}{2}R$$

The quantity $\Delta\overline{KE}$ represents the heat energy required to raise the temperature of 1 mole of a monatomic gas by 1 degree Kelvin at constant volume. This quantity is the molar heat capacity at constant volume, C_v.

On the other hand, suppose it is desired to raise the temperature of the monatomic gas in such a manner that its pressure remains constant. The molar heat capacity under these conditions is C_p. The experimental arrangement might be that shown in Figure 15-5. If the gas pressure is to remain constant during the heating process, provision must be made for an increase in the volume of the gas against an external pressure which exactly balances the pressure of the gas. The gas does work in expanding against the opposing pressure. For 1 mole of monatomic gas, the work, $-w$, done against the external pressure is given by

$$-w = PV_2 - PV_1$$

Before expansion,

$$PV_1 = RT_1$$

After expansion,

$$PV_2 = R(T_1 + 1)$$
$$P(V_2 - V_1) = R(T_1 + 1) - RT_1 = R$$

The molar heat capacity at constant pressure, designated C_p, is greater than that at

TABLE 15-3. Heat Capacities of Several Gases at About Room Temperature

	cal/mole · deg			J/mole · deg		
	C_p	C_v	$C_p - C_v$	C_p	C_v	$C_p - C_v$
He	5.0	3.0	2.0	20.9	12.6	8.3
Ar	5.0	3.0	2.0	20.9	12.6	8.3
H_2	6.8	4.8	2.0	28.5	20.1	8.4
O_2	7.0	5.0	2.0	29.3	20.9	8.4
CO_2	8.9	6.9	2.0	37.2	28.9	8.3
NH_3	8.9	6.9	2.0	37.2	28.9	8.3
C_2H_6	11.6	9.6	2.0	48.5	40.2	8.3
CH_3OCH_3	15.9	13.8	2.1	66.5	57.7	8.8

constant volume, C_v, by R calories per mole per degree (1.987 cal/mole · K) for an ideal gas:

$$C_p = C_v + R$$

When diatomic and polyatomic molecules are heated, energy is absorbed by rotations and internal vibrations of the molecules in addition to that which increases the translational kinetic energy of the molecules. Depending on the structures of the molecules, the additional energy required will vary. However, as shown in Table 15–3, even with polyatomic molecules the difference between C_p and C_v is equal to R.

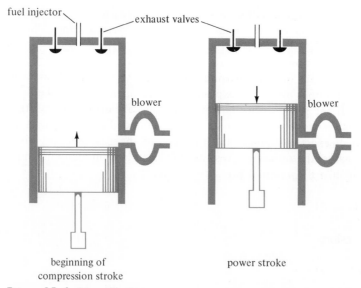

beginning of
compression stroke

power stroke

Figure 15–6. Diesel Engine

The cylinder is filled with air by means of a blower. As the piston rises, the valves close, and the air is compressed to one sixteenth of its original volume. The compression stroke is so rapid that little heat is lost to the surroundings. Because of the work done by the piston, the temperature of the air rises above the ignition temperature of the fuel. At the top of the stroke, fuel is injected into the cylinder and immediate combustion occurs. The piston is driven downward in the power stroke by the expanding gaseous reaction products. When the piston reaches the lowest position, the exhaust gases are blown out and the cylinder refilled with air.

15-8   Adiabatic Processes

Processes which take place in a manner such that no heat energy enters or leaves the system are called **adiabatic processes.** Gases can be expanded or compressed adiabatically, using the apparatus of Figure 15-5, for example, provided that the electric heater is not used and the walls of the container and the piston are insulated so that no heat can be transferred between the system and the surroundings. If the gas within the container is compressed adiabatically by pushing in the piston, energy will be transferred from the piston to the molecules which collide with it, and they will rebound with greater kinetic energy. The temperature of the gas must necessarily rise. This principle is applied in the operation of the diesel engine (Figure 15-6).

The reverse process, in which the gas within the container expands adiabatically and pushes back the piston, will result in a lower temperature of the gas. Since no heat enters the system, the energy used to drive back the piston must come from the kinetic energy of the molecules. In these processes, both ideal gases and real gases experience heating or cooling. If an ideal gas is allowed to expand adiabatically without doing work, for example, by expanding into a vacuum, it would not be cooled because no work is done. However, if a real gas is allowed to expand into a vacuum, some energy must be expended to overcome the attractive forces between the molecules. This energy comes from the kinetic energy of the molecules, and hence the temperature is lowered somewhat, since the average kinetic energy of the molecules is lowered. As shown in Figure 15-7, this effect is used in the liquefaction of gases.

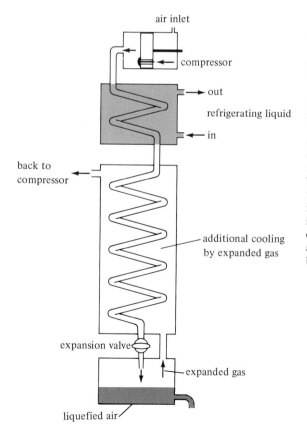

Figure 15-7. **Liquefaction of Air**
Air is pumped by a compressor through pipes leading through a cooling bath containing a refrigerating liquid such as ammonia (normal boiling point, $-33\,°C$). The compressed air is then passed through other cooling coils and finally through an expansion valve. Upon expansion, the air is further cooled because it uses up kinetic energy to push back the atmosphere and to overcome the intermolecular forces. The cold, expanded air is passed over the cooling coils, further cooling the incoming air. Finally, the incoming air is sufficiently cold so that part of it condenses to liquid, which is drawn off at the bottom of the expansion chamber. Liquid air boils at about $-190\,°C$.

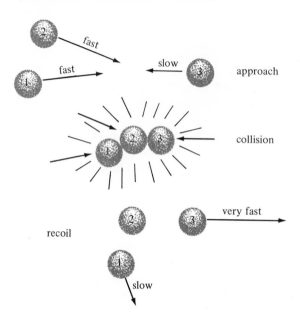

Figure 15–8. **Schematic Diagram of a Molecular Collision**

15-9   The Distribution of Molecular Speeds

At any given instant, the molecules in a sample of gas have a range of speeds. As the molecules collide with each other and with the walls of their container in a perfectly random manner, some of them will have an instantaneous speed of zero owing to simultaneous collisions of more than two molecules in such a manner that all the kinetic energy of one molecule is transferred to the others. Other molecules will have extremely high speeds owing to collisions in which the momentum of several molecules is transferred to one. Such possibilities are represented in Figure 15–8.

While it is not possible to know the speed of any individual molecule in a sample of gas, it is possible to determine the statistical distribution of molecular speeds at a given temperature. Such a distribution is shown graphically in Figure 15–9. Graphs such as this one are known as Maxwell–Boltzmann distributions of molecular speeds. The shape of the curves is predicted theoretically and also can be demonstrated experimentally (Figure 15–10). In Figure 15–9, the total area under the curve is

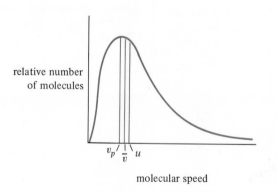

relative number of molecules

molecular speed

Figure 15–9. **Maxwell–Boltzmann Distribution of Molecular Speeds**

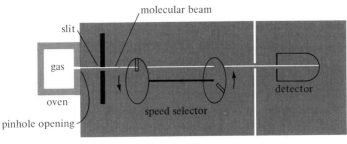

Figure 15-10. **An Experimental Method of Determining Molecular Speeds**

Molecules of a gas heated to a certain temperature in an oven are allowed to escape through a pinhole opening into an evacuated chamber. Those entering the chamber in such a direction that they pass through a collimating slit form a "molecular beam." The beam is directed toward a "speed selector," which is merely a pair of rotating disks connected by a shaft and having slits offset in such a manner that only molecules traveling with a certain speed can pass through both slits and onto a detector. The shaft connecting the disks is rotated at various rates, and the numbers of molecules reaching the detector for each rate of rotation is determined. The results obtained confirm the theoretically predicted Maxwell–Boltzmann distribution of molecular speeds.

proportional to the total number of molecules in the sample. The maximum in the curve corresponds to the most probable speed, v_p. More molecules possess this speed than any other. The average speed, $\bar{v}$, and the root mean square speed, u, are also shown.

In Figure 15-11 are shown the distributions of molecular energies for equal numbers of molecules of a given kind at two different temperatures: T_1 and T_2, where T_2 is greater than T_1. Since the total number of molecules at the two temperatures is the same, the areas under the two curves are equal. However, as may be predicted

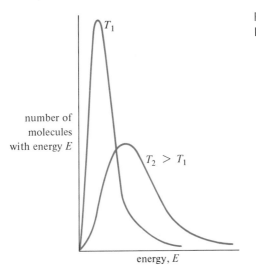

Figure 15-11. **Distributions of Molecular Energies at Two Temperatures**

from the kinetic molecular theory, the maximum in the distribution curve at T_2 is shifted toward higher energy compared to that at T_1. The curve for the higher temperature is also broader, indicating that more molecules possess higher energies and that there is a wider distribution of higher energies. Also, at the higher temperature, there is a proportionate decrease in the number of molecules having energies lower than the most probable value.

15–10   Exercises

Basic Exercises

1. Which postulate(s) of the kinetic molecular theory can be used to justify (a) Dalton's law of partial pressures? (b) Graham's law?

2. What is the relationship between k, the Boltzmann constant, and R, the ideal gas law constant?

3. If separate samples of argon, neon, nitrogen, and ammonia, all at the same initial temperature and pressure, are expanded adiabatically to double their original volumes, which would require the greatest quantity of heat to restore the original temperature?

4. Explain why spark plugs are not necessary in a diesel engine.

5. At what temperature will hydrogen molecules have the same root mean square speed as nitrogen molecules have at 35°C?

6. Calculate the pressure of 0.60 mole of NH_3 gas in a 3.00 liter vessel at 25°C (a) with the ideal gas law, (b) with the van der Waals equation.

7. Calculate the pressure of 15.00 moles of neon at 30°C in a 12.0 liter vessel using (a) the ideal gas law, (b) the van der Waals equation.

8. Calculate the root mean square speed and the average kinetic energy of oxygen molecules at 18°C.

9. Show that change in momentum per unit time has the same dimensions as mass times acceleration.

10. (a) Under what sets of experimental conditions is the van der Waals equation more applicable than the ideal gas law equation? Calculate the pressure of 12.0 moles of CO at 25°C in a 10.0 liter vessel using (b) the ideal gas law, (c) the van der Waals equation. Calculate the pressure of 0.120 mole of CO at 25°C in a 10.0 liter vessel using (d) the ideal gas law, (e) the van der Waals equation.

11. (a) Show that the average of the squares of the following series of numbers is different from the square of the average of the numbers: 5, 10, 15, 20. (b) Which of the results is larger? (c) For molecules having a given mass, why does the root mean square speed, rather than the average speed, have greater physical significance?

12. When a bicycle tire is pumped up rapidly, its temperature rises. Would this effect be expected if air were an ideal gas? Explain.

13. When CO_2 under high pressure is released from a fire extinguisher, particles of solid CO_2 are formed, despite the low sublimation temperature of CO_2 at 1.0 atm pressure (-77°C). Explain this phenomenon.

14. Explain why Boyle's law cannot be used to calculate the volume of a real gas which is changed from its initial state to its final state by an adiabatic expansion.

15. The total kinetic energy of molecules of a given gaseous substance has components of translation, molecular vibration, and molecular rotation. (a) Which one(s) of these components is(are) proportional to the absolute temperature of the gas? (b) Which one(s) is(are) related to the specific heat of the substance? (c) Which one(s) is(are) related to the infrared absorptions of the molecules?

16. A certain saturated hydrocarbon effuses about half as fast as methane. What is the molecular formula of this hydrocarbon?

17. At what temperature will hydrogen molecules have the same kinetic energy as nitrogen molecules have at 35°C?

18. Prove that for a vector u,

$$u^2 = u_x{}^2 + u_y{}^2 + u_z{}^2$$

19. What is the numerical value of n'/n, where n' is the number of molecules in a given sample of gas and n is the number of moles of the gas.

20. Distinguish between the total kinetic energy of a molecule and its translational kinetic energy. For what type of gas molecules are they the same?

21. Avogadro's law states that equal volumes of gases under the same conditions of temperature and pressure contain equal numbers of molecules. Develop from that law the fact that at a given temperature and pressure, the density of a pure gas is proportional to its molecular weight (as stated on page 424).

22. For the following set of speeds, calculate **(a)** the average speed, **(b)** the mean square speed, **(c)** the root mean square speed, **(d)** the most probable speed.

10 meters/sec	15 meters/sec	20 meters/sec	25 meters/sec
25 meters/sec	30 meters/sec		

23. Which of the postulates of the kinetic molecular theory are only approximations when applied to real gases?

24. Compare the average velocity of the molecules in a sample of air at 25°C to its bulk velocity in a 10 miles/hour wind.

General Exercises

25. What volume would 3.00 moles of oxygen occupy at 50.0 atm pressure and 100°C according to **(a)** the ideal gas law? **(b)** the van der Waals equation? *Hint:* Solve by successive approximations, using the volume obtained in part **(a)** in the pressure correction term to calculate a better answer in part **(b)**. Use that answer to compute a still better answer, continuing until the results of two successive steps differ by a negligible amount.

26. According to the van der Waals equation, how many moles of ammonia will occupy 7.00 liters at 20.0 atm and 100°C? (If necessary, see the hint in exercise 25.)

27. For a given number of moles of gas, show that the van der Waals equation predicts greater deviation from ideal behavior **(a)** at high pressure rather than low pressure at a given temperature, **(b)** at low temperature rather than high temperature at a given pressure.

28. At what temperature would the average kinetic energy of gaseous hydrogen molecules equal the energy required to dissociate the molecules into atoms (104 kcal/mole)?

29. Assuming ideal behavior, calculate the volume per mole of gaseous water at 1.00 atm pressure and 100°C. Also calculate the volume of 1.00 mole of liquid water at 100°C (density = 0.958 gram/ml). Assuming that this value is the total volume of the molecules themselves, with negligible space between molecules, calculate the percent of the total volume of gaseous water at 100°C which is "free volume." Compare the volume of 1.00 mole of water molecules at 100°C to the value of the van der Waals constant b for water. Suggest a reason for the difference.

30. At 1200°C, the following equilibrium is established between chlorine atoms and chlorine molecules:

$$Cl_2 \rightleftharpoons 2\,Cl$$

The composition of the equilibrium mixture may be determined by measuring the rate of effusion of the mixture through a pinhole. It is found that at 1200°C and 1.8 torr, the mixture effuses 1.16 times as fast as krypton effuses under the same conditions. Calculate the fraction of chlorine molecules dissociated into atoms.

31. A space capsule is filled with neon gas at 1.00 atm and 290 K. The gas effuses through a pinhole into outer space at such a rate that the pressure drops by 0.30 torr/sec. **(a)** If the capsule were filled with ammonia at the same temperature and pressure, what would be the rate of pressure drop? **(b)** If the capsule were filled with 30.0 mole percent helium, 20.0 mole percent oxygen, and 50.0 mole percent nitrogen at a total pressure of 1.00 atm and a temperature of 290 K, what would be the corresponding rate of pressure drop?

32. Which one(s) of the following represent vector quantities: **(a)** force, **(b)** momentum, **(c)** energy, **(d)** speed, **(e)** velocity, **(f)** the square of velocity, **(g)** pressure.

33. What factors influence the magnitude of the temperature drop on expansion of a pure real gas into a vacuum?

34. Does a molecule of a gas sample which has a kinetic energy equal to the average kinetic energy of the gas sample also have a speed equal to the average speed of the molecules of the sample? Explain.

35. At the start of an experiment, one end of a U tube of 6 mm glass tubing is immersed in concentrated ammonia solution and the other end is immersed in concentrated hydrochloric acid solution. At the point in the tube where vapors of NH_3 and HCl meet, a white cloud of $NH_4Cl(s)$ forms. At what fraction of the distance along the tube from the ammonia solution does the white cloud *first* form?

36. A porous cup filled with hydrogen gas at atmospheric pressure is connected to a glass tube which has one end immersed in water, as shown in the accompanying figure. Explain why the water rises in the glass tube.

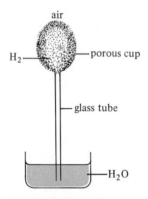

37. Explain in terms of atomic and molecular motions why the heat capacity of CH_3OCH_3 is so much greater than that of He (Table 15–3).

38. Compare the values of the van der Waals constants for NH_3 and N_2. Explain why the value of a is larger for NH_3 but that of b is larger for N_2.

39. A study of automobile speeds on a highway yielded the following data for a 1 hour period. What are the average speed, the root mean square speed, and the most probable speed of this group of cars?

Speed	Number of Cars	Speed	Number of Cars
47	2	56	210
48	7	57	192
49	20	58	144
50	45	59	112
51	102	60	63
52	150	61	26
53	205	62	11
54	220	63	4
55	250	66	2

40. Explain why you would suspect that some data were missing if the second pair of columns of the data in exercise 39 had been accidentally omitted.

41. The absolute zero of temperature may be determined experimentally in several ways. List the various experimental techniques which might be used for this purpose which have been described so far in this book.

Advanced Exercises

42. The separation factor, f, of a process of isotopic separation is defined as the ratio of the relative *concentration* of a given species after processing to its relative *concentration* before processing:

$$f = \frac{n_1'/n_2'}{n_1/n_2}$$

where n_1 and n_2 are the concentrations of two species before processing. For a single stage diffusion process, the maximum separation factor is merely the ratio of the rates of diffusion of the isotopic molecules as determined from Graham's law. Naturally occurring uranium consists of 99.3% by mass of ^{238}U and 0.7% of ^{235}U. **(a)** Calculate the overall separation factor necessary to achieve a product containing 99.7% ^{235}U starting with natural uranium. **(b)** What is the theoretical separation factor for a single diffusion step using isotopic molecules of UF_6? **(c)** How many ideal diffusion steps would be required to produce 99.7% pure ^{235}U? **(d)** Natural hydrogen consists of 99.98% ^1H and 0.02% ^2H by mass. In the manner outlined above, calculate the number of ideal diffusion steps required to produce 99% pure $^1H^2H$ from ordinary gaseous hydrogen.

43. Part **d** of the preceding exercise intimates that ^2H in hydrogen gas prepared from naturally occurring sources of hydrogen is more likely to be in the form of $^1H^2H$ than in the form $^2H_2$. Hence, to obtain pure ^2H by a diffusion method, isotopic *compounds* of hydrogen are used. What characteristics would be desirable in such a compound? What compound(s) can you suggest?

44. A given quantity of real gas is expanded adiabatically into a vacuum, from an initial volume of 1 liter to a final volume of 2 liters, and the temperature of the gas decreases by 1.0°C. If half the given quantity of the same gas were expanded adiabatically into a vacuum from 1 to 2 liters, would the same decrease in temperature be expected? Explain.

45. According to the ideal gas law, at constant volume the pressure of a gas becomes zero at 0 K and is independent of the volume. Solve the van der Waals equation explicitly for pressure. Is the variation of pressure of a real gas with temperature at constant volume independent of the volume? Using the data of Table 15-1, determine the absolute temperature at which 1.0 mole of helium gas in a volume of 1.0 liter would have zero pressure. Repeat the calculation for ethylene, C_2H_4. Compare these temperatures with the normal boiling points of He and C_2H_4.

46. Expand the van der Waals equation (for 1 mole of gas) into a cubic equation with respect to the volume. At the critical temperature, T_c, the pressure of a gas is the critical pressure, P_c, and the volume of 1 mole is the critical volume, V_c. At that temperature, $V - V_c = 0$. Expand the cubic equation $(V - V_c)^3 = 0$. By equating coefficients of corresponding powers of V in the two cubic equations, express a and b in terms of T_c, V_c, and P_c. Experimental values of a and b are determined in this manner.

47. Expand the van der Waals equation into a cubic equation in V. Using appropriate solution methods, for example that given in the *Handbook of Chemistry and Physics,* CRC Publishing Co., solve this equation to calculate the volume occupied by 3.00 moles of oxygen at 50.0 atm pressure and 100°C. Compare the result with that obtained by successive approximations in exercise 25.

16

Liquids and Solutions

In this chapter, the characteristics of liquids and liquid solutions will be described. Like gases, liquids are fluids; that is, the individual molecules have sufficient freedom of motion to enable the bulk material to flow and to diffuse. However, in contrast to gases, which have volumes dependent on both pressure and temperature, a given mass of a liquid at a given temperature occupies a definite volume.

In terms of the kinetic molecular theory, the nature of a liquid is postulated as follows:

1. There are appreciable attractive forces between the molecules of a liquid.
2. The molecules are relatively close together.
3. The molecules are in constant, random motion, as evidenced by such phenomena as diffusion and evaporation.
4. The average kinetic energy of the molecules in a given sample of liquid is proportional to the absolute temperature.

When the temperature is raised, the molecular motion in the liquid is increased, and there is a slight lessening of the influence of the intermolecular forces. The volume of the liquid increases with increasing temperature, and therefore its density decreases.[1]

Because of its definite volume, there is a surface between the bulk of a liquid and the surrounding space. Homogeneous regions of matter, separated from other matter by such definite boundary surfaces, are called **phases.** Within a given system, all regions containing matter in the same state and having the same composition are considered to belong to a single phase. For example, several ice cubes in a glass of water comprise just one solid phase. Liquids and solids are **condensed phases.** A given system may consist of several phases, and at a given temperature the phases will tend to be interconverted until all of the phases are in equilibrium with each other. In this chapter, two approaches to the description of equilibria between phases will be described: the Clausius–Clapeyron equation and the Gibbs phase rule.

16-1 Viscosity and Surface Tension

One consequence of the strong intermolecular attractions in liquids as compared to gases is the greater viscosity of liquids. Viscosity is the resistance of a fluid to flow. Liquids comprised of large, irregularly shaped molecules are generally more viscous

---

[1]Between 0° and 4°C, water is a notable exception (see Section 21–3).

than liquids made up of small, symmetrical molecules. In general, as the temperature of a liquid is increased, the viscosity decreases. Both of these observations can be explained in terms of the kinetic molecular theory. The more nearly the molecules of a substance resemble hard spheres, the more elastic will be the collisions between the molecules. Collisions between irregularly shaped molecules will be less elastic. In collisions of the latter some of the energy of translation is transformed into vibrational and rotational energy. With less translational energy, the molecules have a greater tendency to stick together. As the temperature is increased, the average kinetic energy of the molecules also increases, and the translational kinetic energy becomes relatively more important for all molecules. Two techniques of measuring the viscosities of liquids are noted in Figure 16-1.

Another property of liquids which is related to intermolecular forces is **surface tension,** which is defined as the work (energy) per square centimeter required to increase the surface area. Energy divided by area has the same dimensions as force divided by length; therefore, an appropriate dimension for surface tension is force per unit length of surface, such as dynes per centimeter. A molecule in the bulk of a liquid is attracted more or less equally in all directions by surrounding molecules. However, since molecules at the surface of a liquid have a net attraction toward the interior of the liquid, there is a tendency for the liquid to acquire as small a surface area as possible. Falling water droplets or mercury droplets on glass have nearly spherical shapes. The sphere has the smallest surface per unit volume of any geometric figure; hence a spherical shape gives a lower surface energy than any other shape.

The rise of liquid in a capillary tube is a familiar phenomenon involving surface tension (Figure 16-2). Whether a liquid rises in a capillary depends on the relative attraction of the liquid molecules for each other compared to their attraction for the walls of the capillary. If the attraction between the molecules and the capillary walls is the larger, the liquid is said to "wet" the surface. When a capillary is placed in such a liquid, the liquid wets the walls, and this wetting tends to increase the surface area of the liquid. Surface tension tends to reduce the surface area; consequently, some of the bulk of the liquid rises up into the capillary to a height, h, above the original surface (see Figure 16-2). That is, the liquid rises until the weight of the liquid above the original surface just balances the attractive force between the glass and the liquid.

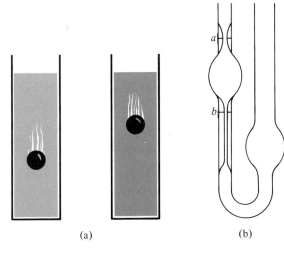

(a)   (b)

Figure 16-1. **Measurement of Viscosity**
(a) The greater the viscosity of a liquid, the slower will be the fall of a spherical ball through it. The viscosity may be calculated using Stokes' law:

$$v = \frac{(m - m_0)g}{6\pi r \eta}$$

where v is the terminal velocity, m is the mass of the sphere of radius r, m_0 is the mass of displaced liquid, η is the coefficient of viscosity (in poise), and g is the acceleration due to gravity.

(b) An Ostwald viscometer. The more viscous the liquid, the longer it takes for its level to fall from line a to line b. Viscosity is determined by comparison with a standard substance.

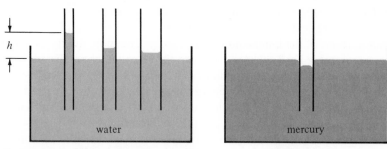

Figure 16-2. **Effects of Surface Tension**

If a liquid wets the walls of the glass tube, it rises to a height which is inversely proportional to the inner radius of the tube. A liquid which does not wet glass, such as mercury, will be depressed in a capillary tube.

If γ (gamma) denotes the surface tension and r is the radius of the capillary, the attractive force of the glass is equal to $2\pi r \gamma$. The weight of the liquid drawn up into the capillary, which just balances this force, is equal to mg, where m is the mass of the liquid above the original surface and g is the gravitational constant. If the density of the liquid, d, is known, the mass of the liquid above the original surface is equal to $\pi r^2 h d$. Hence

$$2\pi r \gamma = (\pi r^2 h d)g$$
$$\gamma = \tfrac{1}{2} r h d g$$

Therefore, the surface tension of a liquid may be determined experimentally by measuring the height to which the liquid rises in a capillary of known radius. Another method of measuring surface tension, known as the falling drop method, is described in Figure 16–3.

Figure 16-3. **Determination of Surface Tension by the Falling Drop Method**

Liquid, allowed to flow from bulb A through a capillary tube, forms drops at the bottom of the apparatus. Several drops are allowed to fall and are collected and weighed precisely. The average mass of a drop and the known density of the liquid allow calculation of the volume and radius of the drop. Just before a drop falls, the total force due to the surface tension and the weight of the drop are equal in magnitude, and the former may be calculated from the equation $\gamma = mg/2\pi r$.

16-2   Vapor Pressure

A substance which is **volatile** is one which vaporizes readily. The volatilities of liquids can be explained by means of the kinetic molecular theory. As in gases, molecules in a liquid have a Maxwell–Boltzmann distribution of kinetic energies ranging from very low values to very high values. Consequently, a certain fraction of the molecules at the surface of a liquid will have such large kinetic energies that they can escape from the condensed phase. If the temperature is kept constant, the remaining liquid will possess the same distribution of molecular energies, and the most energetic fraction will continue to escape from the liquid into the vapor state. If the liquid is in an open vessel, evaporation will continue until ultimately no liquid remains.

If a quantity of a liquid is introduced into a container which is then closed and maintained at a constant temperature, the molecules having sufficient kinetic energies will escape into the vapor phase. Since the temperature is kept constant, evaporation continues at a constant rate. However, in the closed container, as the number of molecules in the gas phase increases, these molecules, too, achieve a Maxwell–Boltzmann distribution, and a fraction will have such low energies that they will condense to liquid when they collide with the liquid surface. The greater the concentration of molecules in the gas phase, the more molecules will so condense. Ultimately, condensation and evaporation occur at equal rates, and there is no net transfer of molecules between the phases. A state of equilibrium exists. The pressure exerted by the molecules in the gas phase under these equilibrium conditions is defined as the **vapor pressure** of the liquid.

The magnitude of the vapor pressure for a given liquid depends only on the temperature, and not on the volume of the liquid or of the space above the liquid. For example, suppose a liquid sample is placed in a container such as shown in Figure 16-4. If the volume of space above the liquid is doubled, the gas pressure is halved (approximately), and the rate of condensation is lowered. Since the temperature is maintained constant, the rate of evaporation is not affected. The pressure of the gas builds up until equilibrium is again established. Since the temperature has not changed, the rate of condensation will be equal to the rate of evaporation only when the pressure is the same as it was before the gas space was doubled. Thus the vapor pressure of the liquid does not depend on the volume of the space above it.

The vapor pressure of a liquid increases with increasing temperature. The magnitude of the increase in vapor pressure with temperature is characteristic of the particular liquid. The variations of vapor pressure with temperature of several liquids

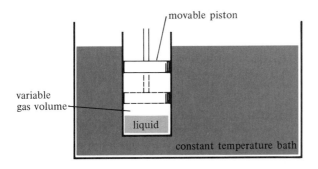

Figure 16-4. **Vapor Pressure of a Liquid Is Independent of the Volume Above the Liquid**

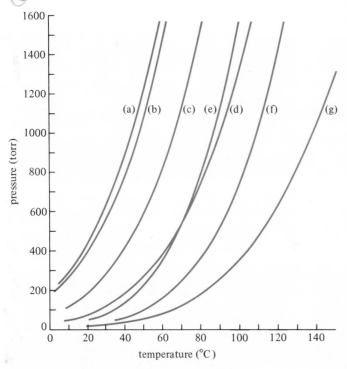

Figure 16-5. Variation of Vapor Pressure with Temperature

(a) Diethyl ether, (b) ethyl bromide, (c) acetone, (d) benzene, (e) ethyl alcohol, (f) water, (g) octane.

are shown in Figure 16–5. Another way of graphing these variations is shown in Figure 16–6: when the logarithm of the observed vapor pressure is plotted against the reciprocal of the absolute temperature, $1/T$, a straight line is obtained. In these cases, the usual equation for a straight line, $y = mx + b$, will have the following form:

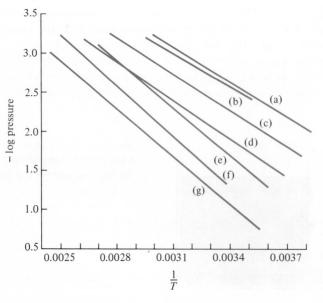

Figure 16-6. Plot of Log P_{vap} Versus $1/T$

(a) Diethyl ether, (b) ethyl bromide, (c) acetone, (d) benzene, (e) ethyl alcohol, (f) water, (g) octane.

$$\log P_{\text{vap}} = -A \left(\frac{1}{T}\right) + B$$

where A and B are constants characteristic of the particular substance.

16-3 The Clausius-Clapeyron Equation

To understand the linear relationship between the logarithm of the observed vapor pressure and temperature, it is necessary to recall some concepts about equilibrium which were presented earlier (Chapters 6 through 8). The equilibrium between a liquid and its vapor at a given temperature can be represented by the equation

liquid $\rightleftharpoons$ gas

The equilibrium constant expression may be written in the customary manner, using square brackets to denote concentrations (activities):

$$K_{\text{eq}} = \frac{[\text{gas}]}{[\text{liquid}]}$$

Since the liquid is pure, its concentration is constant and [liquid] can be represented by a constant, k'. At a given temperature the concentration of a gas is proportional to its pressure:

$$[\text{gas}] = kP_{\text{vap}}$$

Hence the above equilibrium constant expression may be rewritten

$$K_{\text{eq}} = \frac{kP_{\text{vap}}}{k'}$$

Rearranging and collecting constants yields

$$K_{\text{vap}} = P_{\text{vap}}$$

In other words, an equilibrium constant for the vaporization, K_{vap}, is merely the observed vapor pressure of the liquid at the given temperature.

For any process the equilibrium constant and the standard free energy change are related by the following expression (Section 6-5):

$$\Delta G° = -2.30 \, RT \log K$$

where R is the gas constant (1.987 cal/mole $\cdot$ K) and T is the absolute temperature. Therefore, for the vaporization process,

$$\Delta G°_{\text{vap}} = -2.30 \, RT \log P_{\text{vap}}$$

Also, for any vaporization at constant temperature,

$$\Delta G^{\circ}_{vap} = \Delta H^{\circ}_{vap} - T\Delta S^{\circ}_{vap}$$

where ΔH°_{vap} and ΔS°_{vap} are the standard enthalpy of vaporization and standard entropy of vaporization, respectively. Combining these two equations for ΔG°_{vap} gives

$$-2.30\,RT\log P_{vap} = \Delta H^{\circ}_{vap} - T\Delta S^{\circ}_{vap} \quad \text{or} \quad \log P_{vap} = -\frac{\Delta H^{\circ}_{vap}}{2.30R}\left(\frac{1}{T}\right) + \frac{\Delta S^{\circ}_{vap}}{2.30R}$$

If it is assumed that for a given substance, ΔH°_{vap} and ΔS°_{vap} are constants in the temperature range being considered, the last equation is identical in form to the straight line equation relating the experimentally observed vapor pressures with temperature. The term $-A(1/T)$ corresponds to $-(\Delta H^{\circ}_{vap}/2.30R)(1/T)$, and the term B corresponds to $\Delta S^{\circ}_{vap}/2.30R$. In other words, the slopes of the straight lines plotted in Figure 16-6 are proportional to the standard enthalpies of vaporization of the respective liquids. The enthalpies of vaporization of the substances whose vapor pressures are plotted in Figures 16-5 and 16-6 are listed in Table 16-1.

If the enthalpy of vaporization of a substance and its vapor pressure at any temperature are known, the vapor pressure at any other temperature may be calculated. For a given liquid at two different temperatures, T_1 and T_2:

$$\log P_1 = -\frac{\Delta H^{\circ}}{2.30R}\left(\frac{1}{T_1}\right) + \frac{\Delta S^{\circ}}{2.30R}$$

$$\log P_2 = -\frac{\Delta H^{\circ}}{2.30R}\left(\frac{1}{T_2}\right) + \frac{\Delta S^{\circ}}{2.30R}$$

Subtracting the first equation from the second,

$$\log P_2 - \log P_1 = \log\frac{P_2}{P_1} = \frac{\Delta H^{\circ}}{2.30R}\left(\frac{1}{T_1} - \frac{1}{T_2}\right)$$

The last equation is known as the **Clausius–Clapeyron equation.** It may be derived in

TABLE 16-1. Some Enthalpies of Vaporization

Compound	Formula	ΔH°_{vap} at Temperature t		Temperature t (°C)
		kcal/mole	kJ/mole	
diethyl ether	$C_2H_5OC_2H_5$	6.22	26.0	34.6
ethyl bromide	C_2H_5Br	6.53	27.3	38.4
acetone	$CH_3\overset{\overset{\text{O}}{\|\|}}{C}CH_3$	7.23	30.3	56.1
ethyl alcohol	C_2H_5OH	9.40	39.3	78.3
water	H_2O	9.72	40.7	100.0
octane	$CH_3(CH_2)_6CH_3$	8.36	35.0	125.7
benzene	C_6H_6	7.37	30.8	80.2

a more rigorous fashion using calculus. The variation with temperature of *any* equilibrium constant, K, can be represented by an analogous equation:

$$\log \frac{K_2}{K_1} = \frac{\Delta H^\circ}{2.30R} \left(\frac{1}{T_1} - \frac{1}{T_2} \right)$$

Example

Between 20 and 80°C, the enthalpy of vaporization of benzene is 7800 cal/mole. At 26°C, the vapor pressure of benzene is 100 torr. Calculate its vapor pressure at 60°C.

$$\log \frac{100 \text{ torr}}{P_{60^\circ}} = \frac{7800 \text{ cal/mole}}{(2.30)(1.99 \text{ cal/mole} \cdot \text{K})} \left(\frac{1}{333 \text{ K}} - \frac{1}{299 \text{ K}} \right) = -0.58$$

$$P_{60^\circ} = 380 \text{ torr}$$

16-4 Boiling Point

Vaporization of a liquid occurs because some of the molecules have sufficient kinetic energies to escape from the surface of the liquid against the pressure of the surrounding atmosphere. If the liquid is heated, a temperature is eventually reached at which large numbers of molecules have sufficient energies to push back the atmosphere. The free energy of the liquid becomes equal to that of the vapor. As a result, bubbles of vapor are formed throughout the liquid phase. The pressure of the vapor inside the bubbles is at least equal to the pressure of the atmosphere plus the very small pressure due to the weight of the liquid above the bubble. This process of vaporization within the body of the liquid is called ebullition or **boiling**. The temperature at which the vapor pressure of a liquid is equal to the external pressure above the surface of the liquid is called the **boiling point** of the liquid. The temperature at which the vapor pressure of a liquid is exactly 1 atm is called its **normal boiling point.** This is the temperature at which a liquid boils when the surrounding pressure is 760 torr. However, if the pressure above a liquid is adjusted appropriately, the liquid can be made to boil at any temperature between its freezing point and its critical temperature.

Example

Use the Clausius–Clapeyron equation to estimate the boiling point of water at 24 torr pressure. The average ΔH_{vap} over the temperature range is 10.12 kcal/mole.

Since the normal boiling point of water is known to be 100°C and P_{vap} at that temperature is 1.00 atm, the Clausius–Clapeyron equation may be applied to determine the temperature at which the vapor pressure is 24 torr:

At 100°C (373 K): $P_{vap} = 760$ torr

At T: $P_{vap} = 24$ torr

$$\log \frac{760}{24} = \frac{10.12 \times 10^3}{2.30(1.99)} \left(\frac{1}{T} - \frac{1}{373} \right)$$

$$T = 298 \text{ K} = 25°\text{C}$$

To make water boil at 25°C, it must be placed in a closed container from which air is pumped until the pressure of the gas phase is 24 torr. If all the water is to be boiled at this temperature, continuous pumping must be applied so that the gas pressure is maintained at 24 torr.

16-5   Distillation

If a liquid is boiled and its vapor is subsequently condensed into another vessel, the process is called **distillation** (Figure 16-7). This process is often used to purify a liquid and also serves as a means of identifying pure liquids because each substance has a characteristic normal boiling point. However, if a compound happens to be unstable at elevated temperatures or if its normal boiling point is extremely high, the distillation may be effected at a temperature lower than the normal boiling point by **vacuum distillation.** The apparatus used for vacuum distillation is similar to that of Figure 16-7, but all connections are made airtight, and the entire apparatus is evacuated to achieve the desired low pressure.

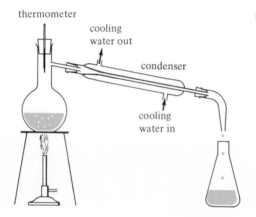

Figure 16-7. **Distillation Apparatus**

16-6   Achieving the Critical Temperature

When a liquid is heated in an evacuated, sealed tube, the vapor pressure of the liquid increases and hence the density of the gas phase increases. Simultaneously, the expansion of the liquid causes its density to decrease. If heating is continued toward the critical temperature of the substance, the densities of the liquid and the vapor approach each other, as shown graphically in Figure 16-8 for the case of CCl_2F_2. The average of the densities of the two phases plotted against temperature yields a straight line. The temperature at which the two curves and the straight line intersect is the critical temperature. As the critical temperature of a liquid is neared, its surface tension approaches zero. At the critical temperature, the meniscus (the surface separating the liquid phase from gas phase) suddenly vanishes. Below the critical temperature, the pressure in the tube is the vapor pressure of the liquid. At the critical temperature, the vapor pressure is, of course, the critical pressure. Above the critical temperature, the pressure of the substance varies with temperature and volume as is expected for a gas under high pressure.

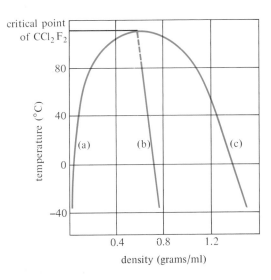

Figure 16-8. **Densities of Liquid CCl₂F₂ and Its Vapor near the Critical Point** (a) Density of the vapor phase, (b) average of the densities of the liquid and vapor phases, (c) density of the liquid phase.

16-7 Freezing Point

If the temperature of a pure liquid is lowered, the average kinetic energy of its molecules decreases, and its vapor pressure also decreases. At a certain temperature, characteristic of the particular liquid, the intermolecular attractive forces predominate, and random translational molecular motions cease. If this were the only effect of lowering the temperature, such cooled liquids could be regarded as being extremely viscous. However, a more profound change usually occurs: not only does the random molecular motion stop but the molecules become fixed in definite geometric arrangements. In short, the liquid is transformed into a **crystalline solid** (see Section 18-1). At a given pressure, this transformation occurs at a definite temperature and is accompanied by a definite enthalpy change, known as the enthalpy of fusion. When the external pressure is 1 atm, the temperature of solidification is called the **freezing point.**

In a closed container at a given temperature, a crystalline solid, like a liquid, can exist in equilibrium with its vapor. The transformation of a substance directly from the solid state into the vapor state is called **sublimation,** and the equilibrium vapor pressure of the solid is called its **sublimation pressure.** The sublimation pressure of a solid substance is plotted against temperature in Figure 16-9, along with a similar plot of the vapor pressure of the same substance as a liquid. It should be noted that there are two separate curves rather than one continuous curve. One curve corresponds to the sublimation of the solid and the other to vaporization of the liquid. At the point of intersection of the two curves, the solid and liquid phases have the same vapor pressure. It is a general principle that when two phases are simultaneously in equilibrium with a third phase, the two phases are also in equilibrium with each other. Therefore, in Figure 16-9, the point of intersection denotes a unique situation in which solid, liquid, and gas are simultaneously in equilibrium, with no other substance present. The temperature at which this situation occurs is called the **triple point.** For water the triple point, at which the equilibrium vapor pressure is 4.58 torr, is *defined* as exactly 0.01 °C. The triple point and the freezing point are not identical.

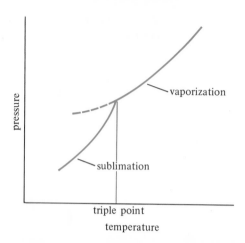

Figure 16-9. **Vapor Pressures of a Solid and Its Liquid; the Triple Point**

The freezing point is the temperature at which the solid and liquid of a substance are in equilibrium under a pressure of 1 atm. For water, this temperature is 0.000°C.

In some cases, particularly when the molecules of the liquid are quite complex and irregularly shaped, the liquid may be cooled to a temperature below the triple point without formation of a solid. This phenomenon, called **supercooling,** represents an unstable state. In Figure 16-9, the dashed line extending beyond the point of intersection represents the vapor pressure of the supercooled liquid.

16-8 Phase Diagrams and the Phase Rule

A given substance may exist in various states depending on the specific conditions which prevail. The behavior of the substance may be summarized on a graph, called a **phase diagram.** Examples of phase diagrams are shown in Figures 16–10 and 16–11. In Figure 16–10, the phase diagram for water, curve *TBC* represents the equilibrium between liquid water and water vapor; along line *TA*, solid and liquid water are in equilibrium; and curve *TS* represents the sublimation equilibrium of solid water. Point *T* is the triple point, point *F* is the freezing point, point *C* is the critical point,

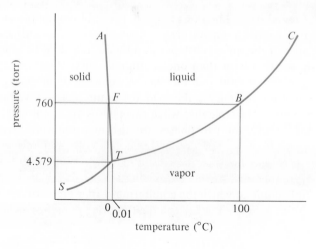

Figure 16-10. **Phase Diagram for Water (not drawn to scale)**

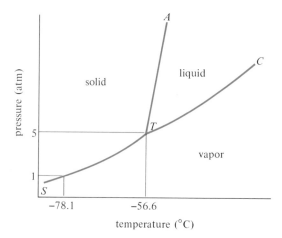

Figure 16-11. **Phase Diagram for Carbon Dioxide (not drawn to scale)**

and point B is the normal boiling point of water. It should be noted in Figure 16-10 that the line TA has a negative slope. With increasing pressure, the melting point of ice is shifted toward lower temperatures. This property of water makes ice skating possible; under the pressure of the skate blade, the ice melts and the skate glides along on a thin layer of water. This behavior of water is unique; the melting points of most substances increase with increasing pressure.

From the phase diagram for carbon dioxide (Figure 16-11) it may be noted that CO_2 has a sublimation pressure of 1 atm at $-78.1°C$, while its triple point is $-56.6°C$. This substance can exist as a liquid only under pressures greater than 5 atm. Solid carbon dioxide in equilibrium with gaseous carbon dioxide at 1 atm pressure maintains a constant temperature of $-78.1°C$. It is used as a refrigerant known as dry ice (so named because it sublimes rather than melts).

Phase diagrams, such as those in Figures 16-10 and 16-11, are readily interpreted in terms of a generalization known as the **phase rule.** The rule states that the number of phases, p, plus the number of variables, also called degrees of freedom, f, exceeds the number of components, c, by 2. Written algebraically the phase rule is

$$p + f = c + 2$$

Variables are conditions, such as temperature or pressure, which might be arbitrarily changed. The number of components is the number of chemically unique species which are present.[2] For example, consider the phase diagram for water, Figure 16-10. There is but one component, H_2O. At the triple point (point T), three phases are in equilibrium; $p = 3$. Therefore, applying the phase rule,

$$p + f = c + 2$$
$$3 + f = 1 + 2$$
$$f = 0$$

The system is invariant; there is only one temperature and one vapor pressure at which the three phases can exist simultaneously at equilibrium.

---

[2] Chemical species whose concentrations are *not* independently variable in the system, such as H_3O^+ and OH^- ions in equilibrium with H_2O, are not considered to be separate components.

Consider the possible equilibria along the curve TC, where liquid and vapor coexist. Since $p = 2, f = 1$. The system can be defined in terms of one variable, either temperature or pressure. If one of these is specified, the other is also fixed. For example, at 25°C the equilibrium vapor pressure of water *must* be 24 torr.

Finally, consider any position not on one of the curves, say in the region of water vapor. Since only the gas phase exists, $p = 1$ and therefore $f = 2$. A stated value of the temperature does not fix the pressure; that is, both temperature and pressure can be independently varied as long as only one phase is present. The phase rule is applicable to all systems containing phases in equilibrium. It is particularly useful in interpreting phase equilibria which involve mixtures (see Section 19–4).

16-9 Cooling Curves

Evaporation cools. During evaporation of a liquid only the more energetic molecules escape into the vapor state. If no heat is supplied from the surroundings, the molecules which remain will have a lower average kinetic energy, and the temperature of the liquid will be lowered. The cooling effect of evaporation is apparent to anyone who has emerged from a swim on a breezy day. A practical application of the cooling effect of evaporation is in the storage of liquefied gases. When kept in an insulated vessel, the evaporation of a small fraction of the liquefied gas causes sufficient cooling to maintain the bulk of the liquid below its boiling point.

On the other hand, if the temperature of the liquid is to remain constant as part of it is vaporized, heat must be added from the surroundings. The quantity of heat required per mole is, of course, the enthalpy of vaporization. If a pure liquid is at its boiling point, all of the added heat goes into the vaporization process. Increasing the rate of heating will increase the rate of boiling, but the temperature of the liquid will remain constant. Similarly, if heat is added to a pure crystalline solid at its melting point, the substance melts at constant temperature. The added heat (enthalpy of fusion) changes the solid to liquid but does not change the temperature. A diagram showing these effects for a given substance is known as a heating (or cooling) curve. In such curves, the temperature of a substance being heated is plotted against the heat added. The heating curve for water is shown in Figure 16–12. The curve shows

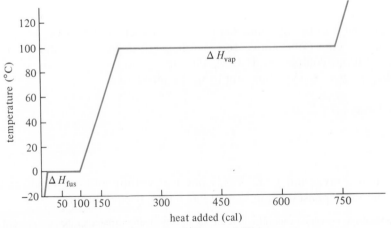

Figure 16-12. **Heating Curve for 1 Gram of Water**

the change in temperature for 1.0 gram of water, initially in the form of ice at $-20\,°C$, as heat is added. At first there is a rapid rise in temperature corresponding to the absorption of 0.49 cal/gram · deg until the melting point is reached. Then upon addition of the next 80 cal, the gram of ice changes to liquid water at constant temperature. When all of the ice has melted, additional heat causes the temperature to rise at an average rate of 1.0 deg/cal until the normal boiling point is reached. Then 540 cal, corresponding to the enthalpy of vaporization, is absorbed at constant temperature until all of the water is converted to its vapor. Thereafter, the temperature changes at a rate corresponding to the heat capacity at constant pressure for water vapor.

16-10   Thermodynamics of Phase Change

Any process occurring under equilibrium conditions is a reversible process, and if two phases of a substance are in equilibrium, the molar free energy of the substance must be the same in each of the phases. Thus for a liquid in equilibrium with its solid form,

$$\text{liquid} \rightleftharpoons \text{solid}$$
$$\Delta G = 0 = \Delta H - T\Delta S$$
$$\Delta S = \frac{\Delta H}{T}$$

In this case, ΔH is the enthalpy of fusion.

Example

The enthalpy of fusion of H_2O at $0\,°C$ is 1.435 kcal/mole. Calculate ΔS for the following process at $0\,°C$: $H_2O(l) \rightleftharpoons H_2O(s)$.

$$\Delta S = \frac{\Delta H}{T} = \frac{-1435 \text{ cal/mole}}{273 \text{ K}} = -5.26 \text{ cal/mole} \cdot \text{K}$$

Note that, as shown in the above example, there is a decrease in entropy in going from liquid to solid. This decrease is expected, since the solid state is much less random than the liquid state.

Often it is desired to calculate changes in thermodynamic properties for reactions which are not reversible. These calculations may be done, provided that the process can be conceived of as occurring by means of a series of reversible steps for which appropriate data are available. For example, the freezing of supercooled water at $-10\,°C$ to ice at $-10\,°C$ is an irreversible process.

$$H_2O(l) \xrightarrow[-10\,°C]{} H_2O(s)$$

However, ΔH, ΔS, and ΔG can be obtained for this process because the enthalpy and entropy changes for the following series of reversible steps can be calculated:

I. $H_2O(1, -10°C) \rightleftharpoons H_2O(1, 0°C)$

II. $H_2O(1, 0°C) \rightleftharpoons H_2O(s, 0°C)$

III. $H_2O(s, 0°C) \rightleftharpoons H_2O(s, -10°C)$

Since heating and cooling can occur reversibly, the enthalpy changes can be calculated as follows: In the range $-10°$ to $0°C$, liquid water has a specific heat of 1.0 cal/gram $\cdot$ deg:

$$\Delta H_{rev} = C_p \Delta t = (18 \text{ grams})(1.0 \text{ cal/gram} \cdot \text{deg})[(0 - (-10)) \text{ deg}] = +180 \text{ cal}$$

An equation to calculate ΔS for processes which involve change in temperature, assuming C_p is constant, is derived using calculus.

$$\Delta S = 2.30 C_p \log \frac{T_2}{T_1}$$

where T_1 is the initial temperature, and T_2 is the final temperature. Applied to this example,

$$\Delta S_I = 2.30(18 \text{ cal/mole} \cdot \text{K}) \log \frac{273}{263} = 0.67 \text{ cal/mole} \cdot \text{K}$$

Similarly, in the temperature range 0 to $-10°C$, ice has a specific heat of 0.49 cal/gram $\cdot$ K, and the enthalpy and entropy changes for step III are thus given by the following equations:

$$\Delta H_{III} = \frac{18 \text{ grams}}{\text{mole}} \left(\frac{0.49 \text{ cal}}{\text{gram} \cdot \text{K}} \right) (263 \text{ K} - 273 \text{ K}) = -88 \text{ cal/mole}$$

$$\Delta S_{III} = 2.30(18 \times 0.49) \log \frac{263}{273} = -0.33 \text{ cal/mole} \cdot \text{K}$$

At $0°C$, the freezing is accompanied by the following enthalpy and entropy changes:

$$\Delta H_{II} = -1440 \text{ cal/mole} \qquad \text{(Table 3-4)}$$

$$\Delta S_{II} = \frac{\Delta H}{T} = \frac{-1440 \text{ cal/mole}}{273 \text{ K}} = -5.27 \text{ cal/mole} \cdot \text{K}$$

Overall, the enthalpy and entropy changes are the sums of these changes for the individual steps:

$$\Delta H = 180 \text{ cal/mole} - 1440 \text{ cal/mole} - 88 \text{ cal/mole} = -1350 \text{ cal/mole}$$

$$\Delta S = 0.67 \text{ cal/mole} \cdot \text{K} + (-5.27 \text{ cal/mole} \cdot \text{K}) + (-0.32 \text{ cal/mole} \cdot \text{K})$$
$$= -4.92 \text{ cal/mole} \cdot \text{K}$$

Finally, at $-10°C$:

$$\Delta G = \Delta H - T\Delta S$$
$$= -1350 \text{ cal/mole} - (263 \text{ K})(-4.92 \text{ cal/mole} \cdot \text{K}) = -60 \text{ cal/mole}$$

Since the free energy change depends only on the initial and the final states, ΔG is the same for both this three-step reversible path and the irreversible path, as well as any other.

16–11   Solutions

Homogeneous mixtures are called **solutions.** Therefore a solution may be defined as a mixture of two or more substances in a single phase. Since they are mixtures, solutions have properties which depend on the properties of their constituents and on the relative quantities of each of the constituents present. For example, the freezing points of 1 M and 2 M sugar solutions in water are different. In the study of solutions, one of the chief objectives is to ascertain how the properties of a solution depend on its constitution and concentration.

It is customary and convenient to regard one constituent of the solution as the **solvent** and all other constituents as **solutes.** Water is perhaps the most widely used solvent, and the study of aqueous solutions is of great importance. Solutions which contain relatively small quantities of solute are said to be **dilute.** Solutions which contain relatively large quantities of solute are said to be **concentrated.**

16–12   Solubility

Unless they react with each other, gases mix readily with each other in any proportions to form gas phase solutions. Some pairs of liquids, such as water and ethyl alcohol, also form solutions regardless of the proportions of the constituents. In contrast, it is more often found that mixtures of gases and liquids, liquids and liquids, and solids and liquids form solutions only over a limited range of compositions. For example, at 20°C, carbon tetrachloride dissolves in water only to the extent of 0.08 gram/100 ml of water. The terms **soluble, slightly soluble,** and **insoluble** are used to describe decreasing tendencies of solutes to dissolve in a particular solvent at a given temperature.

A general rule often useful for predicting solubilities in terms of molecular structure is "like dissolves like." Solvents having molecules with dipole moments, such as water, dissolve solutes with dipolar molecules and solutes which are ionic. Solvents having molecules without dipoles dissolve solutes of a similar nature. Thus carbon tetrachloride dissolves hydrocarbons, iodine, and other nonpolar materials, which generally are insoluble in water.

The process of dissolving may be represented by an equation of the type

$$\text{solute} \;+\; \text{solvent} \;\rightarrow\; \text{solution} \quad (\text{heat evolved} \;=\; -\Delta H_{\text{soln}})$$

The **enthalpy of solution,** ΔH_{soln}, results from the difference between the energies of attraction of solute molecules for solvent molecules and of solute molecules for each other and solvent molecules for each other. In general, the formation of a solution increases the entropy of the system, because the mixture is more random than the separate constituents. Thus substances which dissolve in a solvent with large negative enthalpies of solution (highly exothermic processes) are likely to be quite soluble, because both enthalpy effects and entropy effects are favorable. Substances which dissolve with the absorption of heat (positive enthalpy of solution) will have limited solubilities because only the entropy effects favor the solution process.

The variation of solubility with temperature can be predicted from the enthalpy of solution. If ΔH_{soln} is negative, the solubility of the solute decreases as the temperature increases; if ΔH_{soln} is positive, the solubility increases with temperature. The variations with temperature of the solubilities in water of several substances are plotted in Figure 16-13. Note that the solubility of NaCl does not vary much with temperature, a consequence of the relatively small magnitude of its enthalpy of solution (1.3 kcal/mole). Note also that the solubility of Na_2SO_4 decreases with temperature, indicating that its enthalpy of solution is negative.

In cases where the solute is present in quantity greater than will go into solution, an equilibrium state is established:

$$solute \;+\; solvent \;\rightleftharpoons\; solution$$

The solution in equilibrium with the excess solute is said to be **saturated.** The mixture contains two components—solvent and solute—and three phases—vapor, solution, and undissolved solute. According to the phase rule (Section 16-8),

$$p + f = c + 2$$
$$3 + f = 2 + 2$$

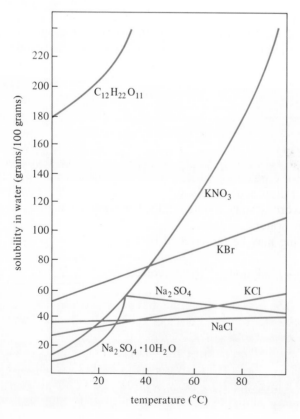

Figure 16-13. **Variation of Solubility with Temperature**

There is one degree of freedom, which customarily is taken as temperature. If the temperature is fixed, all other variables, such as concentration and vapor pressure, must also be fixed. Thus at a given temperature, a given saturated solution has a definite concentration. That concentration is called the **solubility,** s, at that temperature. Solubility may be expressed in a variety of units, including mass per unit volume, moles per unit volume, or moles per unit mass of solvent. Saturated solutions of ionic substances have already been discussed in Chapter 8. For these cases, the solubility is expressed in terms of the solubility product constant, K_{sp}.

16-13 Supersaturated Solutions

A supersaturated solution is one which contains solute at a concentration greater than that of a saturated solution of the substance at the same temperature. In the case of some substances whose solubilities increase with temperature, it is possible to prepare a supersaturated solution in the following manner. First, a saturated solution containing an excess of undissolved solute is prepared at a given temperature. The mixture is then heated until all of the excess solute is dissolved. The solution is allowed to cool slowly to its initial temperature. If no solute precipitates upon cooling, the solution now contains more solute than the equilibrium concentration.

Supersaturated solutions are possible because crystallization always begins at some "nucleus" at which the crystal pattern can be established. Therefore, supersaturated solutions can exist only in the absence of a nucleation site. A minute crystal of solute will cause the excess solute to precipitate. Shaking, scratching, or even adding a slight bit of dust may also induce crystallization in a supersaturated solution. Supersaturated solutions of sodium acetate are easy to prepare and are often used to demonstrate this phenomenon.

16-14 Solubility of Gases—Henry's Law

It is generally observed that the solubility of gases decreases with increasing temperature. For example, water at room temperature usually contains dissolved air. When the water is heated to boiling, the air is expelled. If the water were cooled back to room temperature and a fish were put into it, the fish would have to surface and gulp air in order not to drown.

At a given temperature, the solubilities of gases which do not react extensively with the solvent are directly proportional to the partial pressures of the gases above the solution. This proportionality is known as Henry's law. Mathematically, it can be expressed as follows:

$$[X] = KP_X$$

where [X] is the concentration of dissolved gas and P_X is its partial pressure above the solution. K is the Henry's law constant. If the solubility of a gas is known for one pressure, the Henry's law constant may be calculated and then used to calculate the solubilities under any other gas pressure.

Example

The solubility of N_2 in water is 2.2×10^{-4} gram in 100 grams of H_2O at 20°C when the pressure of nitrogen over the solution is 1.2 atm. Calculate the solubility at that temperature when the nitrogen pressure is 10 atm.

$$[N_2] = KP_{N_2}$$

Per 100 grams of water,

$$2.2 \times 10^{-4} \text{ gram} = K(1.2 \text{ atm})$$

$$K = \frac{2.2 \times 10^{-4} \text{ gram}}{1.2 \text{ atm}} = 1.8 \times 10^{-4} \text{ gram/atm}$$

$$[N_2] = (1.8 \times 10^{-4} \text{ gram/atm})(10 \text{ atm}) = 1.8 \times 10^{-3} \text{ gram}$$

16-15 Temperature-Independent Concentration Units

The term *concentration* is used to describe the quantity of solute in a given quantity of solution. The terms *molarity* and *normality,* introduced earlier, specify the concentrations of solute in a given volume of solution. Since the volume of a solution changes with temperature, for precise work the temperature must be specified when these concentration units are used. Concentration units which are independent of temperature are more useful for specifying the quantities of both solute and solvent for a given solution. Two such ways of expressing concentration are *mole fraction* and *molality.* These units are useful for discussing the properties of the solution as a whole or for discussing how the presence of dissolved solute affects the properties of a *solvent.* They are temperature independent because they involve the mass or number of moles of material rather than the volume. The volume, and therefore the density, of a solution is not independent of temperature.

If a solution is composed of several constituents, A, B, C, . . . , the **mole fraction** of constituent A is denoted N_A and is equal to the number of moles of that component divided by the total number of moles present. No particular component need be labeled solvent in order to use the mole fraction concept. It follows from the definition of a mole that the mole fraction of a constituent is also the ratio of the number of molecules of that substance to the total number of molecules of all species present.

$$N_A = \frac{\text{moles A}}{\text{total moles}}$$

Similarly,

$$N_B = \frac{\text{moles B}}{\text{total moles}}$$

and so on. It follows from the definition of mole fraction that the sum of the mole fractions of all the components of a solution is 1.

$$N_A + N_B + N_C + \cdots = 1$$

Example

Calculate the mole fraction of water in a mixture consisting of 9.0 grams of water, 120 grams of acetic acid, and 115 grams of ethyl alcohol.

The molecular weights of water, acetic acid, and ethyl alcohol are 18, 60, and 46 D, respectively. The mixture thus contains 0.50 mole of water, 2.0 moles of acetic acid, and 2.5 moles of ethyl alcohol, for a total of 5 moles:

$$N_{H_2O} = \frac{0.50}{5.0} = 0.10$$

In other words, one out of each ten molecules in the solution is a water molecule.

The **molality** (*m*) of a solution is defined as the number of moles of solute per kilogram of solvent. It is important to note that the quantity of solvent is fixed, and thus the number of moles of solvent is fixed. Molality is most often used when discussing the properties of the solvent as affected by the solute concentration (Section 16-18).

Example

Calculate the molalities and the mole fractions of acetic acid in two solutions prepared by dissolving 120 grams of acetic acid **(a)** in 100 grams of water and **(b)** in 100 grams of ethyl alcohol.

(a) In water,

$$\text{molality} = \frac{2.00 \text{ moles acetic acid}}{0.10 \text{ kg water}} = 20 \, m$$

$$N_{\text{acetic acid}} = \frac{2.00 \text{ moles}}{(2.00 + 5.55) \text{ moles}} = 0.265$$

acetic acid
$$HC_2H_3O_2$$
$$\overset{1}{2}4$$
$$4$$
$$\overset{3}{3}2$$
$$60$$

(b) In ethyl alcohol, .

$$\text{molality} = \frac{2.00 \text{ moles acetic acid}}{0.10 \text{ kg ethyl alcohol}} = 20 \, m$$

$$C_2H_6O$$
$$24$$
$$\frac{16}{6}$$
$$46$$

$$N_{\text{acetic acid}} = \frac{2.00 \text{ moles}}{(2.00 + 2.17) \text{ moles}} = 0.480$$

The above example shows that although the molalities of the two solutions are identical, their mole fractions are quite different. Comparison of the properties of two such solutions of equal molality has little meaning. However, comparison of the properties of solutions in a given solvent but having different molalities does provide useful information.

In converting from concentration units involving volume, such as molarity, to units not involving volume, the density of the solution is required.

Example

The density of a 2.03 *M* solution of acetic acid in water is 1.017 grams/ml. Calculate the molality of the solution.

One liter of the solution, containing 2.03 moles of solute, has a mass of 1017 grams. Thus the solution contains 2.03 moles $\times$ 60.0 grams/mole = 122 grams of acetic acid and also $1017 - 122 = 895$ grams of water. The molality is

$$\frac{2.03 \text{ moles solute}}{0.895 \text{ kg water}} = 2.27 \ m$$

16–16 Raoult's Law

In a solution, the properties of the solvent, such as its vapor pressure and its boiling point, are modified by the presence of a solute. The modifications depend on the relative numbers of solute and solvent molecules and also on the intermolecular forces between the various molecules. In real solutions the strengths of the intermolecular forces range from weak van der Waals attractions to strong dipole–dipole and ion–dipole interactions.

A hypothetical **ideal solution** is defined as one which can be formed from its components with no evolution or absorption of heat, and whose volume is the sum of the volumes of the individual constituents. These criteria imply that the forces of attraction of solute particles for solvent particles are equal in magnitude to the forces of attraction of solvent molecules for each other and of solute particles for each other. Benzene and toluene form almost ideal solutions. The properties of an ideal solution of a given set of components depend only on its concentration. For example, the vapor pressure, P_A, of a volatile component, A, of such a solution is proportional to its mole fraction. Moreover, the proportionality constant is simply the vapor pressure of the pure component, P_A°, at the same temperature:

$$P_A = N_A P_A^\circ$$

This relationship is known as **Raoult's law,** which describes the behavior of ideal solutions at all concentrations. Real solutions may show large deviations from Raoult's law, but they approximate the law when solute concentrations are small.

The consequences of Raoult's law on the boiling point and the freezing point of a solution containing a nonvolatile solute are shown graphically in Figure 16–14. Since only the solvent is volatile, the vapor pressure is due only to solvent molecules. At any temperature, the vapor pressure of the solution is lower than that of the pure solvent. Therefore, at the temperature at which the pure solvent has a vapor pressure of 1 atm (its normal boiling point), the solution will have a vapor pressure which is lower by an amount approximated by Raoult's law. The solution must be heated to a higher temperature to achieve a vapor pressure of 1 atm. Thus at any pressure, the boiling point of a solution containing nonvolatile solute will be higher than the boiling point of the pure solvent at that same pressure. In Figure 16–14 the elevation of the normal boiling point is designated Δt_b.

In Figure 16–14 it may also be seen that the vapor pressure curve of the solution intersects the sublimation pressure curve of the solid solvent at a temperature which is lower than the triple point of the pure solvent. Accordingly, the triple point and freezing point of a solution from which only the solvent separates as solid are lower than those of the pure solvent. In the figure, the freezing point depression due to the presence of the solute is denoted Δt_f.

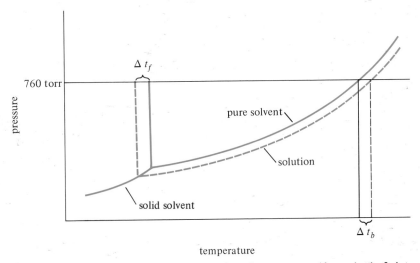

Figure 16–14. **Vapor Pressure of a Solution Containing a Nonvolatile Solute** (not drawn to scale)

16–17   Eutectic Mixtures

It is of interest to apply the phase rule to the freezing of a solution consisting of a solvent and one solute. At the temperature at which the vapor pressure curve of the solution intersects the sublimation pressure curve of the solvent (the freezing point of the solution), crystals of solvent appear. Thus there are two components and three phases in the system—solution, solid solvent, and vapor.

$$3 + f = 2 + 2$$
$$f = 1$$

Hence, specification of only one variable (temperature or vapor pressure or concentration) will define the system. As the temperature is lowered further, crystallization of solvent continues, and the solution phase becomes more concentrated until finally a saturated solution is obtained. At this point, both solute and solvent crystallize separately but simultaneously, and there are four phases—solid solute, solid solvent, solution, and vapor.

$$4 + f = 2 + 2$$
$$f = 0$$

The system is invariant. Simultaneous crystallization of solvent and solute continues at constant temperature until all of the liquid solution is solidified. The mixture which solidifies under these conditions is called a **eutectic mixture,** and the constant temperature at which this solidification takes place is called the **eutectic temperature** or **eutectic point.** These changes are shown in Figure 16–15. The composition and the eutectic temperature are characteristics of the mixture. Several aqueous eutectics are described in the accompanying table. Such mixtures are often used to maintain constant, low temperatures.

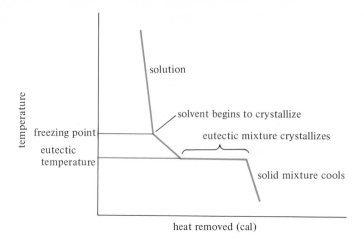

Figure 16-15. **Cooling Curve for a Solution Containing a Single Solute**

Solute	Eutectic Composition (grams/100 grams H_2O)	Eutectic Temperature (°C)
KCl	24.6	−10.7
NH_4Cl	19.7	−15.4
NaCl	35	−21.3
$CaCl_2$	48	−51

16-18   Colligative Properties

As seen in Figure 16–14, the elevation of the boiling point and the lowering of the freezing point of a solution relative to the corresponding points of the pure solvent are consequences of the lowering of the vapor pressure. These properties of solutions are known as **colligative properties.** All three of these phenomena result from the lowering of the "escaping tendency" of the solvent molecules due to the presence of the solute. The chemical nature of the solute does not significantly influence the magnitude of the colligative properties. Regardless of the nature of the solute, the number of solvent particles in a given quantity of solution is a certain fraction of those which would be there if the solvent were pure. The escaping tendency of solvent molecules in the solution is therefore a fraction of the escaping tendency of the molecules in the pure solvent.

The equations expressing the magnitudes of the colligative properties are quite similar in form. For example, consider a solvent, A, containing a solute, B, at a given temperature. The mole fractions of solvent and solute are N_A and N_B, respectively. The lowering of the vapor pressure, ΔP, may be expressed in terms of Raoult's law, as follows:

$$\Delta P = P_A^\circ - P_A = P_A^\circ - N_A P_A^\circ = (1 - N_A)P_A^\circ = N_B P_A^\circ$$

Since at a given temperature, P_A° is a constant, the lowering of the vapor pressure is directly proportional to the mole fraction of solute, N_B. If the solution is dilute, the number of moles of B will be small compared to the number of moles of A, and the mole fraction of solute may be approximated as the number of moles of solute per mole of solvent:

$$N_B = \frac{n_B}{n_B + n_A} \cong \frac{n_B}{n_A} \qquad \text{(dilute solution of B in A)}$$

The number of moles of solvent in 1 kg of solvent is given by

$$n_A = \frac{1000 \text{ grams}}{M_A}$$

where M_A is the molecular weight of the solvent, which gives

$$N_B = \frac{n_B M_A}{1000 \text{ grams}}$$

The number of moles of solute per kilogram of solvent is the solute molality, m. Thus, for dilute solutions the lowering of the vapor pressure can be written in terms of molality, as follows:

$$\Delta P = N_B P_A^\circ = m M_A P_A^\circ$$

Since for a given solvent, $M_A P_A^\circ$ is a constant, k_p,

$$\Delta P = k_p m$$

Example

Estimate the lowering of the vapor pressure due to the solute in a 1.0 m aqueous solution at 100°C.

The normal boiling point of water is 100°C; hence

$$P_{H_2O}^\circ = 1.00 \text{ atm}$$

$$M_{H_2O} = 18.0 \text{ grams/mole}$$

$$\Delta P = m M_A P_A^\circ = \left(\frac{1.0 \text{ mole}}{1000 \text{ grams}}\right)\left(\frac{18.8 \text{ grams}}{\text{mole}}\right)(1.0 \text{ atm}) = 0.019 \text{ atm} = 14 \text{ torr}$$

Thus, k_p for H_2O is 14 torr/m.

The elevation of the boiling point of a solvent, Δt_b, and the lowering of its freezing point, Δt_f, due to the presence of a solute can be shown by analogous derivations to be

$$\Delta t_b = k_b m$$
$$\Delta t_f = k_f m$$

where k_b and k_f are called the molal boiling point elevation constant and the molal freezing point depression constant, respectively. These constants are characteristic of

TABLE 16-2. Boiling Point Elevation and Freezing Point Depression Data

Solvent	Normal Boiling Point (°C)	k_b (°C/m)	Freezing Point (°C)	k_f (°C/m)
acetic acid	118.9	3.1	16.6	3.9
benzene	80.1	2.53	5.5	5.12
chloroform	61.2	3.63	—	—
naphthalene	—	—	80.22	6.85
water	100.0	0.512	0.00	1.86

the particular solvent. Some typical examples are given in Table 16-2. It should be noted that for every solvent, k_f has a value different from k_b.

Example

Calculate **(a)** the freezing point of a solution of 0.0100 mole of glucose dissolved in 100 grams of water, and **(b)** the freezing point of a 0.100 m solution of naphthalene in benzene.

(a) concentration $= \dfrac{0.0100 \text{ mole}}{0.100 \text{ kg}} = 0.100 \; m$

$$\Delta t_f = (1.86°C/m)(0.100 \; m) = 0.186°C$$

Since the freezing point of pure water is 0.000°C, the freezing point of the solution will be −0.186°C.

(b) $\Delta t_f = (5.12°C/m)(0.100 \; m) = 0.512°C$

The freezing point is lowered by 0.512°C, from 5.5 to 5.0°C. (Note the distinction between freezing point and freezing point depression.)

16-19 Molecular Weights from Colligative Properties

Since measurement of a colligative property permits the determination of the effective molal concentration of the solute, convenient methods of determining molecular weights are based on colligative properties. For example, the molality of a solution is obtained from the freezing point depression, as follows:

$$m = \Delta t_f/k_f$$

By dissolving a certain number of grams of solute in a weighed quantity of solvent, one can calculate the molecular weight of solute from the freezing point depression of the solution.

Example

The freezing point of a solution containing 4.80 grams of a compound in 60.0 grams of benzene is 4.50°C. What is the molecular weight of the compound?

$$\Delta t_f = 5.5°C - 4.5°C = 1.0°C$$

$$m = \frac{\Delta t_f}{k_f} = \frac{1.0°C}{5.1°C/m} = 0.20 \; m = 0.20 \; \text{moles/kg}$$

$$M_{\text{compound}} = \frac{4.8 \; \text{grams}/60.0 \; \text{grams benzene}}{0.20 \; \text{moles}/1000 \; \text{grams benzene}} = 400 \; \text{grams/mole}$$

The molecular weight is 400 D.

Example

An aqueous solution containing 288 grams of a nonvolatile compound having the stoichiometric composition $C_n H_{2n} O_n$ in 90.0 grams of water boils at 101.24°C at 1.00 atm pressure. What is the molecular formula of the compound?

$$\Delta t_b = 101.24°C - 100.00°C = 1.24°C$$

$$m = \frac{\Delta t_b}{k_b} = \frac{1.24°C}{0.512°C/m} = 2.42 \; m$$

$$M = \frac{288 \; \text{grams}/90.0 \; \text{grams water}}{2.42 \; \text{moles}/1000 \; \text{grams water}} = 1320 \; \text{grams/mole}$$

The molecular weight is 1320 D. If the value of n were 1, corresponding to CH_2O, the molecular weight would be 30 D. Hence n must be $1320/30 = 44$, and the molecular formula is $C_{44}H_{88}O_{44}$.

16-20   Osmotic Pressure

Certain animal membranes, such as the outer covering of the intestines, are **semipermeable;** that is, they permit the passage of water but not any solutes dissolved in the water. (Artificial semipermeable membranes may also be made, for example, by precipitating copper(II) hexacyanoferrate(III), $Cu_3[Fe(CN)_6]_2$, in the pores of unglazed porcelain.) When a semipermeable membrane is placed between pure water and an aqueous solution, as shown in Figure 16–16, the phenomenon of **osmosis** occurs. Solvent diffuses *into* the solution, and the height of the solution rises above that of the pure solvent. The process continues until the hydrostatic pressure due to the extra height of the solution prevents further osmosis. Because the solution is diluted by the solvent which has passed through the membrane, the concentration after osmosis is not that of the original solution. Therefore the **osmotic pressure** of a solution is defined as the pressure which must be applied above the solution to *prevent* passage of solvent through a semipermeable membrane into the solution.

Quantitative measurements show that the osmotic pressure is proportional to the mole fraction of solute. Osmotic pressure is a colligative property because it depends only on the concentration of solute and not on the nature of the solute particles. The osmotic pressure is the pressure which must be applied to the solution to restore the partial pressure of solvent over the solution to the value of the vapor pressure of pure solvent at that temperature. For dilute *aqueous* solutions, the osmotic pressure, π, of a solution containing n moles of solute in V liters of solution is given by

$$\pi V = nRT$$

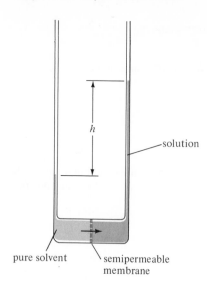

Figure 16-16. Osmosis

The arrow indicates the direction of spontaneous flow of solvent.

where R is the gas constant and T is the absolute temperature. Note that if V is in liters, n/V is molarity.

Example

Calculate the osmotic pressure of a 0.100 M solution of a nonelectrolyte at 0°C.

$$\pi = \frac{n}{V} RT = \left(\frac{0.100 \text{ mole}}{\text{liter}}\right)\left(\frac{0.0821 \text{ liter} \cdot \text{atm}}{\text{mole} \cdot \text{K}}\right)(273 \text{ K}) = 2.24 \text{ atm}$$

This example shows that even for dilute solutions osmotic pressures are very large. Since pressure measurements as low as 1 torr can be made conveniently, osmotic pressure measurements provide an accurate way of determining the molecular weights of very large molecules, such as protein molecules (Section 23–11).

Example

What is the molecular weight, M_A, of a solute, A, if the osmotic pressure of a solution containing 10.0 grams/liter is 10.0 torr at 27°C?

$$\pi V = nRT = \frac{m}{M_A} RT$$

$$M_A = \frac{mRT}{\pi V} = \frac{(10.0 \text{ grams})(0.0821 \text{ liter} \cdot \text{atm/mole} \cdot \text{K})(300 \text{ K})}{[(10.0/760) \text{ atm}](1.00 \text{ liter})} = 18,700 \text{ grams/mole}$$

16–21 Escaping Tendency and Free Energy Change

The escaping tendency of solvent molecules from a solution or from pure solvent can be associated with the change in free energy of a given process. For example, when a semipermeable membrane separates a solution from a sample of the solvent,

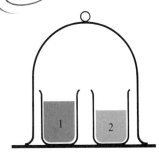

Figure 16–17. **Demonstration of Relative Escaping Tendencies of Solvent in Two Solutions**

Beakers of two aqueous solutions of different concentrations are placed in a bell jar and allowed to stand for a period of time. Water evaporates more rapidly from the more dilute solution (beaker 2), building up a higher vapor pressure than that of the more concentrated solution (beaker 1). Consequently, water vapor condenses in beaker 1, diluting the solution, and water continues to evaporate from beaker 2, causing that solution to become more concentrated. Ultimately, the two solutions attain equal concentrations, and therefore equal vapor pressures.

osmosis occurs spontaneously. Hence the free energy change for this process is negative. The more concentrated the solution, the more negative will be ΔG. It follows that if two *solutions* are separated from each other by a semipermeable membrane, osmosis will proceed by passage of solvent from the more dilute to the more concentrated until equilibrium is established, at which point ΔG will be zero. Another way of stating this principle is that at equilibrium, the escaping tendency of solvent molecules is the same on the two sides of the semipermeable membrane. Another type of experiment which demonstrates this point is shown in Figure 16–17.

The arguments presented above show that osmosis is mainly an entropy effect. The enthalpy change upon addition of solvent to a dilute aqueous solution is negligible. Since in osmosis the process is spontaneous, the free energy change must be negative, and a large negative value of ΔG implies an appreciable increase in entropy in this case. An entropy increase should be expected, since a dilute solution is more random than either a concentrated solution or the pure solvent.

16–22   Ideal Solutions of Volatile Solutes

If two constituents of an ideal solution are both volatile, the vapor pressure of each will be lowered in accordance with Raoult's law. For two such substances, A and B, at a given temperature,

$$P_A = N_A P_A^\circ \qquad \text{and} \qquad P_B = N_B P_B^\circ$$

The vapor pressure of the solution is the sum of the vapor pressures of the two constituents; hence

$$P_{soln} = P_A + P_B = N_A P_A^\circ + N_B P_B^\circ$$

A vapor pressure–mole fraction diagram of benzene–toluene mixtures at 30°C, which illustrates this principle, is shown in Figure 16–18. Benzene–toluene solutions exhibit ideal behavior, and the total vapor pressures of various solutions of the two are intermediate between the vapor pressure of pure benzene and pure toluene, 119 and 37.0 torr, respectively, at 30°C.

In such solutions, the composition of the vapor which is in equilibrium with a given solution will be different from the composition of the liquid phase, as can be shown by the following example.

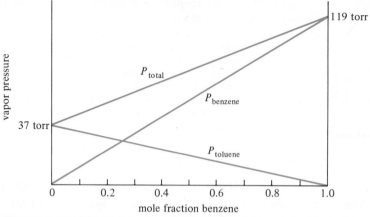

Figure 16-18. **Vapor Pressures of Benzene–Toluene Solutions at 30°C**

Example

What is the composition of the vapor which is in equilibrium at 30°C with a benzene–toluene solution with a mole fraction of benzene of 0.400? with a mole fraction of benzene of 0.600?

$P^{\circ}_{benzene} = 119$ torr

$P^{\circ}_{toluene} = 37.0$ torr

$P_{benzene} = (0.400)(119 \text{ torr}) = 47.6$ torr

$P_{toluene} = (0.600)(37.0 \text{ torr}) = 22.2$ torr

$P_{total} = 47.6 \text{ torr} + 22.2 \text{ torr} = 69.8$ torr

The composition of the vapor is determined by applying Dalton's law of partial pressures (Section 2–17):

$$N_{benzene\ vapor} = \frac{P_{benzene}}{P_{total}} = \frac{47.6 \text{ torr}}{69.8 \text{ torr}} = 0.682$$

$$N_{toluene\ vapor} = \frac{22.2 \text{ torr}}{69.8 \text{ torr}} = 0.318 = 1.000 - 0.682$$

Similarly, for the case of the solution in which the mole fraction of toluene is 0.400,

$$N_{benzene\ vapor} = \frac{71.5}{86.3} = 0.829$$

$$N_{toluene\ vapor} = \frac{14.8}{86.3} = 0.171$$

In both cases cited above, the mole fraction of benzene in the vapor phase is greater than its mole fraction in the liquid phase. This example illustrates the general principle that the vapor phase is always richer in the more volatile component than is the liquid phase with which it is in equilibrium.

Figure 16–19 shows the complete range of liquid–vapor compositions for benzene–toluene solutions at 30°C. In this diagram, the composition of vapor in equi-

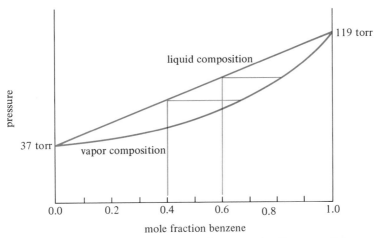

Figure 16-19. **Pressure-Composition Diagram for Benzene-Toluene Solutions at 30°C**

librium with the liquid phase at a given vapor pressure is read from the vapor composition curve at the same pressure. For instance, the compositions of liquid and vapor from the example given above are shown by horizontal lines (tie lines) connecting the respective points. The entire vapor composition curve can be calculated in the same manner.

16-23   Fractional Distillation

In practical work, a more useful diagram is one in which the compositions of vapor and liquid in equilibrium are plotted against the boiling points of the solutions. Such a temperature-composition diagram for benzene-toluene solutions at 1 atm total pressure is shown in Figure 16-20. The vapor of any mixture which boils at a given

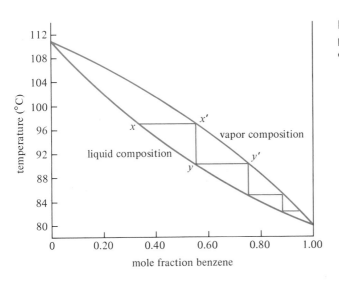

Figure 16-20. **Temperature-Composition Diagram for Benzene-Toluene Solutions at 1 Atm Pressure**

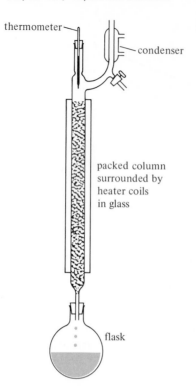

thermometer

condenser

Figure 16–21. **Fractionating Column**

packed column
surrounded by
heater coils
in glass

flask

temperature may be seen to be richer in the more volatile constituent. A procedure for separating mixtures of volatile components based on this phenomenon is called **fractional distillation.** For example, consider the liquid having a composition represented by point x on the diagram. When this liquid is boiled, the vapor will have the composition x', richer in benzene. This vapor can be condensed to give the liquid of composition $y = x'$, which in turn will boil at a lower temperature and give a vapor having composition y', which is even richer in benzene. Every time a fraction of the vapor is condensed and the liquid is reboiled, the new boiling point is nearer that of pure benzene, and the vapor phase is richer in benzene. By successively distilling the fractions, pure benzene will ultimately be obtained.

Instead of repeated evaporations and condensations of numerous fractions, liquid solutions may be conveniently separated by distillation through a fractioning column, such as shown in Figure 16–21. The column is designed so that condensations and evaporations occur along its entire length. The hot vapors moving up the column come into equilibrium successively with the liquid returning down the column, and ultimately the liquid at the top of the column has a composition and boiling point close to that of the pure component of higher vapor pressure. The more volatile component is collected from the top of the column, while the residue in the flask becomes richer in the less volatile component. Efficient fractionating columns are capable of separating mixtures of liquids whose boiling points differ by as little as 1 or 2°C. Huge columns based on this principle are used in the production of gasoline (Figure 16–22).

Figure 16-22. **Commercial Distillation Columns (Courtesy of Texaco, Inc.)**

16-24   Nonideal Solutions

A solution which obeys Raoult's law at all concentrations is said to be ideal. However, most real solutions are not ideal. In cases in which both the solute and the solvent are volatile, the vapor pressures of the constituents can deviate from ideal behavior in one of two ways: The observed partial pressures of the constituents may be greater than calculated from Raoult's law **(positive deviation)**, or they may be less than predicted from the law **(negative deviation).** Diagrams representing these types of behavior are shown in Figure 16-23.

In solutions exhibiting positive deviations from Raoult's law, the attractions of the solvent molecules for the solute molecules are less than the attractions of solute molecules and solvent molecules for like molecules. The formation of such solutions is accompanied by the absorption of heat; thus the solution is in a state of high energy relative to the unmixed constituents. It is not surprising that the escaping tendencies of the molecules are increased over that predicted by Raoult's law. Examples of this type include ethyl alcohol and water solutions as well as carbon disulfide and acetone solutions. Indeed, any solution formed by mixing polar molecules with nonpolar molecules is likely to show positive deviations from Raoult's law.

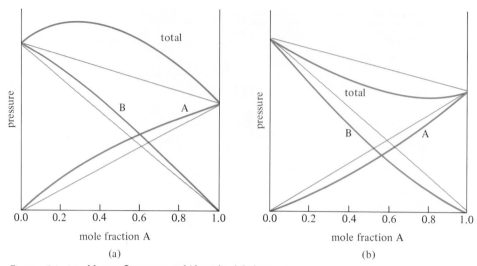

Figure 16-23. Vapor Pressures of Nonideal Solutions
(a) Positive deviation from Raoult's law, (b) negative deviation from Raoult's law.

In solutions showing negative deviations from Raoult's law, the attractions between solute molecules and solvent molecules are large compared to the attractions between like molecules in the mixture. Formation of the solution is accompanied by the evolution of heat. The constituents are therefore in lower energy states than would be the case for an ideal solution, and consequently their escaping tendencies are lower than predicted by Raoult's law. Examples of this type include solutions of formic acid in water and also chloroform–acetone solutions.

For a solution which shows positive deviations from Raoult's law, there is some mixture which has a higher total vapor pressure than the vapor pressure of either pure component; hence this solution will boil at a temperature lower than the boiling

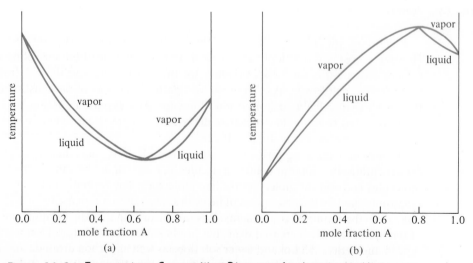

Figure 16-24. Temperature–Composition Diagrams for Azeotropic Mixtures
(a) Positive deviation from Raoult's law, (b) negative deviation from Raoult's law.

TABLE 16-3. Azeotropic Mixtures

Components		Normal Boiling Points (°C)	Normal Boiling Point of Azeotrope (°C)	Mole Fractions
ethyl alcohol	C_2H_5OH	78.5	78.2	0.91
water	H_2O	100.0		0.09
acetone	CH_3COCH_3	56.5	39.3	0.40
carbon disulfide	CS_2	46.3		0.60
ethyl alcohol	C_2H_5OH	78.5	65.0	0.385
carbon tetrachloride	CCl_4	76.8		0.615
benzene	C_6H_6	80.1		0.559
ethyl alcohol	C_2H_5OH	78.5	64.6	0.218
water	H_2O	100.0		0.223
hydrogen chloride	HCl	−83.7	108.6	0.11
water	H_2O	100.0		0.89

point of either pure component. By similar reasoning it may be shown that a solution which shows negative deviations from Raoult's law has a certain composition having a higher boiling point than that of either pure component. Representative diagrams showing the effect of composition on the normal boiling point for these cases are shown in Figure 16-24. At the compositions corresponding to the minimum (or maximum) in the boiling point curves, the compositions of the vapor and the liquid are identical. Therefore, mixtures of these particular compositions boil at a constant temperature and cannot be separated by distillation at 1 atm pressure. Such constant boiling mixtures are called **azeotropes** or **azeotropic mixtures**. Some specific examples of azeotropic mixtures are listed in Table 16-3. When solutions of substances capable of forming azeotropic mixtures are distilled, only one pure component and the azetropic mixture can be obtained. For example, pure water and a 95% solution of alcohol in water can be obtained but both pure water and pure alcohol cannot be obtained by distilling an alcohol–water mixture at 1 atm pressure.

It should be noted that the composition of an azeotropic mixture depends on the pressure. If the pressure is changed, the boiling point of a given azeotropic mixture will change. The vapor will become richer in one of the components until a new azeotropic mixture is obtained. If the pressure is changed enough, it is possible to obtain a good separation of the mixture.

16-25   Exercises

Basic Exercises

1. A liquid of density 0.85 gram/ml, having a surface tension of 55 dynes/cm, will rise how far in a glass capillary of 0.140 cm inside diameter?
2. (a) Distinguish between the triple point and the freezing point of a pure substance.
   (b) For most pure substances, is the triple point or the freezing point apt to be higher?

3. Distinguish **(a)** between the boiling point of a liquid and the normal boiling point of the liquid, **(b)** between freezing point and freezing point depression.

4. All other factors being equal, which will cool to room temperature faster—a closed container of water at 100°C or an open container of water at 100°C? Explain your answer.

5. In the process of **recrystallization,** an impure solid sample is dissolved in a minimum quantity of hot solvent; the solution is filtered and allowed to cool, whereupon purer solute separates. **(a)** Explain why this process removes both soluble and insoluble impurities from the solid sample. **(b)** Which compound, KCl or KNO_3, would be more apt to be easily purified by recrystallization? (Hint: see Figure 16–13.)

6. The density of 10.0% by mass KCl solution in water is 1.06 grams/ml. Calculate the molarity, molality, and mole fraction of KCl in this solution.

7. Show algebraically that the sum of the mole fractions of all components of a solution must equal 1.00.

8. A solution of 10.0 grams of a nonionic solute in 100 grams of benzene freezes at 4.2°C. Calculate the molecular weight of the solute.

9. An astronaut in an orbiting spaceship spilled a few drops of his drink, and the liquid floated around the cabin. In what geometric shape was each drop most likely to be found? Explain.

10. When a liquid in a thermostated chamber is vaporized at constant temperature by the very slow withdrawal of a piston, the vapor is continuously in equilibrium with the liquid. The heat absorbed from the bath in the process is equal to what thermodynamic property of the liquid? Write an expression for calculating the value of $\Delta G°$ for this isothermal process at constant pressure.

11. A certain nonvolatile nonelectrolyte contains 40.0% carbon, 6.7% hydrogen, and 53.3% oxygen. An aqueous solution containing 5.00% by mass of the solute boils at 100.15°C. Determine the molecular formula of the compound.

12. What mass of ammonium chloride is dissolved in 100 grams of water in each of the following solutions? **(a)** 1.10 m NH_4Cl solution. **(b)** a solution which is 75% water by mass, **(c)** a solution with a mole fraction of 0.15 NH_4Cl.

13. At 25°C, the vapor pressure of pure benzene is 100 torr, while that of pure ethyl alcohol is 44 torr. Assuming ideal behavior, calculate the vapor pressure at 25°C of a solution which contains 10.0 grams of each substance.

14. Calculate the freezing point of 0.200 m solutions of fructose, $C_6H_{12}O_6$, **(a)** in water, **(b)** in acetic acid.

15. An aqueous solution of a nonvolatile solute boils at 100.17°C. At what temperature will this solution freeze?

16. When 36.0 grams of a solute having the empirical formula CH_2O is dissolved in 1.20 kg of water, the solution freezes at −0.93°C. What is the molecular formula of the solute?

17. Calculate the osmotic pressure of an aqueous solution which contains 4.00 grams of glucose, $C_6H_{12}O_6$, in 250 ml of solution at 25°C.

18. At 25°C, the vapor pressure of methyl alcohol, CH_3OH, is 96.0 torr. What is the mole fraction of CH_3OH in a solution in which the (partial) vapor pressure of CH_3OH is 23.0 torr at 25°C?

19. Determine the number of degrees of freedom in a system containing a solution of NaCl in water, ice, and solid NaCl, all in equilibrium with each other.

20. Pure water contains H_3O^+ and OH^- in addition to H_2O. How many degrees of freedom are there in pure water at 50°C and 1 atm pressure?

21. Explain on a molecular basis why benzene and toluene form nearly ideal solutions with each other.

22. Camphor, $C_{10}H_{16}O$, which has a freezing point of 174°C, has a freezing point depression constant of 40.0°C/m. Explain the usefulness and the limitations of camphor as a solvent for determination of molecular weights. For what kind(s) of solutes would camphor be especially useful?

General Exercises

23. Solid $Ce_2(SO_4)_3$ has a positive enthalpy of solution—it dissolves in water to a greater extent at lower temperature than at higher. Suggest how this substance can be purified by recrystallization. (*Hint:* see exercise 5.)

24. A clear solution is heated in an open vessel, whereupon a solid separates from the hot solution. Give two possible explanations for this behavior. Devise an experiment which would distinguish between these two possibilities.

25. Describe how and explain why a standard solution (of precisely known concentration) of HCl can be prepared from an aqueous solution of HCl without chemical analysis.

26. Explain why the melting point of a substance gives an indication of the purity of the substance. Explain why one usually recrystallizes an unknown product until successive recrystallized samples show no increase in melting point.

27. Explain under what conditions water would completely fill a fine capillary tube which is open at both ends when one end is immersed in water.

28. The surface tension of liquid mercury is 490 ergs/cm². How far will the mercury level be *depressed* when a glass capillary with 0.40 mm radius is placed in a dish of mercury?

29. In a measurement of surface tension using an apparatus such as shown in Figure 16-3, five drops of a liquid weighed 0.220 gram and occupied a volume of 0.276 ml. Calculate the surface tension of the liquid.

30. A certain liquid has a viscosity of 1×10^4 poise and a density of 3.2 grams/ml. How long will it take for a platinum ball with a 2.5 mm radius to fall 1.0 cm through the liquid? The density of platinum is 21.4 grams/cm³.

31. Knowing that the density of ice is less than that of water, explain why the slope of the line *TA* in Figure 16-10 is in accord with Le Châtelier's principle.

32. The vapor pressure of ethyl acetate is 300 torr at 51°C, and its enthalpy of vaporization is 9.0 kcal/mole. Estimate the normal boiling point of ethyl acetate.

33. Using the data $\Delta H_{fus}^{-24°} = 4.16$ cal/gram and $C_p(s) = C_p(l) = 0.20$ cal/gram · deg, calculate the entropy change and the free energy change for the freezing of 1.00 gram of supercooled $CCl_4(l)$ at $-34°C$ to solid at that same temperature.

34. The vapor pressure of pure benzene, C_6H_6, at 50°C is 268 torr. How many moles of nonvolatile solute per mole of benzene is required to prepare a solution of benzene having a vapor pressure of 167.0 torr at 50°C?

35. Roughly sketch a heating curve between $-40°$ and $100°C$ for 1.0 gram of CCl_4 using the following data: $\Delta H_{fus} = 4.16$ cal/gram, $\Delta H_{vap} = 46.4$ cal/gram, fp $= -24°C$, normal bp $= 76.8°C$, C_p of $CCl_4(l)$ and $CCl_4(s) = 0.20$ cal/gram · deg, and C_p of $CCl_4(g) = 0.13$ cal/gram · deg.

36. The densities of ethyl alcohol liquid and vapor are tabulated below. By means of a graph, estimate the critical temperature of ethyl alcohol.

Temperature (°C)	Density (grams/ml)		Temperature (°C)	Density (grams/ml)	
	Vapor	Liquid		Vapor	Liquid
100	0.004	0.716	220	0.085	0.496
150	0.019	0.649	240	0.172	0.383
200	0.051	0.557			

37. A 250 ml water solution containing 48.0 grams of sucrose, $C_{12}H_{22}O_{11}$, at 300 K is separated from pure water by means of a semipermeable membrane. What pressure must be applied above the solution in order to just prevent osmosis?

38. Calculate the composition of the vapor in equilibrium with an ideal solution of ethyl-benzene and methylbenzene in which the mole fraction of ethylbenzene in the liquid is 0.35:

$$P^\circ_{\text{ethylbenzene}} = 10.0 \text{ torr}$$
$$P^\circ_{\text{methylbenzene}} = 37.0 \text{ torr}$$

Calculate the total vapor pressure of the solution.

39. A solution containing 3.50 grams of solute X in 50.0 grams of water has a volume of 52.5 ml and a freezing point of $-0.86°C$. **(a)** Calculate the molality, mole fraction, and molarity of X. **(b)** Calculate the molecular weight of X.

40. Calculate the molarity, molality, and mole fraction of ethyl alcohol in a solution of total volume 95 ml prepared by adding 50 ml of ethyl alcohol (density $= 0.789$ gram/ml) to 50 ml of water (density $= 1.00$ gram/ml). Calculate the molality of water in alcohol.

41. Calculate the vapor pressure lowering of a $0.100\ m$ aqueous solution at $75°C$.

42. Calculate the molecular weight of a substance which forms a 7.0% by mass solution in water which freezes at $-0.89°C$.

Advanced Exercises

43. The enthalpy change of the reaction $H^+ + OH^- \rightarrow H_2O$ is practically independent of temperature. Calculate the pH of pure water at $100°C$.

44. By means of calculus, derive the expression for the temperature variation of entropy at constant pressure:

$$\Delta S = 2.30 C_p \log (T_2/T_1)$$

What assumption about C_p is made in this derivation?

45. At $50°C$ the vapor pressure of pure CS_2 is 854 torr. A solution of 2.0 grams of sulfur in 100 grams of CS_2 has a vapor pressure 848.9 torr. Determine the formula of the sulfur molecule.

46. In water at $20°C$, the Henry's law constant for oxygen is 4.6×10^4 atm, and for nitrogen it is 8.2×10^4 atm, where the concentrations are expressed as mole fractions. Suggest a method based on these data for preparing 99 mole % pure oxygen from air. How many cycles would be necessary to achieve this result?

47. A certain solution of benzoic acid in benzene has a freezing point of $3.1°C$ and a normal boiling point of $82.6°C$. Explain these observations, and suggest structures for the solute particles at the two temperatures.

48. At $50°C$, the vapor pressure of pure benzene, C_6H_6, is 268 torr, while that of ethylene chloride, $C_2H_4Cl_2$, is 236 torr. Assuming ideal behavior, plot a vapor pressure–composition diagram for solutions of the two at $50°C$. Include on the diagram the compositions of vapor in equilibrium with the solutions.

49. Using Table 16–3, devise a method for preparing practically pure ethyl alcohol from the alcohol–water azeotrope.

50. As supercooled water freezes spontaneously, its temperature rises to $0°C$. What is ΔH for this spontaneous process:

$$H_2O(l)(-10°C) \rightarrow H_2O(s)(0°C)$$

17

Chemical Kinetics and Reaction Mechanisms

Chemical reactions occur at a variety of rates from very rapid to very slow. For example, the paper on which this book is printed reacts extremely slowly with atmospheric oxygen; however, if the temperature is raised sufficiently, rapid reaction results. An explosion is an extremely fast reaction which produces gaseous products; however, not all rapid reactions are explosive.

One might guess that the more the free energy is lowered during a reaction, the faster the reaction would be. However, this guess is not necessarily correct. A comparison of the following two reactions illustrates this point:

$$16\ H^+ \ + \ 2\ MnO_4^- \ + \ 5\ Sn^{2+} \ \rightarrow \ 5\ Sn^{4+} \ + \ 2\ Mn^{2+} \ + \ 8\ H_2O$$
$$\epsilon^\circ = 1.39\ V;\ \Delta G^\circ = -1390\ kcal$$

$$8\ H^+ \ + \ MnO_4^- \ + \ 5\ Fe^{2+} \ \rightarrow \ 5\ Fe^{3+} \ + \ Mn^{2+} \ + \ 4\ H_2O$$
$$\epsilon^\circ = 0.78\ V;\ \Delta G^\circ = -376\ kcal$$

Per mole of permanganate, ΔG° is more negative for the first reaction, yet the iron(II) ion reduces the permanganate ion to manganese(II) much more rapidly than does tin(II) ion.

Another example further emphasizes that the free energy of reaction does not determine the reaction rate. For a given reaction run two different ways, the free energy change must be the same, but the rates of reaction may be quite different. The sugar, glucose, reacts with oxygen according to the following equation:

$$C_6H_{12}O_6(s) \ + \ 6\ O_2(g) \ \xrightarrow[25^\circ C]{} \ 6\ CO_2(g) \ + \ 6\ H_2O(l) \quad \Delta G = -791\ kcal$$

However, pure glucose can be stored in pure oxygen gas at 25°C for indefinite periods of time without undergoing reaction. On the other hand, in cell metabolism, which occurs very close to room temperature, glucose reacts rapidly with oxygen to produce carbon dioxide and water.

In the first example, the reduction of the permanganate ion, thermodynamics allows the prediction that both tin(II) and iron(II) in their standard states would reduce permanganate spontaneously but offers no clue as to why the reaction having the lower potential is the faster. In the second example, thermodynamics predicts that the reaction between glucose and oxygen should be spontaneous at room temperature. But thermodynamics does not explain why no reaction occurs when the reagents are mixed at room temperature, while under the conditions peculiar to a living cell the reaction does go readily as predicted. Thermodynamics is concerned only with the initial and final states of a system. Chemical kinetics, on the other hand, seeks to account for the rates of reactions and to describe the conditions and individual steps by which reactions proceed.

In this chapter the methods for measuring the rates of reactions and for summarizing each in a mathematical equation, called a **rate law,** are given. Also a theory of reaction rates which is known as the **collision theory** is described. The basic assumption is that in order for two molecules to react with each other chemically, they must at least come into contact. The rate of any reaction should therefore depend on the number of collisions occurring per second between the reacting molecules. By no means will every collision be an effective collision—that is, produce a chemical reaction—but the more collisions there are, the more effective collisions can be expected.

When a reaction goes from the original reactants to the final products, the collisions between reacting species may occur in several steps. The detailed sequence of steps is called a **reaction mechanism.** It will be shown how, in some cases, a reaction mechanism can be inferred from the observed rate law. Also some techniques of experimentally verifying proposed reaction mechanisms will be described.

17-1   The Rate Laws

The rate of a chemical reaction is defined as the change in concentration of some reactant or product per unit time. It is convenient to study the influence of concentration on reaction rates by studying reactions between gases or reactions occurring in solution, because in both of these cases initial concentrations can be controlled and changes in concentration may be measured comparatively easily. For example, in a reaction in which a gas is produced from solid materials, such as

$$MgCO_3(s) \rightarrow MgO(s) + CO_2(g)$$

the rate at which CO_2 is formed can be determined by running the reaction in a fixed volume and measuring the rate at which the pressure increases.

Because of the necessity of obtaining accurate analyses after precisely measured intervals of time, considerable ingenuity is often required in designing experiments to determine reaction rates. One such apparatus is shown in Figure 17–1.

The effect of concentration on the rate of reaction is expressed mathematically as a rate law. For a reaction between substances A and B, the rate law has the form

$$rate = k[A]^m[B]^n$$

where k is the **rate constant** and [A] and [B] are the concentrations of the reacting substances A and B. The exponents m and n are numbers, often integers, called the **order** of the reaction with respect to A and B, respectively. The sum $m + n$ is called the overall order of the reaction, or more simply the order of the reaction. The rate constant, k, and the order of the reaction with respect to the individual reactants, m and n, must be experimentally determined. They cannot be predicted theoretically. Once the rate constant and the order with respect to each reactant are known, the rate of reaction can be predicted for any set of concentrations of reactants.

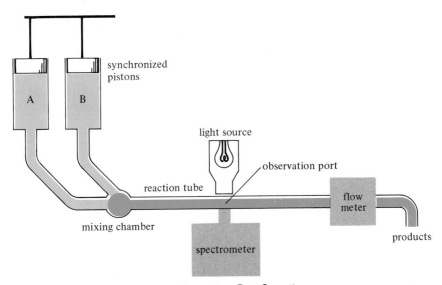

Figure 17-1. Flow Method for Measuring Fast Reactions
Solutions of reactants, A and B, are injected into the mixing chamber and forced through the reaction tube. The reaction is followed by observing some property of the system, such as the absorption of light at a characteristic wavelength, using an instrument placed at a fixed point "downstream" along the tube. Experiments are performed at different known rates of flow and the quantity of reaction occurring in the time required for the mixture to flow from the mixing chamber to the observation port is obtained. Reactions which go to completion in only a few milliseconds (10^{-3} sec) may be studied in this manner.

The order of a reaction with respect to each reactant may be determined by noting the effect of a change in concentration of that reactant on the rate of the reaction. It is easier to interpret the effect of initial concentrations on the initial rate of reaction, because as the reaction proceeds, the concentrations of reactants change. Also as the concentrations of the products build up, the occurrence of a reverse reaction may influence the rate.

Example

Five 5.00 ml samples of 1.0 *M* reagent B were poured into five vessels containing 5.00 ml samples of A, each having the concentration tabulated below. The temperature was 25 °C, and all other conditions were constant. The initial rates are also tabulated. Another set of experiments was performed in which the concentration of B was varied; these results are also tabulated. What is the order of the reaction with respect to A and with respect to B? What is the value of the rate constant?

Conc A (M)	Conc B (M)	Initial Rate (M/sec)	Conc A (M)	Conc B (M)	Initial Rate (M/sec)
1.0	1.0	1.2×10^{-2}	1.0	1.0	1.2×10^{-2}
2.0	1.0	2.3×10^{-2}	1.0	2.0	4.8×10^{-2}
4.0	1.0	4.9×10^{-2}	1.0	4.0	1.9×10^{-1}
8.0	1.0	9.6×10^{-2}	1.0	8.0	7.6×10^{-1}
16.0	1.0	1.9×10^{-1}	1.0	16.0	3.0

It is apparent from the data on the left that, all other factors being held constant, the initial rate of the reaction is proportional to the initial concentration of A. The reaction is first order with respect to A. Note that no conclusions about the order with respect to B can be drawn from the data on the left-hand side of the table. The data on the right show that the initial rate is proportional to the square of the initial concentration of B, and therefore the reaction is second order with respect to B. The reaction is third order overall. The numerical value of the rate constant, k, may be obtained by substituting data for each experiment into the rate law expression:

$$\text{rate} = k[A]^m[B]^n = k[A][B]^2$$

Thus for the data in the first line on the left,

$$1.2 \times 10^{-2} \text{ mole/liter} \cdot \text{sec} = k(1.0 \text{ mole/liter})(1.0 \text{ mole/liter})^2$$
$$k = 1.2 \times 10^{-2} \text{ liter}^2/\text{mole}^2 \cdot \text{sec}$$

Knowledge of the order of a reaction with respect to each reactant involved suggests ways in which the rate of reaction may be controlled by variation of the relative concentrations of reactants. For example, in the reaction between A and B described above, for a given concentration of B, doubling the concentration of A doubles the rate. For a given concentration of A, doubling the concentration of B increases the rate fourfold. It will be shown below how the rate laws provide some insight into the mechanisms of individual reactions.

The order of a reaction with respect to a reagent which is present in large excess is difficult to determine, because its concentration remains essentially constant throughout the reaction. For example, in dilute aqueous solution, water is present in large excess. Even if water is one of the reactants, its concentration will be essentially unchanged throughout the reaction; hence its effect on the overall order of the reaction may not be apparent. By performing the reaction in another solvent and adding water in quantities comparable to those of the other reagents, the effect of the water on the rate can be determined.

Relationships between time and various functions of concentration of reagents and products for reactions of various orders may be derived from the rate law equations. Some results of the derivations, which were obtained by methods of calculus, are listed in Table 17–1. In the table, $[A_0]$ refers to the initial concentration of a reactant, $[A]$ is the concentration of the reactant at time t, and $[X]$ is the concentration of a product of the reaction at time t. In the last column of the table are listed the dimensions in which the rate constants are expressed. It should be emphasized that in giving a rate law expression or in expressing a rate constant, it is often necessary not only to express the concentration units but also to note the substance to which the rate refers.

The mathematical forms of the rate laws given in Table 17–1 provide simple tests for determining the order of a reaction from experimental data. Thus, for a first order reaction, a plot of the logarithm of the concentration of reactant remaining, log [A], or the logarithm of the fraction of reactant remaining, log ([A]/[A_0]), versus time gives a straight line having a slope equal to $-k/2.30$. Similarly, a plot of [A] versus t yields a straight line for zero order reactions,[1] and a plot of 1/[A] versus t yields a straight line for second order reactions.

---

[1] In zero order reactions the rate is independent of the concentration of any reactant. The decomposition of ammonia on a tungsten surface is one example of such a reaction.

TABLE 17-1. Concentration as a Function of Time[a]

Order (n) with Respect to A	Rate Equation in Terms of Reactant Concentration	Rate Equation in Terms of Product Concentration	Dimensions of k
0	$[A_0] - [A] = kt$	$[X] = kt$	$\dfrac{mole}{liter \cdot sec}$
1	$\begin{cases} \log [A_0] - \log [A] = \dfrac{kt}{2.30} \\[2mm] \text{or} \quad 2.30 \log \dfrac{[A]}{[A_0]} = -kt \end{cases}$	$\begin{cases} \log [A_0] - \log ([A_0] - [X]) = \dfrac{kt}{2.30} \\[2mm] \text{or} \quad 2.30 \log \dfrac{[A_0] - [X]}{[A_0]} = -kt \end{cases}$	$\dfrac{1}{sec}$
2	$\dfrac{1}{[A]} - \dfrac{1}{[A_0]} = kt$	$\dfrac{1}{[A_0] - [X]} - \dfrac{1}{[A_0]} = kt$	$\dfrac{liter}{mole \cdot sec}$
3	$\dfrac{1}{[A]^2} - \dfrac{1}{[A_0]^2} = 2kt$	$\dfrac{1}{([A_0] - [X])^2} - \dfrac{1}{[A_0]^2} = 2kt$	$\dfrac{liter^2}{mole^2 \cdot sec}$

[a] rate $= k[A]^n$
$[A]$ = concentration of a reactant at a given time, t
$[A_0]$ = initial concentration of the reactant
$[X]$ = concentration of a product at a given time, t
Time t is measured in seconds for the dimensions listed.

17-2 Half Life

For a first order reaction, the rate of reaction at any time, t, is directly proportional to the concentration of reactant remaining. As the reactant is used up, the rate diminishes. The time required for one half of a sample to react is known as its **half life.** After one half-life period, the rate of reaction is half as great as it was originally. In the succeeding half-life period, half as much will react as reacted in the first half-life period; that is, half of the remaining material will react. In the third half-life period, again one half of the remaining sample will react. For first order reactions only, the half life is independent of initial concentration.

There is a direct relationship between the rate constant of a first order reaction and the half life. If $[A]$ is the concentration of reactant at any time and $[A_0]$ is the original concentration, in one half life t is equal to $t_{1/2}$ and the fraction $[A]/[A_0]$ is equal to 0.500. Therefore in this case the rate equation may be written

$$2.30 \log 0.500 = -kt_{1/2}$$
$$k = \frac{0.693}{t_{1/2}}$$

Example

Sucrose decomposes in acid solution into glucose and fructose according to a first order rate law, with a half life of 3.33 hours at 25°C. What fraction of a sample of sucrose remains after 9.00 hours?

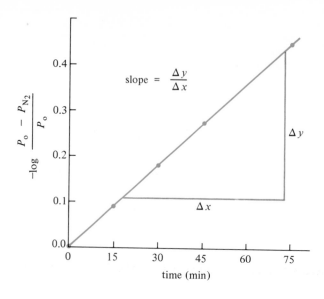

Figure 17-2. **Plot of a First Order Reaction (Decomposition of Azomethane)**

At any time, t, the fraction of sucrose remaining is $[A]/[A_0]$.

$$2.30 \log \frac{[A]}{[A_0]} = -kt = -\left(\frac{0.693}{t_{1/2}}\right)t = -\left(\frac{0.693}{3.33 \text{ hours}}\right)(9.00 \text{ hours})$$

$$\log \frac{[A]}{[A_0]} = -0.814$$

$$\frac{[A]}{[A_0]} = 0.153$$

fraction remaining $= 0.153$

Example

Azomethane, $(CH_3)_2N_2$, decomposes with a first order rate according to the equation

$$(CH_3)_2N_2(g) \rightarrow N_2(g) + C_2H_6(g)$$

The following data were obtained for the decomposition in a 200 ml flask at 300°C:

Time (min)	Total Pressure (torr)
0	36.2
15	42.4
30	46.5
48	53.1
75	59.3

Calculate the rate constant and the half life for this reaction.

It may be seen that for each mole of azomethane which decomposes, 2 moles of gaseous products are obtained. Since in a constant volume at constant temperature the pressure of a gas is directly proportional to the number of moles of gas, the total pressure in the flask will increase as the reaction proceeds. The pressure increase is equal to the partial pressure of N_2 (or C_2H_6). Since the reaction is first order, the rate equation in terms of pressure (concentration) of N_2 is as follows (Table 17-1):

$$\log \frac{P_0 - P_{N_2}}{P_0} = -\frac{k}{2.30} t$$

A plot of the logarithm term against time, t, gives a straight line with a slope equal to $-k/2.30$ (Figure 17-2). In this case, the slope is $-5.67 \times 10^{-3}/\text{min}$, and therefore $k = 1.30 \times 10^{-2}/\text{min}$. The half life is given by $0.693/k = t_{1/2}$; therefore $t_{1/2} = 53.3$ min.

17-3 Catalysts

Substances which accelerate a chemical reaction but which are themselves not used up in the net reaction are called **catalysts.** The ways in which catalysts work are almost as varied as the kinds of chemical reactions involved. However, all catalysts function to provide an additional path or mechanism for the reaction.

At this point it is necessary to point out the difference between the concepts of kinetic stability and thermodynamic stability. Many substances are thermodynamically unstable at ordinary temperatures and exist only because the rate at which they react is low. Such substances are stable in a kinetic sense only, and if some change which increases the rate of reaction is made, the predicted reaction would occur spontaneously. Some examples of thermodynamically unstable substances are hydrogen peroxide, ozone, trinitrotoluene (TNT), and nitroglycerine.

If at a given temperature a system is thermodynamically unstable but reaction is too slow to be observed, a suitable catalyst may be employed to effect the reaction. On the other hand, if the reaction is not possible thermodynamically, no catalyst would be able to make it go.

Catalysts are widely used in the chemical industry and in chemical research. The manufacture of sulfuric acid from sulfur is one example. The sequence of reactions is as follows:

$$S + O_2 \rightarrow SO_2$$
$$2\,SO_2 + O_2 \rightarrow 2\,SO_3$$
$$SO_3 + H_2O \rightarrow H_2SO_4$$

In the absence of a catalyst, the second reaction is very slow. In the older process for the manufacture of sulfuric acid, known as the lead chamber process, the catalysts used were nitrogen monoxide and nitrogen dioxide. The reaction sequence, which takes place in a lead vessel in the presence of the nitrogen oxides, is as follows:

$$NO_2 + NO + 2\,H_2SO_4 \rightarrow 2\,ONOSO_3H + H_2O$$
$$SO_2 + 2\,H_2O + 2\,ONOSO_3H \rightarrow 3\,H_2SO_4 + 2\,NO$$
$$2\,NO + O_2 \rightarrow 2\,NO_2$$

Nitrosyl bisulfate, $ONOSO_3H$, is the reagent which actually oxidizes the SO_2. Part of the NO produced by the reaction reacts with oxygen to yield NO_2, and the resulting mixture reacts with more H_2SO_4 to regenerate more nitrosyl bisulfate.

The more modern method for the manufacture of sulfuric acid is the contact process, in which SO_2 is treated at $400°C$ with an excess of air in the presence of finely divided platinum deposited on asbestos. The resulting SO_3 cannot be treated with pure water, because its reaction with water is so exothermic that a fine mist would be formed. Therefore the SO_3 is absorbed in 96% sulfuric acid, in which it reacts with the small quantity of water present. An excess of SO_3 also dissolves. The resulting "fuming sulfuric acid" is diluted to 96% sulfuric acid by addition to sufficient water. In order not to "poison" the catalyst, very pure sulfur dioxide must be used. As a result, very pure sulfuric acid is produced.

Some reactions are autocatalytic; that is, the reaction is catalyzed by its own products. For example, the reaction of iodate ion with sulfite ion in acid solution produces iodide ion:

$$IO_3^- + 3 H_2SO_3 \rightarrow I^- + 3 SO_4^{2-} + 6 H^+$$

The iodide ion catalyzes further reaction, according to the following equations:

$$IO_3^- + 5 I^- + 6 H^+ \rightarrow 3 I_2 + 3 H_2O$$
$$I_2 + H_2SO_3 + H_2O \rightarrow 2 I^- + SO_4^{2-} + 4 H^+$$

The overall rate is faster because the H_2SO_3 reacts faster with I_2 than it does with IO_3^-.

A very important class of catalysts in biological systems are complex substances known as enzymes. Enzymes are produced by living organisms as catalysts for specific reactions. Thus ptyalin, the enzyme of saliva, promotes the conversion of starch into sugar; and zymase, an enzyme of the yeast plant, promotes conversion of sugars into alcohol and carbon dioxide. A number of enzymes have been isolated in pure crystalline form.

Since catalysts are not consumed by the chemical reactions which they promote, a small quantity is sufficient to catalyze the reaction of a large quantity of reactants. The term **catalytic amount** refers to the relatively small number of moles of catalyst which can effect the reaction of a large number of moles of reagents. For example, 10 grams of the milk protein casein must be boiled with 100 grams of concentrated hydrochloric acid for 20 hours in order to break it down into its constituent amino acids. In contrast, in the presence of 0.001 gram of protease, an enzyme isolated from the stomach lining of an animal, dilute HCl will break down the 10 grams of casein at body temperature in only a few hours.

17-4 Collision Theory of Reaction Rates

In Chapter 16, the liquid state and the evaporation process were explained using the postulates of the kinetic molecular theory. Of more immediate interest is the use of the kinetic molecular theory to develop a theory of reaction rates. This theory, known as the **collision theory,** has as its basic assumption that in order for two molecules to react with each other chemically, they must at least come into contact. The rate of any reaction should therefore depend on the number of collisions occurring per second

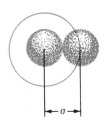

Figure 17-3. Excluded Volume Surrounding a Molecule

between the reacting molecules. By no means will every collision produce chemical reaction, but the more total collisions there are, the more effective collisions can be expected.

The number of collisions per second between molecules in a sample can be calculated by means of the kinetic molecular theory. However, in making such calculations, it is necessary to discard the assumption that molecules are point masses. For simplicity it will be assumed that the molecules are rigid spheres, and the size of a molecule will be defined in terms of its collision diameter, σ. The collision diameter is related to the constant, b, in the van der Waals equation. It is found experimentally that b is four times the actual volume occupied by 1 mole of the molecules, V_M. The factor 4 comes about because there is a volume surrounding each molecule from which the centers of all other molecules are excluded. As shown in Figure 17-3, the centers of two identical molecules cannot get closer than a distance σ to each other. As shown in Figure 17-4, the effective collision diameter of a molecule moving through a mass of gas molecules is 2σ, because other molecules may approach from either side. Hence, a gas molecule moving with an average speed $\bar{v}$ sweeps out a volume $\pi\sigma^2\bar{v}$ in 1 sec, and if there are n' molecules per cubic centimeter in the gas, the number of collisions per second, Z, experienced by the molecule will be

$$Z = \pi\sigma^2\bar{v}n' \quad \text{(one moving molecule)}$$

However, this calculated number of collisions per second is based on the assumption that only one molecule is moving. If the motions of all the molecules of the gas are considered, the total number of collisions per molecule per second is given by

$$Z = \sqrt{2}\pi\sigma^2\bar{v}n' \quad \text{(per molecule)}$$

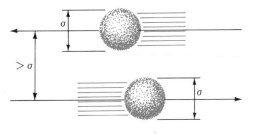
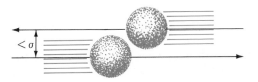

Figure 17-4. Collision Diameters

Two molecules will collide if their centers come within a distance σ, the molecular diameter, of one another. At distances greater than σ, the molecules do not collide.

Since all n' molecules in 1 cm^3 of the gas are undergoing collisions, the total number of collisions per second per cubic centimeter is given by

$$Z = \tfrac{1}{2}\sqrt{2}\pi\sigma^2\bar{v}(n')^2$$

where the factor $\tfrac{1}{2}$ is introduced because each collision involves two molecules.

It can be shown theoretically that the average speed, $\bar{v}$, is $\sqrt{8/3\pi}$ times the root mean square speed, u (see Figure 15–9 for the relative magnitudes of $\bar{v}$ and u):

$$\bar{v} = \sqrt{\frac{8}{3\pi}}\,u = \sqrt{\frac{8}{3\pi}\left(\frac{3RT}{M}\right)} = 2\sqrt{\frac{2RT}{\pi M}}$$

Hence the number of collisions per second per cubic centimeter in any gas is given by

$$Z = \tfrac{1}{2}\sqrt{2}\pi\sigma^2\bar{v}(n')^2 = 2(n')^2\sigma^2\sqrt{\frac{\pi RT}{M}}$$

Example

Calculate the number of collisions per second per cubic centimeter at 0°C in hydrogen gas at a pressure of 1.0 atm.

First, it is necessary to determine the number of molecules per cubic centimeter:

$$n' = Nn = N\frac{PV}{RT} = \frac{(6.02 \times 10^{23} \text{ molecules/mole})(1.00 \text{ atm})(0.00100 \text{ liter})}{(0.0821 \text{ liter} \cdot \text{atm/mole} \cdot \text{K})(273 \text{ K})}$$

$$= 2.69 \times 10^{19} \text{ molecules}$$

Then σ for hydrogen is calculated from the van der Waals constant, $b = 26.6$ cm^3/mole (from Table 15–1).

$$V_M = \frac{b}{4} = \frac{26.6 \text{ cm}^3/\text{mole}}{4} = 6.65 \text{ cm}^3/\text{mole}$$

$$\frac{\text{volume}}{\text{molecule}} = \frac{V_M}{N} = \frac{6.65 \text{ cm}^3/\text{mole}}{6.02 \times 10^{23} \text{ molecules/mole}} = \frac{1.10 \times 10^{-23} \text{ cm}^3}{\text{molecule}}$$

Since $V = \tfrac{4}{3}\pi r^3$ for any sphere, and the radius of the molecule is $\sigma/2$,

$$V = \frac{4}{3}\pi r^3 = \frac{4}{3}\pi\left(\frac{\sigma}{2}\right)^3 = \frac{\pi\sigma^3}{6}$$

$$\sigma = \sqrt[3]{\frac{6V}{\pi}} = \sqrt[3]{\frac{6(1.10 \times 10^{-23}) \text{ cm}^3}{3.14}} = 2.76 \times 10^{-8} \text{ cm}$$

$$Z = 2(2.69 \times 10^{19})^2(2.76 \times 10^{-8})^2\sqrt{\frac{3.14(8.31 \times 10^7)(273)}{2.01}} = \frac{2.08 \times 10^{29} \text{ collisions}}{\text{sec}}$$

17-5 Effective Collisions

As just illustrated, the greater the concentration of molecules, the greater will be the number of collisions per second between the molecules in a given volume. The collision theory thus accounts for the influence of concentration on the rate of a reaction. However, every collision between reacting molecules does not result in reaction. Consider the reaction

$$2 \text{ HI} \rightarrow H_2 + I_2$$

At 500°C with an initial concentration of HI of 1×10^{-3} mole/liter, the rate of decomposition of HI is found to be 1.2×10^{-8} mole/liter · sec. The number of collisions per second between HI molecules under the same conditions is 4×10^{28} collisions/liter · sec. If every collision were effective, the rate of reaction would be

$$\text{rate} = \frac{(4 \times 10^{28} \text{ collisions/liter} \cdot \text{sec})(2 \text{ molecules/collision})}{6.02 \times 10^{23} \text{ molecules/mole}}$$

$$= 1 \times 10^5 \text{ moles/liter} \cdot \text{sec}$$

It is apparent that only a small fraction of the collisions between HI molecules result in the formation of H_2 and I_2.

Two reasons for the very large discrepancy are suggested. In the first place, real molecules, particularly polar diatomic molecules such as HI, are not rigid spheres. It should be expected that in a certain percentage of the collisions, the molecules will be oriented in such a manner that they will not react. A second, more important possibility is that only those collisions which occur with a certain minimum energy will result in reaction. Therefore it is concluded that hydrogen and iodine will be produced only if two HI molecules collide with a proper orientation and with sufficient energy.

Actual calculations for many substances show that in general the orientation factor is somewhat less than unity. (For the hydrogen iodide decomposition reaction, the numerical value is about 0.2.) Therefore this factor alone cannot account for the observed difference between the rate of reaction and the rate of collisions of HI molecules. Consequently, the major factor in determining the rate is the number of collisions which occur with sufficient energy. The minimum energy necessary to cause reaction is called the **activation energy** of the reaction.

17-6   Activation Energy

If two reactant molecules, A_2 and B_2, collide with sufficient energy, they form a kind of transient intermediate which is called an **activated complex.** The activated complex may split apart into the original reactants or it may split in a different way to yield the products AB (Figure 17-5).

In Figure 17-6, the potential energy of a system undergoing reaction is plotted against a function of interatomic distances called the **reaction coordinate.** (For most systems, the reaction coordinate is too complicated to work out in detail.) The difference in potential energy between the reactants and the activated complex is known as the **activation energy,** E_a. This energy is the minimum required to cause the system to undergo reaction through this transition state. If the potential energy of the products is greater than that of the reactants, the energy released in going from the activated complex to products will be less than the activation energy, and the reaction

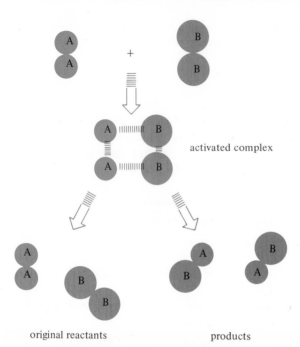

Figure 17–5. Formation and Breakup of Activated Complex

activated complex

original reactants   products

will be endothermic. If the potential energy of the products is less that that of the reactants, the energy difference between the activated complex and the products is greater than the activation energy, and the reaction is exothermic.

Further insight into the nature of the activation energy is provided by a study of the variation of reaction rate with temperature. For many reactions occurring near room temperature, the reaction rate increases by a factor between 1.5 and 3 with a rise of 10°C. The actual increase depends on the specific reaction. However, a simple calculation reveals that the increase in rate is not due to an increase in the average kinetic energy of the molecules. The ratio of kinetic energies at 35° and 25°C is merely the ratio of absolute temperatures:

$$\frac{\overline{KE}_{35}}{\overline{KE}_{25}} = \frac{\frac{3}{2}R(273 + 35)}{\frac{3}{2}R(273 + 25)} = 1.03$$

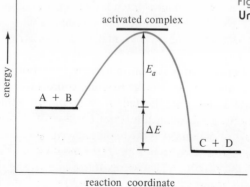

Figure 17–6. Potential Energy of a System Undergoing Reaction

The ratio of average kinetic energies is very much less than the ratio of reaction rates at the two temperatures.

Examination of the change in the Boltzmann distribution of molecular energies with temperature (Figure 15–11) reveals that even for a small temperature rise, the number of molecules having a relatively large energy is very much increased. For a temperature rise from 25 to 35°C, the number of high energy molecules is approximately doubled. Collisions of such high energy molecules are expected to occur proportionately more often. Thus the increase in reaction rate with temperature is attributed to the increase in concentration of high energy molecules.

Even before the development of the kinetic molecular theory to the point of being able to explain reaction rates, Svante Arrhenius had proposed that the rate constant was related to the absolute temperature by the equation

$$k = A(10^{-E_a/2.30RT})$$

where A is a constant characteristic of the particular reaction, E_a is the activation energy, and R is the gas constant. The following equation has since been developed:

$$k = \rho A(10^{-E_a/2.30RT})$$

where ρ is a steric factor related to the shapes and orientations of the reacting molecules. Taking the logarithm of each side of this equation yields

$$\log k = \log \rho A - \frac{E_a}{2.30R}\left(\frac{1}{T}\right)$$

Assuming that E_a is a constant, when $\log k$ is plotted against $1/T$, a straight line should be obtained. The slope of the line is equal to the value of the activation energy divided by $-2.30R$. A typical plot is shown in Figure 17–7.

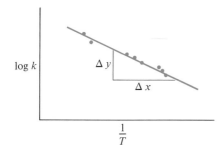

Figure 17-7. **Variation of Rate Constant with Temperature**

Data for the reaction $H_2 + I_2 \rightarrow 2\,HI$

Example

From the following data, estimate the activation energy for the reaction

$$H_2 + I_2 \rightarrow 2\,HI$$

T (K)	$1/T$ (K^{-1})	$\log k$
769	1.3×10^{-3}	2.9
667	1.5×10^{-3}	1.1

$$\text{slope} = \frac{(2.9 - 1.1)}{(1.3 - 1.5) \times 10^{-3}\,\text{K}^{-1}} = -9 \times 10^3\,\text{K}$$

$$E_a = -2.30R(\text{slope}) = -2.30(1.99\,\text{cal/mole} \cdot \text{K})(-9 \times 10^3\,\text{K}) = 4 \times 10^4\,\text{cal/mole}$$

Reactions with large activation energies tend to be slow reactions because only a small fraction of the molecules have enough kinetic energy to undergo effective collisions. In contrast, reactions having low activation energies proceed rapidly since large fractions of the molecules have sufficient energies to undergo effective collisions.

When a catalyst is added to a reaction mixture, it affects the rate of the chemical reaction by changing the path of the reaction. The formation of a new activated complex involving the catalyst results in a lower activation energy. The lower activation energy allows the reaction to proceed at a more rapid rate. This new path is represented in Figure 17–8 by the colored line.

The reaction between hydrogen and oxygen to form water is an interesting example of these principles. At room temperature, the activation energy for this reaction is so great that mixtures of hydrogen and oxygen are stable indefinitely at room temperature. However, if only a few molecules are activated by ultraviolet light or by an electric spark, reaction occurs:

$$2\,H_2(g) \;+\; O_2(g) \;\rightarrow\; 2\,H_2O(g) \qquad \Delta H = -58\,\text{kcal/mole}$$

The reaction of these few molecules produces so much energy that other molecules are activated, and the reaction proceeds with explosive violence. On the other hand, if a mixture of hydrogen and oxygen is passed over a catalyst, such as finely divided platinum maintained at room temperature, the formation of water proceeds by a different mechanism. The activation energy is quite low. Again, the net energy released is still 58 kcal per mole of water produced, but no explosion occurs.

Although it can be rigorously applied only in simple cases, the collision theory does account for the factors which influence reaction rates, as follows:

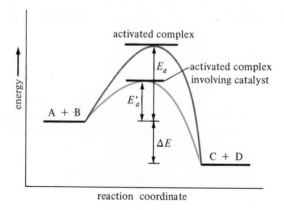

Figure 17–8. **Influence of a Catalyst on Energy of Activation**

1. The rate of reaction depends on the concentration of the reactants because the number of collisions per second increases as the concentrations are increased.
2. The rate of reaction depends on the nature of the reactants because the activation energy and thus the relative number of effective collisions differ from one reaction to another.
3. The rate of reaction depends on temperature because the proportion of molecules having energies sufficient for reaction is a function of the temperature.
4. The effect of catalysts is to change the activated complex and thereby alter the path of the reaction so as to lower the activation energy, thus providing for an increase in the reaction rate.

17-7   Rate Laws and Reaction Mechanisms

It is found experimentally that reactions having similar stoichiometries nevertheless may have very different rate laws. For example

$$H_2 \;+\; I_2 \;\rightarrow\; 2\,HI \qquad \text{rate} = k[H_2][I_2]$$

$$H_2 \;+\; Br_2 \;\rightarrow\; 2\,HBr \qquad \text{rate} = \frac{k[H_2][Br_2]^{1/2}}{1 + [HBr]/k'[Br_2]}$$

Two reactions having different rate laws will have different reaction mechanisms. But the overall order of the reaction ought to be related to the reaction mechanism. The term **reaction mechanism** as used here means the sequence of stepwise reactions by which the overall change occurs. The stepwise reactions are not necessarily observable; they are merely proposed pathways from reactants to products which logically fit all the available experimental data. Each reaction must be studied individually to infer its mechanism from its rate law.

17-8   Elementary Processes

To help keep in mind the distinction between the overall reaction and each proposed step in the mechanism, the proposed steps are called **elementary processes.** For example, the reaction between nitrogen(II) oxide and ozone to form nitrogen dioxide and oxygen may be represented as occurring in one step, as follows:

$$NO \;+\; O_3 \;\rightarrow\; NO_2 \;+\; O_2 \qquad \text{(tentative)}$$

In this case, the overall reaction is an elementary process, since it occurs in one step. Another possibility for the same overall reaction is a sequence of two elementary processes, such as the following:

$$O_3 \;\rightarrow\; O_2 \;+\; O$$
$$NO \;+\; O \;\rightarrow\; NO_2 \qquad \text{(tentative)}$$

An acceptable mechanism is one which predicts the experimentally observed rate law. Therefore, as shown below, the choice of a mechanism is possible only after examination of experimental data.

17-9   Molecularity

Elementary processes are classified according to the number of reactant molecules which are involved. A **unimolecular** elementary process involves only one reactant molecule; a **bimolecular** process involves two reacting molecules. **Termolecular** elementary processes, involving three molecules, are also possible. Elementary processes with a molecularity greater than 3 are not known because the simultaneous collision of more than three particles is extremely unlikely, and in order for many molecules to react in a single step, they must collide with each other at the same time.

The order of an *elementary process* is equal to its molecularity. Moreover, the order with respect to each reactant is equal to the molecularity with respect to the reactant. For example, any bimolecular elementary process must follow a second order rate law.

Example

Write the equation which represents the rate law for each of the following elementary processes:

$$A \; + \; B \; \rightarrow \; C \; + \; D$$
$$2\,A \; \rightarrow \; E \; + \; F$$

The first elementary process is first order with respect to A and with respect to B, and therefore second order overall:

rate $= k[A][B]$

Although the second elementary process involves only one reactant, A, it is a bimolecular process, and therefore is second order:

rate $= k[A]^2$

Although for elementary processes a unimolecular process is first order, a bimolecular process is second order, and a termolecular process is third order, it must be emphasized that the same statements cannot be made about overall reactions in general because the mechanism of a reaction may consist of several elementary steps.

The overall rate of reaction can be no faster than the rate of its slowest step in the sequence of elementary processes. Therefore, if one of the steps is much slower than the others, the rate of the overall reaction will be determined by the rate of this slow step. Hence the slowest elementary process in a sequence is called the **rate-determining step**. As a result, the rate law observed for the overall reaction will be equivalent to that of the slowest step.

17-10   Determination of Reaction Mechanisms

Consider the two step mechanism proposed above for the reaction between NO and O_3. If the slower step were the unimolecular process,

$$O_3 \rightarrow O_2 + O \quad \text{(slow)}$$

then the overall reaction should be first order and independent of the concentration of NO:

$$\text{rate} = k[O_3] \quad \text{(proposed)}$$

It is not likely that the second step,

$$NO + O \rightarrow NO_2$$

is the rate-determining step because it is known that atomic oxygen is an extremely reactive species and, once formed, tends to react rapidly with whatever is available for it to react with. On the other hand, the experimentally observed rate law for the reaction is

$$\text{rate} = k[NO][O_3] \quad \text{(actual)}$$

Consequently, of the two mechanisms proposed, the two-step mechanism cannot be correct. It must be emphasized, however, that the experimental findings do not *prove* that the bimolecular, one step process is correct. Some other mechanism not considered here might lead to the same rate law and might be the correct mechanism. Rate data alone cannot prove a mechanism correct but can exclude mechanisms which predict rate laws other than those found experimentally.

Another example of a complex reaction mechanism is that for the reaction

$$CO + Cl_2 \rightarrow COCl_2$$

with the experimental rate law

$$\text{rate} = k[Cl_2]^{3/2}[CO]$$

Hence the overall order is 5/2. The following three-step mechanism has been proposed to account for these results:

$$Cl_2 \overset{1}{\rightleftharpoons} 2\,Cl \qquad \text{(fast)}$$

$$M + Cl + CO \overset{2}{\rightleftharpoons} COCl + M \quad \text{(fast)}$$

$$COCl + Cl_2 \overset{3}{\rightarrow} COCl_2 + Cl \quad \text{(slow)}$$

The first step is a rapidly established equilibrium reaction, producing chlorine atoms (in small concentration). The second step is a combination of a chlorine atom with a carbon monoxide molecule in the presence of a third body, M, which can remove some excess energy and allow the other reagents to remain together; it is also a rapidly established equilibrium reaction. The third step is the rate-determining step. The rate of this step, and thus of the overall reaction, is given by

$$\text{rate} = k_3[Cl_2][COCl]$$

It is customary to express the rate in terms of the concentrations of initial reactants because the concentrations of intermediates are difficult to measure. Using the equilibrium constants for steps (1) and (2)

$$K_1 = \frac{[Cl]^2}{[Cl_2]} \qquad K_2 = \frac{[COCl]}{[CO][Cl]}$$

the concentration of COCl can be expressed in terms of the concentrations of Cl_2 and CO:

$$[Cl] = (K_1[Cl_2])^{1/2}$$
$$[COCl] = K_2[CO][Cl] = K_2[CO](K_1[Cl_2])^{1/2}$$

Therefore the rate for the overall reaction can be expressed in terms of concentrations of reactant molecules as follows:

$$\text{rate} = k_3[Cl_2][COCl] = k_3[Cl_2] \cdot K_2[CO](K_1[Cl_2])^{1/2} = k_3 K_2 K_1^{1/2}[Cl_2]^{3/2}[CO]$$

The product of the constants is merely another constant, k, and so the rate law can be written as follows:

$$\text{rate} = k[Cl_2]^{3/2}[CO]$$

Thus the same equation as the experimentally determined rate law is deduced from this sequence of elementary processes, and the proposed mechanism is acceptable.

In some cases more than one reaction mechanism is found to give agreement with the experimentally observed rate law. As will be shown below, most of these cases can be resolved by consideration of the chemical properties of the system and the structures of the substances involved, or perhaps by gathering more data from experiments specifically designed to distinguish between the possible alternative mechanisms.

17-11   The Transition State

It is useful to picture the rate-determining step of a reaction as involving the formation of an unstable state, which is called the **transition state** or the **activated complex.** For example, the bimolecular rate-determining step in a reaction between substances A and B may be represented by

$$A + B \rightleftharpoons [AB]^* \rightarrow C + D$$

In this case, the species AB represents the unstable transition state. To emphasize the energetic nature of this state, it is labeled with an asterisk. The activated complex is capable of changing either into the products C and D or back into the reactants A and B.

Transition states have very short lives, and since they possess more energy than any of their possible decomposition products, it is impossible to isolate them. However, it is reasonable to assume structures by analogy with the properties of stable species. For example, for the reaction of H_2O_2 with HBr in aqueous solution:

$$H_2O_2 \; + \; 2\,H^+ \; + \; 2\,Br^- \; \rightarrow \; 2\,H_2O \; + \; Br_2$$

a third order rate law is observed experimentally:

$$\text{rate} = k[H_2O_2][Br^-][H^+]$$

The following mechanism has been proposed:

$$H_2O_2 \; + \; Br^- \; + \; H^+ \; \overset{1}{\rightleftharpoons} \; HOBr \; + \; H_2O$$

$$HOBr \; + \; Br^- \; + \; H^+ \; \overset{2}{\rightarrow} \; Br_2 \; + \; H_2O$$

HOBr is not actually observed in the reaction mixture at any time. Hence it might be concluded that the first elementary process is the rate-determining step, since the second step proceeds so rapidly. The formation of HOBr can be assumed to proceed through the formation of an activated complex such as the following:

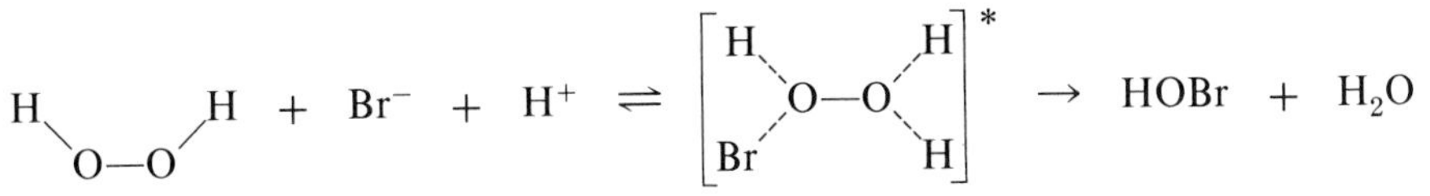

It is known that in acid solution the following reaction occurs rapidly:

$$HOBr \; + \; H^+ \; + \; Br^- \; \rightarrow \; H_2O \; + \; Br_2$$

The proposed mechanism is consistent with the known chemical properties of the substances involved.

Another way of inferring the structure of an activated complex is to use an isotope of some atom in one of the reactants and then to observe the distribution of the isotopic atoms in the products. This technique is called **isotopic labeling.** The distribution of isotopes in the products may possibly be determined in each case by one or more of the following techniques: infrared spectroscopy, mass spectrometry, nuclear magnetic resonance spectroscopy, or radioactivity measurements.

17-12 Exercises

Basic Exercises

1. Distinguish explicitly between the rate of a chemical reaction and the rate constant for the reaction.

2. The half life of a first order reaction is 2.50 hours. Calculate the value of the rate constant in sec^{-1}.

3. What is the half life of a first order reaction for which $k = 7.1 \times 10^{-5}$ sec^{-1}?

4. A reaction between A and B is second order. Write three different rate law expressions which might possibly apply to the reaction.

5. Determine the order of the reaction and the value of the rate constant for the decomposition of N_2O_5 from the following data:

t (min)	$P_{N_2O_5}$ (torr)	t (min)	$P_{N_2O_5}$ (torr)
1.83	0.826	79.50	0.372
16.33	0.708	109.66	0.277
38.50	0.562	149.80	0.184
61.17	0.449		

6. The reaction

$$CCl_3CO_2H(aq) \rightarrow CO_2 + CHCl_3$$

was found to proceed at 70°C according to the data in the following table. Determine the order of the reaction and the value of the rate constant.

Time (hours)	CCl_3CO_2H Concentration (M)	Time (hours)	CCl_3CO_2H Concentration (M)
0.00	0.10000	3.00	0.08314
1.00	0.09403	4.00	0.07817
2.00	0.08842	5.00	0.07351

7. For the reaction of reagent A in solution to give products, the following data were obtained:

t (min)	[A]	t (min)	[A]
0.0	0.583	18.0	0.170
9.0	0.343	20.0	0.133
12.0	0.257	24.0	0.118
14.0	0.223	30.0	0.079

Determine the order of the reaction, and calculate the value of the rate constant.

8. The initial rate of the reaction

$$A + B \rightarrow C + D$$

is doubled if the initial concentration of A is doubled, but is quadrupled if the initial concentration of B is doubled. **(a)** What is the order with respect to each of the reactants? **(b)** What is the overall order of the reaction? **(c)** Could this possibly be a single-step reaction? Explain.

9. Three sets of railroad tracks are to accommodate trains which are 14 feet wide. What must be the minimum distance between the *centers* of the two outermost tracks? Explain why the effective collision diameter of a molecule (Section 17–4) is 2σ.

10. Show that Z, the number of collisions of molecules per cubic centimeter per second, has the dimensions cm$^{-3} \cdot$ sec^{-1}. $Z = 2(n')^2\sigma^2\sqrt{\pi RT/M}$ (*Hint:* If necessary, see Appendix A–3.)

11. Carbon dioxide and helium are stored in separate but identical containers initially at the same temperature and pressure. Using data from Table 15–1, determine **(a)** which gas (if either) would require the greater number of calories to reheat to the original temperature if the volume of each were doubled by sudden expansion into a vacuum, **(b)** which gas, if either, would do more work on the surroundings if the original volume of each gas were doubled by expanding against a constant pressure of 1 atm at constant temperature, **(c)** which gas would show the greater rate of pressure decrease if each gas were allowed to effuse from its container through a very small orifice. **(d)** Molecules of which gas should travel the greater average distance between collisions? **(e)** If a given quantity of heat were added to each gas, which would experience the greater rise in temperature?

12. For the reaction A → C + D, the initial concentration of A is 0.010 M. After 100 sec, the concentration of A is 0.0010 M. The rate constant of the reaction has the numerical magnitude of 9.0. What is the order of the reaction?

13. From the accompanying data for the reaction A $\xrightarrow{\text{catalyst}}$ products, calculate the value of k.

Time (min)	[A]	Time (min)	[A]
0.0	0.100	2.0	0.080
1.0	0.090	3.0	0.070

14. Graph the following data from an experimental determination of the rate of a reaction. From the slope of the line, calculate the order of the reaction and the value of the rate constant.

Time (min)	[Reactant]	Time (min)	[Reactant]
0.00	0.500	15.00	0.236
5.00	0.389	20.00	0.184
10.00	0.303	25.00	0.143

15. What is meant by an "accepted" mechanism? Is there any such thing as a proven mechanism?

16. What is the rate law for the single-step reaction A + B → 2 C?

17. Distinguish between an unstable intermediate and a transition state or activated complex. Select an example of each from the reaction between H_2O_2 and HBr described in this chapter.

18. Explain why transition states cannot be isolated as independent chemical species.

19. Deduce rate law expressions for the conversion of H_2 and I_2 to HI at 400°C corresponding to each of the following mechanisms:

(a) H_2 + I_2 → 2 HI   (one step)

(b) I_2 ⇌ 2 I
       2 I + H_2 → 2 HI   (slow)

(c) I_2 ⇌ 2 I
        I + H_2 ⇌ IH_2
      IH_2 + I → 2 HI   (slow)

(d) Can the observed rate law expression given in the text distinguish among these mechanisms? **(e)** If it is known that ultraviolet light causes the reaction of H_2 and I_2 to proceed at 200°C with the same rate law expression, which of these mechanisms becomes most improbable? Is any of these mechanism proved?

20. Explain why a catalyst which could accelerate the reactions

$$2\,NO_2 \;+\; 4\,CO \;\rightarrow\; N_2 \;+\; 4\,CO_2$$
$$2\,NO \;+\; 2\,CO \;\rightarrow\; N_2 \;+\; 2\,CO_2$$

would be useful to the automobile industry.

General Exercises

21. At 326°C, 1,3-butadiene dimerizes according to the equation

$$2\,C_4H_6(g) \;\rightarrow\; C_8H_{12}(g)$$

In a given experiment, the initial pressure of C_4H_6 was 632.0 torr at 326°C. Determine the order of the reaction and the value of the rate constant from the accompanying data.

Time (min)	Total Pressure (torr)	Time (min)	Total Pressure (torr)
0.00	632.0	24.55	546.8
3.25	618.5	42.50	509.3
12.18	584.2	68.05	474.6

22. For a reaction in which 1 mole of reactant produces 1 mole of product, show that each of the expressions on the left-hand side of Table 17–1 becomes identical to the corresponding expression on the right.

23. At 25°C, the second order rate constant for the reaction $I^- + ClO^- \rightarrow IO^- + Cl^-$ is 0.0606 liter/mole · sec. If a solution is initially $3.50 \times 10^{-3}\,M$ with respect to each reactant, what will be the concentration of each species present after 300 sec?

24. Substance A reacts according to a first order rate law with $k = 5.0 \times 10^{-5}$/sec. **(a)** If the initial concentration of A is 1.00 M, what is the initial rate? **(b)** What is the rate after 1.00 hour?

25. Which of the following will react fastest (produce most product in a given time) and which will react at the highest *rate*? **(a)** 1 mole of A and 1 mole of B in a 1 liter vessel, **(b)** 2 moles of A and 2 moles of B in a 2 liter vessel, **(c)** 0.2 mole of A and 0.2 mole of B in a 0.1 liter vessel.

26. For the nonequilibrium process A $+$ B $\rightarrow$ products, the rate is first order with respect to A and second order with respect to B. If 1.0 mole each of A and B were introduced into a 1.0 liter vessel, and the initial rate were 1.0×10^{-2} mole/liter · sec, calculate the rate when half the reactants have been turned into products.

27. At 25°C, the second order rate constant for the reaction $I^- + ClO^- \rightarrow IO^- + Cl^-$ is 0.0606 liter/mole · sec. If a solution is initially 1.0 M in I^- and $5.0 \times 10^{-4}\,M$ in ClO^-, what additional information is required to determine what the concentration of ClO^- will be after 300 sec?

28. The reaction A $+$ 2 B $\rightarrow$ C $+$ 2 D is run three times. In the second run, the initial concentration of A is double that in the first run, and the initial rate of the reaction is double that of the first run. In the third run, the initial concentration of each reactant is double the respective concentrations in the first run, and the initial rate is double that of the first run. What is the order of the reaction with respect to each reactant?

29. A certain reaction, A $+$ B $\rightarrow$ C, is first order with respect to each reactant, with $k = 1.0 \times 10^{-2}$ liter/mole · sec. Calculate the concentration of A remaining after 100 sec if the initial concentration of each reactant was 0.100 M.

30. A certain reaction, $A + B \rightarrow$ products, is first order with respect to each reactant, with $k = 5.0 \times 10^{-3}$ liter/mole $\cdot$ sec. Calculate the concentration of A remaining after 100 sec if the initial concentration of A was 0.100 M and that of B was 6.00 M. State any approximation made in obtaining your result.

31. Calculate the activation energy for a reaction which doubles in rate when the temperature is raised from 18 to 28°C.

32. The activation energy of the reaction $A + B \rightarrow$ products is 24.6 kcal. At 40°C the products are formed at the rate of 0.133 mole/liter $\cdot$ min. What will be the rate of product formation at 80°C?

33. Write a stoichiometric equation for the reaction whose mechanism is detailed below. Determine the value of the equilibrium constant for the first step. Write a rate law equation for the overall reaction in terms of its initial reactants.

$$A_2 \rightleftharpoons 2A \quad k_1 = 10^{10} \text{ sec}^{-1} \quad \text{(forward)}$$
$$k_{-1} = 10^{10} \ M^{-1} \text{ sec}^{-1} \quad \text{(reverse)}$$
$$A + C \rightarrow AC \quad k_2 = 10^{-4} \ M^{-1} \text{ sec}^{-1}$$

34. Devise a mechanism for the following decomposition which will fit the given rate law:

$$COCl_2 \rightarrow CO + Cl_2 \quad \text{rate} = k[COCl_2][Cl_2]^{1/2}$$

35. Above 500°C, the reaction $NO_2 + CO \rightarrow CO_2 + NO$ obeys the rate law: rate $= k[NO_2][CO]$. Below 500°C, the rate law for this reaction is rate $= k'[NO_2]^2$. Suggest mechanisms for each of these cases.

36. List the ways an activated complex differs from an ordinary molecule.

37. For the reaction $2 Fe^{2+} + H_2O_2 \rightarrow 2 Fe^{3+} + 2 OH^-$ the rate of formation of Fe^{3+} is given by rate $= k[Fe^{2+}][H_2O_2]$. Suggest a mechanism for the reaction, and indicate the probable rate-determining step.

38. For the reaction $A + B \rightarrow C + D$, the rate law equation is rate $= k_1[A] + k_2[A][B]$, where k_1 and k_2 represent two different constants. The products of this reaction are formed by two different mechanisms. What may be said about the relative magnitudes of the rates of the two individual mechanisms?

39. The rate of the following reaction in aqueous solution, in which the initial concentration of the complex compound is 0.100 M, is to be studied:

$$[Co(NH_3)_5Cl]^{2+} + H_2O \rightarrow [Co(NH_3)_5(H_2O)]^{3+} + Cl^-$$

Explain why under these conditions it is possible to determine the order of the reaction with respect to $[Co(NH_3)_5Cl]^{2+}$, but not with respect to water. Support your explanation with an appropriate calculation.

40. For a reaction $A + B \rightarrow C + D$, the rate law is rate $= k_1[A][B] - k_2[A]$. What can be said about the nature of this reaction from this rate law? (*Hint:* See exercise 38.)

41. For the reaction $A + B \rightarrow C + D$, the rate law is rate $= k_1[A][B] - k_2[C][D]$. What can be said about the nature of this reaction? (*Hint:* See exercise 38.)

42. The decomposition of SO_2Cl_2 to SO_2 and Cl_2 is first order. When 0.10 mole of $SO_2Cl_2(g)$ is heated at 600 K in a 1.0 liter vessel, the following data are obtained:

Time (hours)	Pressure (atm)	Time (hours)	Pressure (atm)
0.0	4.91	4.0	7.31
1.0	5.58	8.0	8.54
2.0	6.32	16.0	9.50

(a) Calculate the rate constant for the decomposition of SO_2Cl_2 at 600 K. (b) What is the half life of the reaction? (c) What would be the pressure in the vessel after 20 min? (d) What fraction of the original SO_2Cl_2 remains undecomposed after 20 hours at 600 K?

43. The reaction $2 NO(g) + Cl_2(g) \rightleftharpoons 2 NOCl$ was studied at $-10°C$, and the following data were obtained:

	Initial Concentration (mole/liter)		Initial Rate of Formation of NOCl (mole/liter · min)
Run	NO	Cl_2	
1	0.10	0.10	0.18
2	0.10	0.20	0.35
3	0.20	0.20	1.45

(a) What is the order of reaction with respect to NO and with respect to Cl_2? (b) What is the numerical value of the rate constant at $-10°C$?

Advanced Exercises

44. Using the methods of integral calculus, derive the expressions of Table 17-1 from the corresponding rate law expressions.

45. The rate constant for a certain second order reaction is 8.00×10^{-5} liter/mole · min. How long will it take a 1.00 M solution to be reduced to 0.500 M in reactant? How long will it take from that point until the solution is 0.250 M in reactant? Explain why the term *half life* is not used often for second order reactions.

46. The rate law expression for the reaction of H_2 and Br_2 is given in Section 17-7. Derive a mechanism to fit this rate law. In your derivation, assume that all the steps have rates on the same order of magnitude and that a "steady state" is established—that is, after a short time of concentration buildup, any free atoms react as rapidly as they are formed. Thus throughout most of the reaction, the concentrations of free atoms do not change with time. Show specifically that your mechanism yields a rate law expression which corresponds to the experimental rate law expression.

47. Write a rate law expression for the overall reaction between A and B to yield C and D by the two mechanisms proposed below. What will be the initial rate of conversion of A and B in a solution containing 0.10 mole/liter of each. At what concentration(s) of A and/or B will the inherent rates be equal?

I. $A + B \rightarrow AB^* \rightarrow C + D$ $k_1' = 1 \times 10^{-5}$ liter/mole · sec

II. $\begin{cases} A \rightarrow A^* \rightarrow E \\ E + B \rightarrow C + D \end{cases}$ $k_1 = 1 \times 10^{-4}$ sec^{-1}
                                                   $k_2 = 1 \times 10^{10}$ liter/mole · sec

48. A certain reaction, $A + B \rightarrow C$, is first order with respect to each reactant, with $k = 1.0 \times 10^{-3}$ liter/mole · sec. Using calculus, calculate the concentration of A remaining after 100 sec if the initial concentration of A was 0.100 M and that of B was 0.200 M.

18

Crystalline Solids and Ionic Solutions

In this chapter, the criteria for defining a crystalline solid will be discussed, and the terminology used to describe crystal structure will be presented. It will be shown how crystals are classified on the basis of symmetry. Also, the structures will be rationalized in terms of the orderly stacking of spherical units, giving rise to definite geometric patterns.

At ordinary temperatures, ionic substances are characteristically solids. It will be shown how the cohesive energies in such crystals are calculated, both from theoretical considerations and as the result of experimental measurements. Finally, the characteristic properties of solutions containing ions will be discussed.

18-1 Crystalline and Amorphous Solids

A **crystalline solid** is characterized by three criteria:

1. A crystal lattice.
2. A definite melting point.
3. A definite enthalpy of fusion.

These criteria will be defined and more fully discussed below. Substances which are rigid but which do not conform to these requirements are called **amorphous solids.** Glass, rubber, and solidified glue are typical examples. One cannot always tell by external appearance whether a substance is crystalline. Many substances when ground to fine powder appear to be amorphous, but examination under a microscope will reveal their crystalline nature. Moreover, they will have definite melting points and definite enthalpies of fusion.

The distinction between a crystalline solid and an amorphous solid is demonstrated by their behavior when their respective liquids are cooled. When a liquid which will form an amorphous solid is cooled, the average kinetic energy of its molecules decreases. At a sufficiently low temperature, the intermolecular forces become large in comparison to the kinetic energy of the molecules, and the liquid gradually becomes so viscous that ultimately the entire mass of material becomes rigid. There is no temperature characteristic of the transition from the liquid state to the rigid (solid) state.

The behavior upon cooling of a pure liquid which forms a crystalline solid is diagrammed as a cooling curve in Figure 18–1. If the liquid is cooled rapidly, its

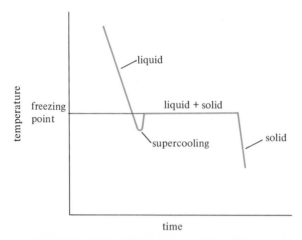

Figure 18-1. **Cooling Curve**

temperature decreases and then suddenly rises to some characteristic temperature known as the freezing point. At this point, crystals of the solid form of the substance appear in the liquid. Both solid and liquid appear together at the freezing point. If cooling is continued, the temperature of the mixture remains constant until the entire mass has solidified. It should be noted that at the freezing point the molecules of the liquid still have appreciable kinetic energies.

As shown in Figure 18-1, the temperature of a liquid may temporarily fall considerably below the freezing point without a solid being produced. This process is called **supercooling.** As the molecules in the supercooled liquid change from their somewhat random orientation to the more ordered arrangement in the solid, a definite quantity of heat is liberated per mole, causing the temperature to rise to the freezing point. The heat associated with the "organization" of the molecules is the negative of the enthalpy of fusion, $-\Delta H_{fus}$. As stated in Section 3-14, the process of going from a disordered state to an ordered state is accompanied by a decrease in entropy:

$$\Delta S = \frac{\Delta H}{T} = \frac{-\Delta H_{fus}}{T}$$

It should be noted that a supercooled liquid is in an unstable state. Any procedure which encourages the formation of an ordered arrangement of its molecules will induce spontaneous crystallization. Such procedures include the addition of a tiny crystal of the solid material or shaking or scratching the container. As mentioned in Section 16-7, liquids having complex and irregularly shaped molecules are most prone to supercooling.

18-2 The Crystal Lattice

The regular geometric pattern of the units making up a crystalline solid is called a **crystal lattice.** A lattice is a regular, three-dimensional arrangement of equivalent points in space. Since crystals are composed of a regular arrangement of atoms, ions, or molecules in space, they may be described in terms of lattice points. **Lattice points**

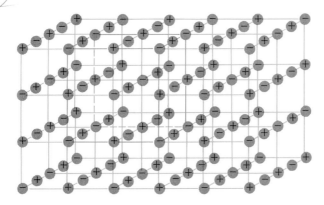

Figure 18-2. **Rock Salt (NaCl) Structure**

are positions within the crystal which all have the same chemical environment. For example, the crystal structure of sodium chloride is shown in Figure 18-2. All of the sodium ions have identical environments of six chloride ions. Similarly, all of the chloride ions have identical environments, which are different from those of the sodium ions. Therefore, in NaCl the lattice points may be chosen at the sites of either the sodium ions or the chloride ions but not at both.

It should be noted that lattice points may or may not be occupied by atoms. For example, in crystalline benzene, the lattice points are best located at the centers of the molecules because this location provides for the greatest symmetry about the lattice points. If the lattice points were chosen at the site of one of the atoms in each benzene molecule, the symmetry about the lattice points would be lower. Therefore, the choice would be less desirable.

(lattice point in color)

18-3 The Unit Cell

A set of angles and a set of lengths characteristic of the lattice as a whole are formed by connecting a lattice point with the three nearest lattice points which are not all coplanar with it, as shown in Figure 18-3(a). The three lengths are denoted a, b, and c. The three angles are called α, β, and γ, where α is the angle which is not bounded by the line of length a, β is the angle which is not bounded by the line of length b, and γ is the angle not bounded by the line of length c. If nine additional lines are now drawn, three parallel to each of those present as shown in Figure 18-3(b), a geometric figure known as the **unit cell** is defined. Movement of this figure along the three primary directions builds up the lattice structure as in Figure 18-3(c). Since, by definition, each and every lattice point has the identical chemical environment, the volume enclosed by the unit cell contains all the components of the crystal as a whole and in their proper stoichiometric ratios.

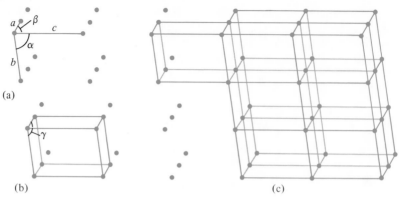

Figure 18-3. **Crystal Parameters and the Unit Cell**

The ideal crystal consists of extended "stacks" of repeating unit cells. The external symmetry of the crystal also reflects the arrangement within the unit cell. For example, in the diagram of the NaCl crystal lattice (Figure 18-2) all the angles in the unit cell are 90°. The characteristic angle between the faces of sodium chloride (rock salt) crystals is also 90°.

Compound unit cells have lattice points in addition to those at their corners. One of the three types of compound unit cells contains a lattice point at the center of the cell and is called a **body-centered** unit cell. The unit cell with lattice points in the center of *each* of its six faces is called **face-centered.** If the centers of only one pair of faces are the sites of lattice points, the unit cell is said to be **end-centered.** The compound unit cells are shown in Figure 18-4. Each of these structures could be regarded as derived from a simple unit cell of lower symmetry, but it is more convenient to describe the lattice in terms of the more symmetrical compound cell.

Example

What type of unit cell best describes the NaCl crystal lattice diagrammed in Figure 18-2?

Either Na^+ or Cl^- ions may be chosen as the lattice points. Choosing Cl^- ions, one sees that the simplest cubic arrangement contains a chloride ion on the center of each face as well as at the corners of the unit cell; hence the unit cell is face-centered cubic.

In making up the entire crystal, a given lattice point may belong to more than one

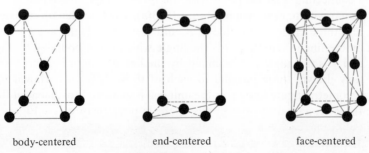

body-centered   end-centered   face-centered

Figure 18-4. **Compound Unit Cells**

unit cell. An atom located at a corner of a three-dimensional unit cell is actually at the corner of seven additional unit cells and therefore should be counted as one-eighth atom in each. Similarly, atoms along the edges of a unit cell are counted as one fourth in each cell, and atoms on the faces of unit cells are counted as one half in each of the two unit cells they link. Only atoms entirely within a unit cell are counted as belonging solely to that unit cell. In this manner the number of atoms in the unit cell can be determined, and the stoichiometry of the crystal can be deduced.

Example

Determine the number of formula units of NaCl in the unit cell (Figure 18-2). Since the cell is face-centered cubic, there are

At the 8 corners:	$8 \times \frac{1}{8} = 1$ Cl$^-$ ion	
At the 6 faces:	$6 \times \frac{1}{2} = 3$ Cl$^-$ ions	
Along the 12 edges:	$12 \times \frac{1}{4} =$	3 Na$^+$ ions
At the cube center:		1 Na$^+$ ion
Total		4 Cl$^-$ and 4 Na$^+$ ions

Hence, the unit cell contains four NaCl units.

18-4 Crystal Classes

Despite the wide variety of unit cell compositions and dimensions, all real crystals can be classified under just seven types of lattices. These are listed in Table 18-1 and are illustrated in Figure 18-5. All crystals are classified in terms of simple or compound unit cells derived from one of these types. Therefore it is necessary to become familiar with this classification. Some examples of substances which crystallize in each type of lattice are also listed in Table 18-1.

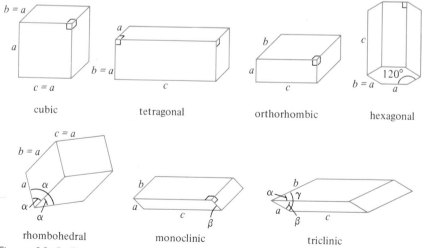

cubic   tetragonal   orthorhombic   hexagonal

rhombohedral   monoclinic   triclinic

Figure 18-5. **The Seven Crystal Systems**

TABLE 18-1. Crystal Classes

Class	Dimensions	Angles	Examples
cubic	$a = b = c$	$\alpha = \beta = \gamma = 90°$	Ag, NaCl
tetragonal	$a = b \neq c$	$\alpha = \beta = \gamma = 90°$	Sn (white), MgF_2
orthorhombic	$a \neq b \neq c$	$\alpha = \beta = \gamma = 90°$	S_8, $HgCl_2$
rhombohedral	$a = b = c$	$\alpha = \beta = \gamma \neq 90°$	Al_2O_3
hexagonal	$a = b \neq c$	$\alpha = \beta = 90°, \gamma = 120°$	Mg, CuS
monoclinic	$a \neq b \neq c$	$\alpha = \gamma = 90°, \beta \neq 90°$	$KClO_3$
triclinic	$a \neq b \neq c$	$\alpha \neq \beta \neq \gamma \neq 90°$	$CuSO_4 \cdot 5H_2O$

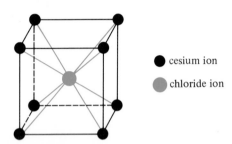

Figure 18-6. **Cesium Chloride Unit Cell**

● cesium ion

● chloride ion

Example

The cesium chloride structure is shown in Figure 18-6. To what system does cesium chloride belong? Is the unit cell compound or simple?

The cesium chloride unit cell is a simple, cubic unit cell. The center of the unit cell is occupied by an ion of opposite charge from the ions at the cell corners and is not the site of a lattice point.

18-5 Packing of Atoms and Spherical Molecules

A simple rationalization of the structures of crystals is provided by analogy with the packing of spherical objects, as, for example, the packing of marbles in a box, as illustrated in Figure 18-7. Figure 18-8 represents several ways in which a number of

Figure 18-7. **Packing of Marbles in a Box**

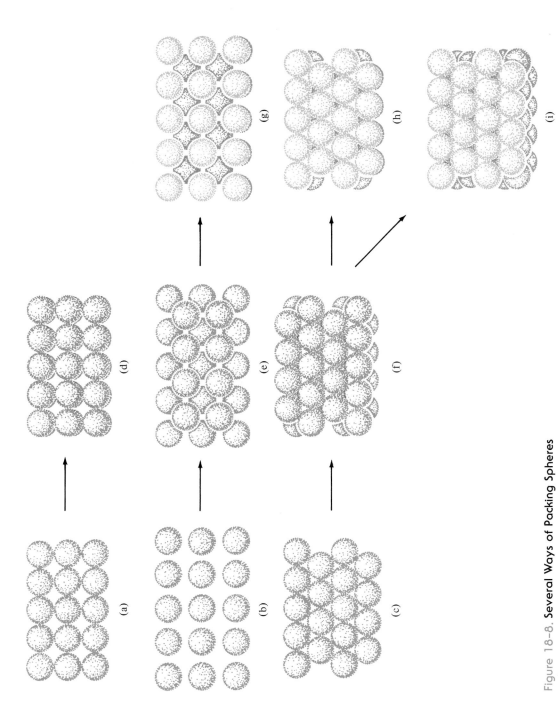

Figure 18-8. Several Ways of Packing Spheres

spheres of equal size can be packed in layers. In (a), (b), and (c) are shown three ways in which the first layer might be arranged. It is easily seen that there is least free space between the spheres in arrangement (c). This arrangement is called closest packed (or close packed), while those of (a) and (b) may be seen to be the beginnings of cubic patterns. The ways of putting additional layers on these first layers are shown in (d) through (i). Starting with the cubic pattern of Figure 18–8(a), the successive layers will generate a simple cubic lattice. Each sphere touches six adjacent spheres. Successive layers of the more open pattern (b) generate a body-centered cubic (bcc) lattice, as shown in (e) and (g). Each sphere touches its eight nearest neighbors at the corners of a cube. Six additional spheres, located at the centers of adjacent cubes, are only slightly farther away.

In the closest packed arrangement, there are two distinct ways of adding successive layers of spheres. These are shown in Figures 18–8(h) and (i). In one case, (h), a **hexagonal closest packed** (hcp) structure is obtained. The spheres of the second layer are placed in the depressions between the first layer spheres, and then the third layer spheres are placed in the depressions of the second layer such that they are directly above the spheres of the first layer. The spheres of the fourth layer are directly above those of the second layer, and each successive layer alternates in position.

In the other closest packed structure (i), the third layer of spheres is placed so that each sphere is *not* directly above a sphere of the first layer. Instead the spheres of the *fourth* layer are above those of the first, and the succession of layers then proceeds by having each layer directly above the spheres three layers below. This pattern leads to the **cubic closest packed** (ccp) structure, in which the unit cell is a face-centered cube (Figure 18–9).

The number of closest neighbors of a given sphere is called its **coordination number.** In both the hcp and the ccp lattices, each sphere has 12 neighboring spheres which it touches directly—6 in the same plane, 3 in the plane above, and 3 in the plane below. Hence, both closest packed structures have atoms characterized by coordination numbers of 12. The body-centered cubic arrangement is less compact; the characteristic coordination number is 8. (Sometimes the slightly more distant atoms are included, and a coordination number of 14 is attributed to this structure.) In the simple cubic arrangement, the coordination number is 6; it is the least compact of those shown.

Many substances whose atoms or molecules are essentially spherical, or whose molecules rotate freely in the lattice producing pseudospherical symmetry, crystallize

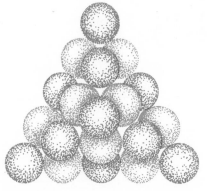

Figure 18–9. Cubic Closest Packing Resulting from Hexagonal Layers
The colored spheres represent the corners of the cubes.

in one or more of these lattice types. For example, hydrogen crystallizes with an hcp lattice, while the noble gases crystallize in the ccp arrangement. In their solid states, the hydrogen halides (HCl, etc.) assume the hcp arrangement. Metal atoms are essentially spherical, and metals crystallize in the bcc, hcp, and ccp structures. Uncharged atoms and molecules never crystallize in the simple cubic arrangement.

18-6   Ionic Solids

Crystals of ionic solids may be regarded as being "derived" from closest packed structures, but there are several important differences from the case of uncharged spheres. In the closest packed arrangements, all of the component units are identical and none has any charge. In contrast, the component units of ionic crystals are at least two different kinds of ions which have positive and negative charges, respectively. In a lattice containing ions, it is geometrically impossible for any ion to have a coordination number greater than 8 if each section of the structure is to be electrically neutral. For example, if both the positive and negative ions have the same charge ($1+$ and $1-$, $2+$ and $2-$, etc.), there must be an equal number of ions of each charge type in the crystal. Ions of each charge type must therefore be surrounded by equal numbers of ions of the opposite charge type. But if each positive ion were surrounded by more than eight negative ions, there simply would be insufficient room to surround each negative ion with the same number of positive ions.

Similarly, if the positive and negative ions are of different charge types ($2+$ and $1-$, for example), their relative numbers must be such as to preserve electroneutrality. The coordination numbers of the respective ions and the type of unit cell in the crystal will be determined by the charges and the sizes of the ions. One set of ions may be regarded as occupying the lattice points of a closest packed structure, while the other ions occupy the interstices or "holes" between them. For example, in the hcp structure, each sphere rests upon three others, as shown in Figure 18-10. Thus their centers locate the apices of a tetrahedron, and the "hole" located between the spheres is called a tetrahedral site. Since each sphere in a closest packed lattice is surrounded by eight tetrahedral sites, each of which is bounded by four spheres, there are two tetrahedral sites per sphere in the lattice as a whole.

Many binary compounds having formulas of the types AB, A_2B, and AB_2 crystallize in the ccp lattice. One set of atoms or ions, say B, occupies the lattice points, and the other set occupies the tetrahedral sites. If the formula is of the type AB, only half the tetrahedral sites are occupied. For example, in the crystal structure of the mineral zinc blende, ZnS, shown in Figure 18-11, the sulfide ions are located at the lattice points, with the zinc ions occupying half the tetrahedral sites.

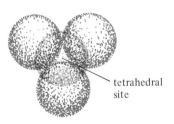

Figure 18-10. **A Tetrahedral Site**

tetrahedral site

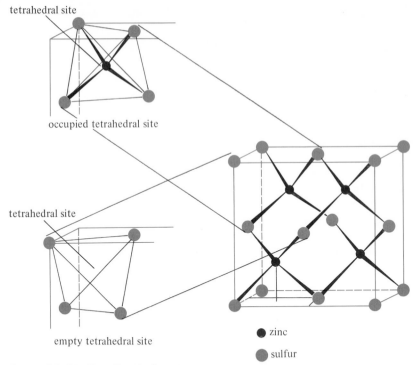

tetrahedral site

occupied tetrahedral site

tetrahedral site

empty tetrahedral site

● zinc

● sulfur

Figure 18-11. Zinc Blende Structure
Alternate tetrahedral sites in the structure are filled.

The mineral fluorite, CaF_2, is typical of a ccp crystal in which all of the tetrahedral sites are occupied. Its structure is diagrammed in Figure 18–12. The calcium ions are located at the lattice points, and the fluoride ions fill all the tetrahedral sites. The opposite situation, in which the anions are located at the lattice points and the cations fill all the tetrahedral holes in a ccp structure, is called the antifluorite structure. Lithium oxide and rubidium sulfide are examples of compounds which crystallize with this structure.

Another type of interstice in close packed structures is the octahedral hole. In a ccp structure, an octahedral site is formed by two sets of three spheres, as shown in Figure 18–13. Three intersecting lines connecting the centers of the spheres in the lower

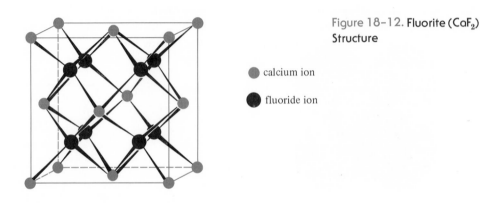

Figure 18-12. Fluorite (CaF_2) Structure

● calcium ion

● fluoride ion

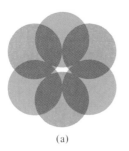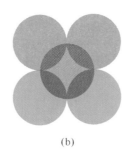

Figure 18–13. An Octahedral Site Viewed from Two Perspectives

(a)   (b)

plane with those in the upper plane would meet at the **octahedral site.** Another way of locating the octahedral sites in the cubic closest packed lattice is to look at these six spheres from a different perspective, as shown in Figure 18–13(b). Recognizing that two of the spheres lie above and below the plane of the paper, one can easily see that the hole between the six spheres is an octahedral site. For example, in the crystal structure of rock salt, NaCl (Figure 18–2), if the chloride ions are considered to be lattice points, then each octahedral site is occupied by a sodium ion. A large number of compounds of the type AB, such as KCl, CaO, TiO, and MnS, crystallize in the rock-salt-type lattice.

In lattices which are not closest packed, sites having a coordination number higher than 6 are possible. For example, the cesium chloride structure can be regarded as a simple cubic lattice. The lattice points are occupied by either cesium ions or chloride ions, with the other ions occupying interstices at the centers of the cubes. The coordination number of each type ion is 8.

18-7 Radius Ratio

The dimensions of the octahedral and tetrahedral sites in closest packed structures are determined by the sizes of the atoms occupying the lattice points. The closest possible approach of like charged ions is such that they touch one another. Figure 18–14 shows the cross section of an octahedral site. If r_1 is the radius of the ions at the lattice points and r_2 is the radius of a smaller ion which just makes contact with those at the lattice points, the radius ratio, r_2/r_1, may be determined by the Pythagorean theorem, as follows:

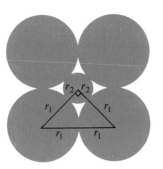

Figure 18–14. **Cross Section of an Octahedral Site**

TABLE 18-2. Radius Ratios and Coordination Numbers in Ionic Lattices

$\dfrac{r_2}{r_1}$	Coordination Number of Smaller Ion	Arrangement of Nearest Neighbors
1.000–0.732	8	corners of a cube
0.732–0.414	6	octahedral
0.414–0.225	4	tetrahedral

$$2(r_1 + r_2)^2 = (2r_1)^2$$
$$r_1 + r_2 = \sqrt{2}r_1 = 1.414r_1$$
$$r_2/r_1 = 0.414$$

An analogous calculation shows that for tetrahedral sites, the ratio of the radius of the ion at the tetrahedral site to that of an ion at a lattice point is 0.225.

The ranges of radius ratio values which correspond to various coordination numbers of the smaller ion are listed in Table 18–2. When the relative sizes of the ions fall within a given range, the most stable geometry of the crystal lattice can be predicted. For example, in ZnS the ratio of the radius of Zn^{2+} to the radius of S^{2-} is found experimentally to be 0.40. Hence, Zn^{2+} will occupy tetrahedral holes in the closest packed lattice of sulfide ions. Similarly, in NaCl the ratio r_{Na^+}/r_{Cl^-} is 0.52, a value in keeping with a coordination number of 6 for the sodium ion. Real atoms and ions cannot be considered as hard spheres, because the probability densities of their surrounding electrons do not end abruptly at a certain distance from their nuclei. The actual distances of closest approach may be somewhat larger than that given by the simple hard sphere model. As a result, in some cases radius ratios exceeding the theoretical limits are encountered. For example, the radius ratio for RbCl is 0.82, and the salt might be expected to have a coordination number of 8, but it actually crystallizes in the rock salt structure, with a coordination number 6.

18-8 Ionic Radii

To predict coordination numbers of ions using the radius ratio concept, it is necessary to have a set of values for the radii of the ions. However, the absolute radius of an ion cannot be defined precisely. Nevertheless, it is reasonable to assume that the internuclear distances in ionic crystals can serve as practical measures of the radii of the respective ions. These internuclear distances can be determined experimentally by X-ray diffraction (see Figure 13–15). A problem arises as to what fraction of the interionic distance between unlike ions should be apportioned to each of them. One method of assigning sizes which yields reasonably consistent results is to assume that when very large anions, such as iodide ions, are packed around a small cation, such as lithium ion, the anions will be in contact with each other (see Figure 18–14, for example). The radius of the large anion is taken as one half the *anion–anion* distance. Once a value for the radius of the anion is established, the radius of a different cation can be determined from the observed distance in another crystal. That cation radius can be used to determine still more ionic radii. For example, the iodide ion radius can

H^- 2.08	Li^+ 0.60	Be^{2+} 0.31		
O^{2-} 1.40	F^- 1.36	Na^+ 0.95	Mg^{2+} 0.65	Al^{3+} 0.50
S^{2-} 1.84	Cl^- 1.81	K^+ 1.33	Ca^{2+} 0.99	Ga^{3+} 0.62
Se^{2-} 1.98	Br^- 1.95	Rb^+ 1.48	Sr^{2+} 1.13	In^{3+} 0.81
Te^{2-} 2.21	I^- 2.16	Cs^+ 1.69	Ba^{2+} 1.35	Tl^{3+} 0.95

	Fe^{2+} 0.75	Fe^{3+} 0.53
Cu^+ 0.96	Cu^{2+} 0.72	Cr^{3+} 0.55
Ag^+ 1.26	Zn^{2+} 0.74	

Figure 18-15. Radii of Some Common Ions (Ångström units)

be used together with the potassium–iodide distance in KI to determine an effective radius for the potassium ion. Radii determined in this manner depend somewhat on the coordination numbers of the respective ions and are most consistent only when applied to lattices of the same type. Some representative values of ionic radii are given in Figure 18–15.

18-9 Lattice Energy

Why should there be exceptions to the radius ratio rules? Why do some ionic solids, such as KCl, show large changes in solubility with temperature, while the solubilities of others, such as NaCl, have relatively small temperature dependence? Why are some other ionic solids, such as AgCl, virtually insoluble in water? These and related questions are answered in part by considerations of the lattice energy.

Lattice energy is defined as the energy change observed when the constituents of a crystal are brought from "infinitely far" apart to their equilibrium positions in the crystal. In general, a large negative value for the lattice energy indicates considerable stability of the crystal.

In ionic crystals, the lattice energy stems mainly from the energy of attraction of oppositely charged ions. Ionic crystals have generally high stabilities, as indicated by their high melting points, low volatilities, and high enthalpies of fusion. Since they consist of oppositely charged ions, it is possible to calculate their lattice energies on the basis of Coulomb's law: $E = q_1q_2/d$ (Section 9–1). It must be noted that in addition to the attractions between oppositely charged ions, there are also repulsions between like charged ions. In principle, every ion in the crystal is acted upon by all of the others, and the magnitudes of the interactions are determined by the geometry of the lattice. Because of the high symmetry of the crystal, all of the electrostatic interactions and the lattice energy can be expressed in terms of the distance, r, between adjacent ions. Using the rock salt, NaCl, structure as an example, the procedure for calculating the lattice energy will be demonstrated.

First, consider a "crystal" of only two ions, at the distance r equal to the shortest interionic distance in solid sodium chloride.

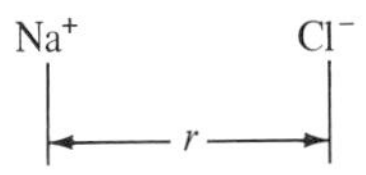

The lattice energy is merely $-e^2/r$, where e is the electronic charge. If the "crystal" consists of four ions arranged in a square,

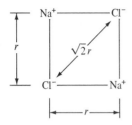

the lattice energy is obtained from six terms: four terms due to the attractions between oppositely charged ions at distance r and two terms due to repulsions between like charged ions at distance $\sqrt{2}r$:

$$-\left[4\left(\frac{e^2}{r}\right) - 2\left(\frac{e^2}{\sqrt{2}r}\right)\right] = -2.59\,\frac{e^2}{r}$$

Hence the lattice energy per ion pair (one sodium ion and one chloride ion) is $-1.29e^2/r$.

If the "crystal" consists of eight ions in a cubic arrangement,

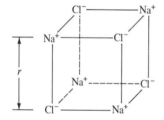

the expression for the lattice energy is

$$-\left[12\left(\frac{e^2}{r}\right) - 12\left(\frac{e^2}{\sqrt{2}r}\right) + 4\left(\frac{e^2}{\sqrt{3}r}\right)\right] = -5.82\,\frac{e^2}{r}$$

The lattice energy per ion pair is $-1.46e^2/r$. As larger and larger segments of crystal are considered, the lattice energy per ion pair approaches a limiting value of $-1.75e^2/r$. The trend toward this value is shown in Table 18–3. After the "crystal" attains a certain size, the coulombic energy per ion pair does not change significantly upon addition of more ion pairs.

Since the limiting value of the coefficient is determined only by the lattice geometry, the lattice energy per mole of ion pairs, U_0, of any substance which crystallizes in the rock salt structure is given by

$$U_0 = -\frac{1.75NZ_1Z_2e^2}{r_0}$$

TABLE 18-3. Lattice Energies of Crystal Fragments in the NaCl Crystal

Number of Ion Pairs	Net Coulombic Energy	Coulombic Energy per Ion Pair
1	$-1.00e^2/r$	$-1.00e^2/r$
2	$-2.59e^2/r$	$-1.29e^2/r$
4	$-5.85e^2/r$	$-1.46e^2/r$
$\vdots$	$\vdots$	$\vdots$
N		$-1.75e^2/r$

where N is Avogadro's number, Z_1 and Z_2 are the magnitudes of the charges on the cation and anion, respectively, and r_0 is the equilibrium distance between adjacent, oppositely charged ions in the crystal. The numerical coefficient, 1.75, is called the Madelung constant. The value of the Madelung constant depends on the geometric arrangement of the ions but is independent of their chemical natures or their charges. Madelung constants for several types of structures are listed in Table 18-4.

It should be mentioned that even for purely ionic crystals, the Madelung constants do not completely determine the lattice energy. The ions in the crystal are not point charges; because of their surrounding electrons they have finite sizes. Therefore, in addition to coulombic forces there are repulsive forces which exist between neighboring ions regardless of their charges. A more complete expression for the lattice energy is given by the Born equation, which includes interionic repulsions between unlike ions:

$$U_0 = -\frac{Z_1 Z_2 N A e^2}{r_0}\left(1 - \frac{1}{n'}\right)$$

where U_0 is the lattice energy, A is the Madelung constant, N is Avogadro's number, and n' is a constant called the Born exponent, which can be obtained experimentally and can also be estimated theoretically. If the electronic configurations of the ions are assumed to be like those of the noble gases, n' has the following values:

Noble Gas Configuration	n'
He	5
Ne	7
Ar	9
Kr	10
Xe	12

TABLE 18-4. Madelung Constants for Common Crystal Lattices

Crystal Type	Example	Madelung Constant
rock salt	NaCl	1.747558
cesium chloride	CsCl	1.762670
wurtzite	ZnS	1.641
fluorite	CaF_2	5.03878

For salts having ions of two different noble gas configurations, the average value of n' is used.

The magnitude of the lattice energy of a substance is thus determined by several factors, including the charges on the ions and the interionic distance (which in turn is related to the sizes of the ions). Any decrease in the size of either anion, cation, or both will result in an increase in the lattice energy.

Example

Calculate the lattice energy of cesium iodide, which crystallizes with the cesium chloride structure and has an interionic distance of 3.95 Å.

The Born exponent for CsI, with two Xe type ions, is 12.

$$U_0 = -\frac{(1)^2(6.02 \times 10^{23})(1.76)(4.80 \times 10^{-10})^2}{3.95 \times 10^{-8}}\left(1 - \frac{1}{12}\right)$$

$$= -5.67 \times 10^{12} \text{ ergs/mole} = -135 \text{ kcal/mole}$$

18-10   Born–Haber Cycle

The lattice energy of an ionic crystal may be regarded as the enthalpy change for the process of combining gaseous anions and cations to produce 1 mole of crystalline solid. For example, the lattice energy of sodium chloride, U_{NaCl}, is approximately the enthalpy change of the following reaction:

$$Na^+(g) + Cl^-(g) \rightarrow NaCl(s) \qquad \Delta H_{298} \cong U_{NaCl}$$

Since it is impossible to obtain even a fraction of a mole of gaseous ions at room temperature, this enthalpy change cannot be determined directly. However, since ΔH is a thermodynamic property, its value depends only on the initial and final states of the system being considered. Consequently, it is possible to obtain the lattice energy by an indirect calculation involving steps whose enthalpy changes at room temperature are known. This procedure uses the **Born–Haber cycle,** so called because the steps may be shown graphically as a cyclic process (Figure 18–16). The enthalpy change of any step in the cycle is the net sum of the enthalpy changes of the remaining steps. To obtain the lattice energy of NaCl, the enthalpies of the following reactions are required:

kcal/mole

$Na(s) + \frac{1}{2}Cl_2(g) \rightarrow NaCl(s)$	$\Delta H = \Delta H_f$	$=$	-98.2
$Na(g) \rightarrow Na(s)$	$\Delta H = -\Delta H_{sub}$	$=$	-25.9
$Cl(g) \rightarrow \frac{1}{2}Cl_2(g)$	$\Delta H = -\frac{1}{2}\Delta H_{diss}$	$=$	-28.9
$Na^+(g) + e^- \rightarrow Na(g)$	$\Delta H \cong -IP$	$=$	-118.4
$Cl^-(g) \rightarrow Cl(g) + e^-$	$\Delta H \cong -(-EA)$	$=$	85.5

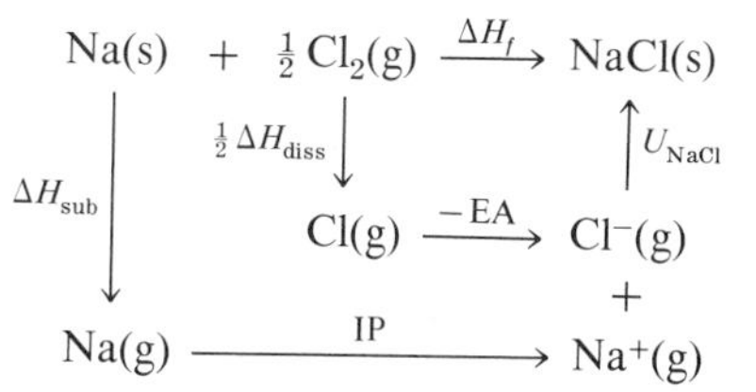

$$Na(s) + \tfrac{1}{2}Cl_2(g) \xrightarrow{\Delta H_f} NaCl(s)$$

Figure 18-16.
Born-Haber Cycle for Sodium Chloride

$$\Delta H_{sub} \downarrow \qquad \tfrac{1}{2}\Delta H_{diss} \downarrow \qquad \uparrow U_{NaCl}$$

$$Cl(g) \xrightarrow{-EA} Cl^-(g)$$
$$+$$
$$Na(g) \xrightarrow{IP} Na^+(g)$$

The net reaction is that of the formation of the NaCl crystal from its gaseous ions, and the net enthalpy change is the lattice energy:

$$Na^+(g) + Cl^-(g) \rightarrow NaCl(s) \qquad \Delta H = U_{NaCl} = -186 \text{ kcal/mole}$$

The enthalpy changes of the steps leading to the net equation can be determined experimentally. The enthalpy change for the first step is the enthalpy of formation of sodium chloride from the elements in their standard states. The second step is the reverse of the sublimation of sodium metal; hence the enthalpy change is $-\Delta H_{sub}$. Similarly, the enthalpy change of the third step is the negative of the enthalpy of dissociation of $\tfrac{1}{2}$ mole of gaseous chlorine. The enthalpy change of the fourth step is merely the ionization potential, IP, of sodium, corrected to a constant pressure process and reversed in sign. The enthalpy change for the fifth step is minus the electron affinity, $-EA$, of chlorine, also corrected to a constant pressure process and reversed in sign.[1] Thus the lattice energy, U_{NaCl}, is determined to be -186 kcal/mole, which compares well with the value -187 kcal/mole calculated from the Born equation.

Since the lattice energy can be calculated using the Born equation, the Born–Haber cycle is often used to determine electron affinities, which in many cases are difficult to measure directly.

Example

Calculate the electron affinity of iodine, given the following data:

$$
\begin{array}{rr}
 & \text{kcal/mole} \\
U_{NaI} = & -165.4 \\
\Delta H_{f(NaI)} = & -64.8 \\
\Delta H_{sub(Na)} = & 25.9 \\
IP_{Na} = & 118.4 \\
\tfrac{1}{2}(\Delta H_{sub(I_2)} + \Delta H_{diss(I_2)}) = & 25.5
\end{array}
$$

The negative of the electron affinity of iodine is the enthalpy of formation of NaI minus the enthalpies of each of the other terms:

$$-EA = \Delta H_f - \Delta H_{sub(Na)} - \tfrac{1}{2}(\Delta H_{sub(I_2)} + \Delta H_{diss(I_2)}) - IP_{Na} - U_{NaI}$$
$$= -64.8 - 25.9 - 25.5 - 118.4 - (-165.4) = -69.2$$
$$EA = 69.2 \text{ kcal/mole}$$

[1] By definition, electron affinity is the energy *released* when an electron is added to a gaseous atom. The energy *absorbed* in this process, ΔH, is denoted $-EA$.

18-11   Solubility of Ionic Solids

Ionic solids are soluble to a significant degree at ordinary temperatures only in solvents having polar molecules. The attractive forces between the ions and the solvent molecules must be rather large in order to be able to overcome the strong attractive forces between the ions in the crystal. In other words, the energy required to break up the crystal must be supplied in great measure by the energy released owing to the attraction of the solvent molecules for the ions in solution. This energy is called the **solvation energy.** The greater the dipole moment of the solvent molecules, the stronger the attraction for an ion, and the greater the solvation energy. For a given solvent, the smaller and the more highly charged the ion (either positive or negative), the greater will be the solvation energy. Water is an excellent solvent for ionic substances because of the polarity of its molecules and because water forms coordinate covalent bonds with many metal ions.

When a salt, MX, dissolves in water, the enthalpy change can be represented by a Born–Haber type of cycle as follows:

$$MX(s) \xrightarrow{\;-U\;} M^+(g) \;+\; X^-(g)$$

$$+ \qquad\qquad + \qquad\qquad +$$

$$H_2O \qquad\qquad H_2O \qquad\qquad H_2O$$

$\Delta H_{soln} \Big|\qquad\qquad \Delta H_{M^+} \Big\downarrow \qquad\qquad \Delta H_{X^-} \Big\downarrow$

$$M^+(aq) \;+\; X^-(aq)$$

and

$$\Delta H_{soln} = \Delta H_{M^+} + \Delta H_{X^-} - U$$

where ΔH_{soln} is the observed enthalpy of solution at infinite dilution, ΔH_{M^+} is the solvation enthalpy of the cation, ΔH_{X^-} is the solvation enthalpy of the anion, and U is the lattice energy.

Example

Calculate the enthalpy of solution of sodium chloride. The lattice energy is -186 kcal/mole, and the solvation enthalpies of the cation and anion, respectively, are -97 and -85 kcal/mole.

$$\Delta H_{soln} = (-97) + (-85) - (-186) = 2 \text{ kcal/mole}$$

The enthalpies of solution of a vast number of ionic crystals are positive, and the high solubilities of some of these salts result from the favorable entropy change involved in the solution process. The dissolved state is much more random than the ordered arrangement of ions in the crystal. However, it should be observed that there is also an ordering effect in solution, as the solvation of the ions tends to orient the solvent molecules. This effect lessens the increase in entropy. Hence, the free energy change of the overall solution process is the result of several factors.

18-12   Conductivity of Electrolytes in Solution

Conductivity measurements can be used to determine the charge type of strong electrolytes. Also, strong and weak electrolytes can be distinguished by measuring their conductivities over a range of concentrations. The experimental arrangement for measuring conductivity was described in Figure 14-2.

The specific conductivity, κ, is defined as the conductivity[2] of a solution between two electrodes each having an area of 1 cm^2 and separated by a distance of 1 cm. By first determining the cell constant, k, any cell regardless of its size and shape can be used to determine κ. The conductivity, L, of a solution of known κ is measured in the cell, and since

$$L = k\kappa$$

$$\kappa = L/k$$

To compare different solutes, it is necessary to define the **equivalent conductivity, Λ,** as the conductivity of a solution containing 1 mole of positive charge (and, of course, 1 mole of negative charge) between electrodes 1 cm apart and large enough to contain all of the solution. (See Section 14-2 for a parallel discussion of molar conductivity.) Λ is related to κ by the equation

$$\Lambda = \frac{\kappa}{C} \times 1000$$

where C is the concentration of the solute in moles of charge per liter (normality). The equivalent conductivities of several electrolytes are listed in Table 18-5. In Figure 18-17 the equivalent conductivities of KCl, a typical strong electrolyte, and acetic acid, a weak electrolyte, are plotted against the square root of concentration. These data show that there is a pronounced difference in the conductivity behavior between strong and weak electrolytes. Strong electrolytes show a comparatively small change in Λ with dilution, and the plot can easily be extrapolated to infinite dilution

TABLE 18-5. Equivalent Conductivity, Λ, of Some Electrolytes in Water at 25°C (cm^2/ohm · equiv)

	Equivalents per Liter				Infinitely Dilute, Λ_0[a]
	1.00	0.10	0.01	0.001	
NaCl		106.74	118.51	123.74	126.45
KCl	111.9	129.0	141.3	147.0	149.9
HCl	332.8	391.3	412.0	421.4	426.2
HC$_2$H$_3$O$_2$		5.2	16.3	49.2	(390.7)
NH$_3$		3.6	11.3	34.0	(271.0)
NaC$_2$H$_3$O$_2$		72.8	83.8	88.5	91.0
NH$_4$Cl		128.8	141.3	146.8	149.7

[a] Extrapolated (or calculated) values.

[2] Actually, the specific resistance, ρ, is determined, and $\kappa = 1/\rho$.

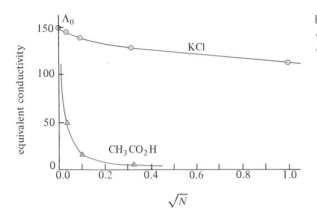

(unlimited solvent). The extrapolated value of the equivalent conductivity at infinite dilution is designated Λ_0. In contrast, weak electrolytes such as acetic acid show very large changes in conductivity near infinite dilution, and it is not possible to obtain a value of Λ_0 by extrapolation of the curve.

18–13 Kohlrausch's Rule

If the conductivities at infinite dilution of several salts of two cations having common anions are compared, the difference in conductivity for each pair is practically constant. An example of this behavior is shown in Table 18–6. This behavior is the basis for **Kohlrausch's rule,** which states that the equivalent conductivity at infinite dilution of any electrolyte can be estimated from an appropriate combination of the known equivalent conductivities at infinite dilution of other electrolytes.

Example

Estimate the equivalent conductivity at infinite dilution for HClO$_4$ using the data of Tables 18–5 and 18–6.

$$\Lambda_{0(HClO_4)} = \Lambda_{0(HCl)} + \Lambda_{0(NaClO_4)} - \Lambda_{0(NaCl)}$$
$$= 426.2 + 117.5 - 126.5 = 417.2 \text{ cm}^2/\text{ohm} \cdot \text{equiv}$$

The equivalent conductivity at infinite dilution for weak electrolytes can also be calculated using Kohlrausch's rule.

TABLE 18–6. Illustration of Kohlrausch's Rule
Λ_0 at 25°C (cm^2/ohm · equiv)

	Anion		
Cation	Cl$^-$	I$^-$	ClO$_4^-$
K$^+$	149.9	150.4	140.0
Na$^+$	126.5	126.9	117.5
difference	23.4	23.5	22.5

Example

Show how the equivalent conductivity at infinite dilution for acetic acid was calculated from other data in Table 18–5.

$$\Lambda_{0(HC_2H_3O_2)} = \Lambda_{0(NaC_2H_3O_2)} + \Lambda_{0(HCl)} - \Lambda_{0(NaCl)}$$
$$= 91.0 + 426.2 - 126.5 = 390.7 \text{ cm}^2/\text{ohm} \cdot \text{equiv}$$

18-14   Arrhenius' Theory

The results of experiments on the conductivity of electrolytes prompted Svante Arrhenius in 1884 to suggest a theory of ionization which at that time was quite novel. The theory has since been shown to be applicable only to weak electrolytes. He proposed that upon dissolving in water, those substances known to be electrolytes spontaneously dissociate into positive and negative ions. This dissociation is not necessarily complete, but as a given solution is diluted, the degree of dissociation increases, approaching 100% at infinite dilution.

At any concentration, the fraction of dissociated electrolyte, α, equals the ratio of the equivalent conductivity at that concentration to the equivalent conductivity at infinite dilution:

$$\alpha = \frac{\Lambda}{\Lambda_0} = \frac{\text{moles ionized}}{\text{total moles}}$$

These proposals were novel in that it was not apparent why oppositely charged particles should spontaneously separate at ordinary temperatures. Arrhenius' contemporaries thought that the separation of substances into ions occurred only after electrodes were introduced into the solution and a potential was applied.

Arrhenius supported his theory with data from experiments on colligative properties, in which no potential was involved. The colligative properties of aqueous solutions of electrolytes do not follow the simple laws observed for nonelectrolytes (Section 16–18). For example, in the case of aqueous solutions of strong electrolytes, the lowering of the freezing point is described by

$$\Delta t_f = 1.86 im$$

where i is a correction factor known as the van't Hoff factor. For nonelectrolytes, this factor is equal to 1. For a given electrolyte, the factor is not constant, but varies with concentration. Some representative values are presented in Table 18–7, where for comparison, data on sucrose, a typical nonelectrolyte, are included.

The data of Table 18–7 show that for strong electrolytes at infinite dilution, the van't Hoff factor approaches the number of ions, ν, expected from the formula of the compound. For a weak electrolyte, such as acetic acid, in moderate concentration, the van't Hoff factor increases very slowly with dilution. In very dilute solutions, the van't Hoff factor increases much more rapidly with dilution, but it is impossible to measure colligative properties precisely in such solutions.

Arrhenius attempted to relate the van't Hoff factor, i, to the degree of dissociation, α. If 1 mole of solute is capable of dissociating into ν moles of ions,

TABLE 18-7. Some Representative Values of the van't Hoff Factor, i

	Molality				Infinite Dilution[a]
	0.10	0.01	0.001	0.00001	
sucrose	1.01	1.00	1.00	1.00	1.00
$HC_2H_3O_2$	1.013	1.043	1.15	1.75	(2.00)
HCl	1.89	1.94	1.98		2.00
KCl	1.85	1.94	1.98		2.00
$MgSO_4$	1.21	1.53	1.82		2.00
K_2SO_4	2.32	2.70	2.84		3.00

[a] Extrapolated (or calculated) values.

$$A_x B_{\nu-x} \rightleftharpoons x\,A^{(\nu-x)+} + (\nu - x)\,B^{x-}$$

an increase of $\nu - 1$ moles of particles is produced by the reaction. The effective number of moles of solute particles in a solution per mole of solute added will be given by

$$1 + (\nu - 1)\alpha = i$$

Hence

$$\alpha = \frac{i - 1}{\nu - 1}$$

Arrhenius showed that the values of α calculated from conductivity measurements and the values calculated from colligative properties were in reasonable agreement with each other.

Example

Calculate the value of α for 0.010 m aqueous acetic acid from the data of Table 18–7 and from the data of Table 18–5.

From Table 18–7: $\alpha_{\text{acetic acid}} = \dfrac{1.043 - 1}{2.00 - 1} = 0.043$

From Table 18–5: $\alpha_{\text{acetic acid}} = \dfrac{16.3}{391} = 0.042$

The equilibrium constant for the reaction of an electrolyte to produce its ions may be expressed in terms of α:

$$HX + H_2O \rightleftharpoons H_3O^+ + X^-$$

$$K = \frac{[H_3O^+][X^-]}{[HX]}$$

In this system,

$$[H_3O^+] = [X^-] = \alpha C$$

where C is the total concentration of electrolyte. Therefore

$$[HX] = (1 - \alpha)C$$

Hence

$$K = \frac{(\alpha C)(\alpha C)}{(1 - \alpha)C} = \frac{\alpha^2 C}{(1 - \alpha)}$$

The above expression predicts the dissociation constants for weak electrolytes adequately. However, it cannot be applied to strong electrolytes.

Example

Using the data of Table 18-7, calculate the value of K_a for a 0.010 M solution of acetic acid. As shown above, the value of α for acetic acid in 0.010 M solution is 0.043:

$$K = \frac{(0.043)^2(0.010)}{0.957} = 1.9 \times 10^{-5}$$

18-15   Debye–Hückel Theory

It is now known that strong electrolytes in aqueous solution are completely ionized. Indeed, salts such as KCl consist of ions even in the solid state. In 1923, Debye and Hückel developed a theory which accounts for the observed behavior of dilute solutions of strong electrolytes in terms of the electrostatic interactions between the oppositely charged ions and between the ions and the dipoles of the solvent molecules.

In concentrated solutions of strong electrolytes there are many ions of opposite charge per unit volume. The solvation sphere of any ion will be surrounded by solvated ions of opposite charge. The higher the charge on a given ion, the stronger will be its attraction for neighboring opposite charges. It may be seen in Table 18-7 that the van't Hoff factors for $MgSO_4$ and K_2SO_4 increase toward the number of moles of ions per mole of salt much more slowly with increasing dilution than do unipositive, uninegative salts.

In a conductivity experiment, an ion moving through a solution will be attracted by the ions of opposite charge, and its motion will be retarded. The more concentrated the solution, the more oppositely charged ions are within interacting distance, and concentrated solutions have lower equivalent conductitivies than dilute solutions of the same material. The effects of interionic attractions on both colligative properties and conductivities led Arrhenius to erroneous conclusions about the nature of concentrated solutions of strong electrolytes.

In contrast, weak electrolytes are only partially dissociated in solution, and in a solution of a given molality the concentration of ions in the solution is comparatively small. Hence, interionic attractions are almost negligible, and dissociation constants can be determined by Arrhenius' method.

18-16   Exercises

Basic Exercises

1. **(a)** Calculate the number of CsCl formula units in a unit cell (Figure 18-6). **(b)** What is the coordination number of each type of ion?

2. How many unit cells are there **(a)** in a 1.00 gram, cube-shaped, ideal crystal of NaCl? **(b)** along each edge of that crystal?

3. Explain why an end-centered unit cell cannot be cubic. What is the highest possible symmetry for this type of unit cell?

4. Explain why uncharged atoms or molecules never crystallize in simple cubic structures.

5. Contrast the visible changes which occur while heat is added to **(a)** an ice cube, **(b)** a bar of chocolate. **(c)** Which type of behavior is characteristic of an amorphous solid? **(d)** List three other common amorphous materials.

6. Calculate the number of formula units in each of the following types of unit cells: **(a)** MgO in a rock salt type unit cell, **(b)** ZnS in a zinc blende structure, **(c)** platinum in a face-centered cubic unit cell.

7. A mineral having the formula AB_2 crystallizes in the ccp lattice, with the A atoms occupying the lattice points. What is the coordination number of the A atoms? of the B atoms? What fraction of the tetrahedral sites is occupied by B atoms?

8. Silver iodide crystallizes in the (ccp) zinc blende structure. Assuming that the iodide ions occupy the lattice points, what fraction of the tetrahedral sites is occupied by silver ions?

9. The intermetallic compound LiAg crystallizes in a cubic lattice in which both lithium and silver atoms have coordination numbers of 8. Sketch a portion of the lattice. To what crystal class does the unit cell belong?

10. Use the Born–Haber cycle and the following data to calculate the electron affinity of chlorine: $\Delta H_{f(RbCl)} = -102.9$ kcal/mole; IP $= 95$ kcal/mole; $\Delta H_{sub (Rb)} = +20.5$ kcal/mole; $D_{Cl_2} = +54$ kcal/mole; $U_{RbCl} = -166$ kcal/mole.

11. State one method discussed in this chapter by which one could distinguish between the compounds $[Co(NH_3)_5SO_4]Br$ and $[Co(NH_3)_5Br]SO_4$. State two methods described in earlier chapters by which this distinction could be made.

12. Explain why $0.100\ m$ NaCl in water does *not* have a freezing point equal to **(a)** $-0.183°C$, **(b)** $-0.366°C$.

13. The equivalent conductivity of a solution containing 2.54 grams of $CuSO_4$ per liter is $91.0\ cm^2/ohm \cdot equiv$. **(a)** Calculate the specific conductivity of the solution. **(b)** What is the resistance of a cubic centimeter of this solution when placed between two electrodes each having an area of $1.00\ cm^2$?

14. Arrange the following aqueous solutions in order of increasing freezing points (that is, lowest first): **(a)** $0.10\ m\ C_2H_5OH$, **(b)** $0.10\ m\ Ba_3(PO_4)_2$, **(c)** $0.10\ m\ Na_2SO_4$, **(d)** $0.10\ m\ KCl$, **(e)** $0.10\ m\ Li_3PO_4$.

15. The equivalent conductivity of a $0.0100\ M$ aqueous solution of ammonia is $10\ cm^2/ohm \cdot equiv$. The equivalent conductivity of ammonia at infinite dilution is calculated to be $238\ cm^2/ohm \cdot equiv$. Calculate the percent ionization of ammonia in $0.0100\ M$ aqueous solution.

16. The freezing point of a solution composed of 10.0 grams of KCl in 100 grams of water is $-4.5°C$. Calculate the van't Hoff factor, i, for this solution.

17. Using data from Figure 18–15, predict the coordination number of the cation in crystals of each of the following compounds: **(a)** MgO, **(b)** MgS, **(c)** CsCl.

18. Calculate the freezing point of the solution containing 15.6 grams of solute per kilogram of benzene for each of the following solutes:

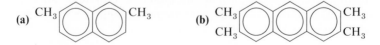

19. For each of the following statements about the nature of aqueous solutions of strong electrolytes **(a)** give the experimental observations which tend to support the statement, **(b)** suggest experimental methods for proving or disproving the statement.
   (i) Ions are formed when charged electrodes are placed in a solution of a strong electrolyte.
   (ii) Ions are formed when strong electrolytes are dissolved in water.
   (iii) Not only is it unnecessary to dissolve an electrolyte in water in order for ions to exist, but in solution there is considerable interionic attraction.

General Exercises

20. **(a)** Calculate the freezing point of the solution containing 24.8 grams of solute per kilogram of water for each of the following solutes: (i) $[Co(NH_3)_3(NO_2)_3]$, (ii) $[Co(NH_3)_4(NO_2)_2][Co(NH_3)_2(NO_2)_4]$, (iii) $[Co(NH_3)_5(NO_2)][Co(NH_3)_2(NO_2)_4]_2$.
   (b) Write empirical formulas for the compounds described in **(a)** and in exercise 18.
   (c) For which set of compounds is the term polymerization isomerism appropriate? Explain.
   (d) Explain why the compounds of one of these sets cannot be distinguished by freezing point depression methods, while those of the other set can be distinguished in this way.
   (e) Suggest an experimental technique which can be used to distinguish between the polymerization isomers.

21. The metal ion–halide ion distances in several alkali metal halides are given in the accompanying table. Suggest why there is such a small difference in internuclear distance between LiI and NaI compared to that between LiCl and NaCl.

	Distance (Å)	
	I^-	Cl^-
Li^+	3.10	2.49
Na^+	3.17	2.79
K^+	3.52	3.14
Rb^+	3.67	3.29

22. As heat is removed from a liquid which tends to supercool, its temperature drops below the freezing point, and then rises suddenly. What is the source of the heat which causes the temperature rise?

23. Locate the 4 three-fold axes, the 3 four-fold axes, and the 8 mirror planes in a cube. How many of these 15 elements of symmetry does a tetragonal unit cell have?

24. Explain why a hexagonal closest packed structure and a cubic closest packed structure for a given element would be expected to have the same density.

25. The following exercise may be performed with spheres of uniform size, such as ping pong or billiard balls or marbles. **(a)** Arrange 15 balls in a triangular manner, as in preparing for a billiards game. What part of Figure 18–8 compares to this arrangement? **(b)** Place a second layer of balls in the depressions of the first layer, then a third layer (one ball) in a depression directly over the center ball of the first layer. What part of Figure 18–8 is represented in this arrangement? **(c)** On the slanting side of the triangle created, identify a square. Find the portion of the face-centered cube of which the top ball forms one corner. Identify as many other corners of the smallest such cube as are visible. **(d)** Explain how a *cubic* closest packed arrangement can start out with a *hexagonal* layer.

26. Arrange uniform balls (such as ping pong or billiard balls) in a rectangular or square arrangement. Pile a second layer of identical balls in the holes of the first layer, and add as many identical balls to the third layer as will fit. Remove balls until you can identify a hexagonal layer of balls slanting up the three layers. Explain how a cubic pattern can generate a closest packed arrangement.

27. Calculate the ionic radius of the fluoride ion using only the following interionic distances and the radius of the iodide ion, 2.19 Å. (Do not use tabulated values from the text.) RbI, 3.67 Å; RbBr, 3.44 Å; KBr, 3.29 Å; KCl, 3.14 Å; NaCl, 2.79 Å; NaF, 2.31 Å.

28. Calculate the lattice energy in terms of e^2/r of a "crystal" consisting of 14 sodium ions and 13 chloride ions arranged in one isolated unit cell, with Na^+ ions at the corners.

29. Calculate the number of octahedral sites per sphere in a cubic closest packed structure.

30. A salt MY crystallizes in the cesium chloride structure. The anions at the corners actually touch each other and the cation in the center. What is the radius ratio, r_+/r_-, for this structure?

31. Estimate ionic radii for M^+, R^+, Q^-, and T^- from the following internuclear distances in NaCl-type crystals:

Salt	Anion–Anion Distance (Å)	Cation–Anion Distance (Å)
MT	2.40	1.70
MQ	1.63	1.15
RT	2.66	1.88
RQ	2.09	1.48

32. Explain why the Madelung constant does not depend on the charges of the ions in a crystal.

33. Using the ionic radii from Figure 18–15, suggest the probable structures of the unit cells of each of the following: **(a)** RbBr, **(b)** MgTe, **(c)** MgO, **(d)** BaO.

34. Calculate the lattice energy of magnesium sulfide using the following energies (all in kilocalories per mole): $\Delta H_{f(MgS)} = -82.2$; $\Delta H_{sub(Mg)} = 36.5$; $IP_1 + IP_2 = 520.6$; $\Delta H_{atom} = 133.2$; $EA_1 + EA_2 = -72.4$.

35. Estimate the equivalent conductivity of Na_2SO_4 in 0.00100 N solutions from the following equivalent conductivities in 0.00100 N solutions: NaCl, 123.7 cm²/ohm · equiv; KCl, 147.0 cm²/ohm · equiv; K_2SO_4, 152.1 cm²/ohm · equiv.

36. Estimate the equivalent conductivity of a 0.0100 M acetic acid solution from its conductivity at infinite dilution, 390.7 cm²/ohm · equiv at 25°, and the value of its dissociation constant, $K_a = 1.8 \times 10^{-5}$.

Advanced Exercises

37. Calculate the value of Avogadro's number from the internuclear distance of adjacent ions in NaCl, 2.82 Å, and the density of solid NaCl, 2.17 grams/cm³.

38. The length of a unit cell in the nickel crystal is 3.52 Å. Diffraction of X rays of 1.54 Å wavelength from a nickel crystal occurs at 22.2°, 25.9°, and 38.2°. Show that these data are consistent with a face-centered cubic crystal structure.

39. Calculate the proton affinity of ammonia, $NH_3(g) + H^+(g) \rightarrow NH_4^+(g)$, given the following data: NH_4F crystallizes in a ZnS (wurtzite) structure; the Born exponent is 8, the ammonium ion to fluoride ion distance is 2.63 Å; the enthalpy of formation of NH_4F is -112 kcal/mole; the enthalpy of formation of ammonia gas is -280 kcal/mole. Use other data from tables as needed.

40. Show that the face-centered cubic lattice of NaCl, shown in Figure 18–2, also contains a smaller, tetragonal unit cell. How many NaCl formula units are there in one such cell?

Outline on a figure of the lattice a still smaller unit cell, containing only one NaCl formula unit per unit cell. Calculate the unit cell lengths and angles of this simple unit cell.

41. Ionic radii have been estimated (by Linus Pauling) by assuming that the internuclear distances between isoelectronic ions is inversely proportional to the effective nuclear charges of the ions. Effective nuclear charges are obtained by subtracting an amount reflecting the shielding of the inner electrons from the actual nuclear charges. For neon-type ions, Na^+ and F^-, for example, the factor is $4.15e$. On this basis, calculate the ionic radii of Na^+ and F^- from the experimentally determined internuclear distance in NaF, 2.31 Å. Compare these values with those given in Figure 18–15.

19
Metals and Metallurgy

The importance of metals for the progress and welfare of mankind is reflected in the fact that the stages of man's evolution from a primitive state are referred to by such terms as "the stone age," "the bronze age," "the iron age," "the age of steel," and so forth. At present, the steel production of a nation is widely regarded as an economic barometer because steel is so essential in the production of other goods and services that the prosperity of a nation can be measured in terms of its use. In this chapter, the structures of metals will be described and a theoretical explanation of the properties of metals will be presented. Also, some metals of commercial importance will be discussed.

The large majority of the chemical elements are metals. On periodic tables, such as shown in Figure 19-1, the division between metals and nonmetals is shown by the heavy, stepped line. The division into these two groups is not sharp, and elements adjacent to the line have several properties characteristic of metals as well as properties more typical of nonmetals.

A characteristic metal is opaque, has a "metallic luster," and is malleable and ductile. A **malleable** substance is easily hammered into thin sheets. **Ductility** is the ability to be drawn or stretched into thin strands. Metals also have relatively high densities, although some, such as sodium, lithium, magnesium, and aluminum, are exceptions. A typical metal is not volatile, but is a good conductor of heat and of

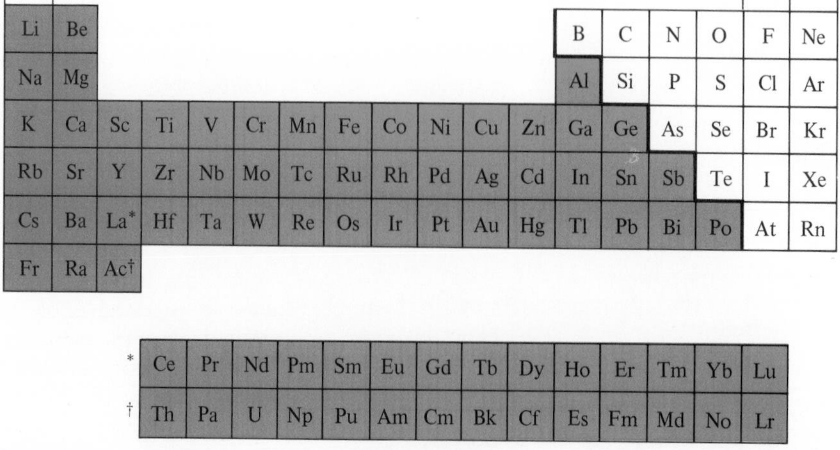

Figure 19-1. Periodic Table Showing Division Between Metals and Nonmetals

		cubic
c		cubic
rho		rhombohedral
tet		tetrahedral
dia		diamond
mon		monoclinic
bcc		body-centered cubic
hcp		hexagonal closest packed
fcc		face-centered cubic

Crystal structures of the metallic elements:

Li bcc	Be hcp															
Na bcc	Mg hcp												Al fcc			
K bcc	Ca fcc	Sc hcp	Ti hcp	V bcc	Cr bcc	Mn c	Fe bcc	Co hcp	Ni fcc	Cu fcc	Zn bcc	Ga c	Ge dia			
Rb bcc	Sr fcc	Y hcp	Zr hcp	Nb bcc	Mo bcc	Te hcp	Ru hcp	Rh fcc	Pd fcc	Ag fcc	Cd hcp	In tet	Sn tet	Sb rho		
Cs bcc	Ba bcc	La hcp	Hf hcp	Ta bcc	W bcc	Re hcp	Os hcp	Ir fcc	Pt fcc	Ag fcc	Hg rho	Tl hcp	Pb fcc	Bi rho	Po mon	
Fr bcc																

Ce fcc	Pr hcp	Nd hcp	Pm hcp	Sm rho	Eu bcc	Gd hcp	Tb hcp	Dy hcp	Ho hcp	Er hcp	Tm hcp	Yb fcc	Lu hcp
Th fcc		U c											

Figure 19-2. Crystal Structures of the Metallic Elements
Only the most important crystal modification is shown for each element.

electricity. Metals are reducing agents, and in reacting they tend to assume positive oxidation states. Thus in many binary compounds with nonmetals, metals exist as cations.

With few exceptions, pure metals crystallize in ccp, hcp, or bcc lattices (Section 18–5). The bcc structure is not a closest packed structure, but the available space is used almost as efficiently as in a closest packed structure. The tendency to assume closest packed structures, along with the compact electronic structures of the transition metals (see Chapter 10), accounts for the relatively high densities of metals. Structures of the metallic elements are shown in Figure 19–2. It should be noted that many exist in more than one crystalline modification, or **allotropic form.**

In the closest packed structures each metal atom has a coordination number of 12. In the bcc structure, each atom has a coordination number of 8, but there are 6 additional atoms sufficiently close to give an effective coordination number of 14. Obviously in a metal such as sodium, which crystallizes in the bcc structure, there are not enough valence electrons to form 14 covalent bonds or even 8 covalent bonds. Consequently, the bonding in a metallic solid must be of another type.

19-1 Metallic Bonding

A theory of bonding is required to explain the properties of metals in the solid state. It is instructive first to recall the formation of molecular orbitals from the atomic orbitals of individual atoms (Sections 11–15 and 11–16). Consider the formation of a diatomic lithium molecule from two lithium atoms. Only the valence shell electrons, one $2s$ electron from each atom, need be considered because the underlying electrons, having very much lower energies, are not involved in the chemical bonding. The two $2s$ orbitals are combined into two molecular orbitals: a σ_{2s} (bonding) orbital and a σ_{2s}^* (antibonding) orbital. The two electrons enter the σ_{2s} orbital and the Li_2 molecule is stable toward dissociation into two lithium atoms. If $2s$ orbitals from four lithium atoms were to be combined, four molecular orbitals would be formed, having relative energies which depend on the geometric configuration of the atoms and whether the orbitals were bonding or antibonding. If a very large number of lithium atoms, say 10^{23}, were brought together in the bcc structure of solid lithium metal, 10^{23} molecular orbitals would be formed. The energies of these molecular orbitals would be so closely spaced that they would form a continuous "band" of energy levels. These possibilities are diagrammed in Figure 19–3. These molecular orbitals are delocalized; that is, they belong to the crystal as a whole. Since each lithium atom contributes only one electron and each orbital can contain two electrons, the energy band is only half filled. It requires very little energy for electrons to change from lower energy orbitals to higher orbitals in the band, and such transitions occur readily. There is little restriction on the movement of the electrons from one orbital to another, and since the orbitals are delocalized, the structure of the metallic crystal is that of positively charged ions in a lattice within a "sea" of relatively mobile electrons. This concept is known as the band theory of metallic bonding.

The "electron sea" model of a solid metal is consistent with the properties of metals. For example, for a metal to be malleable or ductile, the atoms in the lattice structure must be easily displaced with respect to one another without weakening the bonding. Assuming that the electron sea is uniform and nondirectional, the bonding is independent of the identity of, and is not localized between, specific neighboring atoms. In Figure 19–4(a) it is seen that when the planes of atoms in a metal lattice are shifted with respect to each other, the environment about the respective lattice points is essentially unchanged. This is in contrast to the cases of ionic solids or solids in which all the bonds are covalent, as represented in Figure 19–4(b) and (c).

The opaqueness and metallic luster of metals are also explained in terms of the closely spaced energy levels of the sea of electrons. A substance is completely transparent only if it does not absorb any visible light which is passed through it. If an electronic transition within the substance is possible, light having a wavelength

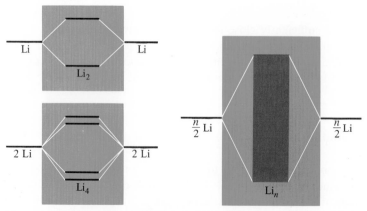

Figure 19–3. **Formation of Delocalized Energy Band from Individual Atomic Orbitals**

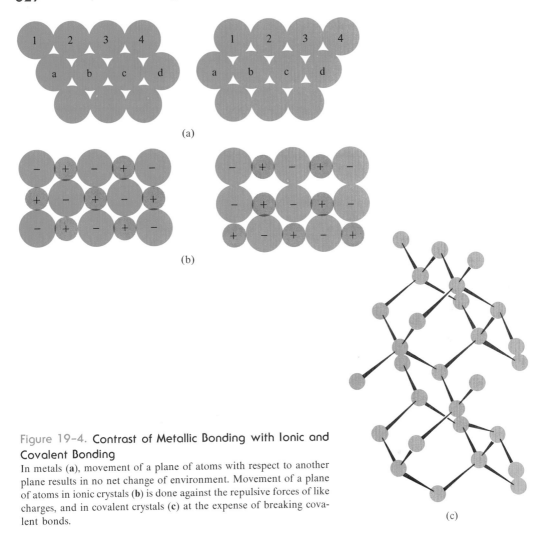

Figure 19-4. **Contrast of Metallic Bonding with Ionic and Covalent Bonding**
In metals (**a**), movement of a plane of atoms with respect to another plane results in no net change of environment. Movement of a plane of atoms in ionic crystals (**b**) is done against the repulsive forces of like charges, and in covalent crystals (**c**) at the expense of breaking covalent bonds.

corresponding to the energy of that transition is absorbed, and the transmitted light will have a color. In metals, the energy levels are so closely spaced that a transition of almost any energy is possible; hence light of any wavelength is absorbed and the metal is opaque. The characteristic metallic luster is due to the re-emission of the absorbed light by the "free" electrons on the surface of the crystal. Hence a smooth or polished metal has a good reflecting surface. In contrast, a finely divided metal powder appears black. In such cases the light reflected by the surface of one microcrystal is out of phase with that reflected by others.

The elements on the left side of the periodic table, groups I A and II A and the transition elements, are typical metals. Such elements, having relatively few electrons in the valence shell, readily assume close packed structures because the large delocalization energy leads to a net bonding. In moving from left to right across any period, metallic properties become less pronounced, and the tendency toward covalently bonded structures increases. In carbon, for example (Figure 11–26), the covalent tetrahedral structure of diamond is energetically more favored than a close packed structure. Graphite consists of carbon atoms covalently bonded to three others in well-separated plane layers. The absence of a fourth covalent bond is compensated for by delocalized π bonding, which effectively makes graphite a "two-dimensional metal." (Indeed, in graphite, electricity is primarily conducted along the planes of atoms.) Silicon, germanium, and gray tin have diamond-like structures, and the metallic allotropic forms of the last two may result from the fact that the metallic closest packed structures and the covalently bonded structures have comparable energies. Arsenic, antimony, and bismuth are similarly borderline cases, tending to form covalent structures, yet showing several characteristic metallic properties.

19-2   Electrical Conductivity of Metals

The band theory accounts for the electrical conductivity of metals. For example, in lithium the valence band is only half filled with electrons which are mobile because of the almost continuous range of energies available to them. An electron is easily excited to the unfilled delocalized orbitals which are part of the same band. The promotion leaves a "hole" in the filled part of the band. If an external source of potential is applied across the crystal, the excited electrons migrate. In addition, an electron adjacent to the hole in the filled part of the band fills the vacancy, leaving a new hole in its former position. In effect, the hole moves across the crystal in the direction opposite to the electron movement until an electron from the external source fills it. On the other side of the crystal, an excited electron leaves the crystal and enters the source of potential. There is no buildup of either charge or mass within the crystal. The net effect is a translation of charge across the crystal at a rate of 3×10^{10} cm/sec—the speed of light.

A metal such as magnesium might be expected to be nonconducting, because the number of available valence electrons is exactly sufficient to completely fill the band of delocalized orbitals obtained from the 3s atomic orbitals. However, the width of a

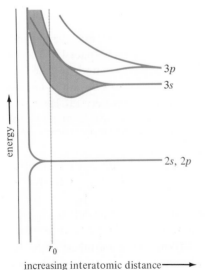

increasing interatomic distance ⟶

Figure 19-5. Energy Bands in Magnesium
At the interatomic distance, r_0, in the magnesium crystal, the filled 3s band overlaps the empty band of 3p orbitals, permitting easy promotion of an electron.

band of orbitals is a function of the interatomic distances, and at the distances in the magnesium crystal, the energy of the band formed from the $3s$ atomic orbitals overlaps the energy of the unfilled band formed from delocalized $3p$ orbitals (Figure 19–5). Hence electron mobility is easily achieved in the magnesium crystal by exciting electrons into the unfilled band.

The electrical conductance of metals decreases as the temperature is increased. As would be predicted by the kinetic molecular theory, when the temperature is raised, the atoms oscillate more energetically about their lattice sites. Electrons passing through the crystal under the influence of an applied potential will frequently collide with the vibrating positive centers, thus encountering an increased resistance.

19-3   Insulators and Semiconductors

A crystal will be an **insulator,** that is, it will not conduct, if the highest energy band which contains electrons is completely filled and if the next higher (empty) band of orbitals is appreciably higher in energy. This situation is diagrammed in Figure 19–6. It is seen in the figure that at the interatomic distances of the crystal lattice, the separation between bands is so great that electrons cannot absorb energy from an applied field and move from their bonded states. The figure also shows that if atoms in the crystal were compressed to smaller internuclear distances, r', than in the normal crystal, the unfilled band would overlap the filled band. In this case, metallic conduction could occur. Indeed, it has been shown experimentally that at extremely high pressures, bonding in all solids tends to become metallic.

If, at the equilibrium distance of the crystal, the energy gap between the filled band and next higher band is small, it may be possible to excite electrons across the energy gap by heating the crystal or by irradiating it with light. Although the crystal is normally an insulator, under these conditions, electrical conductance is possible. Since there are fewer electrons or holes to conduct the current, a smaller conductance is found for this type of material than for a metal. These materials are referred to as **semiconductors.** In contrast to the temperature effect on metallic conduction, the conductance of a semiconductor increases with increasing temperature.

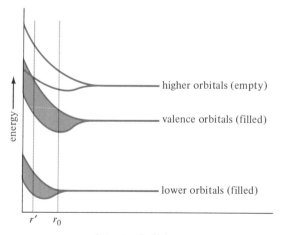

Figure 19-6. Energy Bonds in an Insulator

At the interatomic distance, r_0, there is a large gap between the band of valence shell orbitals and the next higher band of orbitals. However, at smaller distances, r', the filled and empty bands overlap.

Germanium and silicon can be converted into semiconductors by introducing a small number of certain impurity atoms (fewer than 1 in 10^9) into the crystal lattice. For example, if arsenic is added as an impurity to pure silicon, there will be an extra electron at the lattice points occupied by the arsenic atoms compared to the lattice points occupied by silicon atoms. These extra electrons occupy delocalized orbitals lying between the filled and empty bands of the silicon crystal. They can be excited easily into the empty band, and the crystal thereby becomes a semiconductor. In this case, arsenic is called a donor impurity, because it donates an extra electron into the crystal lattice. Crystals which are semiconductors because of the presence of donor impurities are called *n*-type (negative) semiconductors. When an element such as gallium is used as an impurity in the germanium lattice, a *p*-type (positive) semiconductor results. The gallium atom has one fewer electron than germanium, and at the point which it occupies in the lattice, there will be an electron deficiency, or hole. The energies of such holes lie between the energies of the filled and empty bands of the germanium crystal, and electrons can be excited into these holes from the lower band. Under the influence of an applied potential, an electron from an adjacent atom moves into a hole and in turn is replaced by an electron from another atom. Thus the hole moves across the crystal in a direction which is opposite the direction of electron migration. The current can be thought of as being due to the migration of the positive hole—hence the term *p*-type semiconductor. Either *n*-type or *p*-type semiconductors may be used as transistors in electronic circuits.

19-4   Alloys

An **alloy** is a mixture of elements which retains the characteristic properties of the metallic state. Alloys are of immense commercial importance, because it is possible to design them to have useful properties not found in the pure metals. For example, steel, brass, bronze, and pewter are alloys. Pure gold is too soft and deformable to be of practical use for coins or jewelry. Pure copper, in addition to being soft, tarnishes in air and is corroded by acidic gases in the air such as sulfur dioxide. But alloys of gold and copper are quite hard, tough, and corrosion resistant.[1] They have many uses in jewelry, coins, and scientific instruments.

----

[1] The composition of gold alloys is expressed in karats—the number of parts by mass of gold in 24 parts of the alloy. The best jewelry is from 15 to 18 karat gold; gold coins are 21 to 22 karat gold.

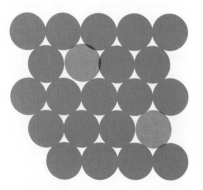

Figure 19-7. **Lattice of a Substitutional Alloy**

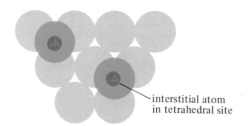

Figure 19-8. **Lattice of an Interstitial Alloy**

interstitial atom
in tetrahedral site

Designing alloys with specific properties is somewhat of an art because the relationship between the composition, structure, and properties of an alloy is not completely understood. Alloys may be classified into several types—solution alloys, eutectic mixtures, and intermetallic compounds. Solution alloys are further divided into **substitutional alloys,** in which the component metals have atoms which are close in size, and **interstitial alloys,** in which the atoms of one of the components are considerably smaller than those of the other. If the atoms of the two metals are very close in size, atoms of one may simply substitute for those of the other in the crystal lattice (Figure 19-7). The substitution is random, but in most cases the crystal structure depends on the composition. For example, up to a zinc/copper ratio of 40/60, alloys of the two metals have a face-centered cubic structure. Alloys containing a higher zinc/copper ratio crystallize in a body-centered cubic structure similar to that of pure zinc.

Interstitial alloys are formed when elements having very small atoms, such as hydrogen, carbon, boron, or nitrogen, are dissolved in transition metals. The small atoms occupy the holes within the close packed structure of the metal crystal (Figure 19-8). Carbon atoms and nitrogen atoms always occupy octahedral holes, while hydrogen atoms always occupy tetrahedral holes. The presence of the interstitial atoms causes the closest packed structure of the metal to be distorted, so that the slippage planes normally present are destroyed (Figure 19-9). As a result, the metal becomes tougher and less malleable.

Interstitial alloys are of great commercial importance. For example, all steels are interstitial alloys of carbon and iron, most having other metals added also. Other elements having small atoms—boron, nitrogen, or silicon—are used to make special purpose steels.

A **eutectic alloy** is a heterogeneous mixture of two metals which has a characteristic melting point lower than that of either pure metal. If two metals, A and B, form solutions when molten but are immiscible in the solid state, the freezing point behavior of their mixture may be represented by a phase diagram of the type shown

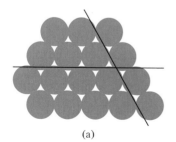

(a)

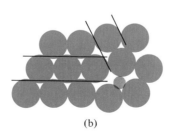

(b)

Figure 19-9. **Effect of Interstitial Atoms on Slippage Planes**
When a malleable metal (**a**) is subjected to stress, the planes of atoms easily slip past each other along slippage planes (black lines). An atom occupying an interstitial site (**b**) deforms the crystal lattice, destroying the slippage planes.

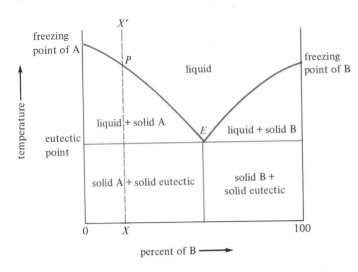

Figure 19-10. **Phase Diagram of Eutectic Alloy of Metals A and B**

in Figure 19–10, in which temperature is plotted against composition. The freezing point of each metal is lowered by the presence of the other. If a molten mixture has a composition represented by the line XX' in the diagram and the mixture is cooled to a temperature corresponding to point P, crystals of metal A separate from the solution. The residual liquid is richer in metal B, and as the temperature is lowered further, additional A solidifies while the freezing point and composition fall along the curve PE. At composition E, both metals crystallize simultaneously. This temperature is called the eutectic point (see Section 16–17). There are four phases present: liquid, solid A, solid B, and vapor. According to the phase rule the number of degrees of freedom, f, is given by the expression

$$f = c + 2 - p = 2 + 2 - 4 = 0$$

There are no possible degrees of freedom. Therefore the eutectic mixture freezes at a constant temperature with a constant composition. The solid is heterogeneous, consisting of crystals of metal A (with a very small quantity of B dissolved in them) and of B (with a slight quantity of A). Figure 19–11 shows a photomicrograph of a eutectic alloy.

Solder is a low melting alloy of tin and lead. A simplified phase diagram for the tin–lead system is shown in Figure 19–12. Pure lead melts at $327°C$; pure tin melts at $232°C$. The eutectic mixture consists of 63% tin, and melts at $181°C$. Some other low melting metal mixtures are listed in Table 19–1.

Intermetallic compounds are alloys which are homogeneous and which have a definite composition and a definite melting point or decomposition temperature. An example of a compound alloy is the gold–copper alloy, Cu_3Au. X-ray studies reveal that this alloy crystallizes in a cubic lattice in which the gold atoms occupy the lattice points at the corners of a cube and the copper atoms occupy the centers of each of the cube faces (Figure 19–13). Compounds of two metals which differ widely in electronegativity are called "salt-like." Examples are $MgCu_2$, Mg_2Pb, and Li_3Bi.

Intermetallic compounds are rarely used as such, but alloys containing crystals of an intermetallic compound distributed throughout a metal are of commercial im-

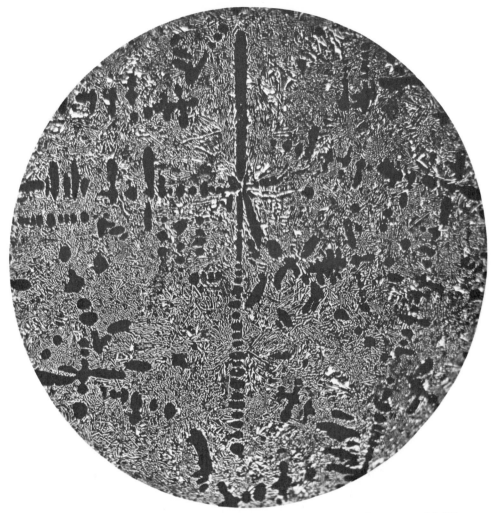

Figure 19-11. **Photomicrograph of Antimony-Lead Eutectic Alloy Containing 11.2%
Antimony (Magnification ~250×, Etched)**
Bright plates are antimony in a dark matrix of lead. There are also bright areas of excess antimony and dark areas of lead dendrites. (Photomicrograph courtesy of the National Lead Company Central Research Laboratory)

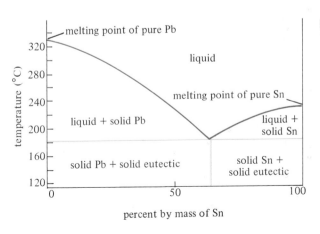

Figure 19-12. **Phase Diagram for Tin-Lead Mixtures**

TABLE 19-1. Low Melting Alloys

	Melting Point (°C)	Composition (% by mass)			
		Bi	Pb	Sn	Cd
Wood's metal	70	50	25	12.5	12.5
Rose's metal	109	50	28	22	
Malotte's metal	123	46.1	19.7	34.2	
solder (eutectic)	183		37	63	

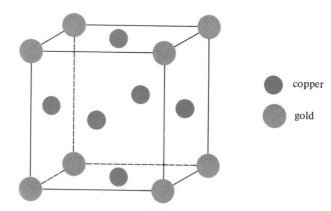

Figure 19-13. **Structure of Cu₃Au, a Compound Alloy**

● copper

● gold

portance. The compound cementite, Fe_3C, is a constituent of some steels. Addition of about 4% copper to aluminum produces an alloy known as duralumin. The compound $CuAl_2$ precipitates in the bulk of the aluminum metal and produces a hard, tough structure. The tensile strength of duralumin is five times that of pure aluminum, making this alloy a useful material for the fabrication of strong but lightweight structures.

19-5   Techniques of Metallurgy

Gold, silver, copper, mercury, and members of the platinum family (Os, Ir, Pt, Rh, and Pd) are sometimes found uncombined in nature. All other metals occur in nature as compounds, mainly in minerals in the form of oxides, sulfides, carbonates, and silicates. Any mineral or mixture of minerals from which it is commercially feasible to extract a metal is called an **ore.** The abundances of some metals in the earth's crust are listed in Table 19-2, along with the limits of composition which would make a given mineral an ore. For example, aluminum is the most abundant metal in the crust of the earth, but the difficulty and relatively high cost of obtaining pure aluminum make it necessary to consider as ores only those materials having a high aluminum content. In contrast, a rare but valuable metal such as gold is worth extracting from ores in which it is present to the extent of only a few thousandths of 1%.

The three steps followed in extracting a metal from its ore are called (1) beneficiation, (2) reduction, and (3) refining. When an ore is mined, usually large quantities of materials which contain none of the metal are also collected. This material is called

TABLE 19-2. Abundances of Some Metals in the Earth's Crust

Metal	Abundance (% by mass)	Minimum Percent to Constitute an Ore
aluminum	8.13	30
iron	4.71	35–65
manganese	0.07	25–50
nickel	0.01	2–5
tin	ca. 10^{-4}–10^{-5}	1.5–3
copper	ca. 10^{-5}	1–10
lead	ca. 10^{-5}	2–2.5
zinc	ca. 10^{-5}	5–25
silver	ca. 10^{-7}	0.03–0.16
gold	ca. 10^{-8}	0.003–0.00016
platinum	ca. 10^{-9}	5×10^{-5}

gangue. The pretreatment of the ore to remove as much gangue as possible is called **beneficiation.** For example, if the mineral is magnetic, the ore may be crushed and poured onto a moving magnetic belt (Figure 19–14). The mineral particles stick to the belt, while the gangue falls off. Panning for gold is another example of beneficiation. The gold is much more dense than the gravel and sand in which it is found. When a mixture of gold and sand is shaken in a flat pan under running water, the sand is washed away, leaving the gold behind.

Flotation is a process for beneficiation of sulfide minerals. Sulfides are more readily wet by oil than by water, while clay and sand exhibit the reverse tendency. The crushed ore is wet with oil and then agitated in a tank of soapy water while air is bubbled through the mixture (Figure 19–15). The air bubbles stick to oil-coated particles, and they float to the surface and are collected as a foam, while the gangue

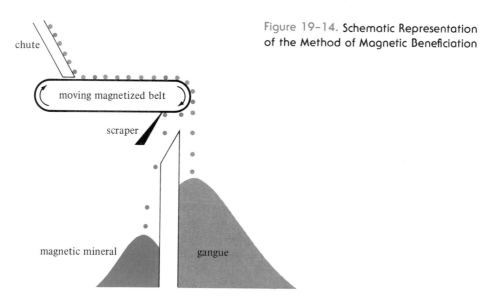

Figure 19-14. Schematic Representation of the Method of Magnetic Beneficiation

chute

moving magnetized belt

scraper

magnetic mineral

gangue

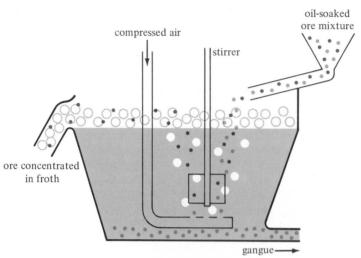

Figure 19-15. Diagram of Flotation Process

settles to the bottom of the tank. The foam on the top of the tank is allowed to flow off, and the mineral is recovered.

Other methods of beneficiation include **leaching.** The ore may be treated with an acid to dissolve the mineral, separating it from the gangue. In the case of bauxite, an aluminum ore, the leaching agent is hot alkali solution. The amphoteric aluminum oxide dissolves in the excess base according to the following equation:

$$Al_2O_3 \ + \ 6\,OH^- \ + \ 3\,H_2O \ \rightarrow \ 2\,[Al(OH)_6]^{3-}$$

After the gangue is filtered, the solution is acidified and heated to give Al_2O_3 suitable for electrolysis (see the Hall process, Section 5-2).

The method used to reduce the concentrated ore to metal depends on several factors, including the ease of reduction, the purity desired, and, of course, the cost. Some examples of reduction processes have already been described. Additional examples will be found in later chapters where specific metals will be discussed. Because of its importance and because the processes illustrate the practices of commercial metallurgy, the reduction of iron ores will be described in some detail.

19-6   Metallurgy of Iron

The principal iron ores contain hematite, Fe_2O_3; limonite, $Fe_2O_3 \cdot 3H_2O$; magnetite, Fe_3O_4; and siderite, $FeCO_3$. In the western hemisphere, vast deposits of hematite are found in Minnesota, New York, Alabama, Michigan, Labrador, Quebec, and Venezuela. The quality of the ores from these locations is such that they require little or no beneficiation. The ore is reduced to iron in **blast furnaces** (Figure 19-16), which are steel structures, lined with silica brick, that stand about 90 feet high with a diameter of 22 feet. A furnace is charged at the top with a mixture of ore; coke, C; limestone, $CaCO_3$; and sand, SiO_2. The impurities in the ore are mainly clay and silicate rocks. The purpose of adding limestone and sand is to convert this gangue to

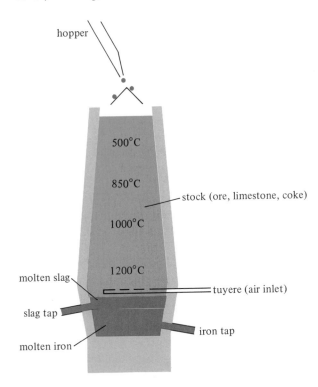

Figure 19-16. **Blast Furnace**

a fusible mixture called **slag.** Typical slag-forming reactions which occur at the elevated temperatures within the furnace are

$$CaCO_3 \rightarrow CaO + CO_2$$
$$CaO + SiO_2 \rightarrow CaSiO_3$$
$$MgO + SiO_2 \rightarrow MgSiO_3$$
$$CaO + Al_2O_3 \rightarrow Ca(AlO_2)_2$$

Preheated, dry air at 425 to 650°C is blown through water-jacketed nozzles called tuyeres into the bottom of the furnace. The hot air oxidizes the coke to carbon monoxide, and the heat liberated is sufficient to raise the temperature above the melting point of iron.

Reduction of the iron oxide proceeds in three steps. Near the top of the furnace where the temperature is about 500°C, hematite is reduced to magnetite:

$$3 Fe_2O_3 + CO \rightarrow 2 Fe_3O_4 + CO_2$$

In the middle of the furnace, at a temperature about 850°C, the magnetite is converted to iron(II) oxide:

$$Fe_3O_4 + CO \rightarrow 3 FeO + CO_2$$

Reduction to the metal takes place near the bottom of the furnace, where the temperature is greater than 1000°C:

$$FeO + CO \rightarrow Fe + CO_2$$

The molten iron flows to the bottom of the furnace, where it is drawn off into insulated cars and transported to steel-making furnaces. The molten slag floats on top of the iron and is drawn off through a separate tap. It is allowed to harden to a rock-like material, which is crushed and used in the manufacture of Portland cement. As iron and slag are drawn off at the bottom of the furnace, fresh charge is added at the top. The blast furnace must be kept in continuous operation, because if the temperature were allowed to fall below the freezing point of the iron and slag, the bottom of the furnace would become a solid mass which could not be remelted by a hot air blast.

The primary product of the blast furnace is known as "pig iron." Pig iron contains considerable quantities of impurity in the form of silicon, carbon, phosphorus, and sulfur, and it is quite brittle. To be used to make tools or materials of construction which must withstand shock, the iron must be refined.

Refined iron which is alloyed with carbon and possibly other metals is called **steel.** To produce steel from pig iron, the silicon, phosphorus, and sulfur must be removed, and the carbon content must be adjusted to something less than 1.5%. In practice, it is easier to remove all the carbon and later add the desired quantity after the purification step. The processes used for the large scale manufacture of steel are (1) the **Bessemer** process, (2) the **open hearth** process, and (3) the **basic oxygen** process. The last two processes now account for more than 90% of the steel production, but the Bessemer process was the first practical method for producing steel in large quantities.

A Bessemer converter is a large, egg-shaped vessel mounted on trunnions so that it can be tipped to a horizontal position (Figure 19–17). The converter is charged with from 15 to 20 tons of molten pig iron, with perhaps some scrap iron added, and a blast of hot air is blown through the bottom of the vessel into the molten metal. During the "blow" the impurities are oxidized, and some metal oxide slag forms above the molten metal. Typical reactions include

$$Si + O_2 \rightarrow SiO_2$$
$$2\,Mn + O_2 \rightarrow 2\,MnO$$
$$2\,C + O_2 \rightarrow 2\,CO$$

The carbon monoxide burns in air, and the end of the oxidation is signaled by the disappearance of the blue flame. The entire process takes less than 20 min. Expert

Figure 19–17. **Bessemer Converter**

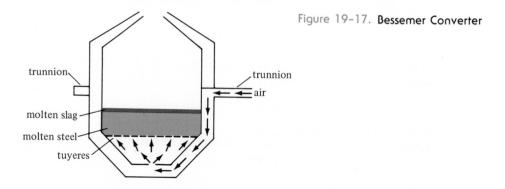

trunnion

trunnion

air

molten slag

molten steel

tuyeres

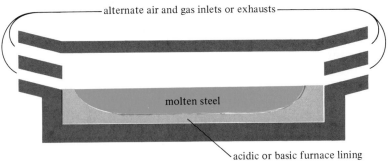

alternate air and gas inlets or exhausts

molten steel

acidic or basic furnace lining

Figure 19-18. **Open Hearth Furnace**

operators are required to judge when all the impurities have been oxidized because there is insufficient time for quality control analysis during the oxidation process. When the oxidation is deemed complete, calculated quantities of carbon and alloying metals are added. Then the steel is poured into molds or transported in the molten state to mills where it is rolled into sheets or drawn into wire or into rails.

In the open hearth process, a charge of 200 to 300 tons of molten pig iron and scrap steel is heated in a shallow hearth furnace, about 40 feet long, 12 feet wide, and 2 feet deep (Figure 19-18). If it is desired to remove impurities which form acidic oxides, such as phosphorus, sulfur, or carbon, a furnace lined with a basic material, such as calcium oxide or magnesium oxide, is used. If the pig iron has a low sulfur and phosphorus content and it is desired to remove metallic impurities, a furnace lined with acidic oxides, mainly SiO_2, is used. Heat is supplied by a gas flame played directly over the surface of the molten metal, for as much as 8 hours. The process is sufficiently slow that several analyses of the material can be made during the process. This quality control results in iron of very high purity. Thereafter, very precise quantities of alloying materials can be added.

Some limitations of these two methods can be avoided in the **duplex** process for refining pig iron. In essence, the Bessemer process is used first to remove most of the impurities. Then the material is transferred to an open hearth furnace for 1 to 2 hours, where the last of the impurities are removed under the slow conditions which facilitate quality control. The extensive heating required in the open hearth process is avoided, and yet a high quality product is obtained.

In the basic oxygen process, pure oxygen from a water-cooled lance is played on the surface of molten pig iron in a basic lined converter of the Bessemer type. The process is very rapid (10 to 20 min), and iron equal in purity to that from the basic open hearth process is produced. The basic oxygen process and its modifications account for the major portion of the steel produced today. This process does not require the use of large quantities of scrap steel, as did the older processes. As a consequence, there has been a decreased demand for scrap metal. Since it is no longer profitable to collect worn out machinery and automobiles, the disposition of these items has become an ecological problem.

19-7   Metallurgy of Magnesium

Magnesium is the sixth most abundant element in the earth's crust, comprising 2.2% of its mass. Magnesium exists in nature as $MgCl_2$, $MgSO_4$, $Mg(OH)_2$, and

$MgCO_3$. Various silicate minerals contain significant quantities of magnesium ion, along with other cations such as potassium, calcium, and aluminum ions. Sea water is an important magnesium "ore"—magnesium constitutes approximately 0.13% by mass of the dissolved solids in sea water. It should be noted also that magnesium is present as the central atom of chlorophyll in all green plants (Figure 14–5).

Commercially, magnesium is often alloyed with aluminum. Such alloys have very high tensile strengths along with very low densities, which makes them excellent materials for the construction of aircraft and space vehicles.

The metallurgy of magnesium provides examples of two techniques which have not been discussed previously: (1) the extraction of minerals from the ocean and (2) the refining of metals by vacuum distillation. The metal is produced either by electrolysis of molten $MgCl_2$ or by the chemical reduction of magnesium oxide with an iron–silicon alloy. In both methods, the original magnesium compound is converted first into MgO.

Magnesium is precipitated from sea water by the addition of lime:

$$Mg^{2+} + CaO + H_2O \rightarrow Mg(OH)_2 + Ca^{2+}$$

The precipitated $Mg(OH)_2$ is filtered and then heated to convert it to MgO. The oxide is mixed with finely divided coke, NaCl, and KCl, and the mixture is heated to 1200 to 1400°C in a steel oven lined with acid-resistant bricks. Gaseous chlorine is passed into the mixture, and ultimately a mixture of molten chlorides containing 10 to 20% $MgCl_2$ is formed. The mixture (melting point, 750°C) is transferred into a cell where it is electrolyzed between graphite electrodes. The magnesium metal, which melts at 650°C, floats on the surface of the melt and is scooped off. The chlorine gas which also is produced during the electrolysis is used to convert another batch of MgO to $MgCl_2$.

In the chemical reduction of magnesium ores, the MgO formed by the thermal decomposition of $MgCO_3$ is ground together with an iron–silicon alloy and then is heated to 1180°C in a vacuum electric furnace in which the pressure is maintained at 10^{-5} torr. At this temperature the oxide is reduced to the metal:

$$2\,MgO(s) + Si(s) \rightarrow 2\,Mg(g) + SiO_2(s)$$

The magnesium vapor distills from the furnace into receivers where it is condensed and solidified. Because the metal has been distilled under vacuum, it has a very high purity.

19–8   Refining Techniques

With notable exceptions, such as magnesium prepared in a vacuum furnace, metals prepared on a commercial scale contain various impurities. Consequently it is necessary to undertake some kind of refining step. One of the most widely used techniques for less reactive metals is the electrolysis of aqueous solutions. The application of this technique to the refining of copper was described in Section 5–15. A variation of the electrolytic method is used in refining the more active metals, such as magnesium. An electrolyte composed of fused alkali metal chlorides, such as the KCl–NaCl eutectic mixture, is used.

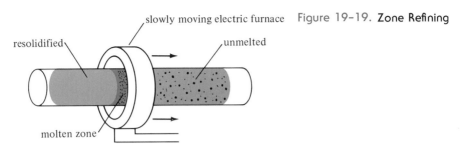

slowly moving electric furnace   Figure 19-19. **Zone Refining**

resolidified
unmelted
molten zone

Vapor phase purification methods include direct distillation of the metal. Magnesium, mercury, and zinc are refined in this manner. The Mond process for the refining of nickel is an example of the formation of a volatile compound of the metal which is distilled from the impurities and then decomposed to recover the metal (Section 14–5). The **van Arkel process** is another variation of this technique. In this process, the crude metal is treated with iodine vapor at 600°C to form a volatile iodide. Vapors of the iodide are passed over tungsten wires heated to 1800°C. The compound decomposes, and pure metal is deposited on the wire. The van Arkel process can be used to purify zirconium, hafnium, silicon, titanium, beryllium, and tungsten.

A technique worthy of special note is **zone refining.** The substance to be refined is placed in a tube equipped with a movable heating coil (Figure 19–19). Sufficient heat is applied to melt a narrow zone of the material, and the coil is gradually moved along the tube so that the molten zone moves with it. The melting point of a mixture is usually lower than that of a pure substance which is its major component, and as the coil is moved, crystals of the pure material form at the trailing edge of the molten zone. The impurities concentrate in the liquid phase and move along with it to the end of the tube. The impurities are eliminated by simply removing the end of the sample. Elements used for semiconductors, gallium, germanium, and silicon, have been purified by zone refining to such an extent that impurities cannot be detected by ordinary analytical methods (less than $10^{-10}\%$ impurities remain).

19-9   Theoretical Principles and Metallurgy

A basic consideration in selection of a method for extracting a metal from its ore is whether the method is thermodynamically feasible. Usually the most desirable process for such extractions is the one which occurs with the largest decrease in free energy. The reduction of oxides to metals will be used to illustrate the principles involved. The more stable the metal oxide, the more difficult will be the reduction to metal. Also, there will be fewer ways of effecting such a reduction.

The ease of reduction of different metal oxides can be quantitatively compared by considering their free energies of formation. The more negative the free energy of formation, the more stable will be the oxide toward reduction. The relationship

$$\Delta G = \Delta H - T\Delta S$$

suggests that if ΔH and ΔS are approximately constant, a straight line will be obtained when ΔG is plotted against the absolute temperature. At 0 K, ΔG is equal to ΔH for the process (in this case the formation of the oxide), and the slope of the line

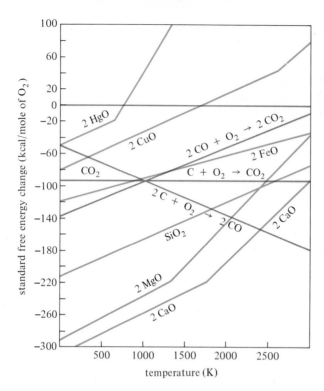

Figure 19-20. **Free Energies of Formation of Oxides Plotted Against Absolute Temperature**

is the negative of the entropy change. Some data for several oxides are shown in Figure 19-20, where the standard free energy change per mole of oxygen gas used is plotted against the absolute temperature. These data show that for each case cited, the formation of the oxide from the element and oxygen is exothermic because at $0 \text{ K}, \Delta G^\circ = \Delta H^\circ$, and all of the ΔG° values at 0 K are seen to be negative. Moreover, the slopes of all the lines (except one) are positive, indicating that there is a decrease in entropy, as would be expected when a solid element reacts with a mole of gaseous oxygen to form a solid oxide. Since a mole of gas is used up, the final state is less random.

Several of the plots show abrupt changes in the slopes. These breaks occur at temperatures at which the metal undergoes a phase transition. For example, the plot for calcium has a break at about 1750 K, corresponding to the normal boiling point of calcium (1490°C). Above this temperature, the solid oxide is formed from 3 moles of gas, which results in an even larger decrease in entropy:

$$2 \text{ Ca(g)} + O_2(g) \rightarrow 2 \text{ CaO(s)}$$

The plot for HgO shows that this compound is thermodynamically unstable above 700 K. Indeed, metallic mercury can be obtained from its oxide merely by heating:

$$2 \text{ HgO} \rightarrow 2 \text{ Hg} + O_2$$

In principle, when the plot for one metal lies below that for another, the first metal is capable of reducing the oxide of the second. Thus, magnesium metal will reduce CuO and FeO but not CaO. Also it is seen that at room temperature (300 K) the order of reducing ability approximates that of the standard electrode potentials (Table 5–2). In Figure 19–20 the plot corresponding to the change

$$C(s) \quad + \quad O_2(g) \quad \rightarrow \quad CO_2(g)$$

is shown by a horizontal line. For this reaction, the enthalpy change is -94 kcal/mole, and the entropy change is $+0.68$ cal/mole $\cdot$ K. ΔS is relatively small because in this case 1 mole of gaseous product is formed while 1 mole of gaseous reactant is used up. $\Delta G°$ for this reaction is almost independent of temperature. The plot for CO_2 is relatively high in Figure 19–20, and at low temperatures carbon will reduce only a few of the metal oxides shown. However, the slopes of the plots for several of the metals are such that they cross the CO_2 plot; hence theoretically these metals can be reduced by carbon at elevated temperatures.

An alternative reaction involving carbon and oxygen is the formation of carbon monoxide:

$$2 \, C(s) \quad + \quad O_2(g) \quad \rightarrow \quad 2 \, CO(g)$$

Since 2 moles of gaseous product is formed from 1 mole of gaseous reactant, this process is accompanied by an increase in entropy. Hence the slope of the corresponding line is negative, as shown by the downward sloping line in Figure 19–20. If the temperature is high enough, carbon should reduce all of the metal oxides, being converted into carbon monoxide.

The plot for the reaction of carbon monoxide with oxygen is also shown in Figure 19–20. It is worth noting that below 973 K, CO is thermodynamically a better reducing agent than carbon. This fact supports the assumption that reduction of iron oxides in the blast furnace is done by carbon monoxide rather than by coke.

Figure 19–20 also shows why silicon can successfully reduce MgO to the free metal. Up to 1380 K, the normal boiling point of magnesium, the $\Delta G°$ plots for the formation of SiO_2 and MgO are parallel. However, above 1380 K the plot for MgO changes slope owing to the increased entropy effect, and above 2400 K the reaction between silicon and the magnesium oxide proceeds with a decrease in free energy. In practice, the reaction is further enhanced by the distillation of magnesium metal from the reaction mixture.

19–10 Exercises

Basic Exercises

1. Would a sample of mercury vapor appear shiny and metallic? Explain.
2. Are diatomic lithium molecules stable toward (a) dissociation into lithium atoms? (b) condensation into solid lithium metal?
3. Explain precisely why the two allotropic forms of carbon differ so markedly in their electrical conductance.
4. A sample of the alloy duralumin contains 4.2% copper by mass. What percent of the alloy consists of the compound $CuAl_2$?
5. Express the composition of the compound alloy Cu_3Au in karats.

6. Name one property of metals for which pure metals are more useful than alloys.

7. The compound alloy Cu_3Au crystallizes in a cubic lattice (Figure 19–13). How many formula units of the compound are there in each unit cell?

8. What type of alloy would most likely be formed by each of the following pairs? Give an example of each pair to illustrate your choice. Discuss cases in which the choice is not clear cut, if any. **(a)** A pair of metals having similar sized atoms, the same number of valence electrons, and the same type of lattice when pure. **(b)** A pair of metals having widely different sized atoms and vastly different electronegativities. **(c)** A pair of metals having atoms widely different in size but similar in electronegativities. **(d)** A pair, one of which is a small, nonmetallic element.

9. Describe the properties of an ore which is to be concentrated by **(a)** leaching with alkali, **(b)** leaching with acid, **(c)** flotation, **(d)** panning.

10. There are many minerals in the earth's crust which contain aluminum, but only bauxite is an important ore of this metal. Explain.

11. Iron oxide is reduced to pig iron with carbon. The excess carbon is oxidized as a part of the steel-making process. Later still, carbon is added to make steel. Explain why the middle step is necessary, and why it is not combined with the first or the third step.

12. In steel manufacture, the word *basic* signifies which of the following? **(a)** fundamental, **(b)** alkaline, **(c)** unalloyed, **(d)** essential.

13. What characteristics are desirable for a furnace lining? State under which circumstances SiO_2 would be preferred over CaO or MgO as a furnace lining.

14. Give an example of at least one metal which is refined by each of the following processes: **(a)** vacuum distillation, **(b)** Mond process, **(c)** electrolysis, **(d)** van Arkel process, **(e)** zone refining.

15. Using Figure 19–20, predict the temperature above which CuO would be reduced by roasting in air.

16. Define each of the following terms: **(a)** karat, **(b)** alloy, **(c)** ore, **(d)** gangue, **(e)** slag, **(f)** steel, **(g)** eutectic point, **(h)** eutectic mixture.

General Exercises

17. Suggest how the properties of liquid metals, such as mercury, support the band theory of metallic bonding.

18. Using data from Figure 10–9, predict on the basis of actual calculations which one(s) of the following sets of elements could form an *undistorted* interstitial alloy: Pd–C, La–B, Ba–Be, Pd–H.

19. Suggest an experimental method of distinguishing between an alloy which is a solid solution from one which is an intermetallic compound.

20. A compound alloy of metals A and B has a unit cell containing A atoms at its corners and center and B atoms at the face centers. What is the formula of the compound?

21. Using Figure 19–12, draw cooling curves and identify the points at which breaks occur as well as the points at which the temperature is constant during the cooling from the molten state for each of the following mixtures: **(a)** 25% tin, **(b)** 75% tin, **(c)** 100% tin, **(d)** the eutectic mixture.

22. Using Figure 19–20, estimate the temperature at which magnesium metal can be prepared by reduction with carbon. What products are expected?

Advanced Exercises

23. Speculate on the types of alloys which might be formed in three-component systems. Describe some properties of each type of alloy.

24. Explain how sodium metal may be prepared from NaOH and iron metal (a preparation reported in the chemical literature).

25. Band theory is only one possible explanation of the structure and properties of metals. Look up the bond theory of metals in L. Pauling, *The Nature of the Chemical Bond*, 3rd ed. (Cornell Univ. Press, Ithaca, N. Y., 1960), and in J. S. Griffith, *Journal of Inorganic and Nuclear Chemistry*, **3**, 15 (1956), and compare and contrast these approaches. What are the advantages of each?

26. W. Hume-Rothery has pointed out that the crystal structures of solution alloys are influenced by the ratio of valence electrons to atoms in the alloy. For example, the crystal structure of solid solutions of zinc in copper have a face-centered cubic structure until the composition approaches the "formula" CuZn, at which point the alloys assume a body-centered cubic structure known as the beta phase. The valence electron to atom ratio in the beta phase is $\frac{3}{2}$ (or $\frac{21}{14}$). A gamma phase and an epsilon phase are also possible at other characteristic electron to atom ratios, where the ratios increase from beta to gamma to epsilon. The "formulas" of alloys listed below correspond to such phases. Deduce the Hume-Rothery ratios, and classify each alloy in the proper phase: $CuZn$, Cu_9Al_4, Cu_9Ga_4, Cu_3Sn, Cu_5Zn_8, Cu_5Al_3, Cu_5Sn, Cu_3Ge, $CuZn_3$, Cu_3Al, $Cu_{31}Sn_8$.

20

Nuclear Chemistry and Radiochemistry

The discovery of radioactivity and some characteristics of radiations from radioactive materials were briefly described in Section 9–5. In this chapter, a more detailed discussion of radioactivity and some concepts of nuclear structures will be presented. Studies of phenomena associated with radioactivity have shed light on many problems in the areas of chemistry, cosmology, energy production, and archeology, among others. Indeed, the subject is so broad that it is convenient to consider it by two separate approaches—nuclear chemistry and radiochemistry.

Nuclear chemistry is concerned with the structure of the nucleus and how this structure influences nuclear stability. In addition, nuclear chemistry deals with phenomena by means of which the nucleus is changed, such as the processes of (natural) radioactivity and of (artificial) nuclear transmutation. The energies involved in nuclear processes are more than a million times greater than the energies released in ordinary chemical reactions because nuclear forces are correspondingly stronger than chemical bonds. The study of nuclear phenomena is a paramount concern of physicists as well as chemists, and the term *nuclear physics* can be considered synonymous with nuclear chemistry.

Radiochemistry is the application of chemical techniques to the study of radioactive substances, and it also includes the study of the chemical effects of the radiations from radioactive materials. In principle, the chief difference between radiochemical and ordinary chemical procedures is the use of radioactivity measurements, including the measurements of the energies of radiations, as a means of following a chemical reaction.

Nuclear Chemistry

20-1 Kinds of Radioactive Transformations

Radioactivity is the result of spontaneous changes occurring in the nucleus of an atom, as a consequence of which the nucleus itself is changed. These changes result in the release of radiations such as alpha, beta, and gamma rays. The nature of these radiations and their behavior in a magnetic field were briefly described in Section 9–5, especially in Figure 9–8. Alpha particles are doubly charged helium ions (helium nuclei). The emission of an alpha particle from the nucleus of an atom results in the production of an atom of a new element, with an atomic number 2 less and mass number 4 less than the parent atom. For example, the emission of an alpha particle from the radium isotope of mass number 224 can be represented

$$^{224}_{88}\text{Ra} \rightarrow \ ^{220}_{86}\text{Rn} \ + \ ^{4}_{2}\text{He}$$

It should be noted that equations representing nuclear processes must be balanced with respect to both nuclear charge and mass number. In the equation given above, the sum of the charges of the radon nucleus and the helium nucleus equals the charge on the nucleus of radium. Similarly, the mass numbers of the products total 224, the mass number of the radium atom. Alpha particles are ejected from a nucleus at speeds about 10^9 cm/sec (approximately 20 million miles/hour). Their kinetic energies are in the range of several million electron volts (MeV).[1] Alpha particles emitted from a given nucleus have a specific energy characteristic of that type of nucleus. Despite their high energy, alpha particles travel only a few centimeters in air and they are stopped by relatively thin layers of solid matter, such as paper or glass. As they pass through air or any other gas, alpha particles ionize the gas molecules in their paths by knocking off electrons. On the average, about 35 eV is required to produce an ionized molecule; hence an alpha particle having about 35 MeV energy will produce at least 10^5 ions before it stops. Since the alpha particle has a relatively short **range** (distance of travel before being stopped), it produces a large number of ions per centimeter.

Beta particles are identical in rest mass to electrons. There are both positively and negatively charged beta particles; the latter are identical to electrons. The emission of a negative beta particle from a nucleus causes an increase of 1 in the atomic number, but no change in the mass number of the nucleus. The emission of a positively charged beta particle, called a **positron,** results in a decrease of 1 in atomic number, with no change in mass number:

$$^{14}_{6}\text{C} \rightarrow \ ^{14}_{7}\text{N} \ + \ ^{0}_{-1}\beta$$
$$^{22}_{11}\text{Na} \rightarrow \ ^{22}_{10}\text{Ne} \ + \ ^{0}_{+1}\beta$$

In contrast to alpha particles, the energies of beta particles from atoms of a given isotope vary from a few to several million electron volts. To explain why beta particle emission is not monoenergetic, it is assumed that another particle, the neutrino, having a rest mass equal to zero but having no charge, is emitted simultaneously with each beta particle. For a given radioactive species, the sum of the energies of the beta particle and the neutrino is assumed to be constant. Beta particles have much greater penetrating power than alpha particles. They travel up to several meters in air but are stopped by relatively thin sheets of metal. Since they are relatively light and since they travel with such tremendous speeds, beta particles produce relatively few ions on passing through air.

Another radioactive process is the capture by the nucleus of an electron from outside the nucleus. The process is called **K capture** because an electron in the $1s$ orbital (K shell) of an atom is most likely to be captured. K capture results in a decrease in atomic number but no change in mass number; in this respect it is similar to positron emission. However, in K capture the nucleus itself emits no particles.

[1] 1 MeV/particle = 1 million eV/particle = 23.06×10^6 kcal/mole = 96.48×10^6 kJ/mole.

There is radiation emitted from the atom because an electron from a higher energy level (shell) will replace the electron captured from the K shell. Hence, an X ray of frequency corresponding to the energy difference between the two shells is emitted:

$$^{133}_{56}\text{Ba} \ + \ e^- \ \xrightarrow{\quad K \text{ capture} \quad} \ ^{133}_{55}\text{Cs} \ + \ \text{X ray}$$

Most cases of alpha particle emission and beta particle emission are accompanied by the emission of gamma rays. Gamma ray emission results in no change in the atomic number or the mass number of the nucleus. Gamma rays associated with a given nuclear transformation have a definite energy, suggesting definite energy levels within the nucleus. In a given nuclear transformation, the product nucleus is formed in an excited state, and in the transition from the excited state to the ground state, gamma rays are emitted. Indeed, some nuclei having metastable excited states are known. For example, the isotope ^{119}Sn can be produced in an excited state, from which it decays with the emission of gamma rays having energies of 0.065 and 0.024 MeV.

$$^{119}_{50}\text{Sn} \ \xrightarrow{\quad \text{internal transition} \quad} \ ^{119}_{50}\text{Sn} \ + \ \gamma$$

Gamma rays are (high energy) photons; hence they travel with the speed of light. They are capable of penetrating several inches of a dense metal, such as lead, and as a corollary to their high penetrating ability, they are very weakly ionizing.

20-2   Nucleons and Nuclides

For convenience, any nuclear species with a specific atomic number and a specific mass number is referred to as a **nuclide.** For example, $^{12}_{6}\text{C}$ is a nuclide. $^{14}_{6}\text{C}$ is an isotopic nuclide of $^{12}_{6}\text{C}$; both have the same atomic number but different mass numbers.

Discussions of nuclear phenomena refer to specific nuclides, and it is important to refer to the mass number, A, as well as the atomic number, Z. The particles making up the nucleus, the protons and neutrons, are called **nucleons.** Hence the mass number is the total number of nucleons within the nucleus.

The nuclides $^{14}_{6}\text{C}$ and $^{14}_{7}\text{N}$ are not isotopic, but they do contain the same number of nucleons. Nuclides related in this way are called **isobars** of each other. The nuclides $^{14}_{7}\text{N}$ and $^{15}_{8}\text{O}$ are neither isotopic nor isobaric, but their nuclei contain the same number of neutrons. Such species are called **isotones** of each other.

20-3   Radioactive Decay

Since radioactivity is a nuclear phenomenon, the process is not altered by changes in the physical or chemical state of the sample. A given radioactive nuclide will emit the same type of radiation at the same rate regardless of the compound it is part of (if any), its temperature, the applied pressure, or the presence of electrostatic, magnetic, or gravitational fields.[2]

---

[2] Some cases of K capture are affected by the chemical environment.

All radioactive atoms in a given sample do not disintegrate simultaneously. The emission of radiation from a given nucleus is a statistical event. When large numbers of nuclei are present, the rate of radioactivity follows a first order rate law (Section 17–1). Consequently, the rate of emission of radiations at any time is proportional to the number of radioactive atoms present. Since this number is continuously decreasing, the process is referred to as **radioactive decay.** The first order rate law is therefore

$$2.30 \log \frac{N}{N_0} = -\lambda t$$

where N is the number of radioactive atoms at any time, t; N_0 is the initial number of such atoms; and λ is called the **decay constant.** In a manner analogous to that shown for first order chemical reactions, it can be shown that the half life ($t_{1/2}$) for radioactive decay is given by

$$t_{1/2} = \frac{2.30 \log 2}{\lambda} = \frac{0.693}{\lambda}$$

The half life is a useful property for the identification of various nuclidic species, because each radioactive nuclide has a characteristic half life. Some typical examples are shown in Table 20–1.

20–4   Activity

The **activity** of a radioactive sample is defined as the number of disintegrations per unit time. The activity, A, is merely the rate of decay and is proportional to the number of atoms present:

$$A = \lambda N$$

The **specific activity** is defined as the number of disintegrations per unit time per

TABLE 20-1. Half Lives of Some Nuclides

Nuclide	Half Life	Radiation[a]
$^{238}_{92}U$	4.5×10^9 years	alpha
$^{237}_{93}Np$	2.2×10^6 years	alpha
$^{14}_{6}C$	5730 years	beta
$^{90}_{38}Sr$	19.9 years	beta
$^{3}_{1}H$	12.3 years	beta
$^{140}_{56}Ba$	12.5 days	beta
$^{131}_{53}I$	8.0 days	beta
$^{140}_{57}La$	40 hours	beta
$^{15}_{8}O$	118 seconds	beta
$^{94}_{36}Kr$	1.4 seconds	beta

[a] In most of these decay processes, gamma radiations are also emitted.

gram of radioactive material. One **curie** (Ci) is defined as an activity of 3.7×10^{10} disintegrations per second (dis/sec). In laboratory work, activities on the order of microcuries (μCi $= 3.7 \times 10^4$ dis/sec) or millicuries (mCi) are used.

Example

A sample of $^{90}_{38}$Sr originally had an activity of 0.500 mCi. **(a)** What is the specific activity of the sample? **(b)** What is the activity of the sample after 30.0 years?

$\lambda = \dfrac{A}{N_0}$

(a) $\qquad A_0 = \lambda N_0 = 0.500 \text{ mCi}$ ✓ $t_{1/2} = 0.693$

$(0.500)(3.7 \times 10^7 \text{ dis/sec}) = \dfrac{0.693}{19.9 \times 365 \times 24.0 \times 3600} N_0$ $\lambda = \dfrac{A}{N_0}$

half life in sec

$N_0 = 1.68 \times 10^{16} \text{ atoms}$ $\dfrac{0.693 N}{A}$

$\text{mass of strontium} = \dfrac{(1.68 \times 10^{16} \text{ atoms})(90.0 \text{ grams/mole})}{6.02 \times 10^{23} \text{ atoms/mole}} = 2.50 \times 10^{-6} \text{ gram}$

$\text{specific activity} = \dfrac{(0.500)(3.7 \times 10^7)}{2.50 \times 10^{-6}} = 7.4 \times 10^{12} \text{ dis/gram} \cdot \text{sec}$

(b) To determine the activity after 30.0 years:

$$2.30 \log \dfrac{N}{N_0} = -\lambda t = -\dfrac{0.693}{19.9 \text{ years}}(30.0 \text{ years}) = -1.04$$

$$\dfrac{N}{N_0} = \dfrac{A}{A_0} = 0.353$$

$$A = 0.353 A_0 = 0.177 \text{ mCi} = 6.55 \times 10^6 \text{ dis/sec}$$

20-5   Successive Radioactive Decay

Some radioactive nuclides decay to stable nuclides, while others decay to give nuclides which are also radioactive. It is customary to refer to the original nuclide as the "parent" and to its decay product as the "daughter." When the daughter nuclide is not radioactive, a plot of the logarithm of the activity of a sample versus time will give a straight line which has a slope equal to $-\lambda$. If both the parent and its daughter are radioactive, the total radioactivity in the sample will be due to both nuclides. The plot of logarithm of activity versus time will not be a straight line; its shape will be governed by the relative magnitudes of the decay constants of the parent and the daughter. Three cases are most often encountered.

Case I. The half life of the parent is shorter than the half life of the daughter. A typical decay curve is shown in Figure 20-1. Assuming that no daughter atoms are present initially, the observed activity is that of the parent only. The number of daughter atoms increases with time (at the same rate that the parent decays), and since they decay more slowly than the parent, the activity of the daughter grows to some maximum value. Thereafter, it decays with its own characteristic half life. The time, t_{max}, at which the activity of the daughter is a maximum is given by

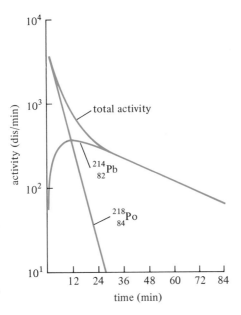

Figure 20-1. **Decay of Parent Nuclide $^{218}_{84}$Po and Long-Lived Daughter Nuclide $^{214}_{82}$Pb**

$$t_{max} = \frac{2.30}{\lambda_d - \lambda_p} \log \frac{\lambda_d}{\lambda_p}$$

where λ_d and λ_p are the decay constants of daughter and parent, respectively.

Case II. The half life of the parent is greater than that of the daughter. This situation is diagrammed in Figure 20-2. If no daughter atoms are initially present, at first the total activity increases and reaches a maximum value. The time of maximum daughter activity is calculated by the equation given above. Beyond t_{max}, the total

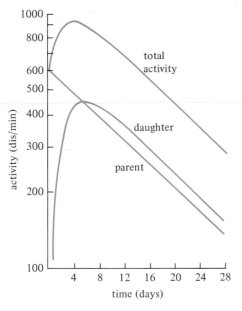

Figure 20-2. **Decay of Parent Nuclide with Half Life Greater Than That of Daughter Nuclide**

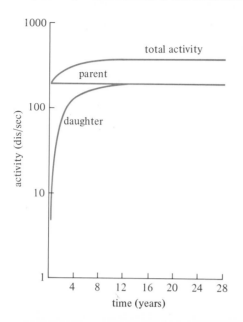

Figure 20-3. Secular Equilibrium: Half Life of Parent Nuclide Very Much Greater Than That of Daughter

activity decays at a rate equal to that corresponding to the half life of the parent. The final decay rate for this steady state process, known as **transient equilibrium,** depends on the parent because the daughter is being used up as fast as it is being formed. After a transient equilibrium has been established, the activity of the daughter, A_d, relative to that of the parent, A_p, is given by the expression

$$A_d = \frac{\lambda_d}{\lambda_d - \lambda_p} A_p$$

Case III. The half life of the parent is extremely long, and also long compared to that of the daughter. In this case, the activity of the parent shows no decrease with time during the period of observation (Figure 20–3). If no daughter atoms are initially present, the total activity first increases and then remains at a constant value. This mixture is said to be in **secular equilibrium.** After the equilibrium is established, the daughter and parent decay at equal rates (the daughter cannot decay at a faster rate than it is formed):

$$A_d = A_p$$

Hence the relative number of daughter atoms, N_d, is given by

$$N_d = \frac{\lambda_p}{\lambda_d} N_p$$

20-6   Natural Radioactivity

Except for ^{40}K, ^{87}Rb, ^{115}In, ^{138}La, ^{144}Nd, ^{147}Sm, ^{176}Lu, ^{187}Re, and ^{190}Pt, each of which has a half life greater than 1 billion years, all of the radioactive elements found

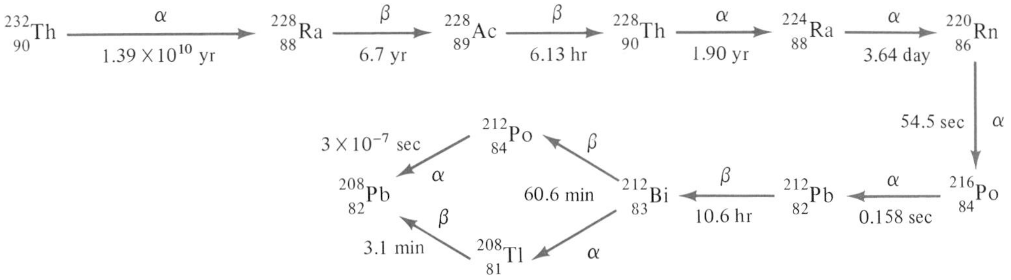

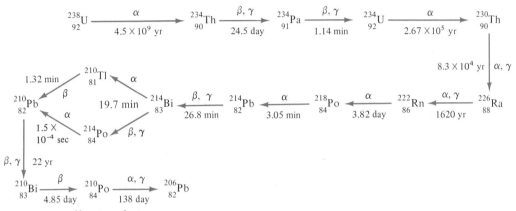

Figure 20-4. **Thorium Series**

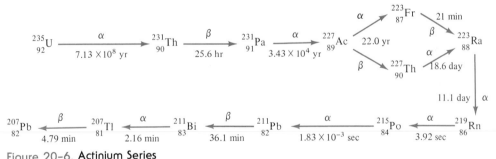

Figure 20-5. **Uranium Series**

Figure 20-6. **Actinium Series**

in nature in appreciable quantity have atomic numbers greater than 82. Only those elements which have long half lives compared to the age of the earth (see below), along with their decay products, are found in nature. Three series of such elements are found. These are described in Figures 20-4, 20-5, and 20-6. In each series, the elements decay successively to another radioactive nuclide until, after a number of steps, a stable nuclide is formed. One series consists of the nuclide $^{232}_{90}$Th and its radioactive "descendants" and is termed the thorium series. The mass number of each member of the series is exactly divisible by 4, and the series is sometimes designated as the $4n$ series. The stable nuclide $^{208}_{82}$Pb ends the series. A second series, the uranium series, begins with $^{238}_{92}$U and decays to $^{206}_{82}$Pb. This series is also known as the $4n + 2$ series. The third series, known as the actinium series, actually has $^{235}_{92}$U as the parent

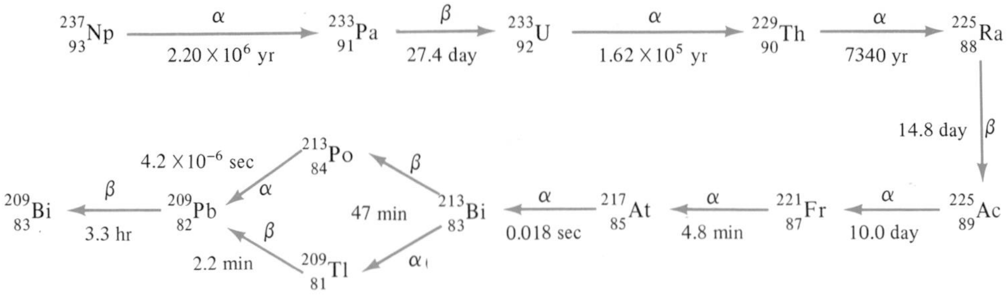

$$^{237}_{93}\text{Np} \xrightarrow[2.20\times10^6\ \text{yr}]{\alpha} \ ^{233}_{91}\text{Pa} \xrightarrow[27.4\ \text{day}]{\beta} \ ^{233}_{92}\text{U} \xrightarrow[1.62\times10^5\ \text{yr}]{\alpha} \ ^{229}_{90}\text{Th} \xrightarrow[7340\ \text{yr}]{\alpha} \ ^{225}_{88}\text{Ra}$$

14.8 day β

$^{209}_{83}\text{Bi} \xleftarrow[3.3\ \text{hr}]{\beta} \ ^{209}_{82}\text{Pb}$

$^{213}_{84}\text{Po}$ 4.2×10^{-6} sec $\ \alpha$

$\xrightarrow{\beta}$ 47 min $^{213}_{83}\text{Bi} \xleftarrow[0.018\ \text{sec}]{\alpha} \ ^{217}_{85}\text{At} \xleftarrow[4.8\ \text{min}]{\alpha} \ ^{221}_{87}\text{Fr} \xleftarrow[10.0\ \text{day}]{\alpha} \ ^{225}_{89}\text{Ac}$

$^{209}_{81}\text{Tl}$ 2.2 min $\ \beta$ $\quad \alpha$

Figure 20-7. **4n + 1 Series**

nuclide, and the terminal member is $^{207}_{82}\text{Pb}$. The series is also known as the $4n + 3$ series.

No evidence has ever been found for a naturally occurring $4n + 1$ series of radioactive elements. A possible explanation is that no member of this series has a sufficiently long half life to have survived since the original formation of the elements. However, a number of nuclides which decay in this series have been produced artificially (as will be described below). One such nuclide is the longest-lived neptunium isotope, $^{237}_{93}\text{Np}$. This nuclide is indeed a parent of a $4n + 1$ series of radioactive elements which successively decay into the stable nuclide $^{209}_{83}\text{Bi}$. This series is diagrammed in Figure 20-7.

The members of each of the respective radioactive series are all in secular equilibrium with the very long-lived parent nuclide. Since they are different elements from the parent, they can be separated by chemical techniques. Mme. Curie separated radium and polonium from uranium ores in this manner.

20-7 Properties of Stable Nuclei

Rutherford's experiments (Section 9-6) in which metal foils were bombarded with alpha particles demonstrated that the radius of an atomic nucleus is about 10^{-13} cm. An estimate of the radius, r, of a given nucleus in terms of its mass number, A, is obtained by means of the empirical expression

$$r = 1.4A^{1/3} \times 10^{-13}\ \text{cm}$$

When several protons are brought together to within a distance corresponding to the nuclear radius, extremely strong binding forces are required to overcome the repulsive force between the positively charged particles. A clue to the nature of these binding forces is found in the neutron/proton ratios of the stable nuclides. A graph showing the number of neutrons versus the number of protons is given in Figure 20-8 for all of the naturally occuring stable nuclides. For all but the lightest elements, the number of neutrons exceeds the number of protons. Moreover, the stable nuclides all lie within a "belt of stability" which rises above the line drawn to represent equal numbers of protons and neutrons.

For a nuclide of given atomic number, if the neutron/proton ratio is above the belt of stability, the nucleus tends to stabilize itself by beta decay. Since beta emission increases the atomic number and decreases the number of neutrons (effectively

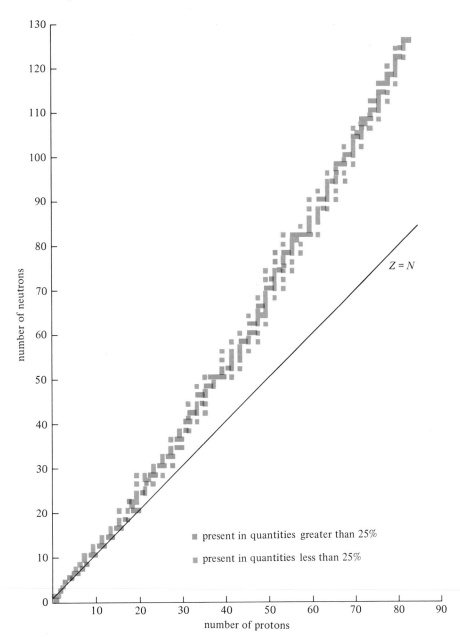

Figure 20-8. **Numbers of Protons (Z) and Neutrons (N) in Stable Nuclides**

converting a neutron to a proton), the neutron/proton ratio is lowered. Conversely, if the neutron/proton ratio is below the belt of stability, the nucleus can achieve stability by positron decay or by electron capture. The net effect of either of these processes is a decrease in the number of protons and an increase in the number of neutrons, which increases the neutron/proton ratio.

Nuclei may be classified according to whether they contain even or odd numbers of each type of nucleon. For example, if there is an even number of protons (even Z)

and also an even number of neutrons (even N), the nucleus is classified as an even-even type. Similarly, if the atomic number is odd and there is an odd number of neutrons, the nucleus is an odd-odd type. Obviously there are four types of nuclei. The numbers of stable nuclides of each type are as follows: even-even, 209, even-odd, 69; odd-even, 61; and odd-odd, 4. The odd-odd classification includes only the very light nuclides ^2_1H, ^6_3Li, $^{10}_5B$, and $^{14}_7N$. The preponderance of stable nuclei having even numbers of each type of nucleon suggests that there is a tendency for like particles to pair up within the nucleus, analogous to the pairing of electrons in atomic and molecular orbitals. The natural abundance of elements having an even atomic number is about ten times greater (on the average) than the natural abundance of elements with an odd atomic number. Moreover, if a given element has an isotope of the even-even type, that isotope will comprise from 70 to 100% of the natural abundance of the element. Two of the most abundant elements on the surface of the earth, oxygen and silicon, illustrate these facts. Oxygen consists of 99.756% $^{16}_8O$, and silicon is 92.27% $^{28}_{14}Si$. It is also worth noting that the two elements having mass numbers below 206 which have *no* stable isotopes, $_{43}Tc$ and $_{61}Pm$, have odd atomic numbers.

20-8   Evidence for Nuclear Shell Structure—The Magic Numbers

Considerable evidence suggests that the nucleus contains energy levels or shells, analogous to the shell structure of extranuclear electrons. For example, alpha radia-

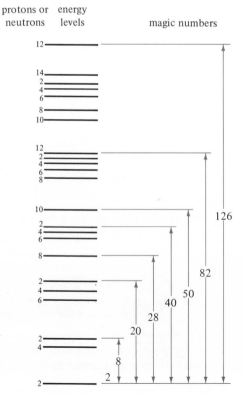

protons or   energy
neutrons   levels   magic numbers

Figure 20-9. Postulated Nuclear Energy Levels

The magic numbers are the sums of the numbers of nucleons up to and including the given level.

tions are monoenergetic, suggesting that their emission corresponds to a definite transition within the nucleus. Also, some nuclei can exist in excited states from which they decay with the emission of gamma radiation of definite energy. Among these is $^{119}_{50}$Sn, which decays to stable $^{119}_{50}$Sn with the emission of 0.089 MeV of energy. Other evidence stems from experiments on the bombardment of nuclei with protons and/or neutrons, from which the yields and stabilities of the products are related to probable energy states. However, because there are at least two types of particles involved, neutrons and protons, it is not possible to define nuclear energy levels with great precision.

A schematic diagram of the postulated nuclear energy levels is shown in Figure 20-9. The numbers of protons or neutrons which can occupy a given shell are correlated with the **magic numbers** corresponding to the nuclides which complete nuclear periods, analogous to the atomic numbers of noble gases which complete the periods in the periodic table. These magic numbers correspond to nuclides having particularly high stability. For example, $^{4}_{2}$He, $^{16}_{8}$O, and $^{40}_{20}$Ca have both N and Z values of 2, 8, and 20, respectively. The element tin (atomic number 50) has the largest number of stable isotopes, 10. The heaviest stable nuclides are $^{208}_{82}$Pb and $^{209}_{83}$Bi, each containing 126 neutrons. Also, all three naturally radioactive series end with stable lead isotopes with atomic number 82.

20-9   Nuclear Binding Energy

It is possible to calculate the **binding energy** of a nucleus from the difference in mass between the separate nucleons and the mass of the composite nucleus. For example, $^{4}_{2}$He is composed of two protons and two neutrons. But the actual mass of $^{4}_{2}$He is less than the sum of the masses of these four nucleons:

$$2 \times \text{mass of proton} = 2(1.00728 \text{ D}) = 2.01456 \text{ D}$$
$$2 \times \text{mass of neutron} = 2(1.00867 \text{ D}) = 2.01734 \text{ D}$$
$$\text{total} = 4.03190 \text{ D}$$
$$\text{mass of } {}^{4}_{2}\text{He} = 4.0015 \text{ D}$$
$$\text{mass "missing"} = 0.0304 \text{ D}$$

The binding energy of the particles in the nucleus is equal to the energy equivalent of this "missing" mass, given by Einstein's equation $E = mc^2$. By use of appropriate conversion factors, it can be shown that 1 D of mass is equivalent to 931 MeV of energy.

Example

How much energy would have to be added to $^{4}_{2}$He in order to make two protons and two neutrons?

To get the $^{4}_{2}$He nucleus to come apart into its four nucleons, energy equivalent to 0.0304 D would have to be added:

$$E = mc^2 = (0.0304 \text{ D}) \left(\frac{1.00 \text{ gram}}{6.02 \times 10^{23} \text{ D}} \right) (3.00 \times 10^{10} \text{ cm/sec})^2$$

$$= 4.54 \times 10^{-5} \text{ erg} \quad \text{(per helium nucleus)}$$

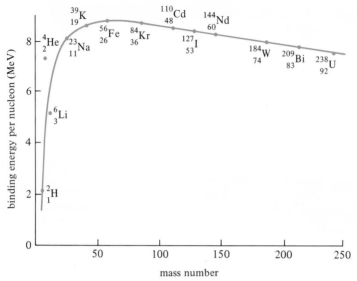

Figure 20–10. Binding Energy per Nucleon Versus Mass Number

For 1 mole of helium nuclei, 653 million kcal is required.

By similar calculations, the binding energies of all the stable nuclei can be determined. For purposes of comparison, it is useful to calculate the binding energy per nucleon. In Figure 20–10, the binding energy per nucleon is plotted against mass number.

The most stable nuclides are those having the greatest binding energy per nucleon. Nuclides of iron, cobalt, and nickel are among the most stable, and it is worth noting that these metals are thought to comprise the core of the earth. Meteorites from outer space are also composed chiefly of iron, cobalt, and nickel.

The shape of the curve in Figure 20–10 suggests that heavy nuclei, such as those of uranium, could be split into lighter nuclei with the release of energy. Also, nuclei of the lightest elements, such as deuterium, ^2_1H, and lithium, ^6_3Li, could be combined to form heavier nuclei with the release of energy. These aspects of conversion of matter to energy by fission and by fusion will be discussed below.

20–10 Nuclear Transmutations

In 1919, Rutherford bombarded nitrogen gas with alpha particles and obtained hydrogen and oxygen:

$$^{14}_7N \ + \ ^4_2He \ \rightarrow \ ^{17}_8O \ + \ ^1_1H$$

This reaction was the first case of **artificial transmutation,** the change of one element into another by human effort. In 1934, Irene Joliot-Curie, the daughter of Mme. Curie, discovered that when aluminum was bombarded with alpha particles from polonium, the metal became radioactive owing to the formation of phosphorus atoms:

$$^{27}_{13}\text{Al} + ^4_2\text{He} \rightarrow ^{30}_{15}\text{P} + ^1_0\text{n}$$

Since that time, hundreds of artificial, radioactive nuclides have been made by a variety of nuclear reactions.

Nuclear reactions are similar to ordinary chemical reactions in some ways and different from them in others. They are alike in that

1. They occur with either the absorption of energy (endoergic) or with the release of energy (exoergic).
2. There is a conservation of charge and of mass-energy.
3. Many reactions have definite energies of activation.

Nuclear reactions differ from ordinary chemical reactions in that

1. Atomic numbers change.
2. Although the sum of the mass numbers is unchanged, the total quantity of matter is changed measurably. (In endoergic reactions, the total quantity of matter in the products is greater than that in the reactants; and in exoergic reactions, the reverse is true.)
3. Quantities are usually defined per particle, rather than per mole, and individual events are considered.
4. Reactions are those of specific nuclides rather than of the mixture of isotopes comprising the element.

Although nuclear reactions can be represented by conventional equations such as those already used, a shortened notation is often used. In nuclear transmutations, a given nuclide is bombarded by a particle, and a particle (or radiation) is usually ejected along with the product nuclide. In the shortened notation, the target nuclide is specified first. Then, in parentheses, are written the projectile particle separated by a comma from the ejected particle(s). Finally, the product nuclide is given. In this notation, the two reactions mentioned above are written as follows:

$$^{14}_7\text{N}(\alpha,\text{p})^{17}_8\text{O} \quad \text{and} \quad ^{27}_{13}\text{Al}(\alpha,\text{n})^{30}_{15}\text{P}$$

The symbols n, p, d, α, e, γ, and X are used in this notation to denote neutron, proton, deuteron (deuterium nucleus), alpha particle, electron, gamma ray, and X ray, respectively.

Projectile particles are of two types: (1) charged particles, including ^1_1H, ^2_1H, ^4_2He, and heavier ions such as $^{12}_6\text{C}$, and (2) gamma rays and uncharged particles, including neutrons. For a positively charged particle to effect a nuclear reaction, it must possess sufficient energy to overcome the coulombic repulsion of the nucleus. For this reason, the first transmutations were done with high energy alpha particles impinging on nuclei having relatively low atomic numbers. However, a number of devices for accelerating charged particles have been designed. Among the earliest of these is the cyclotron, which is described in Figure 20–11. Other accelerators include the synchrotron, the linear accelerator, and various heavy ion accelerators. Details of these very elegant machines can be found in other books.

Neutrons, being uncharged, can approach the nucleus without coulombic repulsion, and accelerating devices are not required for them. Neutrons are obtained as

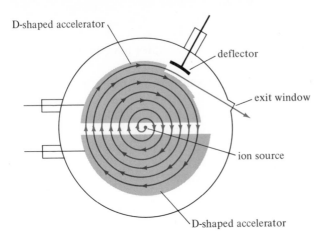

Figure 20-11.

Diagram of a Cyclotron

Positive ions ($^1_1$H$^+$ or $^2_1$H$^+$) are introduced into an evacuated chamber between two hollow, oppositely charged D-shaped accelerators ("dees"). A perpendicular magnetic field (not shown) causes the ions to move in a curved path, and at the same time they are accelerated across the gap from the positively charged dee to the negatively charged dee. Hence the ions travel an increasingly widening curved path. Each time the ion reaches the gap, the polarities on the dees are reversed, and the ion is accelerated up to speeds approaching c. Eventually, the ions pass through the exit window and are available to bombard the target.

products of nuclear reactions and from nuclear reactors, to be described below. A convenient source of neutrons is the reaction on beryllium of alpha particles from radium or polonium:

$$^9_4\text{Be}(\alpha,\text{n})^{12}_6\text{C}$$

20-11 Nuclear Cross Section

A given target nuclide may undergo any of several possible reactions, depending on the nature and energy of the projectile particle. The probability of a given nuclide capturing the projectile particle and undergoing a specific reaction is expressed as the **cross section,** σ, of that nuclide. σ is given in units of area—square centimeters or, conventionally, barns. One **barn** is equal to 10^{-24} cm^2. If the cross section for a particular reaction is known, it is possible to calculate the number of product nuclei, N_{prod}, which will be obtained in time t when a number of target nuclei N_t, are bombarded in a reactor by a known flux, f (particles per second), of projectile particles:

$$N_{\text{prod}} = \frac{N_t \sigma f}{\lambda_{\text{prod}}}(1 - 10^{-\lambda_{\text{prod}}t/2.30})$$

In terms of activity, $A_{\text{prod}} = \lambda_{\text{prod}}N_{\text{prod}}$; hence

$$A_{\text{prod}} = N_t \sigma f(1 - 10^{-\lambda_{\text{prod}}t/2.30})$$

For bombardment times which are large compared to the half life of the product nuclide, the term $10^{-\lambda_{\text{prod}}t/2.30}$ approaches zero, and a "saturation activity" is obtained:

$$A_{\text{prod}} = N_t \sigma f$$

TABLE 20-2. Low Energy Neutron Cross Section

Nuclide	Isotopic Abundance (%)	Cross Section (barns)	Half Life of Product
$^{10}_{5}B$	19	3.99×10^3	stable
$^{23}_{11}Na$	100	5.6×10^{-1}	2.27 min
$^{31}_{15}P$	100	2.3×10^{-1}	14.3 days
$^{55}_{25}Mn$	100	1.34×10^1	2.6 hr
$^{107}_{47}Ag$	51.35	4.4×10^1	2.3 min
$^{109}_{47}Ag$	48.65	1.10×10^2	24 sec
$^{113}_{48}Cd$	12.26	1.95×10^4	stable
$_{48}Cd$	all isotopes	2.4×10^3	
$^{115}_{49}In$	95.8	1.45×10^2	54 min
$^{197}_{79}Au$	100	9.8×10^1	2.7 days

Some cross sections for reactions with low energy neutrons are listed in Table 20-2.

Example

For a target containing 1.00 mg of manganese in a nuclear reactor of flux 1.0×10^{13} neutrons/cm$^2 \cdot$ sec, calculate the activity of $^{56}_{25}Mn$ ($t_{1/2} = 2.6$ hours) formed in 5.2 hours. What would be the activity after 520 hours of irradiation? The isotopic abundance of $^{55}_{25}Mn$ is 100%, and $\sigma = 13.4$ barns.

$$\lambda_{prod} = \frac{0.693}{t_{1/2}} = \frac{0.693}{2.6 \text{ hr}} = 0.267/\text{hr}$$

After 5.2 hours,

$$A_{prod} = N_t \sigma f (1 - 10^{-\lambda_{prod}t/2.30})$$

$$N_t = \left(\frac{1.00 \times 10^{-3} \text{ gram}}{55.0 \text{ grams/mole}}\right)(6.02 \times 10^{23} \text{ atoms/mole}) = 1.09 \times 10^{19} \text{ atoms}$$

$$A_{prod} = (1.09 \times 10^{19})(13.4 \times 10^{-24})(1.00 \times 10^{13})(1 - 10^{-(0.267/\text{hr})(5.2 \text{ hr})/2.30})$$

$$= (1.46 \times 10^9)(1 - 0.25) = 1.1 \times 10^9 \text{ dis/sec}$$

At 520 hours, the exponential term will effectively be zero, and the saturation activity will be 1.46×10^9 dis/sec.

The cross section of a nuclide varies with the energy of the projectile particles; indeed, a given nuclide may undergo different reactions with the same projectile particles, depending on their energy. For example, the possible reactions of the nuclide $^{63}_{29}Cu$ with protons are as follows:

$$^{63}_{29}Cu + ^{1}_{1}H \rightarrow [^{64}_{30}Zn] \begin{cases} \nearrow\ ^{63}_{30}Zn + ^{1}_{0}n & (\sigma = 0.5 \text{ barn for 12 MeV protons}) \\ \rightarrow\ ^{62}_{29}Cu + ^{2}_{1}H & (\sigma = 0.88 \text{ barn for 24 MeV protons}) \\ \searrow\ ^{62}_{30}Zn + 2\,^{1}_{0}n & (\sigma = 0.2 \text{ barn for 24 MeV protons}) \end{cases}$$

20-12 Nuclear Fission

In 1938, O. Hahn and F. Strassmann discovered that when uranium was bombarded with neutrons, elements of much lower atomic number (e.g., barium, lanthanum, rubidium, and strontium) were produced. They correctly concluded that these elements resulted from the splitting, or **fission,** of uranium into two more or less equal fragments. As might be predicted from Figure 20–10, the reaction is exoergic. Further studies of the fission process showed that the nuclide $^{235}_{92}U$ can be induced to fission by low energy neutrons but that $^{238}_{92}U$ cannot. The fission process may occur in many different modes, and a large number of fission products have been isolated ranging from zinc ($Z = 30$) to terbium ($Z = 65$). Also, from one to three neutrons are produced. Examples of typical fission processes are

$$^{235}_{92}U + ^{1}_{0}n \rightarrow ^{140}_{56}Ba + ^{94}_{36}Kr + 2\,^{1}_{0}n$$

$$^{235}_{92}U + ^{1}_{0}n \rightarrow ^{90}_{38}Sr + ^{143}_{54}Xe + 3\,^{1}_{0}n$$

Since the original ^{235}U has a high neutron/proton ratio, the fission products all have neutron/proton ratios far in excess of that corresponding to the belt of stability. As a consequence, they decay by several consecutive beta emissions. Hence, a series of radioactive nuclides all with the same mass number, called a **fission chain,** is observed for each of the original fission products. A plot of fission chain yield against mass number is shown in Figure 20–12. The curve is symmetrical about a minimum at a mass number equal to 233.5/2 rather than 236/2, which indicates that on the average 2.5 neutrons are released per fission event.

With a suitable experimental arrangement, it is possible for the neutrons released in the fission of one nucleus to be absorbed by other ^{235}U nuclei and to induce fission in them. First, the neutrons have to be slowed down so that they may be captured by the target nuclei. Slowing the neutrons is achieved by arranging lumps or rods of

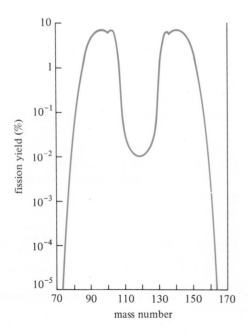

Figure 20–12. **Fission Chain Yield as a Function of Mass Number**

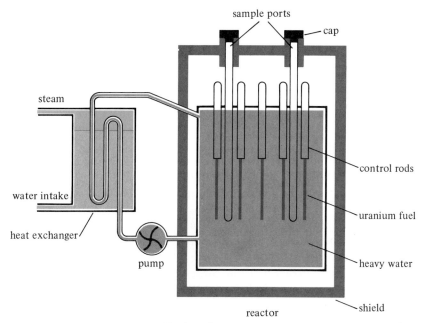

sample ports

cap

steam

water intake

heat exchanger

pump

control rods

uranium fuel

heavy water

shield

reactor

Figure 20-13. **Diagram of a Nuclear Reactor**

uranium in a lattice surrounded by some **moderator,** a material containing atoms of low atomic weight which have low cross sections for neutrons. Suitable moderators are graphite, paraffin, and water, especially D_2O. A schematic diagram of a nuclear reactor containing heavy water as moderator is shown in Figure 20-13. The mass of fuel and moderator is known as a **pile.** The pile must have a certain minimum size, or **critical** size, so that most of the neutrons are moderated and absorbed by the uranium, rather than merely escaping from the device. Fission occurs in a lump of uranium, and neutrons are emitted. They are slowed by collision with the atoms of the moderator and ultimately enter another lump of fuel, where they initiate more reaction. Thus a **chain reaction** is set up, which if left uncontrolled could result in a violent explosion. To prevent such a catastrophe, control rods made of cadmium or boron (substances having very high cross sections for neutrons) are inserted into the pile in such a way that they can be withdrawn to increase the rate of fission or inserted to slow down the chain reaction by absorbing excess neutrons.

Some of the neutrons are absorbed by ^{238}U, which is thereby changed to $^{239}_{94}Pu$ by the following sequence of reactions:

$$^{238}_{92}U(n,\gamma)^{239}_{92}U$$
$$^{239}_{92}U \rightarrow {}^{239}_{93}Np + {}^{0}_{-1}\beta \quad (t_{1/2} = 23 \text{ min})$$
$$^{239}_{93}Np \rightarrow {}^{239}_{94}Pu + {}^{0}_{-1}\beta \quad (t_{1/2} = 2.3 \text{ days})$$

Like $^{235}_{92}U$, the nuclide $^{239}_{94}Pu$ undergoes fission. Hence, a nuclear pile generates some of its own fuel.

Since there is a decrease in the quantity of matter in the fission process, a large quantity of energy (about 200 MeV/event) is liberated, which can be used, for example, to produce steam and ultimately electric power.

20-13   Nuclear Fusion

From the binding energy curve (Figure 20–10), it is apparent that energy would be released if the lightest nuclides are converted into heavier ones. Such **fusion** reactions have been achieved in the "hydrogen bomb." In this device, deuterium atoms are made to react at extremely high temperatures to give heavier nuclides with the release of tremendous quantities of energy. The high temperatures necessary for this reaction are achieved by the explosion of a fission (atomic) bomb, but the quantity of energy released per unit mass of fuel is much greater in the fusion process than in the fission process. The fusion process itself does not yield radioactive products, but because a uranium bomb is required to detonate a hydrogen bomb, the explosion of one of these weapons releases large quantities of radioactive fission products into the atmosphere and menaces the population of the entire world.

Fusion reactions are responsible for the production of energy by the stars. In the process, hydrogen nuclei are transformed into helium nuclei, with the accompanying release of tremendous quantities of energy. One proposed mechanism for the conversion of hydrogen to helium is the carbon cycle, shown in Figure 20–14. The cycle consists of six reactions, in which the net change is the conversion of four protons into one alpha particle, with the release of 30 MeV. In cooler stars such as the sun, the conversion is thought to result from a proton cycle:

$$\mathrm{^1_1H} + \mathrm{^1_1H} \rightarrow \mathrm{^2_1H} + \mathrm{^0_{+1}\beta} + 0.42\ \mathrm{MeV}$$

$$\mathrm{^1_1H} + \mathrm{^2_1H} \rightarrow \mathrm{^3_2He} + \gamma + 5.5\ \mathrm{MeV}$$

$$\mathrm{^3_2He} + \mathrm{^3_2He} \rightarrow \mathrm{^4_2He} + 2\,\mathrm{^1_1H} + 12.8\ \mathrm{MeV}$$

The net reaction is the conversion of four protons into an alpha particle, with the

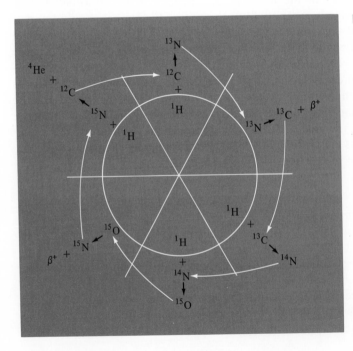

Figure 20–14. Stellar Carbon Cycle

The four proton reactants are shown within the circle. The ^{4}He and the two β^+ particles are the products. The latter react with electrons to produce more energy.

release of 26.6 MeV of energy. This quantity of energy includes that derived from the "annihilation" reaction of the positron with an electron:

$$_{+1}^{0}\beta \ + \ _{-1}^{0}\beta \ \rightarrow \ \text{energy}$$

Radiochemistry

Radiochemistry involves the use of radioactive nuclides in the study of chemical problems. The chemical properties of a radioactive nuclide are identical to the chemical properties of its stable isotope(s), and use of radioactive nuclides provides a convenient means to study a variety of chemical (and biological) reactions because they can be followed by means of the radiations they emit. Such studies include determinations of solubility, chemical equilibrium, reaction kinetics and mechanisms, and structure determinations.

The measurement of radioactivity is critical in the interpretation of radiochemical experiments. For proper measurement, it is necessary to consider the mode of decay of the nuclide, its half life, and the method used for the detection of the radiation. Most artificially produced radioactive nuclides emit beta particles, and a Geiger–Müller counter (Figure 20–15) may be used for their detection. To use the instrument it is necessary to convert the sample to a form suitable for placing beneath the window of the counter. Since all the radiations emitted by the sample do not enter the counter, the observed counting rate must be compared with the counting rate from a standard of known specific activity.

Recently, the more versatile **scintillation counters** are being used for radioactivity measurements. Such counters are based on the principle that ionizing radiations

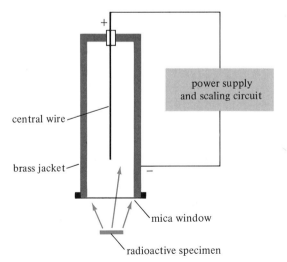

Figure 20-15. The Geiger–Müller Counter
The Geiger–Müller counter consists of a cylinder made of a conducting material which has a potential of about -1000 V with respect to an insulated central wire. The tube contains a mixture of gases, such as argon and ethyl alcohol vapor, under a total pressure of 25 to 50 torr. In the design shown, a thin mica window seals one end of the tube. When ionizing radiation—alpha, beta, or gamma—enters the tube, electrons are knocked from the argon atoms and are accelerated toward the central wire, creating more and more ionization, until an avalanche of electrons and ions spreads along the length of the wire. The positive ions migrate toward the wall in about 10^{-4} sec, accepting the negative charge from the detecting circuit as a pulse of current. The pulse is amplified and registered as a **count.** Photons emitted during recombination of positive ions and electrons also collide with the walls of the tube, causing photoelectrons to be ejected. However, the alcohol molecules combine with these low energy electrons and dissipate their energy so that a secondary avalanche is prevented.

produce flashes of light (scintillations) in crystals of certain materials, such as sodium iodide, zinc sulfide, or anthracene ($C_{14}H_{10}$). The light is detected by sensitive photo-multiplier tubes, which respond to both the number and intensity of the scintillations. Scintillation counters can be adapted to detect radiations in samples in a variety of physical forms.

20-14   Tracers

Radionuclides are sometimes added to a reaction system in order to follow the course of the reaction. Such a sample of radioactive material is called a **tracer.** The nuclide being used as a tracer must be sufficiently long-lived to give an appreciable activity for the duration of the experiment. Unfortunately, both oxygen and nitrogen, two species which would be extremely valuable in tracer studies, do not have long-lived radioactive isotopes. (^{15}O has a half life of only 2 min, while ^{13}N has a half life of 10.1 min.)

Radioactive nuclides for tracer use may be prepared by

1. Extraction from naturally radioactive ores.
2. Production of specific nuclides by bombardment of a suitable target in a nuclear reactor or by use of a particle accelerator.
3. Separation of specific nuclides from the fission products of uranium.

Extraction of suitable isotopes from naturally radioactive sources is limited to a few elements, such as ^{212}Pb (half life 10.5 hours) from ^{232}Th ores, ^{226}Ra (1622 years) from ^{238}U, or ^{210}Pb (22 years) from ^{226}Ra.

Very pure samples of nuclides can be obtained by bombardment techniques. Often the quantity of radionuclide obtained is too small to be weighed, but solutions of very high specific activity can be prepared.

Example

Describe how radiochemically pure $^{62}_{30}Zn$ ($t_{1/2} = 9$ hours) can be prepared.

A suitable reaction would be the irradiation of a copper target with 25 MeV protons:

$^{63}Cu(p,2n)^{62}Zn$

Some radioactive ^{62}Cu will also be formed, as well as ^{63}Zn. The ^{63}Zn, having a half life of 38 min, will decay to ^{63}Cu by electron capture before the separation procedure. The copper produced by these processes will be separated with the bulk of the copper target. About 2 hours after the irradiation (the time lapse to allow the ^{63}Zn to decay), the target is dissolved in nitric acid. The pH is adjusted to about 3, copper is precipitated as CuS, and the solid is filtered. Then the filtrate is boiled to concentrate it and to drive off excess H_2S. The only cations remaining in the solution are $^{62}Zn^{2+}$ and H_3O^+. The solution is electrolyzed between platinum electrodes, and radiochemically pure ^{62}Zn is collected at the cathode.

If a weighable sample of a tracer is desired, nonradioactive atoms may be added before the separation of the radionuclide. For instance, in the example described above, after removal of the copper ions a precipitate of zinc sulfide could have been obtained by adding a few milligrams of zinc ion, adjusting the pH to about 6, and saturating the solution with H_2S. The added, inactive zinc isotope is an example of a

carrier. Carriers are usually used to separate specific nuclides from mixtures of fission products. For example, uranium from a nuclear reactor is dissolved in concentrated nitric acid, and a few milligrams of iodide ion is added. After treatment of the solution with an oxidizing agent, the I_2 produced is extracted with carbon tetrachloride. In this manner the radioactive iodine produced in the fission reaction can be separated from the other fission products.

20-15   Analysis by Means of Radioactivity

Another useful radiochemical technique is **radiometric analysis,** also called **isotope dilution analysis.** To a solution containing an unknown is added a solution containing a known quantity of a radioactive species isotopic with the unknown. Then the desired ion or compound is separated, and the percent of the radioactivity recovered is determined. The nonradioactive material will have been recovered to the same percent; hence its original concentration can be estimated.

Example

To 50.00 ml of a solution containing an unknown concentration of zinc ion was added 0.100 μCi of $^{62}Zn^{2+}$ in 10 ml of solution, and the total volume was diluted to 100 ml with water. Precipitation of a zinc salt yielded 0.2000 gram of zinc in the solid phase, with an activity of 0.0823 μCi. What was the original concentration of the zinc ion?

$$\% \text{ Zn recovered} = \% \ ^{62}\text{Zn recovered} = \frac{0.0823}{0.100} \times 100 = 82.3\%$$

$$\text{total zinc} = \frac{0.2000 \text{ gram recovered}}{0.823 \text{ gram recovered/gram total}} = 0.243 \text{ gram total}$$

The mass of the added $^{62}Zn^{2+}$ is negligible; hence 0.243 gram is the mass of the Zn^{2+} in the original sample.

$$\frac{0.243 \text{ gram}}{(65.37 \text{ grams/mole})(0.0500 \text{ liter})} = 0.0744 \ M \ Zn^{2+}$$

Activation analysis is used to determine traces of impurities of one element in another. For example, it is possible to determine the presence of copper in aluminum by irradiating a sample with neutrons to form 12.8 hour ^{64}Cu:

$$^{63}Cu(n,\gamma)^{64}Cu$$

The sample is irradiated with a known neutron flux until saturation activity is achieved. Then, from the measured activity and the known cross section for the reaction, the number of copper atoms in the original sample can be ascertained.

20-16   Dating

Radioactivity measurements are used to determine the age of minerals and other materials. The process is called radioactive **dating.** The existence of naturally radioactive nuclides on the surface of the earth suggests that their half lives are comparable to the ages of the minerals in which they are found, and these in turn provide an estimate of the age of the earth. Another important use of radioactivity measurement is for radiocarbon dating. The constant bombardment of nitrogen in the atmosphere by cosmic rays produces radioactive ^{14}C:

$$^{14}_{7}N(n,p)^{14}_{6}C$$

In a period of time estimated as about 500 years, this carbon becomes part of the terrestrial carbon cycle and is ingested by plants and animals. Ultimately, a balance is established between the ^{14}C intake and decay in living materials, which results in a natural specific activity of ^{14}C of 15.3 dis/min $\cdot$ gram of carbon. This specific activity appears to have been reasonably constant over at least the past several thousand years. When a living organism dies, the intake of ^{14}C ceases, and the ^{14}C activity decays (half-life = 5730 years). Hence, by measuring the specific activity of a sample of carbonaceous material, its age can be estimated.

Example

   The bones of a prehistoric bison were found to have a ^{14}C activity of 2.80 dis/min $\cdot$ gram of carbon. Approximately how long ago did the animal live?

$$2.30 \log \frac{A}{A_0} = \frac{0.693}{t_{1/2}} t = -\frac{0.693}{5730 \text{ years}} t$$

$$t = -2.30 \left(\log \frac{2.80}{15.3} \right) \left(\frac{5730 \text{ years}}{0.693} \right) = 14,000 \text{ years}$$

20-17   Exercises

Basic Exercises

1. Without consulting tables or other sources of information, explain how one can determine which of the following nuclides is the terminal member of the naturally occurring radioactive series which begins with $^{235}_{92}U$: ^{206}Pb, ^{207}Pb, ^{208}Pb, ^{209}Bi.
2. Using the equation for the radius of the nucleus, calculate the density of the nucleus of $^{107}_{47}Ag$, and compare it with the density of metallic silver (10.5 grams/cm^3).
3. Which one(s) of the following processes—alpha emission, beta emission, positron emission, electron capture—cause **(a)** an increase in atomic number? **(b)** a decrease in atomic number? **(c)** emission of an X ray in every case?
4. Determine the half life of a nuclide which was separated from a cyclotron target and which gave the following readings on a Geiger–Müller counter:

Time (hr)	Counts/Min	Time (hr)	Counts/Min
0	4000	16	1660
2	3660	24	1070
4	3200	40	450
8	2570	60	150

5. Determine the half life of a nuclide which was separated from a cyclotron target and which gave the following readings on a Geiger–Müller counter:

Time (min)	Counts/Sec	Time (min)	Counts/Sec
0.0	300	10.0	66.0
2.5	206	12.0	48.9
5.0	141	15.0	31.2
7.5	96.6	20.0	14.7

6. Prove that $A/A_0 = N/N_0 = m/m_0$, where A and A_0 are activities, N and N_0 are numbers of atoms, and m and m_0 are masses, all of the same decaying nuclide.

7. Show that $1.00 \text{ D} = 931 \text{ MeV}$.

8. If 4.54×10^{-5} erg is needed to produce two protons and two neutrons from one helium nucleus, by use of appropriate conversion factors show that 653×10^6 kcal is required to dissociate one mole of helium atoms into protons and neutrons.

9. Calculate the effective neutron capture radius of a nucleus having a cross section of 1.0 barn.

10. Show that $10^{-\lambda_{\text{prod}} t / 2.30}$ approaches 0 as t becomes very large.

11. When the pure nuclide $^{50}_{24}\text{Cr}$ is bombarded with alpha particles, two reactions occur, and neutrons and deuterons are observed as product particles. Write equations showing the formation of the possible product nuclides.

12. The disintegration of ^{239}Pu is accompanied by the loss of 5.24 MeV/dis. The half life of ^{239}Pu is 24,400 years. Calculate the energy released per day from a 1.00 gram sample of ^{239}Pu, and express the result in million electron volts and kilocalories or kilojoules.

13. The half life of ^{212}Pb is 10.6 hours; that of its daughter ^{212}Bi is 60.5 min. How long will it take for a maximum daughter activity to grow in freshly separated ^{212}Pb?

14. Classify each of the following nuclides as "probably stable," "beta emitter," or "positron emitter": $^{49}_{20}\text{Ca}$, $^{195}_{80}\text{Hg}$, $^{208}_{82}\text{Pb}$, $^{8}_{5}\text{B}$, $^{150}_{67}\text{Ho}$, $^{30}_{13}\text{Al}$, $^{120}_{50}\text{Sn}$, $^{94}_{36}\text{Kr}$.

15. Some radioactive material is decaying with a 30 day half life. Chemical separation procedures yield two fractions. Immediately after separation, one of the fractions decays with a 2 day half life. Would the other fraction show, immediately after separation, constant activity, increasing activity, decay with a 30 day half life, or decay with a 28 day half life? Explain.

16. Select from the following list of nuclides (a) the isotopes, (b) the isobars, (c) the isotones: $^{40}_{18}\text{Ar}$, $^{41}_{19}\text{K}$, $^{40}_{21}\text{Sc}$, $^{42}_{21}\text{Sc}$, $^{90}_{40}\text{Zr}$.

17. Of the three isobars $^{114}_{48}\text{Cd}$, $^{114}_{49}\text{In}$, and $^{114}_{50}\text{Sn}$, which is likely to be radioactive? Explain your choice.

18. Assuming each sample was freshly separated from its decay products, what is the mass of a sample which contains 1.00 mCi of (a) ^{131}I ($t_{1/2} = 8.08$ days), (b) ^{238}U, (c) $^{140}_{56}\text{Ba}$?

General Exercises

19. The cross section for the reaction $^{127}_{53}\text{I}(n,\gamma)^{128}_{53}\text{I}$ is 6.3 barns. The half life of ^{128}I is 25 min. If 12.7 grams of pure ^{127}I is placed in a reactor in which the neutron flux is 2.0×10^5 neutrons/cm² · sec and is bombarded for 25 min, what is the activity of ^{128}I in the sample 100 min *after* the bombardment has stopped?

20. Calculate the cross sectional area of the nucleus of a ^{10}B atom assuming spherical shape. Compare this value to its cross section for neutrons given in Table 20–2. Does the neutron have to make a "direct hit" in order to be captured?

21. Red giant stars, which are cooler than the sun, produce energy by means of the re-action $^9_4Be + ^1_1H \rightarrow ^6_3Li + ^4_2He$ + energy. From the nuclidic masses [^9Be(9.01504) and ^6Li(6.01702)], calculate the energy released in million electron volts, and compare it with the energy released in the carbon cycle and in the solar helium–hydrogen cycle mentioned in this chapter.

22. Potassium (atomic weight = 39.102 D) contains 93.10 atom % ^{39}K, having atomic mass 38.96371 D; 0.0118 atom % ^{40}K, which has a mass of 40.0 D and is radioactive with $t_{1/2} = 1.3 \times 10^9$ years; and 6.88 atom % ^{41}K having a mass of 40.96184 D. Estimate the specific activity of naturally occurring potassium.

23. Russian and American scientists have prepared artificially elements with atomic numbers above 100. To what stable isotope would $^{257}_{103}Lr$ decay after having been produced by artificial means? On what basis can such a prediction be made?

24. Fallout from nuclear explosions contains ^{131}I and ^{90}Sr. Calculate the time required for the activity of each of these isotopes to fall to 1.0% of its initial value. Radioiodine and radiostrontium tend to concentrate in the thyroid and the bones, respectively, of mammals which ingest them. Which isotope is likely to produce the more serious long-term effects?

25. (a) Calculate the neutron capture radius of a $^{113}_{48}Cd$ atom from the data of Table 20–2. (b) Compare this radius to that estimated by means of the equation relating the radius of a nucleus to its mass number on page 556.

26. Antimatter consists of particles having properties and characteristics similar to particles in known atoms, but which have opposite charges. For example, an electron and a positron are two particles identical in every respect except for the sign of their charges. Speculate on the existence of a world composed of negative antiprotons and positrons (along with an array of other antiparticles) and neutrons. Would the chemistry in such a world have formulas similar to those of compounds on earth? What would happen if material from earth encountered material from the "antiworld"?

27. Upon irradiating californium with neutrons, a new nuclide having mass number of 250 and a half life of 0.50 hour is discovered. Three hours after the irradiation, the observed radioactivity due to the nuclide is 10 dis/min. How many atoms of the nuclide were prepared initially?

28. A 7.00 gram sample of nuclidically pure ^{127}I is placed in a nuclear reactor in which the neutron flux is 1.1×10^5 neutrons/cm^2 · sec and is bombarded for 25 min. The half life of the product, ^{128}I, is 25 min. Fifty minutes *after* the bombardment has stopped, the activity in the sample is 9000 dis/sec. What is the cross section, in barns, for the reaction $^{127}I(n,\gamma)^{128}I$?

29. A sample of rock from the Grand Canyon was analyzed for gold content by neutron activation. A 1.00 gram sample was irradiated in a flux of 2.0×10^{12} neutrons/cm^2 · sec for about 2 weeks. Then the sample was removed, dissolved, and some inactive gold carrier was added. Gold was precipitated as the sulfide and was found to have an activity due to ^{198}Au of 50 dis/min, corrected to the time of bombardment. The cross section of ^{197}Au for neutrons is 98 barns. The half life of ^{198}Au is 2.7 days. What mass of gold is there per gram of rock sample?

30. When radioactive sulfur is added to alkaline sodium sulfite solution, radioactive thiosul-fate ion is formed. Upon adding Ba^{2+}, a precipitate of BaS_2O_3 is obtained. The precipitate is filtered and dried and is then treated with acid, producing solid sulfur, SO_2 gas, and water. The SO_2 is *not* radioactive at all. Write equations for the sequence of reactions, and comment on the structure of the thiosulfate ion as elucidated by this experiment.

31. A radionuclide has an initial activity of 2.00×10^6 dis/min, and after 4.0 days its activity is 9.0×10^5 dis/min. Calculate the activity in the sample after 40 days.

32. A sample of radioactive material has an apparently constant activity of 2000 dis/min. By chemical means, the material is separated into two fractions, one of which has an initial activity of 1000 dis/min. The other fraction decays with a 24 hour half life. Estimate the total activity in *both* samples 48 hours after the separation. Explain your estimate.

Advanced Exercises

33. A piece of wood, reportedly from King Tut's tomb, was burned, and 7.32 grams of CO_2 was collected. The total radioactivity in the CO_2 was 10.8 dis/min. How old was the wood sample? Is there a possibility that it is authentic?

34. A mixture of ^{239}Pu and ^{240}Pu has a specific activity of 6.0×10^9 dis/sec. The half lives of the isotopes are 2.44×10^4 and 6.58×10^3 years, respectively. Calculate the isotopic composition of this sample.

35. ^{120}Q, ^{119}R, and ^{120}R are the stable nuclides of elements Q and R. When element Q is bombarded with slow neutrons, radioactive Q is produced, which decays by beta emission with a 3 hour half life to an isotope of R, which is also beta active with an 11 day half life. When Q is bombarded with deuterons, the 11 day activity is observed, but not the 3 hour activity. When element R is bombarded with fast neutrons, the only activities observed are the 11 day R activity and a 2 year Q activity, which decays by beta emission to an inactive daughter. Assign mass numbers to **(a)** the 3 hour Q isotope, **(b)** the 11 day R isotope, **(c)** the 2 year Q isotope. Write equations for the reaction of **(d)** the Q target with slow neutrons, **(e)** the Q target with deuterons, **(f)** the R target with fast neutrons.

36. A 0.20 ml sample of a solution containing 1.0×10^{-7} Ci of ^3_1H is injected into the blood stream of a laboratory animal. After allowing sufficient time for circulatory equilibrium to be established, a 0.10 ml of blood is found to have an activity of 20 dis/min. Calculate the blood volume of the animal.

37. How long must 5.0 grams of the pure nuclide ^{31}P be irradiated with neutron flux 2.0×10^5 neutrons/cm^2 · sec to achieve an activity of ^{32}P of 1.00×10^3 dis/min 100 hours after the bombardment has stopped ($t_{1/2} = 14.3$ days)?

38. A sample of $^{131}_{53}I$, as iodide ion, was administered to a patient in a carrier consisting of 0.10 mg of stable iodide ion. After 4.00 days, 67.7% of the initial radioactivity was detected in the thyroid gland of the patient. What mass of the stable iodide ion had migrated to the thyroid gland? Of what diagnostic value is such an experiment?

39. Devise a set of permitted values for the four nuclear quantum numbers—n, l, j, and m—which describe the shells of nucleons and the stabilities of nuclei with the magic numbers. Compare your values to those of actual nuclear quantum numbers listed in exercise 40, Chapter 10.

40. Chlorine, with an atomic weight of 35.453 D, is a mixture of 75.53% ^{35}Cl and 24.47% ^{37}Cl, having atomic masses of 34.96885 and 36.96590 D, respectively. Suppose that instead of ^{37}Cl, "natural" chlorine contained ^{36}Cl, having an atomic mass of 35.9787 D and a half life of 3.1×10^5 years. Assuming the same atomic weight for this mixture of isotopes, calculate the specific activity of a sample of this "natural" chlorine.

21

Nonmetals

This and the next chapter will cover some topics usually referred to as "descriptive chemistry." It is the purpose of these chapters to demonstrate how the theories and principles taken up earlier in the book can be applied to specific phenomena. Also, examples of the technological applications of the elements and their compounds will be described. In this chapter the chemistry of the nonmetals will be discussed.

As their name implies, nonmetals lack the properties characteristic of metals. Many are volatile; eleven are gases at room temperature. Nonmetals exhibit a variety of colors, and in their pure states most are insulators. Except for hydrogen and helium, atoms of the nonmetals have valence shells containing at least one p-orbital electron. Therefore, nonmetals may react to complete their p subshells and in so doing achieve negative oxidation states. The nonmetallic elements have electronegativity values about 2 or greater on the Pauling scale.

Bonds between like atoms of nonmetallic elements are covalent. Except for boron, carbon, and silicon, in the solid state nonmetals exist as soft and low-melting molecular crystals. Allotropism among nonmetals is usually the result of the element's existing in more than one molecular form; for example, oxygen exists as O_2 and O_3 (ozone). This behavior is in contrast to allotropism in metals, where the difference between allotropes is due to the possibility of more than one close-packed structure. Bonds between unlike nonmetals are essentially covalent, but bonds between nonmetals and metals tend to have ionic character.

The many uses of nonmetals and their compounds include fuels and raw materials for the manufacture of chemicals, agricultural chemicals, and synthetic films and fibers. In densely populated industrial areas, volatile nonmetallic oxides formed in combustion reactions pollute the atmosphere, causing a menace to public health and comfort.

21-1 Hydrogen

Hydrogen is unique. It is the only element which has atoms with incompletely filled $1s$ subshells. All theories of the electronic structure of the atom begin with a consideration of the hydrogen atom, and discussions of chemical bonding begin with discussions of the hydrogen molecule or the hydrogen molecule ion, H_2^+. Hydrogen is the most abundant element in the universe. On the surface of the earth (including the oceans), hydrogen is only the sixth most abundant element in terms of mass, but it is the most abundant in terms of number of atoms. Most of the hydrogen on earth is in the form of water.

Hydrogen exhibits oxidation states of $-1, 0,$ and $+1$, corresponding to the hydride ion, the free element, and the covalently bonded atom or ion, respectively. Since the hydrogen atom has but one electron, it can react to acquire a second electron to achieve an electronic structure analogous to that of helium. In binary compounds with highly electropositive atoms, hydrogen forms the negatively charged hydride ion, H^-, as in LiH; otherwise it achieves the helium configuration by formation of a single covalent bond, as in H_2. The loss of its electron would make the hydrogen atom a free proton. Free protons exist only (1) in a highly evacuated apparatus, such as a canal ray tube (Section 9-4), or (2) at extremely high temperatures such as are observed in the interiors of stars. In ordinary chemical systems protons coordinate with any available electron pair donor; hence, rather than free protons, species such as H_3O^+ and NH_4^+ are observed.

Preparation of Hydrogen

In the laboratory hydrogen gas is conveniently prepared in several ways. Very active metals can reduce water:

$$2\,Na \;+\; 2\,H_2O \;\rightarrow\; H_2 \;+\; 2\,Na^+ \;+\; 2\,OH^-$$

Active metals such as zinc reduce aqueous acids to hydrogen:

$$Zn \;+\; 2\,H_3O^+ \;\rightarrow\; Zn^{2+} \;+\; H_2 \;+\; 2\,H_2O$$

Acids which are strong oxidizing agents, such as HNO_3, cannot be used to prepare hydrogen in this manner. In nitric acid solution it is the nitrate ion which actually undergoes reduction, because it is reduced more rapidly than the hydronium ion:

$$4\,Zn \;+\; 10\,H_3O^+ \;+\; NO_3^- \;\rightarrow\; NH_4^+ \;+\; 4\,Zn^{2+} \;+\; 13\,H_2O$$

Active, amphoteric metals can produce hydrogen from strong aqueous bases:

$$Zn \;+\; 2\,OH^- \;+\; 2\,H_2O \;\rightarrow\; [Zn(OH)_4]^{2-} \;+\; H_2$$

Metal hydrides and complex hydrides react with water or acids to produce hydrogen:

$$BH_4^- \;+\; H_3O^+ \;+\; 2\,H_2O \;\rightarrow\; H_3BO_3 \;+\; 4\,H_2$$

Commercially, hydrogen is prepared by several methods. If cheap electric power is available, the electrolysis of water is likely to be used:

$$2\,H_2O \;\xrightarrow[\text{power}]{\text{electric}}\; 2\,H_2 \;+\; O_2$$

In another commercial preparation, hydrogen is produced along with carbon monoxide in the **water-gas** reaction:

$$C \;+\; H_2O \;\xrightarrow[\text{heat}]{}\; H_2 \;+\; CO$$

If this mixture is used for fuel, no effort is made to separate the two gases. Another method is the reaction of steam with iron at elevated temperatures to yield hydrogen:

$$3\,Fe \ + \ 4\,H_2O(g) \xrightarrow[600°C]{} Fe_3O_4 \ + \ 4\,H_2$$

Binary Compounds of Hydrogen

Hydrogen forms compounds with practically all of the other elements. Its binary compounds are of four types:

1. Saltlike, containing H^- ions.
2. Acidic compounds, in which a proton is readily available for transfer.
3. Nonacidic covalent compounds of discrete or polymeric molecules.
4. Interstitial compounds with transition metals.

The type of binary compound formed with hydrogen can be correlated with the electronegativity of the other element. Elements forming saltlike hydrides have electronegativities no greater than 1.0 on the Pauling scale. The alkali metal hydrides are produced by the direct reaction of the metal and hydrogen, and have the rock salt structure (Figure 18–2):

$$2\,Li \ + \ H_2 \ \rightarrow \ 2\,Li^+H^-$$

The alkaline earth hydrides, CaH_2, SrH_2 and BaH_2, are produced similarly. Their crystals are orthorhombic (Figure 21–1).

Acidic binary hydrides are formed between hydrogen and the elements of groups VI A and VII A in the periodic table. They are prepared by direct combination of the elements or by displacement reactions:

$$H_2 \ + \ I_2 \ \rightarrow \ 2\,HI$$
$$H_2SO_4(conc) \ + \ Cl^- \ \rightarrow \ HCl(g) \ + \ HSO_4^-$$

Nonacidic covalent hydrides are formed between hydrogen and elements (other than transition metals) having electronegativities between 1.0 and 2.5. Examples of this class are CH_4, AsH_3, and SiH_4, which are composed of discrete molecules; and beryllium hydride and aluminum hydride, which are polymeric in nature. These

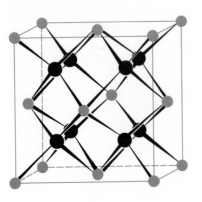

Figure 21-1. **Orthorhombic** CaH_2

● calcium ion

● hydride ion

hydrides are usually prepared from compounds rather than by direct combination of the elements:

$$Al_4C_3 \;+\; 12\,H_2O \;\rightarrow\; 3\,CH_4 \;+\; 4\,Al(OH)_3$$
$$x\,BeCl_2 \;+\; 2x\,LiH \;\rightarrow\; 2x\,LiCl \;+\; (BeH_2)_x$$

Interstitial hydrides are formed by transition metals having incompletely filled d subshells. They are formed by direct combination of the metal with hydrogen gas, and the properties of these hydrides are similar to those of the parent metals. Typical formulas of this type of compound are $TiH_{1.73}$, $ZrH_{1.9}$, $CeH_{2.8}$, and $PrH_{2.70}$. The same compositions are produced as long as the experimental conditions are kept constant, even in the presence of excess hydrogen. However, the formulas do depend on the temperature at which the compound is formed.

The hydride ion, H^-, is analogous to halide ions in several ways, among which is its ability to act as a ligand with metal ions. Important hydride complexes are those with nontransition metal elements in compounds such as $LiAlH_4$ and $NaBH_4$. These compounds are important reducing agents.

21-2   Oxygen

Oxygen is the most abundant element by mass on the surface of the earth, occurring in the form of water and in oxides, silicates, sulfates, and carbonates in the rocky crust of the earth. Also, the free element constitutes 20% of the earth's atmosphere. The supply of atmospheric oxygen is not unlimited but depends on the process of **photosynthesis,** which occurs in green plants:

$$6\,CO_2 \;+\; 6\,H_2O \;\xrightarrow[\text{light}]{\text{chlorophyll}}\; \underset{\substack{\text{glucose}\\\text{(a sugar)}}}{C_6H_{12}O_6} \;+\; 6\,O_2$$

The consumption of oxygen for industrial purposes, such as the production of power, is increasing annually at a rate which threatens to exceed the rate of oxygen production by photosynthesis. Unless appropriate measures are taken, it is quite possible that in the future there will be a shortage of atmospheric oxygen even more critical than the shortage of fresh water which already exists on this planet.

Pure oxygen is an important industrial chemical. Its uses in steelmaking (Section 19-6) and as an oxidizer for fuels (Chapter 5) have already been described. Other uses include life-support systems and the production of high-temperature flames, such as are needed for glass blowing and for welding. Commercially, oxygen is prepared by the fractional distillation of liquid air. Liquid oxygen boils at $-183\,^\circ C$; the two other constituents of air which boil about the same temperature are nitrogen ($-196\,^\circ C$) and argon ($-186\,^\circ C$). Oxygen is also produced along with hydrogen by the electrolysis of water. In the laboratory it can be produced by the thermal decomposition of $KClO_3$. The reaction is catalyzed by MnO_2 or other finely powdered oxides, such as Fe_2O_3 or SiO_2:

$$2\,KClO_3 \;\xrightarrow{\;MnO_2\;}\; 2\,KCl \;+\; 3\,O_2$$

The electronic structure of the oxygen atom is $1s^2 2s^2 2p^4$. In most oxygen compounds, oxygen atoms form two covalent bonds or exist in anions such as OH^- or O^{2-} ions. (In peroxide ion, O_2^{2-}, and superoxide ion O_2^-, the two oxygen atoms are connected to each other by a covalent bond.) As noted in Section 11–15, the structure of the oxygen molecule is best explained in terms of the molecular orbital theory. In an analogous manner, the stabilities of oxygen-containing anions such as CO_3^{2-}, NO_2^-, NO_3^-, and $C_2O_4^{2-}$ can be attributed to bonding via delocalized π orbitals involving all the oxygen atoms.

The bond energy of the oxygen molecule is 119 kcal/mole. This high value accounts for the relatively high percentage of oxygen in the atmosphere. For example, the reaction of oxygen with methane, CH_4, is exothermic at 25°C and proceeds with a decrease in free energy:

$$CH_4(g) \;+\; 2\,O_2(g) \;\rightarrow\; CO_2(g) \;+\; 2\,H_2O(g) \qquad \Delta G = -192 \text{ kcal}$$

However, the bond-breaking steps in both reactants are so endothermic that the reaction has a large activation energy. Therefore a mixture of the two gases is kinetically stable at room temperature.

The other allotropic form of oxygen is the triatomic molecule, O_3, called ozone. It is prepared by passing an electric discharge through oxygen or by the electrolysis of a concentrated aqueous solution of perchloric acid at −50°C between a lead cathode and a platinum anode. In the latter process O_3 is formed at the anode. Ozone is formed in the atmosphere by lightning and by ultraviolet light. The characteristic pungent odor of ozone is often detectable after a thunderstorm. The odor is also detected near electric motors which have been sparking.

Electron diffraction studies show that ozone is a nonlinear molecule. The molecule is diamagnetic, indicating that all of the electrons are paired. The bonding is best regarded as two σ bonds formed with sp^2 hybrid orbitals from the central atom. The two end atoms have a total of five lone pairs of electrons, and the central atom has one. Two electrons are accommodated in a delocalized π orbital.

Ozone is 1.5 times as soluble in water as O_2. When cooled to −111.9°C, O_3 condenses to a deep blue liquid which is dangerously explosive. Ozone is one of the strongest oxidizing agents known, exceeded in oxidizing power only by fluorine, atomic oxygen, and OF_2.

In the upper atmosphere ozone is produced by the reaction of oxygen molecules, O_2, under the influence of ultraviolet radiation of wavelengths between 1600 Å and 2400 Å.

$$3\,O_2 \;+\; h\nu \;\rightarrow\; 2\,O_3$$

In addition to other processes, ozone may be converted back to oxygen atoms and molecules by ultraviolet light of wavelengths 2400 Å to 3600 Å:

$$h\nu \;+\; O_3 \;\rightarrow\; O \;+\; O_2$$

These conversions result in the absorption of much of the ultraviolet light from the sun which otherwise would reach the surface of the earth. It is estimated that an increase of the incidence of ultraviolet radiation on the surface of the earth would cause a significant increase in skin cancer among humans and also produce eye damage in individuals who are especially sensitive to ultraviolet light.

The advent of supersonic aircraft which fly at high altitudes poses a serious threat to the protective ozone layer. The exhaust gases from the engines of these giant planes contain appreciable quantities of NO, which reacts with ozone as follows:

$$NO + O_3 \rightarrow NO_2 + O_2$$

Still another threat to the ozone layer is the accumulation in the upper atmosphere of a class of industrial chemicals called Freons (Section 23-2).

Binary Compounds of Oxygen

Oxygen forms binary compounds with all of the elements except the lighter noble gases—He, Ne, and Ar. In most of these compounds, the oxygen exists in the -2 oxidation state. Notable exceptions are (1) the compound OF_2, where the oxidation state of oxygen is $+2$, and (2) peroxides and superoxides, in which the oxidation number assigned to oxygen is -1 and $-\frac{1}{2}$, respectively.

The normal oxides fall broadly into three classes according to their behavior toward water, acids, and bases. In the first category are the basic oxides (base anhydrides), which react with water to yield a basic solution. Even if they are insoluble in water, they react with acids to give salts:

$$CaO + H_2O \rightarrow Ca^{2+} + 2\,OH^-$$
$$MgO + 2\,H_3O^+ \rightarrow Mg^{2+} + 3\,H_2O$$

The second category of normal oxides includes the acidic oxides (acid anhydrides), which react with water to give acids and, even if insoluble in water, react with bases to give salts:

$$P_4O_{10} + 6\,H_2O \rightarrow 4\,H_3PO_4$$
$$Sb_2O_3 + 6\,OH^- \rightarrow 2\,SbO_3^{3-} + 3\,H_2O$$

The compounds of the third category of oxides, which includes H_2O, N_2O, and CO, are neutral.

This classification of the oxides is diagrammed in Figure 21-2. In the figure oxides adjacent to a dividing line have properties of both classes. For elements which can form more than one oxide, the higher the oxidation state, the more acidic is the oxide. For example, Fe_2O_3 is more acidic than FeO. Cu_2O is weakly basic, but CuO is even less basic, although it does react with strong acids:

$$CuO + H_2SO_4 \rightarrow CuSO_4 + H_2O$$

In reactions with metals having large atoms of very low electronegativity—K, Rb, Cs—oxygen forms superoxides:

$$Cs + O_2 \rightarrow CsO_2$$

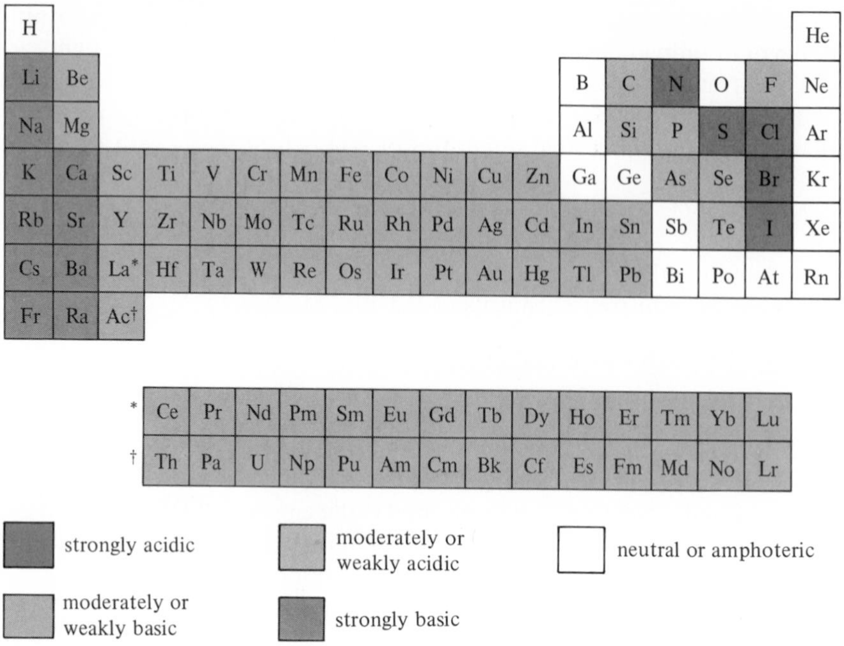

strongly acidic

moderately or weakly acidic

neutral or amphoteric

moderately or weakly basic

strongly basic

Figure 21-2. **Acidities of Oxides**

Oxygen reacts with elements with slightly higher electronegativity (about 0.9) to give peroxides:

$$2\,Na \;+\; O_2 \longrightarrow Na_2O_2$$
$$Ba \;+\; O_2 \longrightarrow BaO_2$$
$$2\,BaO \;+\; O_2 \xrightarrow[\text{high pressure}]{} 2\,BaO_2$$

The superoxide ion and the peroxide ion contain oxygen-to-oxygen covalent bonds. Molecular orbital diagrams analogous to that used to describe the O_2 molecule predict that O_2^- is paramagnetic and that O_2^{2-} is diamagnetic.

Hydrogen peroxide is obtained when Na_2O_2 or BaO_2 is treated with cold, dilute acid:

$$Na_2O_2 \;+\; 2\,H_3O^+ \;\rightarrow\; 2\,Na^+ \;+\; H_2O_2 \;+\; 2\,H_2O$$
$$BaO_2 \;+\; H_2SO_4 \;\rightarrow\; BaSO_4(s) \;+\; H_2O_2$$

Hydrogen peroxide is a colorless, water-like liquid having a melting point when pure of $-0.41\,°C$ and a normal boiling point of $152.1\,°C$. Its density is 1.45 grams/ml at $20\,°C$. These data indicate that H_2O_2 is even more highly associated through hydrogen bonding than is water. The structure of hydrogen peroxide, shown in Figure 21-3, suggests that the molecule has an appreciable dipole moment.

Hydrogen peroxide is prepared commercially by the electrolysis of cold 40% H_2SO_4 solution or of cold ammonium sulfate in sulfuric acid solution. The peroxydisulfate ion, $S_2O_8^{2-}$, is formed and is slowly hydrolyzed to H_2O_2, as shown by the following equations:

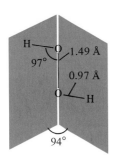

Figure 21-3. **Structure of Hydrogen Peroxide**

cathode: $2 H^+ + 2 e^- \rightarrow H_2$

anode: $2 HSO_4^- \rightarrow S_2O_8^{2-} + 2 H^+ + 2 e^-$

$S_2O_8^{2-} + 2 H_2O \rightarrow 2 HSO_4^- + H_2O_2$

The H_2O_2 is rapidly removed by distillation at low pressure. The dilute aqueous solution so obtained is further concentrated to 90% H_2O_2 by fractional distillation under reduced pressure. Pure H_2O_2 can be obtained by fractional crystallization of the 90% solution. Pure H_2O_2 decomposes explosively at temperatures near its normal boiling point.

H_2O_2 is thermodynamically unstable and capable of disproportionating into H_2O and O_2. It can act as an oxidizing agent as well as a reducing agent. Oxidations by H_2O_2 in acid solution are relatively slow, but in basic solution they are very fast. It follows that peroxides are more stable when kept under acid conditions.

When hydrogen peroxide containing ^{18}O, $^{18}O_2H_2$, undergoes oxidation in aqueous solution, the liberated oxygen gas contains all of the heavy oxygen isotope. This fact suggests that oxidation does not break the oxygen-to-oxygen bond:

$$Cl_2 + {}^{18}O_2H_2 \xrightarrow{{}^{16}OH_2} 2 H^+ + 2 Cl^- + {}^{18}O_2$$

21-3   Water

Of all chemical compounds, water is perhaps the most important and certainly the most versatile. As a chemical reagent, it functions as an acid, a base, a ligand, an oxidizing agent, and a reducing agent. Numerous examples of the chemical properties of water have already been cited throughout this book. Liquid water is a good solvent, especially for substances having polar molecules or ions. For example, substances such as sugars and alcohols dissolve because of the large dipole–dipole attractions and hydrogen bonding, and ionic substances dissolve when the ion–dipole attractions are greater than ion–ion attractions within the crystal. Solubility in water is often accompanied by acid-base reactions, as in the cases of aqueous solutions of HCl, NH_3, or CO_2 or of ions which form hydrates such as $[Cu(H_2O)_4]^{2+}$. However, many simple, nonpolar molecules such as O_2, Cl_2, C_2H_4, and C_6H_6 are soluble in water to some extent.

The physical properties of water are unique. For a substance having such a low molecular weight, water has an extraordinarily high melting point (0°C) and a high normal boiling point (100°C). The enthalpies of fusion and vaporization and the

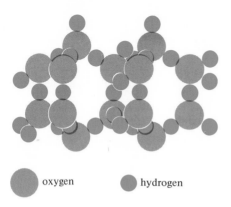

Figure 21–4. **Hydrogen-Bonded Structure of Ice**

oxygen   hydrogen

specific heat of liquid water are all comparatively high. At its melting point, the density of solid water (ice) is less than that of the liquid. All of these physical properties of water can be attributed to the phenomenon of hydrogen bonding (Section 11–8). In solid water the number of hydrogen bonds is a maximum, and in the crystal lattice each oxygen atom is surrounded tetrahedrally by four hydrogen atoms to give the rather open structure diagrammed in Figure 21–4. When the ice melts, the structure collapses, and the density of the liquid is greater than that of the solid. However, hydrogen bonding persists to a significant degree in the liquid. As the temperature rises above the melting point, the hydrogen bonding is lessened, and the density increases still further, reaching a maximum at 3.98°C. Above 3.98°C the density of water decreases with temperature, like the densities of most other liquids.

The physical properties of water are responsible for the world as we know it. On earth, life itself is a consequence of the unique properties of water. The first primitive forms of life probably began in aqueous solution. It is fortuitous that the density of ice is less than that of liquid water. When rivers, lakes, and ponds freeze over, marine life continues beneath the surface. The temperature of the water which remains unfrozen stays close to 0°C, even though the temperature at the surface may be considerably lower. The human body is made up of approximately 75% water. In the body nutrients are transferred to the cells in aqueous solution, and wastes are removed by the elimination of water. Water also functions to regulate the body temperature. Tremendous quantities of water are used daily in industry and in homes. Because of its high specific heat, water is a good heat transfer agent, and water and steam are used to heat buildings. Its large enthalpy of fusion makes ice a good refrigerant.

Water Softening

The purity of the water is of concern in such industrial uses as the generation of steam or as a cleaning agent. Much of the water used for industrial purposes is taken directly from streams or wells, and it contains dissolved impurities, mainly calcium and magnesium salts, which form precipitates when soap is added:

$$Ca^{2+} \quad + \quad 2\,C_{18}H_{35}O_2^{-} \quad \rightarrow \quad Ca(C_{18}H_{35}O_2)_2(s)$$
soap anion

Such water is said to be "hard." There are two kinds of hardness, called temporary and permanent. Temporary hard water contains calcium and/or magnesium ions and a relatively large concentration of hydrogen carbonate ion, HCO_3^- (bicarbonate ion). Temporary hardness can be removed from the water merely by boiling, since the heat shifts the following equilibrium to the right by driving off carbon dioxide:

$$Ca^{2+} + 2\,HCO_3^- \rightleftharpoons CaCO_3(s) + H_2O(l) + CO_2(g)$$

Another way to remove temporary hardness is to add base. Industrially, the cheap base, $Ca(OH)_2$ (limewater), is used:

$$Ca^{2+} + HCO_3^- + OH^- \rightarrow CaCO_3(s) + H_2O$$

In either case, the precipitated calcium carbonate is removed by filtration, leaving water which is suitable for industrial use.

Permanently hard water contains calcium and/or magnesium ions without the hydrogen carbonate ion, and the hardness cannot be removed merely by boiling. One way to soften hard water is to add large quantities of soap to precipitate the alkaline earth metal salts. Another method is to form soluble complexes, using **water softeners** such as EDTA (Table 14–2), tetrasodium pyrophosphate, $Na_4P_2O_7$, or hexasodium hexaphosphate, $Na_6P_6O_{18}$. For example, with the hexaphosphate, calcium ion forms the stable complex ion $[Ca_2P_6O_{18}]^{2-}$, which is soluble even in the presence of soap. Water softeners to which a little perfume is added are packaged and sold as "bath salts." Addition of a detergent along with the perfume makes a "bubble bath."

Ion Exchange

A more elegant way of softening water is by ion exchange. In this process water is percolated slowly through a material which exchanges one ion for another. For example, sodium ion, which gives soluble soaps, might be used to replace the calcium and magnesium ions originally present in hard water. Naturally occurring silicate minerals, called zeolites, were once used for this purpose. An approximate formula for a zeolite is NaH_6AlSiO_7 (abbreviated Na^+Zeo^-). It consists of a giant anionic network of bonded silicon, oxygen, and hydrated aluminum atoms, with sodium ions in the interstices to balance the charge. When hard water is passed through the zeolite (Figure 21–5), an equilibrium between the sodium ions in the zeolite and the calcium ions in the water is established, as represented by

$$Ca^{2+} + 2\,Na^+Zeo^- \rightleftharpoons Ca^{2+}(Zeo^-)_2 + 2\,Na^+$$

The equilibrium constant for this exchange is about 10, and a large quantity of zeolite is required to reduce the concentration of a small quantity of dipositive ion to a negligibly small value. When all of the sodium ions of the zeolite have been replaced by dipositive ions, the exchanger must be regenerated. Regeneration is conveniently done by pouring a concentrated sodium chloride solution through the exchanger tube. The exchange equilibrium is thus shifted in the direction of soluble calcium ion and sodium ion on the zeolite.

In modern ion exchangers zeolite minerals have been replaced by synthetic ion exchangers, which are resinous hydrocarbon networks containing positive or negative charges due to covalently bonded $—NH_3^+$ or $—SO_3^-$ groups, respectively. These resins can be represented by $Res—NH_3^+$ and $Res—SO_3^-$. The negatively charged

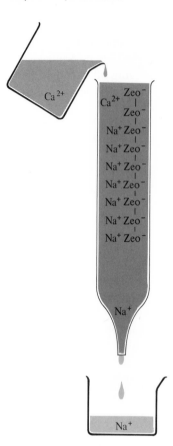

Figure 21-5. Schematic Diagram of Ion Exchange

resin functions as a cation exchanger. It is activated by treatment with an acid to form Res—SO_3H. The positively charged resin is activated by treating it with strong base to form an anion exchanger, Res—$NH_3^+OH^-$. When an aqueous solution is poured through a column of cation exchanger, the cations replace the hydrogen on the resin, and the solution passing through the column is acidic. When this solution is subsequently poured through a column containing anion exchanger, the OH^- on the resin is replaced by the anions in the solution. Since the original solution had equal concentrations of positive and negative charge, the final solution must have acquired equal concentrations of H^+ and OH^-, and pure water results. By use of a combination of the two ion exchange resins, water which is of higher purity than distilled water can be produced.

Ion exchange methods are by no means limited to water softening. This technique can be applied to remove any ionic species from solution and also can be used to separate different ions (see Section 22–17).

Desalinization of Seawater

The demand for pure water in modern society already exceeds that provided by rivers, lakes, and rainfall. Moreover, there are numerous barren, arid regions on the earth which could be made productive if a supply of fresh water were available. There is ample water available in the oceans and seas, but seawater contains too large a percentage of solutes, especially sodium chloride, for industrial use. Considerable research effort is being expended to develop feasible ways for desalinization of seawater on a large scale. It should be noted that the free energy of pure water is greater than that of water in solution. Therefore there is no possibility of finding a process in which seawater changes spontaneously into fresh water. The best compromise is to develop a process which is the least expensive relative to the economic value of the water produced. For example, distillation would not be suitable because the cost of fuel needed to boil the water would far exceed the value of the pure water produced. However, if before being condensed to water the steam from the distillation were first used to drive an electric generator, some of the cost of the fuel would be recovered from the sale of the electric power. Some processes which could be used for the desalinization of sea water are shown schematically in Figure 21-6.

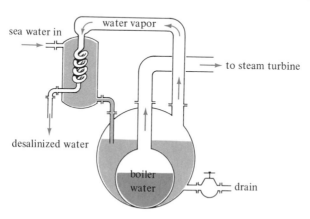

Distillation method. Sea water (in a separate compartment) is heated along with water used to generate steam for the production of electricity. Water vapor condenses in coils which are cooled by the entering sea water. The added cost of producing fresh water along with electric power is much less than the cost of distilling sea water alone.

Freezing process. Water freezes on the refrigerated surface of a rotating drum which is immersed in sea water. The drum then rotates through a melting chamber where the ice is scraped off. The melted ice is desalinized water.

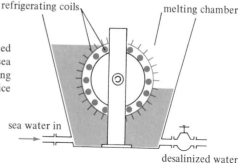

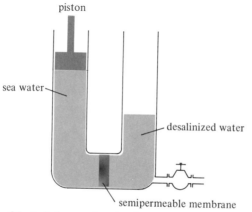

A pressure greater than the osmotic pressure is applied above sea water separated from pure water by a semipermeable membrane. Reverse osmosis occurs, producing desalinized water.

Figure 21-6. Schemes for the Desalinization of Sea Water

21-4   The Halogens

All of the elements in group VII A of the periodic table are nonmetals. Their name, *halogen*, means salt former. Some of their physical and chemical properties are listed in Table 21-1. For the most part, the variation in properties among the members of the group follow regular trends which can be understood in terms of the relative sizes and electronegativities of the elements.

In each halogen atom, X, the configuration of the valence electrons is $ns^2 np^5$. Consequently, they all form the anion X^-, and all of the free elements exist as diatomic molecules, X_2. The trends in the physical properties of the free elements are due to the increasing numbers of electrons in the respective molecules. For example, in going from F_2 to I_2, the trends in melting points and normal boiling points reflect an increase in the van der Waals forces between the molecules.

The bond energy in the fluorine molecule does not follow the trend set by the other elements. The bond is weaker than expected because of repulsion between the non-bonding electrons in this small molecule. Also, the X—X bond energies for Cl_2, Br_2, and I_2 are enhanced by hybridization involving the p and d orbitals of the respective atoms. In addition to the low F—F bond energy, the electron affinity of fluorine is also out of line with the trend.

Fluorine is the most electronegative element and also the strongest chemical oxidizing agent known. All metal fluorides are predominantly ionic, whereas the metal halides of the larger, more polarizable halogens have increasingly more covalent character. For example, although silver fluoride, chloride, and bromide all crystallize in the rock salt structure, silver iodide crystallizes in the zinc blende structure, a structure which is associated with appreciable covalent character.

TABLE 21-1. Some Properties of the Halogens

	Fluorine	Chlorine	Bromine	Iodine	Astatine
atomic number	9	17	35	53	85
electronegativity	4.0	3.0	2.8	2.5	2.2
appearance (at 25°C)	colorless gas	pale green gas	brown liquid	purple solid	
melting point (°C)	−219.6	−101.1	−7.2	113.7	(302)
normal boiling point (°C)	−188.2	−34.7	58	183	
ionization potential (kcal/mole)	402	300	273	241	
(kJ/mole)	1680	1260	1140	1010	
electron affinity (kcal/mole)	79.5	83.3	77.5	70.6	
(kJ/mole)	333	349	324	295	
covalent radius (Å)	0.72	0.99	1.14	1.33	
anionic radius (Å)	1.36	1.81	1.95	2.16	
X—X bond energy (kcal/mole	37.8	57.8	46	36	
(kJ/mole)	158	242	190	150	
$\epsilon°$ ($X_2 \rightarrow 2X^-$) (V)	2.87	1.36	1.09	0.54	
hydration energy of X^- (kcal/mole)	−122	−89	−81	−72	
(kJ/mole)	−510	−370	−340	−300	
ΔH_{vap} (kcal/mole)	0.755	2.44	3.58	5.2	8
(kJ/mole)	3.16	10.2	15.0	22	33
ΔH_{fus} (kcal/mole)	0.061	0.77	1.26	1.87	
(kJ/mole)	0.26	3.2	5.27	7.82	

Occurrence and Preparation of the Halogens

The principal source of fluorine is the mineral fluorspar, CaF_2. This substance is treated with concentrated H_2SO_4 to make HF. Elemental fluorine is prepared by electrolysis of the molten salt, KHF_2, to which an excess of HF is added. The electrolysis produces H_2 and F_2. Chlorine is produced commercially from sodium chloride by electrolysis of the molten salt or by electrolysis of concentrated solutions (see Section 5–2). In the laboratory, chlorine (and bromine and iodine) is prepared by the oxidation of the halide ion with a strong oxidizing agent.

$$MnO_2 + 2\,Cl^- + 4\,H^+ \rightarrow Cl_2 + Mn^{2+} + 2\,H_2O$$
$$2\,MnO_4^- + 10\,Cl^- + 16\,H^+ \rightarrow 5\,Cl_2 + 2\,Mn^{2+} + 8\,H_2O$$

The commercial source of bromine is seawater. The water is first acidified and then saturated with Cl_2 to effect the reaction

$$Cl_2 + 2\,Br^- \rightarrow Br_2 + 2\,Cl^-$$

After the treatment with chlorine, air is bubbled through the solution, sweeping out the bromine, which is then absorbed in aqueous sodium carbonate solution to give a mixture of bromate ion and bromide ion. When this solution is subsequently acidified, bromine is once again produced in more concentrated form:

$$6\,H^+ + BrO_3^- + 5\,Br^- \rightarrow 3\,Br_2 + 3\,H_2O$$

The free element is recovered by distillation.

The chief commercial source of iodine is Chile saltpeter—mainly $NaNO_3$ in which small quantities of $NaIO_3$ are found. Solutions containing the iodate ion are reduced with hydrogen sulfite ion:

$$2\,IO_3^- + 5\,HSO_3^- \rightarrow 3\,HSO_4^- + 2\,SO_4^{2-} + I_2 + H_2O$$

An excess of iodate ion must be used; otherwise, the HSO_3^- will further reduce the iodine to iodide ion.

The fifth halogen, astatine, has no stable isotopes. An isotope $^{211}_{85}At$, having a half life of 7.5 hours, is prepared by bombarding $^{209}_{83}Bi$ with alpha particles. Its chemistry is similar to that of iodine, and it is isolated from the target material using iodine as a carrier element. It forms At^- and AtO_3^- ions.

Hydrogen Halides

The hydrogen halides are volatile, covalent compounds which can be prepared by the direct reaction of the elements or, more conveniently, by the reaction of the respective halide ion with hydrogen ion. HF and HCl are prepared by reaction of the corresponding halide with a strong, nonvolatile acid, such as H_2SO_4. Some properties of the anhydrous compounds are listed in Table 21–2. All of these compounds are Brønsted acids, readily donating protons to Brønsted bases. Thus, with the exception of HF, all are strong acids in aqueous solution. In liquid HF the molecules are

TABLE 21-2. Properties of Anhydrous Hydrogen Halides

	HF	HCl	HBr	HI
molecular weight	20.01	36.46	80.92	127.91
appearance	colorless fuming liquid or gas	colorless gas or liquid	colorless gas or pale yellow liquid	colorless gas or pale yellow liquid
gas density (grams/liter)	$0.991_{(19.5°)}$	$1.00045_{(25°)}$	$3.5_{(0°)}$	$5.66_{(0°)}$
liquid density (grams/ml)	$0.987_{(13.6°)}$	$1.187_{(-84.9°)}$	$2.77_{(-67°)}$	$2.85_{(-4.7°)}$
melting point (°C)	-83.1	-114.8	-88.5	-50.8
normal boiling point (°C)	19.54	-84.9	-67.0	-35.5

strongly hydrogen bonded, and HF has a relatively high boiling point. Vapor density determinations indicate that strong hydrogen bonding persists in gaseous HF.

Hydrogen fluoride undergoes a unique reaction with substances containing silicon-to-oxygen bonds, such as $CaSiO_3$ and SiO_2:

$$SiO_2(s) \;+\; 4\,HF(g) \;\rightleftharpoons\; SiF_4(g) \;+\; 2\,H_2O(g)$$

The high stability of the Si—F bonds and the volatility of SiF_4 are the driving forces of this reaction. HF attacks the silicates in glass and must be stored in other types of containers.

Many nonvolatile, strong acids in concentrated solution, H_2SO_4, $HClO_4$, etc., are also strong oxidizing agents, capable of oxidizing bromide and iodide ions. Therefore HBr and HI are usually prepared from covalent halides:

$$PBr_3(l) \;+\; 3\,H_2O(l) \;\rightarrow\; H_3PO_3(l) \;+\; 3\,HBr(g)$$
$$PI_3(s) \;+\; 3\,H_2O(l) \;\rightarrow\; H_3PO_3(l) \;+\; 3\,HI(g)$$

Binary Halogen-Oxygen Compounds

Some binary compounds of oxygen and the halogens are listed in Table 21-3. All of these compounds are somewhat unstable and may decompose violently when heated

TABLE 21-3. Some Binary Halogen-Oxygen Compounds

Formula	Description	Normal Boiling Point (°C)	Melting Point (°C)
F_2O	colorless gas	-145	
F_2O_2	orange-red solid, decomposes at $-95°C$	-95	
Cl_2O	red liquid, orange gas	2	
ClO_2	red-brown liquid, yellow vapor	11	
Cl_2O_6	red oily liquid	~203	4
Cl_2O_7	colorless oil	82	-92
Br_2O	unstable brown liquid		decomp. > -40
BrO_2	yellow solid		unstable > -80
BrO_3	white powder	unstable	> -80
I_2O_5	white solid		300

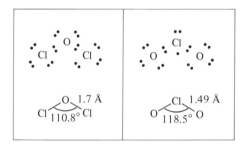

Figure 21-7. **Structures of Cl_2O and ClO_2**

or when they are subjected to mechanical shock. They are vigorous oxidizing agents, and in handling them, one must avoid putting them in contact with grease or any other easily oxidizable material because the heat liberated in the oxidation reaction may trigger an explosion. Several halogen oxides not listed in the table have been observed; however, these substances are so unstable that little is known of their reactions and structures.

Oxygen difluoride is prepared by the reaction of fluorine with dilute sodium hydroxide:

$$2\,F_2 \;+\; 2\,OH^- \;\rightarrow\; 2\,F^- \;+\; OF_2 \;+\; H_2O$$

Dioxygen difluoride, O_2F_2, is made by passing an electrical discharge through a mixture of the elements at liquid air temperature. Chlorine monoxide, Cl_2O, is prepared by passing chlorine and dry air over mercury(II) oxide:

$$2\,Cl_2 \;+\; HgO \;\rightarrow\; Cl_2O \;+\; HgCl_2$$

Chlorine monoxide reacts with water and base to form the hypochlorite ion, ClO^-. It reacts with N_2O_5 to give chlorine nitrate, $ClNO_3$. Chlorine dioxide, ClO_2, is prepared by the reduction of chlorate ion, ClO_3^-, in acid solution:

$$2\,H^+ \;+\; 2\,ClO_3^- \;+\; H_2C_2O_4 \;\rightarrow\; 2\,ClO_2 \;+\; 2\,CO_2 \;+\; 2\,H_2O$$

ClO_2 is paramagnetic. The structures of Cl_2O and ClO_2 are diagrammed in Figure 21-7.

Chlorine hexoxide, Cl_2O_6, is formed when a mixture of ClO_2 and O_2 is irradiated with ultraviolet light or when ClO_2 is treated with ozone:

$$2\,ClO_2 \;+\; O_2 \;\xrightarrow{\;\text{uv light}\;}\; Cl_2O_6$$

$$2\,ClO_2 \;+\; 2\,O_3 \;\longrightarrow\; Cl_2O_6 \;+\; 2\,O_2$$

Molecular weight determinations in carbon tetrachloride solution indicate that the formula is Cl_2O_6, but the paramagnetic behavior of the substance indicates at least partial dissociation to ClO_3:

$$\underset{\text{diamagnetic}}{Cl_2O_6} \;\rightleftharpoons\; \underset{\text{paramagnetic}}{2\,ClO_3}$$

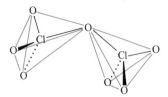

Figure 21-8. **Structure of Cl$_2$O$_7$**

Chlorine heptoxide, Cl$_2$O$_7$, is prepared by the dehydration of perchloric acid with P$_4$O$_{10}$, followed by distillation under reduced pressure:

$$4\,HClO_4 \;+\; P_4O_{10} \;\rightarrow\; 2\,Cl_2O_7 \;+\; 4\,HPO_3$$

The infrared spectrum of Cl$_2$O$_7$ vapor suggests that its structure consists of two chlorine atoms sharing an oxygen atom, with three other oxygen atoms tetrahedrally oriented about each (Figure 21-8). Of all the chlorine oxides, the heptoxide is the most stable, but even this compound detonates when heated or subjected to mechanical shock.

Bromine monoxide, Br$_2$O, is prepared by treating mercury(II) oxide with bromine vapor or by mixing these two reagents in carbon tetrachloride:

$$2\,Br_2 \;+\; HgO \;\rightarrow\; Br_2O \;+\; HgBr_2$$

Bromine dioxide is prepared by passing precooled ozone into a solution of bromine in CF$_2$Cl$_2$ at $-50\,^\circ$C:

$$Br_2 \;+\; 4\,O_3 \;\rightarrow\; 2\,BrO_2 \;+\; 4\,O_2$$

Above $-40\,^\circ$C, BrO$_2$ is unstable and decomposes into Br$_2$O and oxygen.

Bromine trioxide, BrO$_3$, is obtained when bromine vapor reacts with ozone at $0\,^\circ$C or lower. It is unstable above $-70\,^\circ$C unless ozone is present. It is probably polymeric.

Iodine pentoxide, I$_2$O$_5$, is prepared by heating iodic acid to $200\,^\circ$C:

$$2\,HIO_3 \;\xrightarrow[\text{heat}]{}\; I_2O_5 \;+\; H_2O$$

I$_2$O$_5$ is quite stable and can be heated to $300\,^\circ$C without decomposition. However, it is a vigorous oxidizing agent. For example, the reaction

$$I_2O_5 \;+\; 5\,CO \;\rightarrow\; I_2 \;+\; 5\,CO_2$$

proceeds quantitatively at $70\,^\circ$C.

Halogen Oxyacids and Their Salts

Except for fluorine, all of the halogens form oxyacids in which the respective halogen atoms assume positive oxidation states, such as $+1$, $+3$, $+5$, and $+7$. All of these acids as well as their salts are strong oxidizing agents. As the oxidation state of each halogen increases, the corresponding acid (1) increases in thermal stability, (2) increases in acid strength, (3) decreases in rate of oxidation, and (4) generally is slightly lower in reduction potential. Their salts show similar trends in stability and oxidizing properties. The slowness in oxidizing ability of the higher oxidation state compounds suggests that the mechanism of oxidation involves the molecule of acid as a whole. The stronger the acid, the more it exists in ionic form, both in solutions of dilute acid and in solutions as a salt. The weaker acids are already present in molecular form, and so they react faster. Chlorine oxyacids are stronger acids than the corresponding bromine oxyacids, which are stronger than the corresponding iodine oxyacids.

The oxyacids in which the halogens have an oxidation state of $+1$ are known as the hypohalous acids, and the salts are hypohalites. Dilute solutions of HOX are formed when the halogen reacts with water:

$$X_2 + H_2O \rightleftharpoons HOX + X^- + H^+$$

The reactions proceed only to equilibrium, the constants being 4.2×10^{-4}, 7.2×10^{-9}, and 2×10^{-13} for Cl_2, Br_2, and I_2, respectively. The reactions of the halogens with base yield the corresponding salts of the hypohalous acids:

$$X_2 + 2 OH^- \rightarrow X^- + XO^- + H_2O$$

Solutions of sodium hypochlorite are prepared by electrolyzing a solution of sodium chloride while stirring. The chlorine formed at the anode is thus mixed with the hydroxide ion formed at the cathode, and hypochlorite is produced. This solution is sold as a laundry bleach.

The hypohalites disproportionate further in basic solution:

$$3 XO^- \rightarrow 2 X^- + XO_3^-$$

For chlorine, the rate of disproportionation is small at low temperatures, but above $75°C$ the rate increases. BrO^- disproportionates readily even at room temperature, and hypobromite solutions can be prepared and stored only at about $0°C$ or lower. IO^- reacts very rapidly at all temperatures.

Oxyacids in which the oxidation state of the halogen is $+3$ are called halous acids, and the corresponding salts are halites. Chlorous acid is the only known acid of this type. Chlorites are best prepared by the reaction of ClO_2 with peroxides:

$$Na_2O_2 + 2 ClO_2 \rightarrow 2 NaClO_2 + O_2$$

Chlorite ion disproportionates upon heating:

$$3 ClO_2^- \rightarrow 2 ClO_3^- + Cl^-$$

Oxyacids in which the oxidation state is $+5$ are called halic acids; the corresponding salts are halates. Chloric and bromic acids are formed only in aqueous solution, but iodic acid is sufficiently stable to be isolated in the pure state. The halates are prepared by treating the halogens with hot base:

$$3 X_2 + 6 OH^- \rightarrow 5 X^- + XO_3^- + 3 H_2O$$

The reaction of iodine with base is quantitative even at room temperature. Upon heating, chlorates disproportionate to perchlorates and chloride:

$$4\,ClO_3^- \rightarrow 3\,ClO_4^- + Cl^-$$

Halogens have an oxidation number of $+7$ in the perhalic acids and perhalates. Perchloric acid is prepared by mixing potassium perchlorate with concentrated sulfuric acid and distilling the mixture under reduced pressure:

$$KClO_4 + H_2SO_4 \rightleftharpoons HClO_4 + KHSO_4$$

Perchloric acid is a stronger acid than H_2SO_4, and the equilibrium might be expected to lie to the left, but removal of $HClO_4$ by distillation shifts the equilibrium in the desired direction. Perchloric acid reacts explosively with combustible materials and must be handled with extreme caution. Potassium perchlorate, prepared by heating aqueous potassium chlorate, is only slightly soluble in cold water and can be separated upon cooling the solution. In Figure 21–9 the oxidation states of chlorine in acidic and basic media are summarized, along with the relevant standard potential data, in a so-called Latimer diagram.

Perbromates cannot be prepared by disproportionation of bromates. However, BrO_4^- is obtained by oxidation of BrO_3^- with fluorine gas or with XeF_2:

$$BrO_3^- + F_2 + 2\,OH^- \rightarrow BrO_4^- + 2\,F^- + H_2O$$
$$BrO_3^- + XeF_2 + H_2O \rightarrow BrO_4^- + Xe + 2\,HF$$

Perbromic acid is obtained by adding RbF to the reaction mixture to form the slightly soluble $RbBrO_4$. After filtering the salt and dissolving it in a large quantity of water, $HBrO_4$ is obtained by ion exchange.

There are several periodic acids. Paraperiodic acid, H_5IO_6, is made by the action of hypochlorite ion on an iodate:

$$H_2O + ClO^- + IO_3^- + H_3O^+ \rightarrow H_5IO_6 + Cl^-$$

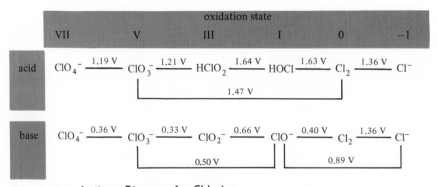

Figure 21-9. Latimer Diagram for Chlorine

Standard reduction potentials are indicated on the lines joining the species in the two oxidation states.

Upon heating or when kept under vacuum with a desiccant, H_5IO_6 loses water to form dimesoperiodic acid, $H_4I_2O_9$, and finally metaperiodic acid, HIO_4. Periodic acids and their salts are strong oxidizing agents, capable of converting manganese(II) ion to permanganate.

Interhalogen Compounds

In view of the similar electronic configurations of the halogen atoms, it is not surprising that atoms of two different halogens combine in 1:1 ratios to form interhalogen compounds, such as ClF. However, it is somewhat surprising that interhalogen compounds with ratios of 1:3, 1:5, and 1:7 are known, where in each case the less electronegative atom is outnumbered by the other. Some interhalogen compounds are described in Table 21-4. It is possible to account for the geometries of these compounds by means of the theory of electron pair repulsion (Section 11-4).

Uses of the Halogens

The halogens, particularly chlorine and fluorine, are commercially very important. Because it reacts with silicates, hydrogen fluoride is used to etch glass objects, such as thermometers and laboratory glassware. Compounds called Freons (CCl_2F_2, $CClF_3$, etc.) are used as refrigerating liquids in refrigerators, air conditioners, and freezers and also as propellant gases in pressurized spray cans for hair sprays, insecticides, and paints. Teflon, a fluorocarbon polymer, is a very tough, high-melting, and chemically inert plastic material. It is used to make "greaseless" cookware, bearings and valves which require no lubrication, and containers for corrosive chemicals.

Chlorine is used as a germicide and disinfectant in the purification of water and as a raw material in the manufacture of other chemicals. Hypochlorites are bleaching agents, and they are used in the manufacture of paper. "Bleaching powder," a mixture of calcium hypochlorite and calcium chloride, is used as a germicide and deodorizer. Chlorates and perchlorates are used in the manufacture of explosives and rocket propellants.

TABLE 21-4. Interhalogen Compounds

Formula	Characteristics	Geometry
ClF	stable gas	linear
ClF_3	stable gas	T-shaped
ClF_5	solid stable below $-195\,°C$	square pyramid
BrF	gas, decomposes above $50\,°C$	linear
BrCl	unstable gas	linear
BrF_3	stable liquid	T-shaped
BrF_5	stable liquid	square pyramid
ICl	solid, decomposes about $97\,°C$	linear
IBr	solid, decomposes about $42\,°C$	linear
IF_3	solid, decomposes above $-35\,°C$	possibly T-shaped
ICl_3	solid, dissociates above $77\,°C$	possibly T-shaped
IF_5	stable liquid	square pyramid
IF_7	stable gas	pentagonal bipyramid
AtI		linear

Bromine, iodine, and their compounds find many uses in photography, as medicinals, and as intermediates in the manufacture of other chemicals. The interhalogen compounds are extremely reactive. ClF_3 and BrF_3 are fluorinating agents which can convert many metals and/or metal oxides to fluorides.

21-5 Group VI A Nonmetals

Of the elements in group VI A, only polonium is a metal. It occurs in nature only as a product of radium disintegration, and its most abundant isotope, ^{210}Po, has a half life of 138.7 days. Tracer studies indicate that it has typical metallic properties.

Some properties of the nonmetals of the group—oxygen, sulfur, selenium, and tellurium—are presented in Table 21-5. The expected trends from nonmetallic character (oxygen and sulfur) toward semimetallic character (selenium and tellurium) are observed in many of the properties of these elements. The chemistry of oxygen has already been discussed in some detail. As is usually the case with second period elements, the properties of oxygen differ significantly from those of the heavier elements of the group. The differences stem from the lower electronegativities of the elements sulfur to tellurium and from the availability in those elements of d orbitals, which allows the formation of a greater number of covalent bonds in some of their compounds.

All of these nonmetals exist in more than one allotropic form. In the case of the heavier elements the allotropism is a variation in crystal structure of the solid element; in the case of oxygen a different molecular form is possible. The phase diagram for sulfur is shown in Figure 21-10. The rhombic modification is stable at

TABLE 21-5. Properties of the Nonmetals of Group VI A

	Oxygen	Sulfur	Selenium	Tellurium
atomic number	8	16	34	52
allotropic forms	O_2, O_3	rhombic, monoclinic	gray, red	gray, black
melting point[a] (°C)	−218.4	122.8	217	452
normal boiling point (°C)	−183.0	444.6	685	1390
electronegativity	3.5	2.5	2.4	2.1
atomic radius (Å)	0.66	1.27	1.40	1.60
anionic radius (Å)	1.32	1.84	1.91	2.11
oxidation states	$-2, -1, -\frac{1}{2},$ $0, +2$	$-2, 0, 4, 6$	$-2, 0, 4, 6$	$-2, 0, 4, 6$
HXH bond angle	105°	93.3°	91°	89.5°
bond length in HXH (Å)	0.958	1.3455	1.47	
ΔH_{vap} (kcal/mole)	0.815	3.01	3.34	11.9
(kJ/mole)	3.41	12.6	14.0	49.8
ΔH_{fus} (kcal/mole)	0.053	0.34	1.25	4.28
(kJ/mole)	0.22	1.4	5.23	17.9
ionization potential				
(kcal/mole)	314	239	225	208
(kJ/mole)	1310	1000	941	870

[a] For allotropic form stable at room temperature.

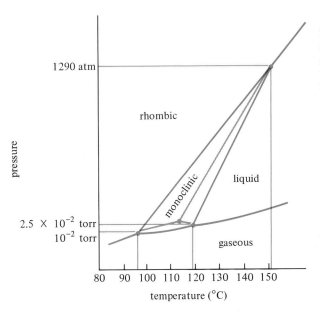

Figure 21-10. **Phase Diagram for Sulfur**
Gray lines within the colored "triangle" indicate the phase changes when rhombic sulfur is heated rapidly. (Not drawn to scale.)

room temperature but undergoes a transition to monoclinic sulfur at 95.5°C. The transition between the solid modifications is slow, but it is possible to obtain the monoclinic form by prolonged heating at 100°C or so. However, if rhombic sulfur is heated rapidly, it melts at 113°C before it gets a chance to change to monoclinic. The equilibrium triple point of monoclinic sulfur is 119°C, and upon cooling of liquid sulfur to that temperature, the monoclinic modification is obtained. Monoclinic sulfur, once formed, can be cooled to room temperature, but upon standing it gradually converts to the rhombic form. Sulfur molecules in both the rhombic and monoclinic forms exist as puckered rings containing eight sulfur atoms (Figure 21-11).

When sulfur is heated to about 200°C and then cooled suddenly by pouring into cold water, a rubber-like product called **plastic** or **amorphous** sulfur is obtained. Plastic sulfur can be drawn into fibers, which X-ray studies reveal consist of long chains of sulfur atoms.

Selenium exists in three forms. The stable gray form has a hexagonal structure consisting of chains of selenium atoms. There are also two red modifications of selenium, both monoclinic, which contain Se_8 molecules in the form of rings.

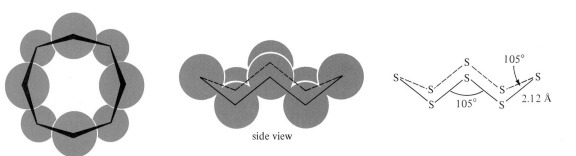

Figure 21-11. **Structure of the S_8 Molecule**

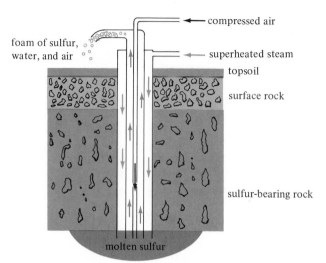

Figure 21–12. Frasch Process
Three concentric pipes are inserted into a hole drilled down to the sulfur deposit. Superheated steam is forced down the outer pipe into the sulfur, causing it to melt. Compressed air is pumped through the inner pipe into the molten liquid, thus forcing a frothy mixture of sulfur and water up the other pipe to the surface.

The two modifications of tellurium are a metallic-appearing hexagonal form containing chains of tellurium atoms and a black, amorphous form.

Sulfur is found in the elemental state in large deposits of volcanic origin in Sicily and Japan, as well as in deep underground beds where it probably resulted from photosynthesis in prehistoric plants. The Frasch process for obtaining the element is diagrammed in Figure 21–12. Selenium and tellurium are much less abundant but frequently occur as impurities in metal sulfide ores. The principal commercial source of these elements is in the anode sludge from the electrolytic refining of copper metal (see Figure 5–17).

Chemical Properties of Sulfur, Selenium, and Tellurium

The reactions of sulfur, selenium, and tellurium and the structures of corresponding compounds of these elements are quite similar. All of the elements form compounds in which they have oxidation numbers of −2, +4, and +6. Each of the elements reacts directly with many metals to form the corresponding compounds:

$$Zn + S \rightarrow ZnS$$
$$2\,Al + 3\,Se \rightarrow Al_2Se_3$$
$$Hg + Te \rightarrow HgTe$$

Sulfides of only the metals having low electronegativity (groups I A and II A) are ionic. Alkali metal sulfides crystallize in the antifluorite structure (cf. Section 18–6), whereas the alkaline earth metal sulfides crystallize in the rock salt structure (Figure 18–2). The so-called ionic sulfides are soluble in water, but the sulfide ion hydrolyzes extensively:

$$S^{2-} + H_2O \rightarrow HS^- + OH^-$$

Most other metal sulfides have considerable covalent character.

Hydrogen sulfide, hydrogen selenide, and hydrogen telluride are colorless, unpleasant-smelling, and highly poisonous gases. (The lethal concentration of H_2S is comparable to that of CO or HCN.) These compounds are produced by the reaction of the respective anion with dilute acid:

$$S^{2-} + 2 H_3O^+ \rightarrow H_2S + 2 H_2O$$

In aqueous solution H_2S, H_2Se, and H_2Te are weak acids. The acid strength increases from H_2S to H_2Te, and the thermal stabilities decrease in that order. H_2Te decomposes slowly into its elements even at $0°C$.

Hydrogen sulfide is used as a reagent in qualitative analysis. For this purpose, it is conveniently prepared by the reaction of thioacetamide with water:

$$\underset{\underset{S}{\|}}{CH_3CNH_2}(aq) + H_2O \rightarrow \underset{\underset{O}{\|}}{CH_3CNH_2}(aq) + H_2S(g)$$

     thioacetamide   acetamide

At room temperature the reaction is slow, but above $60°C$ it proceeds rapidly.

Positive Oxidation States

When sulfur, selenium, or tellurium is burned in air, the corresponding dioxide is formed. Sulfur dioxide is a colorless gas with a very sharp, pungent odor. It liquefies at $-10°C$ under a pressure of 1 atm. It is moderately soluble in water, producing solutions of sulfurous acid, H_2SO_3. This acid is not stable and cannot be isolated in the pure state, but its salts such as sodium sulfite, Na_2SO_3, and sodium hydrogen sulfite, $NaHSO_3$, can be prepared. Because it is easily liquefied, SO_2 is used as a refrigerant. Large quantities of SO_2 are used as a bleaching agent for wool and silk and in paper manufacture. The salts of sulfurous acid are also reducing agents and are used as food preservatives and as bleaching agents.

When an aqueous solution of sulfite is boiled with sulfur, the thiosulfate ion, $S_2O_3^{2-}$, is obtained:

$$8 SO_3^{2-} + S_8 \rightarrow 8 S_2O_3^{2-}$$

In photography, sodium thiosulfate, called *hypo,* is used as a developer. In analytical chemistry laboratories, $S_2O_3^{2-}$ is used as a quantitative reducing agent for the titration of iodine. The tetrathionate ion produced in this reaction has a four-sulfur chain.

$$2 S_2O_3^{2-} \rightarrow \qquad S_4O_6^{2-} \qquad + 2 e^-$$

$$\begin{bmatrix} & \overset{O}{\underset{\|}{}} & & \overset{O}{\underset{\|}{}} & \\ O-\overset{\|}{\underset{\|}{S}}-S-S-\overset{\|}{\underset{\|}{S}}-O \\ & \underset{O}{} & & \underset{O}{} & \end{bmatrix}^{2-}$$

SeO_2 is a white, volatile solid which sublimes at $315°C$. Gaseous SO_2 and SeO_2 are nonlinear molecules in which there is delocalized π bonding. TeO_2 is a nonvolatile, white solid.

Sulfur trioxide is obtained by the reaction of SO_2 with oxygen. The reaction is very slow unless such catalysts as platinum sponge, V_2O_5, or NO are present. (See manufacture of H_2SO_4, Section 17–3.) The gaseous molecule has a planar, trigonal structure involving some delocalized sulfur-to-oxygen π bonding. Sulfur trioxide is the anhydride of sulfuric acid and reacts with water to produce the acid or with metal oxides to produce the corresponding sulfates. SO_3 is both a strong Lewis acid and a powerful oxidizing agent.

SeO_3 is a colorless solid which is produced by passing an electric discharge through a mixture of selenium vapor and oxygen. TeO_3 is an orange solid made by heating telluric acid. Selenic acid is similar to sulfuric acid in that it is a strong acid which forms two series of salts and is a strong oxidizing agent. By contrast, telluric acid, H_6TeO_6, is a very weak acid. It is prepared by the reaction of aqua regia and a chlorate on tellurium. It has an octahedral structure.

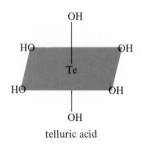

telluric acid

Selenic acid ($\epsilon° = 1.14$ V) is the strongest oxidizing agent among these compounds. H_2SeO_4 is capable of oxidizing chloride ion to chlorine:

$$2\,H^+ \;+\; 2\,Cl^- \;+\; H_2SeO_4 \;\rightarrow\; Cl_2 \;+\; H_2SeO_3 \;+\; H_2O$$

Some Uses of Selenium

Selenium is used for decolorizing glass and in photoelectric cells. As would be expected for a nonmetal, the electrical conductivity of selenium is small (specific conductivity $= 5 \times 10^{-12}$ ohm$^{-1} \cdot$ cm^{-1}), but when a sample is irradiated by light, the conductivity is increased as much as 200 times. The change in conductivity is proportional to the intensity of the light, and selenium is used to make light-measuring devices such as spectrophotometers (Figure 13–4) and automatic exposure (electric eye) cameras.

Another use for selenium is in solar batteries. A cell consists of a sheet of iron covered by a sheet of selenium, which in turn is covered by a virtually transparent layer of another metal. When light strikes the transparent or **collector** electrode, a current flows from it through the external circuit. A number of such cells connected in series and in parallel can be used as a solar battery, converting the energy of sunlight directly into electrical energy.

The iron–selenium junction is also used as a rectifier (device for changing alternating current to direct current) in electronic circuits. When an external source of alternating current is applied across such a junction, current flows more readily from iron to selenium than in the opposite direction.

TABLE 21-6. Properties of the Group VA Nonmetals

	Nitrogen	Phosphorus	Arsenic
atomic number	7	15	33
normal boiling point (°C)	− 195.8	280 (white)	613
melting point (°C)	−210	44.1	817
electronegativity	3.0	2.1	2.0
ΔH_{vap} (kcal/mole)	0.666	2.97	7.75 (subl)
(kJ/mole)	2.79	12.4	32.4 (subl)
ΔH_{fus} (kcal/mole)	0.086	0.15	6.62
(kJ/mole)	0.36	0.63	27.7
ionization potential (kcal/mole)	336	254	231
(kJ/mole)	1410	1060	967
covalent radius (Å)	0.75	1.06	1.19
atomic radius (Å)	0.92	1.28	1.39
anionic radius (Å)	1.71	2.12	2.22
crystal structure	hexagonal	cubic	rhombohedral

21-6   Group VA Nonmetals

Three elements in group VA, nitrogen, phosphorus, and arsenic, are classified as nonmetals. In spite of the fact that many of their compounds have analogous formulas, these elements differ markedly from each other in physical and chemical properties (Table 21-6). These differences result in part from the existence in phosphorus and arsenic of d orbitals which are available for bonding. Nitrogen, on the other hand, has the ability to form π bonds involving atomic p orbitals. Thus nitrogen atoms form multiple bonds with atoms of other second period elements (carbon and oxygen) as well as with other nitrogen atoms.

At ordinary temperatures nitrogen is a diatomic gas, whereas phosphorus and arsenic are solids having more than one allotropic form. The allotropic form of phosphorus known as white phosphorus is a waxy, white material consisting of P_4 molecules (Figure 21-13) in a cubic lattice. This modification has a relatively low density and a low melting point and is soluble in such solvents as carbon disulfide. White phosphorus is extremely reactive, igniting spontaneously in air, and it must be stored under water. Violet (red) phosphorus, a second allotropic form, is less reactive, is less soluble, and sublimes without melting. The solid consists of chains of P_4 tetrahedra (Figure 21-13). A third solid modification, black phosphorus, is formed by heating phosphorus to 200°C under a pressure of 4000 atm. It is similar to graphite in appearance, and like graphite, its crystals are composed of atoms arranged in layers. Also like graphite, this form conducts electricity. It is unreactive at room temperature.

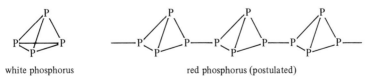

white phosphorus   red phosphorus (postulated)

Figure 21-13. **Structures of White and Red Phosphorus**

Arsenic exists in a stable gray "metallic" form and in a yellow form. The gray modification consists of covalently bonded atoms, and the yellow form consists of crystals of tetrahedral As_4 molecules. Yellow arsenic is even more reactive than white phosphorus. It is quite volatile and is extremely poisonous.

Nitrogen is found in the free state in the atmosphere and also occurs in minerals in the form of nitrates. The large natural deposits of nitrates, such as Chile saltpeter, may have resulted from the decomposition of proteins of prehistoric plants or animals. The chief commercial source of nitrogen is the atmosphere, from which it is obtained by the fractional distillation of liquid air.

Phosphorus occurs mainly in minerals such as calcium phosphate, $Ca_3(PO_4)_2$, and apatite, $Ca_5(F)(PO_4)_3$. The free element is prepared by heating one of these minerals with coke, using sand, SiO_2, as a flux:

$$2\,Ca_3(PO_4)_2 \;+\; 6\,SiO_2 \;+\; 10\,C \;\rightarrow\; P_4 \;+\; 10\,CO \;+\; 6\,CaSiO_3$$

The volatile phosphorus distills from the reaction mixture.

Arsenic is sometimes found in the free state in nature. It is more often found as a sulfide—realgar, As_2S_2; orpiment, As_2S_3; or arsenic pyrites, FeAsS. The element is prepared by converting the sulfide to an oxide, followed by reduction with carbon:

$$2\,As_2S_2 \;+\; 7\,O_2 \;\rightarrow\; As_4O_6 \;+\; 4\,SO_2$$
$$As_4O_6 \;+\; 6\,C \;\rightarrow\; As_4 \;+\; 6\,CO$$

Nitrogen, phosphorus, and arsenic exhibit oxidation states ranging from -3 to $+5$. Compounds in which these elements have negative oxidation states are good reducing agents, with the order of increasing reducing ability $N < P \ll As$. Conversely, compounds in which the elements have a positive oxidation state are oxidizing agents, with the order of increasing oxidizing strength $P \ll As < N$.

Nitrides and Azides

As nonmetals, nitrogen, arsenic, and phosphorus are expected to form compounds in which they have negative oxidation states. Lithium, beryllium, magnesium, and other group IIA elements form ionic nitrides when heated in nitrogen. Sodium, potassium, and rubidium nitrides are prepared by passing an arc between a platinum cathode and an alkali metal anode under liquid nitrogen. The latter compounds are unstable and readily explode at ordinary temperatures.

Covalent nitrides also exist. These include compounds with hydrogen (to be discussed in the next subsection) as well as compounds of the halogens and of carbon. Also included in this class is the network compound boron nitride, BN, which is obtained by the thermal decomposition of many boron–nitrogen compounds. Boron nitride is isoelectronic with graphite. It is chemically inert and closely resembles graphite in its structure (Figure 21–14) and properties. When subjected to very high temperatures and pressures, it is converted into an extremely hard, diamond-like substance.

Interstitial compounds of nitrogen and transition metals are made by heating the powdered metal in nitrogen or ammonia to about 1200°C. The stoichiometries of these compounds, corresponding to Mo_2N, TiN, Fe_4N, and Mn_5N, indicate that they

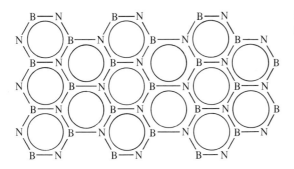

Figure 21-14. **Structure of Boron Nitride**

are more like alloys than ionic compounds. Most of these interstitial nitrides are very hard and high-melting.

When sodium amide, $NaNH_2$, is heated under nitrogen(I) oxide at 190°C, sodium azide is formed:

$$2\,NaNH_2 \;+\; N_2O \;\rightarrow\; NaN_3 \;+\; NaOH \;+\; NH_3$$

The azide ion, N_3^-, is linear; its structure is best described as consisting of a central nitrogen atom which is σ and π bonded to two other nitrogen atoms, as shown in Figure 21-15. Ionic azides are rather stable, decomposing to yield N_2 at temperatures above 200°C. In contrast, covalent azides such as HN_3 and CH_3N_3 are violently explosive. AgN_3, $Pb(N_3)_2$, and $Hg_2(N_3)_2$ detonate when struck.

Hydrogen Compounds of Group VA

The binary hydrogen compounds of nitrogen, phosphorus, and arsenic are described in Table 21-7. Of these, ammonia is the most important, the most polar, and the strongest base. The molecules of ammonia are associated through hydrogen bonding, as revealed by the comparatively high normal boiling point and freezing point of ammonia. Ammonia can be prepared by the action of a strong base on an ammonium salt:

$$NH_4^+ \;+\; OH^- \;\rightarrow\; NH_3 \;+\; H_2O$$

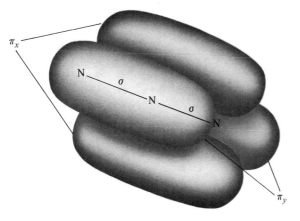

Figure 21-15. **Orbital Structure of the N_3^- Ion**

TABLE 21-7. Binary Hydrogen Compounds of Nitrogen, Phosphorus, and Arsenic

Formula	Name	Melting Point (°C)	Normal Boiling Point (°C)	Oxidation State
NH_3	ammonia	−77.7	−33.3	−3
PH_3	phosphine	−133	−87.7	−3
AsH_3	arsine	−116.3	−55	−3
N_2H_4	hydrazine	1.4	113.5	−2
P_2H_4	diphosphine	<−10	58	−2
HN_3	hydrazoic acid	−80	37	$-\frac{1}{3}$

There is a large demand for ammonia for the production of fertilizers and explosives. In the industrial process for the production of ammonia, known as the Haber process, the principles of kinetics and equilibrium are utilized to maximize the yield of product. As can be seen from the balanced chemical equation,

$$N_2 \ + \ 3\,H_2 \ \underset{\substack{200\ atm, \\ 500°C}}{\rightleftharpoons} \ 2\,NH_3 \ + \ heat$$

increased total pressure favors production of ammonia by shifting the equilibrium to the right. Therefore, the reaction is carried out under a total pressure of 200 atm. On the other hand, high temperatures would favor decomposition of the ammonia to nitrogen and hydrogen. However, the lower the temperature, the lower the rate of production of ammonia from the elements. The operating temperature of 500°C is a compromise, high enough to establish a reasonable rate of reaction and low enough (with the high pressure used) to insure that a reasonable fraction of the reactants is converted to products at equilibrium.

Phosphine, PH_3, and arsine, AsH_3, are prepared by the treatment of a metallic phosphide or arsenide with water:

$$P^{3-} \ + \ 3\,H_2O \ \rightarrow \ PH_3 \ + \ 3\,OH^-$$

Unlike ammonia, phosphine and arsine are only slightly soluble in water, and their solutions are not alkaline. Under certain conditions phosphine does form salts analogous to ammonium salts:

$$PH_3(g) \ + \ HI(g) \ \rightarrow \ PH_4I(s)$$

PH_4Br can also be made in this manner, but PH_4Cl is unstable except at low temperatures under high pressure.

Because of its polar nature, liquid ammonia is a good solvent for ionic and polar substances. Like water, it undergoes self-ionization:

$$NH_3 \ + \ NH_3 \ \rightleftharpoons \ NH_4^+ \ + \ NH_2^- \qquad K_{NH_3} = 10^{-31} \ \text{(at −33°C)}$$

Because ammonia is more basic than water, some Brønsted acids which are weak in water ionize completely in liquid ammonia:

$$HC_2H_3O_2 + NH_3 \rightarrow NH_4^+ + C_2H_3O_2^-$$

Liquid ammonia dissolves alkali metals and alkaline earth metals to give deep blue solutions from which the metal can be recovered unchanged. These solutions contain positive metal ions and solvated electrons as the solutes; therefore, they are very good conductors of electricity.

$$Na \xrightarrow{NH_3} Na^+ + e^-(NH_3)$$

Hydrazine, H_2NNH_2, is prepared by the reaction of ammonia with sodium hypochlorite in the presence of gelatine or glue. The reaction proceeds in two steps, the first of which is rapid:

$$NH_3 + NaOCl \rightarrow NaOH + NH_2Cl$$
$$NH_3 + NH_2Cl + NaOH \rightarrow N_2H_4 + NaCl + H_2O$$

The function of the glue or gelatine is to prevent competing side reactions during the second, slow step. Anhydrous hydrazine is obtained from the reaction mixture by successive distillations (the second distillation from solid NaOH or KOH). Hydrazine, a liquid at room temperature, is a weak base capable of accepting two protons:

$$H_2NNH_2 + H_2O \rightleftharpoons H_2NNH_3^+ + OH^- \qquad K_1 = 8.5 \times 10^{-7}$$
$$H_2NNH_3^+ + H_2O \rightleftharpoons H_3NNH_3^{2+} + OH^- \qquad K_2 = 8.9 \times 10^{-16}$$

Hydrazine has a dipole moment of about 1.85 D; hence it must have a configuration (Figure 21–16) other than *trans,* and there cannot be free rotation about the N—N bond. Derivatives of hydrazine, such as unsymmetrical dimethylhydrazine, $(CH_3)_2NNH_2$, are important rocket fuels.

Other hydrogen compounds of these elements are less important. The phosphorus analog of hydrazine, P_2H_4, is formed in small yield in the preparation of phosphine by hydrolysis of calcium phosphide. Hydrazoic acid, HN_3, is prepared by treating sodium azide with dilute sulfuric acid. Pure hydrazoic acid, a colorless liquid which boils at 37°C, is obtained by distilling the reaction mixture and drying the distillate over calcium chloride. Hydrazoic acid is very explosive, especially in the pure state. It

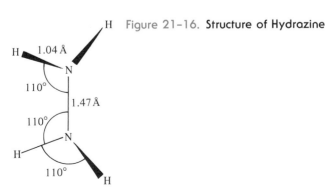

Figure 21–16. **Structure of Hydrazine**

is only slightly dissociated in aqueous solution ($pK_a = 4.7$). It reacts with metals to form nitrogen, ammonia, and the metal azide:

$$\text{Ca} \ + \ 3\,\text{HN}_3 \ \rightarrow \ \text{Ca(N}_3)_2 \ + \ \text{NH}_3 \ + \ \text{N}_2$$

Oxides and Oxyacids of Nitrogen, Phosphorus, and Arsenic

The names and formulas of the oxides and oxyacids of nitrogen, phosphorus, and arsenic are listed in Table 21–8. Several of the nitrogen compounds are of major importance industrially, and the phosphorus compounds also have many uses.

Nitrogen(I) oxide, N_2O, is a colorless gas which is known as "laughing gas." When inhaled, it causes a feeling of light-headedness and gaiety. It is useful as an anesthetic. It is prepared by heating a solution containing ammonium chloride and sodium nitrate:

$$\text{NH}_4^+(\text{aq}) \ + \ \text{NO}_3^-(\text{aq}) \ \xrightarrow{\text{heat}} \ \text{N}_2\text{O} \ + \ 2\,\text{H}_2\text{O}$$

The N_2O molecule is linear and unsymmetrical, having the structure

$$:\!N\!\equiv\!N\!-\!\ddot{O}\!:$$

Nitrogen(II) oxide, NO, is a colorless, reactive gas which can be prepared by the direct reaction of the elements at high temperature, but it dissociates upon cooling. It is produced when an electric spark is passed through a mixture of nitrogen and oxygen. During thunderstorms NO is formed by the action of lightning on the air. NO is an intermediate in the commercial preparation of nitric acid, and for this

TABLE 21–8. Oxides and Oxyacids of Nitrogen, Phosphorus, and Arsenic

Formula	Oxidation Number	Names	Remarks
N_2O	1	nitrogen(I) oxide, nitrous oxide, laughing gas	
NO	2	nitrogen(II) oxide, nitric oxide	
N_2O_3	3	nitrogen(III) oxide, nitrogen sesquioxide	unstable
NO_2	4	nitrogen dioxide	
N_2O_4	4	nitrogen tetroxide, dinitrogen tetroxide	
N_2O_5	5	nitrogen(V) oxide, dinitrogen pentoxide	
HNO_2	3	nitrous acid	unstable
HNO_3	5	nitric acid	
P_4O_6	3	phosphorus(III) oxide, phosphorus trioxide	
PO_2	4	phosphorus dioxide	polymeric
P_4O_{10}	5	phosphorus pentoxide, phosphorus(V) oxide	
H_3PO_3	3	phosphorous acid	diprotic
H_3PO_4	5	phosphoric acid (ortho)	
$H_4P_2O_7$	5	pyrophosphoric acid	
HPO_3	5	metaphosphoric acid	polymeric
As_4O_6	3	arsenic trioxide	
H_3AsO_3	3	arsenious acid	very weak
H_3AsO_4	5	arsenic acid (ortho)	
$HAsO_3$	5	metaarsenic acid	polymeric
$H_4As_2O_7$	5	pyroarsenic acid	

purpose it is prepared by the catalytic oxidation of ammonia (the Ostwald process). NO can be prepared in the laboratory by the reaction of dilute nitric acid with copper metal:

$$8 H^+ + 2 NO_3^- + 3 Cu \rightarrow 2 NO + 3 Cu^{2+} + 4 H_2O$$

As shown by the molecular orbital diagram (Figure 11–22), the NO molecule has one unpaired electron and a bond order of 2.5. However, the molecule does not dimerize to N_2O_2. NO reacts readily with oxygen to form nitrogen dioxide, a red-brown gas:

$$2 NO(g) + O_2(g) \rightarrow 2 NO_2(g)$$

Nitrogen dioxide can be prepared in the laboratory by the decomposition of lead(II) nitrate:

$$2 Pb(NO_3)_2 \xrightarrow{\text{heat}} 2 PbO + 4 NO_2 + O_2$$

The NO_2 molecule contains an odd number of electrons, and the substance is paramagnetic. However, in the gas phase it is in equilibrium with its dimer, N_2O_4. When the red-brown NO_2 gas is condensed, the liquid phase (normal boiling point, 21.3°C) is pale yellow and diamagnetic, indicating that the N_2O_4 predominates at lower temperatures:

$$\underset{\substack{\text{yellow and} \\ \text{diamagnetic}}}{N_2O_4} \quad \rightleftharpoons \quad \underset{\substack{\text{brown and} \\ \text{paramagnetic}}}{2 NO_2}$$

The liquid freezes at −9.3°C to give colorless, solid N_2O_4.

The reaction of NO_2 with water is an essential step in the commercial preparation of nitric acid. In the **Ostwald process** an ammonia and air mixture heated to 300°C is passed over a platinum gauze catalyst maintained at 900° to 1000°C. The reaction is sufficiently exothermic to maintain the catalyst at the required temperature:

$$4 NH_3 + 5 O_2 \xrightarrow[\text{heat}]{\text{catalyst}} 4 NO + 6 H_2O$$

The gas mixture is passed from the catalyst into towers where the following reactions take place:

$$2 NO + O_2 \rightarrow 2 NO_2$$
$$3 NO_2 + H_2O \rightarrow 2 HNO_3 + NO$$

The nitrogen(II) oxide formed in the last step is further oxidized into NO_2, so that eventually all of it is converted into nitric acid. Nitric acid is a strong acid, but its chief uses are as an oxidizing agent and as an intermediate in the preparation of explosives, fertilizers, plastics, varnishes, and dyes.

Nitrogen sesquioxide, N_2O_3, is an unstable compound formed when an equimolar mixture of NO and NO_2 is cooled to −20°C. It is a blue liquid which readily dissociates into NO and NO_2. Formally, nitrogen sesquioxide is the anhydride of nitrous acid:

$$N_2O_3 \ + \ H_2O \ \rightarrow \ 2\,HNO_2$$

Actually, however, its reaction with water is complex, yielding NO, N_2O_4, and HNO_3 as well as HNO_2.

Nitrogen pentoxide, N_2O_5, is a colorless crystalline material which sublimes at 32.5°C. It is prepared by dehydration of nitric acid:

$$4\,HNO_3 \ + \ P_4O_{10} \ \rightarrow \ 4\,HPO_3 \ + \ 2\,N_2O_5(g)$$

In the vapor state or in nonpolar solvents, N_2O_5 molecules have the structure

but in the solid state or in H_2SO_4 solution, N_2O_5 exists as $NO_2{}^+$ and $NO_3{}^-$ ions.

Phosphorus forms three oxides: P_4O_{10}, $(PO_2)_n$, and P_4O_6. The first and last of these are the anhydrides of phosphoric and phosphorous acids, respectively. The second, $(PO_2)_n$, is formed by heating P_4O_6 in a sealed tube above 210°C:

$$n\,P_4O_6 \ \xrightarrow[\text{heat}]{} \ 3\,(PO_2)_n \ + \ \frac{n}{4}P_4$$

It is of minor importance. "Phosphorus trioxide," P_4O_6, is prepared by passing a mixture of nitrogen and oxygen over white phosphorus at 45° to 50°C. "Phosphorus pentoxide," P_4O_{10}, is produced when phosphorus reacts with an excess of oxygen. The molecular structures of the two oxides are diagrammed in Figure 21–17.

Orthophosphoric acid, H_3PO_4, is formed by the reaction of P_4O_{10} with water:[1]

$$P_4O_{10} \ + \ 6\,H_2O \ \rightarrow \ 4\,H_3PO_4$$

This reaction proceeds extremely vigorously, and P_4O_{10} is often used as a drying agent. When phosphoric acid is heated, it loses water to form first the pyrophosphoric

---

[1] With a limited quantity of water, metaphosphoric acid is the product:
$$P_4O_{10} \ + \ 2\,H_2O \ \rightarrow \ 4\,HPO_3$$

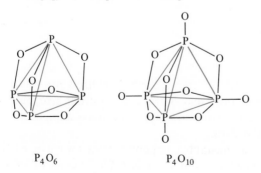

$$P_4O_6 \qquad\qquad P_4O_{10}$$

Figure 21–17. **Structures of P_4O_6 and P_4O_{10}**

acid, $H_4P_2O_7$; stronger heating produces metaphosphoric acid, $(HPO_3)_n$, a glassy polymer:

$$2\,H_3PO_4 \longrightarrow \underset{\underset{OH}{|}}{H-O-\overset{\overset{O}{\|}}{P}-O-\underset{\underset{OH}{|}}{\overset{\overset{O}{\|}}{P}}-O-H} + H_2O$$

$$n\,H_3PO_4 \xrightarrow{\text{heat}} n\,H_2O + \underset{\underset{OH}{|}}{-\overset{\overset{O}{\|}}{P}-O}\left(\underset{\underset{OH}{|}}{\overset{\overset{O}{\|}}{P}-O}\right)_n\underset{\underset{OH}{|}}{\overset{\overset{O}{\|}}{P}-O}-\underset{\underset{OH}{|}}{\overset{\overset{O}{\|}}{P}}-O-P- \cdots$$

Orthophosphorous acid, H_3PO_3, is formed when P_4O_6 reacts with water:

$$P_4O_6 + 6\,H_2O \rightarrow 4\,H_3PO_3$$

It is thermally unstable and decomposes upon heating to give phosphoric acid and phosphine:

$$4\,H_3PO_3 \xrightarrow{\text{heat}} 3\,H_3PO_4 + PH_3$$

It is a diprotic acid because one of the three hydrogen atoms is bonded directly to phosphorus:

$$H-O-\underset{\underset{H}{|}}{\overset{\overset{O}{\|}}{P}}-O-H$$

Phosphorous acid and its salts are strong reducing agents.

Arsenic forms one oxide, As_4O_6, which is structurally similar to P_4O_6. This compound is somewhat amphoteric, reacting with bases to give arsenites and with concentrated HCl to give $AsCl_3$:

$$12\,OH^- + As_4O_6 \rightarrow 4\,AsO_3{}^{3-} + 6\,H_2O$$
$$12\,HCl + As_4O_6 \rightarrow 4\,AsCl_3 + 6\,H_2O$$

A solution of As_4O_6 in water contains arsenious acid, H_3AsO_3, which is a very weak acid having a first ionization constant $K_1 = 6 \times 10^{-10}$. It is also a weak reducing agent:

$$H_3AsO_3 + H_2O \rightarrow H_3AsO_4 + 2\,H^+ + 2\,e^- \qquad \epsilon^\circ = -0.559\,V$$

21-7   Group IVA Nonmetals

In group IVA only carbon and silicon are nonmetals. Carbon is much more versatile in the formation of compounds than silicon, and even in those cases where

TABLE 21-9. Properties of Carbon and Silicon

	Carbon	Silicon[a]
atomic number	6	14
atomic weight (D)	12.0111	28.086
density (grams/cm^3)	2.25 (graphite)	2.42
	3.51 (diamond)	
normal boiling point	4830	2680
melting point	3727 (graphite)	1410
electronegativity	2.5	1.8
ΔH_{vap} (kcal/mole)	171.7 (subl)	(40.6)
(kJ/mole)	717.1 (subl)	(170)
ΔH_{fus} (kcal/mole)		11.1
(kJ/mole)		46.4
ionization potential (kcal/mole)	260	188
(kJ/mole)	1090	787
covalent radius (Å)	0.77	1.11
atomic radius (Å)	0.914	1.32

[a] Values in parentheses are somewhat uncertain.

analogous compounds of the elements exist, there are such vast differences in properties that, rather than look for trends in their properties, it is better to consider each element separately. The underlying reasons for these differences can be deduced from a consideration of the properties of the elements listed in Table 21–9. Carbon can have a maximum covalency of 4; silicon has available $3d$ orbitals and can exhibit a covalency up to 6. Electronegativity values and ionization potentials both indicate that silicon is more likely to form compounds in which it has positive oxidation states. The high stability of the carbon-to-carbon and carbon-to-hydrogen bonds compared to analogous silicon bonds explains why carbon forms hydrocarbons of high molecular weight, whereas silicon forms only silanes (silicon-hydrogen compounds) of low molecular weight. In addition, carbon forms double and triple bonds. On the other hand, silicon forms very strong bonds to oxygen, and this factor strongly influences the chemistry of silicon compounds.

21-8 Carbon

Carbon is found in nature in the forms of coal and diamond, in liquid and gaseous hydrocarbons, and in mineral carbonates—limestone, $CaCO_3$; magnesite, $MgCO_3$; and dolomite, $MgCO_3 \cdot CaCO_3$. Carbon compounds predominate in living creatures, both plant and animal. Two allotropic forms of carbon—diamond and graphite— have already been described in some detail (Section 11–20). It should be noted that at 1 atm pressure, diamond is thermodynamically unstable relative to grȧphite at all temperatures. (At room temperature the free energy change for the reaction of diamond to give graphite is -0.7 kcal/mole.) However, an extremely large activation energy is required for this reaction, and the change is immeasurably slow. Conversely, the conversion of graphite into diamond is not spontaneous under ordinary conditions. The production of synthetic diamonds requires temperatures of 2000°C and pressures of 100,000 atm.

Other familiar forms of carbon—charcoal and carbon black—are merely micro-crystalline forms of graphite. Carbon being used for industrial purposes is usually in the form of coke, a more or less impure form of graphite, which is made by heating coal to drive off volatile impurities and low-melting tars. Charcoal is prepared by heating wood or other materials with a high carbon content, such as sugar, to a temperature high enough to drive off water and other volatile decomposition products. Carbon black is produced by burning natural gas in a limited air supply and collecting the soot on cooled metal plates. Coke is used as a reducing agent in metallurgy, and graphite has many uses, including the manufacture of "lead" pencils, electrodes for arc lamps, and high-temperature crucibles. Carbon black is used as a pigment in ink, but its most important use is in the compounding of rubber for the manufacture of tires.

Oxides of Carbon

Carbon forms three oxides: carbon dioxide, carbon monoxide, and carbon suboxide, C_3O_2. Some properties of these substances are listed in Table 21–10. (In addition, $C_{12}O_9$, the anhydride of mellitic acid, might be considered as an oxide of carbon.) Carbon dioxide is formed whenever carbon or a carbon-containing compound is burned in excess air; it is also a product of the respiration of animals and the decay of living things. It is not toxic but does not support life. Because CO_2 is more dense than air, it collects at the bottoms of wells, caves, or mines. The gas is called "choke damp" by miners, because persons working in mines and tunnels are in danger of being suffocated unless sufficient air circulation is provided.

Solid carbon dioxide is sold under the name "dry ice" and is used to preserve frozen foods and ice cream while they are being transported.

Carbon dioxide is the anhydride of carbonic acid, H_2CO_3. When the gas dissolves in water, the following equilibria are established:

$$CO_2 + H_2O \rightleftharpoons H_2CO_3$$
$$H_2CO_3 + H_2O \rightleftharpoons H_3O^+ + HCO_3^-$$
$$HCO_3^- + H_2O \rightleftharpoons H_3O^+ + CO_3^{2-}$$

At 25°C and 1 atm of CO_2 pressure, the concentration of dissolved gas is about

TABLE 21–10. Properties of Carbon Oxides

	C_3O_2	CO	CO_2
molecular weight	68.03	28.01	44.01
appearance	colorless gas	colorless, odorless, poisonous gas	colorless gas
density			
gas (grams/liter)		1.250 (0°)	1.977 (0°)
liquid (grams/ml)	1.114	0.793	1.101 (−37°)
solid (grams/cm³)			1.56 (−79°)
melting point (°C)	−111.3	−199	−56.6 (5.2 atm)
normal boiling point (°C)	7	−191.5	−78.1 (subl)
structure	linear, symmetrical	linear	linear, symmetrical

0.034 M. If it is assumed that all of the carbon dioxide is converted to carbonic acid, the dissociation constants of the acid are calculated to be $K_1 = 4.3 \times 10^{-7}$ and $K_2 = 4.7 \times 10^{-11}$. However, there is evidence that less than 1% of the dissolved carbon dioxide is in the form of carbonic acid. The structure of carbon dioxide has been described in Section 11–4. Some details of the chemistry of carbonic acid have been given previously.

Carbon monoxide may be formed in several ways. It is produced when either carbon or a hydrocarbon is burned in a limited supply of air. The gas is produced in quantity by the reaction of carbon dioxide and graphite at temperatures about 800°C:

$$CO_2 \;+\; C \;\rightarrow\; 2\,CO$$

Large quantities of CO are produced in the water-gas reaction:

$$C(s) \;+\; H_2O(g) \;\rightarrow\; CO(g) \;+\; H_2(g)$$

A convenient laboratory preparation of carbon monoxide is the dehydration of formic acid by concentrated H_2SO_4:

$$\overset{\displaystyle O}{HC}\!-\!OH \;\xrightarrow{\;H_2SO_4\;}\; H_2O \;+\; CO$$

Carbon monoxide is isoelectronic with N_2, and a molecular orbital treatment shows that carbon monoxide has a bond order of 3. However, the observed bond length is somewhat greater than that expected for a triple bond. Considering the electronegativities of carbon and oxygen, the dipole moment of carbon monoxide is unexpectedly small. Apparently the nonbonding electron pair on the carbon atom contributes a moment in opposition to the C—O bond moment:

$$: \leftarrow C \rightarrow O$$

The poisonous nature of carbon monoxide has already been described in Section 14–5. The gas is not very soluble in water, but it reacts with strong bases under extreme conditions to yield formates—salts of formic acid:

$$CO \;+\; OH^- \;\xrightarrow[200°C]{10\ atm}\; H\!-\!\overset{\displaystyle O^-}{\underset{\displaystyle O}{C}}$$

Carbon monoxide reacts with the halogens to form carbonyl halides. With chlorine, for example, the compound phosgene is produced:

$$CO \;+\; Cl_2 \;\rightarrow\; COCl_2$$
$$\text{phosgene}$$

Phosgene is extremely poisonous; in World War I it was used by troops of both sides as a poison gas.

Commercially, alcohols and hydrocarbons can be formed from hydrogen-enriched water gas in the presence of catalysts:

$$CO \ + \ 2\,H_2 \ \xrightarrow[\text{100 atm}]{\text{ZnO}} \ CH_3OH$$

$$CO \ + \ 3\,H_2 \ \xrightarrow[\text{200 atm}]{\text{Co, Cr}_2\text{O}_3} \ CH_4 \ + \ H_2O$$

The first of these reactions is a major source of methyl alcohol.

Carbon suboxide, C_3O_2, is prepared by the dehydration of malonic acid:

$$HOCOCH_2CO_2H \ + \ P_4O_{10} \ \rightarrow \ 4\,HPO_3 \ + \ O{=}C{=}C{=}C{=}O$$

It is a foul-smelling gas which boils at $7\,°C$. Infrared and Raman studies show that its molecules are linear. C_3O_2 is stable at room temperature, but at higher temperatures it polymerizes to a dark red, water-soluble solid.

21-9   Silicon

Silicon crystallizes in the diamond lattice only; it has no allotropic forms. It is the second most abundant element in the crust of the earth and is found in nature in the form of silica, SiO_2, and in many silicate minerals. Elemental silicon can be obtained by heating silica with carbon or with calcium carbide, CaC_2, in an electric furnace. Very pure silicon is used as a semiconductor material, and for this purpose silicon is purified by zone refining. Other uses of silicon include the preparation of certain alloy steels.

Hydrides and Halides

In contrast to carbon, silicon forms only a limited number of binary hydrogen compounds, having formulas of the type Si_nH_{2n+2}. These are known as silanes. Because of the weakness of the bonds between silicon atoms, the maximum number of such bonds in such a chain is about eight. The silanes all decompose above $300\,°C$. They burn spontaneously in air:

$$Si_5H_{12} \ + \ 8\,O_2 \ \rightarrow \ 5\,SiO_2 \ + \ 6\,H_2O$$

They are attacked by water and dilute bases to form SiO_2 and hydrogen. The strength of the Si—O bond and the availability of d orbitals, which makes possible coordination of water molecules, are two factors which promote the hydrolysis reactions. Silanes are prepared by the partial hydrolysis of magnesium silicide, Mg_2Si, with aqueous acid. The reaction produces a mixture of silanes, which are then separated by fractional distillation. SiH_4 is also prepared by the reduction of silicon tetrachloride with lithium aluminum hydride at $0\,°C$:

$$SiCl_4 \ + \ LiAlH_4 \ \xrightarrow{\text{ether}} \ SiH_4 \ + \ LiCl \ + \ AlCl_3$$

The preparation of SiF_4 by reaction of SiO_2 with HF has already been described. The other tetrahalides are prepared by heating SiO_2 and carbon together with the desired halogen, for example,

$$SiO_2 \ + \ C \ + \ 2\,Br_2 \ \rightarrow \ SiBr_4 \ + \ CO_2$$

Except for SiF_4, the tetrahalides are all volatile, fuming compounds which are readily hydrolyzed by water:

$$SiX_4 \ + \ 2\,H_2O \ \rightarrow \ SiO_2 \ + \ 4\,HX \quad (X \ = \ Cl,\ Br,\ or\ I)$$

Silica and Silicates

The structure of silica was described in Section 11–20. Silica exists in three crystalline forms—quartz, tridymite, and cristobalite—and each of these has a number of modifications. In crystobalite the SiO_4 tetrahedra are arranged in a diamond-like lattice. In the other two forms the arrangement of the tetrahedra has a screwlike pattern. There can be "right-handed" and "left-handed" forms, which are mirror images of each other, and which rotate the plane of polarized light in opposite directions (Figure 21–18).

Pure silica melts at 1710°C, and the highly viscous liquid readily supercools to form quartz glass. This glass is very transparent over a wide range of wavelengths, extending from the infrared region into the ultraviolet region of the spectrum. It is therefore used to manufacture optical instruments, expecially those used to measure ultraviolet light. Also, its low coefficient of thermal expansion and its resistance to thermal shock makes quartz glass very suitable for the construction of laboratory apparatus.

SiO_2 dissolves slowly in strong base to form soluble silicates:

$$SiO_2 \ + \ 2\,NaOH \ \rightarrow \ \underset{\text{sodium metasilicate}}{Na_2SiO_3} \ + \ H_2O$$

$$2\,SiO_2 \ + \ 6\,NaOH \ \rightarrow \ \underset{\text{sodium pyrosilicate}}{Na_6Si_2O_7} \ + \ 3\,H_2O$$

$$SiO_2 \ + \ 4\,NaOH \ \rightarrow \ \underset{\text{sodium orthosilicate}}{Na_4SiO_4} \ + \ 2\,H_2O$$

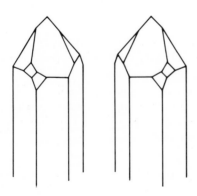

Figure 21–18. Mirror Image Forms of Quartz Crystals

The addition of strong acid to a solution of a soluble silicate results in the formation of a gelatinous precipitate which readily loses water to form partially hydrated silica, $(SiO_2)_m(H_2O)_n$. When the precipitate is dried to a water content of about 5%, a porous material called silica gel is obtained. Silica gel is capable of absorbing large quantities of water, benzene, sulfur dioxide, or other volatile substances. It is therefore useful as a drying agent and absorbent.

Glass

Glass is a mixture of silicates made by melting silica with various oxides and/or carbonates. Common lime glass is made from sodium carbonate, calcium carbonate, and sand, SiO_2. Carbon dioxide is evolved from the melt, and the glass is approximately 75% SiO_2, 12% Na_2O, 12% CaO, and 1 to 2% Al_2O_3. Pyrex glass contains B_2O_3 in place of the CaO.

Silicones

Silicones are organosilicon polymers[2] made by the hydrolysis of organosilicon halides:

$$n\,(CH_3)_2SiCl_2 \;+\; n\,H_2O \;\rightarrow\; 2n\,HCl \;+\; \underset{\underset{CH_3}{|}}{\overset{\overset{CH_3}{|}}{-O-Si}}\left[\underset{\underset{CH_3}{|}}{\overset{\overset{CH_3}{|}}{-O-Si}}\right]_n\underset{\underset{CH_3}{|}}{\overset{\overset{CH_3}{|}}{-O-Si-O-}}$$

The silicones may be obtained as short chain oils or rubber-like, plastic solids. The viscosities of the oils vary little with temperature, and they remain fluid to temperatures as low as $-100°C$. They are also nonvolatile and resistant to chemical attack. They are used as lubricants for both low-temperature and high-temperature applications and are especially suited for use in space vehicles. The solids can be compounded to rubbers which withstand high temperatures and also remain elastic at very low temperatures. By varying the nature of the organic part of the monomer or by using a mixture of monomers, silicone polymers having very specific properties can be designed. The number and uses of such materials are practically limitless.

The Structures of Silicate Minerals

The crust of the earth consists almost entirely of silica and various silicate minerals combined into numerous types of rocks. It is important to distinguish between **minerals**, which are naturally occurring inorganic compounds, and **rocks**, which may be composed of a mixture of several minerals. When it is remembered that 75% of the mass of the crust is composed of oxygen and silicon atoms, the predominance of the silicate minerals is not surprising. The structures of all silicates are based on SiO_4 tetrahedra. In the case of silica, already described, these SiO_4 tetrahedra are condensed, so that each of the four oxygen atoms is shared between two tetrahedra.

---

[2] A polymer is a very large molecule made by the combination of many small molecules, called monomers. Further discussion of polymers will be given in Section 23-10.

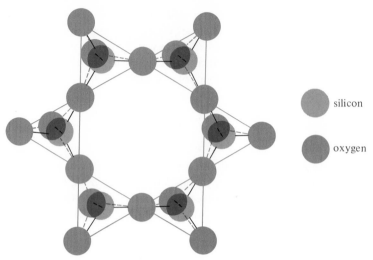

silicon

oxygen

Figure 21-19. **Structure of $Si_6O_{18}^{12-}$**

X-ray crystal structure determinations reveal that there are five types of silicate structures:

1. Simple silicates containing discrete SiO_4^{4-} tetrahedra, along with sufficient positive ions to make the crystal as a whole neutral. Examples are the minerals zircon, $ZrSiO_4$, and olivine, Fe_2SiO_4 or Mg_2SiO_4.
2. Two tetrahedra joined by the sharing of an oxygen atom, or small rings formed by the sharing of two oxygen atoms from each tetrahedron. Crystals containing the $Si_2O_7^{6-}$, $Si_3O_9^{6-}$, and $Si_6O_{18}^{12-}$ ions are in this class. An example is beryl, $Be_3Al_2(Si_6O_{18})$ (Figure 21–19).
3. "Infinitely long" single or double chains of silicate tetrahedra (Figure 21–20). Two or three of the oxygen atoms of each of the tetrahedra are shared between two such tetrahedra. Positive ions which balance the charge are distributed all along the chain. Minerals having this type of structure have long, fibrous crystals. The asbestos minerals, one of which is tremolite, $[Ca_2Mg_5(Si_4O_{11})_2(OH)_2]_n$, are examples of this class of silicates.
4. Two-dimensional sheets resulting from the sharing of three oxygen atoms by each silicate tetrahedron. Positive ions layered between the sheets make the crystals neutral. This type of structure is characteristic of the micas, of which muscovite, $KAl_3Si_3O_{10}(OH)_2$, is an example.
5. Three-dimensional networks arising from the sharing of all the oxygen atoms between tetrahedra. Silica itself is an example of this class. In some minerals aluminum atoms replace some of the silicon atoms of the lattice, producing a negatively charged network. The charge is neutralized by positive ions in the holes. When one out of every four silicon atoms is replaced by an aluminum atom, the resulting mineral is a feldspar, such as $K^+AlSi_3O_8^-$. In this case the negative charge of the network is neutralized by the potassium ions. In addition to the feldspars, the zeolites, $Na_2Al_2Si_3O_{10} \cdot 2H_2O$, have a three-dimensional framework structure. The use of zeolites as ion exchangers has been described earlier.

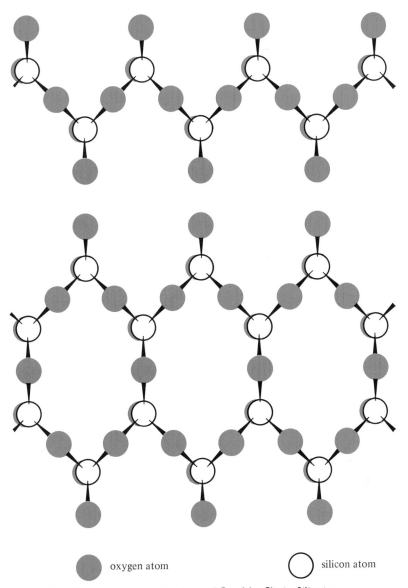

oxygen atom   silicon atom

Figure 21-20. **Structures of Single and Double Chain Silicates**

21-10   Some Aspects of Geochemistry

The earth's crust is in a process of constant change, and the consequences of the change are reflected in the structures of the rocks and the compositions and distribution of the various minerals. Three major processes—vulcanism, diastrophism, and gradation—are responsible for the changes which occur. **Vulcanism** includes processes associated with the formation, flow, and ultimate solidification of molten rock. Active volcanos are examples of vulcanism. **Diastrophism** is the movement of large masses of solid rock. During diastrophism sufficient heat and pressure are generated to cause some rearrangements of the internal structure of rock. Earthquakes are a manifestation of diastrophism. **Gradation** includes the wearing down of rocks and the transfer and deposition of the fragments by such agents as water, ice, wind, and chemical action.

Rocks are classified according to their previous history. **Igneous** rocks are solidified molten rock. Granite and basalt are two examples of igneous rocks, the former containing about 70% SiO_2 plus minerals such as muscovite, biotite, and hornblende, $Ca(Mg,Fe)_3Si_4O_{12}$.[3] Basalt is about half silica and contains olivine, biotite, and hornblende.

Sedimentary rocks result from the cementing together of rock fragments from gradation processes. Minerals deposited between such fragments, which have been submerged under water, bind them together into solid, sedimentary rock, such as shale, sandstone, limestone, magnetite, and gypsum.

Metamorphic rock has been buried and subjected to great heat and high pressures. These stresses cause the minerals to rearrange their crystal structures, and new forms of rock result. Marble, which is metamorphosed limestone, and slate, which is metamorphosed shale, are two typical examples of metamorphic rock.

The degradation of rocks is both a chemical and mechanical process. Rubbing of rock against rock, by such agents as rivers and wind, causes mechanical breakdown. Freezing of water in rock crevices also produces mechanical degradation. Chemical action includes the dissolving and redeposition of limestone by water containing carbon dioxide. Feldspars are also attacked by carbon dioxide in water:

$$2\,KAlSi_3O_8 \; + \; CO_2 \; + \; 2\,H_2O \; \rightarrow \; K_2CO_3 \; + \; Al_2Si_2O_5(OH)_4 \; + \; 4\,SiO_2$$

Olivine is attacked by carbon dioxide and oxygen:

$$Mg_2SiO_4 \; + \; 2\,H_2CO_3 \; \rightarrow \; 2\,MgCO_3 \; + \; SiO_2 \; + \; 2\,H_2O$$
$$3\,Fe_2SiO_4 \; + \; O_2 \; \rightarrow \; 2\,Fe_3O_4 \; + \; 3\,SiO_2$$

Over the several billion years since its origin, the surface of the earth has undergone many changes. Mountains have been created through such processes as volcanic activity (vulcanism) and earthquakes (diastrophism). The primeval rocks have been chemically and mechanically degraded by weathering and erosion. The layers of sediments and metamorphic rocks exposed in canyons, along river valleys, and through excavation provide evidence for the chronological sequence of geological events. Thus it is possible to deduce the history of the earth's surface. However, as a result of the ever-continuing geological processes which have occurred over eons of time, it is impossible to obtain a complete picture of the origin of our planet from a study of its rocks alone.

21–11   Boron

Boron is the only nonmetal in group III A. In the elemental state it is a very hard, high-melting, dark, semiconducting material. It has several crystalline modifications

---

[3] Formulas written with symbols in parentheses and separated by commas denote that either or both of the elements may be present. Thus in the formula given, a total of three moles of Mg and Fe in any ratio is present per mole of Ca.

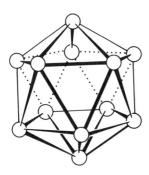

Figure 21-21. The B_{12} Icosahedron

which have structures intermediate between the close-packed structures of metals and the covalent networks of some nonmetals. The structural unit in several of the modifications of boron is an icosahedron (20-sided figure) containing 12 boron atoms at the corners (Figure 21-21).

In nature boron occurs principally in the form of mineral borates. These include borax, $Na_2B_4O_7 \cdot 10H_2O$; kernite, $Na_2B_4O_7 \cdot 4H_2O$; and colemanite, $Ca_2B_6O_{11} \cdot 5H_2O$. The free element is obtained by converting the mineral to the oxide, followed by reduction with magnesium:

$$Na_2B_4O_7 \xrightarrow{HCl} H_3BO_3 \xrightarrow[heat]{} B_2O_3$$

$$B_2O_3 + 3\,Mg \longrightarrow 3\,MgO + 2\,B$$

Very pure boron is made by reducing BBr_3 vapor with hydrogen gas on a hot tantalum filament. Because of its high cross section for neutrons, boron is used to make control rods for nuclear reactors and also to make neutron counters.

Boron Halides

Boron halides can be prepared by the direct reaction of the element with halogens at high temperatures. A more convenient preparation of BF_3 is the treatment of boric oxide with concentrated sulfuric acid and a fluoride:

$$B_2O_3 + 6\,H_2SO_4 + 6\,F^- \rightarrow 2\,BF_3(g) + 6\,HSO_4^- + 3\,H_2O$$

The other boron halides can be prepared by treating BF_3 with the appropriate aluminum halide:

$$BF_3(g) + \tfrac{1}{2}Al_2Cl_6(s) \longrightarrow BCl_3(g) + AlF_3(s)$$

The boron halides are covalent compounds which are monomeric in the vapor state and in benzene solution. They have planar molecules (Section 11-4). They are strong Lewis acids and react readily with electron pair donors to give compounds in which the coordination number of boron is 4. The ability to form such complexes makes BF_3 a useful catalyst for many organic reactions. The BF_4^- ion has a tetrahedral configuration resembling ClO_4^-. It is the anion of tetrafluoroboric acid, HBF_4, which is prepared by treatment of boric acid with hydrogen fluoride:

$$H_3BO_3 \;+\; 4\,HF \;\rightarrow\; HBF_4 \;+\; 3\,H_2O$$

Boron Hydrides and Their Derivatives

A number of boron hydrides (boranes) are known. These compounds are of interest because it has been difficult to account for their compositions and structures in terms of the usual concepts of bonding. Only recently have satisfactory theoretical treatments of their structures been developed. For example, the simplest boron hydride is not BH_3 but its dimer, diborane, B_2H_6. In diborane there are not enough valence electrons to provide seven electron pair bonds, similar to ethane.

$$\begin{array}{cc}
\text{H\ \ H} & \text{H\,H} \\
\text{H:C:C:H} & \text{H:B?B:H} \\
\text{H\ \ H} & \text{H\,H}
\end{array}$$

Structures having one-electron bonds need not be considered because B_2H_6 is diamagnetic. Moreover, infrared and Raman spectroscopy indicate that B_2H_6 has some structural similarity to ethylene, $H_2C{=}CH_2$. Nuclear magnetic resonance studies show the presence of two sets of hydrogen atoms—two of one type and four of another. The experimental evidence suggests a bridged structure, as shown in the accompanying figure. To account for this structure theoretically, it is necessary to

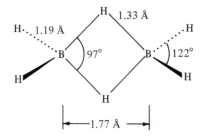

introduce the concept of **three-center "bridge" bonds.** Such bonds may be formed

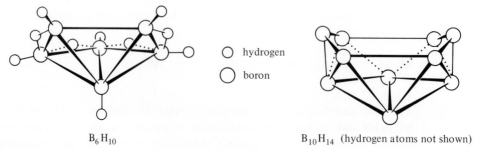

○ hydrogen

○ boron

B_6H_{10} $B_{10}H_{14}$ (hydrogen atoms not shown)

Figure 21–22. **Structures of Two Boron Hydrides**

when there are more orbitals available among several atoms than there are electrons. For example, if in B_2H_6 each boron atom uses sp^3 hybrid orbitals to form σ bonds with two hydrogen atoms (a total of four bonds), there will remain on each boron atom two sp^3 orbitals and one valence electron. The two additional hydrogen atoms have one $1s$ orbital and one electron each; hence there is a total of six available orbitals but only four electrons. The sp^3 orbitals from each boron can overlap with each other and with the $1s$ orbital from a hydrogen atom to form a three-center bond.

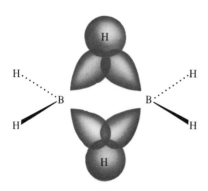

The known boranes include the gases B_2H_6 and B_4H_{10}; the liquids B_5H_9, B_5H_{11}, and B_6H_{10}; and the solid $B_{10}H_{14}$; as well as the anionic species $B_{10}H_{10}^{2-}$ and $B_{12}H_{12}^{2-}$. In addition to three-center bonds involving B—H—B and B—B—B bridging, the molecules of these compounds are characterized by having structures which are fragments of an icosahedral structure (Figure 21-22). The boranes are prepared by the reaction of metal hydrides or complex hydrides with boron halides:

$$6\,LiH \;+\; 8\,BF_3{\cdot}(C_2H_5)_2O \;\longrightarrow\; 6\,LiBF_4 \;+\; B_2H_6 \;+\; 8\,(C_2H_5)_2O$$

$$3\,LiAlH_4 \;+\; 4\,BCl_3 \;\xrightarrow{\text{ether}}\; 3\,LiCl + 3\,AlCl_3 \;+\; 2\,B_2H_6$$

The higher boranes are prepared by the thermal decomposition of diborane. All of the boranes are spontaneously flammable in moist air, possibly because the hydrolysis of B—H bonds is highly exothermic. Diborane reacts with the hydride ion to form the tetrahydroborate ion (also called the borohydride ion):

$$2\,H^- \;+\; B_2H_6 \;\xrightarrow{\text{ether}}\; 2\,BH_4^-$$

Another method of preparing sodium tetrahydroborate is the reaction of trimethoxyboron with sodium hydride:

$$4\,NaH \;+\; B(OCH_3)_3 \;\xrightarrow[250°C]{\text{Lewis base solvent}}\; NaBH_4 \;+\; 3\,NaOCH_3$$

The alkali metal tetrahydroborates are useful as reducing agents and as sources of hydride ions. Because of their very high enthalpies of combustion and their low molecular weights, the boranes and their derivatives have been considered as possible rocket fuels.

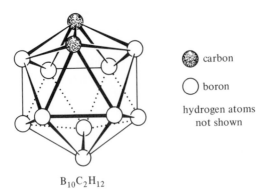

Figure 21-23. **Structure of a Carborane**

⬤ carbon

◯ boron

hydrogen atoms
not shown

$B_{10}C_2H_{12}$

Carboranes

Carboranes are an interesting class of compounds whose molecules contain both carbon and boron in a polyhedral framework. For example, the structure of the carborane $B_{10}C_2H_{12}$ is shown in Figure 21-23. The unusual bond order of 6 for carbon is possible because of the stabilizing influence of the icosahedral structure. This compound is prepared from $B_{10}H_{14}$ by the following sequence of reactions:

$$B_{10}H_{14} + 2(C_2H_5)_2S \xrightarrow{40°C} B_{10}H_{12} \cdot 2(C_2H_5)_2S + H_2$$

$$B_{10}H_{12} \cdot 2(C_2H_5)_2S + HC{\equiv}CH \longrightarrow B_{10}C_2H_{12} + 2(C_2H_5)_2S + H_2$$

The product is a very stable substance. Acids, bases, and oxidizing agents attack the hydrogen atoms bonded to the carbon atoms but leave the boron–hydrogen bonds and the molecular skeleton intact. A number of other carboranes are known, as well as many organic derivatives of carboranes. The study of these substances is currently an active area of chemical research.

21-12   The Noble Gases—Group 0

The elements of Group 0 of the periodic table are all monatomic gases. Some of their properties are listed in Table 21-11. With the exception of helium, all of these elements have filled outer p subshells; helium and neon have filled outer shells of electrons. In every case the filled outer subshell represents a state of high stability, as is reflected in the high ionization potentials and generally low reactivity of these elements.

The noble gases are minor constituents of the atmosphere. Helium is sometimes found in natural hydrocarbon gas deposits and is occluded in certain minerals which contain radioactive elements. In the latter case the helium results from the decay of the radioactive element. All of the isotopes of radon are radioactive. The longest-lived species, ^{222}Rn, has a half life of 3.825 days; consequently, radon is found in nature only in secular equilibrium with its parent nuclide (see Section 20-6). Neon, argon, krypton, and xenon are obtained by the fractional distillation of liquid air.

The noble gases have several commercial uses. Helium, in addition to being used to fill lighter-than-air vehicles, is used as a carrier gas in chromatography and as an inert

TABLE 21-11. Properties of the Noble Gases

	Helium	Neon	Argon	Krypton	Xenon	Radon
atomic number	2	10	18	36	54	86
atomic radius (Å)	0.93	1.31	1.74	1.89	2.09	2.14
ΔH_{vap} (kcal/mole)	0.020	0.431	1.56	2.16	3.02	3.92
(kJ/mole)	0.084	1.80	6.53	9.04	12.6	16.4
ΔH_{fus} (kcal/mole)	0.005	0.080	0.281	0.39	0.55	0.69
(kJ/mole)	0.02	0.33	1.18	1.6	2.3	2.9
ionization potential (kcal/mole)	567	497	363	323	280	248
(kJ/mole)	2370	2080	1520	1350	1170	1040
normal boiling point (°C)	−268.9	−246	−185.8	−152	−108.0	−61.8
melting point (°C)	−269.7	−248.6	−189.4	−157.3	−111.9	−71
atomic weight (D)	4.0026	20.183	39.948	83.80	131.30	(222)[a]

[a] Mass number of most common isotope.

atmosphere in certain experiments where nitrogen gas might react. Neon and argon are used in discharge tubes for advertising display and to provide inert atmospheres in certain gas-filled electric light bulbs, radio tubes, and Geiger counter tubes. Krypton is also used in gas-filled incandescent lamps. Xenon is used in high-intensity flash lamps used for photography. Radon is useful as a therapeutic radiation source in the treatment of cancer.

In the liquid state helium exhibits some rather unusual properties. At its normal boiling point of 4.18 K, helium behaves as a normal liquid, but on further cooling to 2.178 K, a second form of the element called helium(II) is produced. Helium(II) is a "superfluid," having practically zero viscosity. It forms thin films which are only a few atoms thick and which flow without friction. If helium(II) is placed at different heights in two concentric vessels, the liquid spontaneously flows up and over the walls of one into the other until the levels are equal (Figure 21-24). Helium(II) is also an extraordinary conductor of heat.

The apparent lack of chemical reactivity of the noble gases was attributed to the filled outer p subshell of electrons in the atoms of these elements. However, beginning

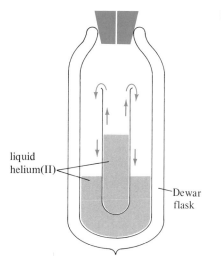

liquid helium(II)

Dewar flask

Figure 21-24. **Superfluidity of Helium(II)**

with argon, these elements should be able to use their outermost d subshells to form chemical compounds. In 1962 Neil Bartlett, at the University of British Columbia, obtained a compound, O_2PtF_6, by the reaction between oxygen and platinum hexafluoride, PtF_6. Reasoning that the ionization potential of xenon is almost identical with that of the O_2 molecule, Bartlett predicted that xenon should react with PtF_6 in an analogous manner. He found that a reaction did occur spontaneously at room temperature to give $XePtF_6$, a red crystalline solid. Subsequently it has been found that thermodynamically unstable fluorides, such as PtF_6, RuF_6, RhF_6, and PuF_6, all react with xenon, whereas stable hexafluorides, such as UF_6 and IrF_6, do not react.

Following Bartlett's discovery, a number of xenon compounds have been prepared as well as a few compounds of krypton and radon. Some of these are listed in Table 21–12. Xenon reacts directly only with fluorine, yielding XeF_2, XeF_4, and XeF_6 under varying conditions of temperature and gas pressures. The reaction of XeF_6 with water vapor yields $XeOF_4$ and XeO_3:

$$XeF_6 \;+\; H_2O \;\rightarrow\; 2\,HF \;+\; XeOF_4$$
$$XeOF_4 \;+\; 2\,H_2O \;\rightarrow\; 4\,HF \;+\; XeO_3$$

$XeOF_4$ is stable in the absence of water, but solid XeO_3 is dangerously explosive. XeO_3 solutions in water are stable but nonconducting. The latter fact suggests that the oxide is not the anhydride of H_2XeO_4, xenic acid. However, salts of this acid are obtained when XeO_3 is dissolved in strong base:

$$XeO_3 \;+\; OH^- \;\rightleftharpoons\; HXeO_4^-$$

$HXeO_4^-$ slowly disproportionates in basic solution to give perxenate solutions:

$$2\,HXeO_4^- \;+\; 2\,OH^- \;\rightarrow\; XeO_6^{4-} \;+\; Xe \;+\; O_2 \;+\; 2\,H_2O$$

Perxenates are powerful oxidizing agents, as is XeO_3.

Xenon difluoride has been found to be an excellent reagent for fluorinating aromatic hydrocarbons:

$$XeF_2 \;+\; C_6H_6 \;\rightarrow\; C_6H_5F \;+\; Xe \;+\; HF$$

TABLE 21–12. Compounds of the Noble Gases

Oxidation State	Compound	Form	Melting Point (°C)	Structure	Remarks
2	XeF_2	colorless crystals	140	linear	
2	KrF_2	white crystals			sublimes below 0°C; very reactive
4	XeF_4	colorless crystals	114	planar	
4	$XeOF_2$	colorless crystals	90		not very stable
6	XeF_6	colorless crystals	47.7	distorted octahedron	
6	$XeOF_4$	colorless liquid	−28	square pyramid	
6	XeO_3	colorless crystals		trigonal pyramid	explosive, stable in solution
8	XeO_4	colorless gas		tetrahedron	explosive

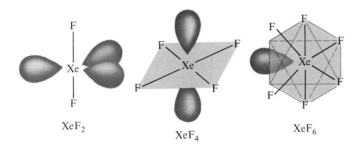

Figure 21-25. **Structures of Some Xenon Fluorides**

Each gray lobe represents a hybrid orbital containing an unshared pair of electrons.

The reaction occurs at room temperature and is apparently catalyzed by HF.

The structures of the rare gas compounds can in general be predicted by the electron pair repulsion theory (Section 11-4). Their structures have been established experimentally by X-ray diffraction and by infrared and Raman spectroscopy. For example, in the XeF_2 molecule there are ten valence electrons about the xenon atom: two shared pairs and three unshared pairs. The minimum repulsion between the unshared pairs is achieved when they are in the equatorial plane of a trigonal bipyramid. Hence the XeF_2 molecule is linear, as illustrated in Figure 21-25. Similar reasoning leads to the conclusion that the XeF_4 molecule is square planar and that XeF_6 has a distorted octahedral structure.

21-13   Exercises

Basic Exercises

1. Suggest three tests which can be used to distinguish between a metal and a nonmetal.
2. Write equations describing the reaction of freshly prepared aluminum oxide with HCl and with NaOH.
3. Classify each of the following oxides as strongly acidic, weakly acidic, neutral, amphoteric, weakly basic, or strongly basic: **(a)** SnO_2, **(b)** SnO, **(c)** CO, **(d)** PbO, **(e)** MnO_2, **(f)** RaO, **(g)** N_2O, **(h)** FeO, **(i)** Ag_2O, **(j)** OsO_4, **(k)** Al_2O_3, **(l)** Fe_2O_3, **(m)** CeO_2, **(n)** CO_2, **(o)** MgO, **(p)** K_2O.
4. Select the strongest and the weakest acid in each of the following sets: **(a)** HBr, HF, H_2Te, H_2Se, H_3P, H_2O, **(b)** HClO, HIO, H_3PO_3, H_2SO_3, H_3AsO_3.
5. Diagram the molecular structure of OF_2.
6. For the reaction (at 25°C)

$$3\,O_2 \;\rightleftharpoons\; 2\,O_3 \qquad \Delta H^\circ = +68\ \text{kcal}$$

   the equilibrium constant is 10^{-54}. Calculate ΔG° and ΔS° for the reaction.
7. Suggest a practical use for water in which advantage is taken of each of the following properties: **(a)** high heat capacity, **(b)** low molecular weight, **(c)** high boiling point, **(d)** polarity of molecules, **(e)** high melting point, **(f)** large enthalpy of fusion, **(g)** large enthalpy of vaporization.
8. Explain why calcium ion makes water "hard," but sodium ion does not.
9. Explain, using equations, how the addition of $Ca(OH)_2$ aids in the removal of calcium ions from temporary hard water. What effect would the addition of too much of the reagent have on the desired process?

10. Write equations showing how each of the following compounds can be prepared, starting with the appropriate elemental halogen: **(a)** $HClO_4$, **(b)** I_2O_5, **(c)** Cl_2O, **(d)** ClO_2, **(e)** $KBrO_3$, **(f)** OF_2, **(g)** BrO_3, **(h)** Br_2O.

11. Comparing manufacturing costs and costs of raw materials, explain why it is cheaper to use H_2SO_4 than HCl as an acid for industrial purposes.

12. Write equations showing a laboratory preparation of each of the following: **(a)** HCl, **(b)** HBr, **(c)** HI, **(d)** HIO_4.

13. Write equations showing a laboratory preparation of each of the following: **(a)** N_2O, **(b)** N_2O_5, **(c)** NO_2, **(d)** H_6TeO_6, **(e)** $COBr_2$.

14. A $0.10\ M$ aqueous solution of which salt in each of the following pairs would have the higher pH? **(a)** $NaNO_2$ or $NaAsO_2$, **(b)** NaF or NaCN, **(c)** Na_2SO_3 or Na_2TeO_3, **(d)** NaOCl or NaOBr.

15. Identify the good oxidizing agent(s), the good reducing agent(s), the good dehydrating agents, and the strong Brønsted acid(s) from among the following substances: H_2SO_3, HNO_3, P_4O_{10}, H_3PO_4, H_2S, H_2SO_4.

16. Write an equation for the disproportionation reaction of P_4 in sodium hydroxide.

17. Suggest tests which could be used to distinguish among NO, N_2O_5, and NO_2.

18. **(a)** Write balanced equations for the reactions of arsenic, antimony, and bismuth with concentrated nitric acid to yield H_3AsO_4, Sb_2O_5, and Bi^{3+} ions, respectively. **(b)** Explain why these three elements do not give analogous products upon reaction with HNO_3.

19. What factors are responsible for the difference in the properties of CO_2 and SiO_2?

20. $B_{10}C_2H_{12}$ is isostructural and isoelectronic with what borane ion, $B_xH_y^{z-}$?

General Exercises

21. Explain why a substance with a formula like $ZrH_{1.9}$ is regarded as a compound rather than a mixture.

22. Draw molecular orbital electronic energy level diagrams for the peroxide and the superoxide ions.

23. Pure $HClO_4$ is a liquid which does not conduct electricity. When melted, the solid hydrate, $HClO_4 \cdot H_2O$, does conduct electricity. Draw possible electron dot structures for both the acid and the acid hydrate. Discuss the importance of hydrogen bonding in these examples.

24. Suggest an explanation as to why the bond angle in Cl_2O is less than that in ClO_2, and why the O—Cl bond is longer in Cl_2O.

25. Tabulate the following data for HCl and HI: **(a)** bond energy, **(b)** percent ionic character, **(c)** dipole moment, **(d)** bond length. Explain why HI is a stronger acid than HCl.

26. Draw an electron dot structure for IF_3. Explain why I_4, a molecule which might have a similar electronic and molecular structure, does not exist.

27. Account for the fact that the formation of Cs_2O from its elements is less exothermic than the formation of ZnO from its elements.

28. State the meaning of each of the following prefixes usually used with oxyacids. Use each one to name a specific compound. **(a)** meta, **(b)** pyro, **(c)** hypo, **(d)** per, **(e)** ortho.

29. List the properties which are desirable for a liquid to be used as a refrigerant.

30. Which of the following should have the greatest enthalpy of combustion at $25\,°C$: rhombic sulfur, monoclinic sulfur, or plastic sulfur?

31. Calculate the equilibrium constant at $25\,°C$ for the disproportionation of 3 moles of aqueous HNO_2 to yield gaseous NO and aqueous NO_3^-. The standard potential for the reduction of HNO_2 to NO is 0.99 V; that for the reduction of NO_3^- to HNO_2 is 0.94 V. Comment on the stability of HNO_2.

32. The formula for the mineral olivine is sometimes written $(Fe,Mg)_2SiO_4$, meaning that 2

moles of any combination of the metal ions is present per mole of orthosilicate ion. Does olivine obey the law of definite proportions? Is olivine a compound or a solid solution?

33. Explain the structure of C_3O_2 in terms of the bonding in the molecule.

34. What is the empirical formula for the anion of beryl? From the number of shared and unshared oxygen atoms in this ion, show how the empirical formula can be deduced without knowing the size of the ring. Determine in this manner the empirical formulas (complete with charges) of the single and double chain silicate minerals illustrated in Figure 21–20.

35. (a) In a regular B_{12} icosahedron, how many boron atoms are equidistant from a given boron atom? (b) How many edges are there? (c) How many valence electrons are there? (d) Can each edge line represent an electron pair bond? (e) Explain the type of bonding involved in elemental boron.

36. Assuming that each has the icosahedral structure, determine how many isomers are possible for the $B_{10}C_2H_{12}$ molecule.

37. Borazene, $B_3N_3H_6$, is isoelectronic and isostructural with benzene. (a) Diagram the borazene structure. (b) Discuss the bonding in borazene. (c) How many isotopic disubstituted borazene molecules, $B_3N_3H_4X_2$, are possible without changing the fundamental ring structure?

38. Diagram possible structures of the B_4H_{10} and B_5H_{11} molecules, showing the existence of three-center bridge bonds.

39. Diborane, B_2H_6, reacts with water to form boric acid and hydrogen. What is the pH of the solution which results when 1.0 gram of B_2H_6 reacts with 100 ml of water? Assume the final volume to be 100 ml.

40. Explain why nuclear magnetic resonance experiments show only one type of hydrogen atom in the $B_{12}H_{12}{}^{2-}$ ion.

41. Show in terms of bond energies and entropies why Cl_2O would be expected to decompose spontaneously into its elements.

42. Account for the bond angles in each of the following on the basis of electron pair repulsion and on the basis of hybridization of the orbitals on the central atom: (a) Cl_2O, (b) ClO_2, (c) Cl_2O_7, (d) $I_3{}^-$.

43. On the basis of the data in Section 21–4, determine which of the following has the greatest affinity for water: P_4O_{10}, Cl_2O_7, I_2O_5.

44. Write electron dot formulas for and explain the geometry of each of the following: (a) ICl_3, (b) XeF_2.

45. Explain why addition of HNO_3 to concentrated H_2SO_4 results in the formation of $NO_2{}^+$ and $NO_3{}^-$ ions. How could one test experimentally to confirm that such ions exist in H_2SO_4 solution?

46. Write other resonance forms for the N_2O molecule, pictured on page 604.

47. In the preparation of P_4O_6 described on page 606, why is a mixture of nitrogen and oxygen used rather than pure oxygen?

48. Explain how the freezing of water in the crevices of rocks causes mechanical degradation.

Advanced Exercises

49. The cyanide ion is often referred to as a pseudo-halide ion. Discuss this concept using as illustrations (a) the cyanogen molecule, C_2N_2; (b) the reaction of cyanogen with OH^-; (c) the formation of cyanide coordination compounds; (d) the properties of hydrogen cyanide; (e) other evidence, if any.

50. Assume that ice has a hexagonal close packed structure with oxygen–oxygen contact and with hydrogen atoms occupying the interstices. (a) What fraction of the tetrahedral holes would be occupied by hydrogen atoms? (b) Estimate the density of ice under such conditions.

51. Show that it is thermodynamically possible to prepare MnO_4^- by the oxidation of Mn^{2+} with sodium perxenate in acid solution. ($\epsilon° = 2.1$ V.)

52. XeF_2 is relatively stable in dilute aqueous HF, despite its reduction potential of approximately 2.2 V. **(a)** Design a galvanic cell in which the following cell reaction occurs:

$$XeF_2 + 2OH^- \rightarrow Xe + \tfrac{1}{2}O_2 + 2F^- + H_2O$$

(b) Predict $\epsilon°$ for this cell. **(c)** Sketch the cell, and state what materials would be used for the electrodes, the containers, the salt bridge, etc. **(d)** Will the H_3O^+ from the ionization of HF in the solution have an appreciable effect on $\epsilon°$? Explain quantitatively.

53. Calculate the pH of 1.0 M $NaHCO_3$, using the fact that in such a solution

$$[H_2CO_3] - [CO_3^{2-}] = [OH^-] - [H_3O^+]$$

54. Calculate the electronegativities of the noble gases using the method of R. S. Mulliken (see exercise 63, Chapter 11). Are these values reasonable? Explain.

55. On the basis of appropriate Born–Haber cycles, show what factor(s) is(are) responsible for the fact that lithium nitride is stable while potassium nitride is unstable.

56. Look up the appropriate cross sections and describe how astatine might be prepared from bismuth. Which astatine isotopes would be produced? Suggest a method of separating the astatine from the target bismuth.

22

Compounds of the Metals

The characteristics which all metals have in common, such as their luster, electrical conductivity, and close-packed structure, were discussed in Chapters 18 and 19. Coordination compounds of the metals were discussed in Chapter 14. In this chapter the chemical properties of the metals will be discussed, and some of the more important compounds of metals will be described. The metals will be divided into (1) main group metals, (2) transition metals, and (3) inner transition metals, with the emphasis on various trends in the properties of the members of each class. It is more efficient to learn generalizations about related metals and their compounds than to attempt to learn all the details of their behavior. Specific facts can be looked up in reference books.

The Main Group Metals

22-1   General Trends

Within each group in the periodic table, similarities exist among the metals, the most obvious of which is their similar oxidation state(s). Each of the metals forms compounds having formulas analogous to those of all the other members of the group. For example, each of the group II metals forms an oxide of the type MO. Among the main group elements, group I A metals show the greatest resemblance to each other. Elements of the succeeding groups show more variation in their chemical and physical characteristics. Thus sodium is more like cesium in its chemical properties than aluminum is like thallium.

In each main group the maximum oxidation state of the metals is equal to the group number; indeed, for many metals the only oxidation state other than 0 is equal to the group number. However, in the heavier metals of groups III A, IV A, and V A the oxidation state two less than the group number becomes increasingly stable as a result of the "inert pair" effect (see Section 10–14). The metals at the bottom left of the periodic table are the strongest reducing agents. Thus cesium or francium might be expected to have the most negative reduction potential, but as a consequence of the large enthalpy of hydration of lithium ion, the reduction potential of lithium is more negative than that of cesium. However, it can be predicted that sodium is a weaker reducing agent than cesium because of its higher position in the periodic table and that thallium is a much weaker reducing agent than cesium because it is located to the right of cesium in the periodic table.

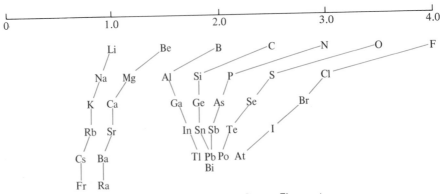

Figure 22-1. **Electronegativities of the Main Group Elements**

Sixth period elements which exhibit more than one oxidation state are more stable in the lower state than the corresponding fifth period elements. Also, the farther to the right the element is in the periodic table, the more stable is the lower oxidation state. Thus lead(II) is less easily oxidized than either tin(II) or thallium(I).

From left to right in the periodic table, and less so from bottom to top, the metal oxides decrease in base strength. Examples of this trend were presented in Section 21–2.

Lithium, beryllium, and boron—elements of the second period—have properties very similar to those of the third period elements of the next higher group. Thus lithium and magnesium have many similar chemical properties, as do beryllium and aluminum as well as boron and silicon. The similarities are due in part to the closeness of their electronegativities (Figure 22–1). Lithium is the only alkali metal which has a carbonate salt less soluble than its hydrogen carbonate salt, a behavior similar to that of magnesium and the heavier alkaline earth compounds. Lithium is also the only alkali metal which reacts directly with atmospheric nitrogen. Magnesium reacts similarly:

$$6 \, Li \; + \; N_2 \; \rightarrow \; 2 \, Li_3N$$
$$3 \, Mg \; + \; N_2 \; \rightarrow \; Mg_3N_2$$

Of the salts of the alkali metals, those of lithium are the least ionic, having about the same covalent character as the corresponding magnesium compounds. For example, the melting point of lithium chloride is more like that of magnesium chloride than like the value extrapolated from the melting points of the other alkali metal chlorides (Table 22–1). Also, lithium is similar to magnesium in that they both form normal oxides (as opposed to peroxides or superoxides). They both form simple carbides, and their phosphates are insoluble in water. Both Li^+ and Mg^{2+} are strongly hydrated in water. Their chlorides are rather soluble in organic solvents.

Beryllium forms compounds with relatively great covalent character, even with such electronegative elements as the halogens. Aluminum has the same tendency, as evidenced by the relatively low melting points of their chlorides (Table 22–1). However, the bonds in both $BeCl_2$ and $AlCl_3$ (Al_2Cl_6) have appreciable ionic character. As shown by the data in the table, the nonmetal chlorides, BCl_3, CCl_4, and $SiCl_4$, are typical covalent compounds.

TABLE 22-1. Melting Points of Some Chlorides (°C)

Compound	Melting Point	Compound	Melting Point	Compound	Melting Point
LiCl	610	$BeCl_2$	405	BCl_3	-107
NaCl	801	$MgCl_2$	714	$AlCl_3$	192
KCl	772	$CaCl_2$	782	CCl_4	-23
RbCl	717	$SrCl_2$	875	$SiCl_4$	-68
CsCl	645	$BaCl_2$	962		

Metals which are good reducing agents, whose oxides form strong bases in water, and whose ions are stable toward hydrolysis are said to be **basic.** The basicity of the main group metals decreases from left to right in the periodic table. Within a group basicity increases from the lighter elements to the heavier.

22-2   Alkali Metals

The electronic configurations of the alkali metals—lithium, sodium, potassium, rubidium, cesium, and francium—consist of one electron outside a noble gas electronic configuration. These metals have lower ionization potentials than any other elements, and they react to form (mainly) ionic compounds. They are all soft, low-melting metals which have high thermal and electrical conductivities. They form alloys with each other and alloys with mercury (amalgams) which are liquid at room temperature. These alloys are often used as reducing agents in place of the solid metals.

The alkali metals dissolve in liquid ammonia at $-33°C$ to give intensely blue solutions which are capable of conducting electricity without any chemical reaction at the electrodes. It is postulated that these solutions actually contain alkali metal ions and solvated electrons:

$$Na \xrightarrow{\text{liquid ammonia}} Na^+ + e^-(\text{solvated})$$

Evaporation of the ammonia at low temperature produces the alkali metal once more. However, heating of the solution causes the formation of a metal amide:

$$2\,Na + 2\,NH_3 \rightarrow 2\,NaNH_2 + H_2$$

The alkali metals are the least electronegative and the most basic elements. They exhibit only one positive oxidation state, and as shown by their reduction potentials, the elements are among the best reducing agents known. Small differences in electronegativity and differences in the sizes of their cations account for some differences in their compounds. Lithium, the smallest, is the only alkali metal which reacts directly with nitrogen gas. it is also the only element in the group which reacts in excess oxygen to form a normal oxide; the other elements form peroxides or superoxides.

The solubilities of some of the alkali metal salts can be related to ionic size. Salts consisting of large cations and large anions or of small cations and small anions are less soluble than salts having ions which differ in size. When the ions are nearly the same size, as in potassium perchlorate, the crystal is more stable, reducing the solubility of the substance in water. Thus lithium perchlorate is soluble, whereas potassium perchlorate is only slightly soluble.

22-3   Alkaline Earth Metals

The group II A metals occur chiefly in the minerals beryl, $Be_3Al_2Si_6O_{18}$; dolomite, $CaCO_3 \cdot MgCO_3$; carnallite, $MgCl_2 \cdot KCl \cdot 6H_2O$; strontianate, $SrCO_3$; and barite, $BaSO_4$. Also their ions occur in seawater in considerable concentrations. The preparation of magnesium from seawater has already been described (Section 19–7). The other group II A elements are prepared from their compounds by electrolysis in a manner analogous to that used for the alkali metals.

Like the alkali metals, the heavier alkaline earth metals dissolve in liquid ammonia to produce intensely blue solutions. In the case of calcium evaporation of the ammonia yields a compound in which the calcium has an oxidation state of 0:

$$Ca(s) \xrightarrow{NH_3(l)} Ca^{2+} + 2\,e^-(solvated) \xrightarrow[evaporation]{} Ca(NH_3)_6$$

Alkaline earth metals are somewhat less basic than the alkali metals. However, the basicity increases with atomic number, so that the heavier elements of the group have reduction potentials comparable to that of sodium. Their only positive oxidation state is +2. Of the group, only barium (and possibly radium) reacts with oxygen to form a peroxide. All the elements except beryllium form insoluble carbonates; calcium, strontium, barium, and radium form insoluble sulfates. The trend in basicity of the elements is illustrated by the facts that beryllium forms no simple carbonate and that magnesium and calcium carbonates lose carbon dioxide on heating, but strontium and barium carbonates do not. Similarly, magnesium and calcium hydroxides can be dehydrated to their oxides, whereas the more basic hydroxides cannot be dehydrated.

An interesting phenomenon resulting from the equilibrium between calcium carbonate, carbonic acid, and hydrogen carbonate ion is the formation of huge limestone caverns containing stalactites and stalagmites, such as the Luray Caverns in Virginia. The original caverns were formed over extensive periods of time through dissolution of the limestone by underground water containing carbon dioxide:

$$CaCO_3(s) + H_2O + CO_2(aq) \rightleftharpoons Ca^{2+}(aq) + 2\,HCO_3^-(aq)$$

Changes in pressure of carbon dioxide in the atmosphere or changes in temperature of the solution cause changes in the concentration of the dissolved carbon dioxide, which shifts this equilibrium. When the CO_2 pressure in the cavern is lowered or the temperature is increased, CO_2 is lost from the aqueous solution of calcium hydrogen carbonate and carbon dioxide as it drips from the ceiling. The equilibrium shifts to the left, causing $CaCO_3$ to precipitate from the drop. Precipitation from countless drops over centuries has built up columns of calcium carbonate suspended from the ceiling (stalactites) or rising from the floor (stalagmites).

22-4   Group III A Metals

The first member of periodic group III A, boron, is a nonmetal, but the other elements are metals. The metallurgy of aluminum has already been described (Section 19–5). The other metallic elements are found in trace quantities with aluminum ores. Their salts are also reduced to the metal by electrolysis. Gallium metal, with a melting point of 29.8°C and a normal boiling point of 2070°C, has the greatest liquid range of any known substance.

Aluminum and gallium exhibit the +3 oxidation state in their compounds. Indium and thallium, in addition to their +3 oxidation state, exhibit a +1 oxidation state in some of their compounds. The group III A metals are less basic than the metals of groups I A and II A. In their +3 oxidation states they form compounds with considerable covalent character. For example, aluminum forms dimeric halides:

$$2\ Al(s) \ + \ 3\ Cl_2(g) \ \rightarrow \ Al_2Cl_6(g)$$

The molecule consists of two tetrahedra sharing one edge (Figure 22–2).

The standard reduction potential of aluminum (-1.66 V) indicates that the metal is extremely reactive and is capable of being oxidized to Al_2O_3 at room temperature. However, Al_2O_3 forms as a thin but tightly adhering coating on the surface of the metal, and further oxidation does not occur even at elevated temperatures. Therefore cooking utensils and other articles made of aluminum are widely used. The rapid reaction of aluminum with air in the absence of the coating can be demonstrated by amalgamating the surface of a sheet of aluminum foil in an acidified solution of mercury(II) ions:

$$2\ Al(s) \ + \ 3\ Hg^{2+}(aq) \ \rightarrow \ 3\ Hg(\text{metal surface}) \ + \ 2\ Al^{3+}(aq)$$
amalgam

The aluminum oxide film does not adhere to the amalgamated surface, and the oxygen in the air attacks the metal vigorously:

$$4\ Al \ + \ 3\ O_2 \ \rightarrow \ 2\ Al_2O_3$$

The chemistry of thallium(I) illustrates the stability of the "inert pair" of $6s$ electrons (Section 10–14). The lower oxidation state thallium(I) ion is more basic than the higher oxidation state thallium(III) ion. Therefore, thallium(I) exists as a simple ion, Tl^+. TlOH is a strong base, and salts such as $TlNO_3$ are neutral. TlCl, TlBr, and TlI are slightly soluble in water. Tl_2S is insoluble ($K_{sp} = 1.2 \times 10^{-24}$). Thallium(I) is isoelectronic with lead(II). In many ways its chemical properties resemble those of silver(I). Thallium(I) is neither easily oxidized to thallium(III) nor reduced to thallium metal.

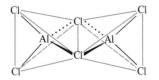

Figure 22–2. **Structure of the Al_2Cl_6 Molecule**

The other elements of group III A are commercially less important than aluminum. The use of gallium as a doping agent for semiconductors has already been mentioned (Section 19–3).

22-5   Group IV A Metals

Germanium, tin, and lead are the metallic elements in group IV A. Germanium scarcely meets this classification because the element crystallizes in the diamond lattice and has semiconducting properties. Tin exists in two allotropic modifications: white tin, which is typically metallic, and gray tin, which crystallizes in the diamond lattice. In contrast to the metallic form, gray tin has a lower density and is very brittle. The transition from white tin to gray tin occurs at 13.2°C, and gray tin is the modification which is stable at low temperatures. The transformation is very slow but is catalyzed by gray tin; hence once some gray tin has been formed, the reaction proceeds more rapidly (Figure 22–3). Lead crystallizes in a metallic form only. All of the metals are obtained by reduction of their oxides with coke. Germanium and tin occur naturally as oxides, but the chief lead ore is galena, PbS. Prior to reduction with coke, galena is **roasted** in air to form the oxide.

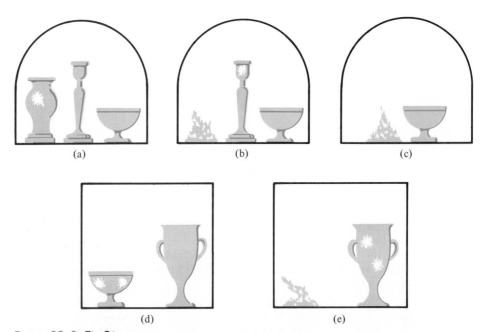

(a)   (b)   (c)

(d)   (e)

Figure 22-3. Tin Disease

The transition from white to gray tin is responsible for the legend of the "tin disease." During a severe winter one of the tin artifacts on display in a museum developed a tiny blemish which gradually spread over its entire surface. Ultimately it crumbled to powder. Another object in the same display case developed blemishes and also crumbled. A third article was apparently unaffected, but when it was transferred to another display case, some of the dust from the previously destroyed objects became attached to it. In the new case the third article began to crumble and "infected" other tin artifacts in the case. The "disease" was arrested merely by moving all of the articles made of tin into a warmer room because at the higher temperature white tin is the stable form.

$$2\,PbS \;+\; 3\,O_2 \;\rightarrow\; 2\,PbO \;+\; 2\,SO_2$$
$$PbO \;+\; C \;\rightarrow\; Pb \;+\; CO$$

The outermost electronic configuration of all of these elements is $ns^2\,np^2$, and their maximum oxidation state is $+4$. However, none forms a simple $4+$ or $4-$ ion. Similar to carbon, all form many compounds having four covalent bonds to the metal atom. Also stemming from the "inert pair" effect, each of these metals exhibits an oxidation state two less than the maximum. In fact, lead is more stable in its lower oxidation state. Tin(II) and germanium(II) are good reducing agents, but lead(II) is not; indeed, lead(IV) is a powerful oxidizing agent.

Compounds of Germanium, Tin, and Lead

The group IV A metals all form volatile, covalent tetrahalides. All of the possible halides have been prepared except PbI_4 and $PbBr_4$, which are not expected to be stable because lead(IV) can oxidize these halides to the free halogens. The preparations of the chlorides are illustrated by the following equations:

$$GeO_2 \;+\; C \;+\; 2\,Cl_2 \;\rightarrow\; GeCl_4 \;+\; CO_2$$
$$Sn \;+\; 2\,Cl_2 \;\rightarrow\; SnCl_4$$
$$PbO_2 \;+\; 4\,HCl \;\rightarrow\; PbCl_4 \;+\; 2\,H_2O$$

The tetrahalides are all Lewis acids, and they hydrolyze readily, possibly through coordination of water:

$$GeCl_4 \;+\; 2\,H_2O \;\rightarrow\; GeO_2 \;+\; 4\,HCl$$

Indeed, when $SnCl_4$ vapor reacts with water vapor, a crystalline hydrate, $SnCl_4 \cdot 5H_2O$, is formed. The ionic character of this hydrate suggests that it contains the complex ion $[Sn(H_2O)_4]^{4+}$. The tetrahalides react further with halide ions to give complex anions:

$$GeF_4 \;+\; 2\,F^- \;\rightarrow\; [GeF_6]^{2-}$$

Germanium dichloride is formed by the reaction of the tetrachloride with germanium:

$$Ge \;+\; GeCl_4 \;\xrightarrow[\text{heat}]{} \; 2\,GeCl_2$$

The dihalides of tin and lead are prepared by the reaction of the metal with the corresponding acid:

$$Sn \;+\; 2\,HX \;\rightarrow\; SnX_2 \;+\; H_2$$

The dihalides of tin and lead are monomeric in the vapor state, but solid $GeCl_2$ has a bridged structure, as shown in the following diagram. The arrows represent coordinate covalent bonds.

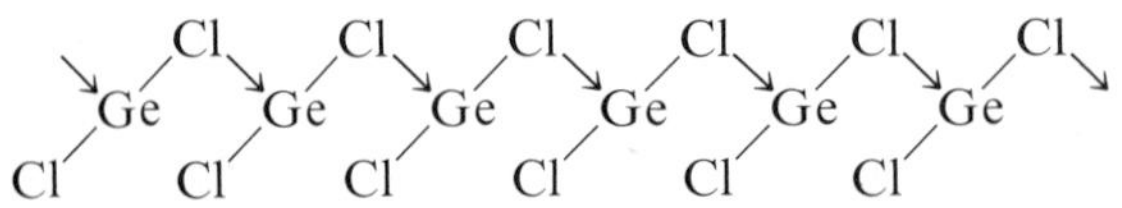

Germanium(IV) oxide and tin(IV) oxide are made by the direct reaction of the metals with oxygen; lead(IV) oxide is prepared by the oxidation of lead(II) ion in basic aqueous solution. Lead dioxide is a strong oxidizing agent. It decomposes upon heating above 300°C:

$$2 \, PbO_2 \xrightarrow[\text{heat}]{} 2 \, PbO + O_2$$

None of the oxides reacts directly with water; however, GeO_2 reacts with alkali metal oxides to form germanates:

$$GeO_2 + Na_2O \rightarrow Na_2GeO_3$$
$$\text{sodium germanate}$$

SnO_2 and PbO_2 are amphoteric. They react with strong acids to give the corresponding salts and also with bases to give stannates and plumbates:

$$SnO_2 + 2 \, H_2SO_4 \rightarrow 2 \, H_2O + Sn(SO_4)_2$$
$$SnO_2 + 2 \, NaOH \rightarrow H_2O + Na_2SnO_3$$
$$PbO_2 + 2 \, KOH \rightarrow H_2O + K_2PbO_3$$

The monoxides, SnO and PbO, react similarly, giving tin(II) salts and lead(II) salts with acids and stannites and plumbites with bases.

All of the group IV A metals form gaseous hydrides, MH_4, the thermal stabilities of which decrease in the order $GeH_4 > SnH_4 > PbH_4$. GeH_4 and SnH_4 are much less readily hydrolyzed by water and base than is SiH_4. The formation of very strong Si—O bonds is probably the driving force in the hydrolysis of SiH_4. The hydrides can be prepared by the reaction of the tetrahalide with $LiAlH_4$:

$$SnCl_4 + LiAlH_4 \xrightarrow{\text{ether}} SnH_4 + LiCl + AlCl_3$$

Higher hydrides of germanium include Ge_2H_6, Ge_3H_8, Ge_4H_{10}, and Ge_5H_{12}. Only one higher hydride of tin, Sn_2H_6, is known; no higher hydrides of lead are known.

Uses of Group IV A Metals

Germanium is used to make transistors and solar batteries. For this purpose the metal must be purified by zone refining (Section 19–8). Tin is used as a protective coating on steel in "tin cans," and low-melting tin alloys are used as solders. Lead metal and also lead glass are used as shields against exposure to nuclear radiation.

Alloys of lead containing antimony and copper are resistant to corrosion and are used for pipes in chemical plants and for sheathing on electric cables. Lead, hardened by alloying with antimony, is used in large quantities in the manufacture of lead storage batteries. Basic lead carbonate, or "white lead," $Pb_3(OH)_2(CO_3)_2$, is used as a white pigment for paint, but such lead-based paints gradually darken with age owing to the formation of black PbS by reaction with H_2S in the atmosphere. Lead paints are poisonous, and in recent years white lead has been replaced largely by titanium dioxide, which is nonpoisonous and also less likely to turn dark. The compound tetraethyllead, $Pb(C_2H_5)_4$, is used as an antiknock agent in gasoline. During the compression stroke of an internal combustion engine the air-gas mixture becomes heated (see the principle of the diesel engine, Figure 15-6), and the fuel may ignite prematurely, causing engine "knock." The tetraethyllead suppresses the premature reaction.

22-6 Antimony and Bismuth

The two metallic elements in periodic group V A, antimony and bismuth, are rather rare. They occur along with lead, copper, and tin, mainly as sulfide ores. The metals are obtained by roasting the ores to form the respective oxides, which are then reduced with carbon:

$$2\,Sb_2S_3 \;+\; 9\,O_2 \;\rightarrow\; 2\,Sb_2O_3 \;+\; 6\,SO_2$$
$$Sb_2O_3 \;+\; 3\,C \;\rightarrow\; 2\,Sb \;+\; 3\,CO$$

Antimony is used mainly as an alloying element with lead, tin, iron, and copper. Alloys of bismuth have low melting points; for example, Wood's metal (50% Bi, 25% Pb, 12.5% Sn, 12.5% Cd) melts at 71°C. Bismuth and many of its alloys expand on solidification, making them useful for casting into printing type.

The elements exhibit oxidation states of $+5$ and $+3$. The trihalides of antimony and bismuth are covalent compounds having pyramidal molecular structures. In water they decompose to form insoluble antimonyl, SbO^+, and bismuthyl, BiO^+, halides:

$$SbCl_3 \;+\; H_2O \;\rightleftharpoons\; SbOCl \;+\; 2\,HCl$$

Addition of excess acid reverses the reaction. Antimony pentachloride is prepared by the chlorination of $SbCl_3$:

$$SbCl_3 \;+\; Cl_2 \;\rightarrow\; SbCl_5$$

No pentahalides of bismuth are known, and antimony pentabromide and pentaiodide cannot be prepared because antimony(V) is too powerful an oxidizing agent. Bismuth(V) is an even more powerful oxidizing agent. Sodium bismuthate, prepared by heating together Bi_2O_3 and Na_2O_2, is able to oxidize manganese(II) to permanganate ion in acid solution:

$$14\,H^+ \;+\; 5\,BiO_3^- \;+\; 2\,Mn^{2+} \;\rightarrow\; 2\,MnO_4^- \;+\; 5\,Bi^{3+} \;+\; 7\,H_2O$$

Transition Metals

22-7   General Trends

Transition metals have one or two electrons in their outermost s subshells, immediately over a partially filled (or completely filled) d subshell. Hence in going across a period their properties do not vary as much as those of the main group elements because the number and kind of valence electrons do not vary much. Also in contrast to the main group elements, sizes in each period vary comparatively little. Major differences in the chemistries of the transition metals stem from differences in the oxidation states of the elements and in the crystal field stabilization energies of their ions (Section 14–7).

The positive oxidation states of the transition metals generally vary from $+2$ ($+1$ for the group I transition metals) up to a maximum oxidation state equal to the group number. Notable exceptions are copper, silver, and gold, which have oxidation states greater than $+1$. Trends in the stabilities of the various oxidation states of the transition metals are shown in Figure 22–4. For example, vanadium(II) and chromium(II) are excellent reducing agents; MnO_4^- is an excellent oxidizing agent; ReO_4^- is very stable; and few compounds exist with tungsten in low oxidation states.

The transition elements are often regarded as consisting of three series of elements. The first series includes the transition elements of period four, Sc through Zn; the second series, the period five transition elements, Y through Cd; and the third series, the transition elements of period six, La and Hf through Hg. Figure 22–4 shows some characteristic differences between the first series and the other two.

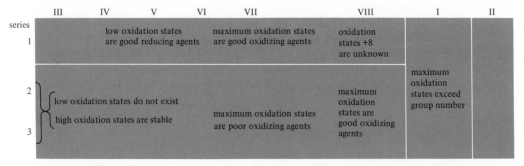

Figure 22-4. **Trends in Stability and Oxidizing Ability of the Transition Metals**

22-8   Group III B

The group III B elements are scandium, yttrium, lanthanum, and actinium. The last two are the precursors of the two series of inner transition elements, known as the lanthanides and actinides, respectively. At this point only the chemistry of scandium will be discussed. This metal has a low electronegativity and is relatively basic. Its characteristic oxidation state is $+3$. In nature it occurs with the lanthanide metal minerals, and in addition it is found in the mineral thortvietite, $Sc_2Si_2O_7$.

Scandium metal is obtained by electrolysis of a fused mixture of $ScCl_3$, KCl, and LiCl. A zinc cathode is used, and the metal is obtained as a zinc–scandium alloy. The zinc is later removed by high-temperature vacuum distillation. Scandium oxide, Sc_2O_3, is more basic than aluminum oxide, but it does not form a definite hydroxide in water. Instead, a hydrous oxide, $Sc_2O_3 \cdot xH_2O$, is formed. The carbonate, oxalate, phosphate, and chloride of scandium are insoluble in water. However, the fluoride is soluble in excess HF, owing to the formation of $[ScF_6]^{3-}$ anions. $ScCl_3$ sublimes and is monomeric in the vapor state. The chemistry of yttrium is very similar to that of scandium and to that of the lanthanide elements as well.

22-9   Group IV B

Of the group IV B elements, only titanium exhibits more than one oxidation state ($+2$, $+3$, and $+4$) in stable compounds. Zirconium and hafnium exhibit mainly the $+4$ oxidation state. Because of the lanthanide contraction (Section 10-11), the atomic radii of zirconium and hafnium are quite similar, as shown.

	Atomic Radius (Å)	Calculated Ionic Radius (Å)
zirconium	1.45	0.74
hafnium	1.44	0.75

The simple tetrapositive ions do not exist in aqueous solution.

Titanium and zirconium are relatively abundant and exist in several ores, of which rutile, TiO_2, and zircon, $ZrSiO_4$, are among the most important. (The first samples of rocks obtained from the moon contained a relatively high percentage of titanium.) TiO_2 cannot be reduced to the free metal using carbon because of the formation of a very stable carbide. Instead, chlorine is passed over a heated mixture of TiO_2 and carbon to form the volatile liquid $TiCl_4$, which is then purified by fractional distillation. Titanium is obtained as a spongy mass from the purified tetrachloride by passing the vapors of $TiCl_4$ over molten magnesium at 800°C in the absence of air. The excess magnesium and the magnesium chloride are removed by vacuum distillation at 1000°C.

$$TiO_2 \; + \; 2\,C \; + \; 2\,Cl_2 \; \rightarrow \; TiCl_4 \; + \; 2\,CO$$
$$TiCl_4 \; + \; 2\,Mg \; \rightarrow \; 2\,MgCl_2 \; + \; Ti$$

Titanium can be prepared in pure form by the decomposition of TiI_4 vapor on a hot wire (see the van Arkel process, Section 19–8). The metal ignites spontaneously in oxygen at high pressure, but at ordinary pressures a passive oxide film forms on the surface of the metal. Titanium has a tensile strength comparable to steel, and because of its low density and resistance to corrosion, the metal is of great importance in the construction of aircraft, space vehicles, and machinery for the chemical industry.

Zirconium metal is obtained in the same manner as titanium, by treatment of the ore with carbon and chlorine, followed by reduction of the chloride by magnesium. Also analogous to titanium, zirconium can be refined by the decomposition of ZrI_4.

Zirconium is much softer than titanium. It is used in the manufacture of alloy steels and also as a "getter" to remove the last traces of oxygen and nitrogen in evacuated tubes. Because the metal has a low cross section for neutrons, very pure zirconium is used in the construction of nuclear reactors.

Compounds of Titanium

Titanium dioxide is insoluble in water but dissolves slowly in sulfuric acid to form the sulfate:

$$TiO_2 + 2 H_2SO_4 \rightarrow Ti(SO_4)_2 + 2 H_2O$$

This salt is hydrolyzed by boiling water, and the product is β-titanic acid, H_4TiO_4. A different crystalline form, α-titanic acid, is precipitated by adding ammonium hydroxide to an *acidified* solution of a titanium(IV) salt. The α-form of the acid is much more reactive than the β-form. It dissolves in concentrated base to give titanates:

$$2 NaOH + H_4TiO_4 \rightarrow Na_2TiO_3 + 3 H_2O$$
$$2 NaOH + 2 H_4TiO_4 \rightarrow Na_2Ti_2O_5 + 5 H_2O$$

Titanium(IV) salts hydrolyze readily in aqueous solution to form titanyl ion, TiO^{2+}:

$$Ti(SO_4)_2 + 3 H_2O \rightleftharpoons TiOSO_4 + 2 H_3O^+ + SO_4^{2-}$$

The anhydrous titanium(IV) halides are volatile, covalent compounds. The preparation of $TiCl_4$ has been described, and the other halides are prepared similarly. $TiCl_4$ hydrolyzes vigorously, with the formation of a dense white cloud of titanium dioxide particles:

$$TiCl_4 + 2 H_2O \rightarrow TiO_2(s) + 4 HCl$$

This reaction was formerly used for the production of smoke screens and is used for sky writing. TiO_2 is used as a pigment in paints; it is very opaque and does not darken with age. Complex anions of the type $[TiCl_6]^{2-}$ and $[TiF_6]^{2-}$ are also known.

Titanium(III) oxide, Ti_2O_3, is made by heating TiO_2 with carbon:

$$2 TiO_2 + C \rightarrow CO + Ti_2O_3$$

This oxide is somewhat more basic than TiO_2 and reacts with dilute sulfuric acid to form the corresponding sulfate. Other titanium(III) salts are prepared by reducing titanium(IV) salts with zinc in an acid solution:

$$Zn + 2 TiO^{2+} + 4 H^+ \rightarrow Zn^{2+} + 2 Ti^{3+} + 2 H_2O$$

There are two isomeric hexahydrates of $TiCl_3$: $[TiCl(H_2O)_5]Cl_2 \cdot H_2O$, a green salt, and $[Ti(H_2O)_6]Cl_3$, a violet salt. Titanium(III) salts are readily oxidized by air.

Titanium monoxide is obtained when a mixture of TiO_2 and titanium metal is heated. It is oxidized by hydrogen ion to give titanium(III):

$$2\,TiO \;+\; 6\,H^+ \;\rightarrow\; 2\,Ti^{3+} \;+\; H_2 \;+\; 2\,H_2O$$

Titanium dichloride is prepared by the reduction of $TiCl_4$ with sodium amalgam. $TiBr_2$ and TiI_2 are prepared in an analogous manner. All three are strong reducing agents, readily oxidized by water.

Compounds of Zirconium and Hafnium

ZrO_2 and HfO_2 are more basic than TiO_2. ZrO_2 does not dissolve in excess strong base, but it does react with hydrochloric acid to form zirconyl chloride, which can be isolated as a hydrate, $ZrOCl_2 \cdot 8H_2O$. The structures of this and analogous zirconyl salts are complex and do not contain the simple ZrO^{2+} ion. In solution both zirconium and hafnium have a marked tendency to form complex ions such as $[ZrF_6]^{2-}$, $[ZrF_8]^{4-}$, $[Zr(C_2O_4)_4]^{4-}$, and analogous hafnium complexes.

22-10   Group V B

Carnotite, $K(UO_2)VO_4 \cdot xH_2O$, and vanadinite, $Pb_5(VO_4)_3Cl$, are the principal ores of vanadium. The metal is obtained by conversion of the vanadium to V_2O_5, followed by reduction with aluminum. Niobium and tantalum occur together in a mineral which contains 1 mole of iron(II) or manganese(II) for every 2 moles of niobate or tantalate. $(Fe,Mn)(NbO_3)_2$ is called columbite; $(Fe,Mn)(TaO_3)_2$ is called tantalate. The mineral, which is a mixture of these, is designated by the name representing the element present in greater quantity. The mineral is treated with K_2CO_3, followed by aqueous carbon dioxide, to convert it to the mixed pentoxides. Treatment with KF and HF allows the separation of the metals as the soluble K_2NbOF_5 and the less soluble K_2TaF_7. The former is reduced to niobium metal with aluminum; the latter is electrolytically reduced to tantalum metal.

Vanadium forms compounds in which its oxidation state ranges from $+5$ to $+2$. Niobium and tantalum are most stable in their $+5$ oxidation states. This trend can be shown by the reactions of the three metals at high temperatures with several nonmetals, the products of which are shown in Table 22–2.

Vanadium(V) is characterized by the amphoteric pentoxide, V_2O_5, which gives the dioxovanadium(V) ion, VO_2^+, in acid and the vanadate ion, VO_3^-, in base. The $+4$ oxidation state can be obtained by reduction of VO_2^+ by such mild reducing agents as sulfur dioxide. The $+4$ oxidation state is characterized by oxovanadium(IV) ion,

TABLE 22–2. Reaction Products of
Group V B Elements with Nonmetals

	Vanadium	Niobium	Tantalum
F_2	VF_5	$(NbF_5)_4$	$(TaF_5)_4$
Cl_2	VCl_4		
Br_2	VBr_3		
I_2	VI_3	NbI_5	
O_2	V_2O_5	Nb_2O_5	Ta_2O_5
N_2	VN	NbN	TaN

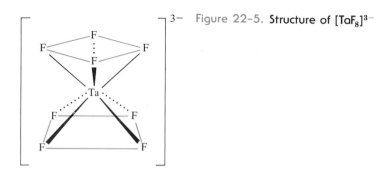

Figure 22-5. **Structure of** $[TaF_8]^{3-}$

VO^{2+}. The $+3$ and $+2$ states, produced electrolytically, are composed of simple hydrated ions, each of which is a good reducing agent. V^{2+} is powerful enough to reduce water.

The $+5$ oxidation states of niobium and tantalum are much more stable than vanadium(V). Niobium(V) and tantalum(V) are reduced only by very strong reducing agents. For example, Nb_2O_5 and Ta_2O_5 can be reduced by strong heating with magnesium metal, yielding Nb_2O_3 and TaO_2, respectively.

Tantalum forms fluoro complexes with as many as eight fluoride ions, $[TaF_8]^{3-}$. The structure of this anion is a square antiprism, shown in Figure 22-5.

22-11 Group VIB

Chromium occurs in the mineral chromite, $FeCr_2O_4$. When the ore is reduced directly, an iron–chromium alloy is produced, which is useful in steelmaking. To produce iron-free chromium, the mineral is oxidized by strong heating in air in the presence of carbonate:

$$4\,FeCr_2O_4 \;+\; 7\,O_2 \;+\; 8\,CO_3{}^{2-} \;\rightarrow\; 8\,CrO_4{}^{2-} \;+\; 8\,CO_2 \;+\; 2\,Fe_2O_3$$

The chromate produced is soluble in water and can be separated from the insoluble iron oxide and then reduced stepwise to the metal with ammonium chloride in acid and then with aluminum.

$$2\,CrO_4{}^{2-} \;+\; 2\,H_3O^+ \;\rightarrow\; Cr_2O_7{}^{2-} \;+\; 3\,H_2O$$
$$Cr_2O_7{}^{2-} \;+\; 2\,NH_4{}^+ \;\rightarrow\; Cr_2O_3 \;+\; N_2 \;+\; 4\,H_2O$$
$$Cr_2O_3 \;+\; 2\,Al \;\rightarrow\; 2\,Cr \;+\; Al_2O_3$$

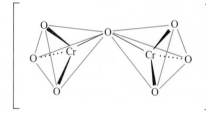

Figure 22-6. **Structure of the Dichromate Ion**

TABLE 22-3. Group VI A Oxidation Products

	Chromium	Molybdenum	Tungsten
O_2	Cr_2O_3	MoO_3	WO_3
Cl_2	$CrCl_3$	$MoCl_5$	WCl_6
Br_2	$CrBr_3$	$MoBr_3$	WBr_6

Molybdenum is obtained from molybdenite, MoS_2. The ore is roasted in air, and the oxide is reduced to the metal with hydrogen, carbon, or aluminum.

Tungsten occurs in a variety of tungstates, including ferberite, $FeWO_4$. For use in alloy steel, this mineral can be reduced directly. To make pure tungsten, such as is used in electric light bulb filaments, the mineral is first treated with carbonate, and insoluble iron compounds are removed by filtration. The filtrate is acidified to yield WO_3. Hydrogen is used to reduce the WO_3 to the metal.

The oxidation states of chromium illustrate the principle that the higher the oxidation state of a given element, the more acidic its behavior. Chromium(II) is basic; it dissolves in acid solution to yield a simple cation, but it does not dissolve in base. Chromium(III) forms a wide variety of coordination compounds, including the tetrahydroxochromate(III). Thus in acid chromium(III) forms either a complex of the anion of the acid or an aquo complex, both of which are fairly inert. In base chromium(III) yields $[Cr(OH)_4]^-$. Chromium(VI) exists as chromate, CrO_4^{2-}, in neutral or basic solution, or as dichromate, $Cr_2O_7^{2-}$, in acid:

$$2\,CrO_4^{2-} \;+\; 2\,H_3O^+ \;\rightleftharpoons\; Cr_2O_7^{2-} \;+\; 3\,H_2O$$

The structure of the dichromate ion consists of two tetrahedra sharing a common oxygen atom at one corner (Figure 22-6).

In a given transition metal group, the higher oxidation states of the heavier elements are the more stable. For example, the products of the reactions of the three group VI B metals with oxygen, chlorine, and bromine (Table 22-3) show that the $+6$ state is favored for tungsten in contrast to the $+3$ state for chromium.

Molybdenum and tungsten form complex polyanions, and even mixed polyanions, such as $Mo_7O_{24}^{6-}$, $SiMo_{12}O_{40}^{4-}$, and $BW_{12}O_{40}^{5-}$.

22-12   Group VII B

Manganese is by far the most important element in group VII B. Technetium is radioactive and does not occur naturally, although gram quantities are available as a by-product of uranium fission. Rhenium, discovered in 1925, has as yet no widespread commercial use.

Manganese occurs naturally as the mineral pyrolusite. This mineral has about 2% less oxygen than is implied by the formula MnO_2. Pyrolusite can be reduced to the metal with aluminum. The principal use of manganese is in the production of alloy steels. It reacts with any sulfur present, and the MnS produced does not adversely affect the properties of the steel. Manganese makes the steel more resistant to chemical attack.

The +3 and +6 oxidation states of manganese are stable only in basic solution, but the +2, +4, and +7 oxidation states are stable in both acid and base. MnO_2 is only marginally stable in acid; almost any reducing agent capable of reducing permanganate ion is also capable of reducing MnO_2 in acid. Permanganate ion is one of the most powerful oxidizing agents capable of existing in water. However, it decomposes on standing over long periods of time. The decomposition is accelerated in the presence of manganese(II) ion.

Rhenium is found as an impurity in molybdenite, MoS_2. The metal can be obtained by the reduction of NH_4ReO_4 with hydrogen at 400°C. Rhenium is much more stable in its +7 oxidation state than is manganese. It does exist in lower oxidation states, but generally in these states rhenium is a reducing agent. The reaction of the metal with fluorine yields ReF_6, which is noteworthy in that reaction of the metal with oxygen gives a higher oxidation state:

$$Re \ + \ 3\,F_2 \ \rightarrow \ ReF_6$$
$$4\,Re \ + \ 7\,O_2 \ \rightarrow \ 2\,Re_2O_7$$

22-13   Group VIII B

Nine metals are included in group VIII B. The first series elements, iron, cobalt, and nickel, resemble each other far more than they resemble the other six (the platinum metals). Iron, cobalt, and nickel do not exhibit an oxidation state as high as the group number. They are all ferromagnetic.

Iron is found in hematite, Fe_2O_3; magnetite, Fe_3O_4; limonite, $Fe_2O_3 \cdot 3H_2O$; and siderite, $FeCO_3$. Roasting of the ores converts them to Fe_2O_3, which is reduced to the metal by carbon monoxide in the blast furnace. The metallurgy of iron has been described in Section 19-6. Nickel occurs as pentlandite, NiS. It is roasted to the oxide in the presence of silica, and the iron impurity is removed as a slag. The nickel oxide is reduced to the metal by water gas ($CO \ + \ H_2$), after which the metal is purified by the Mond process (Section 14-5). Cobalt is found naturally as smaltite, $CoAs_2$, and cobaltite, $CoAsS$. The ore is roasted with silica, which forms a slag with any iron impurity and with the arsenate formed in the roasting process.

$$12\,CoAsS \ + \ 35\,O_2 \ \rightarrow \ 4\,Co_3O_4 \ + \ 6\,As_2O_5 \ + \ 12\,SO_2$$

The Co_3O_4 is reduced to the metallic state with aluminum.

Iron, cobalt, and nickel have sufficiently negative reduction potentials to react with strong acids to liberate hydrogen gas and form the dipositive ions. Iron(III) is formed by reaction with strong oxidizing agents. For example,

$$2\,Fe \ + \ 3\,Cl_2 \ \rightarrow \ 2\,FeCl_3$$

In solution iron exists as the hydrated ion or in complex ions in both the +2 and +3 oxidation states. In its +2 oxidation state cobalt forms simple hydrated ions or complex ions, but its +3 oxidation state is characterized by an extensive variety of complex compounds. The only simple compounds of cobalt(III) are the oxide, Co_2O_3; a double sulfate, $K_2SO_4 \cdot Co_2(SO_4)_3$; and a somewhat unstable fluoride, CoF_3. As an

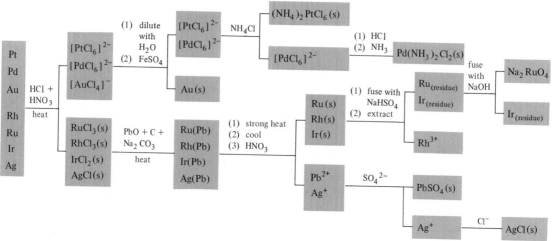

Figure 22-7. Scheme for Separation of Noble Metals

oxidizing agent, Co^{3+} is powerful enough to oxidize water. The chemistry of nickel in solution is mostly that of nickel(II). NiO_2 is the most important compound of nickel(IV). It is insoluble in water and is a powerful oxidizing agent.

The platinum metals, ruthenium, rhodium, palladium, osmium, iridium, and platinum, are quite unreactive. Together they comprise about $2 \times 10^{-5}\%$ of the mass of the crust of the earth. Their scarcity and their nobility make them very valuable. Generally these metals occur together, in ores such as sperrylite, $PtAs_2$; cooperite, PtS; and braggite, a mixture of PtS, PdS, and NiS. Nickel deposits often contain minute quantities of these metals. Osmiridium is a natural alloy of osmium and iridium. Treatment of an ore with hot aqua regia dissolves the platinum and palladium, along with any gold which might be present, but leaves the other platinum metals as insoluble chlorides. The metals can be separated by a type of qualitative analysis scheme (Figure 22-7).

Oxidation states up to $+8$ are observed only in the cases of RuO_4 and OsO_4. Halides of high oxidation state include IrF_6 and RuF_5, but mostly oxidation states from $+2$ to $+4$ are observed. In solution platinum and palladium exist only in $+2$ and $+4$ oxidation states and only as complex compounds.

Platinum and palladium are widely used as catalysts, particularly for hydrogenation reactions. The metals dissolve (absorb) hydrogen extensively, and the absorbed hydrogen is in a form which is more reactive than molecular hydrogen. Platinum is also used to prepare electrodes and to fabricate laboratory apparatus which must withstand chemical attack. Other uses of platinum include the manufacture of jewelry and dental bridgework.

22-14   The Coinage Metals—Group I B

The group I B metals are the only elements, except for the noble gases, which form compounds in which their oxidation numbers exceed their group number. The thermally stable oxidation state for copper is copper(I); but the simple Cu^+ ion disproportionates in aqueous solution, and the chemistry of this oxidation state is not as familiar as that of copper(II).

Copper occurs as copper pyrites, $CuFeS_2$; cuprite, Cu_2O; and malachite, $Cu_2(OH)_2CO_3$. The pyrites is partially oxidized to drive off arsenic impurities, is then roasted with silica to remove the iron as slag, and is fully roasted to produce metallic copper:

$$Cu_2S + O_2 \rightarrow 2\,Cu + SO_2$$

The copper is refined electrolytically, as is described in Section 5–15.

Silver occurs as argentite, Ag_2S; horn silver, $AgCl$; and pyrargyrite, Ag_3SbS_3; as well as an impurity in copper ores. Silver forms a monopositive ion which is not hydrolyzed in water. It is similar to the alkali metal ions in that its compounds are essentially ionic and in that it forms anhydrous compounds. In contrast to alkali metal ions, its oxide is only slightly soluble, but the solubility increases in hydroxide solution. Ag^+ forms insoluble sulfides and halides and has a strong tendency to form complex ions. Silver(II), produced by the reaction of $AgNO_3$ and ozone, is a short-lived species which is a very powerful oxidizing agent.

Gold occurs in nature as the metal; as the telluride, $AuTe_2$; and as an impurity in other ores. The extraction of silver and gold from their ores by the cyanide process was described in Section 14–5. Gold(III) complex compounds are encountered more often than compounds with gold in other oxidation states.

Group I B elements are well known in the form of coins and jewelry. Colloidal gold is used to make red stained glass. Silver compounds are extremely important in photography. Silver metal is the best metallic conductor of electricity known.

22–15   Group II B

Zinc, cadmium, and mercury constitute transition metal group II B. Their electronic configurations are $(n-1)d^{10}\,ns^2$, and in a sense they should not be classified as transition elements because the completely filled d subshell is not involved in bonding. The characteristic oxidation state for all the elements is $+2$, although mercury also exists in the $+1$ state. Compared to copper, silver, and gold, these elements have much higher ionization potentials; but in contrast to the coinage metals, zinc and cadmium are good reducing agents, as shown by their standard reduction potentials (Table 5–2). On the other hand, mercury is surprisingly difficult to oxidize—more difficult than copper and about as difficult as silver. Although having formulas analogous to compounds of the alkaline earth metals, compounds of zinc, cadmium, and mercury have less ionic character. Another difference between the two groups is the tendency of zinc, cadmium, and mercury to form complex compounds.

Zinc metal is obtained from the minerals zinc blende and wurtzite, both having the formula ZnS. The sulfide is converted to the oxide by roasting in air; reduction to the metal is effected with carbon at high temperature. The volatile zinc metal is distilled from the reaction mixture. Cadmium is found in zinc ores, and the metal is obtained along with the metallic zinc. Cadmium has a lower boiling point than zinc and is separated by distillation.

Mercury occurs in the mineral cinnabar, HgS. It is obtained directly from the ore by roasting in air:

$$HgS(s) \; + \; O_2(g) \; \rightarrow \; Hg(l) \; + \; SO_2(g)$$

The metal is purified by distillation.

Zinc and cadmium are soft, white metals. Both crystallize with a distorted hcp lattice. Mercury is a liquid at room temperature and is somewhat volatile. (The vapor pressure at 20°C is 1.3×10^{-3} torr.) The vapors are toxic, and prolonged exposure to the metal should be avoided.

Zinc is used as a protective coating for steel (Section 5-16). It also forms many alloys, the most important of which is brass. Cadmium is also used as an alloying metal. Because of its high cross section for neutrons, control rods in nuclear reactors are made of cadmium. Mercury metal has many uses in scientific work. These include use in thermometers and in barometers and other pressure-measuring devices. Because of its relative inertness and its electrical conductivity, mercury is also used as a flexible electrical contact in laboratory apparatus and in silent switches in the home. The formation of mercury amalgams has already been mentioned.

In addition to the simple dipositive cation, Hg^{2+}, mercury forms the diatomic cation, Hg_2^{2+}. The presence of a mercury-to-mercury bond in mercury(I) salts is indicated by several types of experimental evidence:

1. X-ray crystal analysis of mercury(I) chloride indicates linear structural units, Cl—Hg—Hg—Cl.
2. Conductivities of mercury(I) salts containing univalent anions are characteristic of $(2+, 1-)$ electrolytes.
3. Raman spectra of mercury(I) salts show a line attributed to a mercury–mercury vibration.
4. The observed potential of a concentration cell containing mercury(I) cation can be related to the Nernst equation only if a two-electron change is assumed.
5. Mercury(I) salts are diamagnetic, indicating that the electrons are all paired. The species Hg^+ would have an odd number of electrons, and its salts would be paramagnetic.

Inner Transition Metals

The inner transition metals are those with underlying incomplete f subshells. There are two series of such elements, the first of which is called the lanthanide series. The second series is called the actinide series. The two series differ markedly in their chemical properties, and are therefore best treated separately.

The 14 lanthanide elements, known historically as the "rare earth elements," are not particularly rare. Even the rarest is as abundant as such familiar elements as iodine or bismuth. The most important mineral of the lanthanides is monazite, a heavy, dark sand containing phosphates of thorium, cerium, neodymium, and lanthanum. A source of the heavier lanthanides is the mineral gadolinite, $FeBe_2Y_2Si_2O_{10}$. Thus the chemistries of yttrium and lanthanum are closely associated

TABLE 22-4. Outer Electronic Configurations and Radii of Lanthanide ("Ln") Atoms and Ions

Atomic Number	Name	Symbol	Electronic Configuration				Radius (Å)	
			Ln	Ln^{2+}	Ln^{3+}	Ln^{4+}	Ln	Ln^{3+}
57	lanthanum	La	$5d^1\,6s^2$		[Xe]		1.877	1.061
58	cerium	Ce	$4f^2\,6s^2$		$4f^1$	[Xe]	1.82	1.034
59	praseodymium	Pr	$4f^3\,6s^2$		$4f^2$	$4f^1$	1.828	1.013
60	neodymium	Nd	$4f^4\,6s^2$	$4f^4$	$4f^3$	$4f^2$	1.821	0.995
61	promethium	Pm	$4f^5\,6s^2$		$4f^4$			0.979
62	samarium	Sm	$4f^6\,6s^2$	$4f^6$	$4f^5$		1.802	0.964
63	europium	Eu	$4f^7\,6s^2$	$4f^7$	$4f^6$		2.042	0.950
64	gadolinium	Gd	$4f^7\,5d^1\,6s^2$		$4f^7$		1.802	0.938
65	terbium	Tb	$4f^9\,6s^2$		$4f^8$	$4f^7$	1.782	0.923
66	dysprosium	Dy	$4f^{10}\,6s^2$		$4f^9$	$4f^8$	1.773	0.908
67	holmium	Ho	$4f^{11}\,6s^2$		$4f^{10}$		1.766	0.894
68	erbium	Er	$4f^{12}\,6s^2$		$4f^{11}$		1.757	0.881
69	thulium	Tm	$4f^{13}\,6s^2$	$4f^{13}$	$4f^{12}$		1.746	0.869
70	ytterbium	Yb	$4f^{14}\,6s^2$	$4f^{14}$	$4f^{13}$		1.940	0.858
71	lutetium	Lu	$4f^{14}\,5d^1\,6s^2$		$4f^{14}$		1.734	0.848

with those of the lanthanides. The position of the lanthanides in the periodic table was discussed in Section 10–10. The outer electronic configurations of these elements and of some of their ions are listed in Table 22–4 along with their atomic radii and the radii of their 3+ ions. The trend in ionic radii (Figure 22–8) is what is meant by the term *lanthanide contraction*. The ionic radius of Y^{3+} (0.88 Å) falls on the curve at atomic number 68 (Er^{3+}); hence it is not surprising that the properties of yttrium compounds are similar to those of the compounds of the smaller lanthanides.

The larger lanthanides are more basic, and within the series basicity decreases with decreasing size. The metals are strong reducing agents, comparable to the alkaline earth metals and aluminum. Their oxides range from the very strongly basic La_2O_3 to the insoluble, weakly basic Lu_2O_3.

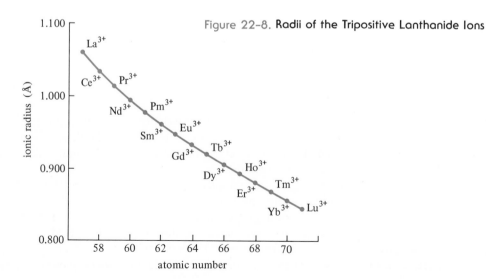

Figure 22–8. **Radii of the Tripositive Lanthanide Ions**

22-16   Electronic Configurations

A comparison of the electronic configurations of the lanthanide elements illustrates the preferred stabilities of half-filled and fully filled subshells. Thus the gadolinium atom has a $5d$ electron above a half-filled $4f$ subshell, and similarly both ytterbium and lutetium exhibit $4f^{14}$ subshell configurations. Lutetium and gadolinium form only tripositive ions, since this state gives them stable electronic configurations. Europium and ytterbium would be expected to form stable dipositive ions in addition to their tripositive ions, as is actually observed. Cerium and terbium attain the f^0 and f^7 configurations, respectively, in their $+4$ oxidation states. Samarium(II), neodymium(II), terbium(II), praseodymium(IV), neodymium(IV), and terbium(IV) states also exist but are of comparatively low stability.

The atomic volumes of the lanthanides are plotted in Figure 22-9. Large deviations from a regular trend occur at europium and ytterbium, the elements which have the greatest tendency to form dipositive ions. All of the lanthanide metals crystallize in the hcp lattice except europium and ytterbium, which crystallize in the bcc and fcc lattices, respectively. Apparently only two electrons from atoms of europium and ytterbium enter the conduction band of the metallic lattice (Section 19-2). Hence, compared to the other lanthanide metals, europium and ytterbium have weaker metallic bonding and therefore less compact crystal lattices.

Those lanthanide ions having partially filled $4f$ subshells are expected to be paramagnetic because of the presence of unpaired electrons. However, the observed paramagnetic moments of lanthanide(III) ions, plotted in Figure 22-10, do not correspond to the predictions of the simple spin-only formula (Section 10-9). Also in contrast to ions of the transition elements, where the magnetic moments are affected by complex formation, the magnetic moments of the lanthanide ions are virtually independent of their chemical environment. Thus it can be concluded that the $4f$ electrons in the ions do not participate in the bonding to any appreciable extent.

Many of the lanthanide ions have characteristic colors which are not influenced by the anion or by any complexing ligands. The colors are due to electronic transitions from one f orbital to another and are virtually independent of the environment of the metal ions. Characteristic colors of the Ln^{3+} ions are shown in Table 22-5. From the data in the table, it is tempting to associate the color with the number of unpaired electrons, but the situation is not that simple. Isoelectronic ions (which have different charges) do not have the same colors. For example, Sm^{2+}, which is reddish, is

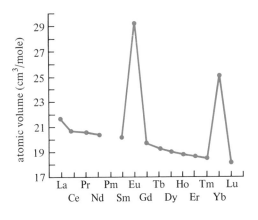

Figure 22-9. **Atomic Volumes of the Lanthanide Elements**

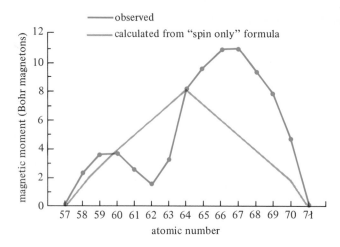

Figure 22–10. **Magnetic Moments of the Tripositive Lanthanide Ions**

isoelectronic with Eu^{3+}, which is colorless. Also, Yb^{2+} is green, whereas the isoelectronic Lu^{3+} is colorless; and Ce^{4+} is orange red in contrast to La^{3+}, which is colorless.

Typical absorption spectra of the lanthanide ions contain one or more very sharply defined absorption bands in the ultraviolet, visible, or near infrared regions. These bands are often so sharp that they resemble the lines of atomic emission spectra more closely than they resemble the typically broad absorption bands of the transition metal ions. The sharp absorptions associated with f electron transitions make the lanthanide ions useful as light filters. For example, glass containing neodymium and praseodymium ions is used to make glassblowers' goggles because wavelengths of the yellow sodium light which is emitted from molten glass correspond to energies of the electronic transitions in these ions.

TABLE 22–5. Colors of Tripositive Lanthanide Ions

Ion	Number of $4f$ Electrons	Number of Unpaired Electrons	Color
La^{3+}	0	0	colorless
Ce^{3+}	1	1	colorless
Pr^{3+}	2	2	green
Nd^{3+}	3	3	blue-violet
Pm^{3+}	4	4	rose
Sm^{3+}	5	5	pale yellow
Eu^{3+}	6	6	colorless (absorbs in uv)
Gd^{3+}	7	7	colorless (absorbs in uv)
Tb^{3+}	8	6	colorless (absorbs in uv)
Dy^{3+}	9	5	pale yellow
Ho^{3+}	10	4	yellow
Er^{3+}	11	3	pink
Tm^{3+}	12	2	green
Yb^{3+}	13	1	colorless
Lu^{3+}	14	0	colorless

22–17   Separation and Purification Techniques

The separation of the lanthanide elements from their ores and from each other is rather involved. In a typical procedure an ore such as monazite is digested with concentrated H_2SO_4 and then extracted with cold water to obtain the soluble sulfates of the lanthanides and thorium. Neutralization of the extract causes precipitation of the thorium and other elements such as zirconium and titanium. After filtration, the lighter lanthanides, La to Eu, are precipitated as a mixture of double sulfates. The heavier elements remain in solution; they are later separated as bromates after concentration of the solution. Cerium is separated from the mixture of lighter elements by oxidizing it to cerium(IV). Then the other lighter lanthanides are separated by fractional crystallization, ion exchange, or solvent extraction. In fractional crystallization, advantage is taken of the very slight differences in solubility of analogous compounds of the lanthanide metals, "Ln," such as the double manganese nitrates, $Ln_2Mn_3(NO_3)_{12} \cdot 24H_2O$. The procedure is very tedious, involving many separate enrichment steps, and it can take many months or even years to separate appreciable quantities of reasonably pure lanthanide compounds. In contrast, ion exchange (Figure 21–5) and solvent extraction procedures can be made practically continuous, and commercial scale separations of the lanthanides can be effected rather quickly. The ion exchange separation is based on the fact that as the lanthanide ions of a given charge decrease in size, they are less strongly held by a cation exchange resin. Thus the lanthanides are absorbed on the resin in order of increasing atomic number. Upon elution from the column, the ions are removed in order of decreasing atomic number. The separation is enhanced if the eluting solution contains a complexing ligand and is maintained at an appropriate pH. Ions possessing smaller radii tend to form stronger complexes and thus enter the eluting phase even more readily. Figure 22–11 shows an elution curve for a typical mixture of lanthanides.

After separation, the metals are obtained from their compounds by the techniques usually employed for active metals. These include (1) electrolysis of the fused chlorides, (2) reduction of anhydrous chlorides with sodium or calcium, and (3) reduction of the anhydrous fluorides with magnesium.

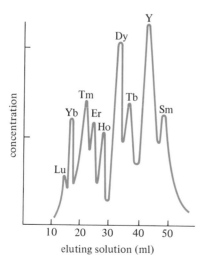

Figure 22–11. Typical Elution Curve for Lanthanide Ions on a Cation Exchange Column Using Buffered Ammonium Citrate Solution

22-18   Properties of the Lanthanides

The lanthanide elements have a silvery white luster but tarnish readily in air. The lighter elements, lanthanum and cerium, are as soft as tin. Hardness of the elements increases rapidly with atomic number, however, and samarium is as hard as steel. The metals react directly with oxygen and with the halogens, slowly at room temperature but vigorously above 150°C. Similarly, they react with sulfur at its normal boiling point to give Ln_2S_3 and with nitrogen above 1000°C to give LnN. They react slowly with cold water, liberating hydrogen:

$$2\,Ln \;+\; 6\,H_2O \;\rightarrow\; 2\,Ln(OH)_3 \;+\; 3\,H_2$$

The metals react exothermically with hydrogen above 300°C to form nonstoichiometric compounds containing LnH_2 and LnH_3 phases. Such formulas as $CeH_{2.8}$ and $PrH_{1.8}$ suggest that these are interstitial compounds; however, they have unusually high stabilities and are saltlike in nature.

22-19   Promethium

The element promethium, Pm, does not exist in nature, except possibly in trace quantities formed by the spontaneous fission of uranium. Indeed, from a consideration of nuclear properties (Section 20–7), it can be deduced that any isotope of promethium will be radioactive, decaying to known stable isotopes of neodymium and samarium. Prior to 1947 claims of the discovery of the element of atomic number 61 all proved to be false. However, in that year there was separated from among the fission products of ^{235}U a radioactive element whose elution behavior in a mixture of known lanthanides corresponded to atomic number 61. The isotope, with a half life of 2.64 years, emitted a low-energy beta particle. It was shown to have a mass number of 147. Its discoverers suggested the name *promethium* (after Prometheus, "born of fire"). Subsequently, $^{147}_{61}Pm$ has been prepared by the irradiation of stable neodymium in a nuclear reactor:

$$^{146}_{60}Nd(n,\gamma)^{147}_{60}Nd \xrightarrow[\text{11 days}]{-\beta^-} {}^{147}_{61}Pm$$

Despite its radioactivity, ^{147}Pm is reasonably easy to work with. Milligram quantities of its salts have been prepared. Solutions of Pm^{3+} are pink.

22-20   Uses of the Lanthanides

The lanthanides are seldom used as the pure metals, but their alloys (known as mischmetals) are used as reducing agents in metallurgy. A typical mischmetal contains about 50% Ce, 25% La, 18% Nd, 5% Pr, and 2% Sm plus other lanthanides. Zirconium–magnesium alloys containing about 3% mischmetal are used to make parts for jet engines. A mixture of 70% mischmetal and 30% iron is pyrophoric (spontaneously combustible in air) and is used to make flints for cigarette lighters.

Mischmetal when alloyed with metals such as aluminum, nickel, copper, vanadium, and steel imparts workability, hardness, and resistance to corrosion to the product.

The use of neodymium and praseodymium oxides in the production of glass filters has already been mentioned. Other lanthanide oxides used in glass manufacture include CeO_2 and La_2O_3. Praseodymium oxide in combination with zirconium oxide is used to make a yellow ceramic glaze.

22-21   The Actinide Series

Actinium, thorium, protactinium, and uranium are the heaviest elements found in nature. All known isotopes of these elements are radioactive, and it is only because the half lives of the isotopes $^{232}_{90}Th$, $^{235}_{92}U$, and $^{238}_{92}U$ are of the same order of magnitude as the age of the earth that they are found at all. Actinium and protactinium have such short half lives that they exist only as daughter elements in the actinium and uranium decay series (Section 20–7).

In 1940 an isotope of the element having the atomic number 93, later called neptunium, Np, was obtained by means of the following sequence of nuclear reactions:

$$^{238}_{92}U(n,\gamma)^{239}_{92}U \xrightarrow[23 \text{ min}]{-\beta^-} {}^{239}_{93}Np$$

Later in that year an isotope of the element having the atomic number 94, subsequently named plutonium, Pu, was prepared by the following sequence of nuclear reactions:

$$^{238}_{92}U(d,2n)^{238}_{93}Np \xrightarrow[2.1 \text{ days}]{-\beta^-} {}^{238}_{94}Pu$$

Comparison of the properties of these new elements and those of actinium, thorium, protactinium, and uranium with the properties of the lanthanides prompted the suggestion that elements 90 through 94 are members of a second inner transition series in which the differentiating electrons are in the $5f$ subshell.

Further nuclear bombardment experiments and separation procedures based on analogy with the lanthanides (to be described later) led to the "discovery" of successive new elements, until a complete actinide series of 14 elements became known. The probable electronic configurations of these actinides, and the radii of their ions, are given in Table 22–6. The actinides have several characteristic oxidation states (Figure 22–12), although in the cases of the heavier members of the series, the +3 state is the most probable. The increased number of oxidation states compared to that observed for the lanthanide elements is attributed to the greater ease of participation of $5f$ orbitals in bonding compared to the $4f$ orbitals.

Paralleling the lanthanide contraction, the radii of both the +3 and +4 actinide ions progressively decrease with increasing atomic number. The atomic volumes of the actinides (Figure 22–13) show a sudden increase at americium, which is analogous to the increase in atomic volume of europium (Figure 22–9). The magnetic susceptibilities of some actinide ions are plotted in Figure 22–14. The variations with the

TABLE 22-6. Outer Electronic Configurations and Radii of Actinide ("An") Atoms and Ions

Atomic Number	Name	Symbol	Electronic Configuration			Radius (Å)	
			An	An^{3+}	An^{4+}	An^{3+}	An^{4+}
89	actinium	Ac	$6d^1\,7s^2$	$[Rn]$		1.11	
90	thorium	Th	$6d^2\,7s^2$		$[Rn]$		0.99
91	protactinium	Pa	$5f^2\,6d^1\,7s^2$ or $5f^1\,6d^2\,7s^2$		$5f^1$		
92	uranium	U	$5f^3\,6d^1\,7s^2$	$5f^3$	$5f^2$	1.03	0.93
93	neptunium	Np	$5f^5\,7s^2$	$5f^4$	$5f^3$	1.01	0.92
94	plutonium	Pu	$5f^6\,7s^2$	$5f^5$	$5f^4$	1.00	0.90
95	americium	Am	$5f^7\,7s^2$	$5f^6$	$5f^5$	0.99	0.89
96	curium	Cm	$5f^7\,6d^1\,7s^2$	$5f^7$	$5f^6$	0.98	0.88
97	berkelium	Bk	$5f^8\,6d^1\,7s^2$ or $5f^9\,7s^2$	$5f^8$			
98	californium	Cf	$5f^{10}\,7s^2$	$5f^9$			
99	einsteinium	Es	$5f^{11}\,7s^2$	$5f^{10}$			
100	fermium	Fm	$5f^{12}\,7s^2$	$5f^{11}$			
101	mendelevium	Md	$5f^{13}\,7s^2$	$5f^{12}$			
102	nobelium	No	$5f^{14}\,7s^2$	$5f^{13}$			
103	lawrencium	Lr	$5f^{14}\,6d^1\,7s^2$	$5f^{14}$			

number of $5f$ electrons is comparable to the variations observed in the magnetic susceptibilities of the tripositive lanthanide ions (Figure 22–10). Moreover, the high value observed for curium(III) suggests the existence of a half-filled $5f$ subshell. Parallels in the absorption spectra of the actinides and the lanthanides are also observed. Species with the same number of f electrons—for example, Pu^{3+} and Sm^{3+},

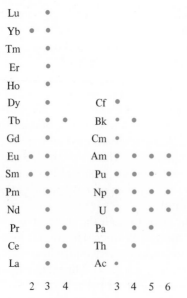

Figure 22-12. **Oxidation States of the Inner Transition Metals**

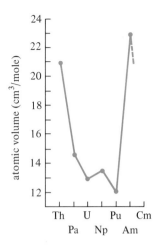

Figure 22-13. **Atomic Volumes of the Actinide Elements**

U^{3+} and Nd^{3+}, and Am^{3+} and Eu^{3+}— have roughly comparable absorption spectra.

The naturally occurring actinides are found in the following abundances in the crust of the earth: Ac, $3 \times 10^{-14}\%$; Th, $1.2 \times 10^{-3}\%$; Pa, $8 \times 10^{-11}\%$; and U, $4 \times 10^{-4}\%$. Thus thorium and uranium are comparable in abundance to zinc. The chief source of thorium is monazite, and the element is separated from the lanthanides as a soluble oxalate. Ultimately, the metal is obtained by reducing $ThCl_4$ with calcium:

$$ThO_2 \;+\; 2\,COCl_2 \;\rightarrow\; ThCl_4 \;+\; 2\,CO_2$$
$$ThCl_4 \;+\; 2\,Ca \;\rightarrow\; Th \;+\; 2\,CaCl_2$$

Highly pure thorium metal can be produced by the thermal decomposition of ThI_4 on a heated filament (the van Arkel process).

Uranium occurs in the minerals carnotite, $KUO_2VO_4 \cdot 1.5H_2O$, and pitchblende, U_3O_8. The ore is fused with Na_2CO_3 and $NaNO_3$, after which it is extracted with dilute H_2SO_4 to obtain the soluble uranyl sulfate, UO_2SO_4. Further purification is achieved by ion exchange, and the uranium is usually recovered in the form of uranyl

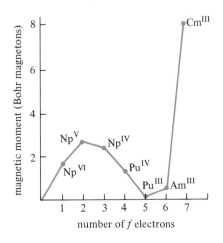

Figure 22-14. **Magnetic Moments of Some Actinide Species**

nitrate, $UO_2(NO_3)_2$. Reduction to the metal is carried out by the following steps:

$$UO_2 \xrightarrow[\text{heat}]{\text{HF}} UF_4 \xrightarrow[\text{heat}]{\text{Mg}} U$$

Very pure metal can be obtained by the van Arkel process. Uranium and plutonium are used in nuclear reactors. Uranium-235 is used as a fuel, and uranium-238 is used as a raw material for the production of plutonium (see Section 20–12). Nonnuclear applications of the elements include the use of thorium in magnesium alloys, gas discharge lamps, and high-intensity lamps. ThO_2 is used as a refractory substance and as a ceramic cathode in heat lamps. Uranium is used as a target in X-ray tubes. Uranium compounds were formerly used in ceramics as coloring agents, but this use has been discontinued in the light of the radioactivity of the element.

Typical nuclear reactions for the production of the transuranium elements (those having atomic numbers over 92) are summarized in Table 22–7. Complete details of their production can be found in other sources. Weighable quantities of elements up to curium have been synthesized. (Plutonium is produced by the ton.) Because of their high radioactivity and tendency toward spontaneous fission, no more than a few atoms at a time of elements heavier than Cm have been made. The separations and characterization of the latter elements have been achieved mainly by use of ion

TABLE 22-7. Synthesis of the Transuranium Elements

Atomic Number	Mass Number	Reaction	Nuclear Device
93	239	$^{238}_{92}U(n, \gamma)^{239}_{92}U \xrightarrow[\text{23 min}]{-\beta^-} {}^{239}_{93}Np$	pile
94	239	$^{239}_{93}Np \xrightarrow[\text{2.3 days}]{-\beta^-} {}^{239}_{94}Pu$	
95	241	$^{239}_{94}Pu(n, \gamma)^{240}_{94}Pu(n, \gamma)^{241}_{94}Pu \xrightarrow[\text{13 yr}]{-\beta^-} {}^{241}_{95}Am$	pile
96	244	$^{241}_{94}Pu(n, \gamma)^{242}_{94}Pu(n, \gamma)^{243}_{94}Pu \xrightarrow[\text{5 hr}]{-\beta^-}$	
		$^{243}_{95}Am(n, \gamma)^{244}_{95}Am \xrightarrow[\text{26 min}]{-\beta^-} {}^{244}_{96}Cm$	pile
97	249	$^{244}_{96}Cm(n, \gamma)^{245}_{96}Cm(n, \gamma)^{246}_{96}Cm(n, \gamma)^{247}_{96}Cm$	
		$^{247}_{96}Cm(n, \gamma)^{248}_{96}Cm(n, \gamma)^{249}_{96}Cm \xrightarrow[\text{65 min}]{-\beta^-} {}^{249}_{97}Bk$	pile
98	250	$^{249}_{97}Bk(n, \gamma)^{250}_{97}Bk \xrightarrow[\text{3.13 hr}]{-\beta^-} {}^{250}_{98}Cf$	pile
99	253	$^{250}_{98}Cf(n, \gamma)^{251}_{98}Cf(n, \gamma)^{252}_{98}Cf(n, \gamma)^{253}_{98}Cf \xrightarrow[\text{20 days}]{-\beta^-} {}^{253}_{99}Es$	pile
100	254	$^{253}_{99}Es(n, \gamma)^{254}_{99}Es \xrightarrow[\text{36 hr}]{-\beta^-} {}^{254}_{100}Fm$	pile
101	256	$^{253}_{99}Es(\alpha, n)^{256}_{101}Md$	cyclotron
102	254	$^{246}_{96}Cm(^{12}_{6}C, 4n)^{254}_{102}No$	cyclotron, HILA[a]
103	257	$^{250}_{98}Cf(^{11}_{5}B, 4n)^{257}_{103}Lr$	cyclotron, HILA[a]

[a] Heavy ion linear accelerator.

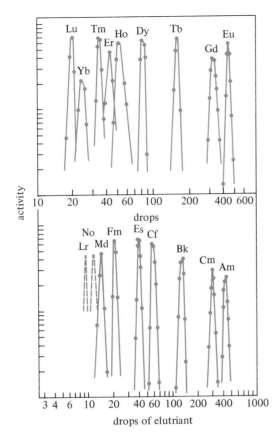

Figure 22-15. **Elution of Lanthanide 3+ Ions and Actinide 3+ Ions from Dowex 50 Cation Exchange Resin**
The predicted positions of No^{3+} and Lr^{3+} are shown by the broken lines. [Reprinted with permission from J. J. Katz and G. T. Seaborg, *The Chemistry of the Actinide Elements* (London: Methuen & Co. Ltd., 1957), p. 435.]

exchange techniques. As shown in Figure 22-15, the parallels between the elution behavior of the lanthanide ions and that of the actinide ions have proved very useful in such separations.

22-22   Exercises

Basic Exercises

1. Define each of the following terms: (a) refractory, (b) galvanized steel, (c) basic metal, (d) engine knock, (e) stalactite, (f) stalagmite.

2. List three properties which illustrate the diagonal relationship between lithium and magnesium, and three properties which illustrate the similarity of lithium to the other alkali metals.

3. Plot the melting points of NaCl, KCl, RbCl, and CsCl versus their formula weights. By extrapolation of the plot, estimate the melting point of lithium chloride. Explain why LiCl actually melts at a much lower temperature.

4. In each set select the substance which is the (a) most basic: Al_2O_3, Tl_2O_3, Tl_2O, (b) lowest melting: $LiBr$, $BeBr_2$, BBr_3, (c) highest in electronegativity: Li, Be, Mg, (d) most stable toward oxidation: $GeCl_2$, $SnCl_2$, $PbCl_2$, (e) strongest oxidizing agent: CrO_4^{2-}, MoO_4^{2-}, WO_4^{2-}.

5. Lead(IV) is a powerful oxidizing agent. What can be said about the reducing ability of lead(II)?

6. Write a balanced chemical equation for the reaction of PbO with NaOH.

7. A solution of $BiCl_3$ in aqueous HCl yields a white precipitate when diluted with pure water. By means of a balanced chemical equation, explain this result.

8. Using tables of reduction potentials, predict the products of the reactions of CrO_4^{2-} with excess **(a)** Fe^{2+}, **(b)** Zn, **(c)** Fe, **(d)** H^-.

9. What properties of tungsten make it suitable for use as filaments in light bulbs?

10. Explain why transition group VIII B includes nine elements, whereas the other transition groups contain three elements each.

11. Using complete equations and stating specific conditions where appropriate, tell how each of the following preparations might be performed: **(a)** $AgNO_3$ from AgCl, **(b)** Hg from Hg_2Cl_2, **(c)** V from $Pb_5(VO_4)_3Cl$, **(d)** pure Ta and pure Nb from an iron, niobium, tantalum oxide.

12. Figure 22–15 shows that the order of elution of the transuranium elements from an ion exchange resin parallels that of the lanthanide elements. Suggest an explanation for this observation.

General Exercises

13. Which would you expect to be more soluble in water, LiI or KI? Explain your choice.

14. Compounds having the formulas $CsBr_3$ and $CsBrCl_2$ are stable below 100°C and crystallize in a cubic lattice. Is the existence of these compounds a contradiction of the statement that alkali metals have only one positive oxidation state? Explain.

15. Solutions of alkali metals in liquid ammonia at −33°C conduct electricity without chemical reaction at the electrodes. Would decreasing the temperature of the solution cause a decrease or increase in conductivity? Explain.

16. Predict the following: **(a)** melting point of francium, **(b)** density of francium, **(c)** type of lattice of francium, **(d)** melting point of FrCl, **(e)** lattice type and density of FrI, **(f)** the product of combustion of francium metal in air, **(g)** the relative enthalpies of formation of FrF and FrI.

17. Magnesium metal burns in air to give a white ash. When this material is dissolved in water, the odor of ammonia can be detected. Suggest an explanation for this observation.

18. Potassium permanganate can be prepared by melting together equimolar quantities of MnO_2, KOH, and KNO_3. The reaction mixture is allowed to solidify, and then the dark green solid, K_2MnO_4, is crushed and treated with dilute H_2SO_4, whereupon a solution of $KMnO_4$ and a precipitate of MnO_2 are obtained. Write all relevant equations for this preparation.

19. What is the oxidation state of iron in pyrites, FeS_2? Is the anion diatomic? What physical measurements could be performed to test this hypothesis?

20. Two iron electrodes connected by a wire are placed in a solution of air-free KCl. When oxygen gas is bubbled around one of the electrodes, the other electrode begins to dissolve, and a current is observed in the wire. The solution around the electrode over which oxygen is being bubbled becomes basic. Explain these observations, write appropriate electrode reactions, diagram the cell, label the electrodes as anode and cathode, and indicate the direction of electron flow in the wire.

21. In steel manufacture manganese metal is used as a *scavenger* to reduce traces of iron oxide and iron sulfide. Suggest two reasons why manganese is effective for this purpose.

22. Osmium forms a +8 oxidation state compound with oxygen but no such compound with fluorine. Explain this behavior.

23. Write complete and balanced equations for reactions between the following substances. If no reaction occurs, so indicate.
    (a) Cu_2O + MnO_4^-
    (b) Cu_2O + H_2SO_4(dil)
    (c) V^{2+} + I_2

(d) WO_3 + I_2

(e) Zn + HCl

(f) ReO_4^- + Bi^{3+}

(g) Sn^{2+} + PbO_2

(h) CrO_4^{2-} + H_3O^+

(i) $Cr_2O_7^{2-}$ + H_2O_2

(j) Cr^{2+} + H_2O_2

24. Calculate ϵ for the cell

$$Pt \mid H_2(1.0\ atm) \mid H^+(1.0\ M) \parallel Hg_2^{2+}(0.10\ M) \mid Hg$$

25. Explain why none of the lanthanide sesquioxides, Ln_2O_3, is as acidic as Al_2O_3.

26. Suggest a reason why, of the Ln^{3+} ions, the magnetic moment of only Gd^{3+} is in agreement with the moment calculated from the spin-only formula (see Figure 22–10).

27. Which tripositive lanthanide ion(s) is(are) most likely to react with chromium(II) chloride? What products are expected? Explain.

28. After consulting the data of Tables 5–2, 8–1, and 8–2, complete and balance the following equations:

(a) Cu_2O + HI →

(b) Cu_2O + $HCl(dil)$ →

(c) Cu_2O + $HCl(conc)$ →

29. Predict whether Tl^+ will disproportionate in aqueous solution, given the following standard reduction potentials:

$$Tl^+ + e^- \to Tl \qquad \epsilon^\circ = -0.34\ V$$

$$Tl^{3+} + 2\,e^- \to Tl^+ \qquad \epsilon^\circ = 1.25\ V$$

30. Explain why aluminum(III) is the only stable oxidation state of aluminum in its compounds, in contrast to thallium, which has states of $+1$ and $+3$.

31. Californium undergoes spontaneous fission as follows: $^{252}_{98}Cf$ → $^{142}_{56}Ba$ + $^{106}_{42}Mo$ + $4\,^1_0n$. With the aid of Figure 20–10, estimate the energy released in this process.

Advanced Exercises

32. Reduction potential measurements show that, thermodynamically, calcium is a more powerful reducing agent in basic solution than barium, while barium is more powerful than calcium in neutral or acidic solution. Explain these observations.

33. Describe, using a diagram, the probable d orbital crystal field splittings in the TaF_8^{3-} ion. Assuming only spin effects, would such an ion be high spin or low spin?

34. What mass of gadolinium will produce the same reduction in neutron flux in a reactor as 1.00 gram of cadmium? The cross section of gadolinium is 46,000 barns/atom.

35. Calculate the concentration of Au^{3+} in 1.0 liter of solution of 0.010 mole of gold(III) and 1.0 mole of chloride ion.

36. The relative magnitudes of the reduction potentials of metals can be estimated by comparison of the net enthalpy changes for the following processes. If ΔH_{ox} is the enthalpy change for the reaction $M(s)$ → $M^{n+}(aq)$ + $n\,e^-$,

$$\Delta H_{ox} = \Delta H_{sub} + \Sigma(IP) + \Delta H_{hyd}$$

where ΔH_{sub} is the enthalpy of sublimation of the metal, $\Sigma(IP)$ is the sum of the successive ionization potentials to go from $M(g)$ to $M^{n+}(g)$, and ΔH_{hyd} is the enthalpy of hydration of the gaseous M^{n+} ion. Obtain appropriate data from this text and/or other sources and show what factors are responsible for the following: (a) ϵ°_{red} for lithium is more negative that that for rubidium, (b) beryllium and magnesium are poorer reducing agents than barium. (c) silver is not as good a reducing agent as is strontium.

23

Organic Reactions and Biochemistry

The chemical properties of organic compounds are determined to a large extent by their functional groups (Section 12–4). Thus alcohols have properties characteristic of the covalently bonded O—H group, while aldehydes and ketones have properties which are due to the presence of the $>C{=}O$ group (Section 12–7). Molecules may contain more than one functional group; indeed many molecules, synthetic as well as naturally occurring, contain several different kinds of functional groups. The synthetic fibers (rayon, nylon, etc.), plastics, synthetic rubber, dyes and pigments, pharmaceuticals, antibiotics, insecticides, and cosmetics contain such polyfunctional molecules. In this chapter, typical reactions of the various types of organic compounds will be described. It will be shown how the properties of relatively complex molecules can be understood in terms of structure and bonding discussed earlier. A number of important applications of organic reactions and organic compounds will be mentioned.

Many characteristic, specialized, and unique chemical reactions occur in living cells. The study of such processes and the compounds involved is called **biochemistry.** Many aspects of biochemistry involve straightforward, typical reactions of organic compounds. However, other reactions occurring in living cells often involve specific catalysts, called **enzymes,** which make it possible for a specific reaction to occur rapidly at relatively low temperatures. Although knowledge of cell processes and their mechanisms is far from complete, enough is known to indicate that heredity, resistance to disease, and indeed many aspects of life itself can be fruitfully studied in terms of the structures and chemistry of the molecules involved.

23-1 Reactions of Hydrocarbons

Hydrocarbons undergo few characteristic chemical reactions. At elevated temperatures in the presence of sufficient oxygen, they all undergo oxidation to carbon dioxide and water. Hence a major use of hydrocarbons is as fuels. Some examples are given in Table 23–1. Hydrocarbons are also useful solvents, particularly for substances having nonpolar molecules. Because of their relative inertness, and their immiscibility with water, and because they wet metal surfaces readily, nonvolatile hydrocarbons are also used as lubricants for machinery.

At ordinary temperatures the alkanes are the least reactive hydrocarbons. They react slowly with Cl_2 or Br_2 to form substituted alkanes and the corresponding hydrogen halide. The reaction is catalyzed by light and proceeds by a chain mecha-

TABLE 23-1. Hydrocarbon Fuels

Name	Composition	Use
acetylene	C_2H_2	fuel for high temperature welding
natural gas	mixture of volatile hydrocarbons, CH_4, C_2H_6, C_4H_{10}	industrial and home use
propane	C_3H_8	bottled fuel gas
butane	C_4H_{10} isomers	bottled fuel gas
gasoline	mixture of C_8H_{18} isomers and other volatile hydrocarbons	internal combustion engines
kerosene	mixture of alkanes with about 12 carbon atoms	jet fuel, home use
paraffin wax	mixture of alkanes with 16 or more carbon atoms	candles

nism. In a **chain reaction** a product which is formed in one of the steps reacts in successive steps to produce more of the reactive species; hence certain steps recur in a cyclic or "chain" manner. For example, the mechanism for the reaction between chlorine and methane is as follows:

$$Cl_2 \xrightarrow{\text{light}} 2\,Cl\cdot \qquad \text{(chain initiation)}$$

$$\left.\begin{array}{l} CH_4 + Cl\cdot \longrightarrow \cdot CH_3 + HCl \\ \cdot CH_3 + Cl_2 \longrightarrow CH_3Cl + \cdot Cl \end{array}\right\} \text{(chain propagation)}$$

The chlorine atoms produced in the third step react with more methane molecules via the second step, and the reaction proceeds over many cycles. The chain process may be terminated before all the reactants are consumed by any of the several steps involving the bonding of two reactive species to form a stable molecule, such as the following:

$$\left.\begin{array}{l} \cdot CH_3 + \cdot Cl \rightarrow CH_3Cl \\ \cdot CH_3 + \cdot CH_3 \rightarrow CH_3CH_3 \\ \cdot Cl + \cdot Cl \rightarrow Cl_2 \end{array}\right\} \text{(alternative chain-termination steps)}$$

The chain might also be broken by the addition of a substance, called an **inhibitor,** which readily combines with chlorine atoms or with methyl free radicals,[1] $\cdot CH_3$, thus breaking the chain. When the hydrocarbon consists of a branched chain of carbon atoms, the hydrogen atoms on the carbon containing the fewest hydrogen atoms are preferentially replaced by the halogen atoms, especially in the case of bromine.

Example

Predict the major product of the reaction of Br_2 with excess $CH_3CH_2CH{-}CH_3$.
$$\qquad\qquad\qquad\qquad\qquad\qquad\qquad\qquad\qquad |$$
$$\qquad\qquad\qquad\qquad\qquad\qquad\qquad\qquad CH_3$$

$$CH_3CH_2\underset{\underset{CH_3}{|}}{C}H{-}CH_3 + Br_2 \rightarrow CH_3CH_2\underset{\underset{CH_3}{|}}{C}Br{-}CH_3 + HBr$$

[1] A **free radical** is a neutral molecule (or atom) which contains an unpaired electron.

Alkenes and alkynes are more reactive than alkanes. A characteristic reaction is addition at the multiple bond:

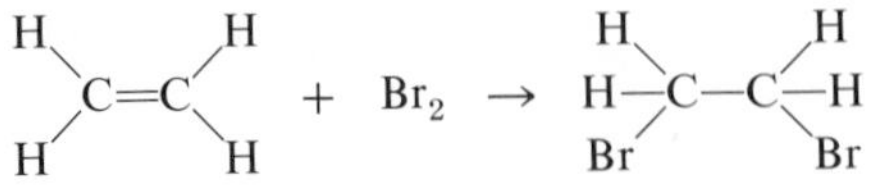

These reactions occur in the dark as readily as in light and are believed to proceed via a mechanism involving coordination of the electrons in the π orbital of the double bond. The electrons in the π orbital polarize the Br_2 molecule, and the positive end of the molecule coordinates with the π electrons. The negatively charged Br^- ion then attacks the complex from the *trans* direction to complete the addition.

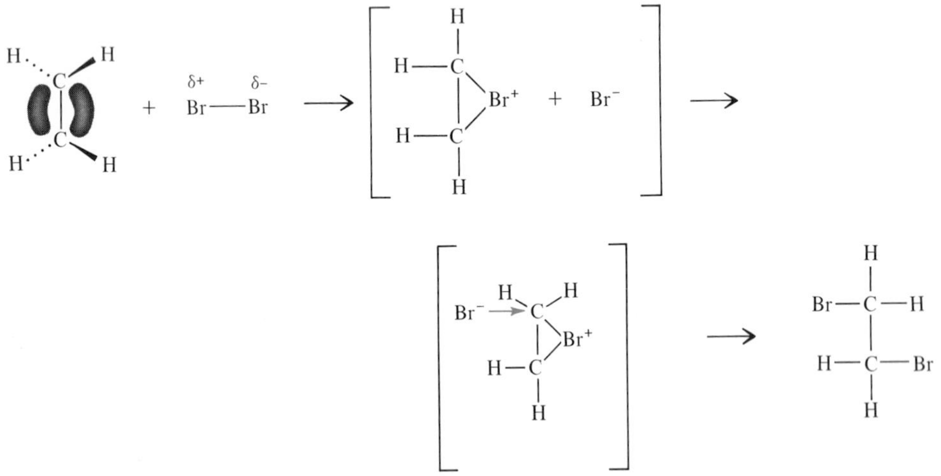

To distinguish between saturated and unsaturated hydrocarbons, a sample is treated with a solution of bromine in carbon tetrachloride. The bromine will react with any carbon–carbon multiple bonds present, and the carbon tetrachloride solution will lose its color. Under these conditions saturated hydrocarbons do not react with bromine.

Hydrogen also may be added to a multiple bond, in the presence of a catalyst:

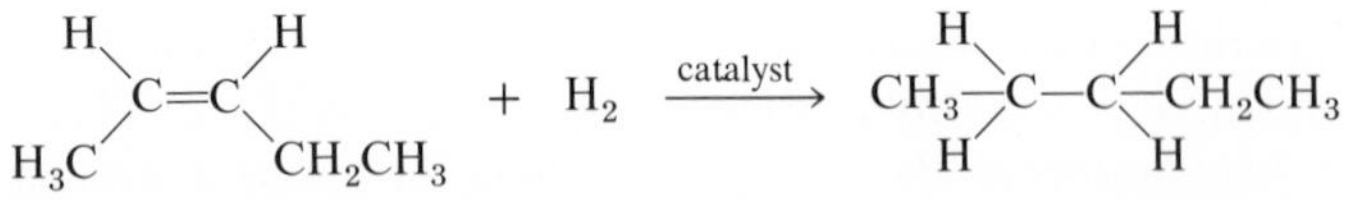

The probable mechanism of the catalyzed addition of hydrogen to double bonds involves the addition of hydrogen *atoms* adsorbed onto the surface of the metal catalyst to the organic molecule, which, through its π orbital, has also been adsorbed onto the surface (Figure 23–1).

The chemical properties of the aromatic hydrocarbons are somewhat more like those of the alkanes than those of the alkenes. For example, the characteristic reaction with halogens is substitution rather than addition.

$$C_6H_6 \;+\; Br_2 \;\rightarrow\; C_6H_5Br \;+\; HBr$$

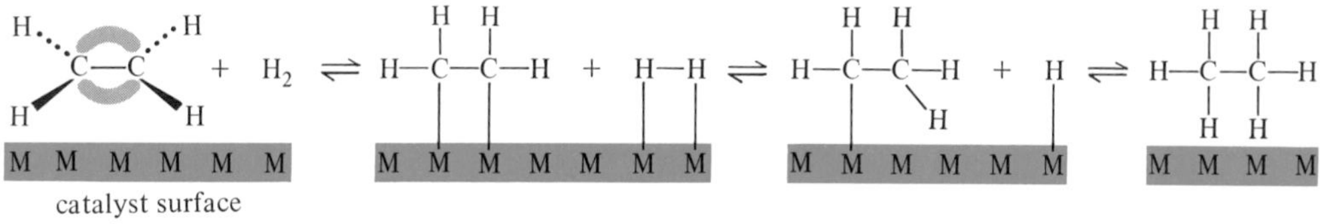

Figure 23-1. Representation of Catalytic Hydrogenation

The C_2H_4 molecule interacts through its π electrons with the atoms (M) in the metal surface. Meanwhile the hydrogen molecule is dissociated by adsorption on the surface. The adsorbed hydrogen atoms add to the adsorbed ethylene one by one in reversible equilibrium processes.

However, in contrast to substitution in alkanes, the substitution reactions of aromatic compounds proceed by ionic rather than free radical mechanisms. The reactions are catalyzed by Lewis acids, which function as "halide ion carriers." A positive halogen ion is formed, which attacks the electrons of the delocalized π orbitals:

$$Br_2 + FeBr_3 \rightarrow \left[Br^+ + FeBr_4^- \right]$$

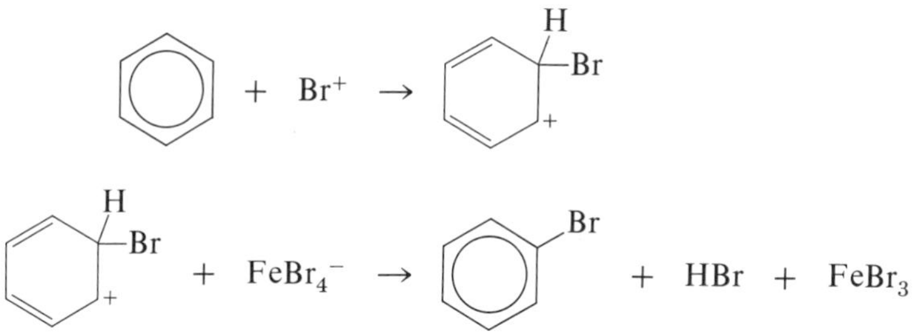

Aromatic hydrocarbons are more reactive than the alkanes, and readily undergo a variety of reactions in which one or more hydrogen atoms are replaced by other groups. In addition to the bromination reaction discussed above and summarized

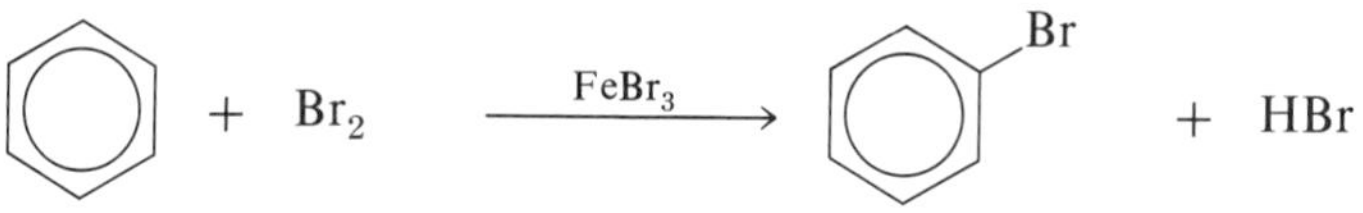

nitration and alkylation reactions can also occur:

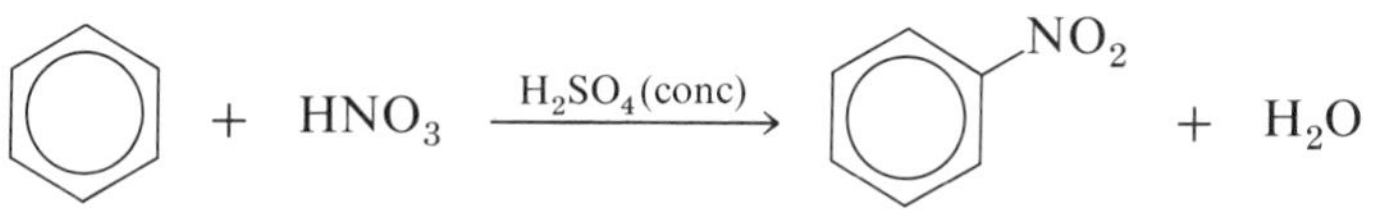

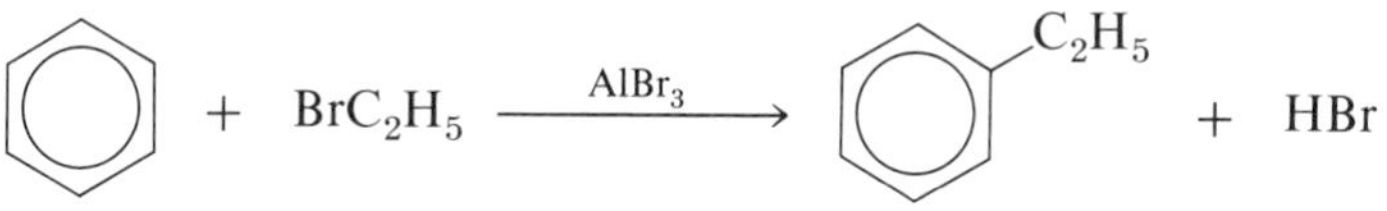

More than one group can be substituted for hydrogen atoms on an aromatic ring. However, the presence of one substituent influences the position at which a second substituent will be attached. For example

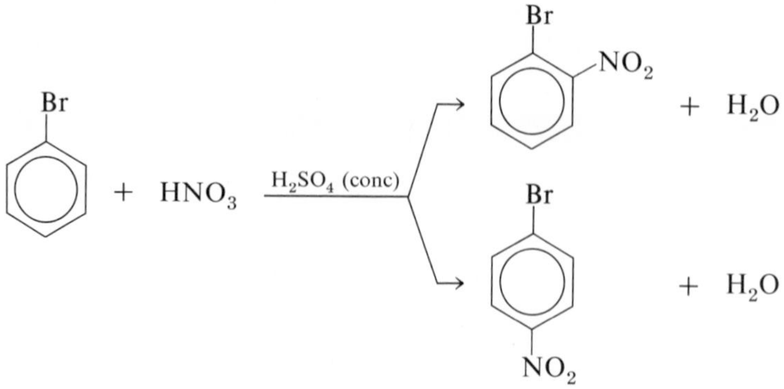

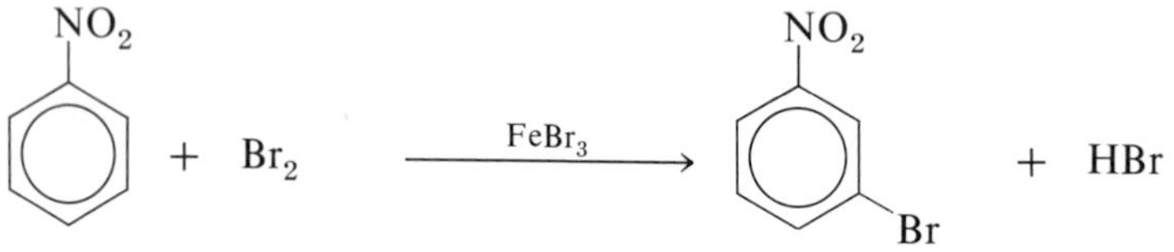

Thus the directive effects of substituents can be used by the organic chemist to synthesize one isomer of a given substance in preference to others.

23-2   Halogenated Hydrocarbons

Halogenated hydrocarbons do not occur in nature. In recent times such compounds have become very important commercial products. Some typical compounds and their uses are listed in Table 23-2.

The chief source of industrial organic chemicals is petroleum, which consists mainly of hydrocarbons. Halogenation is usually the first step in the preparation of other compounds. The halogenation of methane (marsh gas) in the presence of sunlight illustrates the process for all saturated hydrocarbons:

$$CH_4 + Cl_2 \rightarrow CH_3Cl + HCl$$
$$CH_4 + 2\,Cl_2 \rightarrow CH_2Cl_2 + 2\,HCl$$
$$CH_4 + 3\,Cl_2 \rightarrow CHCl_3 + 3\,HCl$$
$$CH_4 + 4\,Cl_2 \rightarrow CCl_4 + 4\,HCl$$

These reactions proceed by free radical mechanisms (Section 23–1). A mixture of products generally results, but the extent of chlorination can be controlled somewhat by use of the proper ratio of reactants. The products can be separated by fractional distillation (Section 16–23).

Some organic halides react with alcoholic silver nitrate to produce an ether and the corresponding silver halide.

$$RBr + Ag^+ + R'OH \rightarrow ROR' + AgBr + H^+$$

This reaction is often used as a means of distinguishing among organic halides. Tertiary halides react rapidly in the cold; secondary halides react upon heating;

TABLE 23-2. Some Commercially Important Halogenated Hydrocarbons

Name	Formula	Description	Applications
carbon tetrachloride	CCl_4	volatile liquid	solvent, fire extinguisher fluid
chloroform	$CHCl_3$	volatile liquid	solvent, industrial raw material
Freon 11 Freon 12 Freon 13	$CFCl_3$ CF_2Cl_2 CF_3Cl	gases	refrigerant, aerosol spray propellant
methyl chloride	CH_3Cl	gas	industrial raw material, refrigerant
1,4-dichlorobenzene	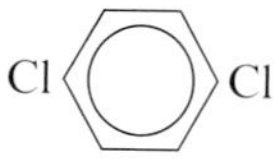	volatile solid	larvicide, moth repellant
hexachlorophene	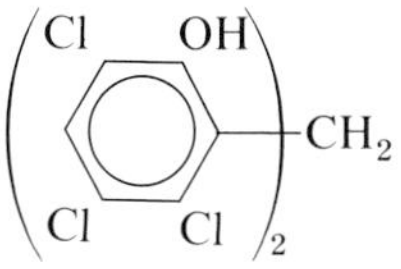	solid	germicide, disinfectant
pentachloronitro-benzene (PCNB)	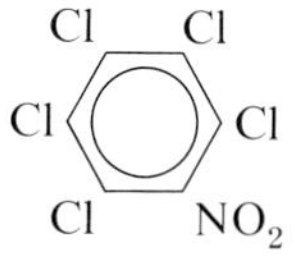	solid	fungicide
2,2-di(4-chloro-phenyl)-1,1,1-tri-chloroethane (dichlorodiphenyl-trichloroethane, DDT)	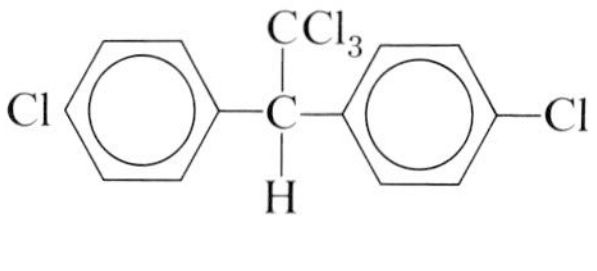	solid	insecticide
chlordane	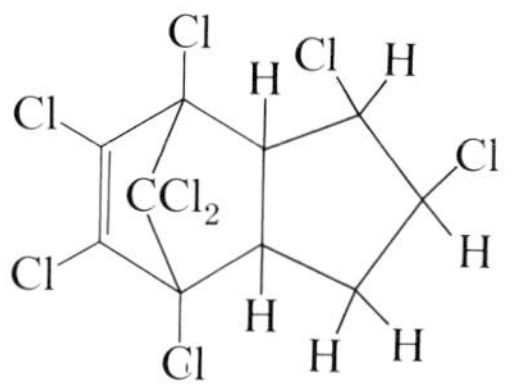	solid	insecticide
gammexane	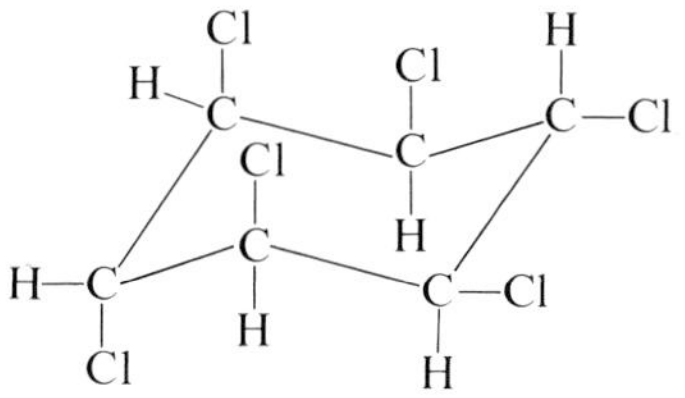	solid	insecticide

primary chlorides and bromides react very slowly if at all; and aromatic halides do not react with this reagent at all.

An important reaction of organic halides is the Grignard reaction, in which the halide is treated with magnesium metal in anhydrous ether:

$$RX \ + \ Mg \ \xrightarrow[\text{ether}]{} \ RMgX \qquad \text{(Grignard reaction)}$$

This exothermic reaction proceeds only if all traces of water are kept out of the system. The **Grignard reagent**, $RMgX$, is then used to prepare a wide variety of other compounds. For example, the length of a carbon chain may be increased, with the production of an acid:

$$2 \, CH_3CH_2MgX \ + \ 2 \, CO_2 \ \rightarrow \ (CH_3CH_2CO_2)_2Mg \ + \ MgX_2$$
$$(CH_3CH_2CO_2)_2Mg \ + \ 2 \, H_3O^+ \ \rightarrow \ 2 \, CH_3CH_2CO_2H \ + \ Mg^{2+} \ + \ 2 \, H_2O$$

Both alkyl and aromatic halides undergo the Grignard reaction; however, the latter compounds react somewhat more slowly.

Uses of Chlorinated Hydrocarbons

Chlorinated hydrocarbons are even more useful as solvents than the hydrocarbons themselves, owing to the polarity of the C—Cl bonds. In the commercial "drycleaning" process, soiled clothing is soaked in volatile chlorinated hydrocarbon solvents, which dissolve much of the grease and oils. The excess solvent is then allowed to evaporate from the cleaned article.

The use of chlorinated hydrocarbons as pesticides is credited with the eradication of many insects and insect-borne diseases. For example, the use of DDT (Table 23–2) has virtually eliminated the mosquito which carries yellow fever in the countries once plagued by this menace. Other epidemic diseases carried by insects, such as malaria and encephalitis, have likewise been brought under control. Extensive crop damage by insects has also been avoided. However, these benefits have cost humanity a price. The chlorinated pesticides are not easily decomposed and they tend to be spread throughout the world by wind, rain, snow, and migrating birds and fish. DDT sprayed upon a lake to control mosquitoes settles on the microscopic plankton in the water, which in turn are ingested by small fish. Birds and larger fish which feed upon the smaller fish accumulate the DDT, so that ultimately the concentration of the compound in tissues of larger animals may reach a level which affects their abilities to reproduce healthy offspring. Possibly as a consequence of DDT ingestion, the populations of several species of birds have been severely depleted.

Over several generations, numerous insect species have developed a resistance to DDT, and other more potent pesticides have been developed. These include chlordane and gammexane. Like DDT, these pesticides tend to persist and accumulate in animal body tissues. Little is known of the ultimate effects on humans of continued and increasing doses of pesticides. At present, the best policy where no biodegradable substitute is available seems to be to use nonbiodegradable pesticides only for essential purposes, such as control of epidemic diseases or prevention of famine from massive crop infestations.

The volatile halogenated hydrocarbons, such as CH_3Cl and the Freons, are also subjects of concern. Widely used as propellants in aerosol sprays, tons of Freons have been released into the atmosphere. Under ordinary conditions, most of these compounds are relatively inert. It is feared that they are accumulating in the atmosphere in such quantities that they may influence the concentration of ozone in the upper atmosphere (Section 21–2). It is postulated that the breakdown of Freons in the upper atmosphere yields chlorine atoms, which react with ozone molecules in a chain reaction involving oxygen atoms present from the decomposition of O_3 by ultraviolet light:

$$Cl + O_3 \rightarrow ClO + O_2$$
$$ClO + O \rightarrow O_2 + Cl$$

23-3 Reactions of Alcohols

Alcohols may be primary, secondary, or tertiary, depending on the location of the OH group on the carbon chain (Section 12–5). Chemically, alcohols exhibit three types of behavior: (1) reactions involving the hydrogen attached to the oxygen, (2) reactions of the entire OH group, and (3) reactions in which dehydration occurs. Alcohols react with very active metals to liberate hydrogen:

$$2\,ROH + 2\,Na \rightarrow H_2 + 2\,Na^+OR^-$$

This reaction is often used to test for the OH functional group in organic molecules.

Example

How could one distinguish by means of a chemical test between $CH_3CH_2OCH_2CH_3$ and its isomer, $CH_3CH_2CH_2CH_2OH$?
Treat samples of the substances with sodium. The ether will give no reaction. The evolution of hydrogen gas will distinguish the alcohol.

Alcohols may be converted to halides, for example according to the following equations:

$$3\,ROH + PBr_3 \rightarrow 3\,RBr + H_3PO_3$$
$$2\,ROH + SOCl_2 \rightarrow 2\,RCl + H_2O + SO_2$$

Alcohols lose water when heated in the presence of concentrated sulfuric acid, a good dehydrating agent. Either ethers or alkenes may be produced, depending on the temperature. The first step is

$$C_2H_5OH + H_2SO_4 \rightarrow C_2H_5OSO_3H + H_2O$$

Heating to different temperatures produces different products:

$$C_2H_5OSO_3H \ + \ C_2H_5OH \ \xrightarrow{140°C} \ C_2H_5OC_2H_5 \ + \ H_2SO_4$$

$$C_2H_5OSO_3H \ \xrightarrow{180°C} \ CH_2{=}CH_2 \ + \ H_2SO_4$$

Primary alcohols may be oxidized to aldehydes or further oxidized to organic acids:

$$R{-}CH_2{-}OH \ \xrightarrow[\text{mild oxidation}]{} \ R{-}\underset{\underset{O}{\|}}{C}{-}H$$

$$R{-}CH_2{-}OH \ \xrightarrow[\text{moderate oxidation}]{} \ R{-}\underset{\underset{O}{\|}}{C}{-}OH$$

Secondary alcohols may be oxidized to ketones:

$$R{-}\underset{\underset{OH}{|}}{\overset{\overset{H}{|}}{C}}{-}R' \ \xrightarrow[\text{moderate oxidation}]{} \ R{-}\underset{\underset{O}{\|}}{C}{-}R'$$

Tertiary alcohols are not oxidized as easily as primary alcohols. Oxidation of a tertiary alcohol would require rupture of a carbon–carbon bond.

23-4 Reactions of Aldehydes and Ketones

Aldehydes are easily oxidized to organic acids, while ketones are oxidized only by very strong oxidizing agents. For example, $[Ag(NH_3)_2]^+$ oxidizes simple aldehydes, being reduced to metallic silver, but ketones are not oxidized by this reagent. A solution containing the complex ion is called Tollens' reagent. It is used to distinguish between simple aldehydes and ketones. Another special reagent for this purpose is Fehling's solution, an alkaline solution of copper(II) tartrate. Aldehydes, but not ketones, are able to reduce the copper(II) complex to copper(I) oxide, which has an easily identifiable red color:

$$2\,[Cu(C_4H_4O_6)_2]^{2-} + \ RCHO \ + \ 5\,OH^- \ \rightarrow$$

$$RCO_2^- \ + \ \underset{\text{red solid}}{Cu_2O} \ + \ 4\,C_4H_4O_6^{2-} \ + \ 3\,H_2O$$

Aldehydes react with Grignard reagents to form alcohols:

$$H{-}\underset{}{\overset{\overset{H}{|}}{C}}{=}O \ + \ RMgX \ \xrightarrow{\text{anhydrous ether}} \ H{-}\underset{\underset{R}{|}}{\overset{\overset{OMgX}{|}}{C}}{-}H$$

$$\underset{\underset{R}{|}}{\overset{\overset{OMgX}{|}}{H-C-H}} + H_2O \longrightarrow \underset{\underset{R}{|}}{\overset{\overset{OH}{|}}{H-C-H}} + \tfrac{1}{2}MgX_2 + \tfrac{1}{2}Mg(OH)_2$$

$$\underset{}{\overset{\overset{H}{|}}{R'-C=O}} + RMgX \xrightarrow{\text{anhydrous ether}} \underset{\underset{R}{|}}{\overset{\overset{OMgX}{|}}{R'-C-H}}$$

$$\underset{\underset{R}{|}}{\overset{\overset{OMgX}{|}}{R'-C-H}} + H_2O \longrightarrow \underset{\underset{R}{|}}{\overset{\overset{OH}{|}}{R'-C-H}} + \tfrac{1}{2}MgX_2 + \tfrac{1}{2}Mg(OH)_2$$

Ketones react similarly with Grignard reagents to form tertiary alcohols.

23-5   Reactions of Carboxylic Acids and Esters

Organic acids, characterized by the functional group —CO_2H, called the carboxyl group, are weak acids, having K_a values about 10^{-5} in water. They undergo the typical reactions of protonic acids, such as the formation of salts upon treatment with active metals or with bases:

$$2\,RCO_2H + Zn \rightarrow Zn^{2+} + 2\,RCO_2^- + H_2$$
$$RCO_2H + OH^- \rightarrow RCO_2^- + H_2O$$

Acids react with alcohols to form esters:

$$\underset{}{\overset{\overset{O}{\|}}{CH_3COH}} + HOCH_2CH_3 \rightleftharpoons \underset{\text{ethyl acetate}}{\overset{\overset{O}{\|}}{CH_3COCH_2CH_3}} + H_2O$$

An ester derives its name from the radical of the alcohol and the anion of the acid from which it was formed. For example, the ester formed from ethyl alcohol and acetic acid is called ethyl acetate. Despite the apparent similarity in name to that of a salt, esters are covalent compounds. Many esters are liquids which have very pleasant odors. The fragrances of many fruits and flowers are due to the esters which they contain. Table 23-3 gives some examples.

The reaction between an acid and an alcohol to form an ester is reversible, and at equilibrium the relative concentration of each of the substances involved is determined by experimental conditions. The formation of the equilibrium mixture, starting with either set of reactants, is catalyzed by dilute mineral acids. Addition of a base causes formation of the salt of the organic acid, and hence the hydrolysis of the ester. The process of hydrolysis of an ester by a base is called **saponification** (Section 23-7).

$$RCO_2R' + OH^- \rightarrow RCO_2^- + R'OH$$

TABLE 23-3. Some Naturally Occurring Esters

Name	Formula	Common Name or Source
methyl salicylate	(benzene ring)—CO_2CH_3 with OH	oil of wintergreen
isoamyl acetate	$(CH_3)_2CHCH_2CH_2OCOCH_3$	banana oil
ethyl butyrate	$CH_3CH_2OCOCH_2CH_2CH_3$	apricot, peach
octyl acetate	$CH_3(CH_2)_7OCOCH_3$	orange
menthyl acetate	(cyclohexane ring structure with $COCH_3$, O, CH_2CH, CH_3, $CH_3—CH$, $CH—CH$, CH_2CH_2, CH_3)	oil of peppermint

Carboxylic acids and their esters can be reduced to alcohols by suitable reducing agents:

$$4\,RCO_2H \;+\; 2\,LiAlH_4 \;\rightarrow\; LiAl(OCH_2R)_4 \;+\; LiOH \;+\; Al(OH)_3$$

$$LiAl(OCH_2R)_4 \;+\; 4\,H_3O^+ \;\rightarrow\; Li^+ \;+\; Al^{3+} \;+\; 4\,HOCH_2R \;+\; 4\,H_2O$$

23-6 Reactions of Amines

Amines can be considered to be derivatives of ammonia in which one or more of the hydrogen atoms of ammonia is replaced by an organic radical (Section 12-9). Many of their reactions are similar to those of ammonia. Like ammonia, amines are Brønsted bases. They accept protons to form the corresponding conjugate acids:

$$R_2NH \;+\; H_2O \;\rightleftharpoons\; R_2NH_2^+ \;+\; OH^-$$

Amines are also Lewis bases. For example, as already mentioned in Chapter 14, amines can act as ligands toward transition metal ions.

An important reaction between carboxylic acids[2] and ammonia, primary amines, or secondary amines is the formation of amides:

$$CH_3CO_2H \;+\; NH_3 \;\underset{heat}{\rightleftharpoons}\; CH_3CONH_2 \;+\; H_2O$$

$$CH_3CO_2H \;+\; R_2NH \;\underset{heat}{\rightleftharpoons}\; CH_3CONR_2 \;+\; H_2O$$

The second of these reactions is analogous to the reaction of carboxylic acids with alcohols. Like esters, amides are covalent compounds. They can be hydrolyzed to form the acid and amine.

---

[2] In actual practice, better yields of amide are obtained by use of reagents which are derivatives of the acids, eg., RCO_2CH_3 or $RCOCl$, rather than the acid itself.

Like ammonia, primary amines react with aldehydes and ketones to form imines:

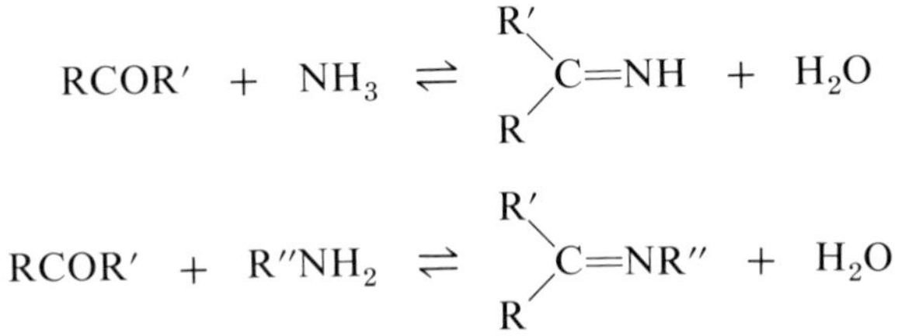

$$RCOR' + NH_3 \rightleftharpoons \begin{matrix} R' \\ \diagdown \\ R \diagup \end{matrix} C{=}NH + H_2O$$

$$RCOR' + R''NH_2 \rightleftharpoons \begin{matrix} R' \\ \diagdown \\ R \diagup \end{matrix} C{=}NR'' + H_2O$$

In the laboratory, amines may be prepared by treatment of an alkyl halide with ammonia, or by the reduction of an imine or amide:

$$RI + 2\,NH_3 \longrightarrow RNH_2 + NH_4I$$
$$2\,RI + 3\,NH_3 \longrightarrow R_2NH + 2\,NH_4I$$
$$3\,RI + 4\,NH_3 \longrightarrow R_3N + 3\,NH_4I$$
$$4\,RI + 4\,NH_3 \longrightarrow R_4N^+ + 3\,NH_4^+ + 4\,I^-$$
$$RCH{=}NH + H_2 \xrightarrow{catalyst} RCH_2NH_2$$
$$4\,RCONH_2 + 2\,LiAlH_4 \longrightarrow LiAl(NHCH_2R)_4 + LiOH + Al(OH)_3$$
$$LiAl(NHCH_2R)_4 + 4\,H_2O \longrightarrow 4\,RCH_2NH_2 + LiOH + Al(OH)_3$$

23-7   Lipids and Waxes

Lipids—fats from animal tissues and oils from plants—are esters of glycerol and long chain carboxylic acids. Glycerol is a trialcohol, $CH_2OHCH(OH)CH_2OH$, and fats are sometimes called triacylglycerols. They have the general formula shown below, where the radicals R, R', and R'' are long chains of carbon atoms ranging in length from 4 to 18 carbon atoms:

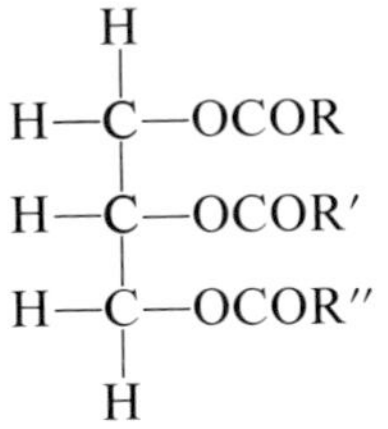

$$
\begin{matrix}
& H & \\
& | & \\
H{-} & C & {-}OCOR \\
& | & \\
H{-} & C & {-}OCOR' \\
& | & \\
H{-} & C & {-}OCOR'' \\
& | & \\
& H &
\end{matrix}
\qquad \text{a triacylglycerol}
$$

Different fats and oils have different R groups. Animal fats are esters of long chain saturated acids, while vegetable oils, such as soybean or peanut oil, have radicals which may contain carbon–carbon double bonds. The liquid vegetable oils, containing unsaturated acids, may be converted into solid fats by hydrogenation. Oleomargarine is produced from vegetable oils in this manner:

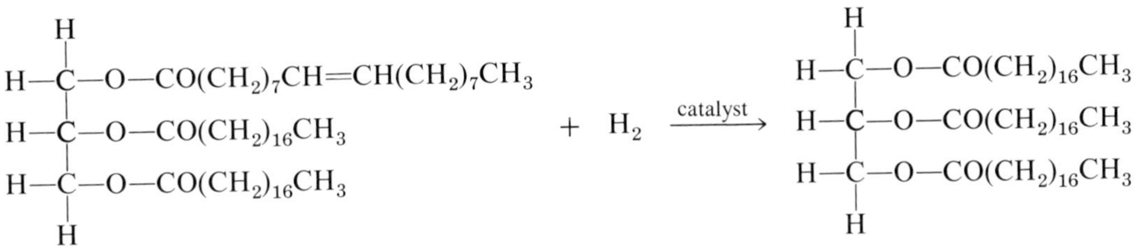

$$
\begin{matrix}
& H & \\
& | & \\
H{-} & C & {-}O{-}CO(CH_2)_7CH{=}CH(CH_2)_7CH_3 \\
& | & \\
H{-} & C & {-}O{-}CO(CH_2)_{16}CH_3 \\
& | & \\
H{-} & C & {-}O{-}CO(CH_2)_{16}CH_3 \\
& | & \\
& H &
\end{matrix}
\; + \; H_2 \xrightarrow{catalyst}
\begin{matrix}
& H & \\
& | & \\
H{-} & C & {-}O{-}CO(CH_2)_{16}CH_3 \\
& | & \\
H{-} & C & {-}O{-}CO(CH_2)_{16}CH_3 \\
& | & \\
H{-} & C & {-}O{-}CO(CH_2)_{16}CH_3 \\
& | & \\
& H &
\end{matrix}
$$

liquid vegetable oil   solid fat

Soaps are salts of long-chain fatty acids. Saponification of a fat with base yields soap and glycerol:

$$
\begin{array}{l}
H-\overset{\displaystyle H}{\underset{\displaystyle H}{\overset{|}{\underset{|}{C}}}}-O-CO(CH_2)_{16}CH_3 \\
H-C-O-CO(CH_2)_{16}CH_3 \\
H-C-O-CO(CH_2)_{16}CH_3
\end{array}
\ + \ 3\,NaOH \ \rightarrow \
\begin{array}{l}
H-C-OH \\
H-C-OH \\
H-C-OH
\end{array}
\ + \ 3\,NaOCO(CH_2)_{16}CH_3
$$

      fat (glyceryl tristearate)　　　　　　　　　　　　　glycerol　　　　sodium stearate (a soap)

An ester of a long chain fatty acid and a long chain alcohol is a **wax.** Such waxes are obtained from both plants and animals. Carnauba wax, obtained from palm trees, is chiefly myricyl cerotate, $C_{25}H_{51}CO_2C_{30}H_{61}$, and is used as a floor polish and an automobile polish. Beeswax is mostly myricyl palmitate, $C_{15}H_{31}CO_2C_{30}H_{61}$. Other animal waxes are found on feathers, skin, and hair, and serve as protective coatings because they are not as easily hydrolyzed as triacylglycerols.

23-8   Soaps and Synthetic Detergents

Soap functions as a **detergent** because the anion of the long chain acid has one end which is essentially a hydrocarbon, while the carboxylate end is negatively charged. When placed in water, soaps form **micelles** in which the hydrocarbon ends cluster together, while the negatively charged carboxylate ends point outward toward the solvent. Oil or grease on an article being washed is dissolved in the interior of the micelle (Figure 23–2) and rendered miscible with water.

Soaps are not effective detergents in hard water because of the formation of insoluble calcium and magnesium salts:

$$
Ca^{2+} \ + \ 2\,CH_3(CH_2)_{16}CO_2^- \ \rightarrow \ Ca(OCO(CH_2)_{16}CH_3)_2
$$

               stearate ion　　　　　　　insoluble calcium stearate

The calcium salt is often more difficult to remove than the dirt originally on the article

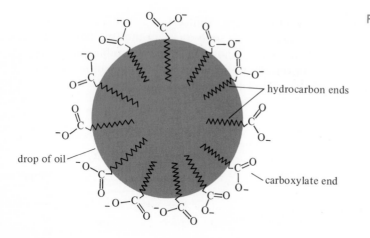

Figure 23-2. **Soap Micelle**

hydrocarbon ends

drop of oil

carboxylate end

being cleaned. To circumvent the problem of precipitation in hard water, synthetic detergents which do not form insoluble calcium and magnesium salts have been developed. Like soaps, synthetic detergent molecules have one portion which is hydrophilic (has a great affinity for water) and another portion which is hydrophobic (has little or no attraction for water). Hence, detergents also form micelles when placed in water. A typical example is a long chain alkylbenzenesulfonate such as

$$CH_3CH_2CH_2\underset{\underset{CH_3}{|}}{C}HCH_2\underset{\underset{CH_3}{|}}{C}HCH_2\underset{\underset{CH_3}{|}}{C}HCH_2 \langle\bigcirc\rangle SO_3^-\ Na^+$$

However, despite their superior performance as cleaning agents, detergents have some drawbacks. Soap is biodegradable; that is, when waste water containing soap is passed into sewage treatment plants or into the soil, the soap molecules are eventually destroyed by naturally occurring bacteria. Some synthetic detergents are not degraded rapidly by bacteria, and as a consequence they accumulate in sewage plants or in lakes and streams with the result that the water becomes covered with a sudsy foam. Biodegradable synthetic detergents, which degrade more rapidly, have been developed and are now in wide use.

23-9   Carbohydrates

Carbohydrates, produced in green plants, are the most abundant class of naturally occurring compounds. This class includes sugars, starches, and cellulose. Simple sugars are produced by photosynthesis:

$$6\,CO_2\ +\ 6\,H_2O \xrightarrow[\text{sunlight}]{\text{chlorophyll}} \underset{\text{a sugar}}{C_6H_{12}O_6}\ +\ 6\,O_2$$

Carbohydrate molecules contain alcohol and aldehyde or ketone functional groups despite the fact that their empirical formulas and their name suggest that they might be complexes of carbon and water. For example, glucose, a common simple sugar, is an aldohexose (a six-carbon sugar containing an aldehyde functional group):

$$
\begin{array}{ccccccc}
& H & H & H & H & H & H \\
& | & | & | & | & | & | \\
H- & C & -C & -C & -C & -C & -C=O \\
& | & | & | & | & | & \\
& O & O & O & O & O & \\
& | & | & | & | & | & \\
& H & H & H & H & H &
\end{array}
$$

An isomer of glucose is fructose, a ketohexose:

$$
\begin{array}{ccccccc}
& H & H & H & H & & H \\
& | & | & | & | & & | \\
H- & C & -C & -C & -C & -C & -C-H \\
& | & | & | & | & \| & | \\
& O & O & O & O & O & O \\
& | & | & | & | & & | \\
& H & H & H & H & & H
\end{array}
$$

It is apparent that branching of the carbon chain and variation in the position of the C=O group can lead to many other possible isomers. In addition, there is the

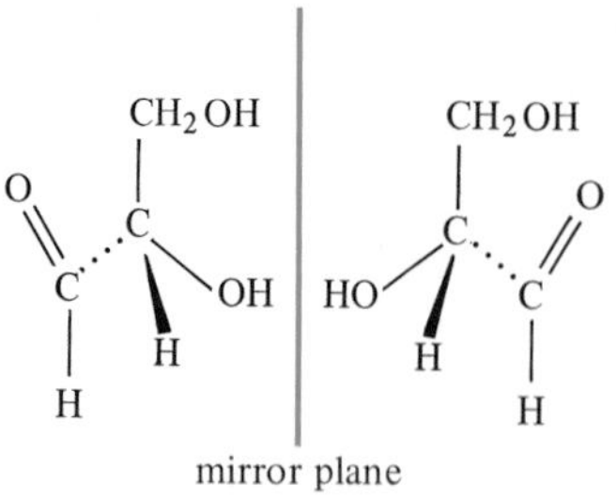

Figure 23-3. Optically Active Forms of Glyceraldehyde, a Compound Containing an Asymmetric Carbon Atom

possibility for optical isomerism (Section 12–11). A carbon atom which has four different groups attached to it is **asymmetric.** If there is only one asymmetric carbon atom, the molecule can have no planes, points, or axes of symmetry. Consequently, compounds containing one asymmetric carbon atom will exist in two optically active forms (Figure 23–3).

Glyceraldehyde, shown in Figure 23–3, is a triose sugar—a three-carbon sugar. It is clear from the figure that there are two isomeric forms which are mirror images. Except for the direction in which they rotate the plane of polarized light, they have the same physical properties. The isomer which rotates the plane of polarized light to the right—the dextro isomer—is called D-glyceraldehyde and the other isomer—the levo isomer—is called L-glyceraldehyde. By convention, their graphic formulas are written as follows:

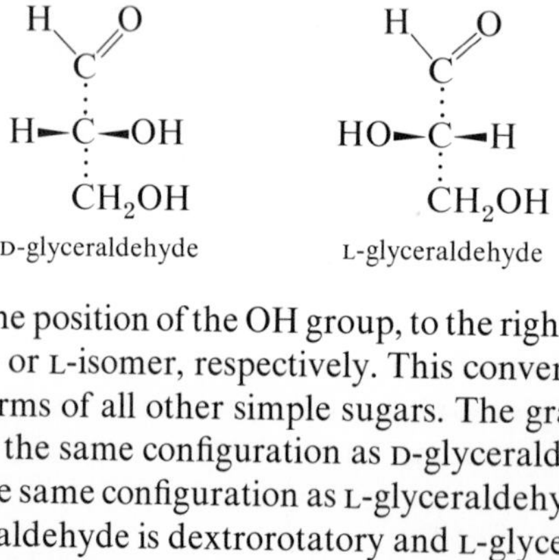

D-glyceraldehyde   L-glyceraldehyde

The position of the OH group, to the right or left of the center carbon atom, denotes the D- or L-isomer, respectively. This convention is the basis for designating the isomeric forms of all other simple sugars. The graphic formulas of all sugars which terminate in the same configuration as D-glyceraldehyde are designated D-sugars; those having the same configuration as L-glyceraldehyde are designated L-sugars. Although D-glyceraldehyde is dextrorotatory and L-glyceraldehyde is levorotatory, it must be emphasized that the prefixes D and L for other compounds serve only to indicate the absolute configuration of the molecule, in contrast to ($+$) and ($-$), which indicate the direction of rotation of plane-polarized light. A D-sugar is one in which the last asymmetric carbon atom (the one farthest from the aldehyde or ketone group) has the same configuration as does the asymmetric carbon atom of D-glyceraldehyde. D-sugars do not necessarily rotate the plane of polarized light to the right (are not necessarily dextrorotatory).

For any organic molecule, the maximum number of optical isomers possible is 2^n, where n is the number of asymmetric carbon atoms (Section 12–11). For example, the glucose molecule has four asymmetric carbon atoms; hence there are 16 optical isomers possible (eight pairs of enantiomers) (Figure 23–4). These isomers have all been observed. Each pair has one ($+$) and one ($-$) form, the designation depending on the direction in which it rotates the plane of polarized light. The ($+$) and ($-$) members of any pair of enantiomers rotate the plane of polarized light through angles which are equal in magnitude but opposite in direction.

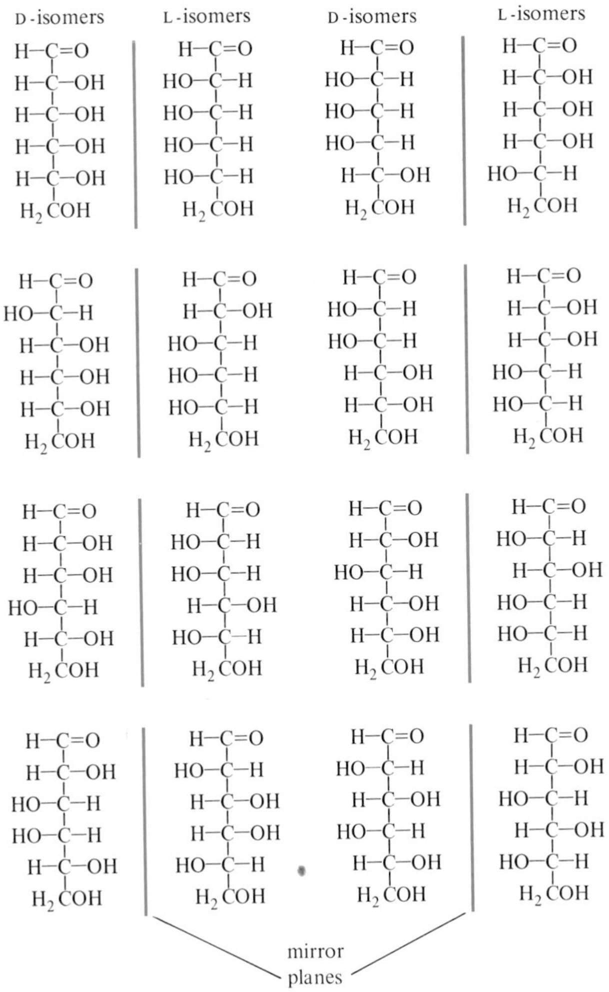

Figure 23-4. **Optical Isomers of Aldohexose**

D-isomers	L-isomers	D-isomers	L-isomers
H—C=O	H—C=O	H—C=O	H—C=O
H—C—OH	HO—C—H	HO—C—H	H—C—OH
H—C—OH	HO—C—H	HO—C—H	H—C—OH
H—C—OH	HO—C—H	HO—C—H	H—C—OH
H—C—OH	HO—C—H	H—C—OH	HO—C—H
H₂COH	H₂COH	H₂COH	H₂COH

H—C=O	H—C=O	H—C=O	H—C=O
HO—C—H	H—C—OH	HO—C—H	H—C—OH
H—C—OH	HO—C—H	HO—C—H	H—C—OH
H—C—OH	HO—C—H	H—C—OH	HO—C—H
H—C—OH	HO—C—H	H—C—OH	HO—C—H
H₂COH	H₂COH	H₂COH	H₂COH

H—C=O	H—C=O	H—C=O	H—C=O
H—C—OH	HO—C—H	H—C—OH	HO—C—H
H—C—OH	HO—C—H	HO—C—H	H—C—OH
HO—C—H	H—C—OH	H—C—OH	HO—C—H
H—C—OH	HO—C—H	H—C—OH	HO—C—H
H₂COH	H₂COH	H₂COH	H₂COH

H—C=O	H—C=O	H—C=O	H—C=O
H—C—OH	HO—C—H	HO—C—H	H—C—OH
HO—C—H	H—C—OH	H—C—OH	HO—C—H
HO—C—H	H—C—OH	HO—C—H	H—C—OH
H—C—OH	HO—C—H	H—C—OH	HO—C—H
H₂COH	H₂COH	H₂COH	H₂COH

mirror planes

Further variations in the structures of carbohydrates result from the fact that many exist as polymeric forms of two or more simpler molecules. For example, sucrose, common table sugar, $C_{12}H_{22}O_{11}$, is a **disaccharide** consisting of two **monosaccharides,** fructose and glucose:

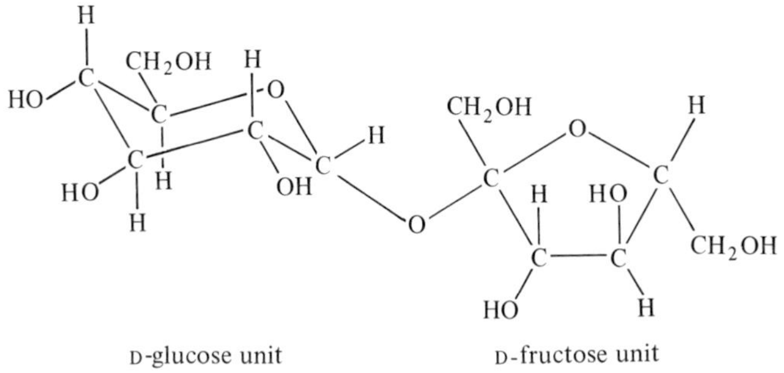

D-glucose unit   D-fructose unit

sucrose

The formula $C_{12}H_{22}O_{11}$ for sucrose, rather than $C_{12}H_{24}O_{12}$, indicates that the dimerization reaction is achieved by loss of one molecule of water between the two simple sugars. Starches and cellulose are examples of **polysaccharides**—polymers of glucose (Section 23–10). The cellulose molecule contains more than 1000 monosaccharide units.

23-10 Polymers

Polymers are large molecules which are made by bonding together a great many small molecules. The small molecules are called **monomers.** Polymers made up from more than one kind of monomer are sometimes called **copolymers.** Starch, cellulose, rubber, and proteins are examples of polymers which occur in nature. Synthetic polymers are of great commercial importance. They can be classified according to their physical behavior as (1) elastomers, which possess the property of exceptional elasticity, (2) thermoplastic polymers, which soften upon heating and may therefore be molded, and (3) thermosetting polymers, which become hard, infusible, and insoluble solids when heated. The synthetic products can be tailor-made to have quite specific properties, which cannot be achieved with natural polymers. For example, synthetic rubber can be designed to withstand very high temperatures without melting or to retain its elasticity at extremely low temperatures, whereas natural rubber melts above 250°C and becomes quite brittle and hard below −30°C. Such commercial polymers as nylon, polyethylene, polystyrene, Teflon, Lucite, and the silicones account for a large fraction of the output of the chemical industry. Synthetic polymers are made into filaments, threads, sheets, and films and are molded into various shapes. In addition, polymerization reactions are used to make tough surface coatings and to make very strong adhesives.

Polymers may also be classified according to their mode of preparation into (1) **addition polymers,** in which small molecules containing double bonds are linked together, and (2) **condensation polymers,** in which there is a loss of small molecules such as water when the monomer molecules link up. The preparation of polyethylene, an addition polymer, proceeds by a chain reaction. A free radical initiator (init·) is used to start the chain. The initiator is often referred to as a catalyst because of the small quantity required (Section 17–3).

$$\text{init·} \ + \ \text{H}_2\text{C}=\text{CH}_2 \ \rightarrow \ \text{init}-\text{CH}_2-\text{CH}_2\text{·}$$
$$\text{init}-\text{CH}_2-\text{CH}_2\text{·} \ + \ \text{H}_2\text{C}=\text{CH}_2 \ \rightarrow \ \text{init}-\text{CH}_2-\text{CH}_2-\text{CH}_2-\text{CH}_2\text{·}$$
$$\text{init}-(\text{CH}_2-\text{CH}_2)_2\text{·} \ + \ \text{H}_2\text{C}=\text{CH}_2 \ \rightarrow \ \text{init}-(\text{CH}_2-\text{CH}_2)_2-\text{CH}_2-\text{CH}_2\text{·}$$

In this type of reaction, ethylene molecules continually add to the unpaired electron on the end of the growing molecule, producing a succession of molecules, each having one more C_2H_4 group on the end and still having an unpaired electron. This process continues until a molecular weight in the range of several hundred thousand is obtained. Finally, addition of another initiator molecule terminates the growth.

$$\text{init}-(\text{CH}_2-\text{CH}_2)_n\text{init} \qquad (n \geqslant 5000)$$

The initiator molecules on the ends of this gigantic molecule do not influence its

TABLE 23-4. Commercial Addition Polymers

Monomer	Plastic	Use
ethylene, $CH_2{=}CH_2$	polyethylene	wrapping film
vinyl chloride, $CH_2{=}CHCl$	poly(vinyl chloride)	raincoats, bottles, etc.
styrene, $C_6H_5CH{=}CH_2$	polystyrene	molded plastic, insulating foam
tetrafluoroethylene, $CF_2{=}CF_2$	Teflon	chemical and heat-resistant coatings
isoprene, $CH_2{=}C(CH_3)CH{=}CH_2$	polyisoprene	synthetic rubber
methyl methacrylate, $CH_2{=}C(CH_3)CO_2CH_3$	Lucite, Plexiglas	transparent sheets

properties significantly, and the material has the characteristics of a huge hydrocarbon. The many polymeric molecules produced in a chain reaction have different chain lengths. The final product contains molecules having a range of molecular weights. In the commercial manufacture of polyethylene, it is possible to control the average molecular weight of the product to obtain various types of materials. Some examples of commercial addition polymers and their uses are listed in Table 23-4.

Condensation polymers result from the reactions of some molecules having more than one functional group. For example, a diamine can react with a dicarboxylic acid to form a polyamide, as in the preparation of nylon. In the first step one amide linkage is formed:

$$HOCO(CH_2)_4CO_2H \quad + \quad NH_2(CH_2)_6NH_2 \quad \rightarrow$$

$$HOCO(CH_2)_4CONH(CH_2)_6NH_2 \quad + \quad H_2O$$

The product of the above reaction still has two functional groups capable of reacting with more of the acid at one end and more of the amine at the other, so the reaction continues until long chains having the formula

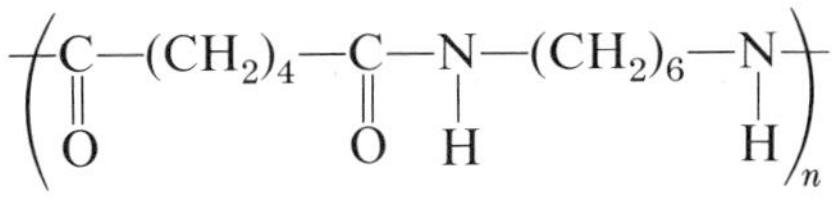

$$\left(\begin{matrix} C-(CH_2)_4-C-N-(CH_2)_6-N \\ \| \qquad\qquad \| \ | \qquad\qquad\quad | \\ O \qquad\qquad O\ H \qquad\qquad\quad H \end{matrix} \right)_n$$

nylon

are built up. In a similar manner, dicarboxylic acids can react with dialcohols to form polyesters. Some commercial condensation polymers are listed in Table 23-5.

The reactions for formation of either addition or condensation polymers can be used to polymerize molecules having branched chains or additional functional groups, and therefore a variety of polymers can be prepared. For example, when propylene, $CH_3CH{=}CH_2$, is polymerized, the resulting polypropylene will have a branched chain structure:

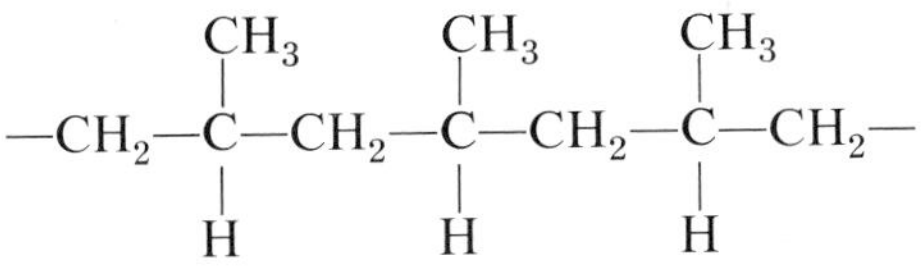

$$\begin{matrix} & CH_3 & & CH_3 & & CH_3 \\ & | & & | & & | \\ -CH_2-&C&-CH_2-&C&-CH_2-&C&-CH_2- \\ & | & & | & & | \\ & H & & H & & H \end{matrix}$$

TABLE 23-5. Some Common Condensation Polymers

Monomer(s)	Plastic
C_6H_5OH + HCHO phenol formaldehyde	Bakelite
C_6H_5OH 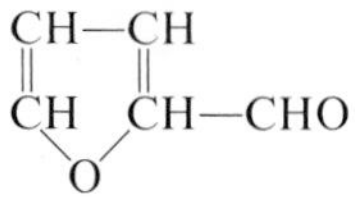 + furfural phenol	Durite
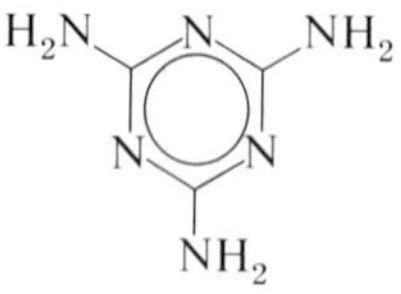 melamine + HCHO formaldehyde	Melmac
$HOC_6H_4SO_3H$ + HCHO phenolsulfonic acid formaldehyde	Amberlite, Dowex
$HOCO(CH_2)_4CO_2H$ + $NH_2(CH_2)_6NH_2$ adipic acid 1,6-diaminohexane	nylon
$(CH_3)_2Si(OH)_2$ dimethyldihydroxysilane	silicone rubber
$HOCOC_6H_4CO_2H$ + $HOCH_2CH_2OH$ terephthalic acid ethylene glycol	Dacron

The methyl groups cause the polypropylene molecule to have a more random orientation than the polyethylene molecule, and therefore the resulting polymer is less crystalline than polyethylene.

Another way of modifying a polymer is by cross-linking between chains. For example, the polymer chain of the synthetic rubber polyisoprene contains double bonds:

$$\left(CH_2-\underset{\underset{CH_3}{|}}{C}=CH-CH_2 \right)_n$$

When the polymer is treated with sulfur, reaction occurs adjacent to some of the double bonds, and sulfur bridges are formed between adjacent chains:

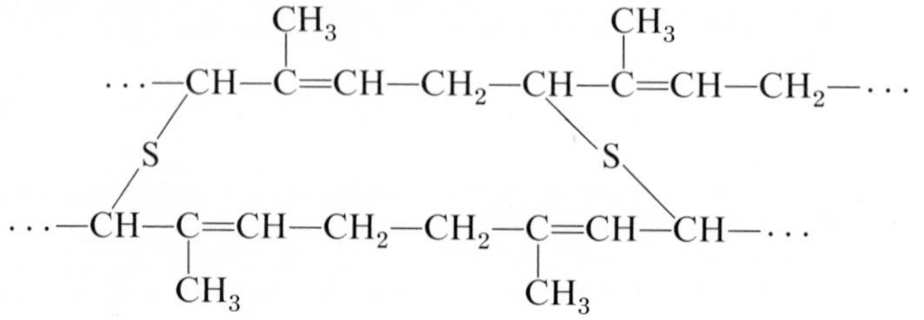

This rubber is said to be **vulcanized** and is harder, less elastic, and much higher melting than the original polymer.

23-11   Amino Acids and Proteins

Amino acids are molecules which contain both an acid functional group and an amine functional group. The simplest such compound is glycine, aminoacetic acid, $NH_2CH_2CO_2H$. Because of the presence of both acidic and basic groups in the same molecule, in aqueous solution amino acids exist as **zwitterions,** ions containing both a positive and a negative charge (at two different places in the molecule). The zwitterion of glycine is $^+NH_3CH_2CO_2^-$.

The amino acids in living organisms contain the amine group on the carbon atom which is attached to the carboxylic acid group—the so-called alpha carbon atom. These can be represented by the general formula

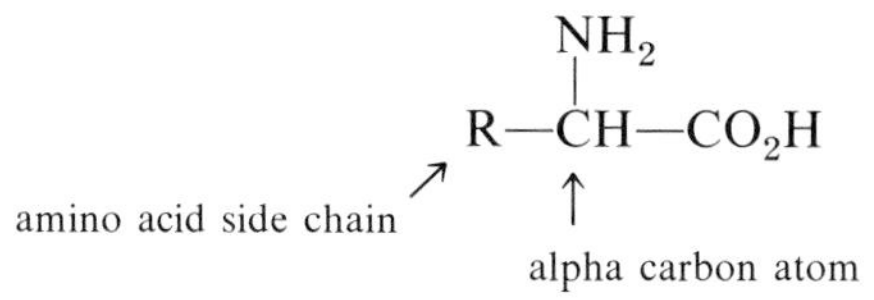

$$\overset{\displaystyle NH_2}{\underset{}{R-\overset{|}{C}H-CO_2H}}$$

amino acid side chain $\nearrow$ $\uparrow$

alpha carbon atom

where R can represent a wide variety of possible groups, not limited only to hydrocarbon radicals. Each different amino acid has unique properties depending on the nature of the R group, which is often referred to as the amino acid side chain. The amino acids are more widely known by their common names rather than by the more cumbersome IUPAC systematic names. The structures of some important amino acids are shown in Table 23-6. The 19 different amino acids found in nearly all protein materials, whether derived from plants or animals, are listed in Table 23-7. Six additional amino acids have been observed in a few proteins.

Except in the case of glycine, in which R represents a hydrogen atom, amino acid molecules contain at least one asymmetric carbon atom and therefore exist in either D or L form. It is noteworthy that with rare exceptions only the L forms of amino acids are found in natural proteins.

Proteins are polyamides, formed by the bonding of an amino group from one amino acid molecule with the carboxylic acid group of another, with the elimination of water as a by-product. The resulting $-\overset{\overset{\displaystyle}{\|}}{\underset{\underset{\displaystyle H}{|}}{C}-N}-$ linkage is called a **peptide** link. The structure of a peptide linkage may be represented in two resonance forms

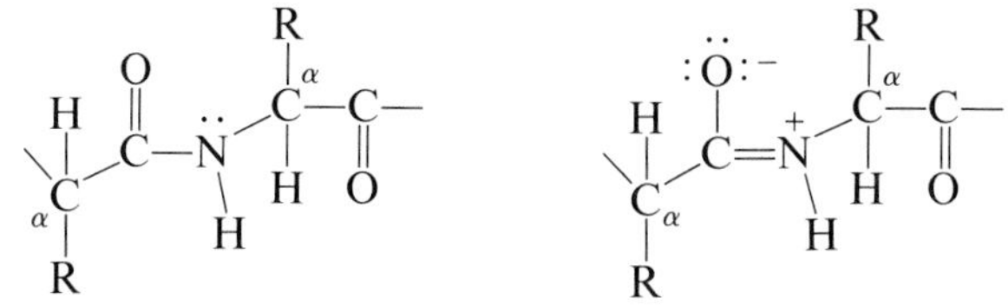

from which it can be seen that the carbon-to-nitrogen bond can have considerable double bond character. Indeed, X-ray studies reveal that the C—N bond length is less than the normal single bond distance of 1.47 Å, averaging about 1.32 Å, and the six

TABLE 23-6. Some Typical Amino Acid Constituents of Protein

Name	Symbol	Formula
glycine	Gly	CH_2—CO_2H │ NH_2
alanine	Ala	CH_3—CH—CO_2H │ NH_2
valine	Val	$(CH_3)_2CH$—CH—CO_2H │ NH_2
leucine	Leu	$(CH_3)_2CH$—CH_2—CH—CO_2H │ NH_2
phenylalanine	Phe	C_6H_5—CH_2—CH—CO_2H │ NH_2
proline	Pro	(cyclic) H_2C—N(H)—CH—CO_2H, H_2C—CH_2
serine	Ser	HO—CH_2—CH—CO_2H │ NH_2
cysteine	Cys	HS—CH_2—CH—CO_2H │ NH_2
methionine	Met	CH_3—S—CH_2—CH_2—CH—CO_2H │ NH_2
aspartic acid	Asp	$\overset{O}{\underset{HO}{\diagup\!\!\diagdown}}C$—$CH_2$—$CH$—$CO_2H$ │ NH_2
lysine	Lys	NH_2—CH_2—CH_2—CH_2—CH_2—CH—CO_2H │ NH_2

atoms adjacent to that bond lie roughly in a plane. A representation of a portion of a protein molecule is

$$-\overset{O}{\overset{\|}{C}}-\overset{H}{\overset{|}{N}}-\underset{R_1}{CH}-\overset{O}{\overset{\|}{C}}-\overset{H}{\overset{|}{N}}-\underset{R_2}{CH}-\overset{O}{\overset{\|}{C}}-\overset{H}{\overset{|}{N}}-\underset{R_3}{CH}-\overset{O}{\overset{\|}{C}}-\overset{H}{\overset{|}{N}}-\underset{R_4}{CH}-\overset{O}{\overset{\|}{C}}-\overset{H}{\overset{|}{N}}-$$

Thus, proteins are condensation polymers of amino acids. The molecular weights of natural proteins vary from a few thousand to over 1 million D. Since R_1, R_2, R_3, R_4,

TABLE 23-7. Amino Acid Content of Some Proteins

| Amino Acid | Symbol | Percent of Amino Acid in Protein from | | | | |
		Cow's Milk	Egg White	Fish Muscle	Whole Wheat	Corn Meal
arginine	Arg	3.5	5.9	6.6	4.3	4.4
histidine	His	2.7	2.5	2.9	1.8	2.4
lysine	Lys	8.0	6.4	10.1	2.5	2.7
tyrosine	Tyr	4.9	4.3	2.4	3.6	4.6
tryptophan	Trp	1.3	1.8	0.9	1.2	0.7
phenylalanine	Phe	5.1	6.0	3.8	4.4	4.5
cystine	Cys	0.9	2.6	1.2	3.3	1.6
methionine	Met	2.4	4.0	2.8	1.2	1.8
serine	Ser	5.2	7.3	3.5	3.8	4.4
threonine	Thr	4.7	4.7	4.5	3.9	4.1
leucine	Leu	9.9	9.0	7.9	6.9	12.7
isoleucine	Ileu	6.5	6.4	5.2	4.4	4.0
valine	Val	6.7	7.8	5.5	4.5	5.3
glutamic acid	Glu	21.7	12.8	12.0	31.4	18.4
aspartic acid	Asp	7.5	7.6	8.6	3.8	12.3
glycine	Gly	2.1	3.7	5.4	3.4	3.5
alanine	Ala	3.6	0.0	6.1	3.0	10.0
proline	Pro	9.2	2.9	5.1	10.3	7.2
hydroxyproline	Hyp	0.0	0.0	0.0	0.0	0.0

etc., each represent a group from any of the amino acids, the number of possible sequences and hence the number of possible kinds of protein is almost limitless. The compositions of proteins from several sources are listed in Table 23-7.

Specific protein molecules have definite biochemical functions. One may be a part of muscle tissue, another a molecule in skin or hair. Still others, such as enzymes or hormones, have specific chemical functions in cell metabolism. The variation in function is related to the sequence of individual amino acids in the molecule. For example, the sequence in the hormone insulin, a protein, is shown in Figure 23-5. The sequence was deduced as the result of painstaking work by F. Sanger of Cambridge University, England. He, with his associates, partially hydrolyzed the protein into molecules containing shorter polyamide chains. Then these fragments were separated by chromatography and again hydrolyzed, until successive steps led to the individual amino acids. Confirmation of the sequence was made in 1964 by P. G. Katsoyannis and his associates at the University of Pittsburgh, who succeeded in synthesizing the insulin molecule from smaller units.

The identities and sequences of amino acids in a protein constitute what is called the **primary structure** of the protein. However, it is known from various physical measurements that each protein has a particular three-dimensional conformation called its **secondary structure.** Linus Pauling and Robert Corey have postulated that some proteins have a special spiral shape much like a spring coiled about an imaginary cylinder (Figure 23-6). The spiral structure assumed by proteins is invariably right handed. The spiral, or helix, is stabilized by hydrogen bonding between the amide hydrogen of one peptide link and the carbonyl oxygen which is located on the next turn of the helix. X-ray data reveal that each turn of the helix consists of about

Figure 23-5. **Sequence of Amino Acids in Insulin**

```
        Phe          Gly
        Val          Ileu
  NH₂—Asp            Val
  NH₂—Glu            Glu
        His          Glu—NH₂
        Leu          Cys———S
        Cys—S—S—Cys
        Gly          Ala
        Ser          Ser
        His          Val
Ala                  Cys———S
Lys     Leu          Ser
Pro     Val          Leu
Thr     Glu          Tyr
Tyr     Ala          Glu—NH₂
Phe     Leu          Leu
Phe     Tyr          Glu
Gly     Leu          Asp—NH₂
Arg     Val          Tyr
Glu—Gly—Cys—S—S—Cys
                     Asp—NH₂
```

Figure 23-6. **α-Helical Conformation of a Protein**

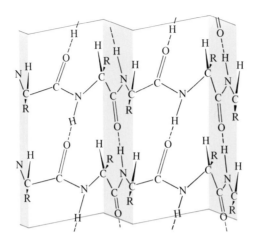

Figure 23-7. **Schematic Diagram of the Pleated Sheet Model of β-Keratin.**

The peptide bonds lie in the plane of the pleated sheet. The side chains lie above or below the sheet and alternate along the chain. The polypeptide chains are held together by hydrogen bonds (shown as dashed lines).

four amino acid units, with the side chains, the R groups, projecting outward from the coiled backbone. This intrapolypeptide hydrogen-bonded structure is called an **alpha helix.**

Not all proteins assume a helical structure. Some, such as the globin portion of hemoglobin and myoglobin, are helical only in certain regions of the polypeptide chain. Others, like gamma globulin, have no helical structure. Still others, like silk fibroin and beta-keratin, are arranged in pleated sheets that are also stabilized by hydrogen bonding (Figure 23-7).

23-12   Enzymes

Enzymes are protein molecules (some of which also contain a nonprotein part) which act as catalysts for specific biochemical reactions. Louis Pasteur (1822–1895) was one of the first scientists to recognize the catalytic activity of enzymes. In 1857, he observed that glucose could be stored indefinitely in sealed sterile containers without undergoing any change. In contrast, addition of yeast cells caused the glucose to be fermented to alcohol and carbon dioxide rather quickly. Thus the name enzyme was derived from the Greek words *en,* "within," and *zyme,* "yeast." It was shown in 1897 by Eduard Buchner that the filtered extract of ground up yeast cells was capable of catalyzing the fermentation reaction; hence the behavior was due to a substance or substances within the cell and not due to the cell itself. However, it was only after 30 years of painstaking attempts by many researchers that a pure crystalline enzyme was isolated.

Practically every chemical reaction occurring within a cell is catalyzed by a specific enzyme. In fact, many reactions occur in several steps, each of which requires its own enzyme catalyst. In addition to reactions within cells, enzymes catalyze specific metabolic reactions such as occur in digestion. For example, the enzyme ptyalin in the saliva catalyzes the hydrolysis of starches to simple sugars. Other enzymes catalyze the conversion of proteins to simple amino acids, while still others catalyze the conversion of fats to glycerol and acids. As a result of research in this area, some 1700 enzymes have been identified and characterized with respect to their function and chemical structure, but there may be as many as 10,000 enzymes at work within a single cell.

TABLE 23-8. Examples of Enzyme Catalysis

Enzyme	Substrate	Type of Reaction	Product(s)
sucrase	sucrose	hydrolysis	fructose, glucose
pepsin	protein	hydrolysis	amino acids
lipase	fats	hydrolysis	fatty acids, glycerol
decarboxylase	$RC-C-OH$ (with $\overset{\|}{O}\,\overset{\|}{O}$)	decarboxylation	$RC-H + CO_2$ (with $\overset{\|}{O}$)
isomerase	$RC-CH_2OH$ (with $\overset{\|}{O}$)	isomerization	$RCHOHCH$ (with $\overset{\|}{O}$)
oxidase	RCH_2OH	oxidation-reduction	$RCHO$

The substance upon which an enzyme acts is known as the **substrate.** An enzyme is often named by appending the ending *ase* to the name of its substrate; in other cases an enzyme is named according to the products formed in the reaction it catalyzes; and in still other cases the enzyme is named according to the type of reaction it catalyzes. In addition, many digestive enzymes are known by names which were given before attempts at systematic nomenclature were made. For example, ptyalin, rennin, trypsin, and pepsin are familiar names of enzymes of the digestive tract. Some examples of enzyme-catalyzed reactions are listed in Table 23-8. The singular specificity of a given enzyme is believed to be due to the conformation (shape) of its molecules and the location of active sites within the molecule. These factors in turn are related to the sequence of amino acids in the protein. The function of any catalyst is to lower the activation energy for the reaction. Apparently the dimensions of an enzyme molecule must exactly match those of its substrate molecule, so that an activated complex (Section 17-6) is easily formed. The activated complex then reacts to form the products.

23-13   The Nucleic Acids

A living system may be described as one which (1) absorbs nutrients from its surroundings, (2) grows by converting the nutrients into its own substance, and (3) reproduces itself. These criteria apply to complex organisms such as humans, trees, and killer whales, as well as to simple one-celled creatures like amoeba. Indeed, they apply to individual cells within a complex organism.

How does the collection of materials in a robin's egg know to develop into a bird which resembles its parents? How do the cells, isolated from outside influence, organize themselves to form a beak, eyes, wings, feathers, a heart, lungs, and so on? What message within the egg determines the red color of the feathers on the breast of the adult male robin? The answers to such questions have long been the concern of that branch of biology called **genetics.** Each living cell contains certain factors called **genes** which determine the characteristic ways in which a given organism will develop. It has been only within the last thirty years or so that evidence was obtained which shows that genes are molecules made up of **nucleic acids.**

Nucleic acids have large molecular weights (about 10^8 D). The individual nucleic acids differ in the sequence of units making up the molecule. They also differ in

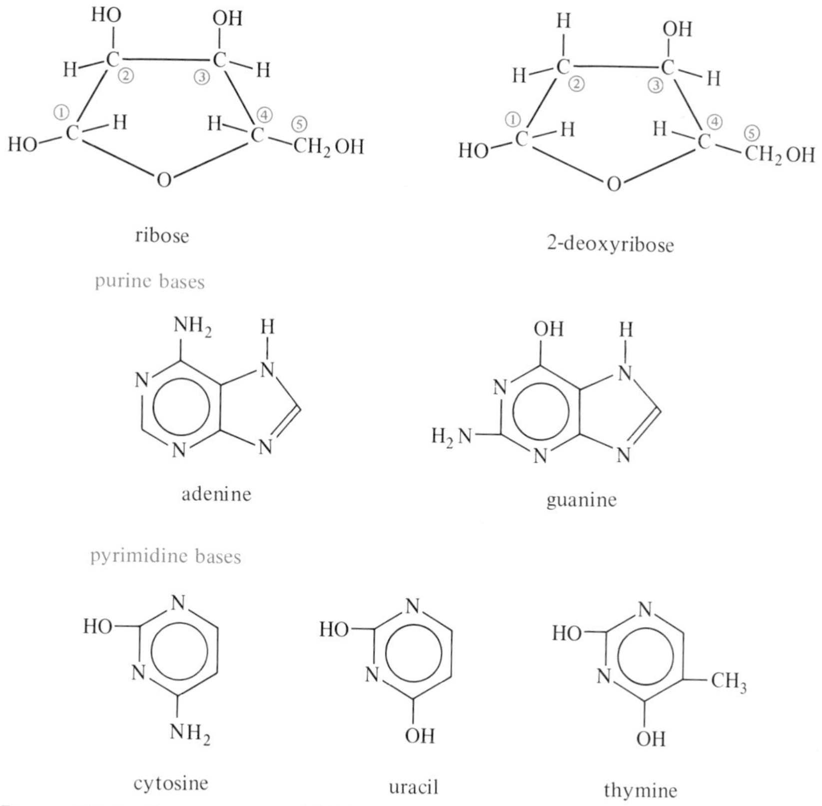

Figure 23-8. **Components of RNA and DNA**

several structural factors, which will be described below. The nucleic acids are of two types, classified on the basis that they contain either of two sugars—ribose or deoxyribose. The corresponding nucleic acids are known as **ribonucleic acid** (RNA) or **deoxyribonucleic acid** (DNA). The basic components of RNA and DNA are shown in Figure 23-8. In addition to the sugar molecule, the nucleic acid contains phosphate, which serves to link two sugars as esters of phosphoric acid, as well as some basic nitrogen-containing constituents—derivatives of purine and pyrimidine. RNA contains adenine, guanine, cytosine, and uracil. In DNA, the uracil is replaced by the structurally similar thymine.

The combination of base, sugar, and phosphate is known as a **nucleotide.** These are the fundamental monomeric units of DNA and RNA. The units are linked in chains, as shown in Figure 23-9, where the shorthand notation on the right is used for a convenient representation of this complex structure.

In 1962 J. D. Watson of Harvard University and F. H. C. Crick of Cambridge University were awarded the Nobel Prize for their demonstration that the structure of the DNA molecule is a double helix, consisting of two strands of nucleotides coiled about one another with the head of one opposite the tail of the other (Figure 23-10). The two strands were shown to be linked by hydrogen bonding between the adjacent nitrogen bases (Figure 23-11).

When a cell divides, the two new cells formed are exact copies of the original. The double helix model of DNA provides an explanation of how this process can occur. It

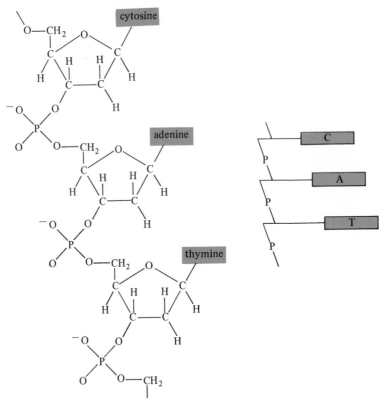

Figure 23-9. **Portion of the DNA Chain**
The representation at the right is a shorthand notation for the chain on the left.

is shown in Figure 23-10 that every cytosine unit in one strand is hydrogen-bonded to a guanine unit in the other strand, and every adenine unit in one strand is hydrogen-bonded to a thymine unit in the other. Only with this pairing can the regular geometry be maintained. Thus the two strands can match only if the sequence of bases in one strand is an exact complement of the sequence of bases in the other. In the process of cell division the two strands separate. Then each strand builds its missing partner from the nutrients present in the cell. Hence two helixes identical with the original are produced. When this task is accomplished for all of the DNA molecules in the cell, complete cell division occurs and two identical new cells result. The process is shown diagrammatically in Figure 23-12.

23-14   Synthesis of Protein—The Genetic Code

Extensive chemical studies combined with X-ray structural determinations have shown that each DNA molecule has its own sequence of bases which effectively stores information about the characteristics controlled by the gene. Within a cell, in the presence of appropriate enzymes, RNA molecules are synthesized by association of purine and pyrimidine bases with the bases in a single strand of DNA in a manner

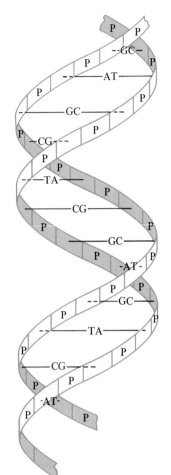

Figure 23-10. **DNA Double Helix**

P = phosphate;   C = cytosine;   A = adenine;   T = thymine;
G = guanine.

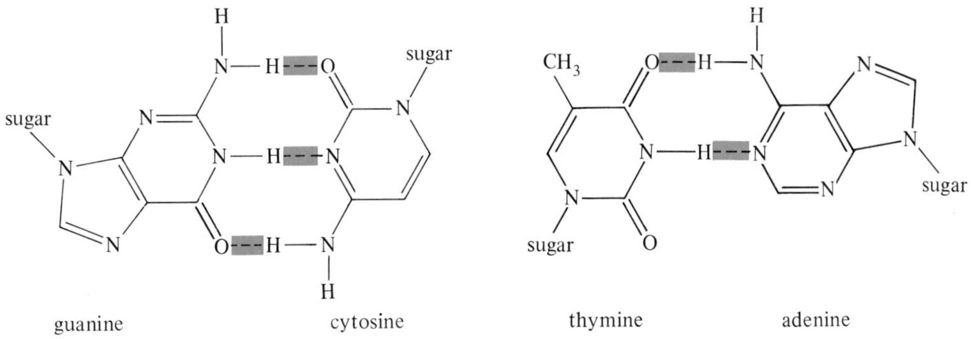

guanine   cytosine   thymine   adenine

Figure 23-11. **Alignment of Nitrogen Bases in a Nucleic Acid by Means of Hydrogen Bonding**

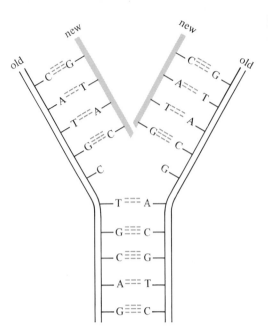

Figure 23-12. **Replication of DNA in Cell Division**

analogous to the formation of new strands of DNA prior to cell division. Unlike DNA, ribonucleic acid, RNA, does not exist in a double strand. It should be recalled that RNA contains the base uracil, rather than thymine. Like thymine, uracil forms two hydrogen bonds to adenine. Therefore whenever the base adenine occurs in the DNA strand, the complementary base in the RNA strand will be uracil. After an RNA strand of appropriate length has been synthesized, it dissociates from the DNA strand which is then free to synthesize another RNA molecule. Each molecule of DNA serves for the synthesis of many molecules of RNA and each RNA molecule reflects the base sequence of its parent DNA strand.

Three types of RNA are involved in the synthesis of polypeptide chains of proteins. The first type is called messenger RNA, mRNA. It is synthesized in conformity with the gene (the DNA molecule) and acts as a template on which are built new protein chains having the specific sequence of amino acids which is required. The second type is called transfer RNA, tRNA. It attaches itself to a particular amino acid and carries that amino acid to the site of protein synthesis. The base sequences of several tRNA molecules have been determined. In all of them the three bases at one end of the molecule are cytosine, cytosine, and adenine, in that order. It is presumed that an amino acid molecule becomes attached to this end of the tRNA molecule. Another portion of the tRNA molecule contains a characteristic sequence of bases which can associate via hydrogen bonding with corresponding bases along the mRNA chain. The third type of RNA is called ribosomal RNA, rRNA. It constitutes about 65% of the ribosome of a cell. Protein synthesis takes place within the ribosomes.

A sequence of three bases on the mRNA strand is required to specify an amino acid for incorporation into a polypeptide chain. Discussion of the process is facilitated by referring to each base by its initial letter, and to each segment of an mRNA molecule by a sequence of three letters corresponding to the bases. The use of such letters makes the discussions and investigations analogous to breaking a code—the

TABLE 23-9. The Genetic Code

SECOND LETTER

		U	C	A	G	
FIRST LETTER	U	UUU ⎤ Phe UUC ⎦ UUA ⎤ Leu UUG ⎦	UCU ⎤ UCC UCA ⎦ Ser UCG	UAU ⎤ Tyr UAC ⎦ UAA xxx UAG xxx	UGU ⎤ Cys UGC ⎦ UGA xxx UGG Trp	U C A G
	C	CUU ⎤ CUC CUA ⎦ Leu CUG	CCU ⎤ CCC CCA ⎦ Pro CCG	CAU ⎤ His CAC ⎦ CAA ⎤ Gln CAG ⎦	CGU ⎤ CGC CGA ⎦ Arg CGG	U C A G
	A	AUU ⎤ AUC ⎦ Ile AUA AUG Met	ACU ⎤ ACC ACA ⎦ Thr ACG	AAU ⎤ Asn AAC ⎦ AAA ⎤ Lys AAG ⎦	AGU ⎤ Ser AGC ⎦ AGA ⎤ Arg AGG ⎦	U C A G
	G	GUU ⎤ GUC GUA ⎦ Val GUG	GCU ⎤ GCC GCA ⎦ Ala GCG	GAU ⎤ Asp GAC ⎦ GAA ⎤ Glu GAG ⎦	GGU ⎤ GGC GGA ⎦ Gly GGG	U C A G

THIRD LETTER

genetic code. Thus the sequence of bases on the DNA molecule (gene) may be regarded as a sequence of three-letter "words" called **codons** formed from a four-letter "alphabet"—adenine, A; thymine, T; guanine, G; and cytosine, C. The mRNA synthesized from the gene will have a corresponding set of codons formed by adenine, A; uracil, U; guanine, G; and cytosine, C. In addition to the sequence of bases which specify the amino acids, the mRNA must also contain codons which start and stop the synthesis of peptide chains. Geneticists have succeeded in determining the codons responsible for each of the twenty amino acids. This genetic code is essentially the same for all organisms. The codons on the mRNA molecule are given in Table 23-9. Some of the codons are redundant; that is, different codons select the same amino acid. Several codons are probably command words or signals for terminating polypeptide chains. In the table, these are designated by the symbol "xxx."

The process of protein synthesis (Figure 23-13) proceeds somewhat as follows:

1. The DNA polynucleotide of a gene transfers its information by synthesizing a messenger RNA molecule.
2. Ribosomes then travel along the mRNA chain (Figure 23-14) and at certain points, corresponding to a given codon, a molecule of tRNA to which a molecule of the proper amino acid has been attached associates with the ribosome.
3. A second tRNA molecule with its attached amino acid attaches itself to an adjacent position on the mRNA and a peptide link is formed between the two amino acids. The first molecule of tRNA then drops off ostensibly to pick up another molecule of its particular amino acid.

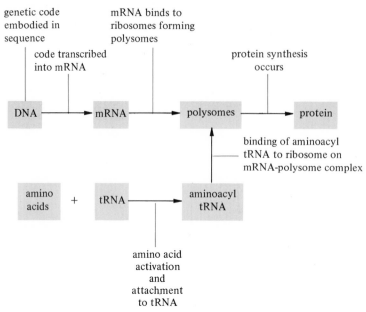

Figure 23-13. **Sequence of Events in Protein Synthesis**

4. As the ribosome moves further down the mRNA chain, a third molecule of tRNA with its amino acid takes an adjacent position, and a second peptide link is formed. The second tRNA molecule drops off and the process continues.

Since each position along the mRNA chain contains specific codons, the tRNA molecules must bring up only the specific amino acid called for. For example, if a segment of the DNA contains the sequence GATGTA. . . , the mRNA will contain the corresponding sequence CUACAU. . . . When a ribosome arrives at this

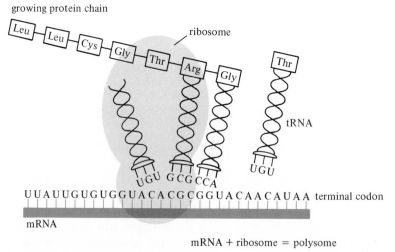

Figure 23-14. **Diagrammatic Representation of Protein Synthesis in a Polysome**

segment of the mRNA strand, the amino acids leucine and histidine will be successively added to the peptide chain. Thus every polypeptide chain formed by this mRNA unit will be identical. Apparently somewhere along the mRNA a codon which signals a termination is encountered.

Literally hundreds of DNA molecules exist within a given cell, each of which gives its information to thousands of mRNA units. Thus all of the characteristic proteins of the cell are generated.

23-15   The Krebs Cycle

The energy utilized by a living animal is obtained from the oxidation of food within its cells. The principal fuel is glucose, which is converted to carbon dioxide and water. (Other nutrients, fats and proteins, are also oxidized, but only after being broken down into much simpler molecules.) The enthalpy of combustion of glucose is -680 kcal/mole, but within the cell this energy is released in a series of steps, each catalyzed by a specific enzyme. Moreover, the energy released by the oxidation is stored by the cell to be used when needed. Thus the two interrelated processes occur: energy production and energy storage. The energy is stored by means of a reaction in which an ester, **adenosine diphosphate** (ADP), is converted into **adenosine triphosphate** (ATP). The structures of these molecules are diagrammed in Figure 23-15, and the reaction is represented by the following equation, where R denotes the rather complex adenosine portion of the molecule:

$$\text{energy} + \begin{bmatrix} & O & & O & \\ \| & & \| & \\ HO-P-O-P-O-R \\ | & & | & \\ O & & O & \end{bmatrix}^{2-} + \begin{bmatrix} & O & \\ \| & \\ HO-P-O \\ | & \\ O & \end{bmatrix}^{2-} \rightleftharpoons$$

ADP

$$\begin{bmatrix} & O & & O & & O & \\ \| & & \| & & \| & \\ HO-P-O-P-O-P-O-R \\ | & & | & & | & \\ O & & O & & O & \end{bmatrix}^{3-} + OH^-$$

ATP

Formation of the triphosphate ester requires about 8 kcal/mole of energy; the reverse reaction releases the same quantity of energy.

As a fuel, glucose is first broken into three-carbon fragments, pyruvic acid and/or lactic acid, in a process called **glycolysis.**

$$C_6H_{12}O_6 \cdot 2\,ATP \;\rightleftharpoons\; 2\,CH_3CHOHCO_2H \;+\; 2\,ATP$$

glucose–ATP complex   lactic acid

Then these acids are converted to carbon dioxide and water by means of a cyclic process known as the **Krebs cycle.** In this cycle pyruvic acid, CH_3COCO_2H, is decarboxylated and then condensed with oxaloacetic acid,

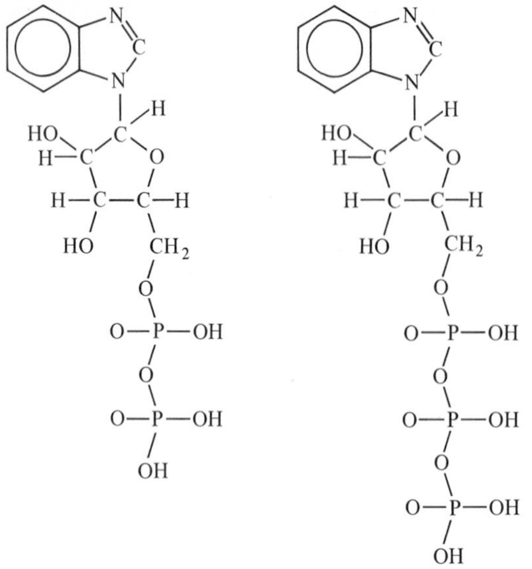

Figure 23-15. **Structures of ADP and ATP**

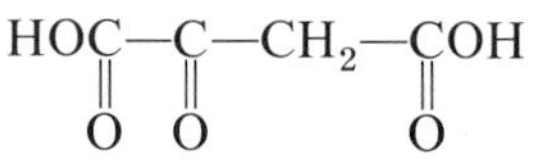

ADP

ATP

$$HOC-C-CH_2-COH$$
$$\ \ \|\ \ \ \|\ \ \ \ \ \ \ \ \|$$
$$\ \ O\ \ O\ \ \ \ \ \ \ \ O$$

to give citric acid,

$$H_2CCO_2H$$
$$HOCCO_2H$$
$$H_2CCO_2H$$

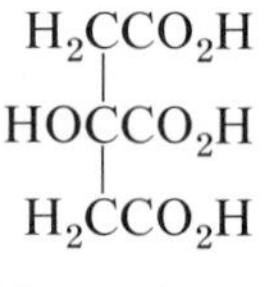

This condensation reaction involves acetylcoenzyme A (coA—SH). This step is then followed by a series of enzyme-catalyzed steps, which produce water, carbon dioxide, and oxaloacetic acid, the compound which began the cycle. The entire oxidation process is shown schematically in Figure 23–16, where pyruvic acid and five half-mole quantities of O_2 are indicated as reactants. More complex oxidizing agents are actually involved, but the overall result is the consumption of $\frac{5}{2}$ moles of O_2.

In the glycolysis step, two ATP molecules are required for the initial phosphorylation of glucose. Then a total of 40 ATP molecules is formed using the energy released in the successive steps of the Krebs cycle. Hence in the complete oxidation of 1 mole of glucose, $38 \times 8 = 304$ kcal of energy is stored for future use.

23-16 Exercises

Basic Exercises

1. Two isomeric compounds, A and B, have the formula C_4H_8. Compound A reacts to decolorize a solution of bromine in CCl_4; compound B does not react with bromine. Diagram possible structures for A and B.

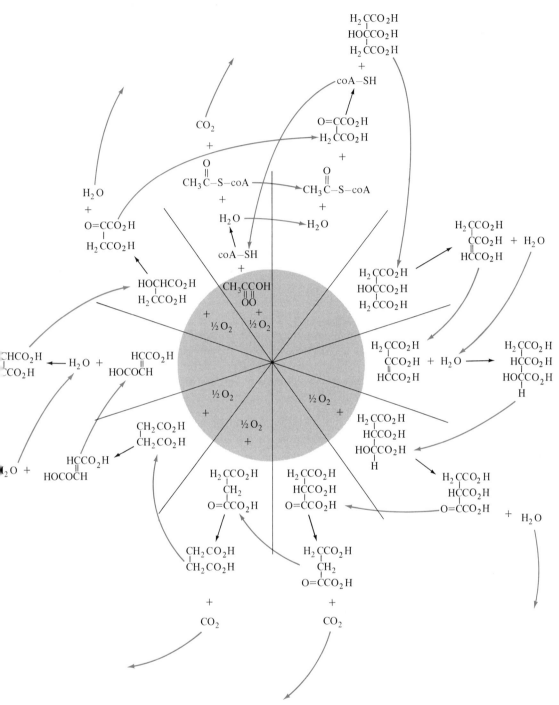

Figure 23-16. **Krebs Cycle (Citric Acid Cycle)**

coA—SH represents coenzyme A, with its —SH group shown explicitly.

2. Using bond energy data, estimate whether CH_4 or C_8H_{18} would produce more heat *per gram* upon complete combustion. Assume all reactants and products to be in the gaseous state.

3. **(a)** Add the equations representing the chain propagation presented on page 659 and determine the net chemical reaction. **(b)** Show how the presence of a small quantity of ethane in the mixture of Cl_2, CH_4, and the chloromethanes, which results from the reaction, supports a chain mechanism (free radical mechanism) for the reaction.

4. Predict the major product of the bromination of a ten-fold excess of $(CH_3)_2CHC(CH_2CH_3)_3$ with Br_2.

5. Unsaturated hydrocarbons are said to "decolorize" bromine water. Write an equation for the reaction and explain why the color disappears.

6. Explain why the use of ultraviolet light to initiate a reaction implies a chain (free radical) mechanism for halogenation of a saturated hydrocarbon.

7. Explain why Br^+, shown in the figure on page 660 and the equations on page 661, is not a stable chemical species.

8. Contrast the experimental conditions necessary to effect the following reactions:
   (a) $CH_4 + Br_2 \rightarrow CH_3Br + HBr$
   (b) $C_2H_4 + Br_2 \rightarrow CH_2BrCH_2Br$
   (c) $C_6H_6 + Br_2 \rightarrow C_6H_5Br + HBr$

9. Which isomer of $C_2H_4Cl_2$ can be prepared by the chlorination of ethylene? Explain. How can the other isomer be prepared?

10. Look up the properties of CCl_4, CCl_2F_2, and CH_4 in a Handbook of Chemistry. Explain what properties of CCl_2F_2 make it effective as a propellant in aerosol cans.

11. What ratio of CH_4 to Cl_2 is best to prepare relatively pure **(a)** CH_3Cl; **(b)** CCl_4?

12. A soap is more soluble in water than the fatty acid from which it is derived. **(a)** How can one prepare a soap from the corresponding acid? **(b)** How can one prepare the acid from the corresponding soap?

13. What class of compounds is represented by the type formula ROR'? Using CH_3 and C_2H_5 as the radicals, write formulas for three compounds which correspond to this type formula.

14. Write an overall equation for the process consisting of reduction of propanoic acid with $LiAlH_4$ followed by treatment of the resulting solution with CH_3OH and then aqueous HCl.

15. **(a)** What is the oxidation state of hydrogen in $LiAlH_4$? **(b)** Write an equation for the reaction of hydrogen in that oxidation state with water. **(c)** What precautions must be exercised in the reduction of organic compounds with excess $LiAlH_4$?

16. What sequence of reactions can be used to convert **(a)** ethanol to ethyl bromide? **(b)** ethyl bromide to propanoic acid? **(c)** propanoic acid to 1-propanol? **(d)** What is the net effect of this sequence of conversions?

17. Write an equation for the reaction with HCl(aq) of **(a)** $CH_3CH_2CH_2NH_2$, **(b)** $(CH_3)_3N$.

18. **(a)** Write equations showing the preparation of acetamide and of ammonium acetate from acetic acid and ammonia. **(b)** What difference in experimental conditions is required in the two cases? **(c)** Distinguish between the properties of the two products.

19. Identify the functional group(s) most characteristic of each of the following: **(a)** lipids, **(b)** soap, **(c)** glycerol, **(d)** wax, **(e)** carbohydrates, **(f)** amino acids.

20. Compare and contrast the properties and structure of a soap with those of an artificial detergent.

21. What ecological problems have been encountered as the widespread use of detergents for laundry and household cleaning increases?

22. Determine the number of optical isomers, including fructose, of 2-ketohexose.

23. D-Glyceraldehyde is (+). Are D-isomers of all carbohydrates (+)? Explain what each of these symbols means in the case of the optical isomers of glucose.

24. Write a balanced chemical equation describing (a) the fermentation of glucose to alcohol and carbon dioxide, (b) the hydrolysis of starch to a simple hexose.

25. Dimerization of six-carbon monosaccharides yields disaccharides, with the general formula $C_{12}H_{22}O_{11}$. Polymerization of such monosaccharides produces starch with what empirical formula?

26. Describe the molecular structure of a polyester fabric.

27. Compare and contrast addition polymers and condensation polymers with respect to (a) by-products produced, (b) number of functional groups necessary in each monomer, (c) ability to incorporate different monomers into one polymer.

28. Write a representation of a portion of a nylon polymer chain (see Table 23–5) containing at least three molecules of each monomer. Draw a similar portion of a Dacron chain.

29. Explain on a molecular level why vulcanization makes rubber harder and less elastic.

30. Referring to alpha helical protein structures, what is meant by the term (a) primary structure? (b) secondary structure? (c) What types of forces are responsible for each of these structures?

31. Draw a structural formula for one loop of an alpha helix, showing the peptide linkages and the hydrogen bonding.

32. What is the genetic code? What is a codon?

33. A segment of the DNA molecule contains the series of bases cytosine, thymine, and guanine. (a) What corresponding segment of an mRNA molecule will be built by this segment? (b) The mRNA segment is capable of specifying which amino acid? (c) What other sequence of bases in mRNA, if any, would specify the same amino acid?

34. The sequence of bases cytosine, uracil, guanine on a mRNA molecule (a) will interact with what series of bases on a tRNA molecule? (b) will be formed from what series of bases of a DNA molecule? (c) will cause addition of what amino acid to a protein chain?

35. Consider the following sequence of amino acids in the insulin molecule (Figure 23–5): Ala–Lys–Pro–Thr–Tyr. . . . Diagram a possible portion of an mRNA molecule responsible for the formation of this sequence of amino acids (as in Figure 23–14).

36. (a) Explain why each of the following equations represents an oxidation of the organic reactant:

$$CH_3CHO \ + \ \tfrac{1}{2}O_2 \ \longrightarrow \ CH_3CO_2H$$

$$CH_3CH_3 \ \xrightarrow{\text{Pt catalyst}} \ CH_2{=}CH_2 \ + \ H_2$$

(b) Which of the oxidation reactions in the Krebs cycle (Figure 23–16) is most similar to the second equation of (a)?

37. Write equations for the ten reactions in the Krebs cycle (Figure 23–16). Add the equations, and determine the overall change. Indicate each reaction in which a new reactant enters the cycle and each reaction in which a product is formed which leaves the cycle.

General Exercises

38. (a) Determine the number of five-membered rings, six-membered rings, and double bonds in chlordane (Table 23–2). (b) Write the molecular formula for the compound and for its parent unsubstituted hydrocarbon. (c) Is the formula consistent with the general formula for a substituted hydrocarbon with that number of rings and double bonds? Explain.

39. Write overall equations for the conversion of ethyl alcohol into (a) ethylene, (b) diethyl ether. (c) What function does the H_2SO_4 serve in these reactions?

40. Describe how each of the following pairs of isomers can be distinguished by at least two instrumental methods: (a) 1-hexene and cyclohexane, (b) 1-butene and 2-butene.

41. Write equations showing how the following acid derivatives react with water:
    (a) acetic anhydride, $CH_3C-O-CCH_3$;   (b) acetyl chloride, CH_3C-Cl.
        $\quad\quad\quad\quad\quad\quad\quad\quad\;\; \| \quad\;\; \| \quad\quad\quad\quad\quad\quad\quad\quad\quad\quad\quad\quad\;\; \|$
        $\quad\quad\quad\quad\quad\quad\quad\quad\;\; O \quad\; O \quad\quad\quad\quad\quad\quad\quad\quad\quad\quad\quad\quad\;\; O$

42. Write equations showing the preparation of the following compounds, starting with ethyl alcohol and any other required chemicals: (a) acetic acid,   (b) ethylene,   (c) diethyl ether,   (d) ethyl acetate,   (e) propanal.

43. Compound A has the formula $C_5H_{12}O$. Reaction of A with Ag_2O, a mild oxidizing agent, yields compound B, which does not reduce Tollen's reagent and which gives an nmr spectrum consisting of a triplet and a quartet. Identify compounds A and B.

44. A solution of bromine in benzene is stable indefinitely, but when an iron nail is put into the solution, bromination of the benzene occurs fairly rapidly. Explain the function of the iron. Write equations for all the reactions.

45. What sequence of reactions can be used to convert   (a) propanoic acid to butanoic acid?   (b) benzene to benzoic acid, $C_6H_5CO_2H$?

46. When 2.86 grams of a mixture of 1-butene, C_4H_8, and butane, C_4H_{10}, was burned in excess oxygen, 8.80 grams of CO_2 and 4.14 grams of H_2O were obtained. Calculate the percentage by mass of butane in the original mixture.

47. The compound C_3H_6ClBrO exists in two optically active forms. Its infrared spectrum shows strong bands at 3300 and 1075 cm^{-1}. Its nmr spectrum consists of three singlets of intensity 1/2/3 from low field to high field, respectively. Reaction with warm, alcoholic $AgNO_3$ gives a precipitate of AgBr. Diagram the structure of this compound.

48. The nmr spectrum of the compound C_3H_7Br shows a septet downfield from a doublet. The intensity of the doublet is six times that of the septet. Would this compound be expected to react with cold, alcoholic $AgNO_3$? Explain.

49. Aromatic alcohols such as phenol, C_6H_5OH, have definite acidic properties in aqueous solution, in contrast with alkyl alcohols. Explain the influence of the aromatic ring on the electrons of the OH functional group.

50. Explain why a solution containing only $K^+ HOCOC_6H_4CO_2^-$ (potassium hydrogen phthalate) in water can serve as a buffer solution.

51. Compound A contains 71.95% carbon, 12.08% hydrogen, and the remainder oxygen. It is treated with aqueous base and then heated to drive off compound B. The residual solution is treated with aqueous H_2SO_4, and compound C precipitates. Compound B reacts with sodium to liberate hydrogen. Its nmr spectrum consists of two peaks, with the downfield peak one third of the intensity of the upfield peak. Compound C contains only carbon, hydrogen, and oxygen. It requires 50.0 ml of 0.200 M NaOH to titrate 1.85 grams of compound C to the phenolphthalein end point. Suggest possible formulas for A, B, and C.

52. Insulin is a rather small protein. Using Figure 23–5 and the covalent radii from Table 11–1, estimate the length the molecule would have if the bonds of the continuous chain were all linear.

53. Throwaway containers made from the polymer poly(vinyl chloride) (PVC) are widely used in packaging consumer goods.   (a) Look up the formula for vinyl chloride, and write the equation for its polymerization.   (b) Write an equation for the combustion of PVC in oxygen.   (c) Suggest why incineration is not a satisfactory method for disposing of used packaging materials made of PVC.

Advanced Exercises

54. In how many ways can three different amino acids be arranged in a three-monomer chain? In how many ways could any three of twenty different amino acids be arranged in a chain containing three different monomers? In how many ways could any of twenty amino acids be arranged in a chain containing three monomers (some may be repeated)? Show how these statistics can be used to explain the vast variety of proteins found in living organisms.

55. Suggest reaction sequences for preparing the following organic compounds starting with graphite as the only source of carbon: **(a)** $C_2H_5OCOC_2H_5$, **(b)** CH_2BrCH_2Br, **(c)** $CH_3OC_2H_5$.

56. In the formation of a phenol–formaldehyde polymer, a water molecule is produced by the reactions of two hydrogen atoms on the aromatic ring and the oxygen atom of formaldehyde. The resulting polymer has the form

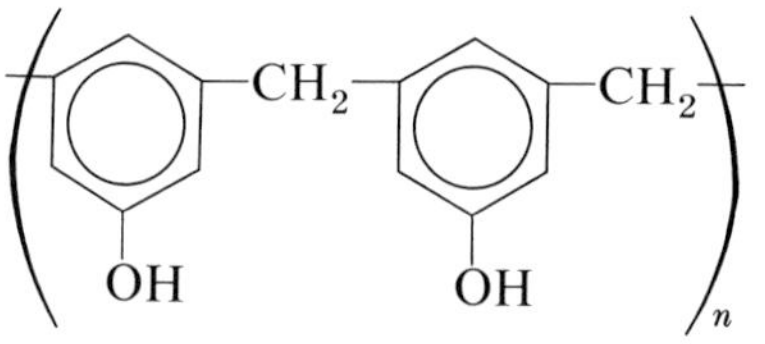

The phenolsulfonic acid–formaldehyde polymer condenses in an analogous manner. Draw an extended structure of the polymer. Explain why this type of polymer might be used as an ion exchange resin.

57. Textbooks represent the formula of a pyrimidine-type base as one of the following:

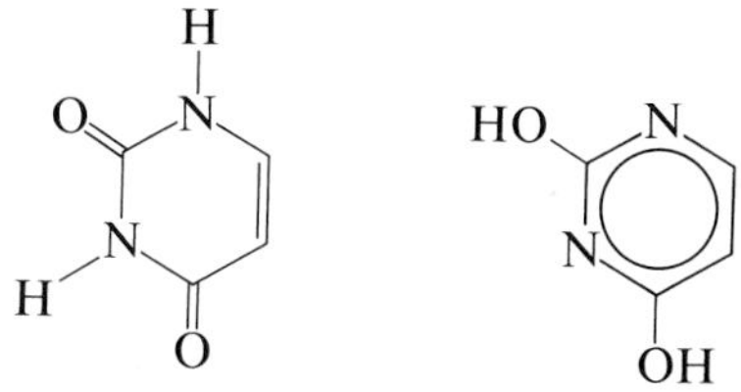

Noting that an OH group attached directly to an aromatic ring is acidic, write the formula for still another possible representation of the base. What relationships exist among these three representations of pyrimidine?

Appendix

A-1   The Metric System

The metric system is used universally in scientific measurements. The units and subunits are interrelated by multiplication or division by powers of 10. The metric standard of length is the meter, which is defined precisely as 1,650,763.73 times the wavelength of the orange-red light emitted by $^{86}_{36}$Kr. The standard of mass is the kilogram, which is defined as the mass of a standard object kept by the International Bureau of Weights and Measures. The unit of volume is the liter, which is now defined as 1000 cubic centimeters. The unit of time is the second.

In practical usage, measurements are commonly expressed as decimal fractions or as multiples of the basic units by using a prefix before the name of the unit. That is, *milli*liter means 0.001 liter. A tabulation of commonly used prefixes is given in Table A-1.

A-2   Concepts from Physics

An understanding of some concepts of elementary physics is necessary for a full understanding of the principles developed in this text. A sketch of some of the important concepts is given here. For fuller explanations see an elementary physics textbook.

Force

A force can be described in an elementary manner as a push or a pull. An unbalanced force on an object will cause the object to accelerate (that is, to change its velocity). If the object has a mass, m, the force, f, is related to the mass and acceleration, a, of the object by Newton's second law of motion:

TABLE A-1. Metric Prefixes

Prefix	Meaning	Multiple	Example
nano	one one-billionth	10^{-9}	1 nanometer (nm) = 0.000000001 meter = 10^{-9} meter
micro	one one-millionth	10^{-6}	1 microgram (μg) = 0.000001 gram = 10^{-6} gram
milli	one one-thousandth	10^{-3}	1 milliliter (ml) = 0.001 liter = 10^{-3} liter
centi	one one-hundredth	10^{-2}	1 centimeter (cm) = 0.01 meter = 10^{-2} meter
kilo	one thousand	10^{3}	1 kilogram (kg) = 1000 grams = 10^{3} grams

$$f = ma$$

The force of attraction of the earth on an object, which is the force of gravity, is called the weight of the object. The acceleration due to gravity, denoted g, is equal to 980 cm/sec^2 on the surface of the earth. Substitution of the quantities specifically related to gravity in the preceding equation yields

$$\text{weight} = mg$$

Forces between two objects resulting from static electric interactions or from magnetic interactions are related to the magnitudes of the charges or of the pole strengths by two similar equations, both of which are attributed to Coulomb.

For electrostatics: $$f = k\frac{q_1 q_2}{d^2}$$

For magnetic interactions: $$f = k\frac{p_1 p_2}{d^2}$$

where q refers to the charge on an object, p to the pole strength of the magnet, and d to the distance between the charged objects or the magnetic poles. In each case the numerical value of the constant, k, depends on the units used.

Energy

Energy exists in many forms. Energy can be expended on an object by the exertion of a force on it over a distance, d. Mechanical energy is the energy due to the motion of the object, called kinetic energy, or to the position of the object, called potential energy. The kinetic energy of an object with mass, m, going at velocity, v, is given by

$$\text{KE} = \tfrac{1}{2}mv^2$$

The potential energy of an object is equal to the force acting on the object times the distance that the force could possibly move the object. For example, a 10.0 gram object on a table at a height, h, 100 cm above the floor has a potential energy with respect to the floor of

$$
\begin{aligned}
\text{PE} &= (\text{force})(\text{distance}) \\
&= (\text{weight})(\text{height}) = mgh \\
&= (10.0 \text{ grams})(980 \text{ cm/sec}^2)(100 \text{ cm}) = 9.80 \times 10^5 \text{ gram} \cdot \text{cm}^2/\text{sec}^2
\end{aligned}
$$

Other forms of energy, such as heat, light, and chemical energy, are generally expressed in calories or joules.

The energy expended in moving a charged object from an infinite distance to a distance, d, away from a second charged object is given by

$$E = k\frac{q_1 q_2}{d}$$

(Note that the energy is inversely proportional to the distance, d, whereas the force is inversely proportional to the square of the distance, d^2.)

A-3   Metric and SI Units and Conversion Factors

For many years, the centimeter, gram, and second have been used, in a system called the cgs system, to define derived units. For example, the dyne is defined as the force required to accelerate a 1 gram mass 1 cm/sec²:

$$1 \text{ dyne} = 1 \text{ gram} \cdot \text{cm/sec}^2$$

A new international system of units, called the *Système Internationale* (SI), has been adopted and is gaining in usage. This system uses the meter, kilogram, and second to define derived units. For example, the unit of force is the newton:

$$1 \text{ newton} = 1 \text{ kg} \cdot \text{meter/sec}^2$$

These systems are compatible and both will be used. A tabulation of the units and conversion factors among them is presented as Table A-2.

A-4   Temperature Scales

On the Celsius (Centigrade) temperature scale, the temperature at which pure ice melts under one atmosphere of pressure is designated as $0\,°C$, while the temperature at which pure water boils under one atmosphere of pressure is designated as $100\,°C$. On the Fahrenheit scale, the same temperature interval is designated as $180°$, and on this scale the melting point of pure ice is $32\,°F$. The temperature in degrees Fahrenheit can be converted into degrees Celsius by means of one of the relationships

$$°C = \tfrac{5}{9}(°F - 32°) \qquad \text{or} \qquad °C + 40° = \tfrac{5}{9}(°F + 40°)$$

On the absolute or Kelvin scale, a one degree interval is identical with that of the Celsius scale, but the absolute zero (0 K) is at $-273.15\,°C$. The relationships among the three scales are illustrated in Figure A-1. The symbol K stands for degrees Kelvin.

A-5   Significant Figures[1]

The numerical value reported for a measurement should reflect the precision with which that measurement was made. For example, the distances between cities are most often reported as certain numbers of miles, although it is hardly possible that all cities are exactly a whole number of miles from each other. As another example, the

[1] Adapted from D. E. Goldberg, *Essentials of Physical Science* (Washington, D.C.: Andromeda Books, 1962), pp. 6–8.

TABLE A-2. Units and Conversion Factors

	CGS and Other Common Units	SI Units
length	1 inch (in.) = 2.540 cm 1 Ångström (Å) = 10^{-8} cm = 10^{-1} nm	1 inch (in.) = 0.02540 meter 1 Ångström (Å) = 10^{-10} meter = 10^{-1} nm
volume	1 U.S. quart = 946.33 ml	1 U.S. quart = 0.94633 liter
mass	1 metric ton = 10^6 grams 1 pound = 453.9 grams 1 dalton (D) = 1.661×10^{-24} gram	1 metric ton = 10^3 kg 1 pound = 0.4539 kg 1 dalton (D) = 1.661×10^{-27} kg
force	1 dyne = 1 gram $\cdot$ cm/sec^2	1 newton (Nt) = 1 kg $\cdot$ meter/sec^2 = 10^5 dynes
energy	1 erg = 1 dyne $\cdot$ cm = 1 gram $\cdot$ cm^2/sec^2 = 1 (esu)^2/cm 1 calorie (cal) = 4.184×10^7 erg = 4.184 joule (J) 1 watt $\cdot$ sec = 1×10^7 erg = 1 J = 1 V $\cdot$ C 1 eV/ion = 23.06 kcal/mole 1 kcal/mole = 350 cm^{-1} 1 dalton (D) = 931 MeV	1 joule (J) = 1 Nt $\cdot$ meter = 1 kg $\cdot$ meter^2/sec^2 = 10^7 erg 1 calorie (cal) = 4.184 J 1 watt $\cdot$ sec = 1 J = 1 V $\cdot$ C 1 eV/ion = 96.48 kJ/mole 1 kJ/mole = 1.46×10^5 meter^{-1} 1 dalton (D) = 931 MeV
pressure	1 atmosphere (atm) = 1.013×10^6 dynes/cm^2 = 760 torr = 760 mm Hg = 14.7 pounds/in.^{2}	1 atmosphere (atm) = 1.013×10^5 Nt/meter^2 = 760 torr = 760 mm Hg = 14.7 pounds/in.^{2} = 760 torr = 760 mm Hg 1 pascal (Pa) = 1 Nt/meter^2
charge	1 coulomb (C) = 3×10^9 esu	1 coulomb (C) = 3×10^9 esu
current	1 ampere (A) = 1 C/sec	1 ampere (A) = 1 C/sec
resistance	1 ohm = 1 V/A	1 ohm = 1 V/A
power	1 watt = 1 A $\cdot$ V	1 watt = 1 A $\cdot$ V
viscosity	1 poise = 1 dyne $\cdot$ sec/cm^2	1 poise = 0.1 Nt $\cdot$ sec/meter^2
potential	1 volt (V)	1 volt (V)
frequency	1 hertz (Hz) = 1/sec	1 hertz (Hz) = 1/sec
activity	1 curie (Ci) = 3.7×10^{10} dis/sec	1 curie (Ci) = 3.7×10^{10} dis/sec

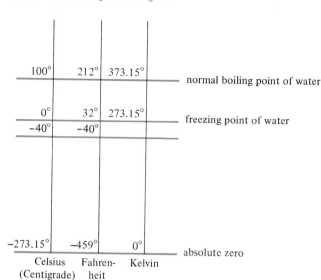

length of a city block can be measured more accurately with a long tape measure than with an automobile mileage indicator. To indicate the accuracy of a measurement, its numerical value must be expressed to show only those digits which are known exactly plus one uncertain digit. These digits are called the **significant digits** or **significant figures.**

Thus the reporting of the number 1.56 means that the measurement (or measurements) upon which the number is based is probably closer to 1.56 than to 1.55 or 1.57; that is, the true value probably lies between 1.555 and 1.565. It is especially important to be aware of significant figures when making calculations based on measurements. No amount of arithmetic can improve the accuracy of the actual measurement; therefore the calculated result must also be expressed with the proper number of significant figures. For example, suppose the area of a rectangle is to be calculated from its measured length and width. If the length is reported as 4.6 cm and the width as 4.1 cm, these figures imply that the measuring instrument used distinguishes between the values reported and those 0.1 cm different. In multiplication by calculator, one obtains for the area 18.86 cm². However, if the numbers are accurate to between 4.05 and 4.15 cm for the width and 4.55 and 4.65 cm for the length, then the true area lies somewhere between the product of the lower limits and the product of the upper limits of measurement, that is, between 18.4275 and 19.2975 cm². The area is certain only to the nearest 1 cm². Therefore the calculator answer should be rounded off so that the accuracy will be reflected in the reported value. The number should be reported as 19 cm², that is, somewhere between 18.5 and 19.5 cm². This analysis is summarized as follows:

	Measurement	Lower Limit	Upper limit
length	4.6 cm	4.55 cm	4.65 cm
width	4.1 cm	4.05 cm	4.15 cm
calculated area	18.86 cm²	18.4275 cm²	19.2975 cm²

reported area: 19 cm²

It is not necessary to perform the kind of analysis described above for every calculation. The number of significant figures to be reported can be estimated rapidly by means of the following rules:

1. In a properly reported number, consider all of the digits significant except for the zeros which serve *only* to locate the decimal point. In 12,300 and 0.000456 there are three significant figures each. In each case, the zeros serve only to fix the magnitude of the number (to locate the decimal point). However, there are five significant figures in both 100.05 and 0.10050.
2. Consider numbers which are not measurements, as, for example, the number of millimeters in a meter or the number of radians in a circle, as exact numbers having an infinite number of significant figures.

Example

How many significant figures are there in each of the following numbers: **(a)** 17, **(b)** 103, **(c)** 1.035, **(d)** 0.0010, **(e)** 1.00×10^6, and **(f)** π.
(a) two, **(b)** three, **(c)** four, **(d)** two, **(e)** three, and **(f)** an infinite number.

3. When adding or subtracting numbers, retain digits to the right only as far as the significant digits of the number which has digits least far to the right.

Example

Perform the following operations:

$$
\begin{array}{r}
12.01 \text{ cm} \\
17.3 \ \text{ cm} \\
+ \ 0.11 \text{ cm} \\
\hline
\end{array}
\qquad
\begin{array}{r}
133 \ \ \text{grams} \\
- \ \ 2.2 \text{ grams} \\
\hline
\end{array}
$$

The answers are 29.4 cm and 131 grams rather than 29.42 cm and 130.8 grams. The 2 in the hundredths column of the sum is farther to the right than the 3 of 17.3, and so it cannot be significant. It is dropped because it is less than 5. The 8 of the second example is not significant for the same reason, but it is over 5, so the answer is rounded off to the next higher integer.

4. When multiplying or dividing, retain only as many significant digits as there are in the measured value with the fewest significant digits.

Example

Perform the following operations: **(a)** $12.7 \times 11.2 = $ **(b)** $108/7.2 = $
In the first calculation, three significant figures may be retained since each factor has three. In the second, only two significant figures are retained in the answer. **(a)** 142, **(b)** 15.

A-6   Standard Exponential Form

Numbers expressed in standard exponential form consist of a decimal part, at least 1 but less than 10, times the appropriate power of 10, as shown in the following examples:

Number	In Standard Exponential Form
1234	1.234×10^3
62.34	6.234×10
0.009239	9.239×10^{-3}
0.7	7×10^{-1}

Very large numbers and very small numbers are frequently encountered in scientific work, and it is convenient to express these in standard exponential form:

$$602,000,000,000,000,000,000,000 = 6.02 \times 10^{23}$$
$$0.00000001 = 1 \times 10^{-8}$$

The significant figures in a large or small number are readily denoted when the number is written in exponential form. For example, the number 12,000 has at least two significant figures. Written as 1.200×10^4, the number is seen to have four significant figures.

To add or subtract numbers written in exponential form, the numbers must be expressed so that all have the base 10 raised to the same power:

$$(1.0 \times 10^4) + (2.0 \times 10^3) = (1.0 \times 10^4) + (0.20 \times 10^4) = 1.2 \times 10^4$$

To multiply numbers written in exponential form, the decimal factors are multiplied in the usual manner, and the exponents are added algebraically:

$$(1.2 \times 10^{-3}) \times (6.02 \times 10^{-5}) \times (5.0 \times 10^2) = (1.2 \times 6.02 \times 5.0) \times (10^{-3-5+2})$$
$$= 36 \times 10^{-6} = 3.6 \times 10^{-5}$$

To divide numbers written in exponential form, the exponents of the base 10 in the divisor are subtracted algebraically from the exponents of 10 in the dividend:

$$\frac{1.2 \times 10^{-3}}{3.6 \times 10^{-5}} = \frac{1.2}{3.6} \times 10^{-3-(-5)} = 0.33 \times 10^2 = 3.3 \times 10$$

A-7 Physical Constants

Avogadro's number	N	6.022×10^{23} molecules/mole
Boltzmann constant	k	1.3806×10^{-16} erg/molecule $\cdot$ K
		$= 1.3806 \times 10^{-23}$ J/molecule $\cdot$ K
charge on the electron	e	1.602×10^{-19} C
		$= 4.806 \times 10^{-10}$ esu
Faraday constant	F	9.6487×10^4 C/mole
rest mass of the electron	m	9.1096×10^{-28} gram
Planck's constant	h	6.6262×10^{-27} erg $\cdot$ sec
universal gas constant	R	8.21×10^{-2} liter $\cdot$ atm/mole $\cdot$ K
		$= 8.3143$ J/mole $\cdot$ K
		$= 1.987$ cal/mole $\cdot$ K
velocity of light in a vacuum	c	2.9979×10^{10} cm/sec
density of mercury at 20°C		13.5939 grams/cm^3
acceleration due to gravity at sea level	g	9.80616×10^2 cm/sec^2

A-8 Vapor Pressure of Water at Various Temperatures

Temperature (°C)	Pressure (torr)	Temperature (°C)	Pressure (torr)
−15	1.436	35	42.175
−10	2.149	40	55.324
−5	3.163	45	71.88
0	4.579	50	92.51
5	6.543	55	118.04
10	9.209	60	149.38
15	12.788	65	187.54
20	17.535	70	233.7
21	18.650	75	289.1
22	19.827	80	355.1
23	21.068	85	433.6
24	22.377	90	525.76
25	23.756	95	633.90
26	25.209	99	733.24
27	26.739	100	760.00
28	28.349	101	787.57
29	30.043	105	906.07
30	31.824	110	1074.56

A-9 Logarithms

Relationship Between Logarithms and Exponents

Any number can be expressed as a power of 10. For example,

$$10 = 10^1$$
$$100 = 10^2$$
$$1 = 10^0$$
$$2 = 10^{0.3010}$$
$$5 = 10^{0.6990}$$

The common logarithm (log) of a number is defined as the power to which 10 is raised to obtain that number. The logarithms of the numbers given above are deduced directly from this definition:

$$\log 10 = 1$$
$$\log 100 = 2$$
$$\log 1 = 0$$
$$\log 2 = 0.3010$$
$$\log 5 = 0.6990$$

Any number greater than 1 but less than 10 can be expressed as 10 raised to a power between 0 and 1; therefore, the logarithm of any number between 1 and 10 is between 0 and 1.

When 10 raised to any power is multiplied by 10 raised to any other power, the product is obtained by merely adding the exponents:

$$(10^x)(10^y) = 10^{(x+y)}$$
$$(10^2)(10^3) = 10^5$$

This relationship suggests an important relation between logarithms given in the next subsection.

Basic Logarithmic Relationships

The logarithm of a product is the sum of the logarithms of the numbers being multiplied:

$$\log(xy) = \log x + \log y$$
$$\log(2.000 \times 10^3) = (\log 2.000) + (\log 10^3) = 0.3010 + 3 = 3.3010$$

The logarithm of a number raised to a power is equal to that power times the logarithm of the number:

$$\log x^n = n \log x$$

If the power, n, happens to be -1, the result is especially important:

$$\log \frac{1}{x} = \log x^{-1} = -1 \log x = -\log x$$

By the use of the two preceding rules, one can simplify the logarithm of a quotient:

$$\log \frac{x}{y} = \log x + \log \frac{1}{y} = \log x - \log y$$

Use of the Logarithm Table

Any number can be written in standard exponential form:

$$10.1 = 1.01 \times 10^1$$
$$65 = 6.5 \times 10^1$$
$$0.123 = 1.23 \times 10^{-1}$$
$$0.000456 = 4.56 \times 10^{-4}$$

From this fact and the rule concerning the logarithm of a product given in the last subsection, it is apparent that a list of the logarithms of the numbers greater than 1 and less than 10 is sufficient to provide the logarithm of any positive number. (The logarithm of a negative number is undefined.) A table of logarithms of these numbers is given on pages 709 and 710.

In the table, the numbers which are listed under N correspond to the integer and one decimal place of the number whose logarithm is to be determined. The second decimal place of the number whose logarithm is to be determined is located from among the numbers across the top of the table. The logarithm of the number is found in the body of the table at the intersection of the row and column thus located. *The body of the logarithm table contains positive decimal fractions only.* To find log 1.23, one looks for 12 in the left hand column under *N;* then one looks to the right of 12 in the column under 3 and finds the number 0899. Therefore

$$\log 1.23 = +0.0899$$

Since the number 1.23 lies between 1 and 10, its logarithm is a positive decimal fraction.

Example

Find the logarithm of each of the following numbers: **(a)** 4.56, **(b)** 1.70, **(c)** 9.75, **(d)** 1.07, **(e)** 3.16, **(f)** 1.00.
(a) 0.6590, **(b)** 0.2304, **(c)** 0.9890, **(d)** 0.0294, **(e)** 0.4997, **(f)** 0.0000.

To find the logarithm of a positive number which is not between 1 and 9.999, the number is first changed into standard exponential form. The logarithm of the decimal part of the number is found in the manner described in the preceding paragraph; the logarithm of the exponential part of the number is merely the exponent. The logarithm of the original number is the sum of these two logarithms. For example, to find the logarithm of 123, one writes this number in the form 1.23×10^2. Then the logarithm of 1.23 is found in the logarithm table, as already described. The logarithm of 10^2 is 2. These two logarithms are then merely added to obtain log 123.

$$\log 123 = \log 1.23 \times 10^2 = (\log 1.23) + (\log 10^2)$$
$$= \quad 0.0899 \quad + \quad 2 \quad = 2.0899$$

Similarly, any positive number less than 1 can be expressed in standard exponential form. Note that, as shown in the following example, a number which is less than 1 has a negative logarithm.

$$\log 0.00246 = \log 2.46 \times 10^{-3} = (\log 2.46) + (\log 10^{-3})$$
$$= \quad 0.3909 \quad + \quad (-3) \quad = -2.6091$$

To take the antilogarithm (antilog) of a positive decimal fraction, that is, to find the number whose logarithm is given, the fraction is found in the body of the table, and the number is determined from the left margin and column head. For example,

$$\text{antilog } 0.6107 = 4.08$$

The 4 and 0 are found at the left of the row in which the 6107 is found, and the 8 is found at the head of the column in which 6107 is found. Since the number corresponding to a logarithm which is a positive decimal fraction is always between 1 and 10, the decimal point is placed to make the number have a value greater than 1 but less than 10. If the exact logarithm given is not found in the table, its antilog can be estimated from the two nearest values given.

Example

Find the antilogarithm of each of the following: **(a)** 0.4502, **(b)** 0.8579, **(c)** 0.7042, **(d)** 0.6080, **(e)** 0.9695.
(a) 2.82, **(b)** 7.21, **(c)** 5.06, **(d)** 4.055, **(e)** 9.322.

If the logarithm whose antilog is sought is a positive number greater than 1, it is converted to the sum of a positive decimal fraction and a positive integer. The antilog of each of these is taken, and the product of the antilogs is the value sought. For example

$$\text{antilog } 1.301 = \text{antilog } (1 + 0.301) = (\text{antilog } 1)(\text{antilog } 0.301)$$
$$= 10^1 \times 2.00 = 20.0$$

Note that the antilogs were multiplied.

If the antilog of a negative number is sought, the number must be converted to the sum of a *positive* decimal fraction and a larger negative integer. To do this transformation, 1 is both subtracted and added to the number, as follows:

$$-2.301 = -2 + (-0.301) = -2 - 1 + (-0.301) + 1 = -3 + 0.699$$

The antilog of the positive decimal fraction can then be found in the logarithm table, and the antilog of an integer is 10 to that power.

$$\text{antilog } -2.301 = \text{antilog } (-3 + 0.699) = 10^{-3} \times 5.00 = 0.00500$$

Note that a positive decimal fraction was obtained.

Example

Determine the antilog of each of the following: **(a)** 2.6170, **(b)** 7.42, **(c)** −2.0057, **(d)** −0.4776.
(a) $4.14 \times 10^2 = 414$, **(b)** 2.6×10^7, **(c)** 9.87×10^{-3}, **(d)** $3.33 \times 10^{-1} = 0.333$.

Exercises

1. Determine the logarithms of 3.00 and 4.00. Add them to check that the sum is equal to log 12 = log (3 × 4).
2. Determine the logarithm of each of the following numbers: **(a)** 1.87, **(b)** 9.36, **(c)** 2.22×10^4, **(d)** 9.41×10^{-8}, **(e)** 0.111, **(f)** 4690, **(g)** 0.03055, **(h)** 12.21, **(i)** 7.87×10^{-8}.
3. Determine the antilog of each of the following: **(a)** 0.6848, **(b)** 0.4346, **(c)** 0.6420, **(d)** 2.6420, **(e)** 7.6873, **(f)** −0.2840, **(g)** −1.6840, **(h)** −3.4982.

Natural Logarithms and Exponents

The natural logarithm of a number (as opposed to the common logarithm) is the power to which the base e ($e = 2.7182$) is raised to obtain that number. Natural logarithms are denoted by scientists by the letters ln. Thus

$$e^{\ln 2} = 2$$

Therefore

$$\ln e = 1.000$$
$$\ln 1 = 0.000$$
$$\ln 2 = 0.6931$$
$$\ln 5 = 1.6094$$
$$\ln 10 = 2.3025$$

The relationships between natural and common logarithms can be summarized by the equations

$$\log x = \frac{\ln x}{\ln 10} = \frac{\ln x}{2.3025}$$
$$\ln x = (\ln 10)(\log x) = 2.3025 \log x$$
$$10^x = e^{x \ln 10} = e^{2.3025x}$$

Exercises

1. Determine the natural logarithms of each of the following three numbers: (a) 56, (b) 0.56, (c) 100. (d) What are the relationships among these three numbers and among their natural logarithms?
2. Determine the value of (a) $e^{2.37}$, (b) $e^{-2.11}$, (c) e^0.
3. Using natural logarithms only, from a table of natural logarithms or an electronic calculator, determine the value of (a) $10^{0.23}$, (b) $10^{-0.73}$, (c) $2.3025 \log 2.222$.

The same rules of logarithmic relationships apply to natural logarithms as to common logarithms, as given above in this section. Thus

$$\ln(xy) = \ln x + \ln y$$

To determine the antilogarithm, given a natural logarithm, simply raise e to the power of the natural logarithm. The number whose natural logarithm is 0.50 is

$$e^{0.50} = 1.65$$

Exercise

1. Determine the number whose natural logarithm is (a) 1.11, (b) -2.34, (c) 50.

N	0	1	2	3	4	5	6	7	8	9
10	0000	0043	0086	0128	0170	0212	0253	0294	0334	0374
11	0414	0453	0492	0531	0569	0607	0645	0682	0719	0755
12	0792	0828	0864	0899	0934	0969	1004	1038	1072	1106
13	1139	1173	1206	1239	1271	1303	1335	1367	1399	1430
14	1461	1492	1523	1553	1584	1614	1644	1673	1703	1732
15	1761	1790	1818	1847	1875	1903	1931	1959	1987	2014
16	2041	2068	2095	2122	2148	2175	2201	2227	2253	2279
17	2304	2330	2355	2380	2405	2430	2455	2480	2504	2529
18	2553	2577	2601	2625	2648	2672	2695	2718	2742	2765
19	2788	2810	2833	2856	2878	2900	2923	2945	2967	2989
20	3010	3032	3054	3075	3096	3118	3139	3160	3181	3201
21	3222	3243	3263	3284	3304	3324	3345	3365	3385	3404
22	3424	3444	3464	3483	3502	3522	3541	3560	3579	3598
23	3617	3636	3655	3674	3692	3711	3729	3747	3766	3784
24	3802	3820	3838	3856	3874	3892	3909	3927	3945	3962
25	3979	3997	4014	4031	4048	4065	4082	4099	4116	4133
26	4150	4166	4183	4200	4216	4232	4249	4265	4281	4298
27	4314	4330	4346	4362	4378	4393	4409	4425	4440	4456
28	4472	4487	4502	4518	4533	4548	4564	4579	4594	4609
29	4624	4639	4654	4669	4683	4698	4713	4728	4742	4757
30	4771	4786	4800	4814	4829	4843	4857	4871	4886	4900
31	4914	4928	4942	4955	4969	4983	4997	5011	5024	5038
32	5051	5065	5079	5092	5105	5119	5132	5145	5159	5172
33	5185	5198	5211	5224	5237	5250	5263	5276	5289	5302
34	5315	5328	5340	5353	5366	5378	5391	5403	5416	5428
35	5441	5453	5465	5478	5490	5502	5514	5527	5539	5551
36	5563	5575	5587	5599	5611	5623	5635	5647	5658	5670
37	5682	5694	5705	5717	5729	5740	5752	5763	5775	5786
38	5798	5809	5821	5832	5843	5855	5866	5877	5888	5899
39	5911	5922	5933	5944	5955	5966	5977	5988	5999	6010
40	6021	6031	6042	6053	6064	6075	6085	6096	6107	6117
41	6128	6138	6149	6160	6170	6180	6191	6201	6212	6222
42	6232	6243	6253	6263	6274	6284	6294	6304	6314	6325
43	6335	6345	6355	6365	6375	6385	6395	6405	6415	6425
44	6435	6444	6454	6464	6474	6484	6493	6503	6513	6522
45	6532	6542	6551	6561	6571	6580	6590	6599	6609	6618
46	6628	6637	6646	6656	6665	6675	6684	6693	6702	6712
47	6721	6730	6739	6749	6758	6767	6776	6785	6794	6803
48	6812	6821	6830	6839	6848	6857	6866	6875	6884	6893
49	6902	6911	6920	6928	6937	6946	6955	6964	6972	6981
50	6990	6998	7007	7016	7024	7033	7042	7050	7059	7067
51	7076	7084	7093	7101	7110	7118	7126	7135	7143	7152
52	7160	7168	7177	7185	7193	7202	7210	7218	7226	7235
53	7243	7251	7259	7267	7275	7284	7292	7300	7308	7316
54	7324	7332	7340	7348	7356	7364	7372	7380	7388	7396
N	0	1	2	3	4	5	6	7	8	9

(continued)

N	0	1	2	3	4	5	6	7	8	9
55	7404	7412	7419	7427	7435	7443	7451	7459	7466	7474
56	7482	7490	7497	7505	7513	7520	7528	7536	7543	7551
57	7559	7566	7574	7582	7589	7597	7604	7612	7619	7627
58	7634	7642	7649	7657	7664	7672	7679	7686	7694	7701
59	7709	7716	7723	7731	7738	7745	7752	7760	7767	7774
60	7782	7789	7796	7803	7810	7818	7825	7832	7839	7846
61	7853	7860	7868	7875	7882	7889	7896	7903	7910	7917
62	7924	7931	7938	7945	7952	7959	7966	7973	7980	7987
63	7993	8000	8007	8014	8021	8028	8035	8041	8048	8055
64	8062	8069	8075	8082	8089	8096	8102	8109	8116	8122
65	8129	8136	8142	8149	8156	8162	8169	8176	8182	8189
66	8195	8202	8209	8215	8222	8228	8235	8241	8248	8254
67	8261	8267	8274	8280	8287	8293	8299	8306	8312	8319
68	8325	8331	8338	8344	8351	8357	8363	8370	8376	8382
69	8388	8395	8401	8407	8414	8420	8426	8432	8439	8445
70	8451	8457	8463	8470	8476	8482	8488	8494	8500	8506
71	8513	8519	8525	8531	8537	8543	8549	8555	8561	8567
72	8573	8579	8585	8591	8597	8603	8609	8615	8621	8627
73	8633	8639	8645	8651	8657	8663	8669	8675	8681	8686
74	8692	8698	8704	8710	8716	8722	8727	8733	8739	8745
75	8751	8756	8762	8768	8774	8779	8785	8791	8797	8802
76	8808	8814	8820	8825	8831	8837	8842	8848	8854	8859
77	8865	8871	8876	8882	8887	8893	8899	8904	8910	8915
78	8921	8927	8932	8938	8943	8949	8954	8960	8965	8971
79	8976	8982	8987	8993	8998	9004	9009	9015	9020	9025
80	9031	9036	9042	9047	9053	9058	9063	9069	9074	9079
81	9085	9090	9096	9101	9106	9112	9117	9122	9128	9133
82	9138	9143	9149	9154	9159	9165	9170	9175	9180	9186
83	9191	9196	9201	9206	9212	9217	9222	9227	9232	9238
84	9243	9248	9253	9258	9263	9269	9274	9279	9284	9289
85	9294	9299	9304	9309	9315	9320	9325	9330	9335	9340
86	9345	9350	9355	9360	9365	9370	9375	9380	9385	9390
87	9395	9400	9405	9410	9415	9420	9425	9430	9435	9440
88	9445	9450	9455	9460	9465	9469	9474	9479	9484	9489
89	9494	9499	9504	9509	9513	9518	9523	9528	9533	9538
90	9542	9547	9552	9557	9562	9566	9571	9576	9581	9586
91	9590	9595	9600	9605	9609	9614	9619	9624	9628	9633
92	9638	9643	9647	9652	9657	9661	9666	9671	9675	9680
93	9685	9689	9694	9699	9703	9708	9713	9717	9722	9727
94	9731	9736	9741	9745	9750	9754	9759	9763	9768	9773
95	9777	9782	9786	9791	9795	9800	9805	9809	9814	9818
96	9823	9827	9832	9836	9841	9845	9850	9854	9859	9863
97	9868	9872	9877	9881	9886	9890	9894	9899	9903	9908
98	9912	9917	9921	9926	9930	9934	9939	9943	9948	9952
99	9956	9961	9965	9969	9974	9978	9983	9987	9991	9996
N	0	1	2	3	4	5	6	7	8	9

A-11   Factor-Label Method

A very useful technique for finding the correct approach to solving a problem in chemistry is called the factor-label method. This method takes advantage of the fact that chemical calculations involve objects and/or quantities, and therefore numbers represent quantities having specific dimensions or units. That is, practically every number has a "label" which tells the units or dimensions of the quantity it represents. In the factor-label method, units and dimensions are treated like algebraic variables and are combined or canceled in such a manner that the correct units or dimensions for the result are obtained. Consider the following example:

> Given that the specific heat of water is 1.00 cal/gram · deg, how many calories are required to raise the temperature of 27 grams of water by 30°C?

The desired result is a number of calories. The data given include a mass (grams) of substance (water), a property of water (its specific heat in cal/gram · deg), and a temperature interval (in degrees). When these "labels" are combined algebraically by multiplication, the result is in calories:

$$(\text{grams})\left(\frac{\text{cal}}{\text{gram} \cdot \text{deg}}\right)(\text{deg}) = \text{cal}$$

(Note that the unit cancels whether it is singular or plural.) Therefore, the correct procedure for solving this problem is to multiply the mass of water by its specific heat and by the indicated temperature change:

$$27 \text{ grams}\left(\frac{1.00 \text{ cal}}{\text{gram} \cdot \text{deg}}\right)(30 \text{ deg}) = 810 \text{ calories}$$

The factor-label method is also useful when it is desired to convert from one set of units to another.

When one converts a measured quantity into a different unit, the quantity can be multiplied by a factor which is a ratio of two equivalent quantities expressed in different units. An example of such a factor is

$$\left(\frac{1 \text{ kg}}{1000 \text{ grams}}\right)$$

which is a ratio relating grams and kilograms. The numerator and denominator of this fraction are equivalent. Another such ratio is

$$\left(\frac{1000 \text{ grams}}{1 \text{ kg}}\right)$$

By considering how the units should combine, one can decide whether a calculation should involve multiplication or division. If one wants to change 565 grams to kilograms, he multiplies the quantity given, 565 grams, by the ratio

$$\left(\frac{1 \text{ kg}}{1000 \text{ grams}}\right)$$

The factor with grams in the denominator is used so that the unit in the denominator will cancel the unit in the quantity given, leaving the unit of the numerator as the unit of the answer:

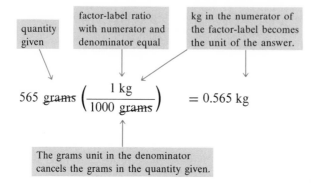

The factor-label method can also be used in percentage problems, or problems in which the quantities of constituents in a mixture are to be determined from the percentages of the constituents. A **percentage** is the number of parts of constituent in 100 parts total of the sample. Thus if a solution is described as 60% antifreeze by mass, it contains 60 grams of antifreeze in 100 grams total. The factor-label ratios which can be used are

$$\left(\frac{60 \text{ grams antifreeze}}{100 \text{ grams total}}\right) \quad \text{and} \quad \left(\frac{100 \text{ grams total}}{60 \text{ grams antifreeze}}\right)$$

Example

How many kilograms of a 30% by mass antifreeze solution can be prepared with 1.2 kg of pure antifreeze?

$$1.2 \text{ kg antifreeze} \left(\frac{100 \text{ kg solution}}{30 \text{ kg antifreeze}}\right) = 4.0 \text{ kg solution}$$

quantity given

Note that the unit in the numerator is the same as the unit in the denominator and that it is a unit of mass.

When the factor-label method is used, actual numerical calculations need not be made until the final step. This is especially useful when one has available a calculating device such as a slide rule, an abacus, an electronic calculator, or a computer. Of course, regardless of the method of computation, the final result must be expressed with the proper number of significant figures.

Example

The density of gold is 19.3 grams/cm^3. Calculate the diameter of a solid gold sphere having a mass of 422 grams.

The result sought is a diameter, D, in centimeters. The data given are a mass of gold (in grams), its density, d (in grams/cm^3), and the shape of the object (spherical). The radius, r, of a sphere is related to its volume, V, by $r = \sqrt[3]{3V/4\pi}$, and the volume of this sphere is 422 grams/(19.3 grams/cm^3). Thus

$$D = 2r = 2\sqrt[3]{\frac{3(422 \text{ grams})}{4\pi(19.3 \text{ grams/cm}^3)}} = 3.47 \text{ cm}$$

Answers to Selected Exercises

Chapter 1

2. (a) 92 protons, 143 neutrons, 92 electrons; **(c)** 1 proton, 1 neutron, 1 electron.
5. 21,450 kg/meter^3. **7.** Its density, 5.00 grams/cm^3, is different.
8. (a) $(NH_4)_2Cr_2O_7$, **(d)** Cu_2O. **9. (a)** Copper(II) chlorite, **(c)** mercury(I) chloride.
10. (a) 1, **(c)** 3, **(e)** 1. **11. (a)** K_3P, ionic, potassium phosphide;
 (e) NF_3, covalent, nitrogen trifluoride. **12.** 182 cm^3.

13. (a) H:C̈:C̈l: **(e)** Na$^+$ [:C̈l:$^-$]
       :C̈l:

14. (c) Ca^{2+} 2[:F̈:$^-$], **(d)** :F̈:F̈:

16. (a) :Ö::S:Ö: **(b)** :Ö:S:Ö: $^{2-}$
18. Seven.

19. ^{13}C   6   6   7   13
   ^{35}Cl   17   17   18   35
   ^{3}H   1   1   2   3

20.

	At. No.	Mass No.	Protons	Neutrons	Electrons
(a)	11	23	11	12	10
(c)	92	238	92	146	88

21. (a) Element only, **(e)** both. **23.** Pure H_2SO_4 is covalent, Na_2SO_4 is ionic.
25. 8.12 grams/cm^3. **26. (a)** Copper(II) sulfate, **(d)** strontium chlorate,
 (h) phosphorus trichloride, **(k)** water. **27. (b)** Both, **(c)** both.
28.

	(a)	**(b)**	**(c)**	**(d)**
metals	+	unusual	1–5, usually 1–3	usual
nonmetals	−	normal	3–8	rare

29. No. **30.** Noble gases, group 0.
31. A, mixture; C, possibly any of the three; D, element. **33. (a)** 1.23,
 (c) -2.8×10^{-3}. **37. (a)** NH_4ReO_4, **(c)** Cu_3As.
38. (21 different structural isomers are possible.)
        H   H   H   H
42. (a) H:C̈:Ö:C̈:H **(c)** H:C̈:C̈:C̈:H
        H   H   H·Ö·H
43. (c) $(CH_3)_3CH$, **(d)** CH_3COCH_3.

Chapter 2

1. (b) 5.927% H, 94.073% O; **(d)** 15.77% Al, 28.11% S, 56.12% O. **2.** $MgSO_3$.
4. All have the same number of atoms; 1 gram of O has the greatest number of
 molecules. **5. (b)** 10.0 moles, **(c)** 12 moles, **(e)** 0.073 mole.

7. **(a)** $NCl_3 + 3 H_2O \rightarrow NH_3 + 3 HOCl$
   (c) $SbCl_3 + H_2O \rightarrow Sb(O)Cl + 2 HCl$
8. **(b)** $BiCl_3 + H_2O \rightarrow Bi(O)Cl + 2 HCl$
9. **(a)** 3.05 grams. **12.** b. **13.** The pressures are the same.
14. **(b)** 1.91 grams. **17.** 3.29 grams. **18.** 0.0231 mole.
19. **(a)** 92.26% C, 7.74% H. **20.** 7.27 liters. **21.** 57.4 D.
22. **(b)** $H^+ + OH^- \rightarrow H_2O$ **(d)** $Mg^{2+} + 2 OH^- \rightarrow Mg(OH)_2$
24. $2.00 \, M$. **25.** 920 ml. **26. (a)** $0.800 \, M \, Na^+$, $0.800 \, M \, Cl^-$;
   (e) $1.8 \, M \, Na^+$, $1.6 \, M \, Ba^{2+}$, $5.0 \, M \, Cl^-$. **28. (a)** 3.95 atm, **(d)** 1.00×10^{-3} atm.
29. $P_{total} = 0.420$ atm. **31.** C_2H_6. **35.** 335 ml. **37.** 2.8 grams.
38. **(a)** 1.59 liters, **(c)** 1.90 liters, **(e)** 1580 torr. **39.** 86.9 ml.
42. ^{12}C is the standard for all. **45.** 8.75 kg (the same as the mass of $H_2C{=}CH_2$).
46. 0.318 liter. **50.** 405 ml. **52.** C_2H_5, C_4H_{10}. **54. (b)** 9.0% Na_2CO_3.
55. $2 Cu^{2+} + 4 I^- \rightarrow 2 CuI + I_2$
56. **(a)** $5 Fe(NO_3)_2 + KMnO_4 + 8 HNO_3 \rightarrow$
   $$5 Fe(NO_3)_3 + Mn(NO_3)_2 + 4 H_2O + KNO_3$$
   or $10 FeSO_4 + 2 KMnO_4 + 8 H_2SO_4 \rightarrow$
   $$5 Fe_2(SO_4)_3 + 2 MnSO_4 + 8 H_2O + K_2SO_4$$
58. $\Delta P = -0.64$ atm. **61. (c)** 3.75 moles of CO and 1.25 moles of CO_2.
63. 9.6 grams of CuS, $0.20 \, M \, H^+$. **64.** 0.83 atm. **65. (a)** 192.2 D,
   (d) 50.42% ^{79}Br. **67. (b)** 136 liters. **73. (a)** 64.9 grams,
   (e) $1.83 \, M \, NaCl$ and $1.42 \, M \, KCl$. **74. (b)** 4.6 mg.

Chapter 3

1. **(a)** Electrical to heat, **(c)** chemical to mechanical, **(i)** potential to kinetic.
2. 300 cal = 1260 J. **5.** 258 cal = 1080 J. **6.** -4.3 kcal = -18 kJ. **7.** 8 grams.
8. 0.03 cal/gram $\cdot$ deg $= 0.13$ J/gram $\cdot$ deg. **9.** 28°C.
11. -146.7 kcal = -613.8 kJ. **12.** $\Delta G° = 0$. **13.** All three.
14. $+41.23$ kcal = 172.5 kJ. **15.** b (oxygen atoms). **19.** Atomic weight $\cong$ 30 D.
20. b. **21.** -540 cal = -2260 J. **22.** $\Delta V = -1.83$ liters, either way.
23. **(b)** $+20.0$ kcal = 83.7 kJ. **25.** 8.80 kcal = 36.8 kJ. **26.** -121 kcal = -506 kJ.
27. **(a)** -0.5 kcal = -2 kJ. **31.** 0.04 kcal = 0.17 kJ. **33. (a)** (i). **34.** 0°C.
37. 99.8 grams. **39.** $-15{,}520$ kcal = -6.494×10^4 kJ. **42.** 27.5 grams O_2.
43. -243 kcal = -1017 kJ. **46. (d)** -46.84 kcal = -196.0 kJ.
49. **(c)** -13.36 kcal = -55.90 kJ. **50.** -117.27 kcal = -490.66 kJ.
51. $+42.92$ kcal = $+179.6$ kJ.

Chapter 4

1. **(a)** -3, **(b)** $-\frac{1}{3}$, **(d)** $+4$, **(k)** -3. **2.** a, c, d, e.
3. **(a)** $3 Sn^{2+} + 8 H^+ + 2 NO_3^- \rightarrow 3 Sn^{4+} + 2 NO + 4 H_2O$
   (c) $Cr_2O_7^{2-} + 6 I^- + 14 H^+ \rightarrow 3 I_2 + 2 Cr^{3+} + 7 H_2O$
5. a and e. In (e), H_2O_2 is both the oxidizing and reducing agent. **6. (a)** $+5$,
   (b) $+\frac{4}{3}$, **(d)** $+6$. **7.** a is an acid and c is a base.
8. **(a)** Dehydrating agent, **(c)** both acid and (mild) oxidizing agent.
9. **(b)** $Cl_2 + 2 NaBr \rightarrow 2 NaCl + Br_2$ **10. (d)** 0, **(e)** -1,
12. S^{2-}, SO_3^{2-}, SO_4^{2-}, $S_2O_3^{2-}$; -2, $+4$, $+6$, $+2$.
14. $ZnCl_2$, NaCl, and $C_{12}H_{22}O_{11}$, respectively, will be recovered upon evaporation.
16. **(a)** 7.18 grams. **17.** 2.97 grams. **18. (a)** $NaCl + Zn$
19. 0.316 mole $FeCl_3$. **20.** b, e, g. **21. (a)** $+1$, **(c)** $+5$.
22. **(a)** Maximum $+5$, minimum 0.

23. (c) $Cr_2O_3 + 2 NaOH + 3 H_2O \rightarrow 2 NaCr(OH)_4$

24. (a) $2 MnO_4^- + 16 H^+ + 5 Sn^{2+} \rightarrow 2 Mn^{2+} + 5 Sn^{4+} + 8 H_2O$

25. (a) $-\frac{1}{3}$, **(c)** $H\!:\!\ddot{N}\!:\!N\!:\!::N\!:$

26. (a) MnO_4^-, Ce^{4+}, $Cr_2O_7^{2-}$, CrO_4^{2-}, HNO_3, F_2.

27. (a) $S_2O_8^{2-} + 9 I^- + 8 H^+ \rightarrow 3 I_3^- + 2 SO_2 + 4 H_2O$
   (d) $3 H_2SO_4 + 2 Br^- \rightarrow Br_2 + SO_2 + 2 H_2O + 2 HSO_4^-$

28. (b) $NaOH + H_3PO_4 \text{ (excess)} \rightarrow NaH_2PO_4 + H_2O$

29. (b) $Cu + Cu^{2+} + 2 OH^- \rightarrow Cu_2O + H_2O$ **30.** 4.67 liters.

31. By any set of assumptions which do not violate rule 1, page 100:

$Hg_2(CN)_2 + 22 Ce^{4+} + 28 OH^- \rightarrow$
$$2 CO_3^{2-} + 2 NO_3^- + 2 Hg(OH)_2 + 22 Ce^{3+} + 12 H_2O$$

33. Ce^{III}.   **35.** 0.110 mole.

37. (c) $5 MnO_4^{2-} + 8 H^+ \rightarrow Mn^{2+} + 4 MnO_4^- + 4 H_2O$ (seven).

38. (d) $Zn + 2 OH^- + 2 H_2O \rightarrow Zn(OH)_4^{2-} + H_2$ **39. (b)** 6.

Chapter 5

1. (a) Watt, **(d)** coulomb, **(g)** volt.   **2.** 0.589 gram Cu.   **3. (a)** $\epsilon° = 0.94$ V.

4. -89 kJ.   **5.** 1.60×10^{-19} C.   **6.** 3.95 grams.   **8. (a)** $2:1$.   **12.** 0.653 V.

13. a.   **19.** 4.47×10^4 kW hr.   **20.** Sn^{II}.   **21.** 0.0248 M.

23. (a) $Hg_2Cl_2 + 2 e^- \rightarrow 2 Hg + 2 Cl^-$
$2 Cl^- \rightarrow Cl_2 + 2 e^-$

28. Cu^+ disproportionates in aqueous solution.   **31. (b)** 1.21 V,   **(d)** 0.344 V.

35. (b) $2 Y^+ \rightarrow Y^{2+} + Y$

38. $\Delta G = \Delta G° + 2.30 RT \log Q$, where Q is the concentration ratio.   **39.** -0.0244 V.

42. $2 Fe^{3+} + Fe \rightarrow 3 Fe^{2+}$

Chapter 6

1. (a) Right,   **(c)** left,   **(e)** no shift.

2. (b) and **(c)** $K = \dfrac{[SO_3]^2}{[SO_2]^2[O_2]}$ **(d)** $K = [CO_2]^2[H_2O]$

4. (a) Shift left.   **5. (a)** 4.79 kJ = 1.15 kcal   **6.** $[XO_2] = 1.4 \times 10^{-2}$ mole/liter

7. 2.4 liters/mole.   **8.** 0.50.   **9.** $[D] = 9.5 \times 10^{-4}$ mole/liter   **10. (a)** Right,
   (e) undeterminable.   **12.** 73.9 kcal = 309 kJ   **13.** c only.   **15. (a)** None.

17. 8.4×10^{-2} liter/mole.   **18.** $P_{total} = 789$ torr   **20.** 0.757 liter/mole.

22. (b) 4.0×10^{-3} mole/liter.   **23. (b)** $K = 0.10$ **29.** 0.42 atm.

Chapter 7

2. $\left[\begin{array}{c} H \\ H\!:\!\ddot{N}\!:\!H \\ H \end{array}\right]^+$ $\left[:\!\ddot{O}\!:\!H\right]^-$

3. b, d, e, f.   **4. (b)** $K = \dfrac{[H_2O][C_2H_3O_2CH_3]}{[HC_2H_3O_2][CH_3OH]}$ **5. (a)** 3.00,   **(d)** 0.82,
   (g) -0.30.   **6. (a)** 1.0×10^{-5},   **(c)** 7.6×10^{-13}.   **7. (b)** 11.00,   **(c)** 11.30.

8. (b) iv.   **9.** $1.0 \times 10^{+6}$.   **11.** 7.30.   **12. (b)** 0.10 mole NaOH,
   (c) 0.10 mole HCl.   **13. (f)** No.   **15. (b)** $[H_3O^+] = 3.6 \times 10^{-5}$ M.

16. (b) 1.0×10^{-6}.   **17.** 11.13.   **19. (a)** 0.32%.   **20.** 11.78.

22. (b) vii, iii, i, iv, v, vi, ii.   **23. (e)** Both hydrolyze to the same extent.

24. a, b, g, h. **25. (b)** K_b and K_w, **(i)** K_w/K_a. **27. (c)** 5.18.

29. (c) $C_2H_3O_2^- + H_3O^+ \rightleftharpoons HC_2H_3O_2 + H_2O$ (greater than 98%).

(d) $OH^- + HC_2H_3O_2 \rightleftharpoons C_2H_3O_2^- + H_2O$ (greater than 98%).

30. 4.00. **31.** 5.13. **32.** $HC_2H_3O_2$ and NH_3. **34.** 1.1×10^{-21}. **36.** 4.35.

Chapter 8

2. (a) Tetrabromopalladate(II) ion, **(e)** hexaamminechromium(III) ion. **3. (b)** 0,

(e) III. **4. (a)** $[Rh(NH_3)_4Cl_2]^+$, **(e)** $[Co(NH_3)_6][Cr(NH_3)_2Cl_4]_3$. **5.** 1×10^{-9} M.

9. 4×10^{-6} M. **11.** 6×10^{-15} M.

12. (a) $Al(OH)_4^- + H_3O^+ \rightleftharpoons Al(OH)_3(s) + 2\,H_2O$ **14.** 7×10^{-6} M.

17. -1.03 V. **19.** 5.8 mg. **21.** 9.04. **24.** 1 mg.

27. (a) NaOH or H_2S plus HCl. **29.** 2×10^{-15} M. **32. (b)** 8×10^{-16} M.

33. 3×10^{-19} M.

36. (e) $NH_3 + H_2O \rightleftharpoons NH_4^+ + OH^-$

$$Mg(OH)_2 \rightleftharpoons Mg^{2+} + 2\,OH^-$$

$$2\,H_2O \rightleftharpoons H_3O^+ + OH^-$$

39. 1.0×10^{-5} mole/liter. **40.** $b > c > d > a$. **41. (c)** $BaSO_4$ is insoluble.

Chapter 9

1. 4.8×10^{-10} esu, 9.6×10^{-10} esu. **2.** 2.2×10^8 dynes. **3. (a)** 90 J, **(c)** 810 J.

4. (a) None. **5.** The atoms have different masses. **8. (a)** 19.9×10^{-17} erg.

9. (a) 7000 Å to 4000 Å, **(f)** 40.8 kcal/mole to 71.5 kcal/mole. **11.** 28:1.

12. (a) Wave, **(c)** particle, **(e)** both. **14.** 8.7×10^{15} Hz **17.** 7.31×10^{14} sec^{-1}.

18. -5.44×10^{-12} erg. **19.** $5 \rightarrow 2$. **20. (b)** 23.0 kcal/mole. **21.** 109,000 cm^{-1}.

24. 4.1×10^{-14} excess electron per atom. **26.** 1.15×10^{-15}.

30. The masses of beta particles are much less than those of alpha particles.

Chapter 10

2. (a) 0, 1, 2, and 3; **(b)** seven; **(c)** 2, 6, 10, 14; **(d)** two; **(e)** four.

4. (a) 32, **(b)** 14, **(c)** 36. **5. (a)** K, **(b)** d. **6. (b)** 18, **(c)** 8.

7. (a) Cl, no ion; **(b)** no atom, Pd^{2+}. **9.** 4.9 B.M.

10. Nuclear properties (*eg* radioactivity) are *not* dependent on electronic configurations; all other properties are. **13. (a)** B, Al, Ga, In, Tl;

(b) H, Li, Na, K, Rb, Cs, Fr, Cu, Ag, Au, Cr, Nb, Mo, Tc, Ru, Rh, Pt.

14. (a) $1s^2\,2s^2\,2p^2$, two; **(b)** $1s^2\,2s^2\,2p^6\,3s^2\,3p^6\,3d^9$; one;

(e) $1s^2\,2s^2\,2p^6\,3s^2\,3p^6\,4s^2\,3d^{10}\,4p^6\,5s^2\,4d^{10}\,5p^6\,4f^7$. **16. (a)** Co, **(b)** S^{2-},

(c) Rb, **(e)** Na, **(f)** La.

17. (a) $1s^2\,2s^2\,2p^6\,3s^2\,3p^6\,4s^2\,3d^{10}\,4p^6\,5s^2\,4d^{10}\,5p^6\,6s^2\,5d^{10}\,4f^{14}$,

(d) ns^2 (where $n = 5$ or 6) has some stability in the ions listed, and is sometimes referred to as the "inert pair."

18. (a) $1s^2\,2s^2\,2p^6\,3s^2\,3p^6\,4s^0\,3d^6$, four, paramagnetic;

(b) $1s^2\,2s^2\,2p^6\,3s^2\,3p^6\,4s^2\,3d^{10}\,4p^6$, zero, diamagnetic. **19.** For aluminum: 53.03 eV; for sodium: 52.20 eV. Since it is more difficult to remove three electrons from aluminum than two from sodium, but there is a stable aluminum(III) oxidation state and no stable sodium(II) oxidation state, it is apparent that some other factors are involved. **20.** The specified jumps occur at the loss of an electron from a complete octet.

21. (b) Fe: $1s^2\,2s^2\,2p^6\,3s^2\,3p^6\,4s^2\,3d^6$; Ni^{2+}: $1s^2\,2s^2\,2p^6\,3s^2\,3p^6\,3d^8$.

22. (c) He.   **23. (a)** Energy is *required* to add an electron to a gaseous atom;
(b) $A + e^- \rightarrow A^-$ (involving a negative ion)
$B^+ + e^- \rightarrow B$ (involving a positive ion).

27. (b) 5.138 eV. A positive sign for EA indicates energy release.

28. (a) $Cl^- < Ar < K^+$; the same electronic configuration, but lower nuclear charge for Cl^-.
(b) $Fe < Fe^{2+} < Fe^{3+}$; the same nuclear charge, but fewer electrons for Fe^{3+}.

30. (a) 1.0×10^{-5}; **(b)** zero.

31. (a) The Cd^{2+} d electrons shield the greater nuclear charge less effectively than the average Sr^{2+} electron shields its nuclear charge. **(b)** As a consequence of the lanthanide contraction, the sizes of the last two are very similar.

33. The inert pair (see exercise 17) is more stable in period VI than in period V, and in lower numbered periods the s^2 pair is not inert at all.

Chapter 11

1. (a) 0.98 Å, **(f)** H—C, 1.05 Å; C≡N, 1.16 Å.

2. (b) C≡C < C=C < C—C.

4. (b) The effects of the various bonds cancel each other.

6. $PH_3 < AsH_3 < SbH_3 < NH_3$.

8. (a) H:O:H⁺, pyramidal; **(e)** H:N:N:H two pyramidal nitrogen atoms in a
H   H
nonplanar molecule. **9.** b. **10. (a)** Trigonal, **(e)** pyramidal.

12. (a) CH_3CH_2OH.   **13.** ca. −90°C.   **15.** The sp hybrid is more directional.

16. (b) HCO_2^-.   **17.** sp^3d.   **18. (b)** sp^3.   **19. (b)** sp^3d or dsp^3.

20. (b) Trigonal, sp^2.   **21.** Yes.

24. B_2

25. (a) Bond order = 1, paramagnetic; **(e)** bond order = 3, diamagnetic;
(g) bond order = 1.5, paramagnetic.   **26.** b.   **27.** e.

29. (e) (i) One, (ii) two, (iii) two; **(g)** (i) two, (ii) two, (iii) four.   **31. (a)** NH_3,
(d) NH_3.   **33.** Longer.

34. (a) H:C:H   H:N:H   **(e)** He: H:⁻
H   H

37. 25%.   **38.** HF is extensively hydrogen bonded.

41. (b) Octahedral, sp^3d^2.   **43. (a)** ↓↑ ↓↑ ↓↑ ↓↑ pyramidal.
sp^3 sp^3 sp^3 sp^3

Chapter 12

1. Hexane, 2-methylpentane, 3-methylpentane, 2,2-dimethylbutane,
2,3-dimethylbutane. None is optically active.   **2. (b)** Octane.   **3. a.**

4. (a) 1-Phenyl-1-propanone, **(c)** propanone, **(f)** methanal.   **5. (b)** Toluene,
(d) dimethyl ether, **(f)** 2-propanol, **(k)** 2-chlorobutane.   **6. (a)** Amines.
7. (b) $CH_3CH_2CH_2CHCH_2CH_2CH_3$,   **8. (b)** and   **(c)** $CH_3CHOHCH_2NHCH_3$.

$$CH_2$$
$$CH_2$$
$$CH_3$$

9.

	(a)		**(b)**	**(c)**
alcohol	CH_3OH	methanol		-2
aldehyde	CH_2O	methanal		0
acid	HCO_2H	methanoic acid		$+2$

10. (b) vii or ix with viii.   **11. (a)** 1,3,5-Hexatriene,
(e) diethylammonium chloride.   **12. (c)** $CH_3COCH_2CH_3$,
(e) $HCO_2CH_2CH_2CH_2CH_3$.   **13. (c)** CH_3CH_2CHO,
(e) $CH_3COCH_2COCH_3$.

14. (b)

20. CH_4.   **24. (c)** C_5H_5N, $(C_2H_5)_2NH$, $C_2H_5OCH_3$.   **27.**

32. Three, no.

Chapter 13

1. Peak at 15 indicates 1,1-dichloroethane, as do nmr doublet and quartet.
3. (a) 7. **(b)** 12.   **4.** The total bond order for nitrogen is an odd number.
5. $(CH_3CH_2)_2O$.   **6.** a, b, d, e.   **7.** One unsplit peak.   **8.** $(CH_3)_2N-C-CH_3$.
                                                                                        O
9. (b) iii, iv, v;   **(e)** i;   **(g)** v.   **11.** Two triplets of equal intensity.
12. (d) 8.28×10^{-14} erg $= 8.28 \times 10^{-21}$ J.   **13.** 36.3%.
14. (b) Microwave, infrared, and ultraviolet, respectively.
15. (b) No net translation is permitted.   **16.** The acetone will have one peak, unsplit; the
aldehyde will have a complex spectrum.
17. H $<$ C $<$ Zn $<$ Sb $<$ U.
24. 15, 35, 37.   **25.** $CH_3CH_2CONHCH_3$.   **26. (a)** The mass spectrum of the first
compound should show a major peak at 16 but not at 29, as opposed to the mass
spectrum of the second. The nmr spectrum of the second compound would show 4
peaks, including the characteristic triplet-quartet combination of a CH_3CH_2- group.
The nmr spectrum of the first compound would show overlapping peaks.
28. SCH_2.   **31.** $CH_2=CHCH_2OH$.

Chapter 14

1. 8, 6, 6.   **2.** b $<$ a $<$ d $<$ c.   **3. (a)** 2, **(d)** 6, **(g)** 6.   **4. (a)** II, **(d)** III,
(g) 0.   **5. (a)** 3$-$, **(d)** 0.   **6. (a)** Dichlorotetraammineplatinum(IV) ion,
(d) carbonatopentaamminecobalt(III) tetrachlorocuprate(II).
7. (b) $[Co(NH_3)_4BrCl]_2SO_4$, **(d)** $[Cren_2Cl_2]_2[PdCl_4]$.
9. (b) Geometric, coordination, linkage. **11.** I is *cis*.

12. (a) i, ii (identical) isomeric with iii; iv, v, vii (identical) isomeric with vi and viii.
13. (b) $1s^2 \, 2s^2 \, 2p^6 \, 3s^2 \, 3p^6 \, 3d^6$, 4 unpaired.   **14.** 3.   **15.** 5.

16. (b)

$$\overbrace{\qquad\qquad\qquad\qquad}^{d^2 sp^3}$$

↑↓ ↑↓ ↑↓   oo oo   oo   oo oo oo

17. (b) $[CoF_6]^{4-}$

$$\overbrace{\qquad\qquad\qquad\qquad}^{sp^3 d^2}$$

↑↓ ↑↓ ↑ ↑ ↑   oo   oo oo oo   oo oo — — —

18. For $[Ag(CN)_2]^-$ **(a)** sp, **(b)** linear, **(c)** 0.   **21. (b)** Mn^{III}.   **23. (a)**

—
↑↓
↑↓
↑↓ ↑↓

25. $-30{,}000 \text{ cm}^{-1}$.   **26.** Highest: $d_{x^2-y^2}$; lowest d_{xz} and d_{yz}.   **35.** ii, v.
38. Geometric, coordination, hydrate.
40. Conducting: 482 D; nonconducting: 241 D.   **45. (b)** ↑   **55. (a)** 8.

↑↓
↑↓
↑↓ ↑↓

56. 28.5 kcal/mole.

Chapter 15

1. (b) The fifth.   **2.** $k = R/N$.   **3.** NH_3.   **5.** 22.0 K.   **6. (b)** 4.8 atm.
8. $u = 4.76 \times 10^4 \text{ cm/sec}$.   **12.** Yes.   **14.** The temperature changes.
15. (a) Translational.   **16.** C_4H_{10}.   **17.** 35°C.
20. They are equal for monatomic gases.   **22. (a)** 21 meters/sec, **(c)** 22 meters/sec,
(d) 25 meters/sec.   **25. (b)** 1.81 liters.   **26.** 4.91 moles.
28. 34,900 K.   **32.** Force, momentum, velocity.

Chapter 16

1. 1.9 cm.   **2. (b)** Freezing point. (Water is a notable exception.)
4. The open container.   **5. (b)** KNO_3.   **6.** 1.42 M, 1.49 m.   **8.** 400 D.
9. Spherical.   **10.** ΔH_{vap}, $\Delta G° = 0$.   **11.** $C_6H_{12}O_6$.   **12. (a)** 5.88 grams.
13. 65 torr.   **14. (b)** 15.8°C.   **15.** $-0.62°C$.   **16.** $C_2H_4O_2$.   **17.** 2.17 atm.
18. 0.240.   **20.** 1.
22. Camphor is good for high molecular weight solutes which will dissolve in it.
24. Positive enthalpy of solution, or evaporating solvent.   **26.** The freezing point,
which is the same as the melting point, depends on the molality of the solutes
(impurities).   **29.** 29.1 dynes/cm.   **30.** 40.5 sec.
32. 350 K (to 2 significant figures).

Chapter 17

2. $7.70 \times 10^{-5} \text{ sec}^{-1}$.
4. Three possible examples: rate $= k[A][B]$, rate $= k[A]^2[B]^0$, rate $= k[A]^0[B]^2$.
5. First order, $k = 1.01 \times 10^{-2} \text{ min}^{-1}$.
8. (a) First order with respect to A; second order with respect to B, **(c)** no.

9. 28 feet.   **11. (a)** CO_2, **(b)** He **(c)** He.   **12.** Second.
14. First order, $k = 0.0500$ min^{-1}.   **16.** Rate $= k[A][B]$.
21. 2.37×10^{-5} torr$^{-1} \cdot$ min^{-1}.   **23.** 3.29×10^{-3} M each.
24. (a) 5.0×10^{-5} mole/liter $\cdot$ sec.   **25.** Fastest, **b**; highest rate, **c**.
26. 1.2×10^{-3} mole/liter $\cdot$ sec.
28. First order with respect to A; zero order with respect to B.
31. 1.15×10^4 cal/mole.   **33.** $K = 1.0$, rate $= k_2[C][A_2]^{1/2}$.
34. One possible mechanism:

$$COCl_2 \rightleftharpoons COCl + Cl \qquad \text{(fast) initiation}$$
$$Cl + COCl_2 \rightarrow COCl + Cl_2 \qquad \text{(slow)}\Big\}\text{chain}$$
$$COCl \rightarrow CO + Cl \qquad \text{(fast)}$$
$$2\,Cl \rightleftharpoons Cl_2 \qquad \text{(fast) termination}$$

Chapter 18

1. (a) One, **(b)** 8.   **2. (b)** 1.37×10^7 units.   **3.** Tetragonal.
5. (d) Examples include glass, wax, butter, plastics.   **6. (a)** 4.   **7.** 8, 4, 100%.
9. Simple cubic (CsCl) structure.   **10.** -79 kcal/mole $= -330$ kJ/mole.
11. Conductivity, infrared spectrum in which sulfate group bands will distinguish, X-ray diffraction, chemical tests, and others.
12. (b) The value of i is somewhat less than 2, because of interionic attractions.
13. (a) 2.89×10^{-3}.   **14.** b $<$ e $<$ c $<$ d $<$ a.   **15.** 4.2%.   **16.** 1.8.
18. (a) $5.0\,°$C.   **20. (a)** All parts: $-0.186\,°$C,
   (d) The value of i increases (almost) proportionally to the formula weight.
21. Iodide to iodide ion contact in LiI.   **22.** ΔH_{fusion}.   **27.** 1.33 Å.   **29.** One.
30. 0.732.   **35.** 128.8 cm^2/ohm $\cdot$ equiv.

Chapter 19

1. No.   **2. (a)** Yes, **(b)** no.   **4.** 7.8%.   **5.** 12.2 karats.
6. Electrical conductivity.   **7.** One.   **8. (a)** Substitutional alloy,
   (b) compound alloy.   **9. (a)** Amphoteric or acidic.   **12.** b.   **14. (a)** Hg,
   (b) Ni, **(c)** Cu, **(d)** W, **(e)** Ge.
15. 1675 K, where it is in equilibrium with O_2 at 1 atm.

Chapter 20

1. Only ^{207}Pb is a member of the $4n + 3$ series.
2. The nucleus is approximately 1.4×10^{13} times as dense as silver metal.
3. (a) β^- emission increases Z, **(b)** α, β^+, and EC decrease Z,
   (c) EC causes X-ray emission.   **4.** 12.6 hours.   **9.** 5.6×10^{-13} cm.
11. $^{50}_{24}Cr(\alpha,n)^{53}_{26}Fe$; $^{50}_{24}Cr(\alpha,d)^{52}_{25}Mn$.
12. 1.03×10^{15} MeV; 2.37×10^{22} kcal/mole or 9.92×10^{22} kJ/mole.
13. 3.78 hours.   **14.** Stable: ^{208}Pb, ^{120}Sn; β^-: ^{49}Ca, ^{30}Al, ^{94}Kr; β^+: ^{195}Hg, ^{8}B, ^{150}Ho.
15. Increasing activity is expected as the daughter builds up again.
16. (a) $^{40}_{21}Sc$ and $^{42}_{21}Sc$, **(b)** $^{40}_{18}Ar$ and $^{40}_{21}Sc$, **(c)** $^{40}_{18}Ar$ and $^{41}_{19}K$.
17. $^{114}_{49}In$, which has odd numbers of both protons and neutrons.
18. (a) 8.1×10^{-9} gram, **(b)** 3.0 kg.   **19.** 2.4×10^3 dis/sec.
20. 2.9×10^{-25} cm^2.   **21.** 3.5 MeV.   **22.** 1.1×10^5 dis/hr.   **23.** ^{209}Bi.
24. ^{131}I: 53 days.   **25. (a)** 7.88×10^{-11} cm, **(b)** 6.8×10^{-13} cm.   **27.** 27,700.
28. 19.7 barns.

30.
$$S^* + SO_3^{2-} \rightarrow S^*SO_3^{2-}$$
$$S^*SO_3^{2-} + Ba^{2+} \rightarrow BaS^*SO_3$$
$$BaS^*SO_3 + 2\,H_3O^+ \rightarrow S^* + SO_2 + 3\,H_2O + Ba^{2+}$$

$$\left[\begin{array}{c} O \\ | \\ O-S-S^* \\ | \\ O \end{array} \right]^{2-}$$

The radioactive sulfur atom and the central sulfur atom, to which it is attached, do not exchange.

32. 2000 dis/min; a case of secular equilibrium.

Chapter 21

3. Neutral: **c**, **e**, and **g**; amphoteric: **a** and **k**.
4. (a) Strongest: HBr; weakest: PH_3.   **6.** $\Delta S^\circ = -18$ cal/K.
11. H_2SO_4 is used to prepare HCl.   **14. (a)** $NaAsO_2$.
16. $P_4 + 2\,H_2O + 4\,OH^- \rightarrow 2\,PH_3 + 2\,HPO_3^{2-}$ **20.** $B_{12}H_{12}^{2-}$.

Chapter 22

3. About 840°C.   **4. (a)** Tl_2O,   **(c)** Be,   **(e)** CrO_4^{2-}.   **13.** LiI.
14. Br_3^-, etc., are interhalogen ions.
17. $3\,Mg + N_2 \rightarrow Mg_3N_2$
$Mg_3N_2 + 6\,H_2O \rightarrow 2\,NH_3 + 3\,Mg(OH)_2$
19. Pyrites consists of Fe^{II} and S_2^{2-} ions.   **24.** 0.76 V.
27. Sm, Eu, and Yb, each of which has a +2 oxidation state.

Chapter 23

1. B could be
$$\begin{array}{c} CH_2-CH_2 \\ | \qquad | \\ CH_2-CH_2 \end{array} \quad \text{or} \quad \begin{array}{c} CH_2 \\ | \!\!\!> CH-CH_3 \\ CH_2 \end{array}$$
3. (a) $CH_4 + Cl_2 \rightarrow HCl + CH_3Cl$ **4.** $(CH_3)_2CBrC(CH_2CH_3)_3$.
5. The color disappears because the Br_2 is used up. The products are colorless.
8. (c) $FeBr_3$ catalyst and heat.   **9.** 1,2-Dichloroethane.
11. (a) A high (*eg.* 5:1) mole ratio.   **13.** CH_3OCH_3, $CH_3OC_6H_5$, and $C_6H_5OC_6H_5$.
14. $CH_3CH_2CO_2H \xrightarrow{\text{LiAlH}_4} \xrightarrow{\text{CH}_3\text{OH}} \xrightarrow{\text{HCl}} CH_3CH_2CH_2OH$
15. (a) -1,   **(b)** $H^- + H_2O \rightarrow H_2 + OH^-$ (explosive violence)
16. (b) $CH_3CH_2Br + Mg \rightarrow CH_3CH_2MgBr$
$2\,CH_3CH_2MgBr + CO_2 \rightarrow Mg(CH_3CH_2CO_2)_2 + MgBr_2$
$Mg(CH_3CH_2CO_2)_2 + 2\,H_3O^+ \rightarrow Mg^{2+} + 2\,CH_3CH_2CO_2H + 2\,H_2O$
17. (b) $(CH_3)_3N + HCl \rightarrow (CH_3)_3NH^+Cl^-$
19. (a) Ester,   **(c)** alcohol,   **(e)** alcohol and aldehyde or ketone.   **22.** 8.
24. (b) $(C_6H_{10}O_5)_n + n\,H_2O \rightarrow n\,C_6H_{12}O_6$ **26.** A polymer of monomers formerly having alcohol and carboxylic acid functional groups, now connected with ester linkages.   **27.** For addition polymers: **(a)** no byproducts, **(b)** one functional group is sufficient,   **(c)** able to copolymerize.
30. (a) The peptide links,   **(b)** the hydrogen bonded helical structure.
33. (a) GAC.   **34. (b)** GAC.   **36. (a)** The oxidation number changes of the carbon atoms are -1 to 0 and -3 to -2, respectively.
38. (a) Three 5-member rings, one 6-member ring, and one double bond,   **(b)** $C_{10}H_6Cl_8$.
39. H_2SO_4 is a dehydrating agent.   **40. (a)** The $C=C$ infrared band is an easy way to identify the hexene. Mass spectra could also be used to distinguish the two.

41. (b) $CH_3\overset{\|}{\underset{O}{C}}-Cl + H_2O \rightarrow CH_3\overset{\|}{\underset{O}{C}}-OH + HCl$

42. (d) $CH_3CH_2OH \xrightarrow{\text{oxidation}} CH_3CO_2H$

$CH_3CH_2OH + CH_3CO_2H \xrightarrow[\text{heat}]{\text{H}_2\text{SO}_4} CH_3CO_2CH_2CH_3 + H_2O$

43. (a) 3-Pentanol. **46.** 39.2% C_4H_8. **48.** $CH_3CHBrCH_3$.

51. C is an acid, $C_{10}H_{21}CO_2H$.

53. (b) $(C_2H_3Cl)_n + \tfrac{5}{2}n\,O_2 \rightarrow 2n\,CO_2 + n\,H_2O + n\,HCl$

Index

The elements are listed in alphabetical order, with additional listings for elements with symbols which do not begin with the initial letter of the element name. Numbers in parentheses refer to the most stable isotopes of radioactive elements.

Element	Symbol	Atomic Number	Atomic Weight	Element	Symbol	Atomic Number	Atomic Weight
Actinium	Ac	89	(227)	Erbium	Er	68	167.26
Aluminum	Al	13	26.9815	Europium	Eu	63	151.96
Americium	Am	95	(243)	Fermium	Fm	100	(253)
Antimony	Sb	51	121.75	Fluorine	F	9	18.9984
Argon	Ar	18	39.948	Francium	Fr	87	(223)
Arsenic	As	33	74.9216	Iron	Fe	26	55.847
Astatine	At	85	(210)	Gadolinium	Gd	64	157.25
Silver	Ag	47	107.868	Gallium	Ga	31	69.72
Gold	Au	79	196.9665	Germanium	Ge	32	72.59
Barium	Ba	56	137.34	Gold	Au	79	196.9665
Berkelium	Bk	97	(249)	Hafnium	Hf	72	178.49
Beryllium	Be	4	9.01218	Helium	He	2	4.00260
Bismuth	Bi	83	208.9806	Holmium	Ho	67	164.9303
Boron	B	5	10.81	Hydrogen	H	1	1.0080
Bromine	Br	35	79.904	Mercury	Hg	80	200.59
Cadmium	Cd	48	112.40	Indium	In	49	114.82
Calcium	Ca	20	40.08	Iodine	I	53	126.9045
Californium	Cf	98	(251)	Iridium	Ir	77	192.22
Carbon	C	6	12.011	Iron	Fe	26	55.847
Cerium	Ce	58	140.12	Krypton	Kr	36	83.80
Cesium	Cs	55	132.9055	Potassium	K	19	39.102
Chlorine	Cl	17	35.453	Lanthanum	La	57	138.9055
Chromium	Cr	24	51.996	Lawrencium	Lr	103	(257)
Cobalt	Co	27	58.9332	Lead	Pb	82	207.2
Copper	Cu	29	63.546	Lithium	Li	3	6.941
Curium	Cm	96	(247)	Lutetium	Lu	71	174.97
Dysprosium	Dy	66	162.50	Magnesium	Mg	12	24.305
Einsteinium	Es	99	(254)	Manganese	Mn	25	54.9380